Advanced Calculus

A Friendly Approach

Advanced Calculus

A Friendly Approach

Witold A. J. Kosmala

Appalachian State University

PRENTICE HALL
Upper Saddle River, New Jersey 07458

Original edition, entitled ADVANCED CALCULUS: A FRIENDLY APPROACH, 1st Edition, 0137379250 by KOSMALA, WITOLD A.J., published by Pearson Education, Inc, publishing as Prentice Hall, Copyright © 1999

KOREA edition published by PEARSON EDUCATION KOREA LTD, Copyright © 2006

This edition is manufactured in KOREA, and is authorized for sale only in KOREA.

ISBN 978-89-450-4512-6
ISBN 89-450-4512-0

To my wife Vanessa,

and my children

Konrad, Doria, and Alina

Contents

[*]Optional sections are marked with an asterisk *.

Preface

1. Purpose and Background

Though designed for a two-semester or three-quarter course in the theory of calculus, the following material may be easily subdivided to suit any length course titled Introduction to Real Analysis, Theoretical Calculus, Honors Calculus, Advanced Calculus, or the equivalent. Intended for undergraduate juniors or seniors who have completed the calculus sequence, linear algebra, and preferably differential equations, this text aids in the pursuit of careers in science, statistics, engineering, computer science, and business. Since branches in these areas possess roots in real analysis, a solid background in undergraduate analysis is essential, and precisely the purpose of this text. Though analysis is an important building block, it is almost universally recognized as one of the toughest undergraduate courses in mathematics. Students are generally not prepared to deal with mathematical proofs. Thus, a new crop of analysis texts have arisen which teach methods of proof together with theory. Unfortunately, such books cover only single variable calculus. This is the first text to cover both single and multivariable calculus at a friendly level. A large portion of the material covered should be somewhat familiar to students from their study of elementary calculus. Here, however, theory and deeper understanding is stressed. Through clear, accurate, and in-depth explanations, ideas are built upon, and the reader is adequately prepared for subsequent new material. A number of goals are achieved. Included with nearly every proof are thorough explanations of what is being done and the desired goal, an essential item in beginning chapters. As the reader develops skill with proofs, the discussion of proof strategies becomes shorter. Curiosity is stimulated through the question "Why?" Questioning encourages the reader to think about what has been said and either promotes logic or requires the reader to look back at previously covered material. The omission of details at appointed places forces the student to read with pencil and paper. *Advanced Calculus: A Friendly Approach* works by pointing out goals, outlining procedures, and giving good intuitive reasons. Detailed and complete graphs are used where necessary, as well as cross-referencing throughout the text. Further goals are to introduce the reader to some important basic concepts and their different terminologies, and various uses of material in other areas. The reader is not expected to provide a proof for every result mentioned. Without taking away a student's mathematical security, a respect for higher mathematics is taught. Historical notes add enrichment and, occasionally, motivation. Also, expressions like "it can be shown," "obviously," and "clearly" are used not to intimidate the reader, but to point out that what follows should be easily grasped by the person reading the text. If not, then review of the necessary material is in order.

Each section in this textbook is followed by a set of exercises usually arranged in the order of the associated text material, and of varied difficulty. Exercises range from routine, to creative and innovative, mixing theory and applications. In addition, excluding Chapter 9, every chapter is followed by a set of review problems of true/false nature. Often, knowing what is true or false is more important than being able to prove a theorem. True statements need to be proven. A counterexample is requested for false statements. And a change is requested for those false statements that become true with a little help. Such problems foster deeper understanding of the material and stimulate curiosity. Students learn to interrelate ideas in this way. Each review section is followed by optional projects designed to improve students' creativity as well as reinforce ideas covered in previous chapters. The projects either continue the study of real analysis or bring out analysis in other areas of mathematics.

For quick reference, all definitions, examples, theorems, corollaries, and remarks are numbered in succession as chapter number, section number, and position in the sequence within each section. Hints and Solutions to Selected Exercises are located at the back of the text. But since most exercises may be completed in a variety of ways, viewing this part of the text after exercises have been attempted is recommended. A convenient Index of Symbols is also included in the back of the book.

2. Design and Organization

Chapter names closely reflect the content of each chapter. **Chapter 1** provides basic necessary terminology and proof techniques. A thorough development of the number system has been omitted due to both the length and nature of the course. Students should be able to cover Chapter 1 with little or no guidance. Of worth noting, however, is that Section 1.4, if covered at great depth, could potentially take a long time. **Chapter 2** provides detailed proof techniques for progressing throughout the text. **Chapter 3** begins with a smooth transition from Chapter 2 by introducing a more general concept of the limit. Thus, the first section, which fills a few holes from elementary calculus, needs only to be read by the student. **Chapter 4** introduces uniform continuity, a new term for the majority of students at this level. Thus, Section 4.4 will require time to gel. **Chapter 5** involves theory through which elementary calculus experiences involving derivatives are explained. The first three sections are a must to cover. Mastery of integration techniques from elementary calculus is helpful in **Chapter 6** where theory and the development of the integral are stressed. Proofs of theorems presented should not be skipped though they may seem unexciting at times. Section 6.5 which covers improper integrals is needed for the integral test in Section 7.2, and thus is not optional. **Chapter 7** covers all of the standard tests and proofs for convergence or divergence of an infinite series of numbers. **Chapter 8** covers sequences and series of functions. The first four sections may present difficulty. **Chapter 9**, on introductory concepts of Fourier series, is optional though useful to many students whose major may not be mathematics. A study in three-space begins in **Chapter 10**. Several topics should be familiar to the reader with proofs likely found in elementary texts. In addition, these sections with familiar topics are marked as optional though their mastery is essential to material which follows. The study of basic topology begins in **Chapter 11**. A thorough coverage of Section 11.1 may cost one some time. A quick reference back to this section is always possible. The rest of Chapter 11 presents the theory of functions in two variables. The transition to functions of n variables is an easy step, and thus omitted. A computer for graphing is highly recommended. **Chapter 12** follows a standard development of the double integral. Line integrals and Green's theorem are essential parts of this chapter. Section 12.8 shows how previous material in Chapter 12 may be generalized into higher dimensions.

Chapters 1 through 8 with optional sections omitted are recommended in a one-semester single variable course. Chapters 1 through 9, with optional sections covered under the discretion of the instructor are recommended for a single variables course consisting of two semesters. Coverage of the entire textbook with the omission of some optional sections is recommended for a typical advanced calculus class. As mentioned in the next subsection, *Advanced Calculus: A Friendly Approach* is accompanied by an *Instructor's Supplement* which covers additional topics. An instructor may opt to place material on reserve in the library for students. Additional topics provided by the *Instructor's Supplement* as well as project sections in the textbook, supplement the material as well as provide learning skills for students. In addition, instructors are moved into a different teaching mode. Coverage of either the project sections in the text or additional topics from the *Instructor's Supplement* is recommended whenever an instructor chooses to do so. Honors classes and classes with guided discovery approaches may especially wish to incorporate either the project sections or additional topics from the *Instructor's Supplement* into their courses.

3. Supplements

Two supplements to the text available only through the publisher are an *Instructor's Solutions Manual*, which contains solutions to most exercises from the text, and an *Instructor's Supplement*, which contains material and exercises for right- and left-hand derivatives, fixed points, natural logarithmic functions, finer tests for infinite series, an approximation of Fourier series and Bessel's inequality, Fourier integrals and Gibbs' phenomenon, curvature, and extrema of functions in $\Re^3$. The *Instructor's Supplement* also contains project sections on Newton–Raphson and secant methods, numerical integration, summable series, zeta functions, generating functions, Bernoulli's and Euler's numbers, parabolas, and Lagrange multipliers.

4. Acknowledgments

Grateful appreciation is expressed to those who reviewed all or part of this manuscript: Frank De Meyer, Colorado State University; Margaret Gessaman, University of Nebraska at Omaha; Suzanne Lenhart, University of Tennessee at Knoxville; Shing S. So, Central Missouri State University; Michael Mossinghoff, University of California at Los Angeles; Ernest Lane at Appalachian State University; and Vanessa Kosmala. I would like to thank George Lobell, the Acquisitions Editor, as well as the staff of Prentice-Hall, Inc., and members of the Department of Mathematical Sciences at Appalachian State University for their assistance and support in the development and production of this manuscript. In addition, I wish to thank Alethea Vitray for her precise illustrations.

Witold A. J. Kosmala

Advanced Calculus

A Friendly Approach

1

Introduction

1.1* Algebra of Sets

Chapter 1 serves as an introduction to this textbook and discusses terminology for the chapters that follow as well as methods of proving mathematical statements. Although the intent is to cover only the topics that are needed in the following chapters, we must realize that any material covered in this text can always be extended indefinitely.

Almost every math textbook begins with a study of sets, since the idea is fundamental to any area of mathematics. By a *set* we mean a collection of well-defined objects called *elements* or *members* of the set. And by *well-defined* we mean that there is a definite way of determining whether or not a given element belongs to the set. To write a set it is customary to use braces { } with elements of the set listed inside them. Lowercase letters are usually used to represent the elements, whereas capital letters denote sets themselves. If an element r belongs to the set A, then we write $r \in A$. And if r is not an element of the set A, then we write $r \notin A$. Thus, if $A = \{1,2,3\}$, then $1 \in A$, but $4 \notin A$.

At times it is difficult to list all members of a set. For example, if the set B consists of all of the counting numbers smaller than 100, we usually write

$$B = \{1,2,3, \ldots ,99\} \text{ or}$$
$$B = \{x \mid x \text{ is a counting number smaller than } 100\}.$$

The second of these is read as: B is the set of all elements x, such that x is a counting number smaller than 100. Thus, we described what is in the set instead of listing the elements. When using the descriptive method, we have to be sure that what we want to include in the set is indeed what we describe and is regarded by everyone in the same way. For example, the set $C = \{\text{all young men in the USA}\}$ is not a well-defined set since the word "young" has different meanings to different people.

There are many ways of describing any one particular set. For example, the set $D = \{2, 3\}$ can also be written as

$$D = \left\{x \middle| x^2 - 5x + 6 = 0\right\},$$
$$D = \left\{x \middle| x \text{ is a prime number less than } 4\right\}, \text{ or}$$
$$D = \left\{x \middle| x \text{ and } x+1, \text{ or } x \text{ and } x-1 \text{ are prime numbers}\right\},$$

where prime numbers are defined below. Sets used throughout this text are

$$N = \textit{natural (counting) numbers} = \{1, 2, 3, \ldots\},$$
$$W = \textit{whole numbers} = \{0, 1, 2, \ldots\},$$
$$Z = \textit{integers} = \{\ldots, -2, -1, 0, 1, 2, \ldots\}, \text{ and}$$
$$Q = \textit{rational numbers} = \left\{x \middle| x = \tfrac{p}{q} \text{ where } p, q \in Z \text{ and } q \neq 0\right\},$$

where $p, q \in Z$ means that $p \in Z$ and $q \in Z$,

$$Q^+ = \left\{x \in Q \middle| x > 0\right\},$$
$$\Re = \Re^1 = \textit{real numbers},$$
$$\Re^+ = \left\{x \in \Re \middle| x > 0\right\}, \text{ and}$$
$$\Re^- = \left\{x \in \Re \middle| x < 0\right\} = \left\{-x \middle| x \in \Re^+\right\}.$$

When the value of x is not specified, assume that $x \in \Re$. Real numbers that are not rational are called *irrational*. In fact, there are more irrational than rational numbers. Elements in $\Re^+$ and elements in $\Re^-$ are called *positive* and *negative* real numbers, respectively. Elements in $\{0, 1, 2, \ldots\}$ and $\{0, -1, -2, \ldots\}$ are called *nonnegative* and *nonpositive* real numbers, respectively. *Prime* numbers are natural numbers divisible by exactly two distinct natural numbers, 1 and the natural number itself. An integer, p, is *even* if p can be written as $p = 2s$ for some integer, s. And an integer, q, is *odd* if it can be written as $q = 2t + 1$ for some integer, t. Sections 1.6 and 1.7 discuss real numbers in greater depth.

The following sets denote *intervals*, where a and b are real numbers with $a < b$.

$$[a,b] = \left\{x \middle| a \le x \le b\right\} \qquad (a,b) = \left\{x \middle| a < x < b\right\}$$
$$[a,b) = \left\{x \middle| a \le x < b\right\} \qquad (a,b] = \left\{x \middle| a < x \le b\right\}$$
$$[a,\infty) = \left\{x \middle| x \ge a\right\} \qquad (a,\infty) = \left\{x \middle| x > a\right\}$$
$$(-\infty,a] = \left\{x \middle| x \le a\right\} \qquad (-\infty,a) = \left\{x \middle| x < a\right\}$$

The symbols ∞ and $-\infty$ are called (*plus*) *infinity* and *minus infinity*, respectively. These symbols do not represent real numbers. Often, ∞ is written as $+\infty$. Note that we can write $\Re = (-\infty, \infty)$ and $\Re^+ = (0, \infty)$.

Definition 1.1.1. If A and B are sets, then A *equals* B, that is, $A = B$, if and only if both sets consist of exactly the same elements.

The expression *if and only if*, commonly written by instructors as "iff," means that two statements are equivalent. This expression always applies to definitions, although it is often not written. Further, $A = B$ if and only if whenever $x \in A$, then $x \in B$, and whenever $x \in B$, then $x \in A$. If $A \neq B$, then the sets A and B are distinct, that is, some element, say x, exists such that $x \in A$ but $x \notin B$, or $x \in B$ but $x \notin A$.

Definition 1.1.2. If A and B are sets such that every element of A is also an element of B, then A is a *subset* of B, meaning A is contained in B and is denoted by $A \subseteq B$. If A is a subset of B but $A \neq B$, then A is called a *proper subset* of B and is denoted by $A \subset B$.

Thus, $A = B$ if and only if $A \subseteq B$ and $B \subseteq A$. If a set has no elements in it, then it is called an *empty* or *null set* and is denoted by ϕ. The vacuous implication of Definition 1.1.2 is that the empty set is a subset of every set. We are now ready to make new sets from given ones.

Definition 1.1.3. If A and B are sets, then

(a) A *intersection* B, denoted by $A \cap B$, is the set of all elements that belong to both A and B. That is, $A \cap B = \{x \mid x \in A \text{ and } x \in B\}$.

(b) A *union* B, denoted by $A \cup B$, is the set of all elements that belong to either A or B or both. That is, $A \cup B = \{x \mid x \in A \text{ or } x \in B\}$.

(c) the *complement* of B in A, also referred to as the complement of B relative to A, or A *minus* B, denoted $A \setminus B$, is the set of all elements in A that are not in B. That is, $A \setminus B = \{x \mid x \in A \text{ and } x \notin B\}$.

(d) A and B are *disjoint* if they have no elements in common. That is, $A \cap B = \phi$.

The shaded regions in Figure 1.1.1 represent Definition 1.1.3, parts (a)–(c).

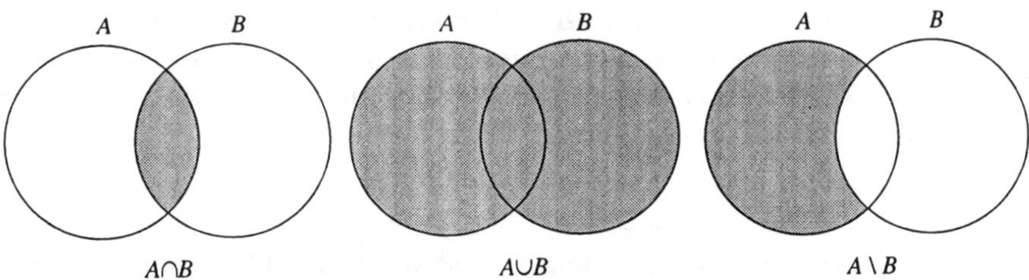

Figure 1.1.1

Example 1.1.4. If $A = \{1,3,5\}$ and $B = \{1,3,8\}$, find the following:

(a) $A \cap B$ **(b)** $A \cup B$ **(c)** $A \setminus B$

Answer. $A \cap B = \{1,3\}$, $A \cup B = \{1,3,5,8\}$, and $A \setminus B = \{5\}$. Elements are never written more than once in a set. ■

The symbol ■ is used to indicate that a proof, answer, or remark is complete. Also, observe that $x \notin A \setminus B$ does not mean that $x \notin A$ but $x \in B$. Why?

THEOREM 1.1.5. *If A, B, and C are sets, then*

(a) $A \cap A = A$ *and* $A \cup A = A$. *(Idempotent property)*
(b) $A \cap B = B \cap A$ *and* $A \cup B = B \cup A$. *(Commutative property)*
(c) $(A \cap B) \cap C = A \cap (B \cap C)$ *and* $(A \cup B) \cup C = A \cup (B \cup C)$. *(Associative property)*
(d) $A \cap (B \cup C) = (A \cap B) \cup (A \cap C)$ *and* $A \cup (B \cap C) = (A \cup B) \cap (A \cup C)$. *(Distributive property)*
(e) $A \setminus (B \cup C) = (A \setminus B) \cap (A \setminus C)$ *and* $A \setminus (B \cap C) = (A \setminus B) \cup (A \setminus C)$. *(De Morgan's[1] laws)*

Proof of first equality in part (d). Since we are to prove that two sets are equal, we need to show that one set is a subset of the other and vice versa. To prove that $A \cap (B \cup C) \subseteq (A \cap B) \cup (A \cap C)$, pick an arbitrary element, say $x \in A \cap (B \cup C)$. We will show that $x \in (A \cap B) \cup (A \cap C)$. By Definition 1.1.3, part (a), $x \in A$ and $x \in B \cup C$. Thus, $x \in A$ and $x \in B$, or $x \in A$ and $x \in C$. Hence, $x \in A \cap B$ or $x \in A \cap C$. And by Definition 1.1.3, part (b), $x \in (A \cap B) \cup (A \cap C)$.

We now need to prove conversely that if $y \in (A \cap B) \cup (A \cap C)$, then $y \in A \cap (B \cup C)$. Since $y \in (A \cap B) \cup (A \cap C)$, then $y \in A \cap B$ or $y \in A \cap C$. Thus, $y \in A$ and $y \in B$, or $y \in A$ and $y \in C$. In other words, $y \in A$, and either $y \in B$ or $y \in C$. Therefore, $y \in A$ and $y \in B \cup C$. Hence, $y \in A \cap (B \cup C)$. And the proof of the first equality in part (d) is complete.

Proof of first equality in part (e). Again, we need to show that one set is a subset of the other, and vice versa, to give equality of the two sets. Let $x \in A \setminus (B \cup C)$. We need to show that $x \in (A \setminus B) \cap (A \setminus C)$. Now since $x \in A \setminus (B \cup C)$, by Definition 1.1.3, part (c), $x \in A$ and $x \notin B \cup C$. Hence, $x \in A$, $x \notin B$, and $x \notin C$. Thus, $x \in A \setminus B$ and $x \in A \setminus C$. Therefore, $x \in (A \setminus B) \cap (A \setminus C)$, and so $A \setminus (B \cup C) \subseteq (A \setminus B) \cap (A \setminus C)$.

To prove inclusion in the other direction, pick an arbitrary element $x \in (A \setminus B) \cap (A \setminus C)$. Then, $x \in A \setminus B$ and $x \in A \setminus C$. Therefore, $x \in A$, $x \notin B$, and $x \notin C$. And so $x \in A$ and $x \notin B \cup C$. Hence, $x \in A \setminus (B \cup C)$. Thus, $(A \setminus B) \cap (A \setminus C) \subseteq A \setminus (B \cup C)$. And the proof of the first equality in part (e) is complete. See the shaded region in Figure 1.1.2. ■

[1] Augustus De Morgan (1806–1871), an English mathematician and logician born in India, is recognized for the development of the foundations of algebra, arithmetic, probability theory, and calculus.

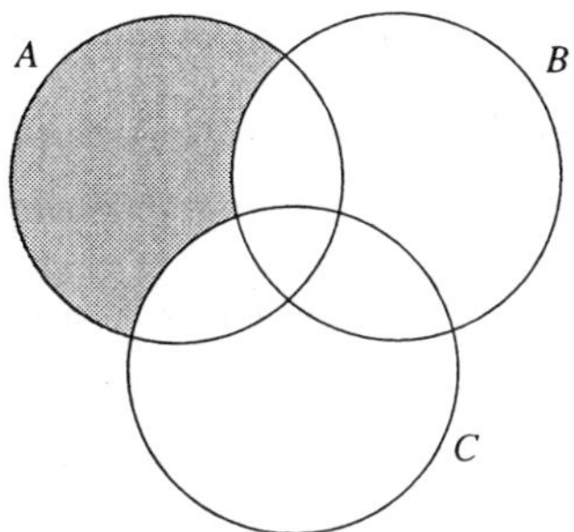

Figure 1.1.2

The diagram in Figure 1.1.2 only illustrates the result and does not constitute a proof. In Exercise 3, you are asked to prove other statements in Theorem 1.1.5. In view of Theorem 1.1.5, part (c), we can write $A \cap (B \cap C) = A \cap B \cap C$ and $A \cup (B \cup C) = A \cup B \cup C$. Furthermore, if $A_1, A_2, \ldots, A_n$ are n sets, where n is some fixed natural number, then

$$A_1 \cap A_2 \cap \ldots \cap A_n = \bigcap_{k=1}^{n} A_k = \{x \mid x \in A_k \text{ for all } k = 1, 2, \ldots, n\}, \text{ and}$$

$$A_1 \cup A_2 \cup \ldots \cup A_n = \bigcup_{k=1}^{n} A_k = \{x \mid x \in A_k \text{ for some } k = 1, 2, \ldots, n\}.$$

Definition 1.1.6. If A and B are two nonempty sets, then the *Cartesian product* (*cross product*) $A \times B$ is the set of all *ordered pairs* (a,b) such that $a \in A$ and $b \in B$. That is,

$$A \times B = \{(a,b) \mid a \in A \text{ and } b \in B\}.$$

Remark 1.1.7. The Cartesian product leads to the *Cartesian* (*rectangular*) *coordinate system*, named after Descartes,[2] where a point, P, in the plane $\Re^2 = \{(x,y) \mid x, y \in \Re\}$ is called an *ordered pair* (a,b). $\Re^2$ is often referred to as the xy-plane. The value a is called the x-*coordinate* (*first coordinate*) whereas the value b is called the y-*coordinate* (*second coordinate*). Often, the value a is also referred to as a point. The same is true of b. $(0,0)$ is called the *origin*. In general, an ordered pair (b,a) is not the same as an ordered pair (a,b). Two ordered pairs (a,b) and (c,d) are equal if and only if $a = c$ and $b = d$. Thus, $(a,b) = (b,a)$ if and only if $a = b$ and $b = a$. Hence, the point represented by this ordered pair must lie on a line passing through the origin and rising at $45°$. ■

---

[2]René Descartes (1596–1650), a French mathematician and philosopher, is known in Latin as Cartesius. Descartes founded analytic geometry, introduced the symbols of equality and square and cube roots, proved how roots of an equation are related to sign changes within it, and did some work in physics.

One of the best known results in mathematics is the Pythagorean[3] theorem.

THEOREM 1.1.8. *(Pythagorean Theorem) A triangle with legs of length a and b and hypotenuse of length c is a right triangle if and only if $a^2+b^2=c^2$.*

In other words, if one angle in a triangle has a measure of $90°$, that is, $\frac{\pi}{2}$ radians, then $a^2+b^2=c^2$. The converse is also true. That is, if $a^2+b^2=c^2$, then the angle opposite the side of length c must be a right angle, that is, it must be of $90°$. For a generalization of the Pythagorean theorem, see the law of cosines in Section 10.3.

We will now assume that the Pythagorean theorem is at hand to easily verify the Euclidean[4] distance formula.

THEOREM 1.1.9. *(Euclidean Distance Formula) If $P_1(x_1,y_1)$ and $P_2(x_2,y_2)$ are two points in the xy-plane, then the distance $d(P_1,P_2)$ between them is given by*

$$d(P_1,P_2)=\sqrt{(x_2-x_1)^2+(y_2-y_1)^2}.$$

Note that $d(P_1,P_2)=d(P_2,P_1)$.

Exercises 1.1

1. Suppose that $A=\{1,3,5\}$, $B=\{1,2,4\}$, and $C=\{1,8\}$. Find the following:
(a) $(A\cup B)\cap C$. **(b)** $A\cup(B\cap C)$. **(c)** $(A\setminus C)\cup B$.
(d) $(A\cap B)\times C$. **(e)** $C\times C$. **(f)** $\{\phi\}\cap A$.

2. Find real numbers x and y so that the ordered pair $(x-2y,-2x+2y)=(-4,2)$.

3. Prove parts (a)–(c) of Theorem 1.1.5.

4. If A, B, and C are sets, prove that
(a) $A\subseteq B$ if and only if $A\cap B=A$.
(b) $A\cap B=A\setminus(A\setminus B)$.
(c) $(A\setminus B)\cup(B\setminus A)=(A\cup B)\setminus(A\cap B)$.
(d) $(A\cap B)\times C=(A\times C)\cap(B\times C)$.
(e) $A\cap B$ and $A\setminus B$ are disjoint, and $A=(A\cap B)\cup(A\setminus B)$.

---

[3]Pythagoras of Samos (c.580–c.500 B.C.), a Greek geometer and philosopher, interpreted all things using numbers. Pythagoras also founded a school, considered a cult, that was political, philosophical, and religious. The cult, also known as the secret society, was closely knit, regulated the diet and way of life of its members, and had a common method of education.

[4]Euclid (c.365 B.C.–300 B.C.), a great mathematician of ancient Greece, received mathematical training from the students of Plato (427–347 B.C.) and founded a school in Alexandria in Egypt. Euclid's book *Elements*, in 13 volumes, compiled the most important mathematical facts available at that time, including a complete solution to the study of Pythagorean triples. For over two thousand years, the first six of Euclid's volumes were students' introduction to geometry. Euclid's *Elements* were first printed in 1482 and had more than one thousand editions thereafter.

5. If the sets A, B, C, and D represent the intervals $[1,3)$, $(1,4)$, $(2,5]$, and $[3,5]$, respectively, shade the region in $\Re^2$ that represents

(a) $(A\times B)\cap(C\times D)$. **(b)** $(A\cap C)\times(B\cap D)$.

(c) $(A\times B)\cup(C\times D)$. **(d)** $(A\cup C)\times(B\cup D)$.

6. One hundred students were surveyed. Thirty of them subscribe to *Time*, 42 to *Business Week*, 51 to *Sports Illustrated*, 15 to both *Time* and *Business Week*, 12 to *Time* and *Sports Illustrated*, 22 to both *Business Week* and *Sports Illustrated*, and 3 to all three magazines. How many subscribe to

(a) only *Time*? **(b)** none of these magazines?

7. Prove Theorem 1.1.9.

1.2* Relations and Functions

In this section we will define what is meant by a function and review some terminology.

Definition 1.2.1. A *relation* from a set A to a set B is a subset of $A\times B$. And the *graph* of a relation, also called a *curve*, is the set of ordered pairs that satisfy the relation.

Example 1.2.2. Sets

$$S_1=\{(0,1),(2,3),(-1,3)\},\qquad S_2=\{(0,1),(0,2),(1,1),(-2,3)\},$$

$$S_3=\left\{(x,y)\,\middle|\,y=2x+1\right\},\qquad S_4=\left\{(x,y)\,\middle|\,x=y^2\right\},$$

$$S_5=\left\{(x,y)\,\middle|\,x^2+y^2=1\right\},\text{ and}\qquad S_6=\left\{(x,y)\,\middle|\,x,y\in N\text{ and }y\le x\right\}$$

are all examples of relations. Figure 1.2.1 shows the graphs of these relations. Note that S_3 can be written as $S_3=\left\{(x,2x+1)\,\middle|\,x\in\Re\right\}$. Recall that lowercase letters are real numbers unless otherwise indicated. ■

Categorizing relations, that is, the sets of ordered pairs, into those that are functions and those that are not is intuitively easy. A given relation is a function if it passes a *vertical line test.* That is, if a vertical line drawn through the relation at any point does not intersect the graph of that relation more than once, then the relation is a function. Thus, in Example 1.2.2 sets S_1 and S_3 are functions, whereas the other sets are not. The set S_5 represents the circle with center at the origin and radius of 1. In general, from the distance formula, the relation $x^2+y^2=r^2$ represents the *circle* with center $(0,0)$ and radius $r>0$. If $r=0$, the circle is *degenerate* and consists of only one point. Also, relations S_3, S_4, and S_5 are defined by equations involving x and y.

Definition 1.2.3. A *function* f, often called a *mapping*, is a relation from a set A to a set B in which no two different ordered pairs have the same first coordinate.

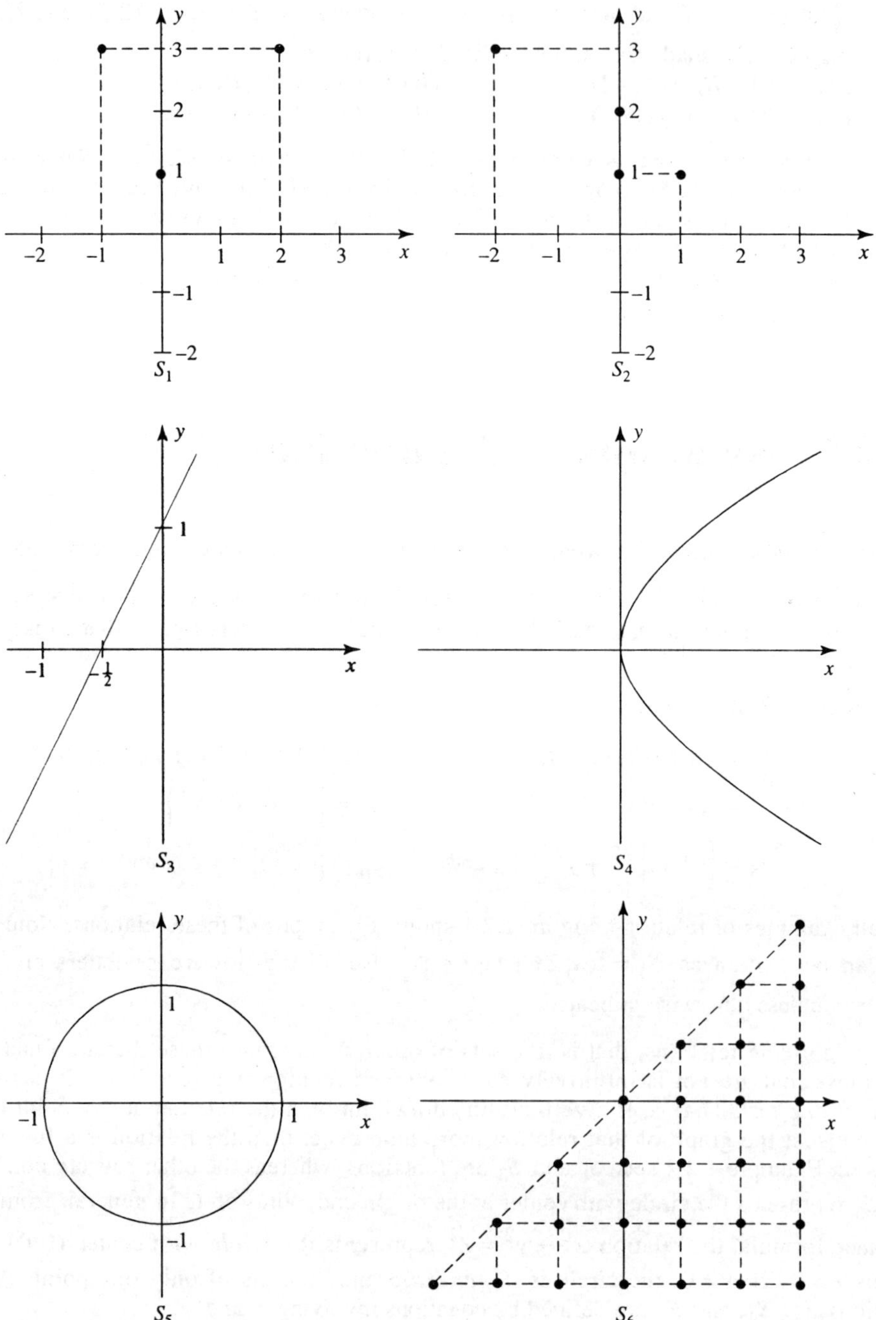
y
3
2
1
-2
-1
1
2
3
x
-1
-2
S_1
y
3
2
1
-2
-1
1
2
3
x
-1
-2
S_2
y
1
-1
$-\frac{1}{2}$
x
S_3
y
x
S_4
y
1
-1
1
x
-1
S_5
y
x
S_6

Figure 1.2.1

The graph of a function f is referred to as "graph(f)." Translating Definition 1.2.3 into different words, a function f from a set $A \subseteq \mathfrak{R}$ to a set $B \subseteq \mathfrak{R}$ is a collection of ordered pairs (a,b) such that

(a) $a \in A$ and $b \in B$,

(b) for every $a \in A$ there exists $b \in B$ such that $(a,b) \in \text{graph}(f)$,

(c) if $(a,c) \in \text{graph}(f)$ and $(a,d) \in \text{graph}(f)$, then $c = d$.

Part (b) says that A is the "domain" of f, a is an *independent variable*, and b is a *dependent variable*, while part (c) relays that f is *well-defined* and *single-valued*, that is, $x_1 = x_2 \in A$ implies that $f(x_1) = f(x_2)$. Equivalently, if $f(x_1) \neq f(x_2)$, then $x_1 \neq x_2$. See Remark 1.2.4 for some discussion of this notation.

Remark 1.2.4. Suppose that f is a function.

(a) Often we write $b = f(a)$ to say that $(a,b) \in \text{graph}(f)$.

(b) The set of all first coordinates of an ordered pair is called the *domain*, D_f, of the function f. And

$$D_f = \left\{a \,\middle|\, \text{there exists } b \text{ such that } (a,b) \in \text{graph}(f)\right\}.$$

If the domain for a function is not specified, then it is the largest possible set of values that make the expression meaningful. The *range*, R_f, of f is the set of all second coordinates, that is, all resulting functional values. So

$$R_f = \left\{b \,\middle|\, \text{for } a \in D_f, (a,b) \in \text{graph}(f)\right\} = \left\{f(a) \,\middle|\, a \in D_f\right\}.$$

The set $f(A)$ with $A = D_f$ is also used to denote R_f. In general, if $E \subseteq D_f$, then $f(E) = \left\{f(x) \,\middle|\, x \in E\right\}$. In chapters 1–9, this textbook considers those functions whose domain and range form a subset of $\mathfrak{R}$.

(c) A function, f, can be thought of as a rule, transformation, mapping, or machine that takes all of the values from D_f and assigns to them new values. We say that f maps set A into set B by writing $f: A \to B$. If $f: A \to B$ or $f: A \to \mathfrak{R}$, then it will be assumed that $A = D_f \subseteq \mathfrak{R}$ and $R_f \subseteq B \subseteq \mathfrak{R}$ and f will be called *real-valued*. (See Figure 1.2.2).

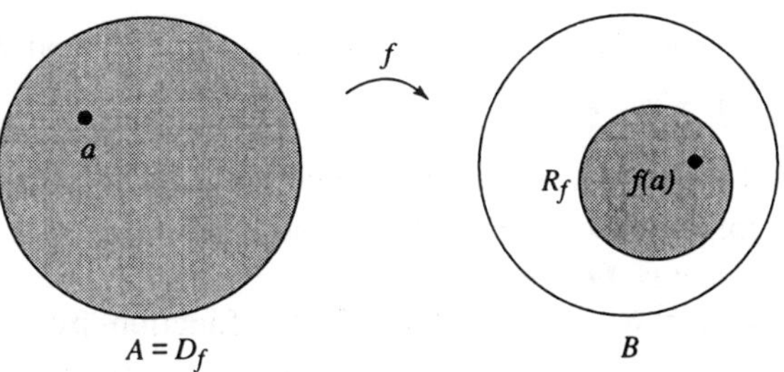

Figure 1.2.2

(d) If $f: A \to B$ and if for some $x_0 \in A$, $f(x_0) = y_0$, then y_0 is called the *image* of x_0 and x_0 is called a *preimage* of y_0. Can y_0 have more than one preimage? If $f(x_0) = y_0$, then we also say that f maps x_0 into y_0. Furthermore, if $x_0 \in A$, then x_0 is in the domain of the function f. Thus, f is *defined* at x_0 and $f(x_0)$ *exists*. If x_0 is not in the domain of f, then we say that f is *not defined* at x_0, or that $f(x_0)$ *does not exist*. Importantly, f is the name of a function, whereas $f(x)$ is the y-coordinate corresponding to some value x. Figure 1.2.3 shows the graph of f. Note the shaded regions represent D_f and R_f. ■

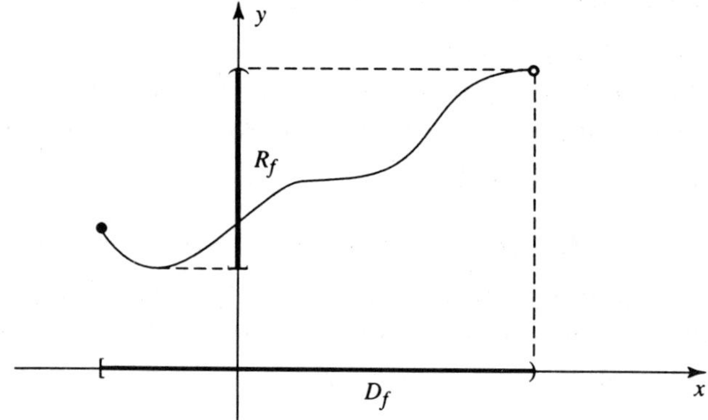

Figure 1.2.3

Example 1.2.5. If $f = \{(x, y) | y = \sqrt{x-1}\}$, find the largest possible D_f and R_f.

Answer. Since we are able to take the square root of only nonnegative real numbers to produce a real number in R_f, we must ensure that $x - 1 \geq 0$. Thus, $D_f = \{x | x \geq 1\}$. After substituting these values for x in the equation $y = \sqrt{x-1}$, we find that $R_f = \{y | y \geq 0\}$. Recall that because the square root of a number is never negative, R_f could not possibly include negative values. ■

Remark 1.2.6. The function f from Example 1.2.5 can be stated in a different and more popular fashion. That is, if $f(x) = \sqrt{x-1}$, find D_f and R_f. ■

Definition 1.2.7. Consider a function f which maps set A into set B.

(a) The function f is *one to one* if for all x_1 and x_2 in A, $f(x_1) = f(x_2)$ implies that $x_1 = x_2$. If f is one to one, then f is called an *injection*, or an *injective function*.

(b) If $B = R_f$, we say that f is a function from A *onto* B and call it a *surjection*, or a *surjective function*. That is, f takes A onto B if for each $b \in B$, there is an $a \in A$ such that $f(a) = b$.

(c) If f is both one to one and onto, f is called a *bijection,* or a *bijective function*.

Compare one to oneness with what is meant by the function being well defined.

Remark 1.2.8. Suppose a function f maps set A into set B.

(a) If f is an injection, then $x_1, x_2 \in A$ with $x_1 \neq x_2$ implies that $f(x_1) \neq f(x_2)$. In addition, any horizontal line drawn through the graph of f will intersect it at most once. Hence, a *horizontal line test* is used to determine whether or not a function is one to one.

(b) If f is a surjection, then every value in B has at least one preimage in A. That is, if $y_0 \in B$, then there exists $x_0 \in A$ such that $f(x_0) = y_0$. ■

Example 1.2.9

(a) The function $f_1:[-1,2] \to [0,4]$, defined by $f_1(x) = \frac{4}{3}(x+1)$, is a bijection. Why?

(b) The function $f_2:[-1,2] \to [0,4]$, defined by $f_2(x) = x^2$, is surjective but not injective. Why?

(c) The function $f_3:[-1,2] \to [0,4]$, defined by $f_3(x) = -x+2$, is injective but not surjective. Why? See Figure 1.2.4. ■

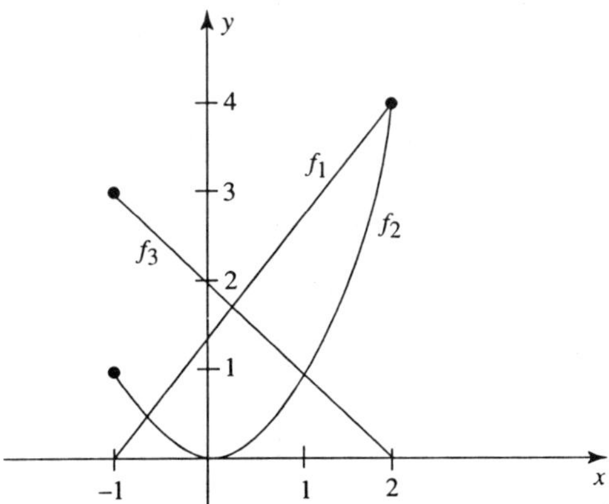

Figure 1.2.4

Definition 1.2.10. A function $f: A \to B$ is *increasing* if and only if for any $x_1, x_2 \in A$ with $x_1 < x_2$, $f(x_1) \leq f(x_2)$. A function f is *strictly increasing* if and only if for any $x_1 < x_2$ in A, $f(x_1) < f(x_2)$.

Decreasing and *strictly decreasing* functions are defined similarly. If a function is either increasing or decreasing, then it is *monotone*. Similarly, a *strictly monotone* function is either strictly increasing or strictly decreasing.

Definition 1.2.11. Consider a function f which maps set A into set B.

(a) f is said to be an *even* function if and only if $f(-x) = f(x)$ for all $x \in A$.

(b) f is said to be an *odd* function if and only if $f(-x) = -f(x)$ for all $x \in A$.

Example 1.2.12

(a) The function $f: \Re \to \Re$, defined by $f(x) = x^2$, is even, since $f(-x) = (-x)^2 = x^2 = f(x)$. However, $f: [-1,3) \to \Re$, defined by $f(x) = x^2$, is not even, since $2 \in [-1,3)$ but $-2 \notin [-1,3)$. The trigonometric function $\cos x$ is another example of an even function. Recall that *trigonometric functions* are functions that can be written in terms of $\sin x$ and $\cos x$. Geometrically, any even function is *symmetric with respect to the vertical axis.*

(b) The function $f: \Re \to \Re$, defined by $f(x) = x^3$, is odd, since $f(-x) = (-x)^3 = x^3 = -f(x)$. Geometrically, any odd function is *symmetric with respect to the origin.* That is, if a straight line drawn through the point $(0,0)$ intersects the function at a point (a,b), then it must also intersect the function at the point $(-a,-b)$ (see Figure 1.2.5). Functions $\sin x$ and $\arctan x$ are other examples of odd functions. In addition, if $x = 0$ is in the domain of an odd function f, then $f(0) = 0$. Why?

(c) The function $f(x) = 2x - 3$ is neither even nor odd on any domain. Why? ■

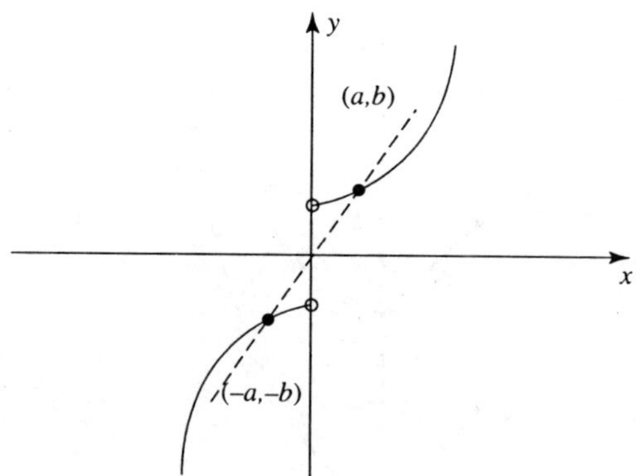

Figure 1.2.5

Definition 1.2.13. The *absolute value* of any real number x, denoted by $|x|$, is a function defined by $|x| = \begin{cases} x, & \text{if } x \geq 0 \\ -x, & \text{if } x < 0. \end{cases}$

For example, $|3| = 3$ and $|-3| = 3$. Therefore, the domain for the absolute value function is the set of all real numbers, whereas the range is all nonnegative real numbers. Note that $|x| = \sqrt{x^2}$ for any real number x. See Exercise 3(c) in Section 1.7.

Definition 1.2.14. Suppose that a function $f: A \to B$. Then f is *bounded* (on A) if and only if there exists a real number M_1 such that $|f(x)| \le M_1$ for all $x \in A$. The number M_1 is called a *bound* for f. If no such M_1 exists, then f is *unbounded*. Furthermore, f is *bounded above* if and only if there exists a real number M_2 such that $f(x) \le M_2$ for all $x \in A$. And f is *bounded below* if and only if there exists a real number M_3 such that $f(x) \ge M_3$ for all $x \in A$. The numbers M_2 and M_3 are called *upper bounds* and *lower bounds* of f, respectively. The smallest of all upper bounds, if such an upper bound exists, is called the *least upper bound*, or *supremum*, for f and is denoted by $\sup f = \sup\limits_{x \in A} f(x)$. And the largest of all lower bounds, if such a lower bound exists, is called the *greatest lower bound*, or *infimum*, for f and is denoted by $\inf f = \inf\limits_{x \in A} f(x)$.

See Theorem 1.6.7 for more about the supremum of a set.

Definition 1.2.15. Consider a function f which maps set A into set B.

(a) A point $x_0 \in A$ is a *root* of f if and only if $f(x_0) = 0$. We say that f *vanishes* at $x_0 \in A$ if and only if x_0 is a root of f. In addition, f is *identically zero*, denoted by $f(x) \equiv 0$, if and only if $f(x) = 0$ for all $x \in A$. An identically zero function is also called a *zero function*.

(b) f has a *maximum* at a value $x_1 \in A$ if and only if $f(x) \le f(x_1)$ for all $x \in A$. The value $f(x_1)$ is the maximum of f and is denoted by $\max f = \max\limits_{x \in A} f(x)$.

(c) If $B = A$ and $f(x) = x$ for all $x \in A$, then f is called an *identity function*.

Minimum is defined similarly and denoted by $\min f$. A function f has an *extremum* at $x = x_1$ if and only if it has a maximum or a minimum at $x = x_1$. If $\max f$ exists and equals a value M, then $\sup f = M$ (see Exercise 5).

Example 1.2.16

(a) If a function $f: (0, \infty) \to \Re$ is defined by $f(x) = \dfrac{x-1}{x}$, then f is bounded above, not bounded below, $\sup f = 1$, $\max f$ does not exist, and $x = 1$ is a root of f (see Figure 1.2.7).

(b) Consider the function $g: \Re \to \Re$ defined by $g(x) = \dfrac{1}{x^2 + 1}$. Graph g and try to convince yourself that g is bounded, $\sup g = \max g = 1$, $\inf g = 0$, and $\min g$ does not exist. In addition, g has no roots. ■

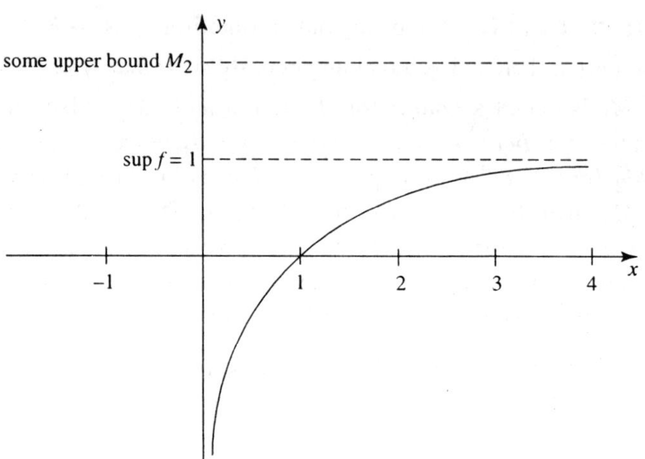

Figure 1.2.7

Definition 1.2.17

(a) A function $f:\Re \to \Re$ is called a *polynomial* if and only if it can be written as

$$f(x) = a_n x^n + a_{n-1} x^{n-1} + \cdots + a_1 x + a_0$$

where n is some fixed whole number called the *degree (order)* of the polynomial, and a_i for $i = 0, 1, 2, \ldots, n$ is a real number called a *coefficient* with $a_n \neq 0$. The number a_n is called the *leading coefficient*. The polynomial $f(x) = c$, where c is any real number, is called a *constant function*, or a polynomial of degree zero. If the degree of a polynomial f is 0 or 1, then the function f is a *linear function*, that is, its graph is a *line*. Part of the line which contains two distinct points and all the points in between is called a *line segment*. If the degree of a polynomial f is 2, then the function f is called a *quadratic function*, that is, its graph is a *parabola*. And if the degree of a polynomial f is 3, then the function f is called a *cubic function*. The quotient of two polynomials is called a *rational function*.

(b) A function $f:A \to \Re$, defined by $f(x) = [x] = \lfloor x \rfloor$, is called the *greatest integer function*, or *integer floor function*, and gives the greatest integer less than or equal to x (see Figure 1.2.8). Symbol $\lfloor \cdot \rfloor$ is referred to as a *floor bracket*.

(c) A function $f:A \to \Re$, defined by $f(x) = \lceil x \rceil$, is called the *integer ceiling function* and gives the least integer greater than or equal to x (see Figure 1.2.9).

(d) A function f is said to be *algebraic* if and only if its formula contains strictly algebraic operations, that is, addition, subtraction, multiplication, division, and radical extraction. Functions that are not algebraic are called *transcendental*. Trigonometric, logarithmic, and exponential functions are familiar transcendental functions. See Remark 3.3.6.

(e) If n^* is some natural number and the set $A = \{x \in N \mid x \geq n^*\}$, then a function $f: A \to \Re$ is called a *sequence*, with $f(n)$ usually denoted by a_n called the nth *term* of the sequence. In number theory, such a function is called *arithmetical* or a *number-theoretic function* (see Definition 2.1.1).

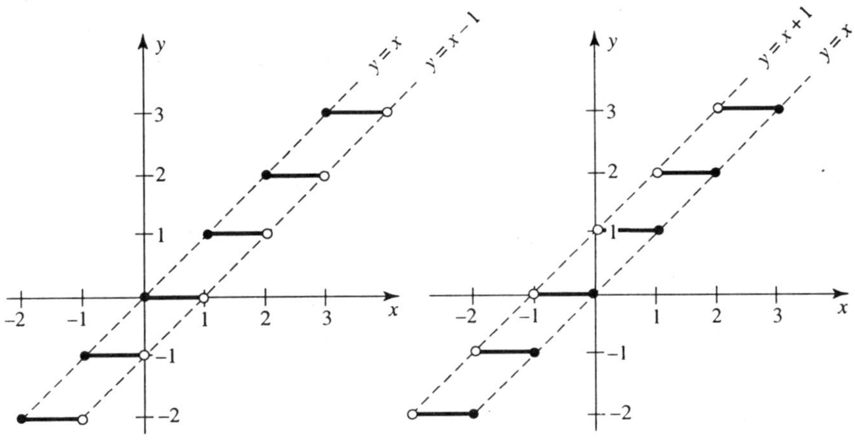

Figure 1.2.8 **Figure 1.2.9**

An important aspect of a linear function is its steepness called the slope. Since two distinct points determine a line, we define the *slope*[5] m of a line by

$$m = \frac{y_1 - y_2}{x_1 - x_2} = \frac{y_2 - y_1}{x_2 - x_1}$$

for any two distinct points $P(x_1, y_1)$ and $Q(x_2, y_2)$ on the line with $x_1 \neq x_2$. Observe that if f is a linear function $f(x) = a_1 x + a_0$, then $a_1 = m$, the slope of the line. Why? If $m > 0$ then the line is increasing, and if $m < 0$ then the line is decreasing. A constant function $f(x) = a_0$ has slope 0. A vertical line is not a function and has no slope. Why? *Parallel* lines have the same slope. And lines which are *perpendicular* have slopes whose product is -1, except when such lines are horizontal and vertical.

The greatest integer function is sometimes called a *postage* or *tax function*. Can you relate to that? Several of the trivial properties for $[x]$, x a real number, are the following:

(a) $x - 1 < [x] \leq x < [x] + 1$. (b) $0 \leq x - [x] < 1$.

(c) $[x + n] = [x] + n$, if $n \in Z$. (d) $[x] + [t] \leq [x + t] \leq [x] + [t] + 1$, $t \in \Re$.

(e) $[x] + [-x] = \begin{cases} 0, & \text{if } x \in Z \\ -1, & \text{otherwise.} \end{cases}$ (f) $\left[\frac{[x]}{n}\right] = \left[\frac{x}{n}\right]$, if $n \in N$.

(g) $x - [x]$ is the fractional part of x.

---

[5]We abbreviate slope by the letter m, which stands for the French word "monter" and means to climb.

(h) $-[-x]$ is the least integer greater than or equal to x.

(i) $[x]=\sum_{1\le k\le x} 1$, if $x\ge 0$, that is, the integer k is incremented, beginning with $k=1$, until k becomes larger than x, and for each such k, a 1 is written. Summing up the 1s we have $[x]$.

Definition 1.2.18. If functions $f, g: A \to B$, that is, functions f and g map A into B, then

(a) $(f \pm g)(x) \equiv f(x) \pm g(x)$. **(b)** $(fg)(x) \equiv f(x)g(x)$.

(c) $\left(\dfrac{f}{g}\right)(x) \equiv \dfrac{f(x)}{g(x)}$, provided $g(x) \neq 0$ for any $x \in A$.

(d) $\sqrt{f}(x) \equiv \sqrt{f(x)}$, provided $f(x) \ge 0$ for all $x \in A$.

(e) $(f \vee g)(x) \equiv \max\{f(x), g(x)\} \equiv \begin{cases} f(x), & \text{if } f(x) \ge g(x) \\ g(x), & \text{if } f(x) < g(x) \end{cases}$

$=\frac{1}{2}(f+g)(x)+\frac{1}{2}|(f-g)(x)|$.

(f) $(f \wedge g)(x) \equiv \min\{f(x), g(x)\} \equiv \begin{cases} f(x), & \text{if } f(x) \le g(x) \\ g(x), & \text{if } f(x) > g(x) \end{cases}$

$=\frac{1}{2}(f+g)(x)-\frac{1}{2}|(f-g)(x)|$.

In parts (e) and (f) the last equalities need to be verified. They are consequences and not definitions. The symbol $\equiv$ means that the expression on the left is defined by the expression on the right. Often, $f^2(x)$ represents $[f(x)]^2$. In general, $f^n(x)=[f(x)]^n$, but not for $n=-1$. The symbol $f^{-1}(x)$ does not represent $[f(x)]^{-1}$. The symbol $f^{-1}(x)$ represents the *inverse* of a function f. See Remark 1.8.2, part (d). And $[f(x)]^{-1}=\dfrac{1}{f(x)}$ represents the *reciprocal* of the function f. Figure 1.2.10 illustrates the meaning of $f \vee g$, where the symbol $\vee$ is read as "vee." Also, note that the symbol $\wedge$ is read as "wedge."

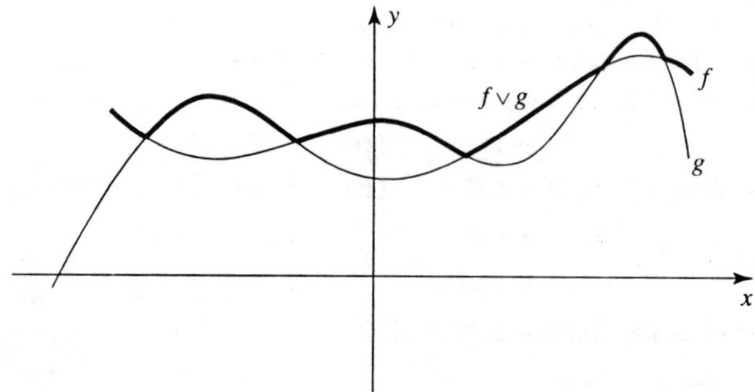

Figure 1.2.10

Definition 1.2.19. Suppose $f: A \to B$ and $g: B \to \Re$. Then the *composition (composite) function* of g and f, $g \circ f$, reads as g circle f, maps A into $\Re$, and is defined by $(g \circ f)(x) = g(f(x))$.

Example 1.2.20. Let $f(x) = x^2 + 1$ and $g(x) = 2x - 3$. Then $(g \circ f)(x) = g(f(x)) = g(x^2 + 1) = 2(x^2 + 1) - 3 = 2x^2 - 1$. ■

In general, $f \circ g$ is not the same as $g \circ f$ (see Exercise 11). A special group of functions for which $(f \circ g)(x) = (g \circ f)(x) = x$ will be discussed in Section 1.8. Figure 1.2.11 demonstrates the meaning of $g \circ f$.

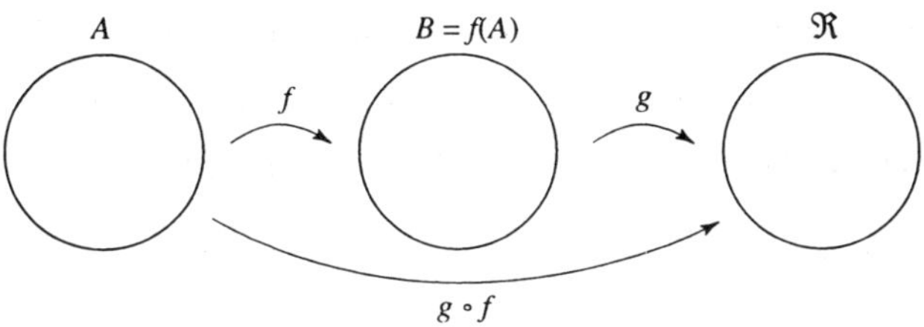

Figure 1.2.11

Many functions can be obtained from simpler functions by shifting and reflecting these simpler functions across the x and y axes.

Exercises 1.2

1. Give two examples of relations that are not functions.

2. For each of the given functions, f, find the largest domain possible.

(a) $f(x) = \dfrac{1-x}{\sqrt{x+1}}$ **(b)** $f(x) = \dfrac{\sqrt{x-1}}{x-2}$

(c) $f(x) = \dfrac{|x|}{x}$ **(d)** $f(x) = \dfrac{x^2 - 1}{x^2 + x - 2}$

3. Determine whether the given function, f, is an injection, surjection, bijection, or none of these. Explain clearly.

(a) $f: \Re \to \Re$, where $f(x) = 2x - 1$.

(b) $f: \{x \mid x \neq 1\} \to \Re$, where $f(x) = \dfrac{x^2 - 1}{x - 1}$.

(c) $f: \Re \to \Re$, where $f(x) = \sqrt[3]{x}$.

(d) $f: [-1, 1] \to [0, 4)$, where $f(x) = x^2$.

4. Determine whether the given function, f, is increasing, strictly increasing, decreasing, strictly decreasing, bounded, bounded above, or bounded below. Also find $\sup f$, $\inf f$, $\max f$, and $\min f$, if they exist.

(a) $f:[0,\infty)\to\mathfrak{R}$, where $f(x)=\dfrac{x}{x+1}$.

(b) $f=\left\{(x,y)\,\middle|\, x\in N \text{ and } y=\dfrac{1}{x}\right\}$.

(c) $f=\left\{(x,y)\,\middle|\, x\in N \text{ and } y=(-1)^x\left(1+\dfrac{1}{x}\right)\right\}$.

(d) $f:\mathfrak{R}^+\to\mathfrak{R}$, where $f(x)=\dfrac{1}{x}$.

5. Prove that if a function f has a maximum, then $\sup f$ exists and $\max f=\sup f$.

6. Determine whether the given function f is even, odd, or neither.

(a) $f(x)=2|x|-x^2$ **(b)** $f(x)=\sqrt{x}+1$

(c) $f(x)=\dfrac{x^3-\sin x}{2x}$ **(d)** $f(x)=\sqrt[3]{x}-x$

7. **(a)** If $f,g:A\to\mathfrak{R}$ are even, prove that $f+g$ is even.

(b) If $f,g:A\to\mathfrak{R}$ are odd, prove that fg is even.

(c) Is there a function f that is both even and odd?

8. Graph the given functions:

(a) $f(x)=\lfloor x\rfloor-\lceil x\rceil$ **(b)** $f(x)=[\![2x]\!]$

(c) $f(x)=2|x|$ **(d)** $f(x)=1-|x-2|$

(e) $f(x)=x-\lfloor x\rfloor$, with $-2\le x\le 2$ **(f)** $f(x)=|x+1|+|x-2|$

(g) $f(x)=\dfrac{x+|x|}{2}$ **(h)** $f(x)=\begin{cases}x, & \text{if } x\in Z\\ 0, & \text{if } x\notin Z\end{cases}$

(i) $f(x)=\lfloor 2x\rfloor-2\lfloor x\rfloor$, with $-2\le x\le 2$

(j) $f(x)=(-1)^n$, where $n\le x<n+1$, $n\in Z$

9. If $f(x)=x^2$ and $g(x)=2-x^2$, find $(f\vee g)(x)$ and $(f\wedge g)(x)$.

10. If $f(x)=x^2-3$ and $g(x)=2x+1$, find $(f\circ g)(-2)$.

11. Find two functions, f and g, so that $f\circ g\ne g\circ f$.

12. Suppose that f is an odd function and g is an even function. Show that the functions $g\circ f$, $g\circ g$, and $f\circ g$ are all even.

13. **(a)** If both functions $f:A\to B$ and $g:B\to C$ are injective, prove that $g\circ f:A\to C$ is injective.

(b) If both functions $f:A\to B$ and $g:B\to C$ are surjective, prove that $g\circ f:A\to C$ is surjective.

14. Define what is meant by a function that is symmetric with respect to the point (a,b).

15. Verify that every function defined on a symmetric interval about the origin can be written as the sum of an even and odd function.

16. If $f(n+1)=\dfrac{3f(n)+1}{3}$ and $f(1)=0$, find $f(79)$.

1.3* Mathematical Induction

Because the development of the set of natural numbers N is a long and rigorous process, we will proceed with the familiarity of the existence of N, thus having at our fingertips all the arithmetic operations of addition and multiplication, as well as the meaning of one number being less than another. Among the axioms used in the development of the natural numbers, the well-ordering principle is a foundation for the natural numbers and a basis for mathematical induction. By an *axiom*, that is, *postulate,* we mean an assumption or property that is built on to obtain other properties.

AXIOM 1.3.1. *(Well-Ordering Principle) Every nonempty subset of N has a smallest element.*

Throughout mathematics we must determine whether or not statements involving natural numbers are correct. For example, consider the following statements:

(a) The sum of two even natural numbers is even.

(b) $1+2+3+\cdots+n=\dfrac{n(n+1)}{2}$, for all $n\in N$.

(c) $n=n+1$, for all $n\geq 3$.

(d) n^2+n+41 is a prime number, for any $n\in N$.

(e) For every $n>2$, there are no nonzero integers x, y, and z such that $x^n+y^n=z^n$.

Our first inclination is to substitute in a few numbers to see whether or not the expression makes sense. If a statement is false for one value, then it is false no matter what happens at the other values, in which case, no proof is necessary. To demonstrate a statement's invalidity, we simply substitute in a value that makes it false. Statement (c) above is false, since if $n=3$, $3\neq 3+1$. Obtaining true statements after many values leads us to suspect that the statement (a) is true for all specified values. However, concluding anything without a valid proof is incorrect. In proving statement (a), we must start with any two arbitrary even natural numbers, say n and m, and show that $n+m$ is even. Proceeding with the proof, since n and m are even, natural numbers r and s exist such that $n=2r$ and $m=2s$. Therefore, $n+m=2r+2s=2(r+s)$. And since 2 is a factor, $n+m$ is divisible by 2 and thus is even. The proof of statement (a) is now complete. Statement (d) can be misleading, since after trying the values 1 through 39, n^2+n+41 continues to result in a prime number. However, substituting the number 40 into the expression gives $40^2+40+41=1681=41^2$, which is certainly not prime. Hence, statement (d) is false. From 1637 until 1995, statement (e) was a conjecture known as Fermat's[6] last theorem. A *conjecture* is a statement that mathematicians feel is true but are unable to prove, as was the case with Fermat's last theorem. No one had been able to verify that a solution of the given expression existed, nor could they prove its impossibility for all $n>2$. Fermat's last theorem finally became a theorem rather than a conjecture in 1995.[7] Recall that if $n=2$, statement (e) reduces to the vast study of Pythagorean triangles. Finally, statement (b), which can be verified in several ways, can be rewritten as

$$\sum_{k=1}^{n} k = \frac{n(n+1)}{2}.$$

This equality is read as "sigma from $k=1$ to n of k equals $\frac{n(n+1)}{2}$." The symbol $\sum$ called *sigma*, represents summation. To expand $\sum_{k=1}^{n} k$, we progressively increment k by 1, beginning with 1 and ending with n, and then add the resulting terms. Clearly, for any two functions $f,g:N\to\Re$ with $m,n\in N$ and $1\le m\le n$, $\sum_{k=1}^{n} cf(k)=c\sum_{k=1}^{n} f(k)$, $\sum_{k=1}^{n}[f(k)\pm g(k)] = \sum_{k=1}^{n} f(k)\pm\sum_{k=1}^{n} g(k)$, and $\sum_{k=1}^{n} f(k) = \sum_{k=1}^{m} f(k) + \sum_{k=m+1}^{n} f(k)$. Example 1.3.3 uses a straightforward, relatively quick and popular method known as mathematical induction to prove statement (b).

THEOREM 1.3.2. *(Principle of Mathematical Induction) If $P(n)$ is a statement for each $n\in N$ such that*

(a) *$P(1)$ is true, and*

(b) *for each $k\in N$, if $P(k)$ is true, then $P(k+1)$ is true,*

then $P(n)$ is true for all $n\in N$.

In part (b), $P(k)$ is called the *induction hypothesis*. The principle of mathematical induction is often referred to as *Mathematical Induction*, or even *Induction* for short. The name "principle" is often used since mathematical induction is equivalent to the well-ordering principle. ,See Example 1.5.4 and Exercise 7 in Section 1.5. For now, let us assume that Theorem 1.3.2 is true and demonstrate its use.

---

[6]Pierre de Fermat (1601–1665) was a French lawyer whose hobby was mathematics. Fermat solved many calculus problems and became best known in number theory. His work, which inspired Newton, was not appreciated by other mathematicians during Fermat's life.

[7]Andrew Wiles, a British mathematician working at Princeton University, presented a "proof" of Fermat's last theorem in 1993, using a 200-page argument. He tried to achieve this result by proving that semistable elliptic curves are modular. The "gap" in his paper was closed by Wiles and Richard Taylor in 1995.

Example 1.3.3. Prove that $\sum_{k=1}^{n} k = \frac{n(n+1)}{2}$ using mathematical induction. See Exercise 10.

Proof. Let $P(n)$ be the statement $1+2+3+\cdots+n = \frac{n(n+1)}{2}$. We need to show that parts (a) and (b) of Theorem 1.3.2 are true. Clearly, $P(1)$ is true, since the left-hand side of the above statement degenerates to 1 and the right-hand side is also $\frac{1(1+1)}{2} = 1$. Next, assume that $P(k)$ is true for some $k \in N$. We will show that $P(k+1)$ is true. Thus, suppose that $1+2+3+\cdots+k = \frac{k(k+1)}{2}$. We need to show that

$$1+2+3+\cdots+(k+1) = \frac{(k+1)(k+2)}{2}.$$

Beginning with the left-hand side of the statement $P(k+1)$, we write

$$\begin{aligned} 1+2+3+\cdots+(k+1) &= (1+2+3+\cdots+k)+(k+1) && \text{(Rewrite)} \\ &= \frac{k(k+1)}{2}+(k+1) && \text{(Use induction hypothesis)} \\ &= \frac{(k+1)(k+2)}{2}. && \text{(Factor and simplify)} \end{aligned}$$

Hence, since $P(k+1)$ is true, we can conclude that $P(n)$ is true for all $n \in N$. ■

If in the statement $P(n)$, the smallest value of n is a natural number other than 1, for generality call it r; then condition (a) in Theorem 1.3.2 is replaced by (a)′, which says that $P(r)$ is true.

Checking condition (a) or (a)′ of Theorem 1.3.2, whichever applies, is essential to the validity of the induction hypotheses $P(k)$. For example, if $P(n)$ is the statement $n = n+1$ for all natural numbers $n \geq 3$, condition (b) of Theorem 1.3.2 would be satisfied, since if $k = k+1$, then certainly $k+1 = k+2$. Therefore, if condition (a)′ is not checked, we would be led to believe things that are false.

Example 1.3.4. Prove that for any fixed real number x and all $n \in N$,

$$x^{n+1} - 1 = (x-1)\left(x^n + x^{n-1} + \cdots + x + 1\right).$$

See Example 1.5.5 for a variation of the proof.

Proof. Although other methods of proof are available, we once again illustrate the use of mathematical induction. Let $P(n)$ be the statement to be proven. $P(1)$ says that $x^2 - 1 = (x-1)(x+1)$, which is true. Next assume that $P(k)$ is true, that is, that

$$x^{k+1} - 1 = (x-1)\left(x^k + x^{k-1} + \cdots + x + 1\right).$$

We will show that $P(k+1)$ is true, that is, that

$$x^{k+2}-1=(x-1)\left(x^{k+1}+x^{k}+\cdots+x+1\right).$$

To this end, we write

$$\begin{aligned}
x^{k+2}-1&=x\left(x^{k+1}-1\right)+x-1 && \text{(Rewrite)}\\
&=x\left[(x-1)\left(x^{k}+x^{k-1}+\cdots+x+1\right)\right]+(x-1) && \text{(Induction hypothesis)}\\
&=(x-1)\left(x^{k+1}+x^{k}+\cdots+x^{2}+x\right)+(x-1) && \text{(Simplify)}\\
&=(x-1)\left[\left(x^{k+1}+x^{k}+\cdots+x^{2}+x\right)+1\right]. && \text{(Factor)}
\end{aligned}$$

Therefore, $P(k+1)$ is true. Hence, $P(n)$ is true for all $n \in N$. ■

We say that an integer b is *divisible* by an integer $a \neq 0$ if there exists some integer c such that $b = ac$. If this is the case, we also say that a is a *divisor* of b, that a is a *factor* of b, or that b is a *multiple* of a. In Example 1.3.4, we proved not only the specified formula but also that $x-1$ is a factor of $x^{n+1}-1$. However, it was not necessary to prove the formula to show that $x-1$ is a factor of $x^{n+1}-1$. See Exercise 2(e). We often write the above proven expression as the *geometric sum*

$$1+x+x^{2}+\cdots+x^{n}=\frac{1-x^{n+1}}{1-x}=1+\sum_{k=1}^{n}x^{k}=\sum_{k=0}^{n}x^{k},\ \text{if } x\neq 1,$$

where in the last expression we assumed that $x^0 = 1$, even if $x = 0$. Why?

Definition 1.3.5. If n is a whole number, then n *factorial*, denoted by $n!$, is defined as follows:

(a) $0! = 1! = 1$, and

(b) $(n+1)! = (n+1)(n!)$, if $n \geq 1$.

In general, from Definition 1.3.5, we can also write that $n! = n(n-1)(n-2)\cdot \cdots \cdot(2)(1)$. If k and n are integers such that $0 \leq k \leq n$, we are often interested in the number represented by $\dfrac{n!}{k!(n-k)!}$. This number, called the *binomial coefficient* and denoted by $\dbinom{n}{k}$, is widely used in probability theory as well as in many other areas of mathematics. Other common notations are ${}_nC_k$ and C_k^n, representing the number of combinations of n objects taken k at a time.

LEMMA 1.3.6. *If k and n are integers such that $0 \leq k \leq n$, prove that*

$$\binom{n}{k}+\binom{n}{k-1}=\binom{n+1}{k}.$$

See Exercise 4(h).

Proof

$$\begin{aligned}\binom{n}{k}+\binom{n}{k-1} &= \frac{n!}{(k!)(n-k)!}+\frac{n!}{(k-1)![n-(k-1)]!} \\ &= \frac{n!}{(k!)(n-k)!}+\frac{n!}{(k-1)!(n+1-k)!} \\ &= \frac{n!(n+1-k)}{k!(n-k)!(n+1-k)}+\frac{n!(k)}{(k-1)!(n+1-k)!(k)} \\ &= \frac{n!(n+1-k)}{k!(n+1-k)!}+\frac{n!(k)}{k!(n+1-k)!} \\ &= \frac{n!}{k!(n+1-k)!}[(n+1-k)+k] \\ &= \frac{(n+1)!}{k!(n+1-k)!}=\binom{n+1}{k}.\end{aligned}$$

■

THEOREM 1.3.7. *(Binomial Theorem[8]) If a and b are any real numbers and $n \in N$, then*

$$(a+b)^n = \sum_{k=0}^{n}\binom{n}{k}a^{n-k}b^k = a^n + na^{n-1}b + \frac{1}{2}n(n-1)a^{n-2}b^2 + \cdots + b^n.$$

Proof. Again proving this theorem by mathematical induction, let $P(n)$ represent the expression

$$(a+b)^n = \sum_{i=0}^{n}\binom{n}{i}a^{n-i}b^i.$$

Obviously, $P(1)$ is true. Next, suppose that $P(k)$ is true, i.e. $(a+b)^k = \sum_{i=0}^{k}\binom{k}{i}a^{k-i}b^i$.

We will prove that $P(k+1)$ is true, that is, $(a+b)^{k+1} = \sum_{i=0}^{k+1}\binom{k+1}{i}a^{k+1-i}b^i$. To this end, we write

$$(a+b)^{k+1} = (a+b)^k(a+b) = \left[\sum_{i=0}^{k}\binom{k}{i}a^{k-i}b^i\right](a+b) \qquad \text{(Induction hypothesis)}$$

---

[8]This theorem is sometimes known as *Newton's Binomial Theorem*. The case $n=2$ of the binomial theorem was known to Euclid in 300 B.C. Other cases were likely discovered by Omar Khayyám (c.1044–1123), an Iranian poet, mathematician, and astronomer, who measured the length of a year.

$$
\begin{aligned}
&= \sum_{i=0}^{k}\binom{k}{i}a^{k+1-i}b^{i} + \sum_{i=0}^{k}\binom{k}{i}a^{k-i}b^{i+1} && \text{(Distribute)}\\
&= \sum_{i=0}^{k}\binom{k}{i}a^{k+1-i}b^{i} + \sum_{j=1}^{k+1}\binom{k}{j-1}a^{k-(j-1)}b^{(j-1)+1} && \text{(Let } j=i+1\text{)}\\
&= \sum_{i=0}^{k}\binom{k}{i}a^{k+1-i}b^{i} + \sum_{i=1}^{k+1}\binom{k}{i-1}a^{k+1-i}b^{i} && \text{(Simplify and rename)}\\
&= a^{k+1} + \sum_{i=1}^{k}\binom{k}{i}a^{k+1-i}b^{i} + \sum_{i=1}^{k}\binom{k}{i-1}a^{k+1-i}b^{i} + \binom{k}{k}b^{k+1} && \text{(Split)}\\
&= a^{k+1} + \sum_{i=1}^{k}\left[\binom{k}{i}+\binom{k}{i-1}\right]a^{k+1-i}b^{i} + b^{k+1} && \text{(Combine)}\\
&= a^{k+1} + \sum_{i=1}^{k}\binom{k+1}{i}a^{k+1-i}b^{i} + b^{k+1} && \text{(Lemma 1.3.6)}\\
&= \sum_{i=0}^{k+1}\binom{k+1}{i}a^{k+1-i}b^{i}. && \text{(Combine)}
\end{aligned}
$$

Thus, $P(n)$ is true for all $n \in N$. ■

The right-hand side of the binomial theorem is the *binomial expression* for $(a+b)^n$. Theorem 1.3.7 can be generalized to powers other than the natural numbers n, to form the so-called binomial series. The coefficients in the expansion of $(a+b)^n$, $n \in N$, can be computed with the use of *Pascal's*[9] *triangle*. See Figure 1.3.1.

In Example 1.3.3, if $a_n \equiv \frac{1}{2}n(n+1)$, the values of a_n are given independently of each other, that is, *explicitly*. However, this is not always the case. Often the first or first few terms are given, with the rest obtainable from those given according to a certain formula called *recurrence*, or a *recursion formula*. If this is the case, then the terms are defined *recursively*, that is, *inductively*, or *implicitly*. For example, suppose that we are given $a_1 = 1$ and $a_{n+1} = a_n + 1$ for all $n \in N$. Then if $n = 1$, the second term $a_2 = a_1 + 1 = 1 + 1 = 2$. The third term $a_3 = a_2 + 1 = 2 + 1 = 3$, and we are led to the conclusion that $a_n = n$. This fact, however, needs to be proven.

---

[9]Blaise Pascal (1623–1662) was a great French geometer, combinatorist, physicist, and philosopher, who laid the foundation for the theory of probability. Blaise's mother died when he was 3. His father Etienne, due to his own views, forbade Blaise to study mathematics before the age of 15. But due to Blaise's geometrical discoveries at the age of 12, which were performed in secret, Etienne changed his educational methods and promoted his son's study of mathematics. When Etienne became a tax collector, Blaise invented the first digital calculator to help his father with his work. In 1647 Blaise proved that the vacuum existed, which fact was not well received by the world at that time. Blaise, a sickly person bedridden at times, had migraine headaches from youth. Blaise lived in great pain with malignant growths in his stomach which killed him at 39 after spreading to his brain. Pascal's triangle appeared in works of Yang Hui (c.1270) and Chu Shih-chie (1303).

$n = 0$						1						
$n = 1$					1		1					Pascal's triangle
$n = 2$				1		2		1				
$n = 3$			1		3		3		1			
$n = 4$		1		4		6		4		1		
$n = 5$	1		5		10		10		5		1	

$\vdots$

Figure 1.3.1

Example 1.3.8. If $a_1 = 1$ and $a_{n+1} = a_n + 1$ for all $n \in N$, prove that $a_n = n$.

Proof. Using mathematical induction, let $P(n)$ be the statement $a_n = n$. Certainly, $P(1)$ is true, since $a_1 = 1$ by assumption. Next, suppose that $P(k)$ is true for some $k \in N$, that is, $a_k = k$. We need to prove that $P(k+1)$ is true, that is, that $a_{k+1} = k+1$. Thus, we write $a_{k+1} = a_k + 1 = k + 1$. Therefore, $P(k+1)$ is true, and so $P(n)$ is true for all $n \in N$. ■

In Example 1.3.8, the recursion was simple since the term a_{n+1} depended on only one previous term. The same was true in Definition 1.3.5. If the term a_{n+1} depends on more than one previous term, then the following second principle of mathematical induction is invoked to carry out the proof. Proof of Theorem 1.3.9 is in Exercise 6 of Section 1.5.

THEOREM 1.3.9. *(Second Principle of Mathematical Induction) If $P(n)$ is a statement for each $n \in N$, and m is some fixed natural number such that*

(a) $P(1), P(2), \ldots, P(m)$ are true, and

(b) for each natural number $k \geq m$, if $P(i)$ is true for each $i = 1, 2, \ldots, k$, then $P(k+1)$ is true,

then $P(n)$ is true for all $n \in N$.

Example 1.3.10. Suppose that $a_1 = 1$, $a_2 = 3$, and $a_{n+1} = 3a_n - 2a_{n-1}$ for all natural numbers $n \geq 2$. Find an explicit formula for a_n, and then prove your assertion.

Proof. We first substitute in values of n and try to find a pattern in the outcome.

$$\begin{aligned}
&a_1 = 1 \\
&a_2 = 3 \\
&a_3 = 3a_2 - 2a_1 = 3(3) - 2(1) = 7 \\
&a_4 = 3a_3 - 2a_2 = 3(7) - 2(3) = 15 \\
&\vdots
\end{aligned}$$

It appears that the resulting values are 1 less than a power of 2. Thus, let $P(n)$ be the statement $a_n = 2^n - 1$ for all $n \in N$, and proceed with the proof using the second principle of mathematical induction with $m = 2$. To this end, the two given values $P(1)$

and $P(2)$ are satisfied. Now suppose $P(k)$ is true for some integer $k > 2$ and for all preceding values up to k. That is, assume that $a_i = 2^i - 1$ for all $i = 1, 2, \dots, k$. We need to show that $P(k+1)$ is true, that is, $a_{k+1} = 2^{k+1} - 1$. So we write

$$\begin{aligned} a_{k+1} &= 3a_k - 2a_{k-1} && \text{(Given)} \\ &= 3\left(2^k - 1\right) - 2\left(2^{k-1} - 1\right) && \text{(Induction hypothesis)} \\ &= 3\left(2^k\right) - 3 - 2\left(2^{k-1}\right) + 2 && \text{(Distribute)} \\ &= 2\left(2^k\right) - 1 && \text{(Collect)} \\ &= 2^{k+1} - 1. && \text{(Rewrite)} \end{aligned}$$

Therefore, $P(k+1)$ is true. Hence, by Theorem 1.3.9, $P(n)$ is true for all $n \in N$. ■

Remark 1.3.11. The sequence $\{a_n\}$, given recursively in Example 1.3.10, can be classified as linear, homogeneous, and with constant coefficients since it can be written in the form

$$a_{n+1} + k_1 a_n + k_2 a_{n-1} = 0,$$

where k_1 and k_2 are real constants not both zero. The recursion formula is homogeneous because the right-hand side of the equality equals zero. This sequence can be explicitly written in the form $a_n = cr^n$, where the nonzero numbers c and r need to be determined. The following example demonstrates this. Observe that the sequence in Example 1.3.8 is not homogeneous. ■

Example 1.3.12. Use the idea from Remark 1.3.11 to find an explicit formula for the sequence $\{a_n\}$, given in Example 1.3.10.

Answer. Set $a_n = cr^n$ in the recursion formula $a_{n+1} = 3a_n - 2a_{n-1}$, and rewrite it to obtain

$$cr^{n+1} - 3cr^n + 2cr^{n-1} = 0.$$

Next, divide both sides by cr^{n-1} to obtain the so-called *characteristic equation* $r^2 - 3r + 2 = 0$. Solutions of this equation are $r = 1$ and $r = 2$. Thus, $a_n = c_1$ or $a_n = c_2 2^n$, where the previous c is now denoted by c_1 or c_2. In fact, a_n can be written as $a_n = c_1 + c_2 2^n$. Why? Next, we need to find c_1 and c_2. By knowing the first two terms of the sequence, we observe that $a_1 = 1$ implies that $c_1 + 2c_2 = 1$, and $a_2 = 3$ implies that $c_1 + 4c_2 = 3$. Solving these two equations simultaneously, we find that $c_1 = -1$ and $c_2 = 1$. Thus, $a_n = -1 + 2^n$. ■

Can the above method be used on a sequence given in Example 1.3.8? Explain. Cases often arise in which the characteristic equation contains more terms, nonhomogeneous terms, or solutions that are not distinct real values. In such cases, lengthy discussions arise. Because this topic is covered in depth in introductory courses on differential equations, elaboration here is unnecessary.

The following theorem is known as *forward and backward induction.*

THEOREM 1.3.13. *If $P(n)$ is a statement for each $n \in N$ such that*

(a) $P(2^m)$ *is true for every $m \in N$, and*

(b) *for each $k \in N$ if $P(k+1)$ is true, then $P(k)$ is true,*

then $P(n)$ is true for all $n \in N$.

That 2^m in part (a) of Theorem 1.3.13 can be replaced by s^m for any positive integer $s \geq 2$ can be verified. In fact, 2^m can be replaced by any bounded strictly increasing sequence of positive integers. See Chapter 2 for a study of sequences. For a proof of Theorem 1.3.13 see Exercise 8 in Section 1.5.

Definition 1.3.14. A function $f:[a,b] \to \Re$ is *(midpoint) convex* on $[a,b]$ if and only if $f\left(\frac{s+t}{2}\right) \leq \frac{f(s)+f(t)}{2}$ for any $s,t \in [a,b]$.

Intuitively speaking, a function f is convex on an interval $[a,b]$ if and only if for any two points $(s, f(s))$ and $(t, f(t))$, the point $\left(\frac{s+t}{2}, \frac{f(s)+f(t)}{2}\right)$ on the line segment joining the points $(s, f(s))$ and $(t, f(t))$ is greater than or equal to the average value $\left(\frac{s+t}{2}, f\left(\frac{s+t}{2}\right)\right)$. See Figure 1.3.2.

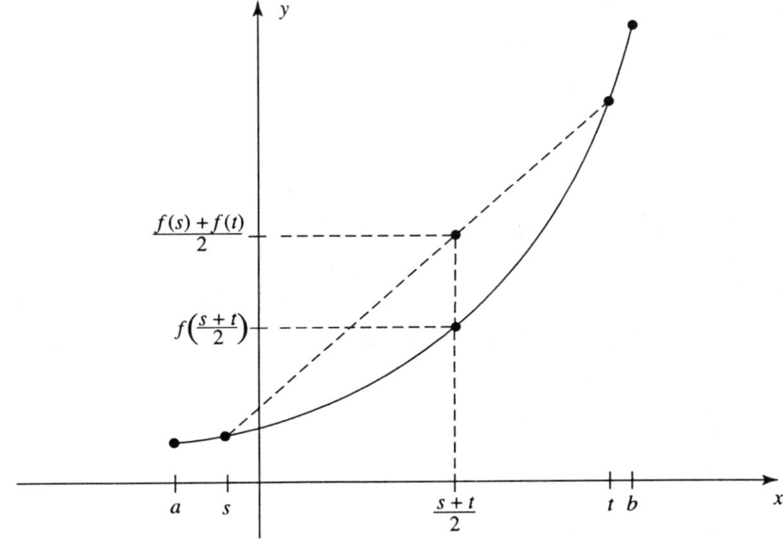

Figure 1.3.2

THEOREM 1.3.15. *(Jensen's[10] inequality) If a function f is convex on $[a,b]$ and $x_1, x_2, \dots, x_n$ are n numbers in $[a,b]$, then*

$$f\left(\frac{x_1 + x_2 + \cdots + x_n}{n}\right) \le \frac{f(x_1) + f(x_2) + \cdots + f(x_n)}{n}.$$

The proof of Jensen's inequality is left as Exercise 8. For further information on convex functions, see Part 5 of Section 5.6. Jensen's inequality appears again in Exercise 27(a) of Section 5.3.

Exercises 1.3

1. **(a)** Prove that the sum of two odd integers is even.
(b) Prove that the product of two odd integers is odd.

2. Use mathematical induction to prove the given statements.

(a) $n^2 + n$ is divisible by 2 for all $n \in N$.

(b) $\displaystyle\sum_{k=1}^{n} k^2 = \frac{n(n+1)(2n+1)}{6}$.

(c) $\displaystyle\sum_{k=1}^{n} k^3 = \left[\frac{n(n+1)}{2}\right]^2$. **(d)** $\displaystyle\sum_{k=1}^{n} (2k-1) = n^2$.

(e) For any fixed integers a and b, $a-b$ is a factor of $a^n - b^n$ for all $n \in N$.

(f) If $x \in (0,1)$ is a fixed real number, then $0 < x^n < 1$ for all $n \in N$.

(g) $2^{n-1} \le n! \le n^n$ for all $n \in N$.

(h) $2^n < n!$ for all natural numbers $n \ge 4$.

(i) $\cos n\pi = (-1)^n$ for all $n \in N$.

(j) $\dfrac{n^5}{5} + \dfrac{n^3}{3} + \dfrac{7n}{15}$ is an integer for every $n \in N$.

(k) $(n-1)(n)(n+1)$ is divisible by 6 for every $n \in N$.

(l) $n^5 - n$ is divisible by 5 for every $n \in N$.

(m) $2^{2n+1} + 1$ is divisible by 3 for every $n \in N$.

(n) If $x_1, x_2, \dots, x_n$ are n real numbers in the interval $[a,b]$, then $a \le \dfrac{x_1 + x_2 + \cdots + x_n}{n} \le b$ for all $n \in N$.

(o) $\dfrac{n}{n+1} \ge \dfrac{1}{2}$ for all $n \in N$.

---

[10] Lohan Jensen (1859–1925), a Danish analyst, algebraist, and engineer, contributed greatly to the study of convex functions. He worked for a telephone company.

(p) $0 < \dfrac{2^n}{n!} \le 2\left(\dfrac{2}{3}\right)^{n-2}$ for all natural numbers $n \ge 3$.

(q) $\displaystyle\sum_{k=1}^{n} k(k!) = (n+1)! - 1.$ **(r)** $\displaystyle\sum_{k=1}^{n} \frac{1}{k(k+1)} = \frac{n}{n+1}$ for all $n \in N$.

(s) $\displaystyle\sum_{k=1}^{n} \frac{1}{\sqrt{k}} > \sqrt{n}$ for all natural numbers $n \ge 2$.

(t) $\displaystyle\sum_{k=1}^{n} \frac{1}{k^2} \le 2 - \frac{1}{n}$ for all $n \in N$.

(u) $\displaystyle\sum_{k=1}^{n} \frac{1}{k^2} \le \frac{7}{4} - \frac{1}{n}$ for all natural numbers $n \ge 2$.

(v) $\displaystyle 2\sum_{k=1}^{n} \frac{1}{k^3} < 3 - \frac{1}{n^2}$ for all natural numbers $n \ge 2$.

(w) $\displaystyle\sum_{k=1}^{n} \frac{1}{(2k-1)(2k+1)} = \frac{n}{2n+1}$ for all $n \in N$.

3. **(a)** If x is some fixed real number, for each $n \in N$ find the sum of $1 - x - x^2 - x^3 - \cdots - x^n$.

(b) Use Exercise 2(b) to find the sum of $\displaystyle\sum_{k=20}^{30} k^2$.

4. **(a)** Compute $\dbinom{6}{4}$.

(b) Verify that $\dbinom{n}{0} = \dbinom{n}{n} = 1$ for all $n \in N$.

(c) Verify that $\dbinom{n}{1} = \dbinom{n}{n-1} = n$ for all $n \in N$.

(d) Verify that $\dbinom{n}{k} = \dbinom{n}{n-k}$ for all integers k and n with $0 \le k \le n$.

(e) Prove that $\displaystyle\sum_{k=0}^{n} (-1)^k \binom{n}{k} = 0.$

(f) Find the coefficient of x^3 in the expansion of $(2-x)^7$.

(g) Use the binomial theorem to prove that $2^n > \dfrac{n(n-1)(n-2)}{6}$ for all $n \in N$.

(h) Prove Lemma 1.3.6 by writing $(1+x)^{n+1} = (1+x)^n(1+x)$ and assuming that the binomial theorem holds.

(i) Find the sum $\displaystyle\sum_{k=1}^{50} 2^k$. **(j)** Verify that $\dbinom{n}{k} \le \dfrac{n^k}{k!}$.

5. **(a)** Use mathematical induction to prove *Bernoulli's*[11] *inequality* that

$$1+nx \le (1+x)^n$$

for all $n \in N$, with $x > -1$ a fixed real number. What if $x = -1$?

(b) Use the binomial theorem to prove Bernoulli's inequality for all $n \in N$ with $x \ge 0$ a fixed real number.

6. Use mathematical induction to prove the given statements.

(a) If $a_1 = 1$, $a_2 = 2$, and $a_{n+1} = a_n + a_{n-1}$ for all natural numbers $n \ge 2$, then $a_n < \left(\frac{7}{4}\right)^n$ for all $n \in N$.

(b) If $a_1 = 1$, $a_2 = 2$, and $a_{n+1} = -a_n + 6a_{n-1}$ for all natural numbers $n \ge 2$, then $a_n = 2^{n-1}$ for all $n \in N$.

(c) If $a_1 = 1$ and $a_{n+1} = \sqrt{3a_n + 1}$ for all $n \in N$, then $a_n < a_{n+1}$ for all $n \in N$.

(d) If $a_1 = 1$ and $a_{n+1} = \sqrt{3a_n + 1}$ for all $n \in N$, then $a_n < \frac{7}{2}$ for all $n \in N$.

7. Write the following recursively written sequences explicitly, and then prove your assertion. (Use patterns or the method used in Example 1.3.12 where applicable.)

(a) $a_1 = 1$, $a_{n+1} = a_n + 2$ for all $n \in N$.

(b) $a_1 = 1$, $a_{n+1} = \frac{n+1}{n} a_n$ for all $n \in N$.

(c) $a_1 = 2$, $a_{n+1} = 2a_n$ for all $n \in N$. (See Exercise 11(f) in Section 2.4.)

(d) $a_1 = 1$, $a_{n+1} = a_n + n + 1$ for all $n \in N$.

(e) $a_1 = 1$, $a_{n+1} = a_1 + a_2 + \cdots + a_n$ for all $n \in N$.

(f) $a_1 = 1$, $a_{n+1} = \left[1 - \frac{1}{(n+1)^2}\right] a_n$ for all $n \in N$.

(g) $a_1 = 1$, $a_{n+1} = 2a_n + 1$ for all $n \in N$.

(h) $a_1 = 1$, $a_2 = \frac{1}{3}$, $a_{n+1} = \frac{5}{6}a_n - \frac{1}{6}a_{n-1}$ for all $n \ge 2$. (See Exercise 11(j) in Section 2.4.)

(i) $a_1 = 1$, $a_2 = 1$, $a_{n+2} = a_n + a_{n+1}$ for all $n \in N$. (See Part 1 of Section 1.10.)

8. Use Theorem 1.3.13 to prove Theorem 1.3.15.

9. Use Theorem 1.3.13 to prove that if $x_1, x_2, \ldots, x_n$ are positive real numbers, then

---

[11]Johann Bernoulli (1667–1748), born to a Swiss family with several generations of outstanding mathematicians, was Euler's teacher and L'Hôpital's tutor. Bernoulli developed a large portion of calculus, parts of which were printed in L'Hôpital's calculus textbook. Optical phenomena, which dealt with reflection and refraction, orthogonal trajectories, and rectification of curves, were just a few of Bernoulli's many other contributions.

$$\sqrt[n]{x_1 \cdot x_2 \cdot \cdots \cdot x_n} \le \frac{x_1 + x_2 + \cdots + x_n}{n}$$

for all $n \in N$. See part (c) of Theorem 1.7.1 and Part 3 of Section 1.10.

10. Reprove the formula in Example 1.3.3 by writing $S = 1+2+3+\cdots+n$, and $S = n+(n-1)+\cdots+2+1$, and adding these expressions together.

1.4* Equivalent and Countable Sets

Definition 1.4.1. Two sets, A and B, are *equivalent*, denoted by $A \sim B$, if and only if a bijection exists from A onto B. See Definition 1.2.7.

Definition 1.4.2

(a) A set, A, is *finite* if and only if $A = \phi$ or $n^* \in N$ exists such that $A \sim \{1, 2, 3, \ldots, n^*\}$. If A is not finite, then it is *infinite*.

(b) A set, A, is *countably infinite* if and only if $A \sim N$.

(c) A set, A, is *countable* if and only if A is finite or countably infinite. If A is not countable, then it is *uncountable*.

Remark 1.4.3. For the case in which A and B are finite, $A \sim B$ if and only if A has the same number of elements as B, but for the case in which both A and B are infinite, counting elements is irrelevant. For example, if $A = W$ (the set of whole numbers) and $B = N$ (the set of natural numbers), then A has one more element than B, but $A \sim B$, because we can pair off every element of W with every element of N by setting up a bijection with the function $f: W \to N$, where $f(x) = x + 1$. ■

THEOREM 1.4.4. *If A, B, and C are sets, then*

(a) *(Reflexive property) $A \sim A$.*

(b) *(Symmetric property) $A \sim B$ if and only if $B \sim A$.*

(c) *(Transitive property) $A \sim B$ and $B \sim C$ implies $A \sim C$.*

THEOREM 1.4.5

(a) *Any subset of N is countable.*

(b) *The Cartesian product of countable sets is countable.*

(c) *The union of a countable number of countable sets is countable.*

(d) *Q is countable.*[12]

---

[12]Georg Ferdinand Ludwig Philipp Cantor (1845–1918) was born in Russia to a Danish family which moved permanently to Germany when he was 11 years old. Cantor introduced the concept of infinite numbers, was the first to prove that the set of rational numbers is countable, and came up with other amazing and revolutionary discoveries. Cantor's findings were supported by Weierstrass, Dedekind, Hilbert, and Zermelo, but very strongly attacked by Cantor's own teacher Kronecker. Cantor was stripped of well-deserved recognition, suffered numerous nervous breakdowns, and died in a mental institution. Ernst Friedrich Ferdinand Zermelo (1871–1951) was a German mathematician best known for his work in axiomatic set theory.

(e) *The interval* $(0,1)$ *is uncountable. (See Theorem 4.6.9, part (d).)*

(f) $\Re$ *is uncountable.*

(g) *The set of all irrational numbers is uncountable.*

Accepting the validity of Theorem 1.4.5, the proofs of this theorem are left as a challenging exercise.

Exercises 1.4

1. Prove that the set of all counting numbers is equivalent to the set of all even counting numbers.

2. Prove Theorem 1.4.4.

3. Show that the given sets are countable.
  (a) all even integers (together with zero)
  (b) $\{2^n \mid n \in N \text{ and } n \geq 3\}$

4. Prove Theorem 1.4.5.

1.5* Proof Techniques

In mathematics we rarely say that some property is true most of the time. An expression is defined, and then the properties created by that expression require proof. Those properties are either right or wrong. If we can demonstrate one situation, called a *counterexample*, in which the property in question is not true, then that property is false. Being unable to find a counterexample leads us to believe that a statement is true, but it remains a conjecture until we prove or disprove it.

Results in mathematics are constructed from two parts. One part is made up of assumptions called *hypotheses* (plural for *hypothesis*). The second part is what must be proven. Of course, all previously proven results can be used to prove the new problem. Results that are proven are usually called *theorems*. If some preliminary steps exist in preparation for the main statement, we call those preliminary results *lemmas*. Thus, we can say that every result can be made up of a sequence of lemmas or steps that require verification. Actually, turning a statement into a sequence of small steps is a common way of proving it. This is what makes mathematical induction so nice. Often, one can draw a few conclusions, called *corollaries*, from the main theorem. A corollary resembles a moral and, like theorems and lemmas, requires a proof.

The term *theory* is misleading. Theory can be expressed in the everyday sense or in the scientific sense. For example, one can have a theory of how an event took place, but it is "only a theory," leading to an impression of uncertainty. In science, theory is not just a statement, but rather a whole body of knowledge dealing with certain ideas, theorems, and conjectures that remain to be proved or disproved and open problems yet to be formulated. Sometimes, however, scientific theory develops from the popular sense of the word theory. In 1905 Einstein[13] developed a theory of relativity. For a while it was "just a theory." His theory, however, is no longer uncertain. It is a firmly established area in physics, and we no longer doubt its results.

Certain terminology is necessary for us to feel more comfortable with theoretical mathematics. The meanings of expressions and symbols, such as $\equiv$, which is used to define some things, as well as a way to handle them, are also important. *Uniqueness*, provided existence, for example, means one and only one. To prove uniqueness, we assume to the contrary that there are two and show that they must be the same.

Suppose that p and q represent statements, and we wish to prove that p implies q, (i.e. assume p and prove q). Written as $p \Rightarrow q$, several ways exist to accomplish this task.

(a) Mathematical induction can be attempted if the statement involves natural numbers.

(b) A direct proof would involve writing the hypotheses in many different ways, writing all the different statements that would yield the conclusion, and linking the ideas together. To successfully complete these tasks, knowledge of the definitions, the meaning of the given statement, and having an intuitive idea of the task are all necessary. So intuition and knowledge play a major part, whereas memorization plays a small one.

(c) An indirect proof, often referred to as a proof by contradiction, involves assuming that p is true and q is not. Since we assume that the opposite of q is true and want to prove that q is true, a contradiction, that is, a statement that we know is false is expected and hoped for. When a contradiction is reached, the proof of $p \Rightarrow q$ is complete.

(d) A contrapositive proof, denoted by $\sim q \Rightarrow \sim p$, involves proving that the opposite of q, that is, not q, implies the opposite of p, that is, not p. This results in a direct proof where the statements $\sim q$ and $\sim p$ are easier to work with.

Often "p implies q" is referred to as "p is sufficient for q" or "q is necessary for p." At times, the converse of a statement is asked for. That is, if given the statement $p \Rightarrow q$, its converse is $q \Rightarrow p$. The proofs of a statement and its converse are different, and often $p \Rightarrow q$ but $q \not\Rightarrow p$. For example, if p is the statement "$x \in N$" and q is the statement "$x \in \Re$," then $p \Rightarrow q$ but $q \not\Rightarrow p$. When $p \Rightarrow q$ but $q \not\Rightarrow p$,we often say that p is *sufficient* to guarantee q, but it is not *necessary*. If the statements p and q are equivalent, then clearly $p \Rightarrow q$ and $q \Rightarrow p$, written $p \Leftrightarrow q$. In other words, p is true if and only if q is true, or p is necessary and sufficient for q, which is often written also as "p iff q." These arguments are usually divided into two parts and proven separately.

---

[13]Albert Einstein (1879–1955), a superb German-American theoretical physicist and philosopher, journeyed to the United States during World War II and took a position at Princeton. Einstein wrote a letter to President F. D. Roosevelt informing him of the dangers of a possible German atom bomb. This letter resulted in the Manhattan Project and in the development of the atom bomb in the United States. Einstein developed relativity theory, creating an increased interest in Riemannian geometry and tensor analysis. In 1921 Einstein won the Nobel Prize for his work in theoretical physics.

That is, to prove that two statements p and q are equivalent, we must prove that both $p \Rightarrow q$ and $q \Rightarrow p$. Examples are used to illustrate given statements, but do not usually prove anything. As stated earlier, if a statement is false, a counterexample is enough. As said earlier, to prove a result it is often easier to prove another equivalent result. For example, to show that two expressions P and R are equal, we may prefer showing that $P - R = 0$, or that $P \geq R$ and $P \leq R$.

Often the negation of a statement is needed and is difficult to write. Here are some examples of a statement p and its negation $\sim p$:

(a) p: every x belongs to A.

$\sim p$: some x does not belong to A, or there is an x that does not belong to A.

(b) p: for every x in A, something happens.

$\sim p$: for some x in A, the something does not happen.

(c) p: there exists $x > 0$ such that $1 \leq f(x) < 5$.

$\sim p$: for every $x > 0$ either $f(x) < 1$ or $f(x) \geq 5$.

(d) p: for all $\varepsilon > 0$ there exists $\delta > 0$ such that whenever $x \in D$ and $|x - a| < \delta$, then $|f(x) - f(a)| < \varepsilon$.

$\sim p$: there exists $\varepsilon > 0$ such that for every $\delta > 0$ there exists $x \in D$ such that $|x - a| < \delta$ and $|f(x) - f(a)| \geq \varepsilon$.

To shorten handwriting, mathematicians use the symbol $\forall$ to mean "for all" and the symbol $\exists$ to mean "there exists."

Example 1.5.1. Prove that the square of an odd integer is odd.

Proof. Using a direct proof, let a be an arbitrary odd integer. Since a is odd, $a = 2t + 1$ for some integer t. We need to show that a^2 is odd. That is, we need to show that $a^2 = 2s + 1$ for some integer s. Using the hypothesis, we write

$$a^2 = (2t+1)^2 = 4t^2 + 4t + 1 = 2\left(2t^2 + 2t\right) + 1.$$

Therefore, since t is an integer, so is $s = 2t^2 + 2t$, and thus a^2 is odd. ■

Example 1.5.2. If b^2 is odd, prove that b is odd.

Proof. Using the contrapositive method of proof, we will prove that if b is not odd, then b^2 is not odd. Thus, suppose that b is even. We will prove that b^2 is even. To this end, b even means that $b = 2t$ for some integer t. Therefore, $b^2 = (2t)^2 = 4t^2 = 2\left(2t^2\right) = 2s$, where $s = 2t^2$. Hence, b^2 is even, completing the proof. ■

Examples 1.5.1 and 1.5.2 can be combined to form an if and only if statement.

Example 1.5.3. Prove that there is no rational number whose square is 2.

Proof. We need to prove that $\sqrt{2}$ is irrational. Using the contradictory method of proof, suppose to the contrary that there exists a rational number x such that $x^2 = 2$. Let

$x = \frac{p}{q}$, where p and q are integers with no common factors, that is, p and q are *relatively prime*, that is $(p,q)=1$. We reach a contradiction with the statement $x^2 = 2$. Using what we have, we write $\frac{p^2}{q^2} = 2$. Hence, $p^2 = 2q^2$ and so p^2 is even. Therefore, p is even, since if it were odd, by Example 1.5.1, p^2 would also be odd, which it is not. Thus, there exists an integer k such that $p = 2k$. This gives $p^2 = 4k^2$. Now, since $p^2 = 2q^2$, combining we have $4k^2 = 2q^2$, which reduces to $q^2 = 2k^2$. Therefore, q^2 is even, implying that q is even. Hence, both p and q have a common factor of 2. But this contradicts the assumption that p and q have no common factors. The proof is thus complete. ■

Example 1.5.4. Prove Theorem 1.3.2.

Proof. Suppose that $P(n)$ is a statement for each $n \in N$ such that

(a) $P(1)$ is true, and

(b) for each $k \in N$ if $P(k)$ is true, then $P(k+1)$ is true.

We need to prove that $P(n)$ is true for all $n \in N$. Proceeding by contradiction, suppose that there exists $n \in N$ for which $P(n)$ is not true. Hence, let S be the set of values n that makes $P(n)$ false. Therefore,

$$S = \{n \in N \mid P(n) \text{ is false}\}.$$

We will assume $S \neq \phi$, which will eventually result in a contradiction. Thus, if $S \neq \phi$, by the well-ordering principle, Axiom 1.3.1, S has the smallest member, say t. Thus, $P(t)$ is false, but $P(t-1)$ is true. Since $P(1)$ is true, t must be greater than 1. Now, from the hypotheses, if $P(k)$ is true, then $P(k+1)$ is also true. Thus, let $k = t-1$. Then, since $P(t-1)$ is true, $P((t-1)+1)$ is also true, but $P((t-1)+1) = P(t)$, which is assumed to be false. Hence, we have a contradiction, and the proof is complete. ■

Often, there are many ways to prove a theorem. We will now reprove Example 1.3.4 without using mathematical induction.

Example 1.5.5. Prove that for any fixed real number x and every $n \in N$,

$$x^{n+1} - 1 = (x-1)\left(x^n + x^{n-1} + \cdots + x + 1\right).$$

Proof. Suppose that $S = x^n + x^{n-1} + \cdots + x + 1$. Then $xS = x^{n+1} + x^n + \cdots + x^2 + x$. Then, subtraction gives $xS - S = x^{n+1} - 1$. Solving for S, we obtain $S(x-1) = x^{n+1} - 1$, which is exactly what we wanted to prove. ■

The following result is very useful in later chapters.

THEOREM 1.5.6. *$x = 0$ if and only if $|x| < \varepsilon$ for any real number $\varepsilon > 0$.*

Proof. ($\Rightarrow$) Trivially, if $x = 0$, then for any $\varepsilon > 0$ we will have $|0| < \varepsilon$. ($\Leftarrow$) Suppose that $|x| < \varepsilon$ for any $\varepsilon > 0$. We need to prove that x must be zero. Using contradiction, assume to the contrary that $x \neq 0$. Since $|x| < \varepsilon$ for any $\varepsilon > 0$, pick $\varepsilon = \frac{|x|}{2}$. Then, $|x| < \frac{|x|}{2}$, and since $x \neq 0$, $|x| \neq 0$. We can divide by $|x|$ to obtain $1 < \frac{1}{2}$. Hence, contradiction, and the implication is proven. ■

In conclusion, whenever possible sketch a graph, illustrate the situation by an example, try to obtain an intuitive feel for the situation, and review other results that deal with similar ideas. Writing is essential for comprehension. Having to reproduce an idea often causes us to realize that we really did not know what we read. Finally, asking ourselves questions helps us adapt old ideas to new ones. Why is this so? Are all of the assumptions necessary? What would happen if a certain assumption was left out or changed? Why is this problem important? What does it do for me? How does it compare with problem solved previously? What are the major steps in the proof? Memorizing does not result in great success in mathematics, but remembering, and most of all understanding, are important. If we do not understand, we should ask our instructor, a tutor, a graduate student, or a friend. Problems in mathematics are like any other social problems. They need to be discussed. When was the last time we talked math with someone, or checked out a mathematics book from the library and looked for other approaches, examples, homework problems, ideas, topics, and so on?

Exercises 1.5

1. Negate the following statements:
   (a) There exists $p > 0$ such that for every x we have $f(x + p) = f(x)$.
   (b) For all $\varepsilon > 0$ there exists $\delta > 0$ such that whenever x and t are in D and satisfy $|x - t| < \delta$, then $|f(x) - f(t)| < \varepsilon$.
   (c) For all $\varepsilon > 0$ there exists $\delta > 0$ such that whenever $x \in D$ and $0 < |x - a| < \delta$, then $|f(x) - A| < \varepsilon$.

2. Prove that if q^2 is divisible by 3, then so is q.

3. Prove that $\sqrt{3}$ and $\sqrt{6}$ are irrational.

4. Prove that $\sqrt[3]{2}$ and $\sqrt{2} + \sqrt{3}$ are irrational. See Example 1.6.12.

5. Consider the statement P: the sum of two irrational numbers is irrational.
   (a) Give an example of a case in which P is true.
   (b) Prove P, or disprove P by giving a counterexample.

6. Prove Theorem 1.3.9.

7. Prove that Theorem 1.3.2 implies Axiom 1.3.1. Then conclude that the principle of mathematical induction is equivalent to the well-ordering principle.

8. Prove Theorem 1.3.13.

1.6* Ordered Field and a Real Number System

One of the most fundamental structures in mathematics is a field, characterized by eight axioms.

Definition 1.6.1. A *field*, F, is a nonempty set together with the operations of addition and multiplication, denoted by + and $\cdot$, respectively, that satisfies the following eight axioms:

(A1) *(Closure)* $a+b$, $a\cdot b\in F$ for any $a,b\in F$. Also, if $a,b,c,d\in F$, with $a=b$ and $c=d$, then $a+c=b+d$ and $a\cdot c=b\cdot d$.

(A2) *(Commutative)* $a+b=b+a$ and $a\cdot b=b\cdot a$ for any $a,b\in F$.

(A3) *(Associative)* $(a+b)+c=a+(b+c)$ and $(a\cdot b)\cdot c=a\cdot(b\cdot c)$ for any $a,b,c\in F$.

(A4) *(Additive identity)* There exists a zero element in F, denoted by 0, such that $a+0=a$ for any $a\in F$.

(A5) *(Additive inverse)* For each $a\in F$, there exists an element $-a$ in F, such that $a+(-a)=0$.

(A6) *(Multiplicative identity)* There exists an element in F, which we denote by 1, such that $a\cdot 1=a$ for any $a\in F$.

(A7) *(Multiplicative inverse)* For each $a\in F$ with $a\neq 0$, there exists an element in F denoted by $\frac{1}{a}$ or a^{-1}, such that $a\cdot a^{-1}=1$.

(A8) *(Distributive)* $a\cdot(b+c)=(a\cdot b)+(a\cdot c)$ for any $a,b,c\in F$.

The symbol 0 in Axiom (A4) was conveniently chosen so that Theorem 1.6.2, part (c), would look familiar. Depending on the field, the additive identity may look very different from the real number 0. Similarly, in view of Theorem 1.6.2, part (d), $-a$ in Axiom (A5) was conveniently chosen to denote the additive inverse of a.

THEOREM 1.6.2. *If F is a field, then*

(a) *both the additive identity and the multiplicative identity are unique.*

(b) *both the additive inverse and the multiplicative inverse of any element in F are unique.*

(c) $0\cdot a=0$ *for any* $a\in F$.

(d) $(-1)\cdot a=-a$ *for any* $a\in F$.

(e) $a\cdot(-b)=(-a)\cdot b=-(a\cdot b)$ *for any* $a,b\in F$.

(f) $-(-a)=a$ *for any* $a\in F$.

(g) *if* $a,b\in F$, *with* $a\cdot b=0$, *then either* $a=0$, $b=0$, *or both.*

Proof of part (a). Proving only the additive case, suppose that F possesses two additive identities 0_1 and 0_2. We will show that $0_1=0_2$. To this end, since 0_1 is an additive identity, by Axiom (A4), $0_2+0_1=0_2$. Similarly, since 0_2 is an additive identity, by Axiom (A4), $0_1+0_2=0_1$. But by Axiom (A2) we have $0_2=0_2+0_1=0_1+0_2=0_1$. Thus, the additive identity is unique.

Proof of part (c). For any $a \in F$, we can write $a = 1 \cdot a = (1+0) \cdot a = 1 \cdot a + 0 \cdot a = a + 0 \cdot a$. Now subtract a from both sides to obtain the desired result. ■

Other parts of Theorem 1.6.2 are left as exercises. Customarily, for $a, b \in F$, we can write $a \cdot b$ as ab, $a(b)$, $(a)b$, or $(a)(b)$, and $a + (-b)$ as $a - b$.

Definition 1.6.3. An *ordered field* is a field, F, together with a relation, $<$, if the following additional "order" axioms are satisfied.

(A9) *(Trichotomy)* For $a, b \in F$, exactly one of the following is true: $a = b$, $a < b$, or $a > b$.

(A10) *(Transitive)* For $a, b, c \in F$, if $a < b$ and $b < c$, then $a < c$.

(A11) For $a, b, c \in F$, if $a < b$, then $a + c < b + c$.

(A12) For $a, b, c \in F$, if $a < b$ and $c > 0$, then $ac < bc$.

Observe that Axiom (A9) also tells us that if $a \in \Re$, then exactly one of the following is true: $a \in \Re^+$, $a = 0$, or $-a \in \Re^+$. In addition, the sets $\Re^+$, $\Re^-$, and $\{0\}$ are disjoint and their union is $\Re$. Moreover, observe that if in Axiom (A12), $a = 0$, it follows that the product of two positive real numbers is also positive.

THEOREM 1.6.4. *Suppose that F is an ordered field.*

(a) *If $a, b \in F$ then $a < b$ if and only if $-a > -b$.*

(b) *If $a, b, c \in F$, $a < b$, and $c < 0$, then $ac > bc$.*

(c) *If $a \in F$ and $a \neq 0$, then $a^2 > 0$.*

(d) *Suppose $a \in F$. If $a > 0$, then $\frac{1}{a} > 0$, and if $a < 0$, then $\frac{1}{a} < 0$.*

Proof of part (a). Suppose that for $a, b \in F$ and $a < b$. Then letting $c = -a - b$ and using Axiom (A11), we obtain

$$\begin{aligned} a + (-a - b) &< b + (-a - b) && \text{(Axiom A11)} \\ a + (-a) - b &< b + (-b) - a && \text{(Axioms A8 and A2)} \\ 0 - b &< 0 - a && \text{(Axiom A5)} \\ -b &< -a && \text{(Axiom A4)} \end{aligned}$$

which is equivalent to $-a > -b$. ■

THEOREM 1.6.5. *Q is an ordered field.*

The proof of Theorem 1.6.5 is left as an exercise. The real numbers, which are a much larger set than the rational numbers, require one additional axiom, known as the completeness axiom.

Definition 1.6.6. Suppose that F is an ordered field. F is a *complete ordered field* if F satisfies the following completeness axiom, sometimes referred to as *Dedekind's*[14] *axiom:*

---

[14]Julius Wilhelm Richard Dedekind (1831–1916), a prominent German contributor to the theory of algebraic functions and number theory, is famous for the "cut" named after him. Dedekind also worked in the theory of irrational numbers.

(A13) *(Completeness axiom)* Every nonempty subset S of F that has an upper bound, has a least upper bound.

Upper and lower bounds, as well as the least upper bound and the greatest lower bound of a set S, are defined parallel to the bounds of a function in Definition 1.2.14. That is, if F is an ordered field, and $S \subseteq F$, then an element $s \in F$ is called an *upper bound* for S if and only if for every $x \in S$, $x \leq s$. Furthermore, t is the *least upper bound* of S, or *supremum* of S, denoted by $\sup S$, if and only if t is an upper bound of S, and for every other upper bound r for S, we have $t \leq r$. Proof of the following theorem is in Exercise 9.

THEOREM 1.6.7. *Suppose S is a nonempty subset of $\Re$ and k is an upper bound of S. The number $k = \sup S$ if and only if for each $\varepsilon > 0$ there exists $s \in S$ such that $k - \varepsilon < s$.*

The set Q is not a complete ordered field. (See Exercise 14.) Intuitively, if we were able to line up the rational numbers, we would see plenty of holes between them. Real numbers, on the other hand, are a complete ordered field, and any other complete ordered field F is *isomorphic* to $\Re$. That is, a bijection from $\Re$ onto F exists. $\Re$ is the underlying number system throughout this textbook, and a thorough development of these numbers would require an ample amount of time and effort. Instead we will take a look at some of their basic properties.

THEOREM 1.6.8. *(Archimedean[15] Order Property of $\Re$) If x is any real number, then a positive integer n^* greater than x exists. See Theorem 2.4.8.*

Proof. Proving by contradiction, suppose that x is an upper bound of N. Since $N \neq \phi$ and $\Re$ satisfies the completeness axiom (A13), there exists $m \in \Re$ such that $m = \sup N$. Thus, $m-1$ is not an upper bound of N. By Theorem 1.6.7 with $\varepsilon = 1$, there exists $n_1 \in N$ such that $m-1 < n_1$. Therefore, $m < n_1 + 1$, where $n_1 + 1 \in N$. But this contradicts that m is the least upper bound of N. Hence, a natural number greater than x must exist. ■

COROLLARY 1.6.9. *The following are equivalent statements:*

(a) *Archimedean order property of $\Re$.*

(b) *For any $x, y \in \Re^+$, there exists $n \in N$ such that $y < nx$.*

(c) *For any $x \in \Re^+$, there exists $n \in N$ such that $\frac{1}{n} < x$.*

(d) *N is unbounded.*

(e) *For any $x \in \Re^+$, there exists $n \in N$ such that $n - 1 \leq x < n$.*

---

[15]Archimedes (c. 287–212 B.C.), born to a Greek family in Sicily, is considered one of the world's greatest mathematicians. Even though algebra and a convenient number system had not been developed, Archimedes devised methods for finding areas, volumes, and square roots. He created a discipline of hydrostatics and used it to find equilibrium positions for certain floating bodies. Archimedes also formulated postulates for mechanics and calculated centers of gravity for solids. When Archimedes was killed during the Roman invasion of Sicily, the practice of great mathematics also disappeared for a very long period of time.

Proofs of Corollary 1.6.9 as well as the next theorem are left as Exercises 10 and 12.

THEOREM 1.6.10. *Every open interval (a,b) contains both a rational number and an irrational number.*

Thus, between any two irrational numbers a rational number exists, and between any two rational numbers, an irrational number exists. In addition, we can conclude that every open interval (a,b) contains infinitely many rational and irrational numbers. Why? Now, if a set $S \subseteq \Re$ satisfies the property that $S \cap I \neq \phi$ for any interval I, then S is called *dense* in $\Re$. Hence, both Q and $\Re \setminus Q$ are dense in $\Re$.

Real numbers can also be broken up into two disjoint sets, algebraic and *transcendental*, that is, not algebraic. An *algebraic* number is a real root of a polynomial equation

$$a_n x^n + a_{n-1} x^{n-1} + \cdots + a_1 x + a_0 = 0$$

with integer coefficients a_i, $i = 0,1,2, \ldots, n$, $n \in N$. A familiar algebraic number is $\sqrt{2}$ since it is the root of $x^2 - 2 = 0$. Examples of transcendental numbers are π, e, e^{π}, and $2^{\sqrt{2}}$, the so-called *Hilbert*[16] *number*. Recall that the smallest positive root of $\sin x$ is called *pi*, denoted by π.[17] Coincidentally, as Cantor proved, the set of transcendental numbers is much larger than the set of algebraic numbers. Great difficulty arises in proving that a certain number is transcendental. The first transcendental number was constructed by Liouville[18] in 1844. In 1873 Hermite[19] proved that e is transcendental

---

[16]David Hilbert (1862–1943) was born in Germany, and is considered by many to be the greatest mathematician of this century. Much of the mathematical research of this century has focused on problems stated by Hilbert. He contributed greatly to many areas of mathematics, but his work in geometry has been more influential than that of anyone since Euclid, who around 300 B.C. provided the sole approach to geometry until the nineteenth-century development of the non-Euclidean geometry.

[17]Johann Heinrich Lambert (1728–1777) was a Swiss-German analyst, number theorist, astronomer, physicist, and philosopher who wrote valuable books on geometry, cartography, and art. Lambert introduced hyperbolic functions, foreshadowed the discovery of non-Euclidean geometry, and proved that π is irrational.

[18]Joseph Liouville (1809–1882), a Frenchman, is remembered for his work in differential equations, number theory, and complex analysis. Liouville, although unable to show that the numbers π and e are transcendental, constructed another transcendental number.

[19]Charles Hermite (1822–1901), an influential French mathematician, taught a number of world-recognized mathematicians. His main areas of expertise were algebra, analysis, and number theory. Among the famous mathematicians Hermite taught was Henri Poincaré (1854–1912) of France, a leading mathematician of his time. Poincaré initiated both algebraic topology and the theory of analytic functions of several complex variables. He also made major contributions to algebraic geometry, number theory, optics, electricity, telegraphy, elasticity, thermodynamics, potential theory, the theory of relativity, celestial mechanics, the theory of light, electromagnetic waves, and so on. Poincaré is considered to be the last of the great "Renaissance" mathematicians.

(see Part 2 of Section 8.8). In 1882, Lindemann[20] proved that π is transcendental. Whether or not the numbers π^π and $e+\pi$ are transcendental is still undetermined.

If $f(x)=ax^2+bx+c$, with a, b, and c real constants, then the roots of f, that is, the values for which $f(x)=0$, or equivalently, the values where f crosses the x-axis, are given by the *quadratic formula*

$$x=\frac{-b\pm\sqrt{b^2-4ac}}{2a},$$

provided $b^2-4ac\geq 0$ (see Exercise 11 in Section 1.7). Thus, if $b^2-4ac\geq 0$, then $f(x)$ can be *factored,* that is, written as a product of polynomials of smaller order, as

$$f(x)=a\left(x-\frac{-b+\sqrt{b^2-4ac}}{2a}\right)\left(x-\frac{-b-\sqrt{b^2-4ac}}{2a}\right).$$

If $f(x)$ is a difference of two squares, then it can easily be factored. Recall that for real numbers A and B, $A^2-B^2=(A-B)(A+B)$, which is a special case of $A^{2n}-B^{2n}=(A^n-B^n)(A^n+B^n)$ for $n\in N$. Also, recall the formulas

$$A^3-B^3=(A-B)(A^2+AB+B^2), \text{ and } A^3+B^3=(A+B)(A^2-AB+B^2),$$

which can be used to factor higher-ordered polynomials. These expressions, as well as Example 1.3.4, can be generalized to yield

$$x^n-a^n=(x-a)\left(x^{n-1}+ax^{n-2}+a^2x^{n-3}+\cdots+a^{n-1}\right)=(x-a)\sum_{k=0}^{n-1}x^{n-1-k}a^k,\ n\in N,$$

$$x^n+a^n=(x+a)\left(x^{n-1}-ax^{n-2}+a^2x^{n-3}-\cdots+a^{n-1}\right), \text{ for odd natural number } n.$$

Can you prove these formulas? Compare the first expression to exercise 2(e) in Section 1.3. If in the first expression $x=1$ and $a=r$, we obtain $1-r^n=(1-r)\sum_{k=0}^{n-1}r^k$, which is equivalent to

$$\sum_{k=0}^{n-1}r^k=\frac{1-r^n}{1-r}, \text{ if } r\neq 1.$$

See Example 1.3.4. When possible, some more complicated polynomials can be factored using the following theorem.

---

[20]Ferdinand Lindemann (1852–1939), a German mathematician, did research on projective geometry and Abelian functions and developed a method of solving equations of any degree using transcendental functions. Lindemann also proved that it is impossible to square the circle using a ruler and compass.

THEOREM 1.6.11. *(Rational Root Theorem) If a rational number* $\frac{p}{q}$ *is a root of a polynomial*

$$f(x) = a_n x^n + a_{n-1}x^{n-1} + \cdots + a_1 x + a_0,$$

where $n \in N$, $a_n \neq 0$, $a_0, a_1, \ldots, a_n \in Z$ *and* $(p,q) = 1$, *then* p *divides* a_0 *and* q *divides* a_n.

Hence, the rational root theorem says that the only choices for the rational solution of

$$2x^4 - 7x^3 + 7x^2 - 7x + 5 = 0$$

are the numbers $\frac{p}{q}$, where p divides 5 and q divides 2. Thus, $\pm\frac{5}{2}$, ± 5, $\pm\frac{1}{2}$, and ± 1 are the only choices for rational solutions of this equation.

The proof of Theorem 1.6.11 is Exercise 16 below. Observe that any polynomial can be written as a product of two nontrivial expressions. For example,

$$x^2 + 1 = \left(x + 1 - \sqrt{2x}\right)\left(x + 1 + \sqrt{2x}\right).$$

However, this is not a factorization of $x^2 + 1$.

Example 1.6.12. Prove that $\sqrt{2} + \sqrt{3}$ is irrational. See Exercise 4 from Section 1.5.

Proof. Let $x = \sqrt{2} + \sqrt{3}$. Then, $x - \sqrt{2} = \sqrt{3}$ and squaring gives

$$x^2 - 2\sqrt{2}x + 2 = 3,$$

which can be written as $2\sqrt{2}x = x^2 - 1$. With the radical isolated again, we square and reduce to obtain

$$x^4 - 10x^2 + 1 = 0.$$

Because of the construction of this equation, we know that the equation has $\sqrt{2} + \sqrt{3}$ as one of its solutions. But since $x = 1$ and $x = -1$ do not satisfy this equation, by Theorem 1.6.11, $f(x) = x^4 - 10x^2 + 1$ has no rational roots. Hence, $\sqrt{2} + \sqrt{3}$ is irrational. ■

Recall from earlier studies that *Descartes' rule of signs*[21] provides information regarding the number of roots a polynomial, p, possesses. In addition, the number of sign variations between $p(x)$ and $p(-x)$ also leads to information regarding the number of positive and the number of negative roots of p. Furthermore, the following particular

[21] A good reference, one of many, is the textbook *Theory of Equations* by J. V. Uspensky, McGraw-Hill 1948.

case[22] of *Gerschgorin's circle theorem*[23] yields a convenient bound on the roots of p.

THEOREM 1.6.13. *Consider a polynomial* $p(x) = a_n x^n + a_{n-1}x^{n-1} + \cdots + a_1 x + a_0$ *with* $a_0, a_1, \ldots, a_n$ *real constants and* $a_n \neq 0$. *If* r *is a root of* p *and* $M = \dfrac{|a_n| + |a_{n-1}| + \cdots + |a_0|}{|a_n|}$, *then* $|r| \leq M$.

The generalized version of this result, the Gerschgorin's circle theorem, can be found in linear algebra, numerical, and/or complex analysis texts when studying "eigenvalues."

Exercises 1.6

1. Prove Theorem 1.6.2.

2. Suppose F is a field and $a, b, c \in F$. Prove that
- **(a)** $(-a)(-b) = ab$.
- **(b)** if $a + b = c + b$, then $a = c$.
- **(c)** $-a - b = -(a + b)$.
- **(d)** if $a \neq 0$, then $\frac{1}{a} \neq 0$ and $\frac{1}{\frac{1}{a}} = a$.

3. Suppose F is an ordered field. Prove that
- **(a)** if $a \in F$ and $a > 0$, then $-a < 0$.
- **(b)** $0 < 1$.
- **(c)** if $a, b \in F$ and $ab > 0$, then either both $a, b > 0$, or both $a, b < 0$. Thus, saying that $ab > 0$ implies that both a and b are of same sign.
- **(d)** if $a, b \in F$ and $ab < 0$, then either $a > 0$ and $b < 0$, or $a < 0$ and $b > 0$. Thus, saying that $ab < 0$ implies that a and b are of opposite sign.
- **(e)** if $0 < a < b$, then $0 < \frac{1}{b} < \frac{1}{a}$.

4. Prove Theorem 1.6.4.

5. Prove that if $r \geq 1$ is a real number, then $r^2 \geq r$ and $\frac{1}{r^2} \leq \frac{1}{r}$.

6. Prove that Q is an ordered field.

7. If $S \neq \phi$ is a subset of real numbers that is bounded below, prove that $\inf S$ exists.

8. Prove that if a set A has a supremum, then $\sup A$ is unique.

9. Prove Theorem 1.6.7.

10. Prove Corollary 1.6.9.

---

[22]This particular case of Gerschgorin's circle theorem was presented by Wade T. Macey, a professor of mathematics at Appalachian State University. Dr. Macey, who has a variety of talents, toured at the age of 17 as a banjo picker with Mac Wiseman, a leading bluegrass artist of the 1950s and 1960s.

[23]Sergi Gerschgorin presented this result in 1931. Gerschgorin, a Russian mathematician, did most of his work in Germany.

11. Prove that the product of a nonzero rational number together with an irrational number is an irrational number.

12. Prove Theorem 1.6.10.

13. If possible, give an example of a nonempty bounded subset of Q that
(a) has a least upper bound and a maximum in Q.
(b) has a least upper bound but no maximum in Q.
(c) does not have a least upper bound in Q.

14. Prove that Q is not a complete ordered field. See Exercise 6.

15. Prove that there exists a real number x such that $x^2 = 2$.

16. Prove the rational root theorem.

17. Suppose $f(x)$ is a polynomial and $c \in \Re$. Prove that $f(c) = 0$ if and only if $x - c$ divides $f(x)$.

18. Find all real solutions of
(a) $x^4 - 4x^3 + 2x^2 + 8x - 8 = 0.$ **(b)** $9x^3 - 30x^2 + 28x - 8 = 0.$

19. Factor, if possible.
(a) $p(x) = x^4 + 4$ **(b)** $p(x) = 4x^2 - 2x - 1$
(c) $p(x) = 2x^4 - 3x^3 - 6x^2 + 13x - 6$ **(d)** $p(x) = x^3 - x^2 + 4x - 4$
(e) $p(x) = x^7 + 1$ **(f)** $p(x) = x^5 - 1$

20. Find all of the real values x that satisfy the given inequalities:
(a) $x^2 - x \geq 6$ **(b)** $x^3 + 2x^2 - 5x - 6 < 0$
(c) $-2 \leq \dfrac{1}{x} < 1$ **(d)** $\dfrac{2x+1}{x-5} \leq 3$
See Exercise 14 in Section 4.3.

21. **(a)** Use Theorem 1.6.11 to verify that $\dfrac{\sqrt{2}}{\sqrt[3]{3}}$ and $\sqrt[3]{2} - \sqrt{3}$ are irrational.
(b) Prove that $\sqrt[3]{2} + \sqrt{3}$ is an algebraic number.

22. If x is rational, then prove that x is algebraic. Is the converse true? Explain.

23. A number of the form $\sqrt[n]{a}$, where $a, n \in N$ is either irrational or it is an integer. If it is an integer, prove that a is its nth power.

24. **(a)** Can you find 2 rational numbers a and b such that a^b is irrational?
(b) Can you find 2 irrational numbers α and β such that α^β is rational?

25. Prove Theorem 1.6.13.

26. A function f that has an unbounded set of roots is said to be *oscillatory*. That is, f *oscillates*.
(a) Is there a polynomial which is oscillatory?
(b) Is there a function defined on an interval $[a,b]$ which is oscillatory?
(c) Give an example of an oscillatory function.
See Exercise 21 in Section 3.3, Exercise 46 in Section 3.4, and Exercise 3 in Section 5.5 for more properties of oscillatory functions.

1.7* Some Basic Properties of a Real Number System

THEOREM 1.7.1. *Suppose $a, b \in \Re$.*

(a) *If $a < b$, then $a < \dfrac{a+b}{2} < b$.*

(b) *If $a, b > 0$, then $a < b \Leftrightarrow a^2 < b^2 \Leftrightarrow \sqrt{a} < \sqrt{b}$.*

(c) *If $a, b \geq 0$, then $\sqrt{ab} \leq \dfrac{a+b}{2}$.* See Part 3 of Section 1.10.

(d) *If $a, b \geq 0$, then $\sqrt{a^2 + b^2} \leq a + b$.*

Proof of part (a). Using the transitive property, Axiom (A10) from Definition 1.6.3, we write

$$a = \frac{a}{2} + \frac{a}{2} < \frac{a}{2} + \frac{b}{2} = \frac{a+b}{2} < \frac{b}{2} + \frac{b}{2} = b.$$ ■

Observe that part (b) of Theorem 1.7.1 can be extended to cover the case when $0 \leq a \leq b$. In that case, we have $a \leq b \Leftrightarrow a^2 \leq b^2 \Leftrightarrow \sqrt{a} \leq \sqrt{b}$. The following are some properties of the absolute value function, $|x|$, which was introduced in Definition 1.2.13.

THEOREM 1.7.2. *Suppose $a, b \in \Re$. Then*

(a) $|a| \geq 0$.

(b) $|a| \leq b$ *if and only if* $-b \leq a \leq b$.

(c) $|a| \geq b$ *if and only if* $a \leq -b$ *or* $a \geq b$.

(d) $|ab| = |a| \cdot |b|$.

(e) *(Triangle Inequality)* $|a + b| \leq |a| + |b|$.

Proof of part (b). ($\Rightarrow$) Suppose that $|a| \leq b$. Since $-|a| \leq a \leq |a|$ and $-b \leq -|a|$ (why?), $-b \leq -|a| \leq a \leq |a| \leq b$, proving the desired inequalities.

($\Leftarrow$) Suppose $-b \leq a \leq b$. If $a \geq 0$, then $|a| = a \leq b$. And if $a < 0$, then $|a| = -a \leq b$. Why? In either case we have $|a| \leq b$. The proof of part (b) is complete.

Proof of part (e). Since we can always write that $-|a| \leq a \leq |a|$ and $-|b| \leq b \leq |b|$, combining these statements, we obtain

$$-|a| - |b| \leq a + b \leq |a| + |b|.$$

But, the left-hand side is $-(|a| + |b|)$. The desired inequality follows from part (b). ■

Note that there is no "best" proof of a given theorem. Each collection of true statements that verifies the desired property makes up a proof. Sometimes a shorter proof is better than a longer one. But often, a longer but straightforward proof is preferred. For example, the following proof of the triangle inequality may be considered better. "Since $0 \le |a+b|^2 = (a+b)^2 = a^2 + 2ab + b^2 \le |a|^2 + 2|a||b| + |b|^2 = (|a|+|b|)^2$, by Theorem 1.7.1, part (b), the triangle inequality follows."

The idea of the absolute value function, $|a|$, becomes much easier when we think of it as the distance from a to 0 (see Figure 1.7.1).

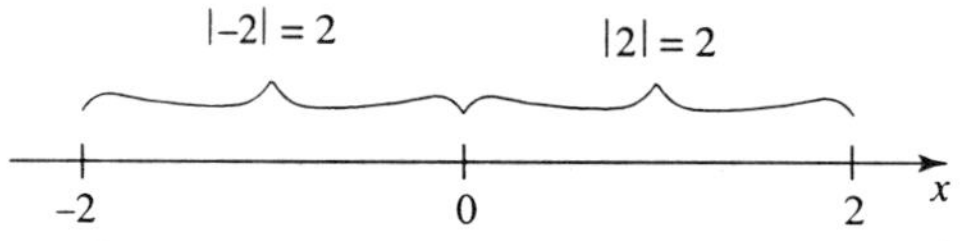

Figure 1.7.1

Now the expression $|x| \le a$ becomes intuitively easy. In other words, we want the values of x that are a units or less away from zero. Recall Theorem 1.7.2, part (b). That is, $-a \le x \le a$, also written as the interval $[-a, a]$ with center at $x = 0$. The expression $|x| < a$ is the interval $(-a, a)$. There is no way to use absolute value in expressing the interval $(-a, a]$. Along these lines, in writing $|x-a|$ or $|a-x|$, we mean the distance from x to a on the x-axis. Hence, $|x-1| \le 3$ means $-3 \le x-1 \le 3$, or equivalently, $-2 \le x \le 4$. The center of this interval is at $x = 1$. Can you explain why the inequality in Theorem 1.7.2, part (e), is called the triangle inequality? When is equality true and when is $<$ true? Compare triangle inequality to the Pythagorean theorem.

Exercises 1.7

1. Prove Theorem 1.7.1, parts (b)–(d).

2. If $a, b \in \Re$ and $a - \varepsilon < b$ for any arbitrary $\varepsilon > 0$, prove that $a \le b$.

3. Suppose that $a, b \in \Re$. Prove that

(a) if $b \in \Re^+$, then $|a| < b$ if and only if $-b < a < b$.

(b) if $b \in \Re^+$, then $|a| > b$ if and only if $a < -b$ or $a > b$.

(c) $|a| = \sqrt{a^2}$. **(d)** $|ab| = |a| \cdot |b|$.

(e) $\left|\dfrac{a}{b}\right| = \dfrac{|a|}{|b|}$, provided $b \ne 0$.

4. Find all real values of x that satisfy the given expressions.

(a) $|x-1| < |x|$ **(b)** $|2x-5| \le |x+4|$

(c) $2|1-3x| > 5$ **(d)** $0 < |x-5| \le 2$

(e) $|2x+1| < -3$ **(f)** $\left|\dfrac{2+x}{3-x}\right| \ge -1$

(g) $|x+6| = |2x+1|$ **(h)** $|x+6| = 2x+1$

5. Use Exercise 3(c) to write part (d) of Theorem 1.7.1 using absolute value notation.

6. Suppose that $a, b \in \Re$. Prove that

(a) $|a-b| \geq |a|-|b|$.

(b) $||a|-|b|| \leq |a-b|$.

7. **(a)** If $a, b, c \in \Re$, prove that $|a-b| \leq |a-c|+|c-b|$.

(b) If $a, b, c \in \Re$ such that $a < b < c$, prove that $|a-b|+|b-c| = |a-c|$.

8. If $|f(x)| \leq M$ for all $x \in [a, b]$, prove that $-2M \leq f(x_1)-f(x_2) \leq 2M$ for any $x_1, x_2 \in [a, b]$.

9. If $a_1, a_2, \ldots, a_n \in \Re$ with $n \in N$, prove $|a_1+a_2+\cdots+a_n| \leq |a_1|+|a_2|+\cdots+|a_n|$. This is a generalization of the triangle inequality.

10. **(a)** *(Cauchy[24]–Schwarz[25] inequality,* also referred to as *Cauchy–Schwarz–Buniakovski[26] inequality)* If $n \in N$ and $a_1, a_2, \ldots, a_n, b_1, b_2, \ldots, b_n$ are real numbers, prove that

$$\left(\sum_{k=1}^{n} a_k b_k\right)^2 \leq \left[\sum_{k=1}^{n} (a_k)^2\right]\left[\sum_{k=1}^{n} (b_k)^2\right].$$

See Exercises 13 in Section 6.3 and 13 in Section 7.2.

(b) *(Minkowski[27] inequality)* If $n \in N$ and $a_1, a_2, \ldots, a_n, b_1, b_2, \ldots, b_n$ are real numbers, prove that

$$\left[\sum_{k=1}^{n} (a_k+b_k)^2\right]^{\frac{1}{2}} \leq \left[\sum_{k=1}^{n} (a_k)^2\right]^{\frac{1}{2}} + \left[\sum_{k=1}^{n} (b_k)^2\right]^{\frac{1}{2}}.$$

This is a generalization of the triangle inequality. Also, see Exercises 15 in Section 6.3 and 13 in Section 7.2.

11. If a, b, and c are real constants with $a \neq 0$, solve $ax^2+bx+c=0$ for x. That is, find roots of the function $f(x)=ax^2+bx+c$, by following the outlined procedure known as "completing the square."

---

[24]Augustin Louis Cauchy (1789–1857), a brilliant French mathematician who led a fascinating life, produced 789 publications. Cauchy introduced rigor into calculus, founded the theory of functions of a complex variable, began the modern era in differential equations, and pioneered work in group theory. In addition to mathematical contributions, Cauchy contributed heavily to the study of surface waves, optical dispersion, fluid flow, and the theory of elasticity.

[25]Hermann Amandus Schwarz (1843–1921), a German mathematician, became famous through his work in complex analysis and calculus of variations.

[26]Victor Jakowlewitsch Buniakovski (1804–1899) was a Russian mathematician.

[27]Hermann Minkowski (1864–1909), a Russian-born mathematician, spent most of his life in Switzerland and Germany, where he was recognized for his work in number theory, analysis, algebra, and geometry.

(a) Divide both sides of the equation by a and subtract the constant term from both sides.

(b) Add $\left(\frac{b}{2a}\right)^2$ to both sides.

(c) Factor the left-hand side and write it as a square.

(d) Take the square root of both sides to obtain an equation involving absolute value. See Exercise 3(c).

(e) Solve for x.

12. If a, b, and c are real constants and $ax^2 + bx + c = 0$, show that

$$x = \frac{-2c}{b \pm \sqrt{b^2 - 4ac}},$$

provided $ac \neq 0$. See the quadratic formula in Section 1.6.

13. Show that $f(x) = ax^2 + bx + c$ attains its maximum value at $x = -\frac{b}{2a}$ if $a < 0$ and its minimum value at $x = -\frac{b}{2a}$ if $a > 0$. The point $\left(-\frac{b}{2a}, f\left(-\frac{b}{2a}\right)\right)$ is called the *vertex* of the parabola f.

14. Show that $f(x) = ax^2 + bx + c$ is symmetrical with respect to the vertical line $x = -\frac{b}{2a}$.

15. Find the center and the radius of the circle $2x^2 + 2y^2 + 4x - 3y - 1 = 0$.

16. $x^2 - 4y^2 + 6x + 8y + 5 = 0$ is an equation of a hyperbola. True or false?

17. If $a^2 = b^2$, then $a = b$. True or false?

1.8* Inverse Functions

Recall Definition 1.2.7, that a function is bijective if it is both one-to-one and onto. A characterization of such functions is of interest to us now.

THEOREM 1.8.1. *The function $f: A \to B$ is a bijection if and only if there exists a unique surjection $g: B \to A$ such that*

(a) $f(g(y)) = y$, *for* $y \in B$, *and*

(b) $g(f(x)) = x$, *for* $x \in A$.

Proof. ($\Rightarrow$) Suppose f is one to one and onto. If $y \in B$, let $x \in A$ be the preimage of y, that is, let x be in A such that $f(x) = y$. Define $g: B \to A$ by $g(y) = x$. Thus, since

$f(x) = y$, it follows that $f(g(y)) = y$. Also, since $g(y) = x$ and $y = f(x)$, we have that $g(f(x)) = x$. Hence, conditions (a) and (b) are satisfied.

($\Leftarrow$) Suppose that there is a function g from B onto A which satisfies conditions (a) and (b). First, we will prove that f is one to one. Thus, choose $x_1, x_2 \in A$ with $f(x_1) = f(x_2)$. From (b) it follows that $x_1 = g(f(x_1)) = g(f(x_2)) = x_2$. Therefore, f is one to one. To prove that f is onto, choose $y \in B$ with $x = g(y)$. Then, from (a) we have that $f(x) = f(g(y)) = y$. And so, f is onto.

To complete the proof we need to show that g is unique. To this end, suppose h maps B onto A. And suppose h satisfies conditions (a) and (b). If $y \in B$, choose an $x \in A$ such that $f(x) = y$. Then (b) implies that $h(y) = h(f(x)) = x = g(f(x)) = g(y)$. Hence, $h(y) = g(y)$ for any $y \in B$. Therefore, g is unique. ■

The function g in Theorem 1.8.1 is called the *inverse function* of f, and is denoted by f^{-1}. If a function f has an inverse, then f is *invertible*. A number of conclusions follow.

Remark 1.8.2. Assume f has an inverse.

(a) If $f: A \to B$ is one to one and onto, then $f^{-1}: B \to A$ is a surjection defined by $f^{-1}(y) = x$, if and only if $f(x) = y$ for any $y \in B$.

(b) The domain of f is the range of f^{-1}, and the domain of f^{-1} is the range of f. Thus, if f maps x into y, then f^{-1} maps y into x.

(c) In order for f to have an inverse, graph of f must pass the *horizontal line test*. That is, for f^{-1} to exist, no horizontal line can cross graph of f at more than one point. See Remark 1.2.8, part (a).

(d) In general, f^{-1} does not represent $\frac{1}{f}$. Can you give an example of a function f for which $f^{-1} \equiv \frac{1}{f}$?

(e) $f^{-1}(f(x)) = x$ for every $x \in A$.

(f) $f(f^{-1}(y)) = y$ for every $y \in R_f$.

(g) If $f: A \to B$ and $D \subseteq B$, then the set $f^{-1}(D) = \{x \in A \mid f(x) \in D\}$ is called the *preimage* of D in A, or f *inverse* of D.

(h) If (a, b) are the coordinates of a point on a function f, then (b, a) are the coordinates of a point on f^{-1}, provided f^{-1} exists. These two points are an equal distance from the line $y = x$.

(i) Functions f and f^{-1} are symmetrical with respect to the line $y = x$. That is, f and f^{-1} will overlap and coincide perfectly with each other if the coordinate plane is folded along the line $y = x$. The function f^{-1} is obtained from f by rotating f counterclockwise 90° and then flipping it over the y-axis.

(j) Functions f and g are *inverses of each other* if $(f \circ g)(x) = x$, for every x in the domain of g, and $(g \circ f)(x) = x$, for every x in the domain of f. Thus, $f(x) = 3x + 1$ and $g(x) = \frac{x}{3} - \frac{1}{3}$ are inverses of each other but $f(x) = x^2$ and $g(x) = \sqrt{x}$ are not. Why? See Figures 1.8.1 and 1.8.2. However, $f(x) = x^2$ with $x \geq 0$ and $g(x) = \sqrt{x}$ are inverses of each other.

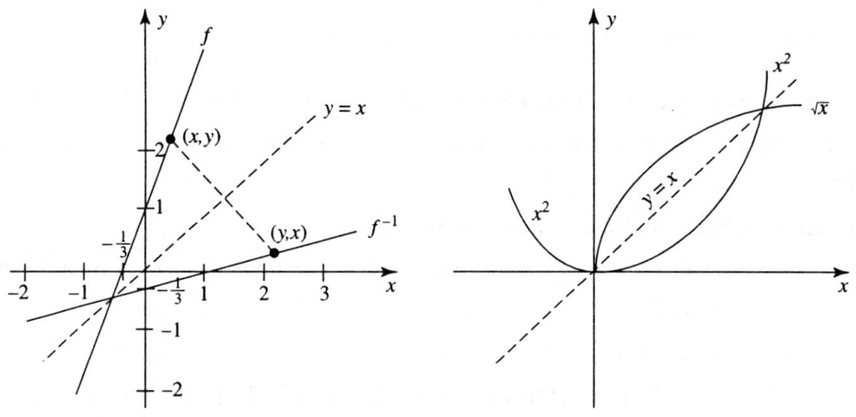

Figure 1.8.1 **Figure 1.8.2**

(k) To compute the inverse of a function f, provided f^{-1} exists, we perform the following steps. Replace $f(x)$ by y, interchange x's and y's, solve for y, and label it $f^{-1}(x)$.

(l) Functions e^x and $\ln x$ are inverses of each other. Thus, $e^{\ln x} = x$ and $\ln e^x = x$. See Remark 3.3.6. Functions $\sin x$, on a restricted domain, and $\arcsin x$ are also inverses of each other. What can be said about $\arcsin\left(\sin\left(\frac{3\pi}{2}\right)\right)$ and $\sin\left(\arcsin\left(\frac{3\pi}{2}\right)\right)$?

(m) $\left(f^{-1}\right)^{-1} = f$.

(n) Even functions have no inverseon their whole domain, but odd functions may. ■

Exercises 1.8

1. Verify that the given functions f and g are inverses of each other.

(a) $f(x) = 2x - 1,\ g(x) = \frac{1}{2}(x+1)$

(b) $f(x) = \sqrt[3]{x-2},\ g(x) = x^3 + 2$

2. Find the unique inverse of the given functions, if one exists.

(a) $f(x) = mx + b,\ m \neq 0$

(b) $f(x) = \frac{2-x}{x+1}$

(c) $f(x) = \frac{x}{x-1}$

(d) $f(x) = 2\arctan(3x+1)$

(e) $f(x) = \sqrt{4 - x^2}$, with $x \in [0,2]$

3. Find two functions f and g for which $(f \circ g)(x) = (g \circ f)(x)$, but for which f and g are not inverses of each other.

4. Find the largest domain for the function $f(x) = \left(x^2+1\right)^2$ on which f is invertible. Find the inverse of f on this domain.

5. If f and f^{-1} are both defined on $\Re$ and f is odd, show that f^{-1} is also odd. Is the statement still true if $\Re$ is replaced by any symmetric interval $(-a, a)$, with $a \neq 0$? How about if f odd is replaced by f even?

6. Determine whether or not the function $f(x) = \frac{1}{x}$ with $x > 0$ is symmetric with respect to the line $y = x$. How about $g(x) = \frac{1}{x^2}$ with $x > 0$? See Exercise 6 in Section 6.5.

7. To convert from x degrees Fahrenheit[28] to y degrees Celsius,[29] we will use the formula $y = f(x) = \frac{5}{9}(x - 32)$. To convert from x degrees Celsius to y degrees Fahrenheit we will use the formula $y = g(x) = \frac{9}{5}x + 32$. Are f and g inverses of each other? Explain.

---

[28]Gabriel Daniel Fahrenheit (1686–1736), a Polish born physicist who later lived in England and Holland, was famous for his Fahrenheit thermometer on which water freezes at 32° and boils at 212°.

[29]Anders Celsius (1701–1744), a Swedish astronomer and physicist as well as a writer of poetry and popular science, is most famous for his temperature scale on which initially 0° represented the boiling point and 100° the melting point. The scale was turned around after his death. In the late 1950s the name "centigrade," which was used for Celsius temperature, became obsolete. "Centigrade" was replaced by "degrees Celsius."

8. If $f:A \to B$ and $g:B \to C$ are bijections, prove that $g \circ f:A \to C$ is a bijection and that $(g \circ f)^{-1} = f^{-1} \circ g^{-1}$.

9. **(a)** Are the functions $f(x) = \sin x$ and $g(x) = \arcsin x$ inverses of each other?
(b) For what values of x is $\sin(\arcsin x) = x$?
(c) For what values of x is $\arcsin(\sin x) = x$?
(d) Do the curves $y_1 = \sin(\arcsin x)$, $y_2 = \arcsin(\sin x)$, and $y_3 = x$ differ in any way?

10. Write each of the following as an algebraic expression.
(a) $\cos(\arcsin x)$ **(b)** $\sin(\arccos x)$
(c) $\sin(\arctan x)$ **(d)** $\sin(2\arcsin x)$

11. **(a)** Find the value of $\arctan\left(\tan\frac{3\pi}{4}\right)$.
(b) For what values of x is $\arctan(\tan x) = x$?
(c) For what values of x is $\tan(\arctan x) = x$?

1.9* Review

Label each statement as true or false. If a statement is true, prove it. If not,
(i) give an example of why it is false, and
(ii) if possible, correct it to make it true, and then prove it.

1. If A and B are any two sets, then $A = (A \cap B) \cup (A \setminus B)$.

2. If A and B are any two sets, then $A \setminus B = A$ if and only if $A \cap B = \phi$.

3. If A and B are any two sets, then $A \setminus B = \phi$ if and only if $A \subset B$.

4. The product of an even function and an odd function is even.

5. $f(x) = x^3 - 1$ is strictly increasing.

6. $f:(-1,2) \to \mathfrak{R}$ with $f(x) = x^3 + 2x$ is an odd function.

7. If functions $f, g:\mathfrak{R} \to \mathfrak{R}$ are odd, then $(fg)(x) \geq 0$ or $(fg)(x) \leq 0$ for all $x \in \mathfrak{R}$.

8. 1 is a prime number.

9. If $a_1 = 4$ and $a_{n+1} = a_n + 1$ for all $n \in N$, then $a_n = n + 3$ for all $n \in N$.

10. If $a_1 = -5$ and $a_{n+1} = a_n + 2$ for all $n \in N$, then $a_n = -7n + 2$ for all $n \in N$.

11. If $a_1 = 1$ and $a_{n+1} = \sqrt{1 + \sqrt{a_n}}$ for all $n \in N$, then $a_{n+1} \geq a_n$ for all $n \in N$.

12. If $a, b \in \mathfrak{R}$ such that $a \leq b + \varepsilon$ for any $\varepsilon > 0$, then $a < b$.

13. If $a, b \in \Re$, then $\left[\frac{1}{2}(a+b)\right]^2 \leq \frac{1}{2}\left(a^2+b^2\right)$.

14. $\sum_{k=1}^{n} ar^{k-1} = \frac{a\left(r^n-1\right)}{r-1}$ for any real numbers a and r with $r \neq 1$. (Assume that $r^0 = 1$ even if $r = 0$.)

15. $1^3 + 2^3 + \cdots + n^3 = (1+2+\cdots+n)^2$ for all $n \in N$.

16. $\sum_{k=1}^{n} \frac{k^2}{(2k-1)(2k+1)} = \frac{n+1}{2(2n+1)}$.

17. $\sum_{k=1}^{n} \frac{1}{(3k-2)(3k+1)} = \frac{n}{3n+1}$.

18. If $A = \{0, \phi, \{\phi\}\}$, then $\phi \in A$ and $\{\phi\} \in A$.

19. If $A = \{0, \phi, \{\phi\}\}$, then $\phi \subset A$ and $\{\{\phi\}\} \subseteq A$.

20. If $a > 1$, then $a^n \geq a$ for all $n \in N$.

21. If $0 < a < 1$, then $a^{n+1} < a^n < 1$ for all $n \in N$.

22. $\left||a| - |b|\right| \leq |a+b|$ for all $a, b \in \Re$.

23. If A, B, C, and D are any sets, then $(A \cap B) \times C = (A \times C) \cap (B \times C)$.

24. If A, B, C, and D are any sets, then $(A \times B) \cap (C \times D) = (A \cap C) \times (B \cap D)$. See Exercises 5(a) and 5(b) in Section 1.1.

25. If A, B, C, and D are any sets, then $(A \times B) \cup (C \times D) = (A \cup C) \times (B \cup D)$. See Exercises 5(c) and 5(d) in Section 1.1.

26. If A and B are any sets and $B \subseteq A$, then $B = A \setminus (A \setminus B)$. See Exercise 4 in Section 1.1.

27. If A and B are any sets, then $A \cap B$ and $A \setminus B$ are disjoint.

28. If $a, b \in Q$ with $ab > 0$ and $a < b$, then $\frac{1}{a} > \frac{1}{b}$.

29. If $a, b, c, d \in \Re^+$ and $\frac{a}{b} < \frac{c}{d}$, then $\frac{a}{b} < \frac{a+c}{b+d} < \frac{c}{d}$.

30. If $a, b \geq 0$, then $\sqrt{a^2+b^2} = a+b$.

31. Suppose F is an ordered field and $a, b, c, d \in F$. If $a < b$ and $c < d$, then $ac < bd$.

32. If $0 < a < 1$, then $\sqrt[n]{a} < 1$ for all $n \in N$.

33. If $a > 1$, then $a^{n+1} > a^n > 1$ for all $n \in N$.

34. If $a, b \in \Re$, then $a^2 + b^2 \geq 2ab$.

35. $-\max\{a, b\} = \min\{-a, -b\}$.

36. If $0 < a < 1$, then $0 \leq a^n \leq a$ for all $n \in N$.

37. $\sum_{k=0}^{n} \binom{n}{k} = 2^n$ and $\sum_{k=0}^{n} \binom{n}{k} 2^k = 3^n$ for all $n \in N$.

38. $(k+1)\binom{n}{k+1}+k\binom{n}{k}=n\binom{n}{k}$ if k and n are integers such that $0 \le k \le n$.

39. $\sum_{k=0}^{n}\binom{n}{k}a^{n-k}b^{k}=\sum_{k=0}^{n}\binom{n}{k}a^{k}b^{n-k}$ for all $n \in N$.

40. The function $f(x)=2x^3+x^2-2x-6$ has three real roots.

41. If a function f has a supremum, then $\max f$ exists and $\max f = \sup f$.

42. If $a_1=1$ and $a_{n+1}=\dfrac{(a_n)^2+2}{2a_n}$ for all $n \in N$, then $a_n \ge \sqrt{2}$ for all natural numbers $n \ge 2$.

43. If $a_1=1$ and $a_{n+1}=\dfrac{(a_n)^2+2}{2a_n}$ for all $n \in N$, then $a_{n+1} \le a_n$ for all $n \ge 2$.

44. If $a_1=1$ and $a_{n+1}=\dfrac{(a_n)^2+2}{2a_n}$ for all $n \in N$, then a_n is rational for all $n \in N$.

45. If $a,b \in \Re$, then $|ab| \le |a|\cdot|b|$.

46. If $a,b \in \Re$, then $|a-b| \le |a|+|b|$.

47. If $a,b \in \Re \setminus \{0\}$, then $\left|\dfrac{a}{b}+\dfrac{b}{a}\right| \ge 2$.

48. If a function f is injective, then it is strictly monotone.

49. If a function f is strictly monotone, then it is injective.

50. If $0 \le a \le b$ and $0 \le c \le d$, then $0 \le ac \le bd$.

51. If $x>0$, then $\sin x < x$.

52. Set of algebraic numbers is countably infinite.

53. Set of transcendental numbers is uncountable.

54. Every rational number is algebraic, but every irrational number is not necessarily algebraic.

55. If $f,g: A \to \Re$ where $A \ne \phi$, and R_f and R_g are bounded, then $\sup_{x\in A}(f+g)(x) \le \sup_{x\in A} f(x)+\sup_{x\in A} g(x)$ and $\inf_{x\in A}(f+g)(x) \ge \inf_{x\in A} f(x)+\inf_{x\in A} g(x)$.

56. For sets A and B, $A \subseteq B$ if and only if $A \cup B = B$. Compare this result to Exercise 4(a) in Section 1.1.

57. The supremum of a set A is unique, if it exists.

58. $|\sin(nx)| \le n|\sin x|$ with $n \in N$ and $x \in \Re$.

59. If f and g are two functions for which $f \circ g$ and $g \circ f$ are defined, then $(f \circ g)(x)=(g \circ f)(x)$ for all $x \in D_f \cap D_g$.

60. If an invertible function f has domain $\Re$, then f^{-1} also has domain $\Re$.

1.10* Projects

Part 1. Fibonacci Numbers

Consider the following sequence of numbers, which begins with two ones and where each successive term is equal to the sum of the preceding pair of terms. These numbers, called *Fibonacci*[30] *numbers*, are arranged in the *Fibonacci sequence*

$$1, 1, 2, 3, 5, 8, 13, 21, 34, 55, 89, 144, \ldots .$$

If $F_1 = 1$, $F_2 = 1$, $F_3 = 2$, $F_4 = 3$, and so on, then this sequence can be written recursively as

$$F_{n+2} = F_{n+1} + F_n \text{ for all } n \in N, \text{ where } F_1 = F_2 = 1.$$

See Exercise 7(i) in Section 1.3. Fibonacci numbers have an incredible number of properties. We will explore a few of the basic ones. To this end, let us relate Pascal's triangle to Fibonacci numbers. Recall Figure 1.3.1. Lining up the numbers of Pascal's triangle on the left and then adding along the dotted lines, we obtain the indicated Fibonacci values (see Figure 1.10.1).

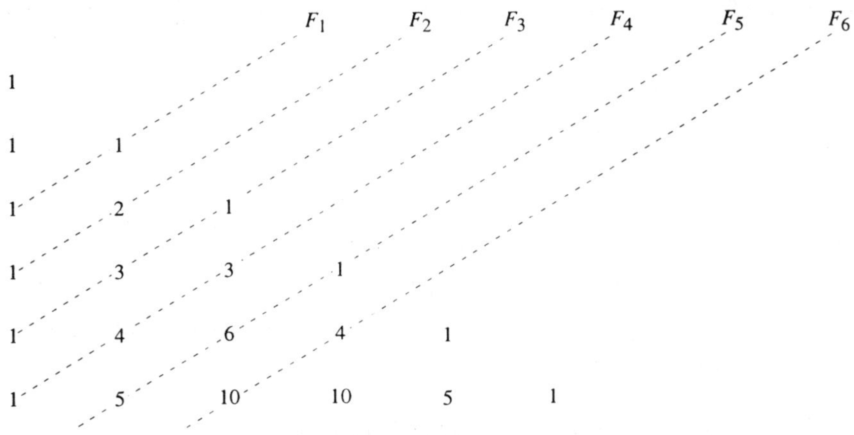

Figure 1.10.1

In general, since Pascal's triangle can be used to obtain binomial coefficients, we can

---

[30]Leonardo Fibonacci (c.1170–1230), also known as Leonardo da Pisa, was an Italian best known for his *Liber Abaci*, which gave a complete and systematic explanation for the use of Hindu and Arabic numbers and a value of positional notation and brought an end to the old Roman system. Fibonacci also worked on Diophantine analysis (named after Greek arithmetician and algebraist Diophantus of Alexandria c. 250 A.D.), wrote on the theory of numbers, applied algebra to solving geometric problems, and did work with trigonometry. Leonardo, the son of Bonaccio is Filius Bonaccii in Latin from which the name Fibonacci is derived.

write that $F_{n+1} = \sum_{k=0}^{\left[\frac{n}{2}\right]} \binom{n-k}{k}$, where $\left[\frac{n}{2}\right]$ is the greatest integer function of $\frac{n}{2}$.

Problem 1.10.1. Compute F_5 and F_8 using the above formula.

One basic identity of Fibonacci numbers is

$$F_1 + F_2 + \cdots + F_n = F_{n+2} - 1 \text{ for all } n \in N.$$

This can easily be derived by writing

$$\begin{aligned} F_1 &= F_3 - F_2 \\ F_2 &= F_4 - F_3 \\ F_3 &= F_5 - F_4 \\ &\vdots \\ F_{n-1} &= F_{n+1} - F_n \\ F_n &= F_{n+2} - F_{n+1}. \end{aligned}$$

Since $F_2 = 1$, in adding the columns we obtain the desired identity.

Problem 1.10.2. Use mathematical induction to prove that $\sum_{k=1}^{n} F_k = F_{n+2} - 1$.

Problem 1.10.3. Prove that

(a) $\sum_{k=1}^{n} (F_k)^2 = F_n F_{n+1}$.

(b) $F_1 + F_3 + F_5 + \cdots + F_{2n-1} = F_{2n}$ for all $n \in N$.

(c) $F_2 + F_4 + F_6 + \cdots + F_{2n} = F_{2n+1} - 1$ for all $n \in N$.

(d) $F_{n-1}F_{n+1} - (F_n)^2 = (-1)^n$ for all $n \in N$.

(e) $F_1 - F_2 + F_3 - F_4 + \cdots + (-1)^n F_n = (-1)^{n+1} F_{n-1} + 1$ for all $n \in N$.

(f) $F_{n+3} = 2F_{n+1} + F_n$ for all $n \in N$.

Consider the following new sequence of numbers with

$$a_1 = \frac{1}{1}, a_2 = \frac{2}{1}, a_3 = \frac{3}{2}, a_4 = \frac{5}{3}, a_5 = \frac{8}{5},$$
$$a_6 = \frac{13}{8}, \ldots, \text{ and } a_n = \frac{F_{n+1}}{F_n} \text{ for all } n \in N.$$

Problem 1.10.4. Prove that if $a_n = \frac{F_{n+1}}{F_n}$, then $a_{n+1} = 1 + \frac{1}{a_n}$ for all $n \in N$. See Exercise 14 in Section 2.5.

Suppose we wish to solve the equation $x = 1 + \frac{1}{x}$. Multiplying both sides by x, we obtain $x^2 = x + 1$, and hence, the equation $x^2 - x - 1 = 0$.

Problem 1.10.5. Show that $\alpha = \frac{1+\sqrt{5}}{2}$ and $\beta = \frac{1-\sqrt{5}}{2}$ are two solutions of $x^2 - x - 1 = 0$. Then conclude that $\alpha^2 = \alpha + 1$ and $\beta^2 = \beta + 1$.

Problem 1.10.6. Verify that $\alpha + \beta = 1$, $\alpha - \beta = \sqrt{5}$, and $\alpha\beta = -1$.

Taking the identities $\alpha^2 = \alpha + 1$ and $\beta^2 = \beta + 1$, and multiplying the first by α^n and the second by β^n, where n is some natural number, we obtain

$$\alpha^{n+2} = \alpha^{n+1} + \alpha^n \text{ and } \beta^{n+2} = \beta^{n+1} + \beta^n.$$

Defining the numbers a_n by $a_n = \frac{\alpha^n - \beta^n}{\alpha - \beta}$ for all $n \in N$ we have that

$$a_{n+2} = \frac{\alpha^{n+2} - \beta^{n+2}}{\alpha - \beta} = \frac{\alpha^{n+1} - \beta^{n+1}}{\alpha - \beta} + \frac{\alpha^n - \beta^n}{\alpha - \beta} = a_{n+1} + a_n$$

$$\text{with } a_1 = \frac{\alpha - \beta}{\alpha - \beta} = 1 \text{ and } a_2 = \frac{\alpha^2 - \beta^2}{\alpha - \beta} = \frac{(\alpha - \beta)(\alpha + \beta)}{\alpha - \beta} = 1.$$

The resulting sequence is the so-called *Binet*[31] *form* for the Fibonacci sequence. Hence,

$$F_n = \frac{\alpha^n - \beta^n}{\alpha - \beta} = \frac{1}{\sqrt{5}}\left[\left(\frac{1+\sqrt{5}}{2}\right)^n - \left(\frac{1-\sqrt{5}}{2}\right)^n\right].$$

The above formula was first derived in 1718 by De Moivre.[32] The number $\alpha = \frac{1+\sqrt{5}}{2}$

[31] Jacques Phillipe Marie Binet (1786–1856), a French mathematician and astronomer, discovered the multiplication rule for matrices in 1812 and investigated the foundations of matrix theory.

[32] Abraham De Moivre (1667–1754), a French Protestant forced to seek asylum in England, was a mathematics tutor who often solved problems presented by wealthy patrons. Problem solving led to the memoir *Doctorine of Chances*, in which De Moivre included a formula known today as Stirling's formula. The Royal Society of London appointed De Moivre to a committee to solve the Newton–Leibniz dispute concerning which of them invented calculus first. De Moivre contributed greatly to statistics, probability, analysis, and analytic geometry. In 1722 he published his famous theorem $(\cos x + i\sin x)^n = \cos nx + i\sin nx$. De Moivre, whose main income was from tutoring mathematics, died in poverty.

is a very special number called the *golden number* or the *golden ratio*.[33]

Problem 1.10.7. Verify that if $\alpha^2 = \alpha + 1$, then

(a) $\alpha^3 = 2\alpha + 1$,

(b) $\alpha^4 = 3\alpha + 2$,

(c) $\alpha^5 = 5\alpha + 3$.

In general, by mathematical induction $\alpha^n = \alpha F_n + F_{n-1}$ for all $n \in N$.

Problem 1.10.8. Verify that

(a) $F_4 = \dfrac{\alpha^4 - \alpha^{-4}}{\sqrt{5}}$.

(b) $\alpha^{n-1} < F_n < \alpha^n$ for all $n > 1$.

Problem 1.10.9

(a) Verify that if $a, b > 0$ and $\dfrac{b}{a}$ satisfies the proportion $\dfrac{b}{a} = \dfrac{a+b}{b}$, then $\dfrac{b}{a} \equiv r = \dfrac{1+\sqrt{5}}{2}$.

(b) Verify that if $a, b > 0$ and $\dfrac{b}{a}$ satisfies the proportion $\dfrac{b}{a} = \dfrac{a}{a+b}$, then $\dfrac{b}{a} \equiv m = \dfrac{\sqrt{5}-1}{2}$.

(c) Show that $m + 1 = r$.

(d) Show that $m(m+1) = 1$ and thus, $m = \dfrac{1}{r}$.

(e) Calculate $\dfrac{1}{r}$ by rationalizing the denominator.

(f) Verify that m may be written as a "continued" fraction given by

$$m = \cfrac{1}{1 + \cfrac{1}{1 + \cfrac{1}{1 + \cfrac{1}{1 + \ddots}}}}.$$

---

[33]The golden ratio, dating back to the fifth century B.C., is the number $\dfrac{b}{a}$, which satisfies the proportion $\dfrac{b}{a} = \dfrac{a+b}{b}$. The golden ratio, $\dfrac{1+\sqrt{5}}{2}$ satisfies the equation $x^2 - x - 1 = 0$. This value plays an important role in a number of areas, including the secant method in approximating roots of functions.

It is customary to say that continued fractions were invented by Bombelli.[34]

(g) Show that

$$\frac{\sqrt{5}+1}{2} = 1 + \cfrac{1}{1+\cfrac{1}{1+\cfrac{1}{1+\cfrac{1}{1+\ddots}}}}.$$

See Problem 3.5.17, part (b).

Part 2. Lucas Numbers

Considering the same recurrence formula used for the Fibonacci sequence with $a_{n+2} = a_{n+1} + a_n$ for all $n \in N$, pick $a_1 = 1$ and $a_2 = 2$. This results in no major change of the sequence; the first term of the Fibonacci sequence is simply omitted. But what if $a_1 = 1$ and $a_2 = 3$? Now there is a real change.

Definition 1.10.10. The *Lucas*[35] *numbers*

$$1, 3, 4, 7, 11, 18, 29, 47, \ldots$$

give rise to the *Lucas sequence*, where $L_1 = 1$, $L_2 = 3$, and $L_{n+2} = L_{n+1} + L_n$ for all $n \in N$.

Problem 1.10.11. Verify that $L_3 = \alpha^3 - \alpha^{-3}$.

Problem 1.10.12. Prove that for all $n \in N$, the given identities are true.

(a) $\sum_{k=1}^{n} L_k = L_{n+2} - 3$ **(b)** $\sum_{k=1}^{n} (L_k)^2 = L_n L_{n+1} - 2$

(c) $F_{2n} = F_n L_n$ **(d)** $L_n = F_{n+1} + F_{n-1}$

(e) $L_n = F_{n+2} - F_{n-2}$ **(f)** $F_n = \frac{1}{5}(L_{n-1} + L_{n+1})$

(g) $5(F_n)^2 = (L_n)^2 - 4(-1)^n$ **(h)** $F_{n+1}L_n - L_{n+1}F_n = 2(-1)^n$

---

[34] Rafael Bombelli (1526–1573), an Italian mathematician, engineer, and architect, wrote two textbooks on algebra, by which he contributed greatly to solving geometric problems algebraically. He also studied solutions of cubic and fourth-degree equations.

[35] Edouard Lucas (1842–1891), a French mathematician, studied primary numbers, figurative arithmetic, and tricircular and tetraspheric geometries. Lucas also made the Fibonacci sequence famous.

Part 3. Mean of Real Numbers

Definition 1.10.13. Let n be some fixed natural number, and suppose that $x_1, x_2, \ldots, x_n$ are some positive real numbers. Then a *mean* of $x_1, x_2, \ldots, x_n$ is a real number that is not less than the smallest and not greater than the largest of these numbers.

Several types of mean exist. We will begin our discussion with means involving two positive real numbers, and later extend to means involving more than two numbers.

Definition 1.10.14. Suppose $a, b \in \Re^+$. Then the

(a) *(Harmonic mean)* $H(a,b) = \dfrac{2ab}{a+b}$, written equivalently as $\dfrac{2}{\frac{1}{a}+\frac{1}{b}}$ or $\left[\dfrac{1}{2}\left(\dfrac{1}{a}+\dfrac{1}{b}\right)\right]^{-1}$.

(b) *(Geometric mean)* $G(a,b) = \sqrt{ab}$.

(c) *(Arithmetic mean)* $A(a,b) = \dfrac{a+b}{2}$, also called an *average.*

(d) *(Quadratic mean)* $Q(a,b) = \sqrt{\dfrac{a^2+b^2}{2}}$, sometimes called *root mean square.*

Problem 1.10.15. If $a, b \in \Re^+$ and $a \le b$, prove that

$$a \le \frac{2ab}{a+b} \le \sqrt{ab} \le \frac{a+b}{2} \le \sqrt{\frac{a^2+b^2}{2}} \le b.$$

Equality is true if and only if $a = b$. Review Theorem 1.7.1.

Problem 1.10.16. Suppose $a, b \in \Re$. Prove that

$$\left(\frac{a+b}{2}\right)^2 \le \frac{a^2+b^2}{2}.$$

Problem 1.10.17. Prove that $n! < \left(\dfrac{n+1}{2}\right)^n$ for all natural numbers $n \ge 2$.

In general, if n is some fixed natural number and $x_1, x_2, \ldots, x_n$ are some positive real numbers, then the *harmonic*, *geometric*, *arithmetic* and *quadratic means* are defined by

$$H(x_1, x_2, \dots, x_n) = \left[\frac{1}{n}\left(\frac{1}{x_1} + \frac{1}{x_2} + \cdots + \frac{1}{x_n}\right)\right]^{-1},$$

$$G(x_1, x_2, \dots, x_n) = \sqrt[n]{x_1 x_2 \cdots x_n},$$

$$A(x_1, x_2, \dots, x_n) = \frac{x_1 + x_2 + \cdots + x_n}{n}, \text{ and}$$

$$Q(x_1, x_2, \dots, x_n) = \sqrt{\frac{(x_1)^2 + (x_2)^2 + \cdots + (x_n)^2}{n}},$$

respectively. In addition, inequalities parallel to the ones in Problem 1.10.15 exist. For related work see Exercises 2(n) and 9 from Section 1.3, Exercises 16 and 17 in Section 2.4, Part 2 of Section 2.8, Exercise 27 in Section 5.3, and Exercise 12 in Section 5.4.

Definition 1.10.18. Let n be some fixed natural number, α any fixed real number, and $x_1, x_2, \dots, x_n$ any fixed positive real numbers. Then, the *exponential mean of order* α for the numbers $x_1, x_2, \dots, x_n$ is given by

$$E_\alpha(x_1, x_2, \dots, x_n) = \sqrt[\alpha]{\frac{(x_1)^\alpha + (x_2)^\alpha + \cdots + (x_n)^\alpha}{n}}.$$

Remark 1.10.19. An exponential mean of order 2 is just the quadratic mean, and an exponential mean of order -1 is the harmonic mean. That is, $E_2 = Q$ and $E_{-1} = H$.■

Problem 1.10.20. Suppose $\alpha \in \Re^-$, $\beta \in \Re^+$, n is some fixed natural number, and $x_1, x_2, \dots, x_n$ are any fixed positive real numbers. Prove that

$$E_\alpha(x_1, x_2, \dots, x_n) \le G(x_1, x_2, \dots, x_n) \le E_\beta(x_1, x_2, \dots, x_n).$$

Problem 1.10.21. Suppose n is some fixed natural number and $x_1, x_2, \dots, x_n$ are any fixed positive real numbers. Prove that

(a) $$n^2 \le (x_1 + x_2 + \cdots + x_n)\left(\frac{1}{x_1} + \frac{1}{x_2} + \cdots + \frac{1}{x_n}\right).$$

(b) $$\frac{x_1 + x_2 + \dots + x_n}{\sqrt{n}} \le \sqrt{(x_1)^2 + (x_2)^2 + \cdots + (x_n)^2} \le x_1 + x_2 + \cdots + x_n.$$

Problem 1.10.22

(a) If $x_1, x_2, \dots, x_n \in \Re^+$, prove that

$$n x_1 x_2 \cdots x_n \le (x_1)^n + (x_2)^n + \cdots + (x_n)^n.$$

In particular, show $2|a||b| \le a^2 + b^2$ for all $a, b \in \Re$.

(b) If $u, v \in \Re^+$ and n_1 and n_2 are fixed natural numbers, prove that

$$\sqrt[n_1+n_2]{u^{n_1} v^{n_2}} \leq \frac{n_1 u + n_2 v}{n_1 + n_2}.$$

Use this result to show that there exists $\alpha, \beta \in \Re$ such that $u^{\alpha} v^{\beta} \leq \alpha u + \beta v$ and $\alpha + \beta = 1$. See Exercise 20(h) in Section 5.2.

Problem 1.10.23. Use an inequality relating the arithmetic mean and the geometric mean to prove that for each $n \in N$,

$$n^{\frac{1}{n}} + n^{\frac{1}{n+1}} + \cdots + n^{\frac{1}{2n-1}} \geq n\sqrt[n]{2}.$$

Looking closer at the arithmetic mean, $A(a,b)$, for two real numbers a and b, the value $A(a,b) = \dfrac{a+b}{2} = \dfrac{1}{2}a + \dfrac{1}{2}b$ is exactly halfway between a and b. That is, a and b are "weighted" equally. In applications, however, the weights of a and b can sometimes vary. For example, a final exam in a course is usually weighted more heavily than the midterm exam. If a midterm exam for a particular class counts $\dfrac{1}{3}$ of the grade, whereas the final exam counts $\dfrac{2}{3}$, then the grade is computed by calculating $\dfrac{1}{3}a + \dfrac{2}{3}b$, where a and b represent grades obtained on the midterm and final exams, respectively. The values 1 and 2 in these fractions are called the *weights*. Here, 2 means that the final exam counted twice as much as the midterm exam. This weighted average of a and b, i.e. $\dfrac{1}{3}a + \dfrac{2}{3}b$ can also be written as $ka + (1-k)b$, with $k = \dfrac{1}{3} \in (0,1)$. So, in general, the following definition applies.

Definition 1.10.24. If $x_1, x_2, \ldots, x_n$ are any fixed real numbers, then the *weighted average* (*weighted mean*) is given by

$$\frac{k_1 x_1 + k_2 x_2 + \cdots + k_n x_n}{k_1 + k_2 + \cdots + k_n},$$

where $k_i \in \Re^+$ with $i = 1, 2, \ldots, n$ are called the *weights*.

Problem 1.10.25. Suppose that a class has a total of four exams, where the first exam counts as 15% of the final grade, the second exam as 30%, the third exam as 20%, and the fourth exam as 35%. If a student scores 95, 70, 80, and 85 on exams, in that order, what is his or her average score?

2

Sequences

2.1 Convergence

This chapter deals with what are regarded as the fundamentals of mathematical analysis. The most basic idea in analysis is the concept of a limit. The simplest version of a limit appears in the study of sequences.

Definition 2.1.1. A *sequence* is a function whose domain is any infinite subset of the natural numbers. See Definition 1.2.17, part (e). Sometimes integral subscripts will also be permitted.

Suppose N, the set of natural numbers, is the domain for the function f. Then by a sequence we mean the values $f(1), f(2), f(3), \ldots$. These values are called *terms* of the sequence. The value, $f(n)$, which denotes the nth-*term* (*general term*) of a sequence, gives the formula by which we evaluate all terms of the sequence, provided such a formula exists. For example, a sequence of primes cannot be represented by a formula. Often we write f_n in place of $f(n)$, where n is referred to as an *index* or a *subscript*. The letters f and n are called *dummy variables*. In other words, a sequence defined by $f(n) = \dfrac{n}{n+2}$, with $n \in N$, is identical to $a(k) = \dfrac{k}{k+2}$ with $k \in N$. In fact, we usually pick lowercase letters to represent sequences. And by convention, we represent a sequence by writing $a(n)$ in braces, as $\{a_n\}_{n=1}^{\infty}$, or simply as $\{a_n\}$ if n begins with 1. However, if the sequence starts with the value $n = i$, where i is a natural number other than 1, then we write that sequence as $\{a_n\}_{n=i}^{\infty}$. In this chapter we will assume that subscripts in a sequence increment by 1. In Section 2.6 subscripts will be incremented in different ways.

In discussing the sequence $\{a_n\}$, with $a_n = \dfrac{n}{n+2}$, it is not recommended that we write $\{a_n\} = \left\{\dfrac{n}{n+2}\right\}$. This notation could lead to confusion. For instance, if we consider the sequence $\{a_n\}$, where $a_n = (-1)^n$, then the notation $\{(-1)^n\}$ could be taken to mean the set $\{-1, 1, -1, 1, \ldots\}$, which one would simply interpret as $\{-1, 1\}$. This, however, is incorrect. Intuitively, one should think of a sequence as an ordered listing of numbers, or as functional values corresponding to each natural number that is in the domain of the sequence.

Many types of sequences will be of interest to us. Those sequences that approach a finite value are defined next.

Definition 2.1.2. A sequence $\{a_n\}$ *converges* to a real number A if and only if for each real number $\varepsilon > 0$, there exists a positive integer n^* such that

$$|a_n - A| < \varepsilon \text{ for all } n \geq n^*.$$

Here A is the *limit* (*limiting value*) of the sequence $\{a_n\}$ and we can write $\lim\limits_{n \to \infty} a_n = A$.

Observe that Definition 2.1.2 is used only to prove that the proposed value is indeed the limit of the sequence. How to actually derive this value is not obtainable from the definition. The idea involved in this definition is actually not as complicated as it may seem. We say that a sequence $\{a_n\}$ converges to A if and only if terms of the sequence are no more than a predetermined value ε away from the point A for all n large enough, that is, if we are far enough in the sequence. Since ε is arbitrary, the terms of the sequence have to get closer and closer to A as ε gets smaller and smaller (examine Figure 2.1.1). In the remainder of this chapter, whenever we refer to the letters n, m, n^*, n_1, n_2, etc., we assume they represent a natural number.

Example 2.1.3. Consider the sequence $\{a_n\}$, where $a_n = \dfrac{n}{n+2}$. If we pick a particular ε, say, $\varepsilon = 0.01$, a so-called *tolerance*, find the values n^* and A such that $|a_n - A| < 0.01$ if $n \geq n^*$.

Answer. From our previous experience with limits, we suspect that the limit of $\{a_n\}$ is 1. Thus, pick $A = 1$ and write

$$|a_n - A| = \left|\frac{n}{n+2} - 1\right| = \left|\frac{n-(n+2)}{n+2}\right| = \frac{2}{n+2}.$$

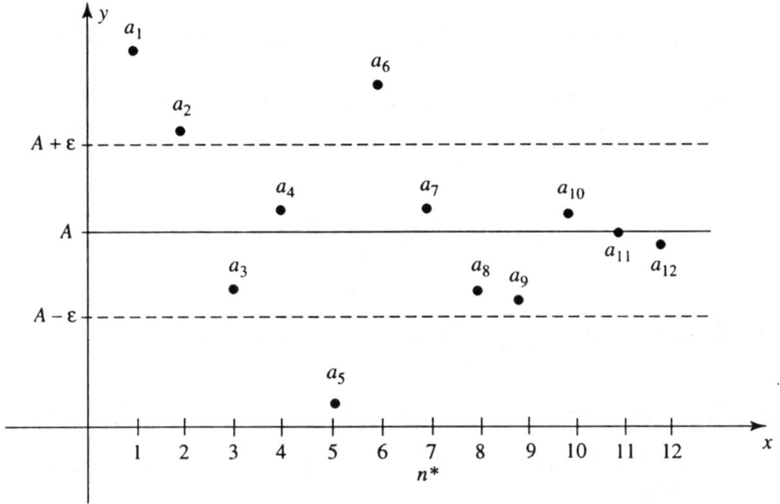

Figure 2.1.1

But $\frac{2}{n+2} < 0.01$, if $2 < 0.01(n+2)$, which is when $n > 198$. Therefore, if $n > 198$ we have $|a_n - 1| < 0.01$. We can, therefore, pick any $n^* \geq 199$. Hence, if $n \geq n^*$, that is, $n \geq n^* \geq 199 > 198$, we have $|a_n - 1| < 0.01$, which was our tolerance. ■

Remark 2.1.4. To prove $\lim_{n\to\infty} a_n = 1$ by the definition, instead of settling for a given tolerance as in the preceding example to prove limits, we use an arbitrary $\varepsilon > 0$. Once this arbitrary $\varepsilon > 0$ is introduced, it stays fixed throughout the problem. Now consider a general case for Example 2.1.3. Begin with an arbitrary $\varepsilon > 0$. We need to find $n^* \in N$ so that

$$|a_n - 1| < \varepsilon \text{ whenever } n \geq n^*.$$

To find this n^*, we write

$$|a_n - 1| = \left|\frac{n}{n+2} - 1\right| = \frac{2}{n+2}.$$

And since we want $|a_n - 1| < \varepsilon$, we consider $\frac{2}{n+2} < \varepsilon$, and solve for n. Hence, $n > \frac{2-2\varepsilon}{\varepsilon}$. Therefore, if we pick $n^* > \frac{2-2\varepsilon}{\varepsilon}$, then for all $n \geq n^*$, we have

$$|a_n - 1| = \left|\frac{n}{n+2} - 1\right| = \frac{2}{n+2} \leq \frac{2}{n^*+2} < \frac{2}{\frac{2-2\varepsilon}{\varepsilon}+2} = \varepsilon.$$

Another possibility for obtaining $|a_n - 1| < \varepsilon$ is to consider $\frac{2}{n+2} < \frac{2}{n}$ and then determine for which n we have $\frac{2}{n} < \varepsilon$. Hence, $n > \frac{2}{\varepsilon}$. Thus, if we pick $n^* > \frac{2}{\varepsilon}$, then for all $n \geq n^*$,

$$|a_n - 1| = \frac{2}{n+2} < \frac{2}{n} \leq \frac{2}{n^*} < \varepsilon.$$ ■

Example 2.1.5. If $a_n = \frac{n^2 - 2}{2n^3 - n - 1}$, prove that the sequence $\{a_n\}$ converges to 0.

Proof. Note that, as we have seen in calculus, $\lim_{n\to\infty} a_n = 0$. Thus, it makes sense to try and prove this assertion. We begin with an arbitrary $\varepsilon > 0$. Again, we will specify the value of n^* later. We write

$$|a_n - 0| = \left|\frac{n^2 - 2}{2n^3 - n - 1}\right| = \frac{|n^2 - 2|}{|2n^3 - n - 1|}.$$

Observe that $n^2 - 2 > 0$ if $n > 1$, and $2n^3 - n - 1 > 0$ if $n > 1$. Therefore,

$$\frac{|n^2 - 2|}{|2n^3 - n - 1|} = \frac{n^2 - 2}{2n^3 - n - 1}$$

if $n > 1$. Although we need to make the preceding expression less then ε, solving for n in

$$\frac{n^2 - 2}{2n^3 - n - 1} < \varepsilon$$

is not trivial. Therefore, we will attempt something similar to the second procedure discussed in Remark 2.1.4. Observe that $n^2 - 2$ can be made a little larger by writing $n^2 - 2 <$ say n^2 for all n. Likewise, $2n^3 - n - 1$ can be made a little smaller by writing $2n^3 - n - 1 >$ say n^3 for $n \geq$ say 100. Actually, $n = 2$ is the smallest value of n for which $2n^3 - n - 1 > n^3$. Thus, we have

$$|a_n - 0| = \frac{n^2 - 2}{2n^3 - n - 1}, \text{ if } n > 1 \text{ and}$$

$$\frac{n^2 - 2}{2n^3 - n - 1} < \frac{n^2}{n^3} = \frac{1}{n}, \text{ if } n \geq 2.$$

Now, $\frac{1}{n} < \varepsilon$ is true if $n > \frac{1}{\varepsilon}$. Thus, $|a_n - 0| < \varepsilon$, if $n \geq 2$ and $n > \frac{1}{\varepsilon}$. Hence, we choose

n^* to be greater than the larger of the two values. So, we write $n^* > \max\left\{2, \frac{1}{\varepsilon}\right\}$. Therefore, if $n \geq n^*$, we have $|a_n - 0| < \varepsilon$. ■

Definition 2.1.6. If for all real numbers A, $\{a_n\}$ does not converge to A, then $\{a_n\}$ *diverges.*

The negation of Definition 2.1.2 can be written as follows. A sequence $\{a_n\}$ does not have a finite limit A, if and only if there exists an $\varepsilon > 0$ such that for every n^* there is an $m > n^*$ with $|a_m - A| \geq \varepsilon$.

Example 2.1.7. Prove that the sequence $\{a_n\}$, where $a_n = \sqrt{n}$, does not converge, that is, it diverges.

Proof. We must begin with an arbitrary real number A and prove that $\lim_{n\to\infty} a_n \neq A$. Let $\varepsilon = 1$. We need to find $n^* \in N$ such that $|a_m - A| \geq \varepsilon$ for all $m \geq n^*$. By Exercise 6(a) from Section 1.7, we can write

$$|a_n - A| = \left|\sqrt{n} - A\right| \geq \sqrt{n} - |A|.$$

But, $\sqrt{n} - |A| \geq 1$, if $n \geq (|A|+1)^2$. Hence, if $m \geq n^* = (|A|+1)^2$, we have $|a_m - A| \geq \varepsilon$.■

Observe that the sequence $\{a_n\}$ defined by $a_n = (-1)^n$ is also a divergent sequence. A formal verification is given in Exercise 2(f).

Remark 2.1.8. Some obvious observations that should be made and proved include the following:

(a) A sequence $\{a_n\}$ is identical to a sequence $\{b_n\}$ if they both attain the same values in exactly the same order.

(b) If $\{a_n\}$ and $\{b_n\}$ differ from each other in only a finite number of terms, then both sequences converge to the same value or they both diverge.

(c) Suppose a_n is defined for all $n \in N$. Then, $\{a_n\}$ converges to A if and only if $\{a_n\}_{n=k}^{\infty}$ for any $k \in N$ converges to A. That is, $\lim_{n\to\infty} a_n = A$ if and only if $\lim_{n\to\infty} a_{n+k} = A$ for $k \in N$. Equivalently, if $\{b_n\}$ is a sequence with $b_n = a_{n+k}$ then $\{b_n\}$ must also converge to A. Therefore, no matter where we start a converging sequence, the new sequence will converge to the same value as the original sequence did.

(d) If the sequence $\{a_n\}$ converges and $a_n = A$ for infinitely many values of n, where A is a constant, then $\{a_n\}$ converges to A. *A constant* means that A is a fixed real number.

(e) If $a_n = A$ for infinitely many values of n, where A is a constant, then $\{a_n\}$ does not necessarily have to converge to A; for example, when $a_n = (-1)^n$.

(f) $\lim_{n\to\infty} a_n = A$ implies that for any $\varepsilon > 0$ there exists $n^* \in N$ such that for all $n \ge n^*$ we have $|a_n - A| < k\varepsilon$ for any fixed $k > 0$.

(g) $\lim_{n\to\infty} a_n = A$ implies that for any constant $K > 0$ there exists $n^* \in N$ such that for all $n \ge n^*$ we have $|a_n - A| < K$.

(h) If $A \ne B$ are real constants and $a_n = A$ and $a_n = B$ for infinitely many values of n, then $\{a_n\}$ diverges. ■

The next few properties are presented as separate theorems.

THEOREM 2.1.9. *Any two limits of a sequence are the same.*

Proof. Suppose a sequence $\{a_n\}$ converges to two values A and B. We will show that A must be identical to B using the triangle inequality, Theorem 1.7.2, part (e).

Let $\varepsilon > 0$ be given. Since $\{a_n\}$ converges to A, there exists $n_1 \in N$ such that

$$|a_n - A| < \boxed{\frac{\varepsilon}{2}}, \text{ for all } n \ge n_1.$$

See Remark 2.1.8, part (f). Also, $\lim_{n\to\infty} a_n = B$ implies that there exists $n_2 \in N$ such that

$$|a_n - B| < \boxed{\frac{\varepsilon}{2}}, \text{ for all } n \ge n_2.$$

Hence, if n is larger than both n_1 and n_2, that is, if $n \ge n^* = \max\{n_1, n_2\}$, we can use both preceding inequalities together and write

$$\begin{aligned} |A - B| &= |A - a_n + a_n - B| && (\text{Add } 0 = a_n - a_n) \\ &\le |A - a_n| + |a_n - B| && (\text{Triangle inequality}) \\ &= |a_n - A| + |a_n - B| && (\text{Rewrite}) \\ &< \boxed{\frac{\varepsilon}{2}} + \boxed{\frac{\varepsilon}{2}} = \varepsilon. \end{aligned}$$

Note that the values in the boxes were chosen so that their sum is equal to a multiple of ε in order to apply Theorem 1.5.6. Hence, $A - B = 0$, which implies $A = B$ and thus the limit, if it exists, is unique. ■

Definition 2.1.10. A sequence $\{a_n\}$ is said to be *bounded* if and only if there exists a positive real constant M such that $|a_n| \le M$ for all $n \in N$. If a sequence is not bounded, then it is said to be *unbounded.*

Intuitively speaking, a sequence is bounded if and only if there exist two horizontal lines $y = M > 0$ and $y = -M < 0$, where all the terms of the sequence lie between them.

At times we talk about the bound of a sequence $\{a_n\}$ on one side only. For instance, if there exists a constant M such that $a_n \leq M$ for all $n \in N$, then $\{a_n\}$ is *bounded above* by M, an *upper bound*. Similarly, if there exists a constant K such that $a_n \geq K$ for all $n \in N$, then $\{a_n\}$ is *bounded below* by K, a *lower bound*. See Definition 1.2.14.

THEOREM 2.1.11. *Any convergent sequence is bounded.*

Proof. Suppose the sequence $\{a_n\}$ converges to A. We need to show that there exists a constant M such that $|a_n| \leq M$ for all $n \in N$. If there were finitely many a_ns, we would just pick the largest (highest) a_n. Since sequences are infinite, there might not be a largest a_n. Why? Hence, bound all values of a_n starting with some a_m, $m \in N$, together with all of its successors. Out of the remaining first finite numbers (exactly $m-1$ of them), we can easily pick the largest element. Thus, since the $\lim\limits_{n\to\infty} a_n = A$, there exists n^*, call it m, such that $|a_n - A| < 1$ for all $n \geq n^* = m$. See Remark 2.1.8, part (g). Adding zero and involving the triangle inequality, we write

$$|a_n| = |a_n - A + A| \leq |a_n - A| + |A| < 1 + |A|$$

for all $n \geq n^* = m$. Hence, $|a_n| \leq M$ for all $n \in N$, if we pick $M = \max\{1+|A|, |a_1|, |a_2|, \ldots |a_{m-1}|\}$. ■

Note that M used in the above proof is not the only possibility available. Also, anything larger than that M would automatically work. Another typical choice for M is

$$M = 1 + |A| + |a_1| + |a_2| + \cdots + |a_{m-1}|.$$

Theorem 2.1.11 is used often, in particular in proving Theorem 2.2.1, part (b). Note that the contrapositive of Theorem 2.1.11 can be used to quickly prove that the sequence given in Example 2.1.7 diverges. The next result is used whenever we need a lower bound. In particular, this result is necessary for the proof of Theorem 2.2.1, part (c).

THEOREM 2.1.12. *Consider a sequence $\{a_n\}$ that converges to a nonzero constant A. Then there exists n^* such that $a_n \neq 0$, for all $n \geq n^*$. In fact, $|a_n| \geq \frac{1}{2}|A|$, for all $n \geq n^*$.*

Proof. Since $\lim\limits_{n\to\infty} a_n = A$, by Remark 2.1.8, part (f), there exists n^* such that

$$|a_n - A| < \boxed{\frac{|A|}{2}}, \text{ if } n \geq n^*.$$

Thus, for $n \ge n^*$, incorporating Exercise 6(a) from Section 1.7, we can write

$$|a_n| = |a_n - A + A| = |A - (A - a_n)| \ge |A| - |A - a_n| > |A| - \frac{|A|}{2} = \frac{|A|}{2} > 0. \qquad \blacksquare$$

Figure 2.1.2 demonstrates Theorems 2.1.11 and 2.1.12. Generalizing Theorem 2.1.12, if $\lim_{n\to\infty} a_n = A > 0$, $K \in [0, A)$, and $L \in (A, \infty)$, then there exists n^* such that $K < |a_n| < L$, for all $n \ge n^*$. We conclude this section by proving a very special limit.

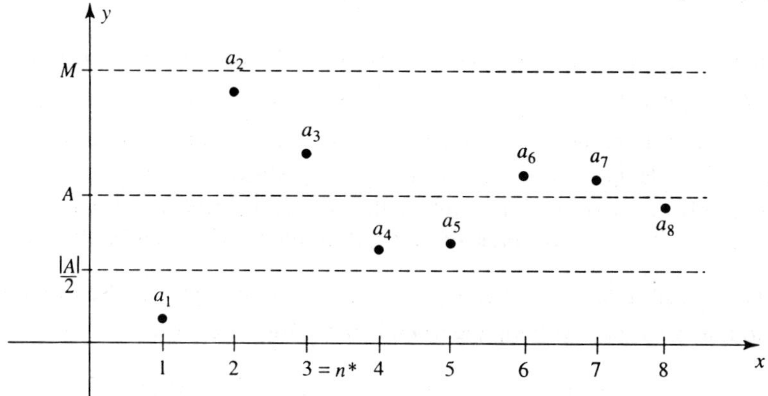

Figure 2.1.2

THEOREM 2.1.13. *If a real number, r, satisfies* $|r| < 1$, *then* $\lim_{n\to\infty} r^n = 0$.

Proof. If $r = 0$, then $r^n = 0$, for any $n \in N$. And so $\lim_{n\to\infty} r^n = \lim_{n\to\infty} 0 = 0$, which is what we wanted to prove. If $r \ne 0$, then we write $|r| = \dfrac{1}{1+b}$, for some $b > 0$. Using the binomial theorem, Theorem 1.3.7, and the fact that $b > 0$, we obtain the expression

$$\begin{aligned}(1+b)^n &= 1 + nb + \frac{n(n-1)}{2}b^2 + \cdots + b^n \\ &\ge 1 + nb \qquad \text{(Bernoulli's inequality; also see Exercise 5 from Section 1.3 and Exercise 20(a) from Section 5.2.)} \\ &> nb.\end{aligned}$$

To prove $\lim_{n\to\infty} r^n = 0$, let us start with an arbitrary $\varepsilon > 0$ and look for n^* so that $|r^n - 0| < \varepsilon$, for all $n \ge n^*$. If we pick $n^* > \boxed{\dfrac{1}{\varepsilon b}}$, then we can write

$$|r^n - 0| = |r^n| = |r|^n = \frac{1}{(1+b)^n} < \frac{1}{nb} < \varepsilon,$$

if $n \ge n^*$. Therefore, pick $n^* > \frac{1}{\varepsilon b}$, since we need $n > \frac{1}{\varepsilon b}$. Hence, $\lim_{n\to\infty} r^n = 0$, provided $|r| < 1$. ■

Exercises 2.1

1. Find $n^* \in N$ so that $\frac{1}{\sqrt{n+1}} < 0.02$, for all $n \ge n^*$.

2. Determine whether the given sequence $\{a_n\}$ converges or diverges with a_n as given. In each case, prove your conclusion.

(a) $a_n = \frac{1}{2n-3}$

(b) $a_n = \frac{n}{n^2-2}$

(c) $a_n = \frac{1}{n^p}$, p a positive constant (If $p = 1$, this sequence is called a *harmonic sequence.*)

(d) $a_n = \frac{n}{2n+\sqrt{n}}$

(e) $a_n = \frac{2n+3}{n^3+1}$

(f) $a_n = (-1)^n$

(g) $a_n = \sqrt{n+1} - \sqrt{n}$

(h) $a_n = (-1)^n \frac{n}{n+1}$

(i) $a_n = \frac{3n}{1-n^2}$

(j) $a_n = \begin{cases} 0, \text{ if } n \text{ is odd} \\ \frac{1}{n}, \text{ if } n \text{ is even} \end{cases}$

(k) $a_n = \begin{cases} 1, \text{ if } n \text{ is odd} \\ \frac{1}{n}, \text{ if } n \text{ is even} \end{cases}$

3. **(a)** Consider the sequence $\{a_n\}$, where $a_n = \frac{1+2+3+\cdots+n}{n^2}$. Show $\{a_n\}$ converges to $\frac{1}{2}$.

(b) Consider the sequence $\{a_n\}$ where $a_n = \frac{1}{n^3} + \frac{2^2}{n^3} + \frac{3^2}{n^3} + \cdots + \frac{n^2}{n^3}$. What value does this sequence converge to?

4. Prove that the sequence $\{a_n\}$ converges to 0 if and only if the sequence $\{|a_n|\}$ converges to 0.

5. Prove that if the sequence $\{a_n\}$ converges to A, then the sequence $\{|a_n|\}$ converges to $|A|$. Is the converse true? Explain. Does this problem differ in any way from the previous problem?

6. Prove that the sequence $\{a_n\}$ converges to A if and only if $\lim_{n\to\infty}(a_n - A) = 0$. Conclude that $\lim_{n\to\infty} c = c$, for any real constant c.

7. If $\{a_n\}$ is a sequence of real numbers, and if $\lim_{n\to\infty} a_{2n} = A$ and $\lim_{n\to\infty} a_{2n-1} = A$, prove that the sequence $\{a_n\}$ converges to A. Is the converse true? Does this problem help you in any way in doing Exercise 2(j)? How about 2(k)? Explain.

8. Suppose $\{a_n\}$ and $\{b_n\}$ are two sequences with $\lim_{n\to\infty} b_n = 0$. If there exist constants A and k and a positive integer n^* such that $|a_n - A| \le k|b_n|$ for all $n \ge n^*$, prove that the sequence $\{a_n\}$ must converge to A.

9. If sequences $\{a_n\}$ and $\{b_n\}$ both converge to A, prove that the *zipper sequence* $a_1, b_1, a_2, b_2, a_3, b_3, \ldots$ also converges to A.

10. Suppose the sequence $\{a_n\}$ converges to A. Define the sequence $\{b_n\}$ by $b_n = \dfrac{a_n + a_{n+1}}{2}$. Does the sequence $\{b_n\}$ converge? If so, specify the value and prove your conclusion. Otherwise, give an example when this is not true.

11. Give an example of a sequence that is bounded but not convergent.

12. If the sequence $\{a_n\}$ converges to a nonzero constant A and $a_n \ne 0$, for any n, prove that the sequence $\left\{\dfrac{1}{a_n}\right\}$ is bounded.

13. Suppose the sequence $\{a_n\}$ converges to a nonzero constant A. Prove that there exists n^* such that if $n \ge n^*$ we have $|a_n| \ge t|A|$, where t is a fixed real number satisfying $0 < t < 1$. Is this statement true if $t = 0$? How about if $t = 1$? Explain.

14. Use the binomial theorem to prove that $\lim_{n\to\infty} \sqrt[n]{c} = 1$, for any constant $c > 0$.

15. Use the binomial theorem to prove that $\lim_{n\to\infty} \sqrt[n]{n} = 1$.

16. Use the binomial theorem to prove that $\lim_{n\to\infty} \dfrac{n^2}{2^n} = 0$.

17. Use the binomial theorem to prove that $\lim_{n\to\infty} nr^n = 0$ if $|r| < 1$.

18. If $\{a_n\}$ is a *geometric sum* defined by $a_n = a + ar + ar^2 + \cdots + ar^n = \sum_{k=0}^{n} ar^k$, a and r are constants, verify that $\{a_n\}$ must converge to $\dfrac{a}{1-r}$ whenever $|r| < 1$.

19. **(a)** Use Exercise 18 to find the limit of the sequence $\{a_n\}$ where

$$a_n = 1 + \frac{1}{2} + \frac{1}{2^2} + \cdots + \frac{1}{2^n}.$$

(b) Use Exercise 18 to verify that the repeating decimal 1.999..., often written as $1.\overline{9}$, is equal to 2. See Section 7.1 for more on this topic.

20. Consider sequences $\{a_n\}$ and $\{b_n\}$ where $b_n = \sqrt[n]{a_n}$.

(a) If $\{b_n\}$ converges to 1, does the sequence $\{a_n\}$ necessarily converge?

(b) If $\{b_n\}$ converges to 1, does the sequence $\{a_n\}$ necessarily diverge?

(c) Does $\{b_n\}$ have to converge to 1?

21. If the sequence $\{a_n\}$ satisfies the property $\lim_{n\to\infty}(a_n - a_{n-2}) = 0$, prove that $\lim_{n\to\infty} \frac{a_n - a_{n-1}}{n} = 0$.

2.2 Limit Theorems

Now that we are better acquainted with sequences and what we mean by their convergence, let us begin a study of their properties. The next result should be somewhat familiar from calculus.

THEOREM 2.2.1. *If sequences $\{a_n\}$ and $\{b_n\}$ converge to A and B, respectively, then*

(a) $\lim_{n\to\infty}(a_n \pm b_n) = A \pm B$.

(b) $\lim_{n\to\infty} a_n b_n = AB$.

(c) $\lim_{n\to\infty} \frac{a_n}{b_n} = \frac{A}{B}$, *if* $B \neq 0$ *and* $b_n \neq 0$ *for any* n.

(d) $\lim_{n\to\infty}(a_n)^p = A^p$ *for all* $p \in N$.

(e) $\lim_{n\to\infty} \sqrt[k]{a_n} = \sqrt[k]{A}$ *if* A *and* a_n *are nonnegative for all* n, *with* $k \in N$.

(f) $A \leq B$ *if* $a_n \leq b_n$ *for all* $n \geq n_1 \in N$.

Observe that part (a) says that sequences $\{a_n \pm b_n\}$ converge to $A \pm B$. We can also write this result as

$$\lim_{n\to\infty}(a_n \pm b_n) = \lim_{n\to\infty} a_n \pm \lim_{n\to\infty} b_n,$$

provided two (in fact, any two) of these limits exist.

Recall that to prove a limit, say, $\lim_{n\to\infty} c_n = C$, we must start with an arbitrary $\varepsilon > 0$ and then find n^* so that

$$|c_n - C| < \varepsilon, \text{ for all } n \geq n^*.$$

Here, we usually obtain n^* in terms of ε at the end of the proof when solving for n in the last expression. However, if we are given that $\lim_{n\to\infty} c_n = C$, then by Remark 2.1.8, part (g), no matter what $K > 0$ is, there always exists n^* such that

$$|c_n - C| < K, \text{ for all } n \geq n^*.$$

Hence, we usually write

$$|c_n - C| < \boxed{}, \text{ for all } n \geq n^*.$$

We fill in the box later with a particular value that works favorably in the proof. The value in the box, which we called K earlier, is usually chosen to be some fraction of a given ε. Recall that throughout proofs involving limits, expressions written in the boxes are computed at the end of the manipulations in the proof.

Proof of part (a) for addition. Let an arbitrary $\varepsilon > 0$ be given. Now we fix it. Since $\lim_{n\to\infty} a_n = A$, there exists n_1 such that for all $n \geq n_1$, we have $|a_n - A| < \boxed{\frac{\varepsilon}{2}}$. Similarly, $\lim_{n\to\infty} b_n = B$ implies that there exists n_2 such that for all $n \geq n_2$ we have $|b_n - B| < \boxed{\frac{\varepsilon}{2}}$. We need to find n^* so that $|(a_n + b_n) - (A + B)| < \varepsilon$ for all $n \geq n^*$. And to use both of the given statements, pick $n^* = \max\{n_1, n_2\}$. Then, for all $n \geq n^*$, we have

$$\begin{aligned} |(a_n + b_n) - (A + B)| &= |(a_n - A) + (b_n - B)| \\ &\leq |a_n - A| + |b_n - B| = |a_n - A| + |b_n - B| < \frac{\varepsilon}{2} + \frac{\varepsilon}{2} = \varepsilon. \end{aligned}$$

Hence, $\lim_{n\to\infty} (a_n + b_n) = A + B$. ■

Next we will attempt to prove Theorem 2.2.1, part (b). Let $\varepsilon > 0$ be given. Since $\lim_{n\to\infty} a_n = A$, there exists n_1 such that for all $n \geq n_1$, we have $|a_n - A| < \boxed{}$. Similarly, $\lim_{n\to\infty} b_n = B$ implies that there exists n_2 such that for all $n \geq n_2$ we have $|b_n - B| < \boxed{}$. We need to find n^* so that $|a_n b_n - AB| < \varepsilon$ for all $n \geq n^*$. Perhaps $n^* = \max\{n_1, n_2\}$ is a good choice. The question remaining is, what expressions are needed inside the empty boxes. After we decide what is needed in the boxes, we will give a formal proof.

Recall that we wish to show that $|a_n b_n - AB| < \varepsilon$. Somehow we need to separate a_n and b_n, and A from B, to use the hypothesis. The trick of adding 0, that is,

$-b_nA + b_nA$, or $-a_nB + a_nB$, can be employed. Picking the first choice and employing the triangle inequality, we write

$$\begin{aligned} |a_nb_n - AB| &= |a_nb_n - b_nA + b_nA - AB| = |b_n(a_n - A) + A(b_n - B)| \\ &\le |b_n||a_n - A| + |A||b_n - B|. \end{aligned}$$

By making each term less than $\frac{\varepsilon}{2}$, the last expression becomes less than ε. From the hypothesis, $|a_n - A|$ and $|b_n - B|$ can be made arbitrarily small. A first attempt might be to write $|a_n - A| < \frac{\varepsilon}{2|b_n|}$ and $|b_n - B| < \frac{\varepsilon}{2|A|}$. Unfortunately, this is not quite acceptable. Note that in the second inequality, A could be zero. This problem, however, can be easily repaired by writing $|b_n - B| < \frac{\varepsilon}{2|A|+1}$, which is, in turn, less than $\frac{\varepsilon}{2|A|}$. The denominator here is certainly never zero. The problem with the first inequality is that $|b_n|$ is not a fixed constant. See Remark 2.1.8, part (f). In addition, $|b_n|$ could very well be zero. To make $\frac{\varepsilon}{2|b_n|}$ smaller and fixed, Theorem 2.1.11 guarantees us a bound for $|b_n|$. Keeping all this in mind, we now present a formal proof of Theorem 2.2.1, part (b).

Proof of part (b). Assume $|b_n| \le M$ for all n, for some $M > 0$, and let $\varepsilon > 0$ be given. Since $\lim_{n\to\infty} a_n = A$, there exists n_1 such that

$$|a_n - A| < \boxed{\frac{\varepsilon}{2M}}, \text{ for all } n \ge n_1.$$

Similarly, since $\lim_{n\to\infty} b_n = B$, there exists n_2 such that

$$|b_n - B| < \boxed{\frac{\varepsilon}{2|A|+1}}, \text{ for all } n \ge n_2.$$

To use both preceding statements, pick $n^* = \max\{n_1, n_2\}$. Then, for all $n \ge n^*$, we have

$$\begin{aligned} |a_nb_n - AB| &= |a_nb_n - b_nA + b_nA - AB| \le |b_n||a_n - A| + |A||b_n - B| \\ &< M\frac{\varepsilon}{2M} + |A|\frac{\varepsilon}{2|A|+1} < \frac{\varepsilon}{2} + \frac{\varepsilon}{2} = \varepsilon. \end{aligned}$$

Therefore, part (b) has been proven.

Proof of part (c). Let $\varepsilon > 0$ be given. We need to find n^* such that

$$\left|\frac{a_n}{b_n} - \frac{A}{B}\right| < \varepsilon, \text{ for all } n \geq n^*.$$

Since B and b_n are not zero, by Theorem 2.1.12, there exists n_1 such that $|b_n| > \frac{|B|}{2}$, if $n \geq n_1$. Since $\lim_{n\to\infty} a_n = A$, there exists n_2 such that

$$|a_n - A| < \boxed{\frac{|B|}{4}\varepsilon}, \text{ for all } n \geq n_2.$$

Likewise, $\lim_{n\to\infty} b_n = B$ implies that there exists n_3 such that for all $n \geq n_3$, we have

$$|b_n - B| < \boxed{\frac{B^2}{4|A|+1}\varepsilon}.$$

To better understand the process, it is wise to leave these boxes blank, proceed with the proof, and decide later what belongs in the boxes. There are a number of possibilities. If we pick $n^* = \max\{n_1, n_2, n_3\}$, then for all $n \geq n^*$, we have

$$\left|\frac{a_n}{b_n} - \frac{A}{B}\right| = \left|\frac{a_nB - b_nA}{b_nB}\right| = \left|\frac{(a_nB - AB) + (AB - b_nA)}{b_nB}\right| \leq \frac{|a_n - A|}{|b_n|} + \frac{|A||b_n - B|}{|b_nB|} \quad \text{(Why?)}$$
$$< \frac{2}{|B|}|a_n - A| + \frac{2}{B^2}|A||b_n - B| < \frac{\varepsilon}{2} + \frac{\varepsilon}{2} = \varepsilon.$$

Hence, the statement is proven.

Proof of part (f). To prove by contradiction, we shall assume $A > B$, and then, once contradiction is reached, the proof will be complete. Since $\lim_{n\to\infty} a_n = A$, there exists n_2 such that for all $n \geq n_2$, we have $|a_n - A| < \boxed{\frac{A-B}{2}}$. In addition, since $\lim_{n\to\infty} b_n = B$, there exists n_3 such that for all $n \geq n_3$, we have $|b_n - B| < \boxed{\frac{A-B}{2}}$. Let $n^* = \max\{n_1, n_2, n_3\}$. Then, using both preceding expressions, we obtain

$$a_n > \frac{A+B}{2} > b_n$$

for all $n \geq n^*$. Why? Therefore, our assumption that $a_n \leq b_n$ is contradicted, and the proof is complete. ■

Although Theorem 2.2.1, part (f) is true if $a_n \leq b_n$, for all $n \geq n_1$, the inequality $\leq$ can be replaced by $\geq$. Note, however, that the inequality cannot be replaced by $<$ nor $>$. This strict inequality is not preserved by limits. See Exercise 8(b). Proof of parts (d) and

(e) of Theorem 2.2.1 are left as exercises. Observe also that Theorem 2.2.1, part (c), and Theorem 2.1.11 can quickly prove Exercise 12 from Section 2.1.

The moral of Theorem 2.2.1 is that the limit of an expression is found by evaluating the limit of each term individually. As the study of the limits develops, other properties distinct from Theorem 2.2.1 will arise that also use this evaluation procedure. Although this procedure simplifies the evaluation of the limits, at the same time it can create confusion if not followed properly. As an illustration, consider the next two examples.

Example 2.2.2. Use Theorem 2.2.1 to evaluate $\lim\limits_{n\to\infty} \dfrac{\sqrt{3n^2+1}}{2n-1}$.

Answer. We write

$$\begin{aligned}\lim_{n\to\infty} \frac{\sqrt{3n^2+1}}{2n-1} &= \lim_{n\to\infty} \frac{\sqrt{n^2\left(3+\frac{1}{n^2}\right)}}{2n-1} = \lim_{n\to\infty} \frac{n\sqrt{3+n^{-2}}}{2n-1} \\ &= \lim_{n\to\infty} \frac{\sqrt{3+\frac{1}{n^2}}}{2-\frac{1}{n}} = \frac{\sqrt{3+\lim\limits_{n\to\infty}\frac{1}{n^2}}}{2-\lim\limits_{n\to\infty}\frac{1}{n}} = \frac{\sqrt{3+0}}{2-0} = \frac{\sqrt{3}}{2}.\end{aligned}$$ ■

Example 2.2.3. Is there anything wrong with writing

$$\lim_{n\to\infty} \sqrt[n]{\frac{1}{n}} = \lim_{n\to\infty} \sqrt[n]{\lim_{n\to\infty}\frac{1}{n}} = \lim_{n\to\infty} \sqrt[n]{0} = \lim_{n\to\infty} 0 = 0?$$

Answer. There certainly is. See Exercise 11(a). Also, see Exercise 17 in Section 2.3. ■

Remark 2.2.4. Two common limits from trigonometry that are used frequently in this textbook are

$$\lim_{n\to\infty} \sin\frac{1}{n} = 0 \quad\text{and}\quad \lim_{n\to\infty} \cos\frac{1}{n} = 1.$$

It is important to be convinced that these limits indeed represent the correct values since a limit and a trigonometric function can change order. We will verify this in Section 4.1 when discussing continuity. We also state without proof that when n is large relative to k, the value of $\sin\frac{k}{n}$, $k \in \Re$, is very much like the value of $\frac{k}{n}$. This means that $\lim\limits_{n\to\infty} \dfrac{\sin\frac{k}{n}}{\frac{k}{n}} = 1$, or, equivalently, that $\lim\limits_{n\to\infty} n\sin\frac{k}{n} = k$. An easy verification can be found in Example 3.3.8 and Exercise 1(e) of Section 5.4. Also see Exercises 10 and 22 of Section 3.3. We will use this result when needed. For instance, to evaluate the $\lim\limits_{n\to\infty}\left(\dfrac{n^2}{3n+1}\sin\dfrac{2}{n}\right)$, we write

$$\lim_{n\to\infty}\left(\frac{n^2}{3n+1}\sin\frac{2}{n}\right) = \lim_{n\to\infty}\left[\left(\frac{n}{3n+1}\right)\left(n\sin\frac{2}{n}\right)\right]$$

$$=\left(\lim_{n\to\infty}\frac{n}{3n+1}\right)\left(\lim_{n\to\infty} n\sin\frac{2}{n}\right)=\left(\frac{1}{3}\right)(2)=\frac{2}{3}. \text{ Why?}$$ ■

The following theorem, which concludes this section, serves as a very useful tool throughout the text. Examples of its application are given as exercises.

THEOREM 2.2.5. *(Sandwich or Squeeze Theorem) Suppose $\{a_n\}$, $\{b_n\}$, and $\{c_n\}$ are three sequences, and suppose that there exists n_1 such that $a_n \le b_n \le c_n$ for all $n \ge n_1$. If $\{a_n\}$ and $\{c_n\}$ both converge to A, then $\{b_n\}$ must also converge to A.*

Proof. Let $\varepsilon > 0$ be given. We need to find n^* such that for all $n \ge n^*$, we have $|b_n - A| < \varepsilon$. Since $\{a_n\}$ converges to A, there must exist $n_2 \ge n_1$ such that $|a_n - A| < \varepsilon$, for all $n \ge n_2$. Likewise, since $\{c_n\}$ converges to A, there exists $n_3 \ge n_1$ such that $|c_n - A| < \varepsilon$ for all $n \ge n_3$. To use both preceding expressions, we need to go past n_2 and n_3. So, we pick $n^* = \max\{n_2, n_3\}$. Then, for all $n \ge n^*$, we can write

$$A-\varepsilon < a_n < A+\varepsilon \text{ and } A-\varepsilon < c_n < A+\varepsilon.$$

Hence, $A - \varepsilon < a_n$ and $c_n < A + \varepsilon$. Thus, since b_n is between a_n and c_n, we have

$$A-\varepsilon < a_n \le b_n \le c_n < A+\varepsilon.$$

Therefore, $-\varepsilon < b_n - A < \varepsilon$, and $|b_n - A| < \varepsilon$, which proves the desired result. ■

As has probably been apparent, proofs of theorems follow a basic procedure. The idea is to write down what we desire to prove, which usually amounts to restating the definition. Then we write the assumptions in equivalent forms. Finally, completion of the proof requires linking the hypothesis with the conclusion. Of course, the more ideas we have, the easier this process becomes, unless we just cannot decide which idea to pursue. The following exercises provide practice in applying the ideas covered in this section.

Exercises 2.2

1. Reprove Theorem 2.2.1, part (c), by first proving that $\lim_{n\to\infty}\frac{1}{b_n}=\frac{1}{B}$ and then employing Theorem 2.2.1, part (b).

2. If the sequence $\{a_n\}$ converges and c is any constant, prove that $\lim_{n\to\infty} ca_n = c\lim_{n\to\infty} a_n$. Do not use Theorem 2.2.1.

3. Prove part (d) and the negative case of part (a) of Theorem 2.2.1.

4. Use Theorem 2.2.1, parts (a) and (d), to prove Theorem 2.2.1, part (b).

5. If sequences $\{a_n\}$ converges to zero, and the sequence $\{b_n\}$ is bounded, prove $\{a_n b_n\}$ converges to zero.

6. Reprove Theorem 2.2.1, part (b), by first proving that $\lim_{n\to\infty}(a_n b_n - AB) = 0$ and then employing Exercise 6 from Section 2.1.

7. Prove Theorem 2.2.1, part (e), for $k = 2$ and $k = 3$.

8. **(a)** Prove Theorem 2.2.1, part (f), directly, that is, without the use of the contradiction argument.

(b) Suppose that the sequences $\{a_n\}$ and $\{b_n\}$ converge to A and B, respectively, and suppose there exists n_1 such that for all $n \geq n_1$, we have $a_n < b_n$. Verify that it is incorrect to conclude that $A < B$.

9. If $\{a_n\}$ converges to A and there exists n_1 such that $a_n > 0$, for all $n \geq n_1$, prove that $A \geq 0$. Do not use Theorem 2.2.1.

10. Use Theorem 2.2.1, part (f), to prove that any converging sequence has a unique limit.

11. Determine whether the given limits exist. Give clear explanations.

(a) $\lim_{n\to\infty} \sqrt[n]{\frac{1}{n}}$ **(b)** $\lim_{n\to\infty} r^{\frac{n+1}{2}}$, if $0 \leq r < 1$

(c) $\lim_{n\to\infty} \frac{1}{2^n}$ **(d)** $\lim_{n\to\infty} \frac{r^n}{n!}$, if $r \in \Re$

(e) $\lim_{n\to\infty} \frac{\sin f(n)}{n^p}$, $p > 0$, a constant, and $f(n)$ represents images of any function $f: N \to \Re$.

(f) $\lim_{n\to\infty} \left(\sqrt{n+1} - \sqrt{n}\right)$ (See Exercise 2(g) in Section 2.1 and Example 2.4.5.)

(g) $\lim_{n\to\infty} \sqrt{n}\left(\sqrt{n+1} - \sqrt{n}\right)$ **(h)** $\lim_{n\to\infty} \left(\sqrt{n^2+1} - n\right)$

(i) $\lim_{n\to\infty} \sqrt[n]{n + \sqrt{n}}$ **(j)** $\lim_{n\to\infty} \sqrt[n]{2^{n+1}}$

(k) $\lim_{n\to\infty} \frac{n^2}{n!}$

12. Consider the sequences $\{a_n\}$ and $\{b_n\}$, where sequence $\{a_n\}$ converges to zero. Is it true that the sequence $\{a_n b_n\}$ converges to zero? Explain. See Exercise 5.

13. Suppose sequences $\{a_n\}$ and $\{a_n b_n\}$ both converge, and $a_n \neq 0$ for any $n \geq n_1$. Is it true that the sequence $\{b_n\}$ must converge? Explain.

14. Use the sandwich theorem to reprove Exercises 14–17 in Section 2.1.

15. Consider the sequence $\{a_n\}$ where $a_n = \dfrac{1}{4n^2 - 1}$. Define the sequence $\{s_n\}$ by $s_n = a_1 + a_2 + \cdots + a_n$. Determine whether or not $\{s_n\}$ converges. If so, find the limit. See Section 7.1 for more on *sequences of partial sums*. Also, see Exercise 2(w) in Section 1.3.

16. Consider the sequence $\{b_n\}$, where $b_n = \dfrac{1}{2} + \dfrac{1}{6} + \cdots + \dfrac{1}{n(n+1)}$. Determine whether $\{b_n\}$ converges. If so, find the limit. This problem reappears in Example 7.1.7. Also, see Exercise 2(r) in Section 1.3.

17. If $0 \le \alpha \le \beta$, define the sequence $\{a_n\}$ by $a_n = \sqrt[n]{\alpha^n + \beta^n}$. Prove that $\{a_n\}$ converges to β.

18. **(a)** Is it possible to have an unbounded sequence $\{a_n\}$ so that $\lim\limits_{n\to\infty} \dfrac{a_n}{n} = 0$?

(b) Prove that if the sequence $\{a_n\}$ satisfies $\lim\limits_{n\to\infty} \dfrac{a_n}{n} = L \ne 0$, then $\{a_n\}$ is unbounded.

19. Consider the sequences $\{a_n\}$ and $\{b_n\}$, where $b_n = \dfrac{a_n + 1}{a_n - 1}$. If $\{b_n\}$ converges to zero, prove that $\{a_n\}$ converges to -1.

20. Consider the sequence $\{a_n\}$, where

$$a_n = \frac{s_p n^p + s_{p-1} n^{p-1} + \cdots + s_1 n + s_0}{t_q n^q + t_{q-1} n^{q-1} + \cdots + t_1 n + t_0},$$

with $p, q \in N$, s_i with $0 \le i \le p$, and t_j with $0 \le j \le q$, are real constants and $s_p t_q \ne 0$. Then prove that $\{a_n\}$ converges to

(a) $\dfrac{s_p}{t_q}$, if $p = q$.

(b) 0, if $p < q$.

This exercise shows that the limit of this type of sequence is determined only by the powers and coefficients of the leading terms. For the case $p > q$, see Section 2.3, Exercise 8.

21. Compute the given limits, if possible.

(a) $\lim\limits_{n\to\infty} \dfrac{1}{n} \sin \dfrac{1}{n}$

(b) $\lim\limits_{n\to\infty} n \sin \dfrac{1}{2n}$

2.3 Infinite Limits

A number of properties pertaining to convergent sequences were introduced in Section 2.2. Most of these properties are not true when sequences diverge. For example, the sequence $\{a_n + b_n\}$ may or may not converge even though both sequences $\{a_n\}$ and $\{b_n\}$ diverge. An example of this situation occurs when $a_n = (-1)^n$ and $b_n = (-1)^{n+1}$. But if $a_n = (-1)^n = b_n$, then $\{a_n + b_n\}$ diverges. Thus, little can be determined from divergent sequences. However, divergent sequences can be subdivided into categories. Some types of divergent sequences have nice properties.

Definition 2.3.1. A sequence $\{a_n\}$ *diverges to* $+\infty$ (*tends to* $+\infty$) if and only if for any $M > 0$, there exists $n^* \in N$ such that $a_n > M$, for all $n \geq n^*$. If this is the case, then the limit exists and we can write $\lim_{n\to\infty} a_n = +\infty$.

The restriction that M is positive is not necessary. Since the definition requires an arbitrary choice of M, we know that for M equal to, say 1, there must exist n^* so that $a_n > 1$ for all $n \geq n^*$. This same n^* would work equally well if M were not chosen to be positive. In Definition 2.3.1, the sequence $\{a_n\}$ tends to $+\infty$ if its terms eventually get larger than any predetermined fixed positive value. This M need not be an integer. Definition 2.3.1 can easily be rephrased for the sequence $\{a_n\}$ *diverging to* $-\infty$. In that case, we would write $\lim_{n\to\infty} a_n = -\infty$ if and only if for any $K > 0$, there exists n^* such that $a_n < -K$, for all $n \geq n^*$. Clearly, if $\{a_n\}$ tends to $+\infty$, then $\{-a_n\}$ tends to $-\infty$, and vice versa. See Theorem 2.3.3 below. Observe that to prove $\lim_{n\to\infty} a_n = A$, A a constant, we need to start the proof with an arbitrary $\varepsilon > 0$. To prove that a sequence *diverges to infinity*, meaning that $\{a_n\}$ has an *infinite limit*, that is, to prove that $\lim_{n\to\infty} a_n = \pm\infty$ (plus or minus infinity), we need to start with an arbitrary $M > 0$, which usually represents a large number. Note that $+\infty$ and $-\infty$ are not real numbers. Also note that if $\{a_n\}$ diverges to infinity, then the *limit exists*, even though its value is $+\infty$ or $-\infty$.

Observe that there are diverging sequences that do not tend to infinity. Those sequences are said to *oscillate*. They are *oscillatory*. The sequence $\{a_n\}$, with $a_n = (-1)^n$, is an example of an oscillatory sequence. Here $\lim_{n\to\infty} a_n$ *does not exist.* Observe that not all oscillating sequences are bounded. In addition, any desired sequence can be written as the sum of two properly chosen oscillating sequences (see Exercise 20). Let us return to sequences that diverge to infinity.

THEOREM 2.3.2. *(Comparison Theorem) If a sequence* $\{a_n\}$ *diverges to* $+\infty$ *and* $a_n \leq b_n$, *for all* $n \geq n_1$, *then the sequence* $\{b_n\}$ *must also diverge to* $+\infty$.

The proof is straightforward and is left as Exercise 1. In addition, rewrite Theorem 2.3.2 with $-\infty$ replacing $+\infty$.

THEOREM 2.3.3. *If a sequence $\{a_n\}$ diverges to $+\infty$ and a sequence $\{b_n\}$ is bounded below by K, then*

(a) $\{a_n + b_n\}$ *diverges to* $+\infty$.

(b) $\{a_n b_n\}$ *diverges to* $+\infty$*, provided* $K > 0$.

(c) $\{ca_n\}$ *diverges to* $+\infty$*, for all positive constants* c.

(d) $\{ca_n\}$ *diverges to* $-\infty$*, for all negative constants* c.

See Exercise 2 for proofs and comments on Theorem 2.3.3. Questions concerning the subtraction and division of sequences $\{a_n\}$ and $\{b_n\}$ are addressed in Exercise 4.

Example 2.3.4. Verify that if $r > 1$, then the sequence $\{a_n\}$, defined by $a_n = r^n$, diverges to $+\infty$.

Answer. First, write $r = 1 + h$, where $h > 0$, and employ Bernoulli's inequality to obtain

$$r^n = (1+h)^n \geq 1 + nh.$$

Since the sequence $\{b_n\}$ with $b_n = 1 + nh$ diverges to $+\infty$ (why?), by Theorem 2.3.2, $\{a_n\}$ diverges to $+\infty$ as well. ■

For a complete understanding of this very important sequence, $\{a_n\}$ with $a_n = r^n$, refer to Exercise 2(f) in Section 2.1 and Theorem 2.1.13. The trivial case of $r = 1$ should also be considered.

Example 2.3.5. Prove that the sequence $\{a_n\}$ with $a_n = \dfrac{-n^3+1}{n^2-n-5}$ diverges to infinity.

Proof. In determining whether to consider $+\infty$ or $-\infty$, writing out a few terms or simply observing that one of the leading terms has a negative coefficient leads us to consider $-\infty$. Let $M > 0$ be given. We want to find n^* so that for all $n \geq n^*$, we will have $a_n < -M$. But, solving $a_n < -M$ for n is not easy. To avoid this task, we need to bound a_n above by something that tends to $-\infty$. Hence, we need to make the numerator larger and the denominator smaller. Although there are many different choices, let us write

$$-n^3 + 1 < -\frac{1}{2}n^3, \text{ for } n \geq 2, \text{ and}$$
$$n^2 - n - 5 < 2n^2, \text{ for } n \geq 1.$$

Therefore, picking $n \geq 2$, and, since the numerator is negative, we write

$$a_n = \frac{-n^3+1}{n^2-n-5} < \frac{-\frac{1}{2}n^3}{2n^2} = -\frac{1}{4}n.$$

But $-\frac{1}{4}n \le -M$ yields that $n \ge 4M$. Thus, picking $n^* \ge \max\{2, 4M\}$, if $n \ge n^*$, we have $a_n < -M$. ■

The preceding lengthy proof can be shortened as shown next. Hopefully, this "behind-the-scenes" proof provided insight.

Pick any $M > 0$. Let $n^* \ge \max\{2, 4M\}$. If $n \ge n^*$, we have

$$a_n = \frac{-n^3+1}{n^2-n-5} < \frac{-\frac{1}{2}n^3}{2n^2} = -\frac{1}{4}n \le -M.$$

Hence, $\lim_{n\to\infty} a_n = -\infty$. It should be noted that perhaps showing that $\{-a_n\}$ tends to $+\infty$ and implementing part (d) of Theorem 2.3.3 would be an easier approach.

See Exercises 5 and 8 for more information concerning rational expressions. There are other ways to determine divergence to infinity. The next result relates ideas from previous sections to the divergence to infinity.

THEOREM 2.3.6. *Consider a sequence $\{a_n\}$, where $a_n > 0$ for all n. Then $\{a_n\}$ diverges to $+\infty$ if and only if the sequence $\left\{\frac{1}{a_n}\right\}$ converges to zero.*

Proof. First, suppose that $\lim_{n\to\infty} a_n = +\infty$, and prove that $\lim_{n\to\infty} \frac{1}{a_n} = 0$. Since we need to prove the existence of a finite limit, pick $\varepsilon > 0$. Then, find n^* such that

$$\left|\frac{1}{a_n} - 0\right| = \frac{1}{a_n} < \varepsilon \text{ for all } n \ge n^*.$$

Since $\{a_n\}$ diverges to $+\infty$, there exists n_1 such that for all $n \ge n_1$, we have $a_n > \frac{1}{\varepsilon}$, where $\frac{1}{\varepsilon}$ is a particular choice of what we denoted M in Definition 2.3.1. So, pick $n^* \ge n_1$. Then, for all $n \ge n^*$, we have $\frac{1}{a_n} < \frac{1}{\varepsilon^{-1}} = \varepsilon$, which is what we needed to show. The converse is similarly shown. ■

Sometimes we have two sequences that diverge to infinity, and we wish to determine which diverges more quickly. For example, if $a_n = n$ and $b_n = 2n$, we can see that $b_n > a_n$, so $\{b_n\}$ tends to infinity more quickly than $\{a_n\}$, but not much more quickly. For instance, if $a_n = n$ and $b_n = n^2$, $b_n > a_n$ and $\{b_n\}$ tends to infinity much more quickly than $\{a_n\}$. Here, we would say that n^2 is of *higher order* than n. The reason is that $\lim_{n\to\infty} \frac{n^2}{n}$ is infinite. In the previous example, where $b_n = 2n$, $\lim_{n\to\infty} \frac{2n}{n} = 2$, a finite nonzero constant. In that case, $\{a_n\}$ and $\{b_n\}$ were of the *same order*. That is, sequences $\{a_n\}$ and $\{b_n\}$ grow at the *same rate* if $\lim_{n\to\infty} \frac{a_n}{b_n} = L \neq 0$. This result also generalizes to functions on larger domains. What if $\lim_{n\to\infty} \frac{a_n}{b_n} = 0$? In that case, by Theorem 2.3.6, the sequence $\{b_n\}$ is of higher order, provided it does not oscillate (see Exercise 9). Thus, to determine which sequence is of higher order, we often combine the two sequences of interest as a quotient and then study the limit of the new combined sequence. Example 2.3.8 demonstrates this idea and also illustrates the use of the next theorem. Also see Definition 2.3.9 and Theorem 3.1.12 for more on this topic.

THEOREM 2.3.7. *Consider a sequence $\{a_n\}$ of nonzero terms such that*

$$\lim_{n\to\infty}\left|\frac{a_{n+1}}{a_n}\right| = \alpha,\ \alpha \text{ a constant.}$$

(a) *If $\alpha < 1$, then $\lim_{n\to\infty} a_n = 0$.*

(b) *If $\alpha > 1$, then $\lim_{n\to\infty} |a_n| = +\infty$. (Sequence $\{|a_n|\}$ diverges to infinity.)*

(c) *If $\alpha = 1$, then $\{a_n\}$ may converge, diverge to infinity, or oscillate.*

Proof of part (a). Suppose $\lim_{n\to\infty}\left|\frac{a_{n+1}}{a_n}\right| = \alpha$ where $\alpha < 1$. Clearly, $\alpha \geq 0$. Thus, there exists n_1 such that for all $n \geq n_1$, we have $\left|\frac{a_{n+1}}{a_n}\right| \leq \beta$ for some $\beta \in (\alpha, 1)$. Why? So, for $n \geq n_1$, we can write

$$\begin{aligned}
|a_n| &= \frac{|a_n|}{|a_{n-1}|}\cdot\frac{|a_{n-1}|}{|a_{n-2}|}\cdot\frac{|a_{n-2}|}{|a_{n-3}|}\cdot\ \cdots\ \cdot\frac{|a_{n_1+1}|}{|a_{n_1}|}\cdot|a_{n_1}| \\
&\leq \beta\cdot\beta\cdot\beta\cdot\ \cdots\ \cdot\beta\cdot|a_{n_1}| = \beta^{n-n_1}|a_{n_1}| = \beta^n\frac{|a_{n_1}|}{\beta^{n_1}} = c\beta^n.
\end{aligned}$$

where c is a constant equal to $\frac{|a_{n_1}|}{\beta^{n_1}}$. Since $0 < \beta < 1$, we have $\lim_{n\to\infty} \beta^n = 0$ (see Theorem 2.1.13). Thus, by the sandwich theorem, the sequence $\{|a_n|\}$ converges to zero.

And, according to Exercise 4 of Section 2.1, sequence $\{a_n\}$ converges to zero, which proves the desired result.

Proof of part (b). Suppose $\lim_{n\to\infty}\left|\frac{a_{n+1}}{a_n}\right| = \alpha$, where $\alpha > 1$. Define a new sequence $\{b_n\}$ with $b_n = \frac{1}{|a_n|}$. Then $\lim_{n\to\infty}\left|\frac{b_{n+1}}{b_n}\right| = \frac{1}{\alpha} < 1$, and so, by part (a), $\{b_n\}$ converges to zero. But, by Theorem 2.3.6, the sequence $\left\{\frac{1}{b_n}\right\}$ diverges to $+\infty$. Hence, so does $\{|a_n|\}$.

Proof of part (c). For the three examples, refer to Exercise 10. ■

Note that if $\lim_{n\to\infty}\left|\frac{a_{n+1}}{a_n}\right| = +\infty$, then for the sequence $\{b_n\}$ with $b_n = \frac{1}{|a_n|}$, we have $\lim_{n\to\infty}\left|\frac{b_{n+1}}{b_n}\right| = 0$, by Theorem 2.3.6. Why? Therefore, by Theorem 2.3.7, part (a), the sequence $\{b_n\}$ converges to zero. Thus, again by Theorem 2.3.6, $\{|a_n|\}$ diverges to $+\infty$.

From calculus and other areas of mathematics, observe that the increasing exponential function grows more quickly than does any increasing polynomial. Using the next example, we can convince ourselves that the sequence $\{a_n\}$, defined by $a_n = b^n$, $b > 1$ constant, is of higher order than the sequence $\{c_n\}$, where $c_n = n^p$, $p > 0$ constant. If $p \le 0$, then the statement is also true, but trivial. Why? To verify this we will form a new sequence $\{d_n\}$, where $d_n = \frac{n^p}{b^n}$, and prove that $\lim_{n\to\infty} d_n = 0$.

Example 2.3.8. If a sequence $\{d_n\}$ is defined by $d_n = \frac{n^p}{b^n}$, for $p > 0$ and $b > 1$ fixed real numbers, then prove that $\{d_n\}$ converges to zero. Compare to Exercise 11(c).

Proof. Using the result of the previous theorem, we write

$$\lim_{n\to\infty}\left|\frac{d_{n+1}}{d_n}\right| = \lim_{n\to\infty}\frac{(n+1)^p}{b^{n+1}}\cdot\frac{b^n}{n^p} = \frac{1}{b}\left(\lim_{n\to\infty}\frac{n+1}{n}\right)^p = \frac{1}{b}\cdot 1^p = \frac{1}{b}.$$

Since $b > 1$, we have $\frac{1}{b} < 1$, and from Theorem 2.3.7, $\lim_{n\to\infty} d_n = \lim_{n\to\infty}\frac{n^p}{b^n} = 0$. ■

Let us now compare the rate of convergence between sequences.

Definition 2.3.9. Suppose $\{a_n\}$ converges to A, $\{b_n\}$ converges to 0, and $b_n \neq 0$ for any $n \in N$. Then $\{a_n\}$ converges to A with a *rate of convergence* $\mathcal{O}(b_n)$ if and only if $\frac{|a_n - A|}{|b_n|} \leq K$ for sufficiently large values of n and K a positive constant. And we write $a_n = A + \mathcal{O}(b_n)$ and say that $\{a_n\}$ converges to A at approximately the same rate as $\{b_n\}$ converges to its limit, and in short $\{a_n\}$ is "*big oh*" of $\{b_n\}$.

Example 2.3.10. Show that if $a_n = \frac{n+4}{n^2}$ and $c_n = \frac{n^2 - n + 3}{n^4}$, then $\{c_n\}$ converges to zero more quickly than $\{a_n\}$.

Answer. Note that $\lim_{n\to\infty} a_n = \lim_{n\to\infty} c_n = 0$. If $b_n = \frac{1}{n}$ and $d_n = \frac{1}{n^2}$, then because of Definition 2.3.9, we have that $a_n = 0 + \mathcal{O}(b_n)$ and $c_n = 0 + \mathcal{O}(d_n)$. Thus, $\{a_n\}$ converges to zero at a rate similar to that of $\{b_n\}$, and $\{c_n\}$ converges to zero at a rate similar to that of $\{d_n\}$. Since the terms of $\{d_n\}$ tend to zero more rapidly than the terms of $\{b_n\}$, the desired conclusion follows. ■

Finally, recall that in calculus we would write that

$$\lim_{n\to\infty} \frac{n^2 - 3n + 1}{2n^3} = \lim_{n\to\infty} \frac{n^2}{2n^3} = \lim_{n\to\infty} \frac{1}{2n} = 0,$$

concluding that if $a_n = \frac{n^2 - 3n + 1}{2n^3}$, then $\{a_n\}$ converges to zero. Now we can add that the rate of convergence of $\{a_n\}$ to zero is very similar to that of $\{b_n\}$, where $b_n = \frac{1}{2n}$. In fact, the rate of convergence of $\{b_n\}$ to zero is very similar to that of $\{c_n\}$, where $c_n = \frac{1}{n}$. Can you verify this using Definition 2.3.9? The rate of convergence of a sequence can be generalized to a function on a larger domain.

Exercises 2.3

1. Prove Theorem 2.3.2.

2. **(a)** Prove Theorem 2.3.3, part (a). Why must the sequence $\{b_n\}$ be bounded below?

(b) Prove Theorem 2.3.3, part (b). Why must the sequence $\{b_n\}$ be bounded below by a positive value? See Exercise 3 below.

(c) Prove Theorem 2.3.3, parts (c) and (d).

3. If the sequence $\{a_n\}$ diverges to $+\infty$, and the sequence $\{b_n\}$ is bounded below by K, with K a negative constant or zero, does the sequence $\{a_n b_n\}$ have to diverge to $-\infty$? Explain.

4. Suppose that the sequence $\{a_n\}$ diverges to $+\infty$. Find the sequence $\{b_n\}$ so that $\left\{\dfrac{a_n}{b_n}\right\}$

(a) diverges to $+\infty$.
(b) converges to any nonzero constant c.
(c) converges to zero.
(d) diverges to $-\infty$.
(e) oscillates.

5. Prove that the given sequence $\{a_n\}$ diverges to infinity. Do not use Exercise 8 below.

(a) $a_n = \dfrac{n^2+1}{n-2}$
(b) $a_n = \dfrac{n^3-n+1}{2n+4}$
(c) $a_n = \dfrac{-n^4+n^3+n}{2n+7}$
(d) $a_n = \dfrac{n^3+5}{-n^2+8n}$

6. Give an example of an oscillating sequence that is
(a) bounded.
(b) bounded above and unbounded below.
(c) bounded below and unbounded above.
(d) unbounded.

7. Consider the sequence $\{a_n\}$, where $a_n = s_p n^p + s_{p-1}n^{p-1} + \cdots + s_1 n + s_0$, with $p \in N$, and s_i, with $0 \le i \le p$, real constants such that $s_p \ne 0$. Prove that $\{a_n\}$ diverges to infinity.

8. Consider the sequence $\{a_n\}$, where

$$a_n = \frac{s_p n^p + s_{p-1}n^{p-1} + \cdots + s_1 n + s_0}{t_q n^q + t_{q-1}n^{q-1} + \cdots + t_1 n + t_0},$$

with $p, q \in N$, s_i with $0 \le i \le p$, and t_j with $0 \le j \le q$, real constants such that $s_p t_q \ne 0$. Prove that if $p > q$, then $\{a_n\}$ diverges to infinity. For other cases, see Section 2.2 Exercise 20. Note that Exercise 7 above is a special case of this problem.

9. If the sequence $\left\{\dfrac{a_n}{b_n}\right\}$ converges to zero with $a_n \ne 0$ for all n, show that the sequence $\{b_n\}$ may oscillate.

10. Find three examples for Theorem 2.3.7, part (c).

11. Determine whether the given sequences $\{a_n\}$ converge, diverge to infinity, or oscillate. If the limit exists, prove it.

(a) $a_n = \dfrac{b^n}{n!}$, where b is a positive constant. **(b)** $a_n = \dfrac{n^n}{n!}$

(c) $a_n = n^k r^n$, where $|r| < 1$ and k are constants. (See Exercise 17 in Section 2.1.)

12. **(a)** Suppose that $\{a_n\}$ and $\{b_n\}$ are sequences of positive terms. And suppose that

$$\lim_{n\to\infty} \frac{a_n}{b_n} = L > 0.$$

Prove that $\lim_{n\to\infty} a_n = +\infty$ if and only if $\lim_{n\to\infty} b_n = +\infty$.

(b) Show that the conclusion of part (a) fails if $\lim_{n\to\infty} \frac{a_n}{b_n} = 0$ or $\lim_{n\to\infty} \frac{a_n}{b_n} = +\infty$.

13. Suppose that $\lim_{n\to\infty} a_n = +\infty$ and $\lim_{n\to\infty} b_n = B > 0$. Prove that $\lim_{n\to\infty} a_n b_n = +\infty$.

14. Consider two sequences $\{a_n\}$ and $\{b_n\}$, where the sequence $\{a_n\}$ diverges to infinity and the sequence $\{a_n b_n\}$ converges. Prove that $\{b_n\}$ must converge to zero.

15. Suppose $\{a_n\}$ and $\{b_n\}$ are sequences of positive terms such that $\lim_{n\to\infty} \frac{a_n}{b_n} = 0$.

(a) Prove that if $\{a_n\}$ diverges to infinity, then so does $\{b_n\}$.

(b) Prove that if $\{b_n\}$ is bounded, then $\{a_n\}$ must converge to zero.

16. If the sequence $\{a_n\}$ diverges to $+\infty$ and α, β, and k are positive constants, prove that $\lim_{n\to\infty} \dfrac{\alpha a_n}{k + \beta a_n} = \dfrac{\alpha}{\beta}$.

17. In view of Exercise 19(a) in Section 2.1, we can write

$$S = 1 + \frac{1}{2} + \frac{1}{4} + \frac{1}{8} + \frac{1}{16} + \cdots$$
$$S = 1 + \frac{1}{2}\left(1 + \frac{1}{2} + \frac{1}{4} + \frac{1}{8} + \frac{1}{16} + \cdots\right)$$
$$S = 1 + \frac{1}{2}S$$
$$\frac{1}{2}S = 1$$
$$S = 2$$

This evaluation of the limit avoids the usage of Exercise 18 from Section 2.1. Mimicking the preceding line of equalities, consider a different situation. Say, we write

$$\begin{aligned} R &= 1+2+4+8+16+\cdots \\ R &= 1+2(1+2+4+8+16+\cdots) \\ R &= 1+2R \\ R-2R &= 1 \\ R &= -1 \end{aligned}$$

Is there anything suspicious in this argument? Is there any relationship with the one above? Explain.

18. Verify that

(a) $\dfrac{n^2+3n-3}{n^3} = 0 + \mathcal{O}\left(\dfrac{2}{n}\right)$.

(b) $\dfrac{\sin n}{n} = 0 + \mathcal{O}\left(\dfrac{1}{n}\right)$.

19. If $a_n = \dfrac{\sqrt{n^2+1}}{n}$ and $b_n = \dfrac{\sqrt[4]{n^4+1}}{n}$, determine which sequence $\{a_n\}$ or $\{b_n\}$ converges to 1 more quickly.

20. Write any converging sequence as a sum of two oscillating sequences.

2.4 Monotone Sequences

Definition 2.4.1. A sequence $\{a_n\}$ is said to be

(a) *increasing* if and only if for all $n, m \in N$, and $n \le m$, we have $a_n \le a_m$.

(b) *eventually increasing* if and only if there exists $n^* \in N$ such that for all $n, m \in N$, and $n^* \le n \le m$, we have $a_n \le a_m$.

(c) *strictly increasing* if and only if for all $n, m \in N$, and $n < m$, we have $a_n < a_m$.

(d) *eventually strictly increasing* if and only if there exists $n^* \in N$ such that for all $n, m \in N$, and $n^* \le n < m$, we have $a_n < a_m$.

In a similar fashion, we can define *decreasing, eventually decreasing, strictly decreasing*, and *eventually strictly decreasing* sequences. See Definition 1.2.10. Often in the literature, an increasing sequence is called *nondecreasing* and a decreasing sequence is called *nonincreasing*. Although Definition 2.4.1 may seem complicated, the idea behind it is quite simple. By a strictly increasing sequence, we mean a sequence for which the next term is always larger then the previous one. By an eventually strictly increasing sequence, we mean a sequence for which the first finite number of terms are any real numbers, provided that after some point such sequence is strictly increasing. By an increasing and eventually increasing sequence, we mean the same as just stated for strictly increasing and eventually strictly increasing sequences, except that the next term could perhaps be equal to the previous one. Note that a (strictly) increasing sequence is also eventually (strictly) increasing, and a strictly increasing sequence is also an increasing sequence. We demonstrate each case in the next example.

Example 2.4.2

(a) The sequence $1,1,2,2,3,3,\ldots$ is an example of an increasing sequence that is not strictly increasing as well as the Fibonacci sequence. See Part 1 of Section 1.10.

(b) The sequence $4,2,1,3,3,5,5,\ldots$ is an example of an eventually increasing sequence that is not eventually strictly increasing.

(c) The Lucas sequence is an example of a strictly increasing as well as an increasing sequence. See Part 2 of Section 1.10.

(d) The sequence $4,3,2,1,2,3,4,5,\ldots$ is an eventually strictly increasing sequence as well as an eventually increasing sequence.

If a sequence is either increasing or decreasing, then we call it *monotone*. Similarly, a *strictly monotone* sequence is strictly increasing or strictly decreasing. We can define *eventually monotone* and *eventually strictly monotone* sequences in a similar manner.

Remark 2.4.3. Observe that the following are all equivalent statements for a sequence $\{a_n\}$, where $a_n > 0$ for all n.

(a) $\{a_n\}$ is increasing. **(b)** $a_n \le a_{n+1}$, for all n.

(c) $\dfrac{a_{n+1}}{a_n} \ge 1$, for all n. **(d)** $a_{n+1} - a_n \ge 0$, for all n. ■

Remark 2.4.3 can be rewritten for decreasing, strictly monotone, and eventually strictly monotone sequences. In Chapter 5, with the use of differentiation, we can form additional equivalent statements pertaining to this topic. Part (c) of Remark 2.4.3 is reminiscent of Theorem 2.3.7.

Before proceeding to consider convergence and/or divergence of monotone sequences, let us discuss the difference between the sequence $\{a_n\}$ and the set of points $S = \{a_n \mid n \in N\}$, the range of the sequence. To form the set S, we simply throw in all values of a_n and form a set containing them. The difference between $\{a_n\}$, sometimes referred to as a "directed set," and S could be tremendous. For example, consider the sequence $\{a_n\}$, where $a_n = (-1)^n$. This sequence consists of the values $-1,1,-1,1,\ldots$. The set S, however, contains only two elements, 1 and -1. As another example, consider the sequence $\{a_n\}$, with

$$a_n = \begin{cases} \dfrac{1}{n}, & \text{if } n \text{ is odd} \\ 1, & \text{if } n \text{ is even.} \end{cases}$$

See Exercise 2(j) of Section 2.1. The values of this sequence are $1,1,\frac{1}{3},1,\frac{1}{5},1,\frac{1}{7},\ldots$.

Here, $S = \left\{1,\frac{1}{3},\frac{1}{5},\frac{1}{7},\ldots\right\}$. S can be graphed on a real line with dots (see Figure 2.4.1). At this point one might consider different known properties of sets. Is this set bounded?

Does it have a largest or a smallest element? What about an infimum and supremum, etc.?

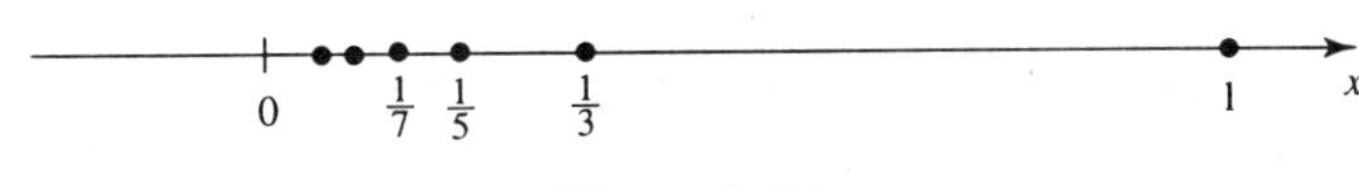

Figure 2.4.1

Having become more accustomed to sequences, let us now think of a sequence $\{a_n\}$ with, say $a_n = 1 - \frac{1}{n}$, not necessarily as points in the two-space $\Re^2$ (see Figure 2.4.2), but rather as points in $\Re^1$ (see Figure 2.4.3). Since it is easier to draw and label points on a horizontal line, the y-coordinates of the sequence are often placed on a horizontal line (see Figure 2.4.4). Thus, sequences are often written as $\{x_n\}$ instead of $\{a_n\}$. For example, compare Figure 2.4.2 with Figure 2.4.3. Both represent the sequence $\{a_n\}$ with $a_n = 1 - \frac{1}{n}$.

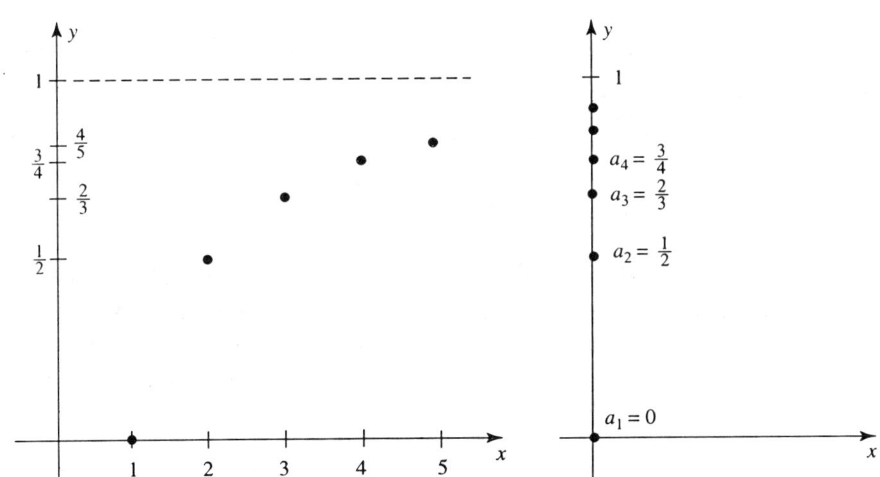

Figure 2.4.2 **Figure 2.4.3**

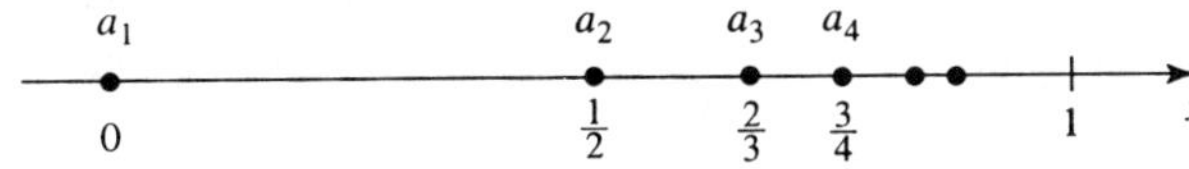

Figure 2.4.4

Next, we present a very important and useful result.

THEOREM 2.4.4. *For an increasing sequence $\{a_n\}$, there are two possibilities:*

(a) $\{a_n\}$ *is bounded above by a constant* M, *in which case* $\lim_{n\to\infty} a_n = L \equiv \sup_{n\in N} a_n \le M$.

(b) $\{a_n\}$ *is unbounded, in which case* $\lim_{n\to\infty} a_n = +\infty$.

Proof of part (a). Suppose $\{a_n\}$ is increasing and bounded above by M. Define the set $S = \{a_n \mid n \in N\}$. Because the sequence is bounded, naturally the set S is bounded, since it contains only values of the sequence. Because these values are real, S has a supremum, call it L. See the completeness property in Definition 1.6.6. Certainly, $L \le M$. To prove $\{a_n\}$ converges to L we start with an arbitrary $\varepsilon > 0$ and look for n^* so that for all $n \ge n^*$, we have $|a_n - L| < \varepsilon$. Since $L = \sup S$, there exists $m \in N$ such that $a_m > L - \varepsilon$. For all $n \ge m$, we have

$$L - \varepsilon < a_m \le a_n \le L < L + \varepsilon.$$

Therefore, $|a_n - L| < \varepsilon$ if $n \ge m$. Hence, if we pick $n^* = m$, the proof is complete.

Proof of part (b). Suppose $\{a_n\}$ is increasing and not bounded above. We will show that $\{a_n\}$ diverges to $+\infty$. Pick any $M > 0$, and show that there exists n^* such that for all $n \ge n^*$, $a_n > M$. Since $\{a_n\}$ is not bounded above, there exists some $m \in N$ such that $a_m > M$. Why? Thus, since $\{a_n\}$ is increasing, $a_n \ge a_m > M$, for all $n \ge m$. So, again pick $n^* = m$ and the conclusion follows. ■

Theorem 2.4.4 is true if $\{a_n\}$ is strictly increasing and eventually increasing. A similar result could be stated for decreasing sequences (see Exercise 5). Note that even though we know by Theorem 2.4.4 that any increasing and bounded sequence converges to a $\sup S$, where $S = \{a_n \mid n \in N\}$, it is not an easy task to actually find this limiting value. Sometimes just proving the existence of the limit suffices.

Example 2.4.5. Determine whether the sequence $\{a_n\}$ with $a_n = \sqrt{n+1} - \sqrt{n}$, converges, diverges, or oscillates. See Exercise 2(g) in Section 2.1 and Exercise 11(f) in Section 2.2.

Answer. Since $\sqrt{n+2} + \sqrt{n+1} > \sqrt{n+1} + \sqrt{n}$, we have that $a_{n+1} < a_n$. Why? Thus, $\{a_n\}$ is decreasing. But, since $\{a_n\}$ is bounded below by zero, it converges. ■

The rest of the section considers sequences that are defined not in terms of n but in terms of previous values of the sequence, that is, recursively (see Section 1.3). For example, take a sequence $\{a_n\}$, where $a_n = n$. This sequence consists of values

$1, 2, 3, 4, \ldots$. Observe that we start with the value of 1 and then, to get the next term, we simply add 1 to the previous term. To generate this same sequence $\{a_n\}$, which is the set of natural numbers, we start with $a_1 = 1$. To get successive terms, we use the recursion formula $a_{n+1} = a_n + 1$ with $n \in N$ (see Example 1.3.8). Similarly, the sequence $\{b_n\}$, where $b_n = 2^n$, is written as $b_1 = 2$ and $b_{n+1} = 2b_n$, $n \in N$. Section 1.3 includes numerous examples of this type. Writing sequences recursively can lead to a successful evaluation of the limit, provided the limit exists. Consider the following example.

Example 2.4.6. Consider the sequence $\{a_n\}$, where $a_n = r^n$ with $0 < r < 1$. In Theorem 2.1.13 we proved that $\{a_n\}$ converges to 0. Here, consider $\{a_n\}$ in a different light. Since $r^{n+1} = r \cdot r^n$ (i.e. $a_{n+1} = ra_n$), we have $r^{n+1} < r^n$ (i.e. $a_{n+1} < a_n$). Therefore, the sequence is strictly decreasing. However, since $r > 0$, we have $r^n > 0$. Thus, $\{a_n\}$ is bounded below. By Theorem 2.4.4, part (a), $\{a_n\}$ converges. To determine the value of the limit, the recursion formula $a_{n+1} = ra_n$ is used. Since $\{a_n\}$ converges, $\{a_{n+1}\}$ must also converge, and the limiting value must be equal in both cases. See Remark 2.1.8, part (c). Call this value A. Taking limits of $a_{n+1} = ra_n$, we obtain $\lim_{n\to\infty} a_{n+1} = r \lim_{n\to\infty} a_n$, resulting in $A = rA$, which implies $A(1-r) = 0$. Since $r \neq 1$, A must be zero. Thus, to prove that the sequence $\{a_n\}$ converges to A, we can write $\{a_n\}$ recursively, then prove that the limit exists using whatever method is applicable, and lastly attempt to compute A by taking the limit on both sides of the recursion formula. This attempt may be at times unsuccessful. See the sequence in Exercise 11(i) in this section and Exercise 16 in Section 2.7. ■

Example 2.4.7. Consider the sequence $\{a_n\}$, defined recursively by

$$a_1 = \sqrt{2} \text{ and } a_{n+1} = \sqrt{2 + a_n} \text{ for } n \in N.$$

Determine convergence or divergence. If $\{a_n\}$ converges, find the limit.

Answer. We will prove that $\{a_n\}$ is strictly increasing, bounded above, and thus convergent. The method of Example 2.4.6 is used to find the limit.

Step 1. Consider a proof by induction that the sequence is strictly increasing. To begin, note that $a_2 = \sqrt{2 + \sqrt{2}} > \sqrt{2} = a_1$. Next, we assume that $a_{k+1} > a_k$, for some $k \in N$, and show that $a_{k+2} > a_{k+1}$. So, we write $a_{k+2} = \sqrt{2 + a_{k+1}} > \sqrt{2 + a_k} = a_{k+1}$. Hence, this sequence is strictly increasing for all $n \in N$.

Step 2. Again using induction, we will prove that $\{a_n\}$ is bounded above by, say, 3. Note that $a_1 = \sqrt{2} < 3$. Suppose $a_k < 3$ and show that $a_{k+1} < 3$, for some $k \in N$. We write $a_{k+1} = \sqrt{2+a_k} < \sqrt{2+3} < 3$. Therefore, $a_n < 3$ for all n.

Step 3. We conclude, by Theorem 2.4.4, part (a), that $\{a_n\}$ converges to some value A. To evaluate A, take the limit of both sides of the recursion formula. Thus, we have $\lim_{n\to\infty} a_{n+1} = \sqrt{2 + \lim_{n\to\infty} a_n}$, yielding $A = \sqrt{2+A}$. This gives us $A^2 - A - 2 = 0$, with the two possibilities that $A = -1$ or $A = 2$. Since the sequence $\{a_n\}$ is certainly positive, we can conclude that $\lim_{n\to\infty} a_n = 2$. ■

Note: steps 1 and 2 in the preceding proof are essential. That is, taking limits in the recursion formula before proving the sequence actually converges can lead to erroneous conclusions. See Exercise 11(d) and 11(e). In addition, as we pointed out at the end of Example 2.4.6, there exist converging sequences whose recursion formulas have limits leading nowhere. Observe also that in Example 2.4.7, we had a sequence of irrational numbers converging to a rational number. Exercises 42–44 from Section 1.9 can be used to show the existence of a sequence of rational numbers that converges to an irrational. Also see exercise 12 below. Another example of the usage of recursively written sequences concludes this section.

THEOREM 2.4.8. *(Archimedean Order Property of $\Re$) Let x be any real number. Then there exists a positive integer n^* greater than x (see Theorem 1.6.8).*

Proof. The proof is by contradiction. Suppose that every positive integer n is less than or equal to x. Then the sequence $\{a_n\}$, where $a_n = n$, is increasing and bounded above by x. Therefore, $\{a_n\}$ converges to, say, A. Yet $a_{n+1} = 1 + a_n$, and taking limits, we have $A = 1 + A$. But this yields 0 = 1, which is certainly not true. Hence, the proof is complete. ■

Exercises 2.4

1. Give an example of a sequence that diverges to $+\infty$ but is not eventually increasing.

2. Give an example of a converging sequence that does not attain a maximum value.

3. Graph each sequence $\{a_n\}$, with a_n as given, in two different ways. In one method, plot the points (n, a_n) in the coordinate plane. In the second method, plot the numbers a_n on a horizontal axis. For each $\{a_n\}$, write the set $S = \{a_n \mid n \in N\}$. Indicate the maximum, minimum, supremum, and infimum values.

(a) $a_n = n - 2$ **(b)** $a_n = \dfrac{1}{n}$ **(c)** $a_n = \dfrac{(-1)^n}{n}$

(d) $a_n = \dfrac{n+1}{n}$ **(e)** $a_n = (-1)^n\left(\dfrac{n-1}{n}\right)$ **(f)** $a_n = 1$

4. Prove that sequences $\{a_n\}$ with a_n as given, are monotone, or eventually monotone.

(a) $a_n = \dfrac{n}{2^n}$ **(b)** $a_n = \dfrac{n^2}{2^n}$

(c) $a_n = \dfrac{3^n}{1+3^{2n}}$ **(d)** $a_n = \dfrac{1\cdot 3\cdot 5\cdot \cdots \cdot(2n-1)}{2^n n!}$

(e) $a_n = \dfrac{n!}{1\cdot 3\cdot 5\cdot \cdots \cdot(2n-1)}$

5. Prove that for an eventually strictly decreasing sequence $\{a_n\}$, there are two possibilities:

(a) $\{a_n\}$ is bounded below by M, in which case $\lim\limits_{n\to\infty} a_n = L \geq M$.

(b) $\{a_n\}$ is unbounded, in which case $\lim\limits_{n\to\infty} a_n = -\infty$.

6. Determine whether the sequence $\{a_n\}$, with a_n as given, is convergent by deciding on monotonicity and boundedness.

(a) $a_n = \dfrac{n+1}{2n+1}$ **(b)** $a_n = \dfrac{n}{2^n}$

(c) $a_n = \dfrac{n^2}{2^n}$ **(d)** $a_n = 1 + \dfrac{1}{2} + \dfrac{1}{4} + \cdots + \dfrac{1}{2^n}$

(e) $a_n = \dfrac{3^n}{1+3^{2n}}$ **(f)** $a_n = \dfrac{n!}{1\cdot 3\cdot 5\cdot \cdots \cdot(2n-1)}$

(g) $a_n = \dfrac{(-1)^{n+1}}{n}$ **(h)** $a_n = \dfrac{2^n}{n!}$

7. Consider the sequence $\{a_n\}$, where $a_n = 1 + \dfrac{1}{2^2} + \dfrac{1}{3^2} + \cdots + \dfrac{1}{n^2}$.

(a) Prove that $\{a_n\}$ converges to A, where $1 \leq A \leq 2$. See Exercise 2(t) from Section 1.3. Is it true that $\dfrac{5}{4} \leq A \leq 2$?

(b) Use Exercise 2(u) from Section 1.3 to prove that $\{a_n\}$ converges to B, where $1 \leq B \leq \dfrac{7}{4}$. It can be proven that $\{a_n\}$ actually converges to $\dfrac{\pi^2}{6}$. See Remark 7.2.2, part (b).

8. Use an idea similar to the one in Example 2.4.6 to prove that the sequence $\{a_n\}$, defined by $a_n = r^n$ with $r > 1$, diverges to $+\infty$.

9. **(a)** Prove that the sequence $\{a_n\}$, with $a_n = \dfrac{3\cdot 5\cdot 7\cdot \cdots \cdot(2n-1)}{2\cdot 4\cdot 6\cdot \cdots \cdot(2n)}$, converges to A, where $0 \le A < \frac{1}{2}$.

(b) Prove that the sequence $\{b_n\}$, with $b_n = \dfrac{2\cdot 4\cdot 6\cdot \cdots \cdot(2n)}{3\cdot 5\cdot 7\cdot \cdots \cdot(2n+1)}$, converges to B, where $0 \le B < \frac{2}{3}$.

(c) Prove that the sequence $\{a_n b_n\}$ converges to zero.

(d) Conclude that $A = 0$.

10. Write the given sequences $\{a_n\}$ recursively.

(a) $a_n = n!$, the *factorial sequence.*

(b) $a_n = ar^{n-1}$, with a and r constants. (This is a *geometric sequence.* Compare with Exercise 18 from Section 2.1 and Example 2.4.6.)

11. For each sequence $\{a_n\}$, defined recursively, determine the limiting value, provided it exists.

(a) $a_1 = \sqrt{6}$ and $a_{n+1} = \sqrt{6 + a_n}$, for all $n \in N$.

(b) $a_1 = 1$ and $a_{n+1} = \sqrt{1 + \sqrt{a_n}}$, for all $n \in N$. (See Exercise 11 from Section 1.9.)

(c) $a_1 = \sqrt{2}$ and $a_{n+1} = \sqrt{2a_n}$, for all $n \in N$.

(d) $a_1 = 2$ and $a_{n+1} = \dfrac{1}{(a_n)^2}$, for all $n \in N$.

(e) $a_1 = 2$ and $a_{n+1} = 3a_n - 1$, for all $n \in N$.

(f) $a_1 = 1$ and $a_{n+1} = 2a_n$, for all $n \in N$. (See Exercise 7(c) from Section 1.3.)

(g) $a_1 = 1$ and $a_{n+1} = 1 + \dfrac{1}{2}a_n$, for all $n \in N$. (See Example 2.5.13 and Exercise 12(b) in Section 2.5.)

(h) $a_1 = 1$ and $a_{n+1} = \left[1 - \dfrac{1}{(n+1)^2}\right]a_n$, for all $n \in N$. (See Exercise 7(f) from Section 1.3.)

(i) $a_1 = 1$, $a_2 = 3$, and $a_{n+1} = 3a_n - 2a_{n-1}$, for all $n \ge 2$. (See Example 1.3.12.)

(j) $a_1 = 1$, $a_2 = \dfrac{1}{3}$, and $a_{n+1} = \dfrac{5}{6}a_n - \dfrac{1}{6}a_{n-1}$, for all $n \ge 2$. (See Exercise 7(h) in Section 1.3.)

(k) $a_1 = 1$, $a_2 = \dfrac{1}{3}$, and $a_{n+1} = \dfrac{5}{3}a_n - \dfrac{4}{9}a_{n-1}$, for all $n \ge 2$.

(l) $a_1 = 1$ and $a_{n+1} = \dfrac{1}{3}a_n$, for all $n \in N$.

12. Suppose A is any positive real number. Define the sequence $\{a_n\}$ recursively by

$$a_{n+1} = \frac{1}{2}\left(a_n + \frac{A}{a_n}\right),$$

with $n \in N$ and with a_1 an arbitrary positive number. Prove that $\{a_n\}$ converges to $\sqrt{A}$. See Exercises 42–44 from Section 1.9. Note that if we write the above a_{n+1} as

$$a_{n+1} = a_n - \frac{(a_n)^2 - A}{2a_n},$$

we obtain the formula for *Newton*[1]*–Raphson*[2] *method* used in calculus when approximating $\sqrt{A}$.

13. Suppose B is any positive real number. Define the sequence $\{b_n\}$ recursively by

$$b_1 = 1 \text{ and } b_{n+1} = \frac{b_n\left[3B + (b_n)^2\right]}{3(b_n)^2 + B}, \text{ for all } n \in N.$$

Prove that $\{b_n\}$ converges to $\sqrt{B}$.

14. Decide what happens to the sequence $\{a_n\}$, defined recursively as

$$a_{n+1} = a_n - \frac{(a_n)^3 - a_n}{3(a_n)^2 - 1}, \text{ for all } n \in N,$$

with the initial value as given. As in Exercise 12, does this expression for a_{n+1} look familiar?

---

[1]Isaac Newton (1642–1727), an English mathematician and astronomer, discovered calculus and laid out the law of gravitation, which he used to explain movement of the planets and tides. Newton initiated theories of light and hydrodynamics and constructed a contemporary reflecting telescope. Although he hesitated often in publishing his findings, Newton's book, *The Mathematical Principles of Natural Philosophy*, is considered to be the most important and influential scientific masterpiece.

Of notable interest is that Newton was born on Christmas day in 1642 under the Julian calendar (introduced by Julius Caesar in 46 B.C.) Christmas day in 1642 was equivalent to Jan. 4, 1643 under the Gregorian calendar, which was what most of Europe used at that time. England adopted the Gregorian calendar in 1752. The Soviet Union, today's Russia, did not adopt the Gregorian calendar until 1916. The Gregorian calendar was put into effect by Pope Gregory XIII, who modified the Julian calendar because, by the 16th century, considerable inaccuracy had accumulated.

[2]Joseph Raphson (1648–1715), an English mathematician, published the work containing the method of approximating the roots of an equation in 1690. Though Newton knew about the method in 1671, it was 46 years before Newton's related published result in 1736.

(a) $a_1 = 2$ (b) $a_1 = \frac{\sqrt{3}}{3}$ (c) $a_1 = 1$

(d) $a_1 = \frac{1}{2}$ (e) $a_1 = 0.1$ (f) $a_1 = \frac{\sqrt{5}}{5}$

15. **(a)** In Problems 1.10.5 and 1.10.9, part (a), we verified that the golden number $\alpha = \frac{1+\sqrt{5}}{2}$ is a solution to the equation $r^2 - r - 1 = 0$. Show that $\alpha = \sqrt{1+\sqrt{1+\sqrt{1+\cdots}}}$.

(b) If $a_1 = 1$ and $a_{n+1} = \sqrt{1+a_n}$ for all $n \in N$, show that the sequence $\{a_n\}$ converges to α. Compare this to Exercises 11(a) and 15(a) above.

(c) Show that $\beta = \frac{\sqrt{5}-1}{2}$ is a solution of $m^2 + m - 1 = 0$. Show that $\beta = \sqrt{1-\sqrt{1-\sqrt{1-\cdots}}}$.

(d) If $b_1 = \frac{1}{2}$ and $b_{n+1} = \sqrt{1-b_n}$ for all $n \in N$, show that the sequence $\{b_n\}$ converges to β even though the sequence is not monotone.

(e) How does the sequence $\sqrt{1}, \sqrt{1-\sqrt{1}}, \sqrt{1-\sqrt{1-\sqrt{1}}}, \ldots$ compare to that in part (d)?

16. Consider sequences $\{a_n\}$ and $\{b_n\}$ which satisfy

$$0 < b_1 < a_1,\ a_{n+1} = \frac{a_n + b_n}{2},\ \text{and } b_{n+1} = \sqrt{a_n b_n},\ \text{for all } n \in N.$$

(a) Show that $b_n < b_{n+1} < a_{n+1} < a_n$.

(b) Show that $0 < a_{n+1} - b_{n+1} < \frac{a_1 - b_1}{2^n}$.

(c) Deduce that $\{a_n\}$ and $\{b_n\}$ converge to the same value called the *arithmetic-geometric mean.*

17. Consider sequences $\{a_n\}$ and $\{b_n\}$, which satisfy

$$0 < b_1 < a_1,\ a_{n+1} = \frac{a_n + b_n}{2},\ \text{and } b_{n+1} = \frac{2a_n b_n}{a_n + b_n},\ \text{for all } n \in N.$$

(a) Prove $\{a_n\}$ and $\{b_n\}$ converge to the same value, called the *arithmetic-harmonic mean.*

(b) Prove that $\lim_{n\to\infty} a_n = \sqrt{a_1 b_1}$.

18. Prove that the following statements are equivalent.

(a) Completeness axiom. That is, every nonempty set of real numbers that is bounded above has a least upper bound. See Definition 1.6.6.

(b) Every monotone sequence that is bounded must converge. See Theorem 2.4.4.

19. Since a number of exercises involve products of only even or odd expressions, we see it fit to introduce a *double factorial* in order to simplify the notation. We define

$$n!! = \begin{cases} 1, \text{ if } n = -1 \text{ or } 0 \\ n(n-2)(n-4)\cdots(5)(3)(1), \text{ if } n \text{ is odd} \\ n(n-2)(n-4)\cdots(6)(4)(2), \text{ if } n \text{ is even.} \end{cases}$$

(a) If $n = 0, 2, 4, \ldots$, verify that $n!! = 2^{\frac{n}{2}}\left(\frac{n}{2}\right)!$.

(b) If $n = 1, 3, 5, \ldots$, verify that

$$n!! = \frac{n!}{2^{\frac{n-1}{2}}\left(\frac{n-1}{2}\right)!}.$$

(c) Is it true that $n!! = (n!)!$? Explain.

20. If $a_1, a_2, a_3 \in \Re$, then $\sqrt[3]{a_1 a_2 a_3} \le \frac{1}{3}(a_1 + a_2 + a_3)$, true or false? Explain. See Exercise 16.

2.5 Cauchy Sequences

We begin this section by introducing accumulation points. Then, after presenting the Bolzano[3]–Weierstrass[4] theorem for sets, we consider Cauchy sequences with Theorem 2.5.9 as the key result. The proof involves knowledge of either subsequences or the Bolzano–Weierstrass theorem for sets. Since subsequences are still foreign to us, the latter is used in the proof. Application of the Cauchy sequence is then demonstrated in the contraction principle, which is widely used in numerical analysis.

---

[3]Bernhard Placidus Johann Nepomuk Bolzano (1781–1848) was born in Prague, Bohemia, known today as the Czech Republic. Although an ordained Roman Catholic priest, Bolzano studied mathematics extensively and in 1824, because of his great human compassion and willingness to stand firm in his beliefs, was forced to retire from his position as chair of Philosophy at the University of Prague. Despite his contributions to the study of force, space, and wave propagation, Bolzano is well known for establishing the importance of rigorous proof versus intuition. He used what we now call a Cauchy sequence four years before Cauchy introduced it. Also, Bolzano's theories of infinity preceded those of Cantor.

[4]Karl Weierstrass (1815–1897), a German mathematician, received international recognition after publishing his research paper in 1854. Opon obtaining an honorary doctorate, Weierstrass quit secondary school teaching to begin a career at the University of Berlin. Recognized as an outstanding teacher, Weierstrass became the world's leading mathematical analyst. Together with his brightest students, Weierstrass paved the road to modern mathematical analysis.

Definition 2.5.1. Let $\varepsilon > 0$ be given. By a *neighborhood* for a real number s, we mean a set $N_\varepsilon(s) \equiv N_\varepsilon = \{x \in \Re \,|\, |x-s| < \varepsilon\}$. A *deleted neighborhood* of s is a set

$$N_\varepsilon{}^*(s) \equiv N_\varepsilon{}^* = \{x \in \Re \,|\, 0 < |x-s| < \varepsilon\}.$$

To better understand the preceding definition, observe that a neighborhood of s is an interval $(s-\varepsilon, s+\varepsilon)$ for any $\varepsilon > 0$. A deleted neighborhood of s is a neighborhood of s that does not contain the central value s. Thus, s is removed (see Figure 2.5.1).

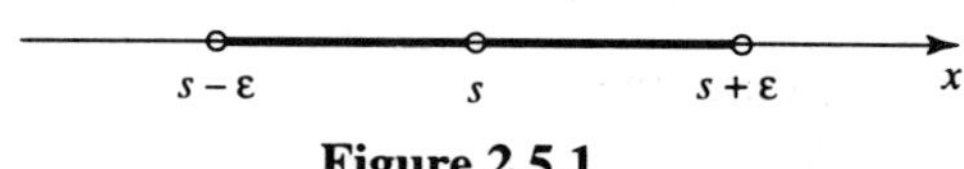

Figure 2.5.1

Definition 2.5.2. Let S be a set of real numbers. A real number s_0 is an *accumulation point* of S if and only if for any $\varepsilon > 0$, there exists at least one point t of S such that $0 < |t - s_0| < \varepsilon$.

An accumulation point is sometimes referred to as a *cluster point.* s_0 is an accumulation point of S if every deleted neighborhood of s_0 contains a point of S. In other words, every neighborhood of s_0 must contain at least one point of S other than s_0 itself. In fact, if s_0 is an accumulation point of S, then any neighborhood of s_0 will have to contain infinitely many points of S (see Exercise 1). Note also that an accumulation point of S need not be an element of S. Consider the following examples.

Example 2.5.3

(a) Any finite set has no accumulation points. For example, consider the set $S = \{1,2,3\}$. A neighborhood of 1 with $\varepsilon = \frac{1}{2}$, that is, an interval $\left(1-\frac{1}{2}, 1+\frac{1}{2}\right) = \left(\frac{1}{2}, \frac{3}{2}\right)$, contains only one point of S, namely, 1 itself. Thus, 1 is not an accumulation point. The same kind of argument applies for the other elements of S.

(b) 0 is an accumulation point of $S \equiv \left\{\frac{1}{n} \,\middle|\, n \in N\right\}$ since the neighborhood $(-\varepsilon, \varepsilon)$ contains infinitely many points of S. Note that zero is not in S. Recall that the sequence $\{a_n\}$ with $a_n = \frac{1}{n}$ converges to zero, so zero is where the sequence "piles up." In fact, zero is the only accumulation point of S. To verify this, pick any $t \in \Re$ with $t \neq 0$. If $t < 0$, then the neighborhood $\left(t - \frac{|t|}{2}, t + \frac{|t|}{2}\right) \cap S = \phi$. If $t > 0$, then pick natural numbers

m_1 and m_2 such that $\dfrac{1}{m_1}$ and $\dfrac{1}{m_2}$ are the closest of all members of S to t, but not equal to t. If $\varepsilon = \min\left\{\left|t - \dfrac{1}{m_1}\right|, \left|t + \dfrac{1}{m_2}\right|\right\}$, then the neighborhood $(t-\varepsilon, t+\varepsilon)$ contains at most one point of S. Why?

(c) The set $S = \{0\} \cup \left\{\dfrac{1}{n} \mid n \in N\right\}$ has only one accumulation point, namely, 0, that is in S.

(d) The interval $I = (0,1]$ has infinitely many accumulation points. In fact, every point of the interval $[0,1]$ is an accumulation point of I. ■

We need to be careful not to confuse an accumulation point of a set S with $\max S / \min S$ or $\sup S / \inf S$. For example, if $S = (0,1] \cup \{2\}$, then $\max S = 2$, $\min S = \text{none}$, $\sup S = 2$, $\inf S = 0$, and the accumulation points form the set $[0,1]$.

THEOREM 2.5.4. *(Bolzano–Weierstrass Theorem for Sets) Every bounded, infinite set of real numbers has at least one accumulation point.*

Proof. Let S be a bounded, infinite set. Since S is bounded, there exists some interval $[a_1, b_1]$ so that $S \subset [a_1, b_1]$. Let

$$c_1 = \frac{1}{2}(a_1 + b_1),$$

that is, the midpoint of $[a_1, b_1]$. Since S is an infinite set, at least one of the intervals $[a_1, c_1]$ or $[c_1, b_1]$ contains infinitely many points of S. We call that interval $[a_2, b_2]$. Note that $a_1 \le a_2 \le b_2 \le b_1$ and $b_2 - a_2 = \dfrac{1}{2}(b_1 - a_1)$, that is, the distance between a_2 and b_2 is one half of the distance between a_1 and b_1. Continuing this process, we get a sequence of intervals $[a_n, b_n]$, each containing infinitely many points of S such that

$$a_1 \le a_2 \le a_3 \le \cdots \le a_n \le \cdots \le b_n \le \cdots \le b_3 \le b_2 \le b_1$$

and

$$b_n - a_n = \frac{1}{2^{n-1}}(b_1 - a_1).$$

Does this remind you of Exercise 16 in Section 2.4? By Theorem 2.4.4, since the sequence $\{a_n\}$ is increasing and bounded above and the sequence $\{b_n\}$ is decreasing and bounded below, they both must converge. So there exist real numbers A and B such that $\lim\limits_{n\to\infty} a_n = A$ and $\lim\limits_{n\to\infty} b_n = B$. Since

$$0 \le b_n - a_n = \frac{1}{2^{n-1}}(b_1 - a_1) \to 0 \text{ as } n \to \infty,$$

by the sandwich theorem, $\{b_n - a_n\}$ converges to zero, which implies that $A = B$. We claim that A is an accumulation point of S. Suppose $\varepsilon > 0$ is given. Then, if n is large enough, we have $A - \varepsilon < a_n \le b_n < A + \varepsilon$. Why? Between a_n and b_n there are infinitely many distinct elements of S, so the interval $(A - \varepsilon, A + \varepsilon)$ contains points of S that are not equal to A. Hence, A is an accumulation point of S. ■

Remark 2.5.5. In Bolzano–Weierstrass theorem for sets, the fact that the set is infinite is essential, because a finite set has no accumulation points. Why? In addition, even if the set is infinite, the existence of an accumulation point does not follow without boundedness. For instance, the set N of natural numbers is infinite with no accumulation points. Thus, an infinite unbounded set may or may not have any accumulation points. Observe that the set $S = \left\{ \frac{1}{n} \middle| n \in N \right\}$ is bounded and infinite, and thus we are guaranteed at least one accumulation point. As in Example 2.5.3, part (b), zero is the accumulation point for set S, $0 \notin S$. In addition, consider the interval $I = (0,1]$ as given in Example 2.5.3, part (d). I is bounded, infinite, and thus contains at least one accumulation point. In fact, I has infinitely many accumulation points. Can you find them? Bolzano-Weierstrass theorem for sets does not necessarily hold if we change $\Re$ to some other ordered field, say Q. For example, the infinite bounded set $\{a_n \mid n \in N\}$ with $\{a_n\}$ as given in Exercise 12 of Section 2.4 has no accumulation points in Q. Why? ■

Definition 2.5.6. A sequence $\{a_n\}$ is called a *Cauchy sequence* if and only if for each $\varepsilon > 0$ there exists $n^* \in N$ such that $|a_m - a_n| < \varepsilon$, for all m and n $\ge$ n*.

There is very little difference in the appearance of this definition and the definition of the convergence of the sequence $\{a_n\}$ given by Definition 2.1.2. However, the meaning and the consequences of the two are quite different. One can see this especially when studying more advanced real analysis. At our level, that is, in the context of real numbers, Cauchy sequences are equivalent to convergent sequences. This equivalency will be proven in Theorem 2.5.9. Note that knowledge of the limit of the sequence is unnecessary in determining whether or not the given sequence is Cauchy. We need to show only that the difference between any two terms of the sequence tends to zero. It is, however, not enough to show that the difference between the successive terms tends to zero. In other words, for an arbitrary $\varepsilon > 0$, even if there exists n^* such that $|a_{n+1} - a_n| < \varepsilon$, for all $n \ge n^*$, the sequence $\{a_n\}$ may diverge. Exercise 8 is an example of this situation. It is important to realize that to show that a given sequence is Cauchy, the indices m and n must be kept entirely independent of one another. On the other hand, to prove that a given sequence is not Cauchy, we need to exhibit a particular relationship between m and n that contradicts the condition $|a_m - a_n| < \varepsilon$. For example, the sequence $\{a_n\}$ with $a_n = (-1)^n$ is not Cauchy since in the case where $m = n+1$, for all n, we have

$$|a_m - a_n| = |a_{n+1} - a_n| = 2.$$

Here, $|a_m - a_n|$ is certainly greater than some positive ε, say $\varepsilon = 1$. Further examples are given in Exercise 7.

Example 2.5.7. Suppose that the sequence $\{a_n\}$ satisfies $|a_{n+1} - a_n| \le \frac{1}{2^n}$. Show that $\{a_n\}$ is Cauchy.

Proof. Let m and n be two distinct natural numbers. Without the loss of generality, suppose that $m > n$. Consequently, we have

$$\begin{aligned} |a_m - a_n| &\le |a_m - a_{m-1}| + |a_{m-1} - a_{m-2}| + \cdots + |a_{n+2} - a_{n+1}| + |a_{n+1} - a_n| \\ &\le \frac{1}{2^{m-1}} + \frac{1}{2^{m-2}} + \cdots + \frac{1}{2^{n+1}} + \frac{1}{2^n} \\ &= \frac{1}{2^n}\left(\frac{1}{2^{m-n-1}} + \frac{1}{2^{m-n-2}} + \cdots + \frac{1}{2^2} + \frac{1}{2} + 1\right) \\ &< \frac{1}{2^n} \cdot 2 = \frac{1}{2^{n-1}}, \end{aligned}$$

using Exercise 19(a) from Section 2.1 in the last step above. Let $\varepsilon > 0$ be given. Since $\lim_{n\to\infty} \frac{1}{2^{n-1}} = 0$, there exists n^* such that $\frac{1}{2^{n-1}} < \varepsilon$, for all $n \ge n^*$. Hence, $|a_m - a_n| < \varepsilon$, for all $m, n \ge n^*$. Therefore, $\{a_n\}$ is Cauchy. ■

Now let us look at a few important properties of Cauchy sequences. Refer to the exercises for additional properties.

THEOREM 2.5.8. *Every Cauchy sequence is bounded.*

The proof of this theorem is so similar to the proof of Theorem 2.1.11 that we have left it as an exercise. The next very important result is not true in the more abstract spaces discussed in higher mathematics. Thus, this theorem cannot be treated as an equivalent definition for convergence. However, as stated earlier, in the context of real numbers, Cauchy sequences are equivalent to convergent sequences. With the use of Theorem 2.5.9, we can determine convergence of the sequence without actually knowing its limit. The same has been done with bounded monotone sequences.

THEOREM 2.5.9. *In $\Re$, a sequence is Cauchy if and only if it is convergent.*

Proof. ($\Leftarrow$) Suppose that the sequence $\{a_n\}$ converges to A. We will prove that it must be Cauchy. Let $\varepsilon > 0$ be given. We want to show that there exists $n^* \in N$ such that for all $m, n \ge n^*$, we have $|a_m - a_n| < \varepsilon$. Since $\lim_{n\to\infty} a_n = A$, there exists n_1 such that

$|a_n - A| < \boxed{\frac{\varepsilon}{2}}$, for all $n \ge n_1$. If we pick $n^* = n_1$, then for all $n \ge n^*$, we have

$$|a_m - a_n| = |a_m - A + A - a_n| \le |a_m - A| + |A - a_n| < \frac{\varepsilon}{2} + \frac{\varepsilon}{2} = \varepsilon.$$

Therefore, the sequence $\{a_n\}$ is Cauchy.

($\Rightarrow$) Suppose that the sequence $\{a_n\}$ is Cauchy. We will prove it must converge. Consider the set $S = \{a_n \mid n \in N\}$. By Theorem 2.5.8, this set is bounded. We need to consider two possibilities: S is finite or S is infinite.

Case 1. Suppose that S is finite. Then the sequence $\{a_n\}$ must be constant from some point on. This is not trivial because the sequence $1, -1, 1, -1, 1, \ldots$ has a finite range but is not constant from any point on. See Exercise 10 for a complete explanation as to why $\{a_n\}$ must be constant from some point on. Therefore, $\{a_n\}$ converges.

Case 2. Suppose that S is infinite. By the Bolzano–Weierstrass theorem for sets, S has an accumulation point, call it A. We will prove that $\{a_n\}$ converges to A. Let $\varepsilon > 0$ be given. We need to find n^* so that $|a_n - A| < \varepsilon$ for all $n \ge n^*$. Since $\{a_n\}$ is Cauchy, there exists n_1 such that for all $m, n \ge n_1$ we have $|a_m - a_n| < \boxed{\frac{\varepsilon}{2}}$. Moreover, since A is an accumulation point of S, any neighborhood of A, in particular, an interval $\left(A - \frac{\varepsilon}{2}, A + \frac{\varepsilon}{2}\right)$, contains infinitely many points of S that are members of $\{a_n\}$. Therefore, there exists an integer $M > n_1$ such that a_M is in this interval. Note that a_M cannot be more than $\frac{\varepsilon}{2}$ away from A. Thus, $|a_M - A| < \frac{\varepsilon}{2}$. Now if $n^* = n_1$, then for all $n \ge n^*$, we have

$$|a_n - A| = |a_n - a_M + a_M - A| \le |a_n - a_M| + |a_M - A| < \frac{\varepsilon}{2} + \frac{\varepsilon}{2} = \varepsilon.$$

Consequently, $\{a_n\}$ converges to A. ■

Observe that in Q the sequence $\{a_n\}$, with $a_n = \left(1 + \frac{1}{n}\right)^n$, is Cauchy but not convergent in Q. See Part 1 of Section 2.8. For a similar reason, in $\Re^+$, the sequence $\{b_n\}$ with $b_n = \frac{1}{n}$ is Cauchy but not convergent. Note that neither $\{a_n\}$ nor $\{b_n\}$ given above diverge in $\Re$. We conclude this section with a method for determining whether a given sequence converges by studying the behavior of the difference between successive terms of the sequence. Again, no limiting value is involved. Review Exercise 8 in this section before continuing.

Definition 2.5.10. A sequence $\{a_n\}$ is said to be a *contractive sequence* if and only if there exists a constant k, with $k \in (0,1)$, such that

$$|a_{n+2} - a_{n+1}| \leq k|a_{n+1} - a_n|$$

for all $n \in N$.

Can you tell what happens to $\{a_n\}$ when $k = 0$ or $k = 1$?

THEOREM 2.5.11. *(Contraction Principle) Every contractive sequence is a Cauchy sequence, and hence, convergent.*

Proof. Let the sequence $\{a_n\}$ be a contractive sequence, as given in Definition 2.5.10. Since the inequality can be used repeatedly, we have

$$|a_{n+2} - a_{n+1}| \leq k|a_{n+1} - a_n| \leq k^2|a_n - a_{n-1}| \leq \cdots \leq k^n|a_2 - a_1|.$$

Without loss of generality, assume that $m > n$ and write

$$\begin{aligned}
|a_m - a_n| &\leq |a_m - a_{m-1}| + |a_{m-1} - a_{m-2}| + \cdots + |a_{n+2} - a_{n+1}| + |a_{n+1} - a_n| \\
&\leq \left(k^{m-2} + k^{m-3} + \cdots + k^{n-1}\right)|a_2 - a_1| && \text{(Why?)} \\
&= k^{n-1}\left(k^{m-n-1} + k^{m-n-2} + \cdots + 1\right)|a_2 - a_1| \\
&= k^{n-1}\left(\frac{1-k^{m-n}}{1-k}\right)|a_2 - a_1| < k^{n-1}\left(\frac{1}{1-k}\right)|a_2 - a_1|. && \text{(See Example 1.3.4.)}
\end{aligned}$$

Since $\left(\frac{1}{1-k}\right)|a_2 - a_1|$ is a constant and $\lim_{n\to\infty} k^{n-1} = 0$ since $0 < k < 1$, the last expression is eventually smaller than any predetermined $\varepsilon > 0$. Review Theorem 2.1.13. Therefore, $\{a_n\}$ is Cauchy, and hence, convergent. ■

COROLLARY 2.5.12. *Suppose $\{a_n\}$ is a contractive sequence. If $\{a_n\}$ converges to A and $k \in (0,1)$, then*

(a) $|a_n - A| \leq \left(\frac{k^{n-1}}{1-k}\right)|a_2 - a_1|$, *for all $n \in N$.*

(b) $|a_n - A| \leq \left(\frac{k}{1-k}\right)|a_n - a_{n-1}|$, *if $n > 1$.*

The proof of this corollary is left as an exercise. Corollary 2.5.12 is covered because of its beauty. Its real meaning will come to life when we study "fixed points" and rates of convergence to the fixed points.

Example 2.5.13. Consider the sequence $\{a_n\}$, defined recursively by $a_1 = 1$ and $a_{n+1} = 1 + \frac{1}{2}a_n$, for all $n \in N$. Verify that $\{a_n\}$ is a contractive sequence. Also, use the contraction principle to show that $\{a_n\}$ converges. See Exercise 11(g) from Section 2.4.

Proof. We write

$$|a_{n+2} - a_{n+1}| = \left|\left(1 + \frac{1}{2}a_{n+1}\right) - \left(1 + \frac{1}{2}a_n\right)\right| = \frac{1}{2}|a_{n+1} - a_n|.$$

Since $0 < k = \frac{1}{2} < 1$, the sequence $\{a_n\}$ is contractive. And by Theorem 2.5.11, it must converge. ■

Intuitively, a sequence $\{a_n\}$ is contractive if the successive difference between the terms is small "enough." This "enough" is represented by k in Definition 2.5.10. Using the definition to show $\{a_n\}$ is Cauchy is often time consuming and tedious. The contraction principle serves to alleviate this problem.

Exercises 2.5

1. Let s_0 be an accumulation point of S. Prove that the following two statements are equivalent.
  (a) Any neighborhood of s_0 contains at least one point of S different from s_0.
  (b) Any neighborhood of s_0 contains infinitely many points of S.

2. Find the set of all accumulation points for the given set S.
  (a) $S = \{x \mid 0 < |x-1| < 3\}$ **(b)** $S = \{x \mid x \in (-\infty, 2)\}$
  (c) $S = \{a_n \mid n \in N\}$, where the sequence $\{a_n\}$ is defined by

$$a_n = \begin{cases} 0, & \text{if } n \text{ is odd} \\ \dfrac{n}{n+1}, & \text{if } n \text{ is even.} \end{cases}$$

  (d) $S = \{a_n \mid n \in N\}$, where the sequence $\{a_n\}$ is defined by $a_n = 2$ for all n
  (e) $S = \{x \mid x \in [0,1] \text{ and } x \text{ is rational}\}$

3. **(a)** Give an example of a sequence $\{a_n\}$ for which the set $S = \{a_n \mid n \in N\}$ has exactly two accumulation points.
  (b) Give an example of a set S that contains infinitely many points, but not every point of S is an accumulation point of S.

(c) Give an example of a set S, where both $\sup S$ and exactly one accumulation point exist, but the values are not equal.

(d) Give an example of a set S, where $\inf S$ and $\sup S$ are in S, but the accumulation point, or points, is (are) not.

4. **(a)** Prove that a real number s_0 is an accumulation point of a set S if and only if there exists some sequence $\{a_n\}$ in S such that $a_n \neq s_0$ for every $n \in N$ and $\lim_{n\to\infty} a_n = s_0$.

(b) Is the restriction that $a_n \neq s_0$ in part (a) necessary? Explain.

5. Suppose S is a nonempty set of real numbers. Prove that $M = \sup S$ if and only if M is an upper bound of S and there exists a sequence $\{a_n\}$ in S, with $\lim_{n\to\infty} a_n = M$.

6. Let S be a bounded, infinite set of real numbers.

(a) Prove that either $\sup S = \max S$ or $\sup S$ is an accumulation point of S. In other words, either $\sup S \in S$ or $\sup S$ is an accumulation point of S.

(b) Give an example where one of the possibilities in part (a) is true but not the other.

(c) Give an example where $\sup S = \max S$ and where $\sup S$ is also an accumulation point of S.

7. Determine which of the sequences $\{a_n\}$ with a_n, as given, are Cauchy. Explain. Do not use Theorem 2.5.9.

(a) $a_n = \dfrac{n+1}{n}$

(b) $a_n = \dfrac{1}{2} + \dfrac{1}{6} + \cdots + \dfrac{1}{n(n+1)}$ (See Exercise 16 in Section 2.2.)

(c) $a_n = 1 + \dfrac{1}{2^2} + \dfrac{1}{3^2} + \cdots + \dfrac{1}{n^2}$ (See Exercise 7 in Section 2.4.)

(d) $a_n = 1 + \dfrac{1}{2} + \dfrac{1}{3} + \cdots + \dfrac{1}{n}$ (Compare with Exercise 2(c) in Section 2.1.)

(e) $a_n = \dfrac{1}{1!} - \dfrac{1}{2!} + \dfrac{1}{3!} - \dfrac{1}{4!} + \cdots + \dfrac{(-1)^{n+1}}{n!}$

(f) $|a_{n+1} - a_n| < r^n$ for some constant $r < 1$. Note that this inequality further restricts r to $0 < r < 1$. Compare this sequence to the one in Example 2.5.7.

8. **(a)** Give an example of a sequence $\{a_n\}$, where $\lim_{n\to\infty} |a_{n+1} - a_n| = 0$, but the sequence diverges. Beware of this possibility when using $|a_{n+1} - a_n| <$ given tolerance, as a stopping criteria in computer programming.

(b) Prove that if the sequence $\{a_n\}$ converges, then $\lim_{n\to\infty} |a_{n+1} - a_n| = 0$. See Exercise 20 in Section 7.4.

9. Prove Theorem 2.5.8.

10. Verify that if $\{a_n\}$ is a Cauchy sequence and the set $S=\{a_n \mid n \in N\}$ is finite, then $\{a_n\}$ is constant from some point on.

11. Prove that if $\{a_n\}$ and $\{b_n\}$ are two Cauchy sequences, then so are $\{a_n+b_n\}$ and $\{a_n b_n\}$. Do not use Theorem 2.5.9.

12. **(a)** Prove Corollary 2.5.12.

(b) What is the importance of the value a_1 in Example 2.5.13? What is the behavior of the sequence in Example 2.5.13 if a_1 is changed to 2 or 4? Explain.

13. Define the sequence $\{a_n\}$ by $a_{n+1}=(a_n)^2$ for all $n \in N$ where $0<a_1 \le \frac{1}{3}$. Prove that $\{a_n\}$ is contractive. See Exercise 12 in Section 4.4.

14. Consider the sequence $\{a_n\}$, defined recursively by

$$a_1 = 1 \text{ and } a_{n+1} = 1+\frac{1}{a_n}, \text{ for all } n \in N.$$

(a) Verify that $\{a_n\}$ is not monotone.

(b) Use the contraction principle to prove that $\{a_n\}$ converges.

(c) Show that this sequence converges to $\dfrac{1+\sqrt{5}}{2}$. See Problem 1.10.4.

15. Suppose $\{a_n\}$ is the Fibonacci sequence. Define the sequence $\{b_n\}$ by $b_n=\dfrac{a_{n+1}}{a_n}$ for all $n \in N$. Prove that $\{b_n\}$ converges to the golden number $\dfrac{1+\sqrt{5}}{2}$. See Problem 1.10.9.

16. Suppose the sequence $\{a_n\}$ is defined recursively by $a_{n+2}=\dfrac{a_{n+1}+a_n}{2}$, for all $n \in N$ with a_1 and a_2 any two real numbers.

(a) Prove that $\{a_n\}$ converges to the value given by $\dfrac{a_1+2a_2}{3}$.

(b) Prove that $\{a_n\}$ is Cauchy without using Theorem 2.5.9.

17. Consider the sequence $\{a_n\}$, where an equally weighted average defines $a_{n+2}=\dfrac{2a_{n+1}+a_n}{3}$, with $n \in N$ and a_1 and a_2 any real numbers.

(a) Prove $\{a_n\}$ converges to $\frac{1}{4}a_1+\frac{3}{4}a_2$. Compare this sequence with the one in exercise 16.

(b) If the term a_n is weighted heavier than a_{n+1}, that is, if

$$a_{n+2}=\frac{2}{3}a_n+\frac{1}{3}a_{n+1},$$

will the sequence still converge? Will it converge to the same value as in part (a)? Explain.

18. Consider the sequence $\{a_n\}$, where $a_1 = 1$ and $a_{n+1} = 1 + \dfrac{1}{1+a_n}$, for all $n \in N$.

(a) Is $\{a_n\}$ monotone?

(b) Use the contraction principle to show that $\{a_n\}$ converges.

(c) Verify that $\{a_n\}$ converges to $\sqrt{2}$. Compare with Exercises 12 and 13 in Section 2.4.

(d) Convince yourself that $\{a_n\}$ represents the continued fraction given by

$$\sqrt{2} = 1 + \cfrac{1}{2 + \cfrac{1}{2 + \cfrac{1}{2 + \cfrac{1}{2 + \ddots}}}}.$$

See Part 2 of Section 3.5.

19. Define the sequence $\{a_n\}$ recursively by $a_{n+1} = a_n + \cos a_n$, for all $n \in N$ with $a_1 = 1$. Write out enough terms to convince yourself that $\{a_n\}$ converges. What does it converge to? Explain your reasoning.

20. Define the sequence $\{a_n\}$ recursively by $a_{n+1} = \cos a_n$, for all $n \in N$ with $a_1 = 1$. Write out enough terms to convince yourself that $\{a_n\}$ converges. What does it converge to? Explain your reasoning. The limiting value is called a fixed point of the function $f(x) = \cos x$.

2.6 Subsequences

Knowledge of subsequences will provide a powerful tool in discussing convergence of sequences. The idea of a subsequence, which was previously previewed in Exercise 7 of Section 2.1 is easy. Unfortunately, formal proofs might at first create some problems. Let us consider a sequence $\{a_n\}$. To obtain a subsequence of $\{a_n\}$, we simply omit some terms from $\{a_n\}$ without changing the order of the remaining terms. The resulting sequence is called a subsequence of $\{a_n\}$. For example, all even terms of $\{a_n\}$ and all odd terms of $\{a_n\}$ form two subsequences of $\{a_n\}$. In addition, the sequence $\{a_n\}$ is a subsequence of itself. To write a definition of this idea, consider that we have a sequence $\{a_n\}$ in which we delete certain terms and then renumber the remaining terms to get a new sequence $\{b_n\}$. For instance, suppose we have a sequence $a_k, a_{k+1}, a_{k+2}, a_{k+3}, a_{k+4}, a_{k+5}, \ldots$, and we wish to consider only, say, $a_{k+1}, a_{k+2}, a_{k+5}, \ldots$. On relabeling the terms under consideration by $b_i, b_{i+1}, b_{i+2}, \ldots$, there must be a function relating the subscripts $k+1$ to i, $k+2$ to $i+1$, $k+5$ to $i+2$, etc. Here is a formal definition.

Definition 2.6.1. The sequence $\{b_n\}_{n=i}^{\infty}$ is a *subsequence* of $\{a_n\}_{n=k}^{\infty}$, with $i, k \in N$, $i \geq k$, if and only if there exists a strictly increasing function f where $f: \{m \in N \mid m \geq i\} \rightarrow \{m \in N \mid m \geq k\}$ and $b_n = a_{f(n)}$, for all $n \in N$.

Example 2.6.2. Consider the sequence $\{a_n\}$, where a_n is as given in the following. Observe that the sequence $\{b_n\}$, with b_n as given, is a subsequence of $\{a_n\}$. The function relating the subscripts is also given. All these sequences begin with $n = 1$.

(a) Let $a_n = n$, $b_n = 2n$, and $f(n) = 2n$. To see that $\{b_n\}$ is a subsequence of $\{a_n\}$ we write out some terms.

$$a_1 = 1, a_2 = 2, a_3 = 3, a_4 = 4, a_5 = 5, a_6 = 6, \ldots$$

$$\downarrow \qquad \downarrow \qquad \downarrow$$

$$b_1 = 2, \qquad b_2 = 4, \qquad b_6 = 6, \ldots$$

Therefore, $b_n = a_{2n}$, for all $n \in N$, and clearly, f is strictly increasing.

(b) Let $a_n = \dfrac{1}{n}$, $b_n = \dfrac{1}{n^2}$, and $f(n) = n^2$. Again, $\{b_n\}$ is a subsequence of $\{a_n\}$. To visualize this we write

$$a_1 = \frac{1}{1}, a_2 = \frac{1}{2}, a_3 = \frac{1}{3}, a_4 = \frac{1}{4}, a_5 = \frac{1}{5}, \ldots, a_9 = \frac{1}{9}, \ldots$$

$$\downarrow \qquad \downarrow \qquad \downarrow$$

$$b_1 = \frac{1}{1}, \qquad b_2 = \frac{1}{2^2}, \qquad b_3 = \frac{1}{3^2}, \ldots$$

Therefore, $b_n = a_{n^2}$, for all $n \in N$. Thus, $b_n = a_{f(n)}$.

(c) Let $a_n = (-1)^{n+1}$, $b_n = 1$, and $f(n) = 2n - 1$. We write

$$a_1 = 1, a_2 = -1, a_3 = 1, a_4 = -1, a_5 = 1, a_6 = -1, a_7 = 1, \ldots$$

$$\downarrow \qquad \downarrow \qquad \downarrow \qquad \downarrow$$

$$b_1 = 1, \qquad b_2 = 1, \qquad b_3 = 1, \qquad b_4 = 1, \ldots$$

Therefore, $b_n = a_{2n-1}$, for all $n \in N$.

(d) Let $a_n = \begin{cases} 1, \text{ if } n \text{ is odd} \\ \dfrac{1}{\text{n}}, \text{ if } n \text{ is even,} \end{cases}$ $b_n = \dfrac{1}{n}$, and $f(n) = 2n$. Clearly $b_n = a_{2n}$, for all $n \in N$. ■

It is customary to write a subsequence $\{b_n\}$ obtained from the sequence $\{a_n\}$ as $b_n = a_{n_k}$, since the terms of the subsequence come from a_n. Recall that only certain values of n are used, say, $n_1, n_2, n_3, \ldots$. These values are in ascending order with their own subscripts indicating the position in the subsequence $\{b_n\} = \{a_{n_k}\}$. The term a_{n_1} is the first term of the subsequence, a_{n_2} is the second term, and so on. Thus, k is the index for the subsequence $\{a_{n_k}\}$. Clearly, the notation $\{a_{n_k}\}$ is easier to use than is $\{a_{f(n)}\}$, for some function f. If $\{a_{n_k}\}$ is a subsequence of $\{a_n\}$, then $n_k \geq k$, for each $k \in N$. Can you prove this statement by induction?

As we observed in parts (c) and (d) of Example 2.6.2, there are sequences that diverge yet contain subsequences that converge. The limiting value of subsequences is called a *subsequential limit point.* Some authors refer to these points as simply *limit points*.

Definition 2.6.3. The real number α is a *subsequential limit point* of a sequence $\{a_n\}$ if and only if there exists a subsequence of $\{a_n\}$ that converges to α. Furthermore, if T is the set of all subsequential limits of $\{a_n\}$, then the *limit superior* (*upper limit*) of $\{a_n\}$ is called $\sup T$ and

$$\limsup_{n\to\infty} a_n \equiv \overline{\lim_{n\to\infty}}\, a_n = \sup T.$$

Similarly, the *limit inferior* (*lower limit*) of $\{a_n\}$ is defined by

$$\liminf_{n\to\infty} a_n \equiv \underline{\lim_{n\to\infty}}\, a_n = \inf T.$$

Do not confuse subsequential limit points with accumulation points. For example, if

$$a_n = \begin{cases} 1, \text{ if } n \text{ is odd} \\ \dfrac{1}{\text{n}}, \text{ if } n \text{ is even,} \end{cases}$$

then $\{a_{2n}\}$ and $\{a_{2n-1}\}$ are two subsequences of $\{a_n\}$ that converge to the values 0 and 1, respectively. Thus, 0 and 1 are the subsequential limit points of this sequence. But 0 is the only accumulation point of $S = \{a_n \mid n \in N\}$. Moreover, $\limsup_{n\to\infty} a_n = 1$, and $\liminf_{n\to\infty} a_n = 0$. Can you find yet another subsequence of $\{a_n\}$ that also converges to zero? A different example would occur if we consider a sequence $\{a_n\}$ with

$a_n = n\cos\frac{n\pi}{2}$. The subsequential limits are $-\infty, 0, +\infty$. Also, $\limsup_{n\to\infty} a_n = +\infty$, and $\liminf_{n\to\infty} a_n = -\infty$. The set $S = \{a_n \mid n \in N\}$ has no accumulation points. Why? Also, note the difference between lim sup and sup. If $c_n = \frac{1}{n}$ for n odd, and $c_n = \frac{1}{n} - 1$ for n even, then $\limsup_{n\to\infty} c_n = 0$, but $\sup\{c_n \mid n \in N\} = 1$. See Section 8.8 for a more involved study of the limit superior and limit inferior.

The following theorem is actually a corollary to Theorem 2.5.4.

THEOREM 2.6.4. *(Bolzano–Weierstrass Theorem for Sequences) Any bounded sequence must have at least one convergent subsequence.*

Proof. Suppose that $\{a_n\}$ is a bounded sequence and define the set $S = \{a_n \mid n \in N\}$. Consider two cases: S is finite or S is infinite.

Case 1. Suppose that S is finite. There must be infinitely many values of n, say, $n_1, n_2, n_3, \ldots$, for which $a_{n_k} = s$, where s is a member of S. The subsequence $\{a_{n_k}\}$ is constant and thus converges.

Case 2. Suppose S is infinite. Then by Theorem 2.5.4, there exists an accumulation point s_0 of S. The existence of a subsequence that converges to s_0 then follows from Exercise 4 in Section 2.5. Why? ■

THEOREM 2.6.5. *A sequence converges to A if and only if each of its subsequences converge to A.*

Proof. ($\Leftarrow$) If every subsequence of the sequence $\{a_n\}$ converges to A, then $\{a_n\}$ converges to A, since it is a subsequence of itself.

($\Rightarrow$) Suppose that the sequence $\{a_n\}$ converges to A. Pick an arbitrary subsequence $\{a_{n_k}\}$ and prove that $\{a_{n_k}\}$ must also converge to A. Thus, we need to pick an arbitrary $\varepsilon > 0$ and look for n^* so that

$$|a_{n_k} - A| < \varepsilon, \text{ for all } n_k \geq n^*.$$

Since the sequence $\{a_n\}$ converges to A, there exists m_1 such that $|a_n - A| < \boxed{\varepsilon}$ for all $n > m_1$. Since $\{a_{n_k}\}$ is a subsequence of $\{a_n\}$, we have $n_1 < n_2 < \ldots$, with $i \leq n_i$, $i \in N$. The equality is true if $\{a_{n_k}\}$ is a subsequence of itself. Thus, for $k \geq m_1$, we have $n_k \geq m_1$. So pick $n^* = m_1$. Then

$$|a_{n_k} - A| < \varepsilon, \text{ for all } n_k \geq n^*.$$

Therefore, $\{a_{n_k}\}$ converges to A. Since $\{a_{n_k}\}$ was arbitrary, every subsequence of $\{a_n\}$ must converge to A. ∎

Theorem 2.6.5 is very powerful and has many consequences. A sequence $\{a_n\}$ converges to a subsequential limit α of $\{a_n\}$ if and only if every subsequence of $\{a_n\}$ converges to α. Theorem 2.6.5 can be used to prove divergence. If some subsequence of $\{a_n\}$ diverges, then so does $\{a_n\}$; all it takes is one such subsequence. Furthermore, if we can exhibit two subsequences of $\{a_n\}$ that converge to two different values, then $\{a_n\}$ must diverge. Thus, to prove that the sequence $\{a_n\}$, with $a_n = (-1)^n$ diverges, simply observe that $\{a_{2n}\}$ converges to 1, $\{a_{2n-1}\}$ converges to -1, and both of these are subsequences of $\{a_n\}$. See Exercise 2(f) from Section 2.1.

Using subsequences to prove convergence is impossible because we would have to prove convergence of every imaginable subsequence, meaning the sequence itself. However, if we know that a sequence converges, we can use subsequences to actually find the limiting value. Recall from previous sections that this task was not always easy.

Example 2.6.6. Consider the sequence $\{a_n\}$ with $a_n = \sqrt[n]{c}$, for any constant $c > 1$. Prove that $\{a_n\}$ converges to 1 (see Exercise 14 in Section 2.1).

Proof. If we were to write out a few terms, we would suspect that $\{a_n\}$ is decreasing. To prove this, using Remark 2.4.3, part (d), we write

$$a_n - a_{n+1} = \sqrt[n]{c} - \sqrt[n+1]{c} = \sqrt[n+1]{c}\left(\sqrt[n(n+1)]{c} - 1\right) > 0.$$

Therefore, $\{a_n\}$ is decreasing and clearly bounded below by 1. Thus, using Theorem 2.4.4, $\{a_n\}$ converges to, say, A. Subsequences are used to evaluate A. First, observe that by Theorem 2.2.1, part (e), we have $\{\sqrt{a_n}\}$ converging to $\sqrt{A}$. Here, $\sqrt{a_n} = \sqrt{\sqrt[n]{c}} = \sqrt[2n]{c}$. But, $\{\sqrt[2n]{c}\}$ is a subsequence of $\{\sqrt[n]{c}\}$. Thus, $\{\sqrt{a_n}\}$ must converge to A. Therefore, $\sqrt{A} = A$. Why? The two choices for A are 0 and 1. But $a_n > 1$, for all $n \in N$, and thus zero is not a possibility. Hence, $\lim_{n\to\infty} \sqrt[n]{c} = 1$. ∎

Exercises 2.6

1. Determine whether the sequence $\{b_n\}$ is a subsequence of $\{a_n\}$, with a_n and b_n as given.

(a) $b_n = -1$ and $a_n = (-1)^n$ **(b)** $b_n = \frac{1}{\sqrt{n}}$ and $a_n = \frac{1}{n}$

(c) $b_n = \frac{1}{3n^2}$ and $a_n = \frac{1}{n^2}$

2. Verify that the sequences $\{a_n\}$ with a_n as given diverge. Find all of the subsequential limit points α, $\limsup_{n\to\infty} a_n$, and $\liminf_{n\to\infty} a_n$.

(a) $a_n = \frac{1+(-1)^{n+1}}{2}$ **(b)** $a_n = \sin\frac{n\pi}{2}$

(c) $a_n = r^n$, $r \geq 0$ **(d)** $a_n = (-1)^n\left(\frac{n-1}{n}\right)$

3. Let the sequence $\{a_n\}$ be defined by $a_n = 1 + \frac{1}{2} + \frac{1}{3} + \cdots + \frac{1}{n}$. See Exercise 7(d) from Section 2.5 and Example 7.1.13. Prove that $\{a_n\}$ is unbounded by showing that there exists some subsequence that is unbounded.

4. Prove that the following two statements are equivalent.

(a) Any bounded sequence must have a converging subsequence (see Theorem 2.6.4).

(b) Every bounded monotone sequence must converge (see Theorem 2.4.4, part (a)).

Compare this exercise to Exercise 18 from Section 2.4.

5. Prove that every unbounded sequence contains a monotone subsequence that diverges to infinity.

6. Reprove Theorem 2.5.9 using the Bolzano–Weierstrass theorem for sequences rather than the Bolzano–Weierstrass theorem for sets.

7. Let $\{a_n\}$ be a sequence and let set $S = \{a_n \mid n \in N\}$. If s_0 is an accumulation point of S, prove that there exists a subsequence of $\{a_n\}$ that converges to s_0 (see Exercise 4 from Section 2.5).

8. Suppose that the sequence $\{a_n\}$ converges to A, and suppose that B is an accumulation point of the set $S = \{a_n \mid n \in N\}$. Prove that $A = B$.

9. Use subsequences to find the limit of the sequence $\{a_n\}$ with a_n as given.

(a) $a_n = \sqrt[n]{c}$, with any constant $c \in (0,1)$. (See Exercise 14 from Section 2.1.)

(b) $a_n = r^n$ with any constant $r \in (0,1)$. (See Exercise 2(c) above and Theorem 2.1.13.)

(c) $a_1 = 1$ and $a_{n+1} = \sqrt{2a_n}$ for all $n \in N$. (See Exercise 11(c) of Section 2.4.)

2.7* Review

Label each statement as true or false. If a statement is true, prove it. If not,

(i) give an example of why it is false, and

(ii) if possible, correct it to make it true, and then prove it.

1. If $\lim_{n\to\infty} \frac{a_n}{n} = A$, with $A \neq 0$ a constant, then the sequence $\{a_n\}$ diverges to infinity.
2. The sequence $\{a_n\}$, given recursively by $a_1 = 1$ and $a_{n+1} = \sqrt{1+a_n}$, for all $n \in N$, converges to the golden ratio.
3. If the sequence $\{a_n\}$ is bounded, then it is Cauchy.
4. If $\{a_n\}$ is a bounded sequence with exactly one subsequential limit α, then $\{a_n\}$ converges to α.
5. If $\{a_n\}$ is a subsequence of $\{b_n\}$ and $\{b_n\}$ is a subsequence of $\{a_n\}$, then $\{a_n\} = \{b_n\}$.
6. If a sequence $\{a_n\}$ is bounded and all of its convergent subsequences have the same limit, then $\{a_n\}$ converges.
7. The sequence $\{a_n\}$ with $a_n = \frac{(-1)^n}{n}$ is an oscillating sequence.
8. If $\{a_n\}$ and $\{b_n\}$ are sequences of positive terms such that $\lim_{n\to\infty} \frac{a_n}{b_n} = +\infty$, and if $\{a_n\}$ is bounded, then $\{b_n\}$ converges to zero.
9. The sequence $\{a_n\}$ with $a_n = \frac{2^n}{1+2^n}$ is monotone.
10. If the sequence $\{a_n\}$ converges to a constant A, then there exists n^* such that $\frac{|A|}{3} \leq |a_n| \leq |A| + \frac{1}{3}$ for all $n \geq n^*$.
11. If sequences $\{a_n\}$, $\{b_n\}$, and $\{c_n\}$ converge to A, B, and AB, respectively, then there exists n^* such that for all $n \geq n^*$, we have $c_n = a_n b_n$.
12. The sequence $\{a_n\}$ converges if and only if it is monotone.
13. If any neighborhood of a constant A contains infinitely many terms of a sequence $\{a_n\}$, then $\{a_n\}$ converges to A.
14. Suppose that the sequence $\{a_n\}$ converges to A. Define a new sequence $\{b_n\}$ by $b_n = \frac{2}{3} a_n + \frac{1}{3} a_{n+2}$, for all $n \in N$. Then $\{b_n\}$ converges to A.

15. Suppose that $\{a_n\}$ and $\{b_n\}$ are two sequences such that $\{a_n + b_n\}$ converges to A and $\{a_n - b_n\}$ converges to B. Then the sequence $\{a_n b_n\}$ converges to $\frac{1}{4}\left(A^2 - B^2\right)$.

16. If $a_1 = \frac{3}{2}$ and $a_{n+1} = \frac{1}{5}\left[(a_n)^2 + 6\right]$ for all $n \in N$, then the sequence $\{a_n\}$ converges to 2.

17. If $\lim\limits_{n\to\infty} a_n = A \neq 0$, then the sequence $\{(-1)^n a_n\}$ diverges.

18. If k is any natural number, then $\lim\limits_{n\to\infty} \sqrt[n]{n^k} = 1$.

19. If k is any natural number, then $\lim\limits_{n\to\infty} \sqrt[n]{kn} = 1$.

20. If the sequence $\{a_n\}$ is defined by $a_{n+1} = \dfrac{\alpha a_n + \beta a_{n+1}}{\alpha + \beta}$, where α and β are positive constants, and $n \in N$, with a_1 and a_2 any real numbers, then

$$\lim_{n\to\infty} a_n = \frac{\alpha a_1 + (\alpha + \beta) a_2}{\alpha + 2\beta}.$$

21. Suppose that $\{a_n\}$ and $\{b_n\}$ are two sequences such that $\{a_n\}$ and $\left\{\dfrac{a_n}{b_n}\right\}$ both converge. Then $\left\{\dfrac{b_n}{a_n}\right\}$ or $\{b_n\}$ must converge.

22. If a sequence of positive numbers is unbounded, then the sequence diverges to $+\infty$.

23. If $\{a_n\}$ is a sequence and A is an accumulation point of the set $S = \{a_n \mid n \in N\}$, then $\{a_n\}$ must converge to A.

24. If S is a set of real numbers and $A = \sup S$, then $A - \varepsilon$ must be in S for any $\varepsilon > 0$.

25. If $\{a_n\}$ is a converging sequence with $a_n \neq 0$ for any $n \in N$, then $\lim\limits_{n\to\infty} \left|\dfrac{a_{n+1}}{a_n}\right| < 1$.

26. The sequence $\{a_n\}$ converges to zero if and only if $\{|a_n|\}$ converges to zero.

27. If $\lim\limits_{n\to\infty} a_n = +\infty$, then $\lim\limits_{n\to\infty} \dfrac{a_n}{2 + a_n} = \dfrac{1}{2}$.

28. If $\{a_n\}$ and $\{b_n\}$ are two sequences, where $\{a_n\}$ converges to 0 and $\{b_n\}$ is bounded, then the sequence $\{a_n b_n\}$ converges to zero.

29. If $\{a_n\}$ and $\{b_n\}$ are two sequences such that $\lim\limits_{n\to\infty} a_n = \lim\limits_{n\to\infty} b_n$, then there exists $n^* \in N$ such that $a_n = b_n$ for all $n \geq n^*$.

30. If $\{a_n\}$ is a sequence that converges to A, then the set $S=\{a_n \mid n \in N\}$ has an accumulation point.
31. If $\{a_n\}$ is a sequence such that the set $S=\{a_n \mid n \in N\}$ is finite, then $\{a_n\}$ converges.
32. If $a_n = \dfrac{1\cdot 4\cdot 7\cdot \cdots \cdot(3n-2)}{2\cdot 5\cdot 8\cdot \cdots \cdot(3n-1)}$, then the sequence $\{a_n\}$ converges to A, where $0 \le A \le \frac{1}{2}$.
33. If A is an accumulation point of the set S, then $A=\sup S$, provided it exists.
34. The set $S=\{0,1\}$ has exactly two accumulation points.
35. If S is a set, $s=\sup S$, and $s \notin S$, then s must be an accumulation point of S.
36. If $\{a_n\}$ and $\{b_n\}$ are two sequences such that $b_n = \dfrac{a_1+a_2+\cdots+a_n}{n}$ for all $n \in N$, then $\{a_n\}$ converges to A if and only if $\{b_n\}$ converges to A. See summable sequences in Section 2.8.
37. Any sequence has a unique limit.
38. If $\{a_n\}$ and $\{b_n\}$ are two sequences such that $\{a_n\}$ and $\{a_nb_n\}$ both converge, then $\{b_n\}$ converges.
39. If the sequence $\{a_n\}$ is defined recursively by $a_{n+2} = \dfrac{1+a_{n+1}}{a_n}$, for $n \in N$ with $a_1=1$ and $a_2=1$, then $\{a_n\}$ converges to the golden number.
40. There exists a sequence of rational numbers that has an irrational limit, and there exists a sequence of irrational numbers that has a rational limit.
41. If $\{a_n\}$ diverges to $+\infty$ and $\{b_n\}$ is any sequence, then $\{a_n+b_n\}$ diverges to $+\infty$.
42. If $\{a_n\}$ and $\{b_n\}$ both diverge, then $\{a_n+b_n\}$ diverges.
43. If $\{a_n\}$ converges and $\{b_n\}$ diverges, then $\{a_n+b_n\}$ diverges.
44. If $\{a_n\}$ and $\{b_n\}$ both diverge to $+\infty$, then $\{a_n-b_n\}$ diverges.
45. If $\{a_n\}$ and $\{b_n\}$ are sequences so that $\{a_n\}$ and $\{a_nb_n\}$ both converge, then $\{b_n\}$ converges.
46. If $\{a_n\}$ and $\{b_n\}$ are sequences so that $\{a_n\}$ and $\{a_nb_n\}$ both diverge, then $\{b_n\}$ diverges.
47. If $\{a_n\}$ and $\{b_n\}$ both diverge, then $\{a_nb_n\}$ diverges.
48. If all the convergent subsequences of $\{a_n\}$ have the same limit, then $\{a_n\}$ converges.
49. If $\{a_n\}$ is convergent, then it is contractive.
50. The value of $\sqrt{2+\sqrt{2+\sqrt{2+\cdots}}}$ is 2. (See Exercise 15 in Section 2.4.)
51. Every bounded sequence has a monotone subsequence.

2.8* Projects

Part 1. The Transcendental Number e

The transcendental number "e," often called *Euler's*[5] *number*, was first printed in 1733 and is among the most famous of irrational numbers. The other two prominent irrational numbers are π and the golden number. Transcendental numbers such as π and e are never roots of any polynomial equation with integer coefficients. In fact, in Section 8.8 we shall prove that e is indeed irrational. The limiting value of the sequence $\{e_n\}$, where $e_n = \left(1+\frac{1}{n}\right)^n$, is the irrational number e. The fact that this sequence actually converges follows from the next exercise. Also, see Exercise 10(a) in Section 3.1. One should beware of taking the limit inside of the parentheses to obtain 1^n and then claiming the value of the limit is 1. Also, the symbol $\ln x$ represents the *natural logarithm* $\log_e x$, where x is any positive real number.

Problem 2.8.1. Use the following outline to prove that the sequence $\{e_n\}$, with $e_n = \left(1+\frac{1}{n}\right)^n$, converges. Verify that $\left(1+\frac{1}{n}\right)^n < e < \left(1+\frac{1}{n}\right)^{n+1}$ and that $2 \le e < 3$, for all $n \in N$.

(a) Use the binomial theorem to write out $n+1$ terms of e_n using the fact that terms such as, say, $\frac{n(n-1)(n-2)}{n^3}$ can be written as $\left(1-\frac{1}{n}\right)\left(1-\frac{2}{n}\right)$.

(b) Do the same to $n+2$ terms of e_{n+1} as in (a).

(c) Conclude that $e_n < e_{n+1}$ for all $n \in N$.

(d) Use $n-k < n$ for $k > 0$, $n^n \ge n! \ge 2^{n-1}$ for all $n \in N$, and Exercise 19(a) from Section 2.1, to prove that $e_n < 3$ for all $n \in N$.

(e) Conclude that $2 \le e < 3$. (In fact, $e \approx 2.71828$.)

---

[5]Leonhard Euler (1707–1783), a Swiss mathematical genius who spent most of his life in St. Petersburg, Russia, was prominent in every area of mathematics as well as in the theory of optics, general mechanics, planetary motion, and electricity. At a very young age Euler was tutored by Johann Bernoulli, and at the age of 18 he began mathematical research. Because of his phenomenal memory, Euler could solve unbelievably complicated problems, such as some in lunar motion. He made mathematics more systematic and was the first mathematician to solve physical problems using calculus and to use the notations e for the base of the natural logarithm (in 1727), $f(x)$ for functions (in 1734), Σ for summation (in 1755), i for the square root of -1 (in 1777), and π for pi. In 1748, Euler gave the formula $\exp(ix) = \cos x + i \sin x$, which was equivalent to that of Roger Cotes given earlier. The most mathematically productive years of Euler's life were the last 17 years, during which he was blind. Does this make you think of Ludwig van Beethoven, the celebrated German composer, who was deaf during his most productive time?

(f) In a similar fashion to the preceding, show that $\{d_n\}$, with $d_n = \left(1+\frac{1}{n}\right)^{n+1}$, decreases to e.

Problem 2.8.2. Prove that the sequence $\{f_n\}$, with $f_n = 1+\frac{1}{1!}+\frac{1}{2!}+\cdots+\frac{1}{n!}$, converges to e by following the suggested steps.

(a) Verify that $\{f_n\}$ is increasing and bounded above by 3.

(b) Conclude that $\{f_n\}$ converges, and let f be the limiting value.

(c) With e_n defined as in the previous problem, verify that $e_n \le f_n$, yielding $e \le f$.

(d) Rewrite the binomial expansion of e_n to obtain $e \ge f$.

(e) Draw a final conclusion.

(f) Compare the sequence $\{f_n\}$ with the one from Exercise 7(e) of Section 2.5.

Problem 2.8.3. Evaluate the given limits, if possible.

(a) $\lim_{n\to\infty}\left(1+\frac{1}{2n}\right)^{2n}$ **(b)** $\lim_{n\to\infty}\left(1+\frac{1}{n+1}\right)^{n}$

(c) $\lim_{n\to\infty}\left(1+\frac{2}{n}\right)^{n}$ **(d)** $\lim_{n\to\infty}\left(1+\frac{1}{2n}\right)^{3n}$

(e) $\lim_{n\to\infty}\left(\frac{n+1}{n}\right)^{n+1}$ **(f)** $\lim_{n\to\infty}\left(1+\frac{1}{n^2}\right)^{n}$

(g) $\lim_{n\to\infty}\left(\frac{n+3}{n+1}\right)^{n+4}$ **(h)** $\lim_{n\to\infty} r^n n^n$, $r>0$ is a real constant.

Problem 2.8.4. Label as true or false: Suppose that $\{a_n\}$ is a sequence and p is a real number. If $\lim_{n\to\infty} n^p a_n = 0$, then there exists $\varepsilon > 0$ such that $\lim_{n\to\infty} n^{p+\varepsilon} a_n$ is finite.

Part 2. Summable Sequences

In Exercise 36 of Section 2.7, we pointed out that there are divergent sequences $\{a_n\}$, which produce convergent *arithmetic average sequences* $\{\alpha_n\}$, with $\alpha_n = \frac{a_1+a_2+\cdots+a_n}{n}$. For example, if $a_n = (-1)^n$, then the sequence $\{a_n\}$ diverges, even though the average size of the terms tends to zero. Compare this idea to a sequence $\{a_n\}$ that diverges to infinity and never possesses the concept of a converging average.

Definition 2.8.5. Let $\{a_n\}$ be a sequence of real numbers and let

$$\alpha_n = \frac{a_1 + a_2 + \cdots + a_n}{n},$$

for all $n \in N$. Then we say that $\{a_n\}$ is $(C,1)$ *summable* or *Cesàro*[6] *convergent* to A if and only if $\{\alpha_n\}$ converges to A.

From Exercise 36 of Section 2.7 we have the following theorem. Also, see Exercise 10 from Section 2.1.

THEOREM 2.8.6. *If a sequence $\{a_n\}$ converges to A, then $\{a_n\}$ is $(C,1)$ summable to A.*

The converse of Theorem 2.8.6 is false. Note that sequences diverging to infinity are not the only sequences that are not $(C,1)$ summable. Some oscillating sequences are also not $(C,1)$ summable. Consider the following example.

Problem 2.8.7. Use Definition 2.8.5 to show that the sequence $\{a_n\}$, where

$$a_n = \begin{cases} \frac{n+1}{2}, & \text{if } n \text{ is odd} \\ -\frac{n}{2}, & \text{if } n \text{ is even,} \end{cases}$$

is not $(C,1)$ summable.

Problem 2.8.8. Suppose that a sequence $\{a_n\}$ is $(C,1)$ summable. Prove that $\lim_{n\to\infty} \frac{a_n}{n} = 0$. Use this result to verify that the sequence $\{a_n\}$, with a_n as given in Problem 2.8.7, is not $(C,1)$ summable.

Problem 2.8.9. If sequences $\{a_n\}$ and $\{b_n\}$ are $(C,1)$ summable to A and B, respectively, prove that the sequences $\{a_n \pm b_n\}$ are $(C,1)$ summable to $A \pm B$, respectively.

Problem 2.8.10. Prove or disprove the following statement. Let $\{a_n\}$ be a sequence of real numbers and let $\alpha_n = \frac{a_1 + a_2 + \cdots + a_n}{n}$, for all $n \in N$. If $\{a_n\}$ is monotone and bounded, then so is $\{\alpha_n\}$.

---

[6]Ernesto Cesàro (1859–1906) was a prominent Italian geometer and analyst. $(C,1)$ summability plays an important role in the study of Fourier series.

Definition 2.8.11. Let $\{a_n\}$ be a sequence of real numbers, and let

$$\beta_n \equiv \frac{na_1 + (n-1)a_2 + (n-2)a_3 + \cdots + 2a_{n-1} + a_n}{1+2+3+\cdots+n} = \frac{2[na_1 + (n-1)a_2 + (n-2)a_3 + \cdots + a_n]}{n(n+1)},$$

for all $n \in N$. Then $\{a_n\}$ is $(C,2)$ *summable* to A if and only if $\{\beta_n\}$ converges to A.

Example 2.8.12

(a) If $a_n = 1$, then $\{a_n\}$ is $(C,2)$ summable.

(b) If $a_n = (-1)^n$, then $\{a_n\}$ is $(C,2)$ summable.

(c) If a_n is as given in Problem 2.8.7, then $\{a_n\}$ is $(C,2)$ summable. ■

Verify these examples using Definition 2.8.11. Thus, we see that the fact that the sequence $\{a_n\}$ is $(C,2)$ summable does not imply that $\{a_n\}$ is $(C,1)$ summable. However, the converse is true. This yields the following result.

THEOREM 2.8.13. *If $\{a_n\}$ converges to A, then $\{a_n\}$ is $(C,2)$ summable to A.*

Problem 2.8.14. Prove Theorem 2.8.13.

3

Limits of Functions

3.1* Limit at Infinity

We will now generalize the concept of a limit to functions with a domain and a range in the real numbers $\Re$. Several types of limits are covered. First, from Definition 2.1.2 consider $a_n = f(n)$, with $n \in N$. In the following definition, any x from the domain of a function f is acceptable, not just $x = n \in N$.

Definition 3.1.1. Let f be a function with domain $D \subseteq \Re$, which contains arbitrarily large values. This function *f has a limit as x approaches (tends to) plus infinity*, that is to ∞, also denoted $+\infty$, if and only if there exists a real number L such that for every $\varepsilon > 0$, there exists a real number $M > 0$ such that

$$|f(x) - L| < \varepsilon, \text{ if } x \geq M \text{ and } x \in D.$$

If such an L exists, we say that L is the *limit of the function f as x tends to* ∞, or simply, L is the *limit of f at plus infinity*, or L is the *limiting value at* ∞, and we write $\lim_{x \to \infty} f(x) = L$, where x is a *dummy variable*.

Definition 3.1.2. A property, P, is true *eventually* if and only if there exists $M > 0$ such that P is true for all $x \geq M$.

Whenever we represent a limit by a letter, we assume that this letter represents a real number. Intuitively, $\lim_{x\to\infty} f(x) = L$ if after a certain point (we called it M in Definition 3.1.2) all the values $f(x)$ are within ε of L for any $\varepsilon > 0$. This idea is demonstrated in Figure 3.1.1. The line $y = L$ is called a *horizontal asymptote* for the function f provided $f(x) \neq L$ for all $x \geq M^*$, where M^* is some positive real number. In other words, $f(x)$ is never eventually equal to the proposed horizontal asymptote, that is, the limiting value at $+\infty$.

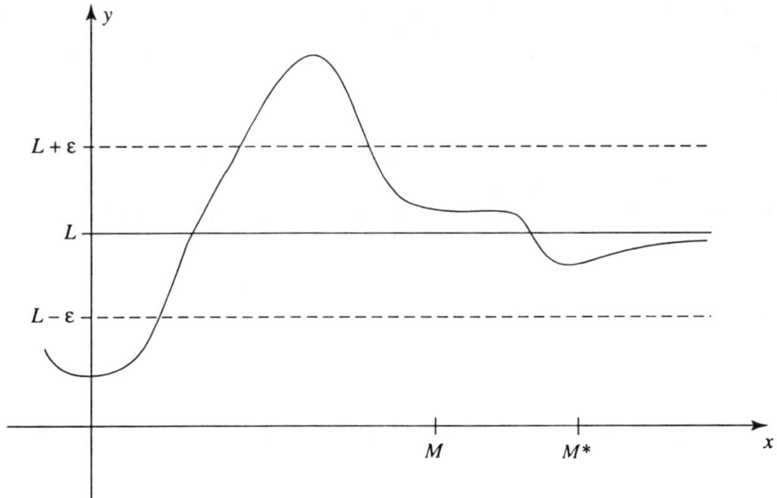

Figure 3.1.1

Example 3.1.3. Verify that if $f(x) = \dfrac{x}{x+2}$ with $x \in Q^+$, the positive rational numbers, then $\lim_{x\to\infty} f(x) = 1$, that is $\lim_{x\to\infty} \dfrac{x}{x+2} = 1$.

Proof. Observe first that if the domain D of the function f is N, then we are considering a sequence that converges to 1 (see Remark 2.1.4). Although $D = Q^+$, the procedure is still similar to the one in Remark 2.1.4. We begin with an arbitrary but fixed $\varepsilon > 0$, and then verify that Definition 3.1.1 is true with $L = 1$. We need to find a positive real number M so that $|f(x) - 1| < \varepsilon$ whenever $x \geq M$, provided $x \in D = Q^+$. To find this M, we write

$$|f(x) - 1| = \left|\frac{x}{x+2} - 1\right| = \frac{2}{x+2} < \frac{2}{x},$$

if $x > 0$, say. Since we want $|f(x) - 1| < \varepsilon$, consider $\dfrac{2}{x} < \varepsilon$ and solve for x. Hence,

$x > \frac{2}{\varepsilon}$ and $x > 0$. Hence, pick any $M > \frac{2}{\varepsilon}$. Then, for all $x \ge M$ with $x \in D$, we have

$$|f(x) - 1| = \frac{2}{x+2} < \frac{2}{x} \le \frac{2}{M} < \varepsilon.$$

Observe that if D were bounded, then $\lim_{x \to \infty} f(x)$ would not exist. Why? ■

At times we will write the symbol $\lim_{x \to \infty} f(x)$ even though this limit may not exist for various reasons, meaning Definition 3.1.1 fails. This concept is formulated in the next remark.

Remark 3.1.4. A function with an unbounded above domain $D \subseteq \Re$ has no limiting value at plus infinity, that is, f has no finite limit at ∞, if and only if for every real number L there exists $\varepsilon > 0$ such that for every $M > 0$ there exists an $x \ge M$ with $x \in D$, where $|f(x) - L| \ge \varepsilon$. ■

Example 3.1.5. Verify that $f(x) = \sin x$ with $x \in D = \Re^+$ has no finite limit at plus infinity.

Answer. Observe that for all $n \in N$ we have $\sin\left(\frac{\pi}{2} + 2\pi n\right) = 1$ and $\sin\left(\frac{3\pi}{2} + 2\pi n\right) =$ -1. Thus, pick $\varepsilon = \frac{1}{2}$, say, and fix any positive real number M. Clearly values of $x \in \Re^+$ exist that satisfy $|f(x) - L| \ge \frac{1}{2}$ no matter what L is. Why? Hence, by Remark 3.1.4, f has no finite limit at infinity. ■

As we have already observed, the limit at plus infinity in the new context of functions is very similar to that for limits of sequences. In fact, one is a generalization of the other. That is, if $\lim_{x \to \infty} f(x) = L$, then $\lim_{n \to \infty} f(n) = L$. This is a consequence of the next result. The converse, however, is false.

THEOREM 3.1.6. *Suppose $D \subseteq \Re$ is an unbounded above domain of the function f; that is, D contains arbitrarily large values. Then $\lim_{x \to \infty} f(x) = L$ if and only if for every sequence $\{x_n\}$ in D that diverges to plus infinity, that is, $\lim_{n \to \infty} x_n = \infty$, the sequence $\{f(x_n)\}$ converges to L.*

Before giving a proof of this theorem, let us consider a few of its consequences. Recall the key words "for every sequence." Even though there may exist some sequence $\{x_n\}$

diverging to plus infinity for which $\{f(x_n)\}$ converges to L, we cannot assume that $\lim_{x\to\infty} f(x) = L$. For example, if $f(x) = \sin \pi x$, with $x \in \Re$, from Example 3.1.5, we found that $\lim_{x\to\infty} f(x)$ does not exist. However, if x is restricted to only the natural numbers, then $f(n) = \sin \pi n = 0$, and so $\{f(x_n)\}$ converges to 0, where $x_n = n$ for all $n \in N$ (see Figure 3.1.2).

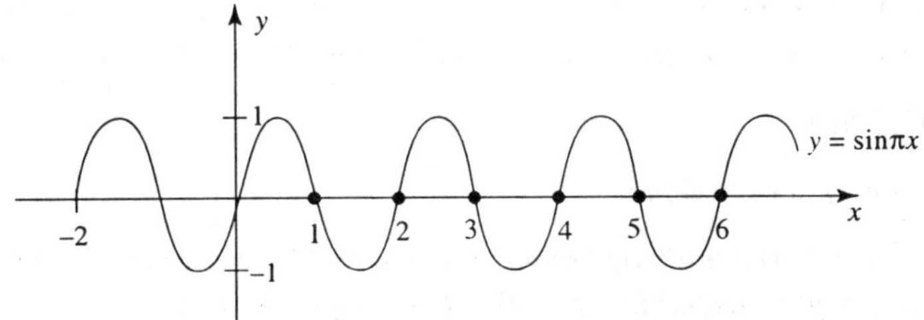

Figure 3.1.2

Now, if $\lim_{x\to\infty} f(x) = L$, then this limit holds for any $x \in \Re$ which tends to infinity, in particular, for $x = n$. Therefore, $\lim_{x\to\infty} f(x) = L$ implies that $\lim_{n\to\infty} f(n) = L$. Therefore, to evaluate limits of sequences, we can use techniques for evaluating more general limits of functions, but not conversely. However, to prove that $\lim_{x\to\infty} f(x)$ does not exist, we often show that $\lim_{n\to\infty} f(n)$ does not exist. Often this can be achieved by finding two subsequences that converge to two different values. This procedure could have been invoked in Example 3.1.5.

Proof of Theorem 3.1.6. ($\Rightarrow$) Suppose that $\lim_{x\to\infty} f(x) = L$. Let $\{x_n\}$ be any sequence in D diverging to plus infinity, and consider the sequence $\{f(x_n)\}$. We will prove that $\{f(x_n)\}$ converges to L. Let an arbitrary $\varepsilon > 0$ be given. We need to find $n^* \in N$ so that $|f(x_n) - L| < \varepsilon$ whenever $x_n \geq n^*$. Since $\lim_{x\to\infty} f(x) = L$, there exists $M > 0$ such that $|f(x) - L| < \boxed{\varepsilon}$ if $x \geq M$ and $x \in D$. So pick $n^* \geq M$. Then $x_n \geq M$. Hence, $|f(x_n) - L| < \varepsilon$, which is what we wanted to prove.

($\Leftarrow$) Suppose that for every sequence $\{x_n\}$ in D that diverges to infinity, the sequence $\{f(x_n)\}$ converges to L. We will prove that $\lim_{x\to\infty} f(x) = L$ by assuming the contrary and thus reaching a contradiction. To this end, there exists an $\varepsilon > 0$ such that for every $M > 0$ there exists an $x \geq M$ with $x \in D$ such that $|f(x) - L| \geq \varepsilon$ (see Remark 3.1.4). In particular, for each integer n there exists an $x_n \in D$ such that $|f(x_n) - L| \geq \varepsilon$

for $x_n \geq n$. However, since the terms x_n tend to $+\infty$, the sequence $\{f(x_n)\}$ converges to L by the hypothesis. But this contradicts the fact that $|f(x_n)-L| \geq \varepsilon > 0$ for all $n \in N$. Hence, the proof is complete. ■

Limits of functions at plus infinity, like sequences, satisfy a number of special properties and some of them are presented in the next theorem.

THEOREM 3.1.7. *Suppose that the functions f, g, and h are defined on $D \subseteq \Re$, which is unbounded above, with $\lim_{x\to\infty} f(x) = A$, $\lim_{x\to\infty} g(x) = B$, and $\lim_{x\to\infty} h(x) = C$. Then*

- ***(a)*** *$\lim_{x\to\infty} f(x)$ is unique.*
- ***(b)*** *f must eventually be bounded above and below, where the bound on f is as given in Definition 2.1.10 with N replaced by D.*
- ***(c)*** $\lim_{x\to\infty} [f(x)-A] = 0$.
- ***(d)*** $\lim_{x\to\infty} |f(x)| = \left|\lim_{x\to\infty} f(x)\right| = |A|$.
- ***(e)*** $\lim_{x\to\infty} [f(x)]^n = \left[\lim_{x\to\infty} f(x)\right]^n = A^n$, *for all $n \in N$.*
- ***(f)*** $\lim_{x\to\infty} \sqrt[n]{f(x)} = \sqrt[n]{\lim_{x\to\infty} f(x)} = \sqrt[n]{A}$, *if $A \geq 0$ and $f(x) \geq 0$ for all $x \in D$, with $n \in N$.*
- ***(g)*** $\lim_{x\to\infty} (f \pm g)(x) = \lim_{x\to\infty} f(x) \pm \lim_{x\to\infty} g(x) = A \pm B$.
- ***(h)*** $\lim_{x\to\infty} (fg)(x) = \left[\lim_{x\to\infty} f(x)\right]\left[\lim_{x\to\infty} g(x)\right] = AB$.
- ***(i)*** $\lim_{x\to\infty} \left(\dfrac{f}{g}\right)(x) = \dfrac{A}{B}$, *if $B \neq 0$ and g not identically 0, eventually.*
- ***(j)*** *$A \leq B$, if $f(x) \leq g(x)$ eventually, for $x \in D$.*
- ***(k)*** *$A \leq B \leq C$, if $f(x) \leq g(x) \leq h(x)$ eventually, for $x \in D$. This property is called the sandwich theorem.*

Theorem 3.1.7 follows from Theorem 3.1.6 and properties of sequences in Sections 2.1 and 2.2.

To refresh your memory concerning what is meant by an eventually monotone function, review Definitions 1.2.10 and 3.1.2, or simply extend Definition 2.4.1 to the appropriate domain. The proof of the next result, which should sound somewhat familiar, is left as an exercise.

THEOREM 3.1.8. *If the function f is defined on an unbounded above domain $D \subseteq \Re$ and is eventually monotone and eventually bounded, then $\lim_{x\to\infty} f(x)$ is finite.*

Next, we extend Definition 2.3.1 to our present context.

Definition 3.1.9. Let f be a function with domain $D \subseteq \Re$, which contains arbitrarily large values. We say that f *tends to plus infinity as x tends to* $+\infty$ if and only if for any real $K > 0$, there exists a real number $M > 0$ such that $f(x) > K$ provided $x \geq M$ for $x \in D$. Whenever this is the case, we write $\lim_{x \to \infty} f(x) = +\infty$.

We define what is meant by a *function tending to* $-\infty$ in a similar fashion. If a function f tends to either $+\infty$ or $-\infty$, we say that f has an *infinite limit.* If $\lim_{x \to \infty} f(x)$ has values of L, $+\infty$, or $-\infty$, we say that the *limit exists*. For example, $\lim_{x \to \infty} x = \infty$ and $\lim_{x \to \infty} \left(-x^2\right) = -\infty$. Theorems 2.3.2, 2.3.3, and 2.3.6 trivially extend to cover functions with domains other than natural numbers. We now turn to another type of limit not discussed in Chapter 2, namely, the *limit at minus infinity*.

Definition 3.1.10. Let f be a function with domain $D \subseteq \Re$, which contains arbitrarily large negative values. Then, $\lim_{x \to -\infty} f(x) = L$ if and only if for every $\varepsilon > 0$ there exists a real number $M > 0$ such that $|f(x) - L| < \varepsilon$ if $x \leq -M$ and $x \in D$.

The terminology in Definition 3.1.10 resembles that in Definition 3.1.1. Similarly, Definition 3.1.9, with $x \geq M$ replaced by $x \leq -M$, defines what is meant by $\lim_{x \to -\infty} f(x) = +\infty$. Moreover, all of the results covered thus far in this section can easily be extended to the case where x tends to minus infinity. Whenever $\lim_{x \to \pm\infty} f(x)$ is written, we mean two limits, one at $+\infty$ and one at $-\infty$. The limiting values for both of these limits do not always coincide. For instance, $\lim_{x \to \pm\infty} \frac{|x|}{x} = \pm 1$ means that at $+\infty$ the limit is 1 and at $-\infty$ the limit is -1. Hence, a function may very well have two different horizontal asymptotes. What are the horizontal asymptotes for $f(x) = \frac{\sqrt{x^2 + 4}}{x}$? Moreover, functions such as $f(x) = \frac{1}{\sqrt{4 - x}}$ and $g(x) = \frac{1}{\sqrt{x - 4}}$ have asymptotes only at $-\infty$ and $+\infty$, respectively. Why? Also,

$$\lim_{x \to \infty} a^x = \begin{cases} 0, \text{ if } 0 < a < 1 \\ +\infty, \text{ if } a > 1. \end{cases}$$

This limit was verified for $x = n \in N$ in Theorem 2.1.13 and Example 2.3.4. The above limit, for a general $x \in \Re$, is left to the reader to formally verify in Exercise 38 of Section 5.2. See also Remark 5.4.7. In the case $a = e$, see also Exercise 12 of Section 4.3 and Remark 3.3.6. Before concluding this section with a discussion of rational functions and oblique asymptotes, we make a few observations dealing with polynomials.

Remark 3.1.11. Clearly, $\lim_{x\to\pm\infty}\frac{1}{x}=0$. Why? Thus, $\lim_{x\to\pm\infty}\frac{k}{x^p}=0$ for any fixed constants $k\in\Re$ and $p\in\Re^+$. We can now verify that the leading term of any polynomial determines its course, that is, if

$$f(x)=a_nx^n+a_{n-1}x^{n-1}+\cdots+a_1x+a_0$$

then $\lim_{x\to\pm\infty}f(x)=\lim_{x\to\pm\infty}a_nx^n$, which is equal to plus or minus infinity, depending on the sign of a_n and whether the value of n is even or odd. The details are left to Exercise 4. The idea can just as well be extended to cover other functions. For example, $\lim_{x\to\infty}\left(3x^3+2\sqrt{x}-1\right)=\lim_{x\to\infty}3x^3$, $\lim_{x\to-\infty}5x\sqrt{x^2+4}=\lim_{x\to-\infty}5x|x|$, and $\lim_{x\to\infty}\frac{x^2-1}{x-1}=\lim_{x\to\infty}(x+1)=\lim_{x\to\infty}x$. However, we should always be cautious about the domain of the function involved. For example, $\lim_{x\to\infty}\left(x^2+\sqrt{x}\right)=\lim_{x\to\infty}x^2=+\infty$, but $\lim_{x\to-\infty}\left(x^2+\sqrt{x}\right)$ does not exist. Why? ■

THEOREM 3.1.12. *Suppose a function, f, is defined by $f(x)=\frac{p(x)}{q(x)}$, where*

$$p(x)=a_nx^n+a_{n-1}x^{n-1}+\cdots+a_1x+a_0 \text{ and}$$
$$q(x)=b_mx^m+b_{m-1}x^{m-1}+\cdots+b_1x+b_0,$$

with $m,n\in N$; a_i with $0\le i\le n$, and b_j with $0\le j\le m$, real constants; and $a_nb_m\ne 0$. Then,

(a) *if $n<m$, then $\lim_{x\to\pm\infty}f(x)=0$.*

(b) *if $n=m$, then $\lim_{x\to\pm\infty}f(x)=\frac{a_n}{b_n}$.*

(c) *if $n>m$, then $\lim_{x\to\pm\infty}f(x)$ is infinite.*

To prove Theorem 3.1.12, multiply f by $1=\frac{x^{-m}}{x^{-m}}$ in each case, and use the theorems from this section. The details are left to the reader. Part (a) of Theorem 3.1.12 indicates that p is of smaller order than q, whereas in part (c), p is of higher order than q. Recall this terminology from Section 2.3. See Remark 5.4.7 for more on this topic. Also see Exercise 20 from Section 2.2 and Exercises 7 and 8 from Section 2.3. Rational functions of the type described in Theorem 3.1.12, parts (a) and (b), have identical horizontal asymptotes at $+\infty$ and $-\infty$. This verifies that $f(x)\equiv\arctan x$ is not a rational function. Why? However, part (c) of that result yields the fact that a function made up of a higher degree polynomial in the numerator than in the denominator has no horizontal asymptotes. Keep in mind that any such polynomial can perhaps be divided to yield a polynomial with a remainder, which, in fact, must tend to zero. Why? The resulting polynomial, excluding the remainder portion, is called an *oblique asymptote* for the original function, unless on division the remainder for the original function is 0.

Example 3.1.13. Find oblique asymptotes for the function $f(x)=\dfrac{x^2-1}{2x+4}$, if any exist.

Answer. Since the degree of the numerator for f is larger than the degree of the denominator, the existence of an oblique asymptote is likely. Now, using long or synthetic division, or a clever rearrangement, we obtain $f(x)=\dfrac{1}{2}x-1+\dfrac{3}{2x+4}$. The line $y=\dfrac{1}{2}x-1$ is an oblique asymptote for f since $\lim_{x\to\pm\infty}\left[f(x)-\left(\dfrac{1}{2}x-1\right)\right]=0$. This intuitively means that $y=\dfrac{1}{2}x-1$ is the "dominant" term for f when the values for $|x|$ are very large (see Figure 3.1.3). ■

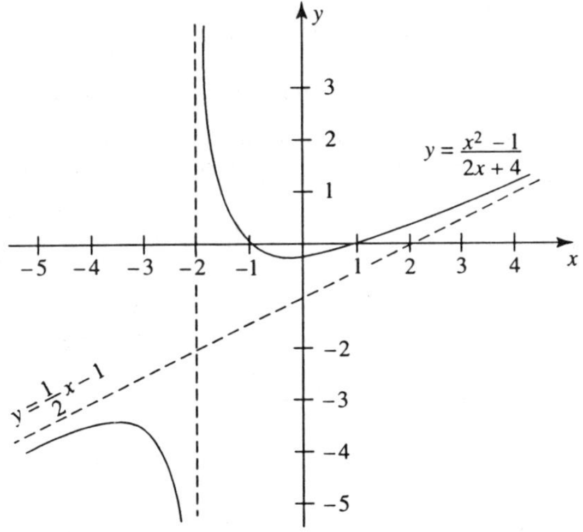

Figure 3.1.3

We conclude this section with a word of caution. Using the comments from Remark 3.1.11 for f, as given in Example 3.1.13, we can write

$$\lim_{x\to\pm\infty}\frac{x^2-1}{2x+4}=\lim_{x\to\pm\infty}\frac{x^2}{2x}=\lim_{x\to\pm\infty}\frac{1}{2}x.$$

Note that the first two functions in these limits are not equal. Even though these two functions diverge to $+\infty$ at the same rate, $y=\frac{1}{2}x$ is not an equation for the oblique asymptote of f. The reason for this is that $\lim_{x\to\pm\infty}\left[f(x)-\frac{1}{2}x\right]\neq 0$. Why?

Exercises 3.1

1. Define $\lim_{x\to\infty} f(x)=-\infty$ and $\lim_{x\to-\infty} f(x)=-\infty$.

2. Use Definitions 3.1.1 and 3.1.9 to prove the following limits.

(a) $\lim_{x\to\infty} f(x)$ with $f(x)=\frac{2x}{x-3}$ and x irrational (b) $\lim_{x\to\infty}\frac{1-x^2}{x-2}$

(c) $\lim_{x\to-\infty} f(x)$ with $f(x)=\frac{x^2+1}{x-2}$ and x rational (d) $\lim_{x\to-\infty}\frac{-1}{x+1}$

3. Prove Theorem 3.1.8.

4. Verify that $\lim_{x\to\pm\infty}\left(a_nx^n+a_{n-1}x^{n-1}+\cdots+a_1x+a_0\right)=\lim_{x\to\pm\infty}a_nx^n$. See Exercise 7 from Section 2.3. Does this prove that $\sin x$ is not a polynomial?

5. Determine whether or not the given limits exist. Find the values for those that do. Verify your assertion in each case, using definitions or theorems from this section.

(a) $\lim_{x\to\infty}\frac{-x^2}{2x^2-3}$ (b) $\lim_{x\to\infty}\frac{3x}{x^2-\sqrt{2x}}$

(c) $\lim_{x\to\infty}\frac{-2x}{3\sqrt{x^2-1}}$ (d) $\lim_{x\to-\infty}\frac{2x}{3\sqrt{x^2+1}}$

(e) $\lim_{x\to\infty}\frac{\sin x}{x}$ (See Exercise 18(b) from Section 2.3.)

(f) $\lim_{x\to\infty}\cos x$ (g) $\lim_{x\to\infty}\sqrt{x}$

(h) $\lim_{x\to-\infty}\sqrt{2+x}$ (i) $\lim_{x\to-\infty}\frac{x-3}{|x-3|}$

(j) $\lim_{x\to\infty}\frac{2^x}{3^{2x+1}}$ (k) $\lim_{x\to\infty}e^{-x}\sin x$

6. (a) Give an example of a function, f, that is unbounded on $\Re$, yet the $\lim_{x\to\infty} f(x)$ is finite.

(b) Prove that $\lim_{x\to\infty} f(x)=\lim_{t\to-\infty} f(-t)$, provided one of these limits exists.

(c) Find $\lim_{x\to-\infty} 2^x e^{-x}$.

7. Explain what is meant by each of the following statements.

(a) $\lim_{x\to\infty} f(x)$ is infinite. **(b)** $\lim_{x\to\infty} f(x)$ does not exist.

(c) $\lim_{x\to\infty} f(x)$ is undefined. **(d)** $\lim_{x\to\infty} f(x)$ is finite.

(e) $\lim_{x\to\infty} f(x)$ exists.

8. Find the horizontal and/or oblique asymptotes for each of the given functions.

(a) $f(x)=\dfrac{3x^2-x+1}{4-2x^2}$ **(b)** $f(x)=\dfrac{2x^2-1}{x+1}$

(c) $f(x)=\dfrac{x^2-1}{x-1}$ (See Exercise 6 of Section 3.2.)

(d) $f(x)=\dfrac{x^4+1}{x^2}$ **(e)** $f(x)=\begin{cases}\dfrac{1-2x}{3x^2}, & \text{if } x<0\\[2mm] \dfrac{x^4+x^3+1}{x+1}, & \text{if } x\ge 0\end{cases}$

9. Suppose that f is an even function defined on $(-\infty,\infty)$. Prove that $\lim_{x\to\infty} f(x)=\lim_{x\to-\infty} f(x)$, provided one of these limits exists. What can be said in the case where f is an odd function? Explain.

10. **(a)** Prove that $\lim_{x\to\infty}\left(1+\dfrac{1}{x}\right)^x=e$. See Example 5.4.11.

(b) Evaluate $\lim_{x\to\infty}\left(\dfrac{x}{x+1}\right)^x$. **(c)** Evaluate $\lim_{x\to\infty}\left(1+\dfrac{1}{\sqrt{x}}\right)^x$.

11. **(a)** Verify that $\dfrac{x-1}{x}<\dfrac{\lfloor x\rfloor}{x}\le 1$ for $x>0$.

(b) Evaluate $\lim_{x\to\infty}\dfrac{\lfloor x\rfloor}{x}$.

(c) Evaluate $\lim_{x\to-\infty}\dfrac{\lfloor x\rfloor}{x}$.

12. If the functions f and g are defined on (a,∞) with $a\in\Re$, $\lim_{x\to\infty} f(x)=L$, and $\lim_{x\to\infty} g(x)=+\infty$, prove that $\lim_{x\to\infty}(f\circ g)(x)=L$.

13. Prove that if f and g are polynomials and $\lim_{x\to\infty}\dfrac{f(x)}{g(x)}=L$, with $L\neq 0$, then f and g are of the same degree.

14. If f and g are polynomials of the same degree, prove that $\lim_{x\to\infty}\dfrac{f(x)}{g(x)}=$

$\lim\limits_{x \to -\infty} \frac{f(x)}{g(x)}$. This means that the function $h(x) = \frac{f(x)}{g(x)}$ has exactly one horizontal asymptote. Is this statement true if the degree of f is smaller than the degree of g? Explain.

15. If $S = \left\{ \frac{x}{x+1} \,\middle|\, x \geq 0 \right\}$, find $\sup S$, $\inf S$, $\max S$, and $\min S$. See Exercise 4(a) in Section 1.2.

16. Prove that a hyperbola $\frac{(x-h)^2}{a^2} - \frac{(y-k)^2}{b^2} = 1$, where a, b, h, and k are real numbers with $a \neq 0$ and $b \neq 0$, has oblique asymptotes given by $y = k \pm \frac{b}{a}(x-h)$.

3.2 Limit at a Real Number

In this section we study yet another type of limit that it was not possible to introduce in Chapter 2. Before going on to the next definition, we recommend a quick review of neighborhoods and deleted neighborhoods covered in Definition 2.5.1.

Definition 3.2.1. Suppose that a function $f: D \to \Re$, with D a subset of $\Re$. And suppose a is an accumulation point of D. The function f *has a limit as* x *approaches, tends to,* a if and only if there exists a real number L such that for every $\varepsilon > 0$ there exists a real number $\delta > 0$ such that

$$|f(x) - L| < \varepsilon, \text{ provided } 0 < |x - a| < \delta \text{ and } x \in D.$$

Whenever this is the case, we write $\lim\limits_{x \to a} f(x) = L$.

Note that if the domain of the function f, labeled D, consists of a deleted neighborhood of $x = a$, then a is automatically an accumulation point of D and the definition is identical to the one given in basic calculus texts. It is not essential that the function f be defined at $x = a$. But even if f is defined for all $x \in \Re$, we are concerned only with what happens to f in its domain "near" the value $x = a$. Thus, in order for f to attain a finite limit at $x = a$, an arbitrary $\varepsilon > 0$ is chosen. This ε is usually a very small value. Corresponding to this ε, we must be able to find a small $\delta > 0$ so that whenever $x \in (a - \delta, a) \cup (a, a + \delta)$ and x is in the domain D of f, we will have $f(x) \in (L - \varepsilon, L + \varepsilon)$ for some real number L. Again, the value of f at $x = a$, if any, does not influence the value of L or its existence. Also, observe that δ is actually a function of ε, since it changes as ε changes. Figure 3.2.1 attempts to demonstrate the previous context pertaining to Definition 3.2.1. Observe that in Figure 3.2.1 it is essential that the neighborhood of a is deleted. Also note that Definition 3.2.1 proves merely whether or not a given L is the correct limit. Thus, the value of L must be predetermined.

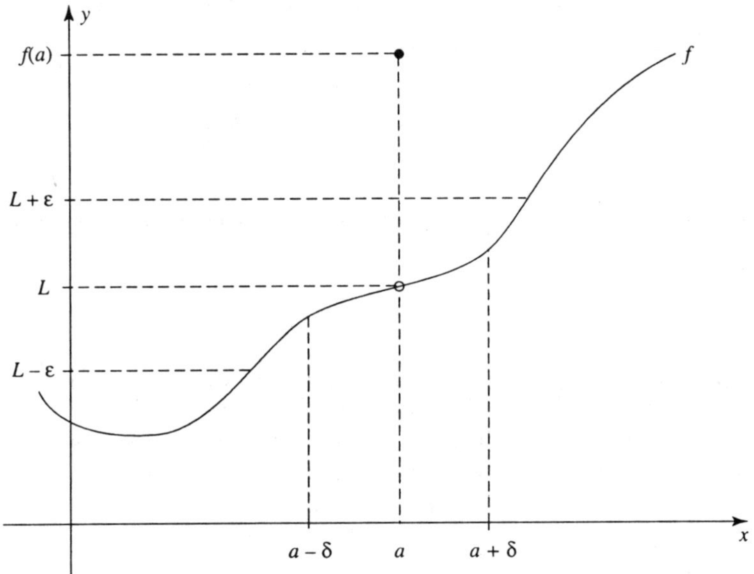

Figure 3.2.1

Example 3.2.2. Prove that $\lim_{x \to 3}(4x-7) = 5$.

Proof. Let $\varepsilon > 0$ be given. We need to find $\delta > 0$ such that if $0 < |x-3| < \delta$, then $|(4x-7)-5| < \varepsilon$. Thus, we write

$$|(4x-7)-5| = |4x-12| = 4|x-3|.$$

And since $4|x-3| < \varepsilon$ whenever $|x-3| < \frac{1}{4}\varepsilon$, we need to pick $\delta = \frac{1}{4}\varepsilon$. Hence, if $0 < |x-3| < \delta = \frac{1}{4}\varepsilon$, then $|(4x-7)-5| = 4|x-3| < 4 \cdot \frac{1}{4}\varepsilon = \varepsilon$. ■

The process of proving a limit such as the previous one involves bounding the expression $|f(x)-L|$ by a constant multiple of $|x-a|$.

Example 3.2.3. Prove that $\lim_{x \to 2} x^2 = 4$.

Proof. Let $\varepsilon > 0$ be given. We need to find $\delta > 0$ such that whenever $0 < |x-2| < \delta$, we will have $|x^2 - 4| < \varepsilon$. So we write $|x^2 - 4| = |x-2| \cdot |x+2|$. Now, saving the term $|x-2|$ with values of close to 2, we will try to find an upper bound for $|x+2|$. To this end, we observe that $|x+2| = |x-2+4| \le |x-2| + 4$, and hence restricting δ to be no more than 1, we have $|x+2| < 5$ for $|x-2| < 1$. Thus,

$$|x^2 - 4| = |x-2| \cdot |x+2| < 5|x-2| < 5 \cdot \frac{1}{5}\varepsilon = \varepsilon,$$

if $|x-2| < 1$ and if $|x-2| < \frac{1}{5}\varepsilon$. Hence, for both restrictions to be true we pick $\delta = \min\left\{\frac{1}{5}\varepsilon, 1\right\}$. ■

Replacing 2 by some constant a would actually prove that $\lim_{x \to a} x^2 = a^2$. Thus, simply replacing x by a in the function $f(x) = x^2$ would allow us to arrive at a proper limiting value, L. Whenever f is "continuous" at $x = a$, the limiting value L can be found in this way. Chapter 4 covers more on this topic.

Example 3.2.4. If $f(x) = \frac{8}{x-3}$, with $x \in Q \setminus \{3\}$, prove that $\lim_{x \to 7} f(x) = 2$. What would the value of this limit be if the domain of f was changed to the set of irrational numbers?

Proof. Let $\varepsilon > 0$ be given. With $x \in Q \setminus \{3\}$ we can write $\left|\frac{8}{x-3} - 2\right| = |x-7| \cdot \frac{2}{|x-3|}$. In restricting x to satisfy $|x-7| < 1$, we can bound $|x-3|$ below by observing that $|x-7| < 1$ implies $3 < |x-3| < 5$. Why? Therefore, we have

$$\left|\frac{8}{x-3} - 2\right| = |x-7| \cdot \frac{2}{|x-3|} < |x-7| \cdot \frac{2}{3} < \frac{3}{2}\varepsilon \cdot \frac{2}{3} = \varepsilon,$$

if $|x-7| < 1$ and if $|x-7| < \frac{3}{2}\varepsilon$. Hence, we pick $\delta = \min\left\{\frac{3}{2}\varepsilon, 1\right\}$. Therefore, if rational x satisfies $0 < |x-7| < \delta$, then $\left|\frac{8}{x-3} - 2\right| < \varepsilon$. Note that $\lim_{x \to 7} f(x) = 2$, provided $x = 7$ is an accumulation point of the domain, D, of the function. Hence, $\lim_{x \to 7} f(x) = 2$ for the domain D equal to the set of irrational numbers. ■

THEOREM 3.2.5. *Suppose functions f, g, $h: D \to \Re$, with $D \subseteq \Re$, a is an accumulation point of D, and $\lim_{x \to a} f(x) = A$, $\lim_{x \to a} g(x) = B$, and $\lim_{x \to a} h(x) = C$. Then all of the conclusions for Theorem 3.1.7 are true with ∞ replaced by a and with "eventually" replaced by "near $x = a$."*

THEOREM 3.2.6. *Let the function f be defined on some deleted neighborhood D of the real number a. The following two statements are equivalent.*

(*a*) $\lim_{x \to a} f(x) = L$.

(*b*) *For every sequence $\{x_n\}$ converging to $x = a$, with $x_n \in D$ and x_n eventually not identically a, the sequence $\{f(x_n)\}$ converges to L.*

Proofs of the above two theorems are left as Exercises 2 and 4.

Example 3.2.7. Consider the function $f(x) = 2x$ with the value of a equal to 3. Clearly, $\lim_{x \to 3} 2x = 6$. Next, pick a sequence $\{x_n\}$ that converges to 3, say, $x_n = \dfrac{3n}{n+1}$. Note that $f(x_n) = \dfrac{6n}{n+1}$ and $\lim_{n \to \infty} f(x_n) = 6$. In fact, by Theorem 3.2.6, the sequence $\{f(x_n)\}$ will converge to 6 no matter what the sequence $\{x_n\}$ converging to 3 is. ■

Theorem 3.2.6 also works well when showing that certain functions do not possess a specified limit. We demonstrate this idea in the next example.

Example 3.2.8. Consider $f(x) = \sin\dfrac{1}{x}$. Does $\lim_{x \to 0} \sin\dfrac{1}{x}$ exist? Explain.

Answer. No. To verify that this limit does not exist we will use two procedures.

Procedure 1. Pick a sequence $\{x_n\}$ of distinct values that converges to $x = 0$ but for which $\{f(x_n)\}$ diverges. Then, by Theorem 3.2.6, $\lim_{x \to 0} \sin\dfrac{1}{x}$ does not exist. To this end, let $x_n = \dfrac{2}{(2n-1)\pi}$. Then, since

$$f(x_n) = \begin{cases} 1, & \text{if } n \text{ is odd} \\ -1, & \text{if } n \text{ is even,} \end{cases}$$

the sequence $\{f(x_n)\}$ diverges.

Procedure 2. Exhibit two sequences $\{x_n\}$ and $\{t_n\}$, both distinct from $x = 0$ and both converging to $x = 0$ but whose functional values, $\{f(x_n)\}$ and $\{f(t_n)\}$, converge to two different limits. Then, using Theorem 3.2.5, part (a), and Theorem 3.2.6, we conclude that the limit does not exist. To this end, pick $x_n = \frac{1}{n\pi}$ and $t_n = \frac{2}{(4n+1)\pi}$. Then, $f(x_n) = \sin n\pi = 0$ and $f(t_n) = \sin\frac{(4n+1)\pi}{2} = 1$. Hence, $\{f(x_n)\}$ and $\{f(t_n)\}$ converge to 0 and 1, respectively. ■

Example 3.2.9. Consider the function f defined by

$$f(x) = \begin{cases} 2x-1, & \text{if } x \text{ is rational} \\ 5-x, & \text{if } x \text{ is irrational.} \end{cases}$$

Prove that

(a) $\lim_{x\to 2} f(x) = 3$. **(b)** $\lim_{x\to a} f(x)$ does not exist, if $a \neq 2$.

Proof of part (a). Let $\varepsilon > 0$ be given, and observe that every real number is an accumulation point of the domain $D = \Re$ of the function f. We need to find $\delta > 0$ such that for all x satisfying $0 < |x-2| < \delta$, we have $|f(x)-3| < \varepsilon$. Thus, we write

$$|f(x)-3| = \begin{cases} |(2x-1)-3|, & \text{if } x \text{ is rational} \\ |(5-x)-3|, & \text{if } x \text{ is irrational} \end{cases} = \begin{cases} 2|x-2| < \varepsilon, & \text{if } x \text{ is rational and } |x-2| < \frac{1}{2}\varepsilon \\ |2-x| < \varepsilon, & \text{if } x \text{ is irrational and } |x-2| < \varepsilon. \end{cases}$$

Therefore, if $\delta = \frac{1}{2}\varepsilon$, then for $0 < |x-2| < \delta$, we have $|f(x)-3| < \varepsilon$ whether x is rational or irrational.

Proof of part (b). Let $\{x_n\}$ be a sequence of rational values distinct from a that tends to a and let $\{t_n\}$ be a sequence of irrational values distinct from a that also tends to a. Then, the sequences $\{f(x_n)\}$ and $\{f(t_n)\}$ tend to $2a-1$ and $5-a$, respectively. If $2a-1 \neq 5-a$, that is, if $a \neq 2$, then because of Theorem 3.2.5, part (a), the function f has no limit as x tends to a. ■

Example 3.2.10. Consider *Dirichlet's*[1] *function* $f:(0,1) \to \Re$, defined by

---

[1] Peter Gustav Lejeune Dirichlet (1805–1859), a German mathematician who studied in Paris under Cauchy, taught Riemann, the successor of Gauss, and was the husband of the sister of the composer Mendelsohn. Dirichlet is best known in number theory for his unit theorem, his proof that certain arithmetic sequences contain infinitely many primes, and his work on binary quadratic forms. In Fourier series, Dirichlet is best known for the first general proof of convergence.

$$f(x) = \begin{cases} \dfrac{1}{q}, & \text{if } x = \dfrac{p}{q} \text{ in lowest terms with } p, q \in N \\ 0, & \text{if } x \text{ is irrational.} \end{cases}$$

See Figure 3.2.2. Show that $\lim_{x \to a} f(x) = 0$ for every $a \in (0,1)$. Dirichlet's function will be mentioned again in a few of the following chapters. See Exercise 2(e) of Section 4.2, Exercise 6 of Section 5.1, and Exercise 8 of Section 6.2.

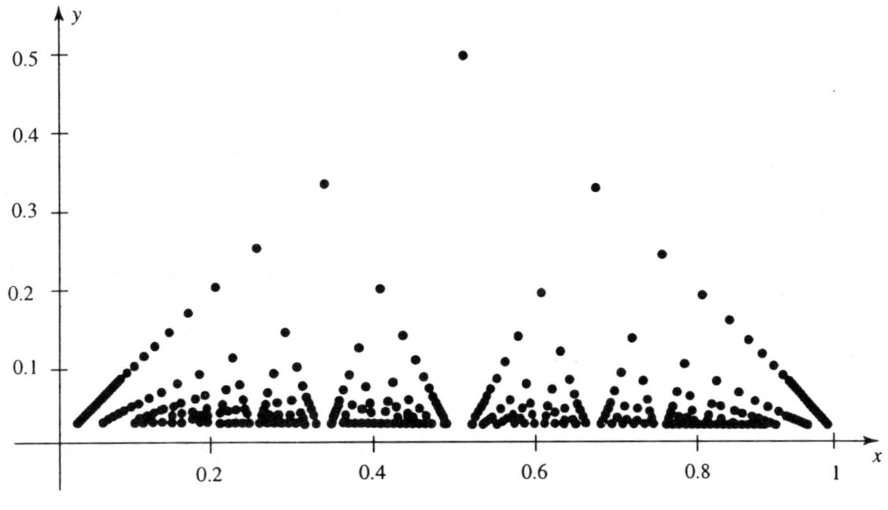

Figure 3.2.2

Proof. Let $\varepsilon > 0$ be given, and let q be a positive integer such that $\dfrac{1}{q} < \varepsilon$. Observe that if any horizontal line is drawn above the x-axis, say $y = \dfrac{1}{q}$, then there are at most a finite number of values of f that are on or above this line. In fact, it can be proven, there are at most $\dfrac{q(q-1)}{2}$ values that satisfy $f(x) \geq \dfrac{1}{q}$. In any event, let δ be the shortest distance from $x = a$ to a value that makes the y-coordinate on or above the line $y = \dfrac{1}{q}$. Then, for $0 < |x - a| < \delta$, we have $|f(x) - 0| = |f(x)| = f(x) < \dfrac{1}{q} < \varepsilon$. Hence, $\lim_{x \to a} f(x) = 0$. ■

Remark 3.2.11. In elementary calculus, through the use of triangles, a sector of a circle, and the sandwich theorem, it was proved that $\lim_{x \to 0} \dfrac{\sin x}{x} = 1$. This result is of great importance throughout the text. See Exercise 5(e) in Section 3.1, Exercise 2(c) in Section

4.1, and Exercise 15 in Section 5.4. Also, we will assume that $\lim_{x\to 0} \sin x = 0$ and $\lim_{x\to 0} \cos x = 1$. These limits will be formally verified in Exercise 22 of Section 3.3. ■

We close this section with a discussion of limits where x tends to a real number $x = a$ but whose corresponding functional values tend to plus infinity or minus infinity.

Definition 3.2.12. Suppose that the function $f: D \to \Re$, with D a subset of $\Re$, and a an accumulation point of D. Then the function f *tends to plus infinity as x approaches, tends to, a* if and only if for any given real number $K > 0$, there exists $\delta > 0$ such that $f(x) > K$, provided $0 < |x - a| < \delta$ and $x \in D$. Whenever this is the case, we write $\lim_{x\to a} f(x) = +\infty$.

Observe that $\lim_{x\to 0} \frac{1}{x^2}$ exists, and we write $\lim_{x\to 0} \frac{1}{x^2} = +\infty$. In a similar fashion, we can write what is meant by $\lim_{x\to a} f(x) = -\infty$. Do not confuse $\lim_{x\to a} f(x) = \pm\infty$ with situations where the limit does not exist (see Example 3.2.8). The next section covers these limits in more detail.

Example 3.2.13. Verify that $\lim_{x\to 0} \frac{1}{x}$ does not exist.

Answer. Pick two sequences, $\{x_n\}$ and $\{t_n\}$, with $x_n = \frac{1}{n}$ and $t_n = -\frac{1}{n}$. Because of the uniqueness of limits, since $\{f(x_n)\}$ and $\{f(t_n)\}$ tend to two different values, $+\infty$ and $-\infty$, respectively, the given limit does not exist. ■

THEOREM 3.2.14. Let the functions f and g be defined on some deleted neighborhood of $x = a$. If $\lim_{x\to a} f(x) = L > 0$ and $\lim_{x\to a} g(x) = +\infty$, then $\lim_{x\to a} (fg)(x) = +\infty$.

Proof. Let $K > 0$ be given. We need to find $\delta > 0$ such that if $0 < |x - a| < \delta$, then $(fg)(x) > K$. Since $\lim_{x\to a} f(x) = L > 0$, by Theorem 3.2.5, part (b), there exists $\delta_1 > 0$ such that $f(x) > \boxed{\frac{L}{2}}$ if $0 < |x - a| < \delta_1$. Also, $\lim_{x\to a} g(x) = +\infty$ implies that there exists $\delta_2 > 0$ such that $g(x) > \boxed{\frac{2K}{L}}$ if $0 < |x - a| < \delta_2$. Let $\delta = \min\{\delta_1, \delta_2\}$. Then, if $0 < |x - a| < \delta$, we have $f(x)g(x) > \frac{L}{2} \cdot \frac{2K}{L} = K$. ■

The importance of L being positive in Theorem 3.2.14 can be found in Theorem 2.3.3 and Exercise 3 of Section 2.3. Also see Exercise 12 at the end of this section.

Exercises 3.2

1. Evaluate the given limits and prove your conclusion using Definition 3.2.1.

(a) $\lim_{x\to 0}(x+1)^3$ **(b)** $\lim_{x\to 1}\frac{x^2-x-2}{2x-3}$ **(c)** $\lim_{x\to 2}\frac{x^2+4}{x+2}$

(d) $\lim_{x\to 0}\frac{x^2}{|x|}$ **(e)** $\lim_{x\to -1}\frac{1}{\sqrt{x^2+1}}$ **(f)** $\lim_{x\to 1}\frac{1-x}{1-\sqrt{x}}$

2. Prove all the properties given in Theorem 3.2.5. Also see Exercise 3.

3. Use mathematical induction to prove that $\lim_{x\to a} x^n = a^n$.

4. Prove Theorem 3.2.6.

5. **(a)** Determine what is wrong with the following argument:

$$\lim_{x\to 0}\left(x\sin\frac{1}{x}\right)=\left(\lim_{x\to 0} x\right)\left(\lim_{x\to 0}\sin\frac{1}{x}\right)=0\cdot(\text{anything})=0.$$

(b) Use a correct method to find $\lim_{x\to 0}\left(x\sin\frac{1}{x}\right)$.

6. Consider the function $f(x)=\frac{x^2-1}{x-1}$.

(a) Evaluate $\lim_{x\to 1} f(x)$, if possible. Explain clearly.

(b) Graph the function f. See Exercise 8(c) of Section 3.1.

7. Use Remark 3.2.11 to evaluate the given limits.

(a) $\lim_{x\to 0}\frac{\sin 5x}{x}$ **(b)** $\lim_{x\to 0}\frac{\sin 3x}{\sin 2x}$

(c) $\lim_{x\to 0}\frac{\tan x}{x}$ **(d)** $\lim_{x\to 0}\frac{1-\cos x}{x}$

(e) $\lim_{x\to 0}\frac{\sin(\sin x)}{x}$ (Compare to Exercise 11(e) of Section 2.2.)

8. Consider the function $f:[-1,1]\to\Re$, defined by

$$f(x)=\begin{cases}0, & \text{if } x=\pm\frac{1}{n} \text{ with } n\in N\\ 1, & \text{otherwise.}\end{cases}$$

Find the given limits and then prove that your results are correct.

(a) $\lim_{x\to\frac{3}{8}} f(x)$ **(b)** $\lim_{x\to-\frac{1}{3}} f(x)$ **(c)** $\lim_{x\to 0} f(x)$

9. Define $\lim_{x\to a} f(x)=-\infty$.

10. Suppose that $f:D\to\Re$, a is an accumulation point of D, $\lim_{x\to a} f(x)=0$, and $f(x)\neq 0$ for any $x\in D$ in some neighborhood of a. Prove that $\lim_{x\to a}\frac{1}{|f(x)|}=\infty$.

11. Suppose $f,g:D\to\Re$, with $D\subseteq\Re$ and a is an accumulation point of D. Prove that if $\lim_{x\to a} f(x)=+\infty$ and $\lim_{x\to a} g(x)=L$, then $\lim_{x\to a}(f+g)(x)=+\infty$.

12. Determine whether Theorem 3.2.14 is true even if $L<0$. How about if $L=0$? Explain.

13. Suppose $f,g:D\to\Re$, with $D\subseteq\Re$ and a is an accumulation point of D. If $\lim_{x\to a} f(x)=L$ and $\lim_{x\to a} g(x)=+\infty$, show that $\lim_{x\to a}\left(\frac{f}{g}\right)(x)=0$. Is this result true if $\lim_{x\to a} f(x)=+\infty$? Explain.

14. Assuming that all of the limits involved exist, prove or disprove the given statements.

(a) $\lim_{x\to 3a} f(x)=3\lim_{x\to a} f(x)$. **(b)** $\lim_{x\to a} f(3x)=3\lim_{x\to a} f(x)$.

(c) $\lim_{x\to 3a} f(x)=\lim_{x\to a} f(3x)$.

15. If $D=\left\{\frac{1}{n}\,\middle|\, n\in N\right\}$ and $f(x)=2x+1$ for all $x\in D$, evaluate $\lim_{x\to 0} f(x)$ and $\lim_{x\to 1} f(x)$, if possible.

16. Verify that if $\lim_{x\to a} f(x)$ exists, then $\lim_{x\to a} f(x)=\lim_{h\to 0} f(a+h)$.

3.3 Sided Limits

The limits at infinity covered in Section 3.1 are just one type of *sided limits* that we are already familiar with. They are called "sided" because they approach infinity from only one side. From Section 3.2, limits at a real number do not necessarily have this property. In $\lim_{x\to a} f(x)$, unless indirectly affected by the domain of f, x can approach a from any direction. If we intentionally restrict x to approach a from only one side, then we have created a sided limit.

Definition 3.3.1. Suppose that the function $f:D\to\Re$, with D a subset of $\Re$ and a an accumulation point of D and $a\neq\sup D$. Then the function f *has a right-hand limit (limit from the right) as* x *approaches, tends to,* a if and only if there exists a real number L such that for every $\varepsilon>0$ there exists a positive real number $\delta>0$ such that

$$|f(x)-L|<\varepsilon,\text{ provided } 0<x-a<\delta \text{ and } x\in D.$$

Whenever this is the case, we write $\lim_{x\to a^+} f(x) = L$, often denoted $f(a^+)$.

Therefore, according to Definition 3.3.1 and Exercise 16 of Section 3.2, we can write $f(a^+)$ as $\lim_{h\to 0^+} f(a+h)$. The *left-hand limit* of a function f at $x = a$, often denoted by $f(a^-)$, is defined similarly. In the one-sided limits defined in Definition 3.3.1, the value of δ depends on the size of ε. Writing $\lim_{x\to a^+} f(x)$ does not imply that this limit actually exists. Depending on the domain of the function f, $\lim_{x\to a^+} f(x)$ may equal $\lim_{x\to a} f(x)$ even though $\lim_{x\to a^-} f(x)$ does not exist. The function $f(x) = \sqrt{x}$ at $x = 0$ is an example of such a function. Thus, it is probable that one of the sided limits exists and the other does not. Also, both sided limits may exist without being equal. However, if $\lim_{x\to a^+} f(x) = \lim_{x\to a^-} f(x)$, then $\lim_{x\to a} f(x)$ will be equal to that same value even when that value is infinite. Why? See Definition 3.3.2 and Exercise 1. Examples of these possibilities are given later in the section and in the following exercises. First, we define what is meant by an infinite sided limit.

Definition 3.3.2. Suppose that the function $f: D \to \Re$, with D a subset of $\Re$ and a an accumulation point of D and $a \neq \sup D$. Then the function f *tends to infinity as x approaches, tends to, a from the right* if and only if for any given real number $K > 0$, there exists a positive $\delta > 0$ such that $f(x) > K$, provided $0 < x - a < \delta$ and $x \in D$. Whenever this is the case, we write $\lim_{x\to a^+} f(x) = +\infty$.

The other three limits, namely, $\lim_{x\to a^-} f(x) = +\infty$, $\lim_{x\to a^+} f(x) = -\infty$, and $\lim_{x\to a^-} f(x) = -\infty$, are defined in a similar fashion. Observe that $\lim_{x\to 0^+} \frac{1}{x} = +\infty$ and $\lim_{x\to 0^-} \frac{1}{x} = -\infty$. This does not trivially follow from Example 3.2.13. Why? Before we proceed, as in Theorem 3.2.6, each of the limits discussed thus far in this section can be written in terms of sequences.

Example 3.3.3. Consider the function $f: D \equiv (-3,-1) \cup (-1,0) \cup (0,\infty) \to \Re$ defined by

$$f(x) = \frac{(x-1)(x+2)}{x^2(x+1)}.$$

Find $f(-3^-), f(-3^+), f(-1^-), f(-1^+), f(0^-), f(0^+)$, and $\lim_{x\to\infty} f(x)$.

Answer. $\lim_{x\to-3^+} f(x) = -\frac{2}{9}$, but $\lim_{x\to-3^-} f(x)$ does not exist. Why? To compute $\lim_{x\to-1^-} f(x)$, observe that $-1 \notin D$ and that an evaluation of f at $x=-1$ yields $\frac{(-2)(1)}{1^2(0)}$, which is undefined. However, if we pick a value of x to the left of but very close to -1, say $x=-1.1$, then $f(-1.1) = \frac{(-2.1)(0.9)}{(1.21)(-0.1)} \approx 15.62$, leading us to believe that $\lim_{x\to-1^-} f(x) = +\infty$, which is indeed the case. Similarly, $f(-1^+) = -\infty$. In generalizing, if a is an accumulation point of the domain D of f which produces $f(a) = \frac{L}{0}$, where $L \neq 0$, then $\lim_{x\to a^+} f(x)$ or $\lim_{x\to a^-} f(x)$ is $+\infty$ or $-\infty$, depending on the signs of the terms involved. We demonstrate this idea on the following two limits.

$$f(0^-) = \lim_{x\to0^-} f(x) = \lim_{x\to0^-} \frac{\overset{-1}{\overset{\uparrow}{(x-1)}}\overset{2}{\overset{\uparrow}{(x+2)}}}{\underset{\underset{0^+}{\downarrow}}{x^2}\underset{\underset{1}{\downarrow}}{(x+1)}} = -\infty,$$

where 0^+ represents values very close to 0 on the positive side. Similarly,

$$f(0^+) = \lim_{x\to0^+} f(x) = \lim_{x\to0^+} \frac{\overset{-1}{\overset{\uparrow}{(x-1)}}\overset{2}{\overset{\uparrow}{(x+2)}}}{\underset{\underset{0^+}{\downarrow}}{x^2}\underset{\underset{1}{\downarrow}}{(x+1)}} = -\infty.$$

Since the power 2 of x^2 is even, $\lim_{x\to0^-} f(x) = \lim_{x\to0^+} f(x)$, and so $\lim_{x\to0} f(x) = -\infty$. See Definition 3.2.12. Note that $\lim_{x\to-1^-} f(x) = -\lim_{x\to-1^+} f(x)$ because of the odd power of $x+1$ in the denominator.

From Theorem 3.1.12, part (a), $\lim_{x\to\infty} f(x) = 0$. The graph of the function f is Figure 3.3.1. ■

Definition 3.3.4. If the limit from the right or from the left at $x=a$ of a function f is infinite, meaning $+\infty$ or $-\infty$, then the line $x=a$ is called a *vertical asymptote*.

To elaborate on vertical asymptotes, if $\lim_{x\to a^+} f(x)$ and $\lim_{x\to a^-} f(x)$ are both infinite but opposite in sign, then $x=a$ is called an *odd vertical asymptote*. If both limits are infinite and equal, $x=a$ is an *even vertical asymptote*. However, if only one of these limits is infinite and the other is finite or does not exist, then $x=a$ is still a vertical asymptote, but it is neither odd nor even. Consider the following examples. The line

$x = 0$ is an odd vertical asymptote for the previously mentioned function $f(x) = \frac{1}{x}$. Also, at $x = 1$, there is a vertical asymptote for the function $f(x) = \frac{-1}{\sqrt{x-1}}$ since $f(1^+) = -\infty$ and $f(1^-)$ does not exist. When discussing vertical asymptotes for rational functions, it is simply not enough to set the denominator equal to 0. Observe that if $f(x) = \frac{x-1}{x^2-1}$ and $g(x) = \frac{1}{f(x)}$, then f has exactly one odd vertical asymptote, namely, $x = -1$, and g has no vertical asymptotes (see Exercise 8). Also, if $f(x) = \frac{p(x)}{q(x)}$ is in a reduced form, where p and q are polynomials and q contains the term $(x-a)^n$ with $n \in N$ even, then $x = a$ is an even vertical asymptote for f, and if $n \in N$ is odd, then the vertical asymptote $x = a$ for f is odd. This idea can also be extended to other types of functions involving quotients.

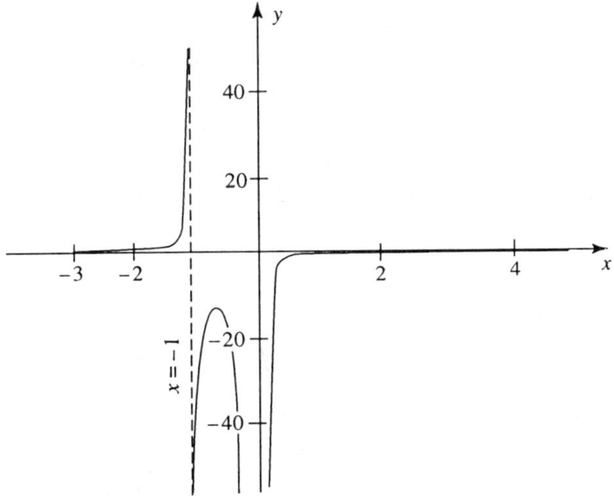

Figure 3.3.1

Remark 3.3.5. The function $f(x) = \frac{x^4+1}{x^2} = x^2 + \frac{1}{x^2}$, which was introduced in Exercise 8(d) of Section 3.1, has an even vertical asymptote at $x = 0$ since $\lim_{x\to 0} f(x) = +\infty$. Here, observe the "dominant" term for the function f. When $|x|$ is very large, $\frac{1}{x^2}$ tends to 0, making $y = x^2$ the dominant term. In Section 3.1, $y = x^2$ was called an oblique asymptote. Near 0, $\frac{1}{x^2}$ is the dominant term since $\lim_{x\to 0} x^2 = 0$ and $\frac{1}{x^2}$ is very large near 0 (see Figure 3.3.2). ■

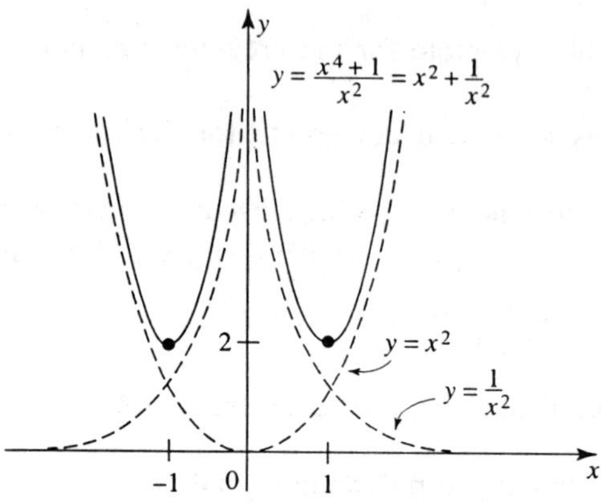

Figure 3.3.2

Remark 3.3.6. Recall the *exponential function* $f(x) = e^x$, with e as given in Exercise 10(a) in Section 3.1, often denoted $\exp(x)$. Although the behavior of e^x is not formally verified until Exercise 12 of Section 4.3, the graph is presented in Figure 3.3.3 along with the inverse function of f, namely the *logarithmic function* $g(x) = \ln x$, so that we can better visualize ideas involving these functions. We will assume familiarity with the properties of exponents and logarithms. Observe that g has a vertical asymptote $x = 0$, which is neither even nor odd. See Exercise 18. The functions f and g transcend —that is, go beyond—purely algebraic methods, and thus are called *transcendental.* See Definition 1.2.17, part (d). ■

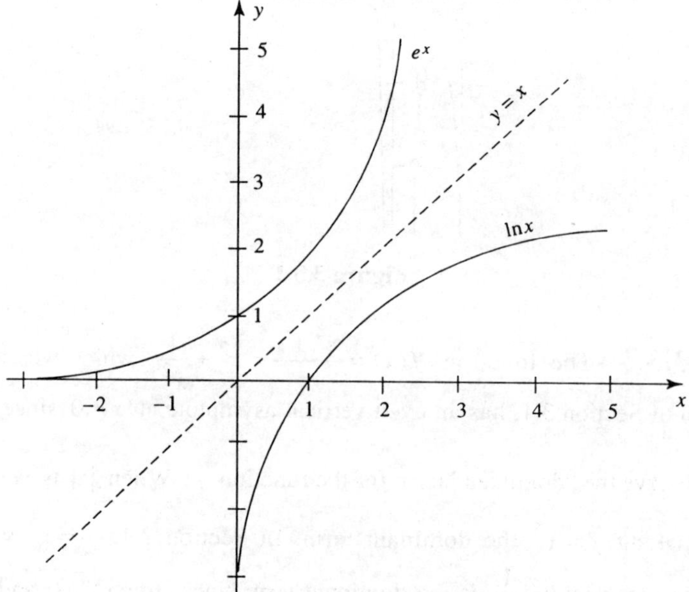

Figure 3.3.3

THEOREM 3.3.7. *Let a function f be defined for $x \in (0,a)$, with $a > 0$ a real number. If either of the limits*

$$\lim_{x \to 0^+} f(x) \text{ or } \lim_{t \to \infty} f\left(\frac{1}{t}\right)$$

exists, then both limits have the same value.

Proof. We will prove Theorem 3.3.7 in the case where the first limit is finite. The converse and all other possibilities are left as an exercise. Cases are to be taken for infinite limits as well.

Suppose $\lim_{x \to 0^+} f(x) = L$. Let $\varepsilon > 0$ be given. We need to find $M > 0$ such that if $t \geq M$, we will have $\left|f\left(\frac{1}{t}\right) - L\right| < \varepsilon$, provided $\frac{1}{t} \in (0,a)$. Since $\lim_{x \to 0^+} f(x) = L$, there exists $\delta > 0$, with $\delta < a$, such that if $0 < x < \delta$, we have $|f(x) - L| < \boxed{\varepsilon}$. So let $x = \frac{1}{t}$, then $\left|f\left(\frac{1}{t}\right) - L\right| < \varepsilon$ for $0 < \frac{1}{t} < \delta$, which is equivalent to $t > \frac{1}{\delta}$. Thus, if we pick $M > \frac{1}{\delta}$, $t \geq M$ implies $\left|f\left(\frac{1}{t}\right) - L\right| < \varepsilon$. ■

Example 3.3.8. If $f(x) = x \sin \frac{1}{x}$, evaluate $\lim_{x \to \infty} f(x)$. Graph the function f by evaluating $\lim_{x \to 0} f(x)$ and computing those values that make $f(x) = 0$ and those that make $f(x) = x$. See Remark 2.2.4, Exercise 5 in Section 3.2, Exercise 1(e) in Section 5.4, and Exercise 6(c) in Section 8.6.

Answer. We write

$$\begin{aligned} \lim_{x \to \infty} x \sin \frac{1}{x} &= \lim_{x \to \infty} \frac{\sin \frac{1}{x}}{\frac{1}{x}} = \lim_{t \to 0^+} \frac{\sin t}{t}, \text{ if } t = \frac{1}{x} \\ &= 1, \text{ by Remark 3.2.11.} \end{aligned}$$

Observe that $f(x) = x \sin \frac{1}{x}$ is an even function. Hence, $\lim_{x \to -\infty} x \sin \frac{1}{x} = 1$ as well. Thus, f has one horizontal asymptote, $y = 1$.

Next we find $\lim_{x \to 0} f(x)$. Since $\left|\sin \frac{1}{x}\right| \leq 1$, we have $0 \leq \left|x \sin \frac{1}{x}\right| \leq |x|$, and the sandwich theorem gives $\lim_{x \to 0} \left|x \sin \frac{1}{x}\right| = 0$. Hence, by Exercise 4 of Section 2.1, we have

$\lim_{x\to 0} x\sin\frac{1}{x} = 0$. A different argument that gives this conclusion can also be used. Observe that since $\sin\frac{1}{x}$ is bounded near $x = 0$ and $\lim_{x\to 0} x = 0$, by Exercise 5 of Section 2.2, we can conclude that $\lim_{x\to 0} x\sin\frac{1}{x} = 0$.

Since f is even, we will find only positive values of x that make $f(x) = 0$ or $f(x) = x$. So, $f(x) = 0$ when $x = \frac{1}{n\pi}$, $n \in N$, with $x = \frac{1}{\pi}$ as the largest root. To find where f touches the line $y = x$, we find x for which $f(x) = x$. This gives the equation $\sin\frac{1}{x} = 1$, with the solution $x = \frac{2}{(4n-3)\pi}$, $n \in N$. The largest of these values is $x = \frac{2}{\pi}$ (see Figure 3.3.4). ■

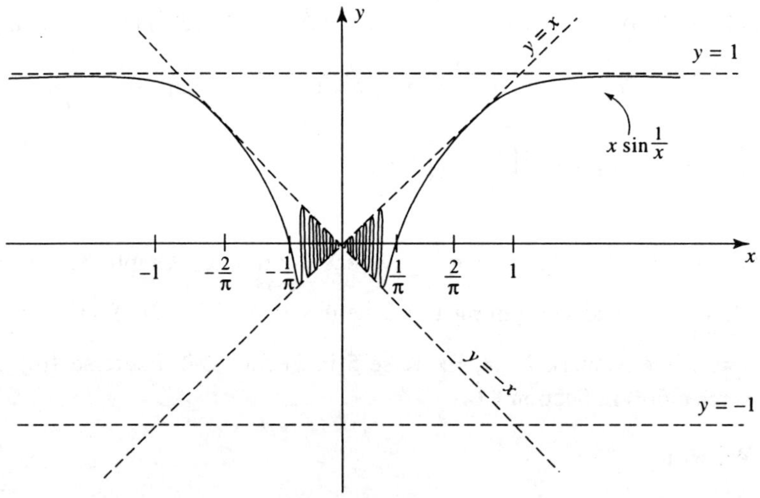

Figure 3.3.4

Example 3.3.9. In view of Example 3.3.8, observe that the odd function $f(x) = x^2 \sin\frac{1}{x}$ has an oblique asymptote, $y = x$ (see Figure 3.3.5). ■

Remark 3.3.10. On completion of the following exercises, observe that often when $\frac{f(x)}{g(x)}$ is evaluated at $x = a$, then $\lim_{x\to a} \frac{f(x)}{g(x)}$ creates the expressions $\frac{0}{0}$, $\frac{\infty}{\infty}$, and $\frac{L}{0}$, with $L \neq 0$. Certainly, if $x = a$ is not an accumulation point of domain D for $\frac{f}{g}$, then this limit does not exist. Suppose a is an accumulation point of D. Then the *indeterminate forms* $\frac{0}{0}$ and $\frac{\infty}{\infty}$ tell us nothing about the value of the limit. Why? Other computations and methods are necessary. Factoring, dividing, multiplying by a fancy form of 1, rationalizing the numerator or denominator, using methods discussed in Section 5.6, and using geometry, are all possibilities that enable us to evaluate such limits. However, the situation $\frac{L}{0}$, with $L \neq 0$, tells us that the value of the limits $f(a^+)$ and $f(a^-)$ is infinite if it exists. Why? All that needs to be determined is the sign of the limit, as we observed in Example 3.3.3. Sections 5.4 and 8.6 study other indeterminate forms of the limit. For $\lim\limits_{x \to \pm\infty} \frac{f(x)}{g(x)}$, meaning $\lim\limits_{x \to \infty} \frac{f(x)}{g(x)}$ or $\lim\limits_{x \to -\infty} \frac{f(x)}{g(x)}$, a similar discussion is involved. ■

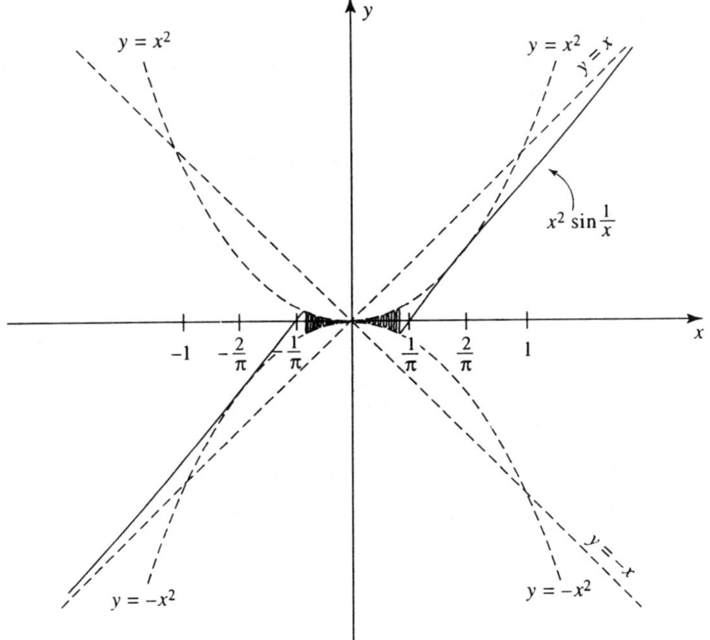

Figure 3.3.5

Exercises 3.3

1. Define each of the given limits.

(a) $\lim\limits_{x \to a^-} f(x) = L$ **(b)** $\lim\limits_{x \to a^-} f(x) = +\infty$

(c) $\lim\limits_{x \to a^+} f(x) = -\infty$ **(d)** $\lim\limits_{x \to a^-} f(x) = -\infty$

2. Verify that if $\lim_{x \to a^+} f(x) = \lim_{x \to a^-} f(x)$, then $\lim_{x \to a} f(x)$ will be equal to the same value without the converse necessarily being true.

3. Use Definition 3.3.2 to prove that $\lim_{x \to 1^+} \dfrac{x}{x-1} = +\infty$.

4. Evaluate the given limits. Use a convincing argument without necessarily using a definition.

(a) $\lim_{x \to 2^+} \dfrac{|x-2|}{x-2}$ **(b)** $\lim_{x \to -1^-} \lfloor x \rfloor$

(c) $\lim_{x \to 0^+} \exp\left(-\dfrac{1}{x}\right)$ **(d)** $\lim_{x \to 0^-} \exp\left(-\dfrac{1}{x}\right)$

(e) $\lim_{x \to 0^+} x^{\frac{1}{x}}$ **(f)** $\lim_{x \to \infty} x^{-x}$

(g) $\lim_{x \to 0^+} \left[\exp\left(\dfrac{1}{x}\right) + \sin\left(\dfrac{1}{x}\right)\right]$ **(h)** $\lim_{x \to 0^-} \left[\exp\left(\dfrac{1}{x}\right) + \sin\left(\dfrac{1}{x}\right)\right]$

5. Find all of the vertical asymptotes and label them as either even, odd, or neither. Graph each function. Also, in parts (a)–(d), find the "dominant" term near the vertical asymptote.

(a) $f(x) = \dfrac{x^2+1}{x}$ **(b)** $f(x) = \dfrac{x^2-1}{x}$

(c) $f(x) = \dfrac{2x^3 - x^2 + 1}{x-1}$ **(d)** $f(x) = \dfrac{x^2+1}{(x+1)^2}$

(e) $f(x) = \dfrac{x^4}{(x-1)(x+1)^2\sqrt{x^2+1}}$ **(f)** $f(x) = \dfrac{x}{\sqrt{4-x^2}}$

6. Prove all the remaining parts of Theorem 3.3.7.

7. Rewrite Theorem 3.3.7 in the case of $\lim_{x \to 0^-} f(x)$.

8. Verify that $f(x) = \dfrac{x-1}{x^2-1}$ has only one vertical asymptote, namely, $x = -1$, and that $g(x) = \dfrac{1}{f(x)}$ has no vertical asymptotes. See Exercise 8(c) from Section 3.1 and Exercise 6 from Section 3.2.

9. Use limits, asymptotes, and roots to graph the given functions.

(a) $f(x) = \dfrac{x-1}{x^2(x-2)}$ **(b)** $f(x) = \begin{cases} \exp\left(\frac{1}{x}\right), & \text{if } x \neq 0 \\ 1, & \text{if } x = 0 \end{cases}$

(c) $f(x) = \dfrac{\exp\left(\frac{1}{x}\right)}{\exp\left(\frac{1}{x}\right)+1}$ **(d)** $f(x) = \begin{cases} \dfrac{1}{\exp\left(\frac{1}{x}\right)+1}, & \text{if } x \neq 0 \\ \frac{1}{2}, & \text{if } x = 0 \end{cases}$

(e) $f(x) = \dfrac{\ln x}{x}$, $x > 0$ (See Remark 5.4.7.)

(f) $f(x) = \dfrac{2x}{\sqrt{x^2 - 1}}$

(g) $f(x) = \dfrac{x^3 + 1}{x}$

(h) $f(x) = \dfrac{x^3 - x^2}{x^2 - 4}$

(i) $f(x) = \dfrac{3x^2}{x^2 - 2x - 3}$

10. (a) Evaluate $\lim_{x\to\infty} \sin\dfrac{1}{x}$.

(b) Conclude that when $|x|$ is "large" the function $f(x) = \sin\dfrac{1}{x}$ "looks very much like" the function $g(x) = \dfrac{1}{x}$, that is, $\sin\dfrac{1}{x} \approx \dfrac{1}{x}$ eventually, and thus has a horizontal asymptote $y = 0$.

(c) Find the values of x for which $\sin\dfrac{1}{x} = 0$, that is, roots of f, and the values of x for which $\sin\dfrac{1}{x} = 1$. What is the largest such value? What is the largest root? Is $\sin\dfrac{1}{x}$ an even or odd function? (See Example 3.2.8.)

(d) Graph the function f.

(e) Graph $h(x) = \sqrt{x}\sin\dfrac{1}{x}$ on $\Re^+$.

11. Prove or disprove.

(a) $\lim_{x\to 0} f(x) = L$ implies $\lim_{x\to\infty} f\left(\dfrac{1}{x}\right) = L$.

(b) $\lim_{x\to\infty} f\left(\dfrac{1}{x}\right) = L$ implies $\lim_{x\to 0} f(x) = L$.

12. Prove that if f is an even function, then $\lim_{x\to 0} f(x) = L$ if and only if $\lim_{x\to 0^+} f(x) = L$. This result also follows if we replace L by $+\infty$ or $-\infty$.

13. (a) Prove that if f is an odd function and $\lim_{x\to 0} f(x) = L$, then $L = 0$.

(b) Is part (a) true if $\lim_{x\to 0} f(x) = L$ is changed to $\lim_{x\to 0^+} f(x) = L$? Explain.

14. Let the function f be defined by $f(x) = \lim_{n\to\infty} x^n$, with $x \in [0,1]$.. Evaluate the given limits.

(a) $\lim_{x\to 1^-} f(x)$ **(b)** $\lim_{x\to 0^+} f(x)$

(c) $\lim_{x\to a} f(x)$, with $a\in(0,1)$ **(d)** $\lim_{x\to 0} f(x)$

15. Let the function f be defined by

$$f(x)=\lim_{n\to\infty}\frac{x^n}{2+x^n},\text{ with } x\in\Re.$$

Evaluate the given limits. See Exercise 27 from Section 2.7 and Exercise 2(p) from Section 4.2.

(a) $\lim_{x\to -1^-} f(x)$ **(b)** $\lim_{x\to -1^+} f(x)$

(c) $\lim_{x\to 1^-} f(x)$ **(d)** $\lim_{x\to 1^+} f(x)$

16. Determine whether or not the given string of equalities is correct.

(a) $\displaystyle\lim_{x\to 1^-}\frac{x^{\frac{3}{2}}-4x^{\frac{1}{2}}}{x^{\frac{1}{2}}}=\lim_{x\to 1^-}\frac{x^{\frac{1}{2}}(x-4)}{x^{\frac{1}{2}}}=\lim_{x\to 1^-}(x-4)=-3$

(b) $\displaystyle\lim_{x\to 0^-}\frac{x^{\frac{3}{2}}-4x^{\frac{1}{2}}}{x^{\frac{1}{2}}}=\lim_{x\to 0^-}\frac{x^{\frac{1}{2}}(x-4)}{x^{\frac{1}{2}}}=\lim_{x\to 0^-}(x-4)=-4$

17. Verify that the number e can also be written as

$$e=\lim_{x\to 0^+}(1+x)^{\frac{1}{x}}.$$

See Exercise 10 in Section 3.1.

18. Prove that $f(x)=\ln|x|$ has a vertical asymptote at $x=0$. That is, prove that $\lim_{x\to 0}\ln|x|=-\infty$. You can assume that $\ln x$, with $x>0$, is an increasing function.

19. If $a_n=\left(\frac{1}{n}\right)^{\frac{1}{\ln n}}$, determine whether or not the sequence $\{a_n\}$ converges.

20. Show that there exists only one value $k\ge 1$ for which

(a) $\displaystyle\lim_{x\to 0}\frac{\sin x}{x^k}$ is finite. See Exercise 7(a) in Section 3.2.

(b) $\displaystyle\lim_{x\to 0}\frac{\sin(\sin x)}{x^k}$ is finite.

21. The function $f(x)=\sin\frac{1}{x}$ is oscillatory, true or false? See Exercise 26 in Section 1.6.

22. **(a)** Prove that $\lim_{x\to 0}\sin x=0$.

(b) Prove that $\lim_{x\to 0}\cos x=1$. See Remark 2.2.4.

3.4* Review

Label each statement as true or false. If a statement is true, prove it. If not,

(i) give an example of why it is false, and

(ii) if possible, correct it to make it true, and then prove it.

1. $\lim_{x \to \pm\infty} c = c$ for any $c \in \Re$.
2. If $\lim_{x \to \pm\infty} f(x) = L$ and c is any real constant, then $\lim_{x \to \pm\infty} cf(x) = cL$.
3. If $y = L$ is a horizontal asymptote for the function f, then f cannot cross this line.
4. If $\lim_{x \to \infty} |f(x)| = |L|$, then $\lim_{x \to \infty} f(x) = L$.
5. If $\lim_{x \to \infty} [f(x)]^2 = L^2$, then $\lim_{x \to \infty} f(x) = L$. (How does this differ from Theorem 3.1.7, part (f), if $n = 2$?)
6. If n is odd then $\lim_{x \to a} \sqrt[n]{x} = \sqrt[n]{a}$ for all $a \in \Re$.
7. If $S = \left\{ \frac{1}{x^2 + 1} \middle| x \in \Re \right\}$, then $\sup S = 1$, $\inf S = 0$, and $\min S$ does not exist.
8. The function $r(x) = \sqrt{x^2 + 4}$ is not a rational function.
9. The function $r(x) = \sqrt{x + 4}$ is a rational function.
10. The function $f(x) = \sqrt{x^2 + 4}$ has an oblique asymptote $y = |x|$.
11. Functions f exist that satisfy $\lim_{x \to 0} f(x) = \lim_{t \to \infty} f\left(\frac{1}{t}\right)$.
12. A function f exists for which $\lim_{x \to a} f(x) = L$, $\lim_{x \to b} f(x) = L$, but $a \neq b$.
13. If f is an odd function, then $\lim_{x \to 0} f(x) = 0$.
14. If $\lim_{x \to 2} f(x) = \lim_{x \to 2} g(x)$, then $f(x) \equiv g(x)$.
15. If $\lim_{x \to a} f(x) = A$, $\lim_{x \to a} g(x) = B$, and $f(x) < g(x)$ for all $x \in \Re$, then $A < B$.
16. If f is a rational function whose denominator equals 0 at $x = a$, then $x = a$ is a vertical asymptote for f.
17. If the domain for the function f is $\Re$, and if there exists a sequence $\{x_n\}$ such that $\lim_{n \to \infty} f(x_n) = +\infty$, then $\lim_{x \to \infty} f(x) = +\infty$.
18. If $\lim_{x \to \pm\infty} f(x) = \pm\infty$, then $\lim_{x \to \pm\infty} \frac{1}{f(x)} = 0$.
19. $\lim_{x \to \infty} f(x) = +\infty$ if and only if $\lim_{x \to 0^+} \frac{1}{f\left(\frac{1}{x}\right)} = 0$.

20. A function exists that is defined on $\Re$, but unbounded in every neighborhood of $x = a$ for any $a \in \Re$.

21. If f is an odd function, then $\lim_{x \to 0^+} f(x) = 0$ implies that $\lim_{x \to 0} f(x) = 0$.

22. If $\lim_{n \to \infty} f(n) = L$ with $n \in N$, then $\lim_{x \to \infty} f(x) = L$.

23. If $\lim_{x \to 0} (fg)(x) = 0$, then both $\lim_{x \to 0} f(x)$ and $\lim_{x \to 0} g(x)$ exist and one of them must equal 0.

24. If the functions f and g are defined on some deleted neighborhood of $x = a$ but $\lim_{x \to a} f(x)$ and $\lim_{x \to a} g(x)$ do not exist, then $\lim_{x \to a} (f + g)(x)$ does not exist.

25. $\lim_{x \to 1} \sqrt{1 - x^2} = 0$.

26. If the function f is defined by $f(x) = \begin{cases} x^2, & \text{if } x \text{ is rational} \\ x, & \text{if } x \text{ is irrational,} \end{cases}$ then $\lim_{x \to 0} f(x) = 0$.

27. $\lim_{x \to \infty} \sin\left(\frac{1}{x}\right) = 1$.

28. If $\lim_{x \to a} (f + g)(x) = L$, then $\lim_{x \to a} f(x) + \lim_{x \to a} g(x) = L$.

29. Let f be defined on some deleted neighborhood of $x = a$, and suppose that the sequences $\{x_n\}$ and $\{t_n\}$ converge to a. If the sequences $\{f(x_n)\}$ and $\{f(t_n)\}$ both converge to some real number L, then $\lim_{x \to a} f(x) = L$.

30. If the functions f and g are defined on some neighborhood of $x = a$ and $\lim_{x \to a} f(x) = -\infty$, then $\lim_{x \to a} (f + g)(x) = -\infty$.

31. Suppose that f is defined on a neighborhood D of $x = a$. If the sequence $\{x_n\}$ converges to $x = a$ and $x_n \in D$ for all $n \in N$, then the sequence $\{f(x_n)\}$ must converge to $f(a)$.

32. Suppose $\lim_{x \to \infty} f(x) = L > 0$ and $\lim_{x \to -\infty} g(x) = +\infty$, then $\lim_{x \to \infty} \left(\frac{f}{g}\right)(x) = 0$.

33. If $0 < f(x) < g(x)$ for all x in some deleted neighborhood of $x = a$, with $\lim_{x \to a} f(x) = A$ and $\lim_{x \to a} g(x) = B$, then $0 < A \le B$.

34. If the function f is defined on $(0, \infty)$ and $\lim_{x \to 0^+} f(x) = L$, then $\lim_{x \to \infty} f\left(\frac{1}{x}\right) = \frac{1}{L}$.

35. Suppose $f : D \to \Re$ with $D \subseteq \Re$ and a is an accumulation point of D. If $\lim_{x \to a} f(x) = L$, then f is bounded.

36. If f is an odd function, then $\lim_{x \to 0} f(x) = f(0)$.

37. If $f(x)=\begin{cases} x^3, & \text{if } x \neq 1 \\ 0, & \text{if } x=1, \end{cases}$ then $\lim_{x\to 1} f(x)=1$.

38. If f is bounded in some neighborhood of $x=a$, then $\lim_{x\to a} f(x)$ is finite.

In Exercises 39 and 40 consider the function f, defined by $f(x)=2$, where $x\in[0,1]$ and $x\neq \frac{1}{n}$, $n\in N$.

39. $\lim_{x\to 1} f(x)$ does not exist.

40. $\lim_{x\to 0^+} f(x)$ does not exist.

In Exercises 41–43 consider the function $f:[0,1]\to\Re$, defined by

$$f(x)=\begin{cases} 2, & \text{if } x\neq \frac{1}{n},\ n\in N \\ 0, & \text{otherwise.} \end{cases}$$

41. $\lim_{x\to 1} f(x)$ does not exist.

42. $\lim_{x\to 0^+} f(x)$ does not exist.

43. $\lim_{x\to \frac{1}{2}} f(x)=2$.

44. There exists a function $f:\Re\to\Re$ that has two horizontal asymptotes.

45. If $\lim_{x\to\infty} f(x)=\infty$, then $\lim_{x\to\infty}\sqrt{f(x)}=\infty$.

46. Oscillatory functions may have a horizontal asymptote.

47. For $-\infty\leq a\leq\infty$, if $\lim_{x\to a} g(x)=a$ and $\lim_{x\to a} f(x)=A$, then $\lim_{g(x)\to a} f(g(x))=A$.

3.5* Projects

Part 1. Monotone Functions

The definition and several examples of monotone functions were given in Chapter 1. Here, we present a few more important results dealing with these functions.

THEOREM 3.5.1. *If $f:(a,b)\to\Re$ is increasing, then for each $c\in(a,b)$, $\lim_{x\to c^+} f(x)$ is finite.*

Proof. Since f is increasing, for every $x\in(c,b)$ we have $f(x)\geq f(c)$. Therefore, the set $S=\{f(x)\mid x\in(c,b)\}$ is bounded below by $\alpha=\inf S$. Our goal is to show that

$\lim_{x \to c^+} f(x) = \alpha$. Thus, let $\varepsilon > 0$ be given. We need to find $\delta > 0$ such that $|f(x) - \alpha| < \varepsilon$, provided $0 < x - c < \delta$ with $x \in (c,b)$. Now, since $\alpha + \varepsilon > \alpha = \inf S = \inf_{x \in (c,b)} f(x)$, there exists $d \in (c,b)$ such that $f(d) < \alpha + \varepsilon$. So, pick $\delta = d - c$. Then, for $0 < x - c < \delta$, which is equivalent to $c < x < c + \delta$, we have $f(x) \le f(d)$, yielding

$$|f(x) - \alpha| = f(x) - \alpha \le f(d) - \alpha < \varepsilon.$$

Hence, the proof is complete. ■

Problem 3.5.2. Restate Theorem 3.5.1 for a decreasing function and then prove it.

Problem 3.5.3. Prove that if $f:(a,b) \to \Re$ is monotone, then for each $c \in (a,b)$, the $\lim_{x \to c^-} f(x)$ is finite.

Problem 3.5.4. Prove that if $f:(a,b) \to \Re$ is monotone, then for each $c \in (a,b)$, we have

$$\lim_{x \to c^-} f(x) = \sup_{x \in (a,c)} f(x) \le f(c) \le \inf_{x \in (c,b)} f(x) = \lim_{x \to c^+} f(x).$$

Problem 3.5.5

(a) Suppose $f:(a,b) \to \Re$ is monotone, then clearly f is bounded at each $c \in (a,b)$. Give an example of a monotone function $f:(a,b) \to \Re$ that is unbounded.

(b) If $f:(a,b) \to \Re$ is increasing, verify that $\lim_{x \to c^+} f(x) \le \lim_{x \to d^-} f(x)$ whenever $a < c < d < b$.

THEOREM 3.5.6. *If $f:(a,b) \to \Re$ is an increasing function, then $\lim_{x \to c} f(x) = f(c)$ for all except perhaps countably many points $c \in (a,b)$.*

Problem 3.5.7. Restate and prove Theorem 3.5.6 for a decreasing function.

Problem 3.5.8. Show by an example that Theorem 3.5.6 is false when f is not monotone.

When looking for an example for Problem 3.5.8, we should ask ourselves if it is possible to find a function $f:(a,b) \to \Re$ that is not monotone but where $\lim_{x \to c} f(x)$ is finite and not equal to $f(c)$ for uncountably many points $c \in (a,b)$. Unfortunately, such a function does not exist. We state it formally as follows.

THEOREM 3.5.9. *If $f:(a,b)\to\Re$, there exist only countably many points $c\in(a,b)$ for which $\lim_{x\to c} f(x)$ is finite, but $\lim_{x\to c} f(x)\neq f(c)$.*

THEOREM 3.5.10. *If $f:(a,b)\to\Re$, there exist only countably many points $c\in(a,b)$ for which $\lim_{x\to c^+} f(x)$ and $\lim_{x\to c^-} f(x)$ are finite, but $\lim_{x\to c} f(x)$ does not exist.*

The proofs of the last three theorems are left to ambitious students.

Part 2. Continued Fractions

As seen in Problem 1.10.9 and in Exercise 18(d) from Section 2.5, $\frac{\sqrt{5}+1}{2}$ and $\sqrt{2}$ can be written as continued fractions. In fact, this is true for any real number. But to begin with, there are a number of types of continued fractions. We will briefly study *simple continued fractions*, that is, expressions of the type

$$x_0+\cfrac{1}{x_1+\cfrac{1}{x_2+\cfrac{1}{x_3+\ddots}}},$$

denoted by $\langle x_0,x_1,x_2,\ldots\rangle$, where $x_0\in Z$ and $x_n\in N$ for $n\in N$. If $x_n=0$ for all $n>m\in N$ and $x_m\neq 0$, then the simple continued fraction is said to be *finite*, is denoted by $\langle x_0,x_1,x_2,\ldots,x_m\rangle$, and represents a rational real number. A simple continued fraction that is not finite, is said to be *infinite*. And a converging infinite continued fraction is always represented by an irrational number (see Definition 3.5.14). $\langle\cdot\rangle$ are referred to as *angle brackets*. The following two theorems are always found in every number theory book.

THEOREM 3.5.11. *(Division Algorithm) If a and b are positive integers, $b\neq 0$, then there exist unique integers q and r with $0\le r<b$ such that $a=bq+r$.*

THEOREM 3.5.12. *(Euclidean Algorithm) If a and b are positive integers, $b\neq 0$, and*

$$\begin{aligned}
a&=bq+r, && 0\le r<b\\
b&=rq_1+r_1, && 0\le r_1<r\\
r&=r_1q_2+r_2, && 0\le r_2<r_1\\
&\vdots && \vdots\\
r_k&=r_{k+1}q_{k+2}+r_{k+2}, && 0\le r_{k+2}<r_{k+1},
\end{aligned}$$

then there exists $k=t\in N$ for which $r_{t-1}=r_tq_{t+1}$, and r_t divides both a and b.

A repeated application of the division algorithm leads to a finite simple continued fraction, since if $a, b \in N$, $b \neq 0$ and if $r_1 \neq 0$, then $a = bq + r_1$ implies

$$\frac{a}{b} = q_1 + \frac{r_1}{b} = q_1 + \frac{1}{\frac{b}{r_1}}.$$

Now, if $r_1 \neq 0$, $b = r_1 q_2 + r_2$, which implies that $\frac{b}{r_1} = q_2 + \frac{r_2}{r_1}$, and so if $r_2 \neq 0$, then

$$\frac{a}{b} = q_1 + \cfrac{1}{q_2 + \cfrac{r_2}{r_1}} = q_1 + \cfrac{1}{q_2 + \cfrac{1}{\cfrac{r_1}{r_2}}}.$$

Now, if $r_2 \neq 0$, $r_1 = r_2 q_3 + r_3$, which implies that $\frac{r_1}{r_2} = q_3 + \frac{r_3}{r_2}$, and so if $r_3 \neq 0$, then

$$\frac{a}{b} = q_1 + \cfrac{1}{q_2 + \cfrac{1}{q_3 + \cfrac{1}{\cfrac{r_2}{r_3}}}}.$$

Because of the Euclidean algorithm, this process will eventually stop.

Problem 3.5.13

(a) Expand $\frac{17}{3}$ and $\frac{3}{17}$ into finite simple continued fractions.

(b) Convert $\langle -3, 2, 4 \rangle$ into a rational number.

Definition 3.5.14. The value of $\langle x_0, x_1, x_2, \ldots \rangle$ is defined to be $\lim\limits_{n \to \infty} \langle x_0, x_1, x_2, \ldots, x_n \rangle$, provided this limit is finite.

Problem 3.5.15

(a) Prove that the value of all infinite simple continued fractions is irrational.

(b) If $r = \langle x_0, x_1, x_2, \ldots \rangle$, then $x_0 = [r]$. Moreover, if $r_1 = \langle x_1, x_2, x_3, \ldots \rangle$, then $r = x_0 + \frac{1}{r_1}$.

(c) Prove that the values of two distinct infinite simple continued fractions are different.

Example 3.5.16. Evaluate $\langle 1,1,1,\ldots\rangle$.

Answer. Since the continued fraction $\langle 1,1,1,\ldots\rangle$ has a value r, by Problem 3.5.15, part (b), we have that $r = 1+\dfrac{1}{r}$. Thus, $r^2 - r - 1 = 0$, with roots $\dfrac{1\pm\sqrt{5}}{2}$. Because $\langle 1,1,1,\ldots\rangle$ represents a positive value, we can conclude that $\langle 1,1,1,\ldots\rangle = \dfrac{\sqrt{5}+1}{2}$ (see Problem 1.10.9). Name the continued fraction represented by $\dfrac{\sqrt{5}-1}{2}$. ■

Problem 3.5.17

(a) Solve $x^2 + x - 1 = 0$.

(b) Write the above equation as $x = \dfrac{1}{1+x} = \dfrac{1}{1+\dfrac{1}{1+x}} = \cdots$. See Problem 1.10.9, part (d).

(c) Verify that $\dfrac{\sqrt{5}-1}{2} = \langle 0,1,1,\ldots\rangle$.

(d) Compare the continued fraction in part (c) to $\langle 1,1,1,\ldots\rangle$.

(e) Evaluate $\langle 2,1,1,\ldots\rangle$ and $\langle 1,2,1,2,1,\ldots\rangle$.

Problem 3.5.18

(a) Develop a procedure that will expand any irrational number into a simple continued fraction.

(b) Write $\sqrt{2}$ and $\sqrt{2}+\sqrt{3}$ as simple continued fractions. See Exercise 18(d) from Section 2.5.

Problem 3.5.19. Verify that $f(x) = \dfrac{2x}{6+2x-x^2}$ can be "formally" written as

$$f(x) = \cfrac{x}{3+\cfrac{x}{1+\cfrac{x}{2-x}}}.$$

Evaluate $f(2)$ in both cases.

Example 3.5.20. Several of the more common continued fractions are listed next.

(a) $\tan x = \cfrac{x}{1-\cfrac{x^2}{3-\cfrac{x^2}{5-\cfrac{x^2}{7-\ddots}}}}$, due to Lambert.

(b) $\tanh x = \cfrac{x}{1+\cfrac{x^2}{3+\cfrac{x^2}{5+\cfrac{x^2}{7+\ddots}}}}$, due to Gauss.[2]

(c) $\arctan x = \cfrac{x}{1+\cfrac{1^2x^2}{3+\cfrac{2^2x^2}{5+\cfrac{3^2x^2}{7+\ddots}}}}$, due to Lambert and Lagrange[3]

$= \frac{\pi}{4} + \cfrac{x-1}{2+\cfrac{x-1}{1-\cfrac{x-1}{6-\cfrac{x-1}{\frac{1}{4}+\ddots}}}}$, due to Thiele's[4] expansion.

---

[2]Karl Friedrich Gauss (1777–1855), a German mathematician, is considered, along with Archimedes and Newton, to be one of the three greatest mathematicians of all times. Many of his important discoveries remained unpublished until years after his death. Gauss discovered non-Euclidean geometry some 30 years before the ideas were published by others. He made important contributions to algebra, geometry, analysis, number theory, numerical analysis, probability, statistics, and complex analysis, and established their role in mathematics as well as in astronomy and physics.

[3]Joseph Louis Lagrange (1736–1813), born to a French-Italian family, was the youngest of eleven children and the only one to survive beyond infancy. Lagrange and Euler were the two greatest mathematicians of the eighteenth century. At age 19, Lagrange became a professor at the Royal Artillery School in Turin. By the age of 25, after monumental contributions to the area today known as calculus of variation, Lagrange was regarded as the greatest living mathematician. Due to Euler's recommendation, Lagrange succeeded Euler as director of the Berlin Academy. There, Lagrange distinguished himself in celestial mechanics, partial differential equations, the theory of numbers, and many other areas. Twenty years later, Lagrange moved to Paris, where he was treated with great honor by King Louis XVI and Napoleon. Lagrange, a shy and modest man, was buried with honor in the Pantheon.

[4]Thorvald Nicolai Thiele (1838–1910), known for both his work in astronomy and an interpolation formula named after him, was a mathematician and actuary from Denmark.

(d) $$\ln(1+x)=\cfrac{x}{1+\cfrac{1^2x^2}{2+\cfrac{1^2x^2}{3+\cfrac{2^2x^2}{4+\cfrac{2^2x}{5+\cfrac{3^2x}{6+\ddots}}}}}}.$$

(e) $$\ln\left(\frac{1+x}{1-x}\right)\equiv 2\tanh^{-1}x=\cfrac{2x}{1-\cfrac{x^2}{3-\cfrac{(2x)^2}{5-\cfrac{(3x)^2}{7-\ddots}}}},$$ due to Lagrange. See Exercise 14(d) from Section 8.5. ■

Different representations give different "rates of convergence." One can study the applications of periodic continued fractions in number theory, approximation theory, and many other areas of mathematics. Continued exponents and continued radicals are also of interest. See Exercises 15(e), 19(b), 15,18(d), and 16 in Sections 5.4, 7.3, 2.4, 2.5, 4.1, respectively.

4

Continuity

4.1 Continuity of a Function

Definition 4.1.1. *(Local)* Suppose a function $f: D \to \Re$, with D a subset of $\Re$. Then f *is continuous at* $x = a \in D$ if and only if for any given $\varepsilon > 0$, there exists $\delta > 0$ such that $|f(x) - f(a)| < \varepsilon$, for all $x \in D$ such that $|x - a| < \delta$.

Definition 4.1.2. *(Global)* A function $f: D \to \Re$ *is continuous on a set* $E \subseteq D$ if and only if f is continuous at each point (value of x) in E. If f is continuous at every point in its domain, D, we simply say f is *continuous*.

The main difference between these definitions and those occurring in basic calculus texts is that there D is assumed to be an interval I, and thus every value of I is an accumulation point of I. If D is an interval, then Definition 4.1.1 reduces to the familiar statement f is continuous at $x = a$ if and only if $\lim_{x \to a} f(x) = f(a) = f\left(\lim_{x \to a} x\right)$. Observe that $\lim_{x \to a} f(x) = f(a)$ means not only that a is an accumulation point of the domain D but also $a \in D$, $\lim_{x \to a} f(x)$ is finite and moreover equals $f(a)$. Note that in Definition 3.2.1 we were not interested in what happened when $x = a$. Here we are. Thus, the deleted neighborhood used in Definition 3.2.1 has to be replaced by a neighborhood of a, and L replaced by $f(a)$. The resulting condition is equivalent to the one used in

Definition 4.1.1. If a is an *isolated point*, that is, if a not an accumulation point of D but $a \in D$, then f is automatically continuous at $x = a$. In such a case the point $(a, f(a))$ is called a *singleton*. That f is continuous at $x = a$ is trivially true by Definition 4.1.1, since if a were not an accumulation point of D, then there would exist $\delta > 0$ such that only a is in the neighborhood $(a - \delta, a + \delta)$. Thus, no matter what $\varepsilon > 0$ is, we would have

$$|f(x) - f(a)| = |f(a) - f(a)| = 0 < \varepsilon.$$

For example, consider $f(x) = x^2$ with $x \in D = \{1\} \cup (2,4] \cup [7, \infty)$. Even though $\lim_{x \to 1} f(x)$ does not exist, f is continuous at $x = 1$ since $1 \in D$. In fact, f is continuous at every point of D. Why? As we can see from this example the word "continuous," which suggests the concept of drawing a function in a continuous stroke without lifting the pen from the paper, is not valid. In fact, even if a were an accumulation point for the domain D of the function f and f were continuous at a and D was an interval, we might not be able to draw f. Here is an example.

Example 4.1.3. Consider the function $f:[-1,1] \to [-1,1]$, which consists of straight line segments connecting the point $\left(\frac{1}{2k}, 0\right)$ with the two points $\left(\frac{1}{2k+1}, \frac{1}{2k+1}\right)$ and $\left(\frac{1}{2k-1}, \frac{1}{2k-1}\right)$, where $k = \pm 1, \pm 2, \pm 3, \ldots$, and $f(0) = 0$. The function f is continuous on the entire domain. Why? There is, however, no way of graphing f without lifting the pen from the paper. We can verify this for ourselves by starting at the point $(1,1)$. After several line segments we realize there is no way of getting to the origin, much less to the other end of the curve. An even worse result would occur if we attempted to graph any portion of f beginning at the origin (see Figure 4.1.1). ■

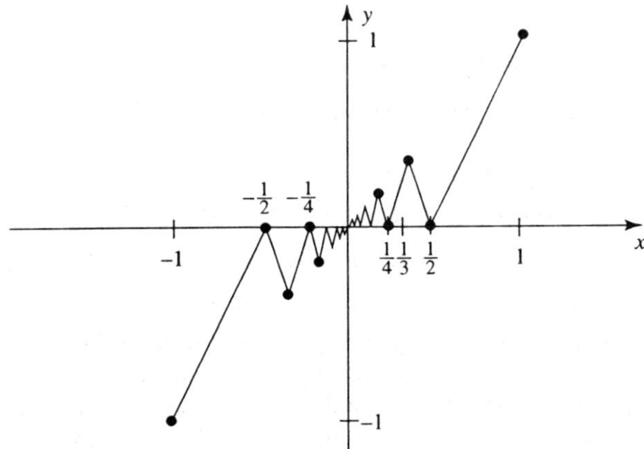

Figure 4.1.1

Example 4.1.4. Verify that every polynomial

$$p(x) = c_n x^n + c_{n-1} x^{n-1} + \cdots + c_1 x + c_0,$$

where $n \in N \cup \{0\}$, $c_i \in \Re$, for $i = 0, 1, 2, \ldots, n$, is continuous, that is, p is continuous at every point of $\Re$.

Proof. Pick any value $a \in \Re$. We will show that p is continuous at $x = a$. Using Definition 3.2.1, we can easily verify that $\lim_{x \to a} x = a$. Thus, by Theorem 3.2.5, part (e), and Exercise 3 from Section 3.2, $\lim_{x \to a} x^n = a^n$ for any $n = 0, 1, 2, \ldots$. So, we can write

$$\begin{aligned} \lim_{x \to a} p(x) &= \lim_{x \to a} \left(c_n x^n + c_{n-1} x^{n-1} + \cdots + c_1 x + c_0 \right) \\ &= c_n \lim_{x \to a} x^n + c_{n-1} \lim_{x \to a} x^{n-1} + \cdots + c_1 \lim_{x \to a} x + c_0 \\ &= c_n a^n + c_{n-1} a^{n-1} + \cdots + c_1 a + c_0 = p(a). \end{aligned}$$

Hence, p is continuous on $\Re$. ■

Remark 4.1.5. Recall from previous experience that trigonometric functions like $\sin x$, $\cos x$, $\arctan x$, rational functions, a^x with $a > 0$, and $\log_a x$ with $a > 0$ and $a \neq 1$, are continuous at every value in their domain. We will use this statement without complete proof. See Exercise 11 in this section, and Exercise 19 in Section 4.3. ■

Definition 4.1.6. Suppose a function $f: D \to \Re$ with $D \subseteq \Re$. Then f *is right continuous at* $x = a$, meaning f *is continuous from the right at* $x = a$ if and only if for any given $\varepsilon > 0$ there exists $\delta > 0$ such that $|f(x) - f(a)| < \varepsilon$, provided $0 \leq x - a < \delta$ and $x \in D$.

If, in fact, a is an accumulation point of D and $a \neq \sup D$, then f is right continuous at a if and only if $\lim_{x \to a^+} f(x)$, which we previously denoted by $f(a^+)$, is actually equal to $f(a)$. If f is right continuous at a, then f does not necessarily have to be continuous at a. Why? The left continuity and $f(a^-)$ can be similarly defined. Certainly, if $f(a^+) = f(a^-) = f(a)$, then f is continuous at $x = a$. Why? When $f(a^+)$ and $f(a^-)$ are finite, but $f(a^+) \neq f(a^-)$, then f is not continuous at a. Observe that if f is defined on $[a, b]$, then continuity at $x = a$ is equivalent to right continuity at $x = a$. A similar statement can be made concerning the continuity at $x = b$ and left continuity. Furthermore, from Definitions 4.1.1 and 4.1.6, if a is not an accumulation point of D, but $a \in D$, then continuity, right continuity, and left continuity at $x = a$ are all equivalent.

Example 4.1.7. Consider

$$f(x) = \begin{cases} 2x+1, & \text{if } 0 \le x \le 2 \\ x^2, & \text{if } x < 0, \end{cases}$$

graphed in Figure 4.1.2. Determine whether f is continuous on the given intervals.

(a) $[0,1]$

(b) $[1,2]$

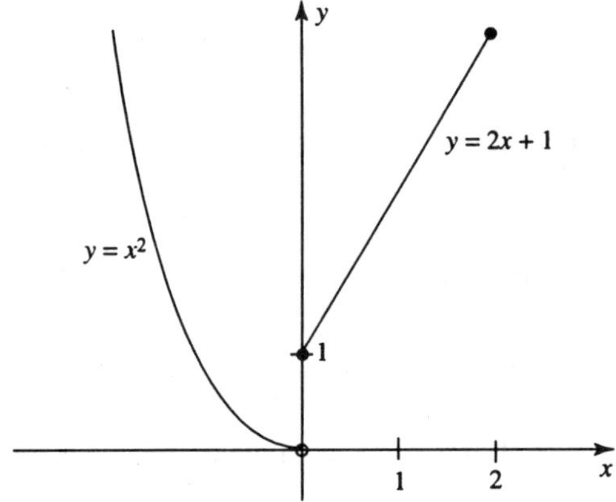

Figure 4.1.2

Answer to part (a). By Example 4.1.4, f is certainly continuous at every x satisfying $x \in (-\infty,0) \cup (0,2)$. So, here all we need to determine is whether f is continuous at $x = 0$. Since $f(0^+) = 1$ and $f(0^-) = 0$, f is not continuous on any set containing 0. Hence, f is not continuous on $[0,1]$. Note that f is right continuous on $[0,1)$ and left continuous on $(0,1]$.

Answer to part (b). Here we only need to determine whether or not f is continuous at $x = 2$. Since f is undefined for all $x > 2$ and $x = 2$ is an accumulation point of the interval $[1,2]$, we have that $\lim_{x \to 2} f(x) = \lim_{x \to 2^-} f(x) = 5 = f(2)$. Thus, f is continuous at $x = 2$ and thus on the entire interval $[1,2]$. ■

THEOREM 4.1.8. *Suppose functions $f, g : D \to \Re$ with $D \subseteq \Re$ are continuous at $x = a$. Then*

(a) *$f \pm g$ are continuous at $x = a$.*

(b) *fg is continuous at $x = a$.*

(c) *$\dfrac{f}{g}$ is continuous at $x = a$, provided $g(a) \ne 0$.*

Proofs of Theorem 4.1.8 are straightforward extensions of those for Theorem 3.2.5 and should be written out in detail. See Exercise 13. It should be noted that part (d) of Theorem 4.1.14 needs to be proven before part (c) of Theorem 4.1.8. Let us now consider continuity for the composition of two continuous functions.

THEOREM 4.1.9. *Consider functions $f: A \to \Re$ and $g: B \to \Re$ with $A, B \subseteq \Re$ such that $f(A) \subseteq B$. If f is continuous at some $x = a \in A$ and g is continuous at $b = f(a) \in B$, then the function $g \circ f$ is continuous at $x = a$.*

Proof. If $x = a$ is an accumulation point of A, we have to show that $\lim_{x \to a} g(f(x)) = g(f(a))$. Let $\varepsilon > 0$ be given. We need to find $\delta > 0$ such that

$$|g(f(x)) - g(f(a))| < \varepsilon, \text{ provided } |x - a| < \delta$$

and x is in the domain of the function $g \circ f$. Since g is continuous at b, say, at $t = b$, that is, $\lim_{t \to b} g(t) = g(b)$, we know that there exists $\delta_1 > 0$, such that

$$|g(t) - g(b)| < \boxed{\varepsilon}, \text{ if } |t - b| < \delta_1$$

and t is in the domain of g. If $t = f(x)$, then the preceding inequalities can be rewritten as

$$|g(f(x)) - g(f(a))| < \boxed{\varepsilon} \text{ whenever } |f(x) - f(a)| < \delta_1.$$

Recall that $f(A) \subseteq B$ and that f is continuous at $x = a$. Thus, there exists $\delta_2 > 0$, such that

$$|f(x) - f(a)| < \boxed{\delta_1} \text{ if } |x - a| < \delta_2.$$

Note that δ_1 is a particular choice of "ε" made when writing down the definition for the limit. Hence, combining the preceding information with $\delta = \delta_2$, we have $|f(x) - f(a)| < \delta_1$, if $|x - a| < \delta$. Thus, $|g(f(x)) - g(f(a))| < \varepsilon$, which is what we needed to show. ∎

The idea behind this proof is actually quite simple. The reader should write out the above proof replacing symbols with verbal descriptions of what is actually taking place. Perhaps the representation given in Figure 4.1.3 will be helpful.

Remark 4.1.10. Theorems 4.1.8, 4.1.9, and Exercise 4 actually play a very important role in evaluating many complicated limits. For example, if f is a composition of functions that satisfy conditions of Theorem 4.1.9, then to evaluate $\lim_{x \to a} f(x)$ we simply evaluate f at $x = a$. Thus,

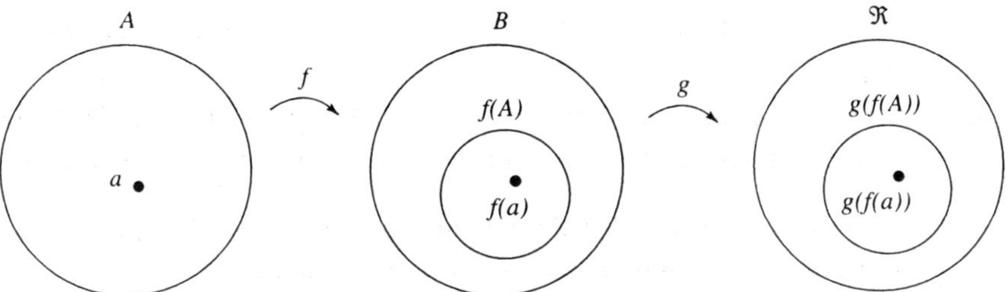

Figure 4.1.3

$$\lim_{x\to 1}\left(\frac{3\sqrt{x^2+3x-1}}{\sqrt{|x-2|}\cos x}\right)^3 = \frac{81\sqrt{3}}{\cos^3 1}.$$

This method was commented on in the paragraph just preceding Example 3.2.4. Do not confuse taking limits of each continuous term all in one step with taking limits of the continuous terms in more than one step. This type of mistake was illustrated in Example 2.2.3 and on the beginning of Part 1 from Section 2.8, where the chain of steps

$$\lim_{n\to\infty}\left(1+\frac{1}{n}\right)^n = \lim_{n\to\infty}(1+0)^n = \lim_{n\to\infty}1^n = \lim_{n\to\infty}1 = 1$$

was incorrect. ■

Remark 4.1.11. As we have seen, there are functions that are *discontinuous* at some value $x = a$, or perhaps, at many values. That is, Definition 4.1.1 is not true. There are several methods for proving discontinuity at a point $x = a$, where a is an accumulation point of D, the domain of f.

(a) (Similar to Remark 3.1.4) There exists $\varepsilon > 0$ such that for any $\delta > 0$ one can find some $x \in D$ satisfying $|x-a| < \delta$ and $|f(x)-f(a)| \geq \varepsilon$.

(b) (Another version of part (a)) Find $\varepsilon > 0$ and a sequence $\{x_n\}$ in D converging to $x = a$ such that $|f(x_n)-f(a)| \geq \varepsilon$ eventually. That is, we need to find $\varepsilon > 0$, $n^* \in N$, and a sequence $\{x_n\}$ in D for which $|f(x_n)-f(a)| \geq \varepsilon$, if $|x_n - a| < \frac{1}{n}$ for all $n \geq n^*$. The value $\frac{1}{n}$ can be replaced by any expression that tends to 0.

(c) (Contradiction to uniqueness) Find two sequences $\{x_n\}$ and $\{t_n\}$ in D, both converging to $x = a$, but whose functional values $\{f(x_n)\}$ and $\{f(t_n)\}$ converge to two different values, or not at all.

(d) (Contradiction to Exercises 5(k) and 5(l) of this section) Find a sequence $\{x_n\}$ of distinct points in D, which converges to $x = a$, but whose functional values $\{f(x_n)\}$ either diverge or converge to something other than $f(a)$.

(e) (Obvious situations) Verify one of the following approaches to showing f is discontinuous at $x=a$: $\lim_{x\to a} f(x)$ is finite but not equal to $f(a)$, $f(a^+)\neq f(a^-)$, a sided limit is infinite, or a sided limit does not exist. Certainly, f is discontinuous at $x=a$ if a is not in D. ■

Also, watch for the terminology. The function $f(x)\equiv\dfrac{1}{x}$ is a continuous function, but f is discontinuous at $x=0$. But,

$$f(x)\equiv\begin{cases}\dfrac{1}{x}, & \text{if } x\neq 0,\\ 2, & \text{if } x=0\end{cases}$$

is not a continuous function. It is discontinuous at $x=0$. Some of our previous work dealt with bounded functions. However, the next definition is presented as a reminder.

Definition 4.1.12. The function $f:D\to\Re$ with $D\subseteq\Re$ is said to be *bounded* on D if and only if there exists a positive constant M such that $|f(x)|<M$, for all $x\in D$.

A function f is bounded if and only if $|f|$ is bounded above.

Remark 4.1.13. If the function $f:D\to\Re$ with $D\subseteq\Re$ is not bounded on D, then we say that f is *unbounded*. To prove that f is unbounded we need to contradict Definition 4.1.12. The negation of the expression "there exists" is "for every." Thus, no matter what positive real constant M one picks, $f(x)$ must be larger than M for at least one value in D. Therefore, we must be able to find a sequence $\{x_n\}$ in D so that $|f(x_n)|>(\text{say})n$, for each $n\in N$. Observe that if D is a *bounded set*, meaning that there exists $L>0$ such that $D\subseteq(-L,L)$, and f is unbounded on D, then $\{x_n\}$ is a bounded sequence and $\{|f(x_n)|\}$ is unbounded. ■

THEOREM 4.1.14. *Suppose D is the domain of f.*

(a) *If f is continuous at a point $x=a$, then there exists $\delta>0$ such that f is bounded, for all $x\in(a-\delta,a+\delta)\cap D$.*

(b) *If f is right continuous at a point $x=a$, then there exists $\delta>0$ such that f is bounded, for all $x\in[a,a+\delta)\cap D$.*

(c) *If f is left continuous at a point $x=a$, then there exists $\delta>0$ such that f is bounded, for all $x\in(a-\delta,a]\cap D$.*

(d) *If f is continuous at a point $x=a$ and $f(a)>0$, then there exists $\delta>0$*

such that $f(x) > \frac{1}{2} f(a)$, *for all* $x \in (a-\delta, a+\delta) \cap D$. *A similar statement is true for* $f(a) < 0$.

Proofs of these statements run parallel to the proofs of Theorem 3.2.5 and are left to the reader. See Exercise 13. We close this section with a brief discussion of sets.

Definition 4.1.15. A set $E \subseteq \Re$ is said to be *closed* if and only if every accumulation point of E is in E.

Examples of closed sets are intervals $[-1,2]$, $[3,\infty)$, $(-\infty,5]$, $\Re$, a set S where $S = \{x \mid x \in [1,2] \cup \{3\}\}$, etc.

Definition 4.1.16. A set $E \subseteq \Re$ is said to be *open* if and only if for each $x \in E$ there exists a neighborhood I of x such that I is entirely contained in E.

Examples of open sets are intervals $(-1,2)$, $(3,\infty)$, $(-\infty,5)$, $\Re$, etc. Also, the interval $[-1,2)$ is not closed and not open. However, the interval $(0,\infty)$ is open but not closed. The sets $\Re$ and ϕ are both open and closed. Are *singletons*, that is sets that contain only one element, closed or open?

THEOREM 4.1.17. *A set* $E \subseteq \Re$ *is closed if and only if* $\Re \setminus E$ *is open.*

Exercises 4.1

1. Find all singletons for the function f, where

$$f(x) = \begin{cases} 1, & \text{if } x = -1 \\ 0, & \text{if } x = 0 \\ 1, & \text{if } x > 0. \end{cases}$$

2. Determine where the given functions are continuous. Explain clearly.

(a) $f(x) = |x-1|$

(b) $f(x) = \frac{|x|}{x}$, where $x = \pm\frac{1}{n}$, $n \in N$

(c) $f(x) = \begin{cases} \frac{\sin x}{x}, & \text{if } x \neq 0 \\ 1, & \text{if } x = 0 \end{cases}$

This function is often called the *sinc function*, and is denoted by $\operatorname{sinc} x$. See Exercise 15 in Section 5.4.

(d) $f(x)=\begin{cases} x\sin\dfrac{1}{x}, & \text{if } x\neq 0 \\ 0, & \text{if } x=0 \end{cases}$ (See Example 3.3.8.)

(e) $f(x)=\begin{cases} x^2\sin\dfrac{1}{x}, & \text{if } x\neq 0 \\ 0, & \text{if } x=0 \end{cases}$ (See Example 3.3.9.)

(f) $f(x)=\dfrac{x+2}{x-1}$

(g) $f(x)=[x]$ (Recall that $[x]=\lfloor x\rfloor$.)

(h) $f(x)=\begin{cases} 1, & \text{if } x>0 \\ 0, & \text{if } x=0 \\ -1, & \text{if } x<0 \end{cases}$

Is the point $(0,0)$ a singleton? This function is often called the *sign function*, and is denoted by $\operatorname{sgn} x$. Compare with part (b).

(i) $f(x)=\dfrac{1}{\sqrt{x}}$

(j) $f(x)=\begin{cases} 1, & \text{if } x \text{ is rational} \\ -1, & \text{if } x \text{ is irrational} \end{cases}$

(k) $f(x)=\begin{cases} \exp\left(-\dfrac{1}{x^2}\right), & \text{if } x\neq 0 \\ 0, & \text{if } x=0 \end{cases}$ (See Exercise 6 of Section 5.3.)

(l) $f(x)=2$ with $x\in[0,3)\cup\{4\}$

3. Determine whether or not the functions f and g defined by

$$f(x)=\begin{cases} x, & \text{if } x=\dfrac{1}{n} \text{ and } n\in Z\setminus\{0\} \\ 1-x, & \text{otherwise} \end{cases} \quad \text{and} \quad g(x)=\begin{cases} x, & \text{if } x \text{ is rational} \\ 1-x, & \text{if } x \text{ is irrational} \end{cases}$$

are continuous at the indicated points.

(a) $x=0$ **(b)** $x=\dfrac{1}{2}$

4. Prove that if the functions $f,g:D\to\Re$ are continuous at $x=a$, then the given functions are also continuous at $x=a$.

(a) $|f|$

(b) $\sqrt{f}$, provided $f(x)\geq 0$ for all $x\in D$

(c) $\max\{f,g\}$ (See Definition 1.2.18, part (e).)

(d) $\min\{f,g\}$

(e) $[f(x)]^n$, $n\in N$

5. Prove or find a counterexample to the following statements in which we assume that f and g are functions defined on the indicated intervals.

(a) f bounded on $[a,b]$ implies f is continuous on $[a,b]$.

(b) f continuous on (a,b) implies f is bounded on (a,b).

(c) $[f(x)]^2$ continuous on (a,b) implies f is continuous on (a,b).

(d) f and g not continuous on (a,b) implies fg is not continuous on (a,b).

(e) f and g not continuous on (a,b) implies $f+g$ is not continuous on (a,b).

(f) f and g not continuous on (a,b) implies $f \circ g$ is not continuous on (a,b).

(g) $f+g$ and f continuous on (a,b) implies g is continuous on (a,b).

(h) fg and f continuous on (a,b) implies g is continuous on (a,b).

(i) $|f|$ continuous on (a,b) implies f is continuous on (a,b).

(j) $f(x)=\frac{1}{x}\sin x$ and $g(x)=\frac{1}{x}$ implies $\frac{f}{g}$ is continuous for all $x \in \Re$.

(k) f continuous at $x=c \in [a,b]$ implies that for any sequence $\{x_n\}$ in $[a,b]$ converging to c, the sequence $\{f(x_n)\}$ converges to $f(c)$, that is, $\lim_{n\to\infty} f(x_n) = f\left(\lim_{n\to\infty} x_n\right)$. See Exercise 31 in Section 3.4.

(l) f defined on $[a,b]$ and $\{f(x_n)\}$ converges to $f(c)$ for any sequence $\{x_n\}$ in $[a,b]$ converging to $c \in [a,b]$ implies f is continuous at $x=c$.

(m) f continuous on D and $\{x_n\}$ in D is a converging sequence, then $\{f(x_n)\}$ converges.

(n) If f is continuous at $x=a$, then $f(a^+)=f(a^-)$.

6. Determine whether or not the function $f(x)=x^x$ is continuous at $x=a>0$.

7. **(a)** Suppose a function f is continuous on (a,b) and $f(r)=c$, for all rational r in (a,b), and c is some fixed real constant. Prove that $f(x)=c$ for all $x \in (a,b)$.

(b) Suppose $S=\{x \mid x \in (-2,3),\ x \text{ rational}\}$ and $f: S \to \Re$ is defined by $f(x)=5$. Is f continuous on the interval $(-2,3)$? Explain.

8. Suppose that a function f is continuous at $x=c \in (a,b)$ and $f(c)>0$. Prove there exists $\delta>0$ such that for all $x \in (a,b)$ with $|x-c|<\delta$, we have $f(x)>0$. Is the continuity assumption necessary? Explain.

9. Suppose functions f and g are continuous at $x=c \in (a,b)$ and $f(c)>g(c)$. Prove there exists $\delta>0$ such that for all $x \in (a,b)$ with $|x-c|<\delta$, we have $f(x)>g(x)$.

10. Give examples of the following requested functions, if possible.

(a) function f defined on $\Re$ but not continuous at any point of $\Re$

(b) function f defined on $\Re$ but continuous at exactly one point of $\Re$

(c) function f defined on $[a,b]$ but continuous at exactly two points of $[a,b]$

(d) function f defined on $[a,b]$ but continuous at only countably many points of $[a,b]$

(e) function f defined on $[0,\infty)$ with unbounded number of countably many points of continuity

(f) function f defined on $[a,b]$ with infinitely many points of both continuity and discontinuity

11. (a) Prove that the exponential function $f(x) = b^x$ with $b > 0$ a real constant, is continuous on $\Re$. That is, if $a \in \Re$, prove that $\lim\limits_{x \to a} b^x = b^a$. Evaluate $\lim\limits_{x \to 2} \dfrac{3^{4x-1}}{5^{3x+1}}$. See Exercise 5(j) in Section 3.1.

(b) Prove that $\sin x$ and $\cos x$ are continuous on $\Re$. See Exercise 22 in Section 3.3.

12. Consider the positive sequences $\{a_n\}$ and $\{b_n\}$ with

$$b_n = \sqrt[n]{a_1 \cdot a_2 \cdot \cdots \cdot a_n},$$

the *geometric mean sequence.*

(a) Prove or disprove. If $\{a_n\}$ converges to A, then $\{b_n\}$ converges to A.

(b) Prove or disprove. If $\{b_n\}$ converges to B, then $\{a_n\}$ converges to B.

For related problems see Exercise 27(c) from Section 5.3 and Exercise 12 from Section 5.4.

13. Prove Theorem 4.1.14 and Theorem 4.1.8.

14. Prove Theorem 4.1.17.

15. If $a_n = \ln n - \ln(n+1)$, determine whether the sequence $\{a_n\}$ converges or diverges. See Example 7.1.5.

16. Suppose $b \ge 1$ and the sequence $\{a_n\}$ is defined recursively by $a_1 = b$ and $a_{n+1} = b^{a_n}$ for all $n \in N$. For a related problem, see Exercise 19 in Section 7.3.

(a) Prove that $\{a_n\}$ is an increasing sequence.

(b) Prove that $\{a_n\}$ is bounded above if $1 \le b < \sqrt[3]{3}$.

(c) Conclude that if $1 \le b < \sqrt[3]{3}$, then $\{a_n\}$ converges to some value A, which satisfies $A = b^A$.

(d) Evaluate the *continued power* $\sqrt{2}^{\sqrt{2}^{\sqrt{2}^{\cdot^{\cdot^{\cdot}}}}}$. See Exercise 19 in Section 7.3.

17. Prove that an interval (a,b) is an open set and an interval $[a,b]$ is a closed set.

4.2* Discontinuity of a Function

Section 4.1 included continuity and an introduction to discontinuity of functions. Functions are discontinuous at points for various reasons. Function $f(x)=\frac{1}{\sqrt{x}}$ is discontinuous at $x=0$ since it is not defined there, function $g(x)=\operatorname{sgn} x$ is discontinuous at $x=0$ since $g(0^+)\neq g(0^-)$, function

$$h(x)=\begin{cases}1, & \text{if } x \text{ is rational}\\ 0, & \text{if } x \text{ is irrational}\end{cases}$$

is discontinuous at $x=0$ since neither $h(0^+)$ nor $h(0^-)$ exist, function

$$k(x)=\begin{cases}\frac{1}{x}, & \text{if } x\neq 0\\ 0, & \text{if } x=0\end{cases}$$

is discontinuous at $x=0$ since $k(0^+)$ and $k(0^-)$ are infinite, function

$$p(x)=\begin{cases}\exp\left(\frac{1}{x}\right), & \text{if } x\neq 0\\ 0, & \text{if } x=0\end{cases}$$

is discontinuous at $x=0$ since $p(0^+)=+\infty$ and $p(0^-)=0$, and function

$$q(x)=\begin{cases}x, & \text{if } x\neq 0\\ 1, & \text{if } x=0\end{cases}$$

is discontinuous at $x=0$ since even though $\lim_{x\to 0} q(x)$ exists, its value of 0 is not equal to $q(0)$. For all these functions $x=0$ is an accumulation point of the domain. In this section we will attempt to clarify and distinguish among the different types of discontinuities.

Definition 4.2.1. A function $g:E\to\mathfrak{R}$ with $E\subseteq\mathfrak{R}$ is an *extension* of the function $f:D\to\mathfrak{R}$ provided $D\subset E$ and $f(x)=g(x)$ for all $x\in D$. If g is continuous, then g is called a *continuous extension* of f.

Thus, if g extends f, then all the functional values of g agree with the functional values of f, and g has some additional values. Extensions are not unique. For example, $g(x)=\operatorname{sgn} x$ and $h(x)=\frac{|x|}{x}$ extend

$$f(x)=\begin{cases}1, \text{ if } x=\dfrac{1}{n} \text{ with } n\in N\\[2ex] -1, \text{ if } x=-\dfrac{1}{n} \text{ with } n\in N.\end{cases}$$

Notice that there is no continuous extension of this function f. Why?

Example 4.2.2. Suppose $f(x)=x\sin\dfrac{1}{x}$. If possible,

(a) find an extension g of f.

(b) find a continuous extension h of f.

Answer to part (a). Since $D=\mathfrak{R}\setminus\{0\}$ is the domain of f, to obtain an extension g of f we must define g at $x=0$. Thus,

$$g(x)=\begin{cases}x\sin\dfrac{1}{x}, \text{ if } x\neq 0\\[2ex] k, \text{ if } x=0\end{cases}$$

is an extension of f, where k is any real constant.

Answer to part (b). As in part (a), we need to define $h(x)$ by $x\sin\dfrac{1}{x}$ for $x\neq 0$. Note that h is continuous at every point except zero. To extend f continuously, since 0 is an accumulation point of D, define $h(0)$ by $\lim\limits_{x\to 0} h(x)=\lim\limits_{x\to 0} x\sin\dfrac{1}{x}$, which is zero. Why? Hence, the desired h is

$$h(x)=\begin{cases}x\sin\dfrac{1}{x}, \text{ if } x\neq 0\\[2ex] 0, \text{ if } x=0.\end{cases}$$

See Exercise 2(d) of Section 4.1. Notice that h is unique in this case. Why? ■

The function f in Example 4.2.2 has a discontinuity at $x=0$. But since f can be extended continuously, we say that this discontinuity is "removable." This is better explained by the next definition.

Definition 4.2.3. A function $f: D\to\mathfrak{R}$, with $D\subseteq\mathfrak{R}$ and a an accumulation point of D, has a *removable discontinuity* at $x=a$ if either

(a) $a\notin D$ and $\lim\limits_{x\to a} f(x)$ is finite, or

(b) $a\in D$ and $\lim\limits_{x\to a} f(x)=L\neq f(a)$.

Basically, if a function f is discontinuous at a point $x=a$ in its domain, and redefining f at this one value $x=a$, we create a new function that is continuous at a, then f has a removable discontinuity at $x=a$. In addition, if $a \notin D$ and $f: D \to \Re$ with $\lim_{x \to a} f(x)$ finite, then f has a removable discontinuity at $x=a$. Since removable discontinuity of f at $x=a$ with $a \in D$ ensures that $\lim_{x \to a} f(x)$ is finite, as noted earlier, it is possible to extend f by a function g, g continuous at $x=a$, if we ensure that $\lim_{x \to a} f(x) = g(a)$. As we have seen in Example 4.2.2, $f(x) = x \sin \frac{1}{x}$ has a removable discontinuity at $x=0$. Any function can have at most countably many removable discontinuities. See Theorem 3.5.9.

Example 4.2.4. Consider the function $f(x) = \frac{x-1}{x^2-1}$, which we worked with in Exercise 8 of Section 3.3. Let us discuss the discontinuity of f. According to Theorem 4.1.8, part (c), Remark 4.1.5, or simply Definition 4.1.1, f is not continuous at $x=-1,1$. Note that these two values are not in the domain of f. But, since

$$\lim_{x \to 1} f(x) = \lim_{x \to 1} \frac{x-1}{x^2-1} = \lim_{x \to 1} \frac{1}{x+1} = \frac{1}{2},$$

Definition 4.2.3 tells us that the discontinuity of f at $x=1$ is removable. The graph of f has a hole at the point $\left(1, \frac{1}{2}\right)$. We can extend the function f at $x=1$ to a function g, g continuous at $x=1$, by defining g as $g(x) = \frac{1}{x+1}$. Now, let us look at the discontinuity of f at $x=-1$. Observe that $\lim_{x \to -1} f(x)$ does not exist since $\lim_{x \to -1^-} f(x) = -\infty$ and $\lim_{x \to -1^+} f(x) = +\infty$. Therefore, the discontinuity at $x=-1$ is not removable. In fact, even if both limits were $+\infty$ or $-\infty$, the discontinuity would not be removable. Why? The function f is graphed in Figure 4.2.1. ■

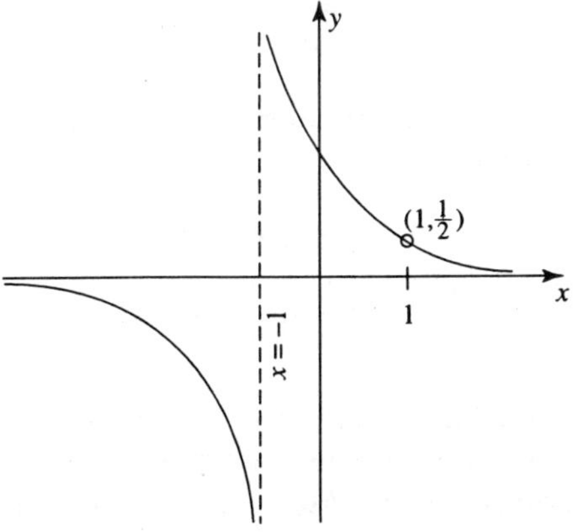

Figure 4.2.1

If $f: D \to \Re$ is continuous at a point $x = a \in D$ but we wish to remove this value from the graph of f, then either we redefine f by $g: E \to \Re$, where $E = D \setminus \{a\}$ and $g(x) = f(x)$ for all $x \in E$, or we multiply f by $\dfrac{x-a}{x-a}$, removing $x = a$ automatically from the domain of f. After performing this multiplication, the resulting function would not be an extension of f. Why? According to Example 4.2.4, discontinuities exist that are not removable. They are referred to as *nonremovable discontinuities* of a function. If $x = a$ is a point of nonremovable discontinuity for the function f, then there is no way to obtain a continuous extension for f. The next three definitions classify the many different types of nonremovable discontinuities. We begin with the most familiar type.

Definition 4.2.5. Suppose a function $f: D \to \Re$ with $D \subseteq \Re$ and a an accumulation point of D. Suppose also that one of the following three conditions is satisfied.

(a) $\lim_{x \to a^+} f(x) = L$ and $\lim_{x \to a^-} f(x) = M$.

(b) $\lim_{x \to a^+} f(x)$ does not exist since f is undefined in that region, but $a \in D$ and $\lim_{x \to a^-} f(x) = M$. In this case, L will denote the value of $f(a)$.

(c) $\lim_{x \to a^-} f(x)$ does not exist since f is undefined in that region, but $a \in D$ and $\lim_{x \to a^+} f(x) = L$. In this case, M will denote the value of $f(a)$.

Suppose further that $L \neq M$. Then f has a *jump discontinuity* at $x = a$. The value $J_a(f) = L - M$ is called the *jump of f at $x = a$*.

Note that if D represents an interval, then parts (b) and (c) of Definition 4.2.5 occur only at the endpoints. Both removable and jump discontinuities may exist at the same point. For example the function

$$f(x) = \begin{cases} 0, & \text{if } x = 0 \\ 1, & \text{if } x > 0 \end{cases}$$

has both removable and jump discontinuities at the point $x = 0$, the endpoint of the interval. Note also that the function

$$r(x) = \begin{cases} 0, & \text{if } x \leq 0 \\ \dfrac{1}{x} \sin \dfrac{1}{x}, & \text{if } x > 0 \end{cases}$$

has a discontinuity at $x = 0$, which is neither removable nor jump. Why?

If at $x = a$ there exists a removable and/or jump discontinuity of the function f, then $x = a$ is referred to as a point of *simple discontinuity* of f, or a *discontinuity of the first kind.* All other discontinuities are *discontinuities of the second kind.*

Example 4.2.6. Consider $f(x) = 2x - \lfloor 2x \rfloor$ with $x \in [-1, 1]$.

(a) Determine where f is discontinuous.

(b) Classify these discontinuities as removable or nonremovable. Determine whether the nonremovable discontinuity, if it exists, is jump or not.

Functions of this type are called *sawtooth functions*.

Answer. The function $f(t) = \lfloor t \rfloor$ is continuous whenever t is not an integer. Why? Thus, to determine the points of discontinuity consider the situation when t, which in our case is $2x$, is an integer. This occurs when $x = -1, -\frac{1}{2}, 0, \frac{1}{2}$, and 1. When $a = -\frac{1}{2}$, 0, and $\frac{1}{2}$, we have $f(a^-) = 1$ and $f(a^+) = 0$. Why? Therefore, by Definition 4.2.5, part (a), f has nonremovable jump discontinuities at $a = -\frac{1}{2}, 0$, and $\frac{1}{2}$ with jump values of -1. If $x = a = -1$, then $f(-1^-)$ does not exist. Why? However, $\lim_{x \to -1^+} f(x) = \lim_{x \to -1^+} (2x + 2) = 0$. And since $f(-1) = 0$ as well, f is continuous at $x = -1$. If $x = a = 1$, then $f(1^+)$ does not exist. But, $\lim_{x \to 1^-} f(x) = \lim_{x \to 1^-} (2x - 1) = 1$. And since $f(1) = 0$, part (b) of Definition 4.2.5 is satisfied resulting in a jump discontinuity at $x = 1$ with a jump of -1. The function f is graphed in Figure 4.2.2. ■

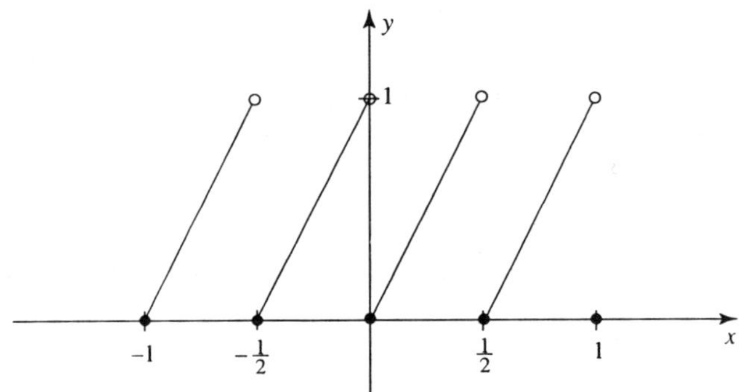

Figure 4.2.2

Next, note that the function

$$k(x) = \begin{cases} \frac{1}{x}, & \text{if } x \neq 0 \\ 0, & \text{if } x = 0 \end{cases}$$

is discontinuous at the point $x = 0$ and fits neither Definition 4.2.3 nor Definition 4.2.5 since $k(0^-) = -\infty$ and $k(0^+) = +\infty$. Suspecting an infinite jump at $x = 0$ can lead to the conclusion that this discontinuity is an "infinite" discontinuity. This is precisely true. We generalize the idea of this discontinuity of the second kind next.

Definition 4.2.7. Suppose a function $f:D \to \Re$ with $D \subseteq \Re$ and a an accumulation point of D. Suppose also that one of the following three conditions is satisfied.

(a) $\lim_{x\to a^+} f(x)$ and $\lim_{x\to a^-} f(x)$ are both infinite.

(b) One of the limits $\lim_{x\to a^+} f(x)$ or $\lim_{x\to a^-} f(x)$ is infinite and the other is finite.

(c) One of the limits $\lim_{x\to a^+} f(x)$ or $\lim_{x\to a^-} f(x)$ is infinite and the other does not exist since f is undefined in that region.

Then, f has an *infinite discontinuity* at $x = a$.

To analyze this concept, consider the next example.

Example 4.2.8

(a) Functions $f(x) = \frac{1}{x}$ and $k(x)$, given above, have an infinite discontinuity at $x = 0$ since $\lim_{x\to 0^+} f(x) = \lim_{x\to 0^+} k(x) = +\infty$ and $\lim_{x\to 0^-} f(x) = \lim_{x\to 0^-} k(x) = -\infty$.

(b) The function r preceding Example 4.2.6 does not satisfy the conditions of Definition 4.2.7, and thus r does not have an infinite discontinuity at $x = 0$. The same is true for functions

$$f(x) = \begin{cases} \frac{1}{x}, \text{ if } x = \pm\frac{1}{n} \text{ and } n \in N \\ 0, \text{ otherwise} \end{cases} \quad \text{and} \quad g(x) = \begin{cases} \frac{1}{x}, \text{ if } x > 0 \\ 1, \text{ if } x \le 0, \ x \text{ rational} \\ \text{-}1, \text{ if } x < 0, \ x \text{ irrational.} \end{cases}$$

(c) The function $g(x) = \frac{1}{x^2}$ has an infinite discontinuity at $x = 0$ since $g(0^+) = g(0^-) = +\infty$.

(d) The function $h(x) = \exp\left(\frac{1}{x}\right)$ has an infinite discontinuity at $x = 0$ since $h(0^+) = +\infty$ and $h(0^-) = 0$. See Exercise 4, parts (c), (d), (g), and (h), from Section 3.3.

(e) The function $f(x) = \frac{1}{\sqrt{x^2 - 4}}$ has infinite discontinuities at $x = -2$ and $x = 2$ due to condition (c) of Definition 4.2.7. ■

One has to be careful with terminology. According to Remark 4.1.5 and Definition 4.1.2, the function $f(x) = \frac{1}{x}$ is continuous. However, $f(x) = \frac{1}{x}$ has an infinite discontinuity at $x = 0$.

If a function $f: D \to \Re$, with $D \subseteq \Re$ and a an accumulation point of D, has a nonremovable discontinuity at $x = a$, which is neither jump nor infinite, then we say f has an *oscillating discontinuity* at $x = a$. Sometimes oscillating discontinuities are broken up into several classes. The reason for the name is obvious if we consider the discontinuity at $x = 0$ of the function $f(x) = \sin \frac{1}{x}$. See Exercise 10 of Section 3.3 and Exercise 1(i) at the end of this section. The functions from Example 4.2.8, part (b), as well as the function $h:(0,1] \to \Re$, defined by

$$h(x) = \begin{cases} 1, & \text{if } x = \frac{1}{n},\ n \in N \\ 0, & \text{otherwise,} \end{cases}$$

also have oscillating discontinuities at $x = 0$.

Before closing this section, we define what is meant by a periodic function, a periodic extension, and a piecewise continuous function.

Definition 4.2.9. The function $f: D \to \Re$ with $D \subseteq \Re$ is *p-periodic*, or just *periodic* (on D) if and only if there exists a number $p > 0$ such that $f(x+p) = f(x)$ for all $x \in D$. The constant p is called a *period* for f. If there is a smallest such number, then this number is called the *fundamental period* for f.

For example, $\sin x$, $\cos x$, and $\tan x$ are periodic on $\Re$ with a period of 2π. Notice that 2π is not the fundamental period for $\tan x$. Why?

Definition 4.2.10. Let f be defined on an interval I of length p. A *periodic extension* of f is the function g which satisfies $g(x+kp) = f(x)$ for each integer k and all $x \in I$.

This definition can be generalized to functions f having domains of only a subset of the interval I of length p. Note also that not every function defined on I has a periodic extension, but that if the function does, then the fundamental period of the periodic extension is no more than p.

Example 4.2.11

(a) If $f(x) = \sin x$ with $x \in [0, \pi]$, then $g(x) = |\sin x|$ with $x \in \Re$ is a periodic extension of f. Actually, π is the smallest period for g.

(b) If $f(x) = \sin x$ with $x \in (0, \pi)$, then $g_1(x) = |\sin x|$ with $x \in \Re$ is a periodic extension of f. Moreover,

$$g_2(x) = \begin{cases} |\sin x|, & \text{if } x \neq n\pi,\ n \in Z \\ 1, & \text{otherwise,} \end{cases}$$

is also a periodic extension of f. Thus, a periodic extension need not be unique.

(c) The function $f_1(x) = \sin x$ with $x \in \left[0, \frac{\pi}{2}\right]$ does not have any periodic extension. Why? But, the function $f_2(x) = \sin x$ with $x \in \left[0, \frac{\pi}{2}\right)$ does.

(d) The function $g(x) = \sin x$ with $x \in \Re$ is an extension of f_2, given in part (c), which is periodic, but g is not a periodic extension of f_2. Why? ■

Definition 4.2.12. The function f is *piecewise continuous* on $D \subseteq \Re$ if and only if there exists finitely many points $x_1, x_2, \dots, x_n$ such that

(a) f is continuous on D except at $x_1, x_2, \dots, x_n$, and

(b) f has simple discontinuities at $x_1, x_2, \dots, x_n$.

As commented just following Definition 4.2.5, the simple discontinuities, that is, those of the first kind, are considered removable or jump discontinuities. Note that if D is an open interval (a,b), then for $x_1, x_2, \dots, x_n$ as given in Definition 4.2.12, we have

$$\lim_{x \to x_i^+} f(x) \text{ and } \lim_{x \to x_i^-} f(x)$$

both of which are finite for each $i = 1, 2, \dots, n$. In this definition, f is not required to be defined at any of the points $x_1, x_2, \dots, x_n$.

Exercises 4.2

1. Determine if $x = 0$ is a point of removable discontinuity for the given functions. If it is, what should $f(0)$ be defined to be to make f continuous at $x = 0$?

(a) $f(x) = \operatorname{sgn} x$ (See Exercise 2(h) of Section 4.1.)

(b) $f(x) = \dfrac{\sin x}{x}$ (See Exercise 2(c) of Section 4.1.)

(c) $f(x) = \dfrac{\sin(\sin x)}{x}$ **(d)** $f(x) = \dfrac{\sin\left(\sin x^2\right)}{x}$

(e) $f(x) = \lfloor x \rfloor - \lceil x \rceil$

(f) $f(x) = \dfrac{x + |x|}{2}$ (See Exercise 8(g) of Section 1.2.)

(g) $f(x) = \begin{cases} x, \text{ if } x = n,\ n \in Z \\ 1 - x, \text{ otherwise} \end{cases}$

(h) $f(x) = \begin{cases} x, \text{ if } x = \dfrac{1}{n},\ n \in Z \\ 1 - x, \text{ otherwise} \end{cases}$ (See Exercise 3 of Section 4.1.)

(i) $f(x) = \sin \dfrac{1}{x}$ (See Example 3.2.8.)

(j) $f(x) = x\sin\frac{1}{x}$ (See Exercise 2(d) of Section 4.1 and Example 4.2.2.)

(k) $f(x) = \begin{cases} x, \text{ if } x = \pm\frac{1}{n},\ n \in N \\ x\sin\frac{1}{x}, \text{ otherwise} \end{cases}$

2. For each given function locate and classify all the points of discontinuity. Then graph the functions.

(a) $f(x) = \lfloor x \rfloor,\ x \in [-4,3]$ (See Figure 1.2.8.)

(b) $f(x) = \frac{|x|}{x},\ x = \pm\frac{1}{n}$ and $n \in N$ (See the discussion preceding Example 4.2.2.)

(c) $f(x) = \frac{\lfloor x \rfloor + x}{2},\ x \in \left(-\frac{3}{2}, 1\right]$

(d) $f(x) = (-1)^n,\ x \in [n, n+1)$ and $n \in N$

(e) $f(x) = \begin{cases} \frac{1}{q}, \text{ if } x = \frac{p}{q} \text{ in lowest terms with } p, q \in N \\ 0, \text{ if } x \text{ is irrational} \end{cases}$ (See Example 3.2.10.)

(f) $f:[-1,1] \to \Re$ defined by $f(x) = \begin{cases} 0, \text{ if } x = \pm\frac{1}{n},\ n \in N \\ 1, \text{ otherwise} \end{cases}$

(See Exercise 8 from Section 3.2.)

(g) $f(x) = \begin{cases} \frac{1}{x}, \text{ if } x = \pm\frac{1}{n},\ n \in Z \\ 0, \text{ otherwise} \end{cases}$ (See Example 4.2.8, part (b).)

(h) $f(x) = \begin{cases} \frac{1}{x}, \text{ if } x > 0 \\ 1, \text{ if } x \le 0,\ x \text{ rational} \\ -1, \text{ if } x < 0,\ x \text{ irrational} \end{cases}$ (See Example 4.2.8, part (b).)

(i) $f(x) = \begin{cases} \exp\left(-\frac{1}{x}\right), \text{ if } x > 0 \\ 0, \text{ if } x \le 0 \end{cases}$ (See Exercise 4(c) of Section 3.3.)

(j) $f(x) = \begin{cases} \dfrac{\exp\left(\frac{1}{x}\right)}{1 + \exp\left(\frac{1}{x}\right)}, \text{ if } x \ne 0 \\ \frac{1}{2}, \text{ if } x = 0 \end{cases}$ (See Exercise 9(c) of Section 3.3.)

(k) $f(x) = \exp\left(\frac{1}{x}\right) + \sin\frac{1}{x}$, $x > 0$ (See Exercise 4(g) of Section 3.3.)

(l) $f(x) = \exp\left(\frac{1}{x}\right) + \sin\frac{1}{x}$, $x < 0$ (See Exercise 4(h) of Section 3.3.)

(m) $f(x) = \exp\left(\frac{1}{x}\right) + \sin\frac{1}{x}$, $x \in \Re$

(n) $f(x) = \begin{cases} x, \text{ if } x = \pm\frac{1}{n}, \ n \in N \\ x^2, \text{ otherwise} \end{cases}$ **(o)** $f(x) = \frac{-1}{\sqrt{x-2}}$

(p) $f(x) = \lim_{n\to\infty} \frac{x^n}{2 + x^n}$ (See Exercise 15 of Section 3.3 and Exercise 1(i) of Section 8.1.)

3. Determine whether or not the given statements are true. Explain yourself.

(a) If $f : D \to \Re$ is continuous, then any periodic extension of f must also be continuous.

(b) If $f : \Re \to \Re$ is periodic, then f is bounded.

(c) Every function has infinitely many periodic extensions.

(d) A periodic extension is an extension of a periodic function.

(e) The function $f(x) = |\sin x|$ with $x \neq n\pi$ and $n \in Z$ is piecewise continuous.

(f) Every periodic function has the smallest period.

(g) If f is periodic, then $\lim_{x\to\infty} f(x)$ does not exist.

4. **(a)** If $f : D \to \Re$ is p-periodic, verify that $f(x + kp) = f(x)$ for all $x \in D$ and k any nonzero integer.

(b) Verify that $\sin nx$ and $\cos nx$ with $n \in N$ are 2π-periodic.

(c) If f and g are p-periodic, show that a *linear combination* $af(x) + bg(x)$ with $a, b \in \Re$ is also p-periodic.

(d) Find the fundamental period for $\sin\frac{x}{2}$.

(e) Give an example of a p-periodic function which is not defined on all of $\Re$.

4.3 Properties of Continuous Functions

One of the many important properties that a continuous function may have is boundedness. As observed from Exercise 5(b) of Section 4.1, a continuous function need not be bounded. However, if the domain is a closed and bounded interval, then continuity implies boundedness. We record this as follows.

THEOREM 4.3.1. *If a function f is continuous on an interval $[a,b]$, then f is bounded on $[a,b]$.*

Proof. We will prove this theorem by assuming to the contrary that f is not bounded on $[a,b]$. If f is not bounded on $[a,b]$, by Remark 4.1.13 there exists a bounded sequence $\{x_n\}$ in $[a,b]$ such that $|f(x_n)| > n$ for all n. However, by Bolzano-Weierstrass theorem for sequences, $\{x_n\}$ has a convergent subsequence. So, let $\{x_{n_k}\}$ be this subsequence, which converges to, say, c. Now, since $[a,b]$ is a closed set, and thus contains all of its accumulation points, $c \in [a,b]$ (see Definition 4.1.15). But f is continuous at c. Thus, by Exercise 5(k) of Section 4.1, we have $\lim_{k\to\infty} f(x_{n_k}) = f(c)$. This, however, contradicts the fact that $|f(x_{n_k})| > n_k$. Why? Hence, the proof is complete. ■

It is important to realize that the assumptions in Theorem 4.3.1 are sufficient but not necessary. See Exercise 5(a) from Section 4.1. In addition, the result is not necessarily true if any one of the conditions in Theorem 4.3.1 is not satisfied. For further details see Exercise 1. The next theorem is among the most powerful results in mathematics.

THEOREM 4.3.2. *(Extreme Value Theorem) If f is a continuous function on an interval $[a,b]$, then f attains its maximum and minimum values on $[a,b]$.*

Proof. A popular method that mathematicians use to prove this result is by contradiction. See Exercise 2 for an alternative proof. Let

$$m = \inf\{f(x) \mid x \in [a,b]\} = \text{greatest lower bound of } f \text{ on } [a,b], \text{ and}$$
$$M = \sup\{f(x) \mid x \in [a,b]\} = \text{least upper bound of } f \text{ on } [a,b].$$

We assume that there is no value $c \in [a,b]$ for which $f(c) = M$. Therefore, $f(x) < M$ for all $x \in [a,b]$. Now define a new function g by

$$g(x) = \frac{1}{M - f(x)}.$$

Why this choice? Because it works. Observe that $g(x) > 0$ for every $x \in [a,b]$ and that by Theorem 4.1.8, g is continuous on $[a,b]$. By Theorem 4.3.1, g is bounded on $[a,b]$. Thus, there exists $K > 0$ such that $g(x) \le K$, for every $x \in [a,b]$. Since for each $x \in [a,b]$, $g(x) = \dfrac{1}{M - f(x)} \le K$ is equivalent to $f(x) \le M - \dfrac{1}{K}$, we have contradicted the fact that M was assumed to be the least upper bound of f on $[a,b]$. Hence, there must be a value $c \in [a,b]$ such that $f(c) = M$.

In proving the second part of this theorem, the case for attaining the minimum value, we either rewrite the preceding proof, or we apply the completed argument to the function $h(x) = -f(x)$. Details are left to the reader. ■

THEOREM 4.3.3. *(Bolzano's Intermediate Value Theorem) If a function f is continuous on $[a,b]$ and if k is a real number between $f(a)$ and $f(b)$, then there exists a real number $c \in (a,b)$ such that $f(c) = k$.*

Just because f is continuous on $[a,b]$ and thus attains maximum and minimum values, M and m, respectively, in $[a,b]$, it is not trivial that every value between m and M is also attained. In other words, it is not obvious that $f:[a,b] \to [m,M]$ is an onto function, that is, a surjection. The geometric meaning of Bolzano's intermediate value theorem is quite simple. If we pick any value k between $f(a)$ and $f(b)$ and draw a horizontal line through the point $(0,k)$, then this line will intersect the function f at a point whose x-coordinate is between a and b. There may very well be more than one such value. This idea is stated formally in Definition 4.3.4 and demonstrated in Figure 4.3.1 where $f(a)$ was chosen to be smaller than $f(b)$.

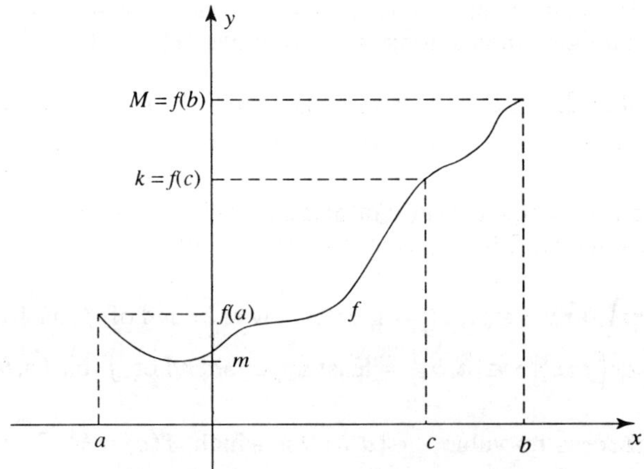

Figure 4.3.1

Definition 4.3.4. A function $f: D \to \mathfrak{R}$ with $D \subseteq \mathfrak{R}$ satisfies the *intermediate value property* on D if and only if for every $x_1, x_2 \in D$ with $x_1 < x_2$ and any real constant k between $f(x_1)$ and $f(x_2)$, there exists at least one constant $c \in (x_1, x_2)$ such that $f(c) = k$.

Clearly, by Bolzano's intermediate value theorem, any continuous function possesses the intermediate value property. However, there are certainly functions that are not continuous and yet satisfy the intermediate value property. See Exercise 3.

Proof of Theorem 4.3.3. There are a number of different ways one can prove this theorem and there are two cases to consider, $f(a) < f(b)$ and $f(b) < f(a)$. The first case is considered here, and the second is left to the reader. We begin with $f(a) < k < f(b)$ and define a set E by

$$E = \{x \in [a,b] \,|\, f(x) < k\}.$$

Note that E might very well be made up of several pieces (i.e., E need not be "connected"). Also, $E \neq \phi$ since $a \in E$. E is bounded and b is its upper bound. Hence, E has a supremum, say, $\sup E = c$. We will show that $f(c) = k$. According to Exercise 5 of Section 2.5, we can choose sequences $\{x_n\}$ in E and $\{t_n\}$ in $[a,b] \setminus E$, both converging to c. That is, $\lim_{n\to\infty} x_n = \lim_{n\to\infty} t_n = c$. Now, since f is continuous on $[a,b]$, f is continuous at $x = c$. Thus, using Exercise 5(k) of Section 4.1, we can write

$$\lim_{n\to\infty} f(x_n) = \lim_{n\to\infty} f(t_n) = f\left(\lim_{n\to\infty} t_n\right) = f(c).$$

But, since $x_n \in E$ and $t_n \notin E$, we also have $f(x_n) < k \le f(t_n)$. Therefore, by the sandwich theorem, $f(c) = k$. ■

An immediate application of this theorem is the following.

COROLLARY 4.3.5. *Any polynomial of odd degree has at least one real root.*

Proof. Let

$$p(x) = a_n x^n + a_{n-1}x^{n-1} + \cdots + a_1 x + a_0$$

be any polynomial, where n is an odd positive integer. Without loss of generality, assume that $a_n > 0$. Then, by Example 4.1.4, p is continuous for all $x \in \Re$, $\lim_{x\to\infty} p(x) = +\infty$, and $\lim_{x\to-\infty} p(x) = -\infty$. See Theorem 3.1.12, part (c). Hence, there exist values a and b such that $p(a) < 0 < p(b)$. Therefore, since $0 \in (p(a), p(b))$, by Bolzano's intermediate value theorem there exists $c \in (a,b)$ such that $p(c) = 0$. ■

In general, if a function f is continuous on $[a,b]$ with $f(a)f(b) < 0$, then the existence of a $c \in (a,b)$ such that $f(c) = 0$ is guaranteed by *Weierstrass intermediate value theorem,* which is a special case of Bolzano's intermediate value theorem when k is chosen to be 0.

COROLLARY 4.3.6. *If a function $f:[a,b] \to \Re$ is nonconstant and continuous, then the range of f is an interval $[c,d]$ with $c,d \in \Re$. Thus, a continuous function maps closed and bounded intervals onto closed and bounded intervals.*

Proof. Suppose $f:[a,b] \to \Re$ is nonconstant and continuous. By the extreme value theorem, f must attain its minimum and maximum values at some points x_1 and x_2, respectively. Let $c = f(x_1)$ and $d = f(x_2)$. Then, $f(x_1) \le f(x) \le f(x_2)$ for all $x \in [a,b]$. Now, pick any $k \in [c,d]$. By Bolzano's intermediate value theorem there exists $x_0 \in (a,b)$ such that $f(x_0) = k$. And since this is true for any arbitrary k, the function f maps $[a,b]$ onto $[c,d]$. Where did we use the fact that f is not constant? ■

Another useful application of Bolzano's intermediate value theorem is solving inequalities. Recall that in Exercise 20 from Section 1.6, we considered several cases in finding values that satisfy a given inequality. Now, using continuity, we can speed up the process. For example, in solving $x^2 - x \ge 6$, we write that $x^2 - x - 6 \ge 0$ and let $f(x) = x^2 - x - 6 = (x-3)(x+2)$. The function f has roots at $x = -2$ and 3. Therefore, due to the continuity of f and since $f(-3) > 0$, $f(x) > 0$ for each $x < -2$. Similarly, since $f(0) < 0$, $f(x) < 0$ for each value of $x \in (-2,3)$. Likewise, $f(4) > 0$ implies that $f(x) > 0$ for each value $x > 3$. Thus, the solution of the inequality $f(x) \ge 0$ consists of $x \in (-\infty,-2) \cup [3,\infty)$. Generalizing, for a continuous function f with the domain D_f, if we can find the roots of f we are equipped to solve $f(x) \ge 0$, $f(x) \le 0$, $f(x) > 0$, or $f(x) < 0$. Indicate D_f and the roots of f on the x-axis. Pick one convenient value in each region and determine whether or not it satisfies the inequality. If it does, then every point in that region satisfies the inequality. If the chosen point does not satisfy the inequality, then none of the points in that region will satisfy it. Endpoints need to be tested separately.

Fixed points are values where the function f crosses the diagonal line $y = x$. Finding these values requires solving the equation $f(x) = x$. Sometimes this equation has no solutions, a unique solution, or perhaps many solutions. Here, we will address the existence of the solution.

THEOREM 4.3.7. *(Brouwer's[1] Fixed Point Theorem) If a function $f:[a,b] \to [a,b]$, is continuous, then f has at least one fixed point, that is, there exists at least one real number $p \in [a,b]$ such that $f(p) = p$.*

Proof. Observe that if $f(a) = a$ or $f(b) = b$, then the existence of a fixed point is obvious. Thus, suppose $f(a) \ne a$ and $f(b) \ne b$. Then it must be true that $f(a) > a$ and $f(b) < b$ (see Figure 4.3.2). Now, define a new function g by $g(x) = f(x) - x$, which must be continuous on $[a,b]$ and satisfy $g(a) > 0$ and $g(b) < 0$. Why? Hence, by Bolzano's intermediate value theorem we know that there exists a real number $p \in (a,b)$ such that $g(p) = 0$. Thus, $g(p) = f(p) - p = 0$ implies that p is a fixed point of f. ■

We close this section with a result involving inverse functions.

---

[1]Luitzen Egbertus Jan Brouwer (1881–1966), a Dutch mathematician, is best known for his invariance and fixed point theorems. In addition, Brouwer is regarded as one of the founders of modern topology.

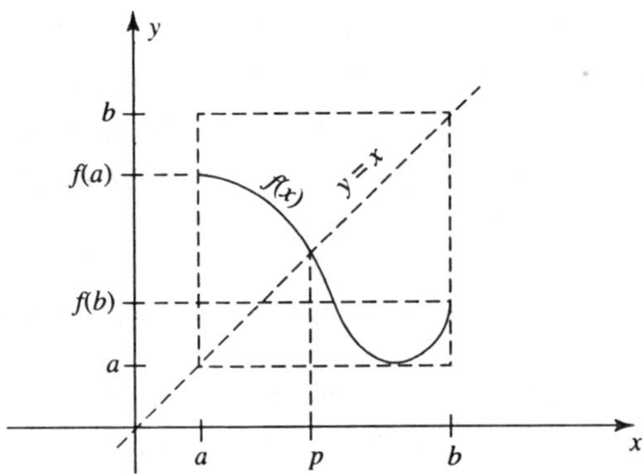

Figure 4.3.2

THEOREM 4.3.8. *If a function $f: D \to \Re$ is a continuous injection and D is closed and bounded, then $f^{-1}: R_f \to D$ is continuous.*

Proof. Choose any $y_0 \in R_f$. And suppose that $\{y_n\}$ is any sequence in R_f converging to y_0, where $y_n \neq y_0$, for any $n \in N$. If we prove that $\lim_{n\to\infty} f^{-1}(y_n) = f^{-1}(y_0)$, then according to Theorem 3.2.6, f^{-1} is continuous. So let $x_n = f^{-1}(y_n)$ for all $n = 0, 1, 2, \ldots$. Then $y_n = f(x_n)$. And since D is bounded, by Bolzano-Weierstrass theorem for sequences, $\{x_n\}$ has at least one converging subsequence. Let $\{x_{n_k}\}$ be any such subsequence that converges to, say, x^*. Since $x^* \in D$, (why?), f is continuous at x^*. Therefore, $\lim_{k\to\infty} f(x_{n_k}) = f(x^*)$. But $\{y_{n_k}\}$, with $y_{n_k} = f(x_{n_k})$, is a subsequence of $\{y_n\}$. Hence, $\{y_{n_k}\}$ converges to $y_0 = f(x_0)$. In addition, since f is an injection, $f(x_0) = f(x^*)$ implies that $x_0 = x^*$. Therefore, by Exercise 6 of Section 2.7 and Theorem 2.6.5, the sequence $\{x_n\}$ converges to x_0. Thus, $\lim_{n\to\infty} x_n = \lim_{n\to\infty} f^{-1}(y_n) = x_0 = f^{-1}(y_0)$, which is what we needed to prove. ■

The preceding proof is very tedious. Knowing a little amount of topology would simplify the proof tremendously. See Section 11.1.

Exercises 4.3

1. **(a)** Give 3 functions for which exactly 2 out of the 3 assumptions of Theorem 4.3.1 are satisfied but the conclusion does not follow.

(b) Give 3 functions for which exactly 2 out of the 3 assumptions of Theorem 4.3.1 are satisfied and the conclusion follows.

(c) Give an example of a bounded function $f: D \to \Re$ with $D \subseteq \Re$ closed and bounded, which does not attain its maximum value in D.

2. (a) Prove, without using the extreme value theorem, that if $f: D \subseteq \Re \to \Re$ is continuous and an interval $[a,b] \subseteq D$, then $f([a,b])$ is closed and bounded. Compare this result with Corollary 4.3.6.
   (b) Prove the extreme value theorem using a direct proof.

3. Give an example of a noncontinuous function f that is defined on $[a,b]$ and
   (a) satisfies the intermediate value property.
   (b) does not satisfy the intermediate value property.

4. Show by an example that Corollary 4.3.5 is not true for polynomials of even degree.

5. If

$$p(x) = a_n x^n + a_{n-1} x^{n-1} + \cdots + a_1 x + a_0$$

with $a_0 < 0$, $a_n > 0$, and n an even positive integer, then prove that p has at least two distinct roots. Is this true if n is odd? Explain.

6. Prove that every positive real number has a unique positive nth root, where $n = 2$, 3, … .

7. Suppose $f:[a,b] \to Q$ is continuous on $[a,b]$. Prove that f is constant on $[a,b]$. Compare this exercise with Exercise 7(a) of Section 4.1.

8. If f is continuous and one-to-one on an interval D, prove that f is strictly monotone. See Exercise 48 in Section 1.9.

9. Give an example of a function $f:[a,b] \to \Re$ that is not continuous but whose range is
   (a) an open and bounded interval.
   (b) an open and unbounded interval.
   (c) a closed and unbounded interval.

10. Which is a stronger assumption on the function f, continuity or the intermediate value property? Explain.

11. Prove that a continuous function that does not vanish on the interval $[a,b]$ is *bounded away from zero* on D. That is, (a) and (b), or (a) and (c) must hold, with (a), (b), and (c) as given as follows.
    (a) If $f:[a,b] \to \Re$ is continuous and $f(x) \neq 0$ for any $x \in [a,b]$, then either $f(x) > 0$ for all $x \in [a,b]$ or $f(x) < 0$ for all $x \in [a,b]$.
    (b) If $f(x) > 0$ for all $x \in [a,b]$, then there exists $\delta_1 > 0$ such that $f(x) \geq \delta_1$ for all $x \in [a,b]$.
    (c) If $f(x) < 0$ for all $x \in [a,b]$, then there exists $\delta_2 > 0$ such that $f(x) \leq -\delta_2$ for all $x \in [a,b]$.

12. Prove that if $f(x) = e^x$, then
    (a) f is strictly increasing for all $x \in \Re$.

(b) $\lim_{x \to \infty} e^x = \infty$. (See Exercise 20(b) in Section 5.2.)

(c) $\lim_{x \to -\infty} e^x = 0$.

13. Suppose the function f is continuous on $[a,b]$, x_1 and x_2 are in $[a,b]$, and k_1 and k_2 are positive real constants. Prove that there exists c between x_1 and x_2 for which

$$f(c) = \frac{k_1 f(x_1) + k_2 f(x_2)}{k_1 + k_2}.$$

14. Use Bolzano's intermediate value theorem to solve given inequalities.

(a) $x^2 + x - 2 \le 0$ **(b)** $\left(x - \sqrt{3}\right)\left(x - \sqrt{3.1}\right) > 0$

(c) $x^3 - x^2 \le 0$ **(d)** $x^3 - x^2 < 0$

(e) $\frac{x-1}{x^2} \le 0$ **(f)** $x^3 - 2x + 1 \ge 0$

(g) $x^3 + 2x^2 - 5x - 6 < 0$ (See Exercise 20(b) from Section 1.6.)

(h) $\frac{2x+1}{x-5} \le 3$ (See Exercise 20(d) from Section 1.6.)

(i) $-2 \le \frac{1}{x} < 1$ (See Exercise 20(c) from Section 1.6.)

15. Show that the equation $2^x = 3x$ has a solution $x = c$ for some $c \in (0,1)$ using

(a) Bolzano's intermediate value theorem.

(b) Brouwer's fixed point theorem.

16. What is wrong with the following argument? To find a fixed point for a continuous function g, simply set up a sequence $x_n = g(x_{n-1})$ and pick an initial value x_0. This sequence converges to a fixed point $x = p$ since

$$\lim_{n \to \infty} x_n = \lim_{n \to \infty} g(x_{n-1}) = g\left(\lim_{n \to \infty} x_{n-1}\right).$$

Therefore, $p = g(p)$.

17. Give an example of a function f that is an injection, but whose inverse function f^{-1} is not continuous.

18. If I is an interval and a function $f : I \to \Re$ is strictly increasing (decreasing), with $f(I)$ an interval, prove that

(a) f is continuous on I.

(b) f^{-1} is strictly increasing (decreasing).

(c) f^{-1} is continuous on $f(I)$.

19. Prove that $\arctan x$ and $\ln x$ are continuous on $\Re$. See Remark 4.1.5.

4.4 Uniform Continuity

Recall that Definition 4.1.2 said that a function $f: D \to \Re$ is continuous on a set $E \subseteq D \subseteq \Re$ if and only if for any given $\varepsilon > 0$ and for every point $a \in E$, there exists $\delta > 0$ such that for all points $x \in E$ satisfying $|x-a| < \delta$, we have $|f(x) - f(a)| < \varepsilon$. This δ mentioned in the previous sentence could vary depending on the choice of a and ε.

Example 4.4.1. Consider the everywhere continuous function $f(x) = x^2$ with domain $D \subseteq \Re$. For a formal proof of this statement we begin with an arbitrary $\varepsilon > 0$ and any real number $x = a$. We need to find $\delta > 0$ so that whenever $|x-a| < \delta$, we will have $|x^2 - a^2| < \varepsilon$. Although different approaches to the method of proof exist, we choose to argue by writing

$$\begin{aligned} |x^2 - a^2| &= |x-a||x+a| \le |x-a|(|x-a| + 2|a|) \\ &< |x-a|(1+2|a|), \text{ if } |x-a| < 1, \\ &< \left(\frac{\varepsilon}{1+2|a|}\right)(1+2|a|), \text{ if also } |x-a| < \frac{\varepsilon}{1+2|a|}, \\ &= \varepsilon. \end{aligned}$$

Hence, pick $\delta = \min\left\{1, \dfrac{\varepsilon}{1+2|a|}\right\}$. Thus, whenever $|x-a| < \delta$, we have $|f(x) - f(a)| < \varepsilon$. Observe that for this particular function, δ is a real number which depends not only on ε but on $x = a$ as well. The larger the number a, the smaller must be the chosen δ. For example, suppose that $\varepsilon = \dfrac{1}{4}$. If $a = 3$, then the largest δ that can be chosen is $\delta = \dfrac{1}{28} \approx 0.0357$. But if $a = 300$, then the largest δ that can be chosen is $\delta = \dfrac{1}{2404} \approx 0.000416$. Why? Finally, notice that as one goes farther out on the x-axis, the x-coordinates that are close together produce functional values far apart. Consider two points close together, say, 1000 and 1000.01. They are within 0.01 units of each other. However, $|f(1000) - f(1000.01)| \approx 20$. ■

Example 4.4.2. For contrast, consider the same function $f(x) = x^2$, but on a different domain. That is, consider the continuous function $f(x) = x^2$ on domain $D = (-2, 1]$. To prove formally, using Definition 4.1.2, that f is continuous at every point $x = a$ in its domain, pick an arbitrary $\varepsilon > 0$ and find $\delta > 0$ such that $|x^2 - a^2| < \varepsilon$ whenever $|x-a| < \delta$ with $x \in (-2, 1]$. To this end, we write

$$\left|x^2 - a^2\right| = |x-a||x+a| \le |x-a|(|x|+|a|) < |x-a|(4) < \frac{\varepsilon}{4}(4) = \varepsilon,$$

whenever $x \in (-2,1]$ and $|x-a| < \frac{\varepsilon}{4}$. Thus, pick $\delta = \frac{\varepsilon}{4}$, guaranteeing continuity of f on this set. Note that here, δ depends only on ε and not on the location of point a. ■

Consider the following definition to help distinguish between the previous two examples involving a continuous function on two different domains.

Definition 4.4.3. A function $f: D \to \Re$ is *uniformly continuous on a set* $E \subseteq D \subseteq \Re$ if and only if for any given $\varepsilon > 0$ there exists $\delta > 0$ such that $|f(x) - f(t)| < \varepsilon$ for all $x, t \in E$ satisfying $|x - t| < \delta$. If f is uniformly continuous on its domain D, we simply say that f is *uniformly continuous*.

The continuous function in Example 4.4.1 is not a uniformly continuous function. However, the continuous function in Example 4.4.2 is uniformly continuous. Also, observe that it makes no sense to talk about uniform continuity at a point, unless it is a singleton. In looking at a uniformly continuous function, whenever two points are within δ the images are within ε of each other. This δ does not depend on the choice of x or t, whereas the δ in Definition 4.1.2 may. That is, if a function is continuous but not uniformly continuous, then the choice of δ will depend not only on ε but on x or t as well. In other words, a function that is continuous but not uniformly continuous on a domain D means that for any chosen point $a \in D$ and any arbitrary $\varepsilon > 0$ we can only find a $\delta > 0$ that depends on both a and ε. This seemingly minor difference in definitions creates a substantially different concept. To prove that $f(x) = x^2$ from Example 4.4.1 is not uniformly continuous, a negation of Definition 4.4.3 needs to be written and then successfully used on the function f. Later, knowing more theory will allow us to complete this task more readily.

Remark 4.4.4. The negation of Definition 4.4.3 is carried out in a way similar to the way the negation of continuity in Remark 4.1.11, part (b), was done. Here, however, two sequences are involved instead of one, since a fixed value for $x = a$ does not exist. Thus, to prove that a given function $f: D \to \Re$ is not uniformly continuous on a set $E \subseteq D \subseteq \Re$, we need to find some particular $\varepsilon > 0$ and two sequences $\{x_n\}$ and $\{t_n\}$ in E that will eventually satisfy $|x_n - t_n| \le \frac{1}{n}$ but $|f(x_n) - f(t_n)| \ge \varepsilon$. Note that the expression $\frac{1}{n}$ here could be changed to any other expression tending to 0 as n becomes arbitrarily large. ■

Example 4.4.5. Prove that $f(x) = x^2$ is not uniformly continuous.

Proof. Here the domain of f is assumed to be all of $\Re$. Restricting the domain may very well lead to a uniformly continuous function. See Example 4.4.2. Remark 4.4.4 will be used in proving that f is not uniformly continuous. Later, other methods will be considered.

We begin with a particular ε, say, $\varepsilon = 1$, and two sequences $\{x_n\}$ and $\{t_n\}$ with $x_n = n$ and $t_n = n + \frac{1}{n}$. Notice that the two sequences, although close together, involve large values. These particular sequences were chosen with the suspicion that the differences in their functional values would be far apart. This would not be the case in a domain near zero. So we have

$$|x_n - t_n| = \left|n - \left(n + \frac{1}{n}\right)\right| \le \frac{1}{n}, \text{ but } |f(x_n) - f(t_n)| = \left|2 + \frac{1}{n^2}\right| \ge 1.$$

Hence, by Remark 4.4.4, $f(x) = x^2$ is not uniformly continuous. ■

Unbounded intervals should not be associated with not uniformly continuous functions, nor bounded intervals with uniformly continuous functions. The function $f(x) = \frac{1}{x}$ with $x \in (0,1]$ is not uniformly continuous even though its domain is bounded. And the function $f(x) = \sin x$ is uniformly continuous on any domain $D \subseteq \Re$. Intuitively, uniformly continuous functions do not rise or fall "too quickly" over their domains. If they do rise or fall "very quickly," it is for only a very short period of time. Linear functions serve as an example of the first situation. Quadratic functions on unbounded intervals are not uniformly continuous because eventually they rise or fall too quickly. The uniformly continuous function $f(x) = \sqrt[3]{x}$ from Exercise 1(d) increases extremely quickly as it passes through the origin, but then it quickly levels off. Also, see Exercise 3(e).

The theory behind uniformly continuous functions is abundant, and the remainder of this section is devoted to such. Uniform continuity is more restrictive than continuity. That is, not every continuous function is uniformly continuous, as verified in Example 4.4.5. Every uniformly continuous function, however, must be continuous as well. Why?

THEOREM 4.4.6. *If $D \subset \Re$ is a closed and bounded set, and a function $f : D \to \Re$ is continuous, then f is uniformly continuous.*

Proof. The fact that D is a closed and bounded domain is essential to this theorem. If this domain is altered in any way, then the result need not be true. Why? We will prove this theorem by assuming that f is not uniformly continuous and then reach a contradiction. To this end, pick $\varepsilon > 0$, and two sequences $\{x_n\}$ and $\{t_n\}$ in D such that $|x_n - t_n| \le \frac{1}{n}$, but $|f(x_n) - f(t_n)| \ge \varepsilon$. By Bolzano–Weierstrass theorem for sequences, there exists subsequence $\{x_{n_k}\}$ that converges to, say, $\alpha \in D$. Since f is continuous we have that

$$\lim_{k\to\infty} f\left(x_{n_k}\right) = f(\alpha).$$

Furthermore,

$$\left|t_{n_k} - \alpha\right| \le \left|t_{n_k} - x_{n_k}\right| + \left|x_{n_k} - \alpha\right| \le \frac{1}{n_k} + \left|x_{n_k} - \alpha\right|.$$

Thus, we can conclude that the subsequence $\{t_{n_k}\}$ converges also to α. Again, due to continuity of f we have that

$$\lim_{k\to\infty} f\left(t_{n_k}\right) = f(\alpha).$$

Hence,

$$\lim_{k\to\infty}\left|f\left(x_{n_k}\right) - f\left(t_{n_k}\right)\right| = 0,$$

contradicting the fact that $\left|f(x_n) - f(t_n)\right| \ge \varepsilon$. The proof is complete. ■

THEOREM 4.4.7. *If a function* $f:(a,b)\to\Re$ *is uniformly continuous, then* $f(a^+)$ *and* $f(b^-)$ *are both finite.*

Observe that if a function $f:(a,b)\to\Re$ is uniformly continuous, then f is continuous on (a,b). And if a function f is continuous and $f(a^+)$ and $f(b^-)$ are both finite, we can conclude that an extension g of f defined by

$$g(x) = \begin{cases} f(a^+), \text{ if } x = a \\ f(x), \text{ if } a < x < b \\ f(b^-), \text{ if } x = b, \end{cases}$$

is continuous on $[a,b]$. The proof of Theorem 4.4.7 is left as an exercise. Note that this theorem does not hold if f is continuous but not uniformly continuous on (a,b). Functions $h(x) = \frac{1}{x}$ on $(0,1)$ and $k(x) = \sin\frac{1}{x}$ on $(0,1)$ are examples. Here, $\lim_{x\to 0^+} h(x) = +\infty$ and $\lim_{x\to 0^+} k(x)$ is undefined. See Exercises 1(a) and (e). In addition, the converse of Theorem 4.4.7 is false. Why?

THEOREM 4.4.8. *If a function* $f:[a,\infty)\to\Re$ *is continuous with* $\lim_{x\to\infty} f(x)$ *finite, then* f *is uniformly continuous on* $[a,\infty)$.

Proof. Let $\varepsilon > 0$ be given. We need to find $\delta > 0$ such that $|f(x) - f(t)| < \varepsilon$, provided $|x - t| < \delta$ and $x, t \in [a, \infty)$. Since $\lim_{x \to \infty} f(x)$ is finite and equals, say, L, there exists $M \ge a$ such that

$$|f(x) - L| < \boxed{\frac{\varepsilon}{2}} \quad \text{if } x \ge M.$$

Since f is continuous on $[a, M+1]$, a compact interval, by Theorem 4.4.6, f is uniformly continuous there. Hence, there exists $\delta_1 > 0$ such that

$$|f(x) - f(t)| < \varepsilon \text{ whenever } |x - t| < \delta_1 \text{ and } x, t \in [a, M+1].$$

Now, let $\delta = \min\{\delta_1, 1\}$. The δ chosen in this way implies that whenever $|x - t| < \delta$, then either x and $t \in [a, M+1]$ or both x and t are greater than M, since the difference between them is less than one. If x and $t \in [a, M+1]$, then clearly $|f(x) - f(t)| < \varepsilon$. And if x and $t > M$, we can write

$$|f(x) - f(t)| \le |f(x) - L| + |f(t) - L| < \frac{\varepsilon}{2} + \frac{\varepsilon}{2} = \varepsilon.$$

In either case, the desired result is obtained. ■

Definition 4.4.9. A function $f: D \to \Re$ with $D \subseteq \Re$ is *Lipschitzian* if and only if there exists $L > 0$, called a *Lipschitz*[2] *constant*, such that

$$|f(x) - f(t)| \le L|x - t|$$

for all x and $t \in D$.

Often, a Lipschitzian function is referred to as a *Lipschitz function* or a function that satisfies the *Lipschitz condition*. Furthermore, if a function f is Lipschitzian with a Lipschitz constant $L \in (0,1)$, then f is a *contraction*, sometimes called a *contractive function*. See Definition 2.5.10. Thus intuitively, with contractive images for two x-coordinates are closer together than those x-coordinates. Next, observe that the inequality in Definition 4.4.9 can be written as

$$\left| \frac{f(x) - f(t)}{x - t} \right| \le L, \text{ provided } x \ne t.$$

Therefore, if the slope of the line segments joining $(x, f(x))$ and $(t, f(t))$ is bounded, then the function f is Lipschitzian. Compare this with Example 4.4.11.

---

[2]Rudolf Otto Lipschitz (1832–1903) was a German mathematician who contributed to mathematics mainly in differential equations, differential geometry, algebra, and number theory.

THEOREM 4.4.10. If a function is Lipschitzian, then it is uniformly continuous.

The proof of this result is straightforward and is left as an exercise. Notice that the converse of Theorem 4.4.10 is not true. See Exercise 8.

In Exercise 13 of Section 5.2, we will verify that if a function f has a bounded "derivative" then it is uniformly continuous, but not conversely. Even though the definition of the derivative of a function is not introduced until Chapter 5, we can prove the preceding statement by writing it as follows.

Example 4.4.11. Let I be an interval and consider a function $f: I \to \Re$. If a set S defined by

$$S = \left\{ \frac{f(x) - f(t)}{x - t} \mid x, t \in I \text{ and } x \neq t \right\}$$

is bounded, prove that f is uniformly continuous on I.

Proof. Suppose that $K > 0$ is an upper bound of S. Then

$$\left| \frac{f(x) - f(t)}{x - t} \right| \leq K \text{ for all } x, t \in I \text{ and } x \neq t.$$

Hence, $|f(x) - f(t)| \leq K|x - t|$ for all $x, t \in I$. Thus, the function f is Lipschitzian and hence, uniformly continuous. Verification that the converse of this theorem is false is left to the reader. ■

Exercises 4.4

1. Use Definition 4.4.3 or Remark 4.4.4 to determine whether or not the given functions are uniformly continuous.

(a) $f(x) = \dfrac{1}{x}$ with $x \in (0,1]$ **(b)** $f(x) = x^3$ with $x \in [0,2)$

(c) $f(x) = \dfrac{x}{x+4}$ with $x \in [0,2)$ **(d)** $f(x) = \sqrt[3]{x}$

(e) $f(x) = \sin\dfrac{\pi}{x}$ with $x \in (0,1)$ **(f)** $f(x) = \dfrac{1}{x^2}$ with $x \in [1,\infty)$

2. Prove or find a counterexample to the following statements in which we assumed that f is a function defined on the indicated interval.

(a) f continuous on (a,b) implies that for any sequence $\{x_n\}$ in (a,b) converging to a, the sequence $\{f(x_n)\}$ converges. See Exercise 5(k) of Section 4.1.

(b) f uniformly continuous on (a,b) implies that for any sequence $\{x_n\}$ in (a,b) that converges, the sequence $\{f(x_n)\}$ converges.

(c) f uniformly continuous on (a,b) implies that for any sequence $\{x_n\}$ in (a,b) that converges, the sequence $\{f(x_n)\}$ is Cauchy.
(d) f continuous on (a,b) and without a finite limit at $x=a$, implies that f is not uniformly continuous on (a,b).
(e) f uniformly continuous on D implies that f is bounded on D.
(f) f uniformly continuous on (a,b) implies that f is bounded on (a,b).
(g) f bounded on D implies that f is uniformly continuous on D.
(h) f bounded and continuous on D implies that f is uniformly continuous on D.
(i) f uniformly continuous on (a,b) and on (b,c) implies that f is uniformly continuous on (a,c).
(j) f uniformly continuous on $[a,b]$ and on $[b,c]$ implies that f is uniformly continuous on $[a,c]$.

3. (a) Prove Theorem 4.4.7.
(b) If a $f:(a,b)\to\Re$ is continuous, then prove that f has a continuous extension to $[a,b]$ if and only if both $f(a^+)$ and $f(b^-)$ are finite.
(c) If a function f is continuous on (a,b) and $f(a^+)$ and $f(b^-)$ are both finite, then prove that f is bounded on (a,b). Explain why the converse is not true.
(d) Prove that a function $f:(a,b)\to\Re$ is uniformly continuous on (a,b) if and only if it can be continuously extended to $[a,b]$.
(e) Use the result of part (d) to prove that $f(x)=\sin\frac{1}{x}$ is not uniformly continuous on $(0,1)$ but $g(x)=x\sin\frac{1}{x}$ is uniformly continuous on $(0,1)$.

4. If a function $f:\Re\to\Re$ is continuous and periodic, prove that f is uniformly continuous on $\Re$. Conclude that $f(x)=\sin x$ is uniformly continuous on $\Re$. See Exercise 3 from Section 4.2.

5. If functions f and g are uniformly continuous on their common domain D, prove that
(a) $f\pm g$ is uniformly continuous on D.
(b) cf is uniformly continuous on D for any real constant c.
(c) f is bounded on D if D is bounded. (See Exercise 3(c) above.)
(d) fg is uniformly continuous on D if f and g are both bounded on D.
(e) fg is uniformly continuous on D if D is compact.
(f) $\frac{f}{g}$ is uniformly continuous on D if $g(x)\neq 0$ for any $x\in D$ and D is closed and bounded.

6. Give an example of a function f_1, f_2, f_3, and f_4 that satisfies each of the given conditions.

(a) f_1 continuous but not uniformly continuous on $[0,\infty)$ with $\lim_{x\to\infty} f_1(x) = -\infty$.

(b) f_2 continuous but not uniformly continuous on $[0,\infty)$ where $\lim_{x\to\infty} f_2(x)$ does not exist.

(c) f_3 uniformly continuous on $[0,\infty)$ with $\lim_{x\to\infty} f_3(x) = -\infty$.

(d) f_4 uniformly continuous on $[0,\infty)$ where $\lim_{x\to\infty} f_4(x)$ does not exist.

7. Determine whether or not the given functions are Lipschitzian. Explain.

(a) $f(x) = x^2$ with $x \in (-2,1]$ (b) $f(x) = x^2$

(c) $f(x) = \sqrt[3]{x}$ (d) $f(x) = \sin\frac{1}{x}$ with $x \in (0,1]$

(e) $f(x) = x\sin\frac{1}{x}$ with $x \in (0,1]$ (See Problem 5.6.19, part (i).)

(f) $f(x) = \sqrt{x}\sin\frac{1}{x}$ with $x \in (0,1]$ (See Exercise 10(e) from Section 3.3.)

8. Give an example of a function that is uniformly continuous but not Lipschitzian.

9. Prove that if $f:[a,b] \to [a,b]$ is a contraction, then f has a unique fixed point. See Brouwer's fixed point theorem, Theorem 4.3.7.

10. Show by counterexamples that if a function $f: D \to \Re$ with $D \subseteq \Re$ is Lipschitzian with Lipschitz constant $L \geq 1$, then

(a) f might have no fixed points.

(b) f might have infinitely many fixed points.

(c) f might have a unique fixed point.

11. Suppose a function $f: D \to D$ with $D \subseteq \Re$ satisfies

$$|f(x) - f(t)| < |x - t|$$

for all $x, t \in D$ with $x \neq t$.

(a) Prove that f has at least one fixed point if D is an interval $[a,b]$.

(b) Prove that f has at most one fixed point.

(c) Give an example of f that has no fixed points.

(d) Give an example of f that is not contractive even if D is closed and bounded.

12. Suppose the function $f(x) = x^2$ is defined for $x \in \left(0, \frac{1}{3}\right]$.

(a) Show that $f: \left(0, \frac{1}{3}\right] \to \left(0, \frac{1}{3}\right]$.

(b) Show that f is contractive.

(c) Show that f has no fixed points.

Does this exercise contradict Theorem 4.3.7? Also, see Exercise 13 in Section 2.5.

13. Prove that the function $f:[1,\infty)\to\Re$ defined by $f(x)=\frac{x}{2}+\frac{1}{x}$ is a contraction. See Exercise 12 in Section 2.4.

4.5* Review

Label each statement as true or false. If a statement is true, prove it. If not,
 (i) give an example of why it is false, and
 (ii) if possible, correct it to make it true, and then prove it.

1. If $0<b<1$ is a real constant, then $\lim_{x\to\infty} b^x=0$.

2. If $f:[a,b]\to\Re$ is one to one and satisfies the intermediate value property, then f is continuous on $[a,b]$.

3. The function $r(x)=\lfloor x\rfloor$ is a rational function. See Exercise 8 of Section 3.4 and Exercise 50 of Section 5.5.

4. A function f is continuous at $x=a$ if and only if $\lim_{x\to a} f(x)=f(a)$.

5. The function $f(x)=\left(1-2^{\frac{1}{x}}\right)^{-1}$ has a jump discontinuity at $x=0$.

6. The function $f(x)=\frac{1}{2}\left(|\sin x|-\sin x\right)$ is continuous and periodic on $\Re$.

7. Functions $f(x)=\sin x$ and $g(x)=\cos x$ are uniformly continuous on $\Re$.

8. If a function f is uniformly continuous on every bounded interval, then f is uniformly continuous on $\Re$.

9. If $\lim_{h\to 0} f(x+h)=f(x)$ for all $x\in[a,b]$, then the function f is uniformly continuous on $[a,b]$.

10. There exists a function $f:\Re\to\Re$ such that the equation $f(x)=c$ has exactly two solutions for any arbitrary real constant c.

11. If a function $f:D\to\Re$ with $a\in D\subseteq\Re$ is continuous and $\{x_n\}$ is a sequence of points in D such that $\{f(x_n)\}$ converges to $f(a)$, then $\{x_n\}$ converges to a.

12. If a function $f:D\to\Re$ with $D\subseteq\Re$ is continuous and the sequence $\{x_n\}$ is Cauchy in D, then $\{f(x_n)\}$ is Cauchy.

13. If a function f is bounded and continuous on D, then f is uniformly continuous on D.

14. If f is contractive, then f is continuous.

15. If a function g is a continuous extension of the function f, then g is unique.

16. If a function f is continuous at $x=a$, then $\lim_{x\to a^+} f(x)=f(a)$ and $\lim_{x\to a^-} f(x)=f(a)$.

17. If functions f and g are continuous on $\Re$ and $f(x)=g(x)$ for every $x\in Q$, then $f(x)=g(x)$ for all $x\in\Re$.
18. A function f exists that is uniformly continuous on (a,∞) and for which $\lim_{x\to\infty} f(x)=\infty$.
19. If a function $f:A\to B$ is uniformly continuous on A, and a function $g:B\to C$ is uniformly continuous on B, then $g\circ f$ is uniformly continuous on A.
20. The function $f(x)=x^2\sin\frac{1}{x}$ is uniformly continuous on $(0,1]$.
21. If functions f and g are both uniformly continuous on D and if f or g is bounded on D, then fg is uniformly continuous on D.
22. If functions f and g are both uniformly continuous on D and $g(x)\neq 0$ for any $x\in D$, then $\frac{f}{g}$ is uniformly continuous on D.
23. If $|f(x)-f(t)|\le(x-t)^2$ for all $x,t\in\Re$, then f is uniformly continuous on $\Re$.
24. If $f:\Re\to\Re$ is periodic, then f is uniformly continuous.
25. If a function f is continuous on both $[0,1]$ and $[1,2]$, then f is continuous on $[0,2]$.
26. The function $f(x)=\begin{cases} x\sin\frac{1}{x}, & \text{if } x \text{ is rational} \\ 0, & \text{if } x \text{ is irrational,}\end{cases}$ is continuous at the origin.
27. Suppose that functions $f,g:[a,b]\to\Re$ are continuous, satisfy $f(a)\le g(a)$, and $f(b)\ge g(b)$. Then there exists a real number $c\in[a,b]$ such that $f(c)=g(c)$.
28. If a function f is continuous at a real number $x=a$, then there exists $\varepsilon>0$ such that f is continuous on $(a-\varepsilon,a+\varepsilon)$.
29. If functions f and g are not continuous on (a,b), then $f\circ g$ is not continuous on (a,b).
30. If a function f is continuous on (a,b) but a function g is not, then fg is not continuous on (a,b).
31. If a function f is continuous at a real number $x=a$, then $[f(x)]^2$ is also continuous at $x=a$.
32. A function f exists that maps a closed interval to an open interval.
33. If $f(x)\le g(x)$ for all real x in an interval I and g is a continuous function, then the function f is uniformly continuous on I.
34. If $\lim_{x\to a} f(x)$ exists, then f has a removable discontinuity at $x=a$.
35. The function $f(x)=\begin{cases} x^2, & \text{if } x \text{ is rational} \\ 4, & \text{if } x \text{ is irrational and positive} \\ 1, & \text{if } x \text{ is irrational and negative}\end{cases}$ is continuous at only 1 and 4.
36. No polynomial of degree greater than 1 is Lipschitzian on $\Re$.

37. If a function f is continuous on a bounded set, then f is bounded.

38. If a function f is not left continuous at a real number $x=a$, then f is not continuous at $x=a$.

39. If a function f is not continuous at a real number $x=a$, then f is not left continuous at $x=a$.

40. Suppose a function f is continuous on an interval I. If $a,b \in I$ such that $f(a)f(b)<0$, then there exists $c \in (a,b)$ such that $f(c)=0$.

41. The function $f(x)=x+\frac{1}{x}$ is uniformly continuous on $[1,\infty)$.

42. If a function f is monotonic on $[a,b]$, then the set of discontinuities of f is countable.

43. If f is uniformly continuous on $(a,b]$ and on $[b,c)$, then f is uniformly continuous on (a,c).

44. If f is uniformly continuous on $[a,b)$ and on $[b,c]$, then f is uniformly continuous on $[a,c]$.

45. The set $\{1,2,3\}$ is closed.

46. There is no $f:\Re \to \Re$ such that f is continuous at each rational, but discontinuous at each irrational number.

47. There is no $f:\Re \to \Re$ that is continuous and such that

$$f(x)=\begin{cases}\text{irrational, if } x \text{ is rational}\\ \text{rational, if } x \text{ is irrational.}\end{cases}$$

48. If f satisfies the intermediate value property on an interval $[a,b]$ and $[c,d] \subset [a,b]$, then f satisfies the intermediate value property on $[c,d]$.

49. A composition of two continuous functions is continuous.

50. A function $f:D \subseteq \Re \to \Re$ is continuous at $x=c \in D$ if and only if for any sequence $\{x_n\}$ in D converging to c, the sequence $\{f(x_n)\}$ converges to $f(c)$.

4.6* Projects

Part 1. Compact Sets

Definition 4.6.1. An *open cover* of a set E is a collection $\{G_\alpha\}$ of open sets in $\Re$ such that $E \subseteq \bigcup_\alpha G_\alpha$, that is, E is completely covered by all of the sets G_α.

Example 4.6.2. A collection of intervals $G_1=(0.2)$, $G_2=(1,3)$, $G_3=(2,4)$, ... covers the interval $(0,\infty)$. ■

Definition 4.6.3. A set E is *compact* if and only if every open cover of E has a finite subcover. That is, if $\{G_\alpha\}$ is any open cover of E, then there exists finitely many α's, say $\alpha_1, \alpha_2, \dots, \alpha_k$ such that $E \subset \bigcup_{i=1}^{k} G_{\alpha_i}$.

Example 4.6.4

(a) The interval $(0,1]$ can be covered by the collection $G_3 = \left\{\left(-1,\frac{1}{2}\right), \left(-\frac{1}{2},2\right), (0,3)\right\}$. Note that $(0,1] \subset \left(-1,\frac{1}{2}\right) \cup \left(-\frac{1}{2},2\right) \cup (0,3)$. Also this open cover of $(0,1]$ has finite subcovers, namely G_3 itself or $\left\{\left(-1,\frac{1}{2}\right), (0,3)\right\}$. But, this fact does not make the interval $(0,1]$ a compact set. The key word in Definition 4.6.3 is "every." Consider an open cover of $(0,1]$ to be H where $H = \left\{\left(\frac{1}{n},3\right) \middle| n \in N\right\}$. The set H has no finite subset that will entirely cover the interval $(0,1]$. And thus, H has no finite subcover. Hence, $(0,1]$ is not compact.

(b) The open cover of $(0,\infty)$ as given in Example 4.6.2 has no finite subcover. Therefore, $(0,\infty)$ is not compact. ■

Problem 4.6.5. Determine which of the given sets is compact.

(a) $(-\infty,0]$ **(b)** $[-1,2] \cup \{3\}$ **(c)** N

(d) $\left\{\frac{1}{n} \middle| n \in N\right\}$ **(e)** $\{0\} \cup \left\{\frac{1}{n} \middle| n \in N\right\}$

THEOREM 4.6.6. *(Heine[3] – Borel[4] Theorem) A set $E \subset \Re$ is compact if and only if E is closed and bounded.*

In the context of real numbers, closed and bounded sets are equivalent to compact. In more abstract spaces, this is not the case. It should be noted that the Heine–Borel theorem is equivalent to Bolzano–Weierstrass theorem for sets, as stated in Theorem 2.5.4.

THEOREM 4.6.7. *A finite union of compact sets is also compact.*

---

[3]Edward Heine (1821–1881), a German analyst, was best known for the study of compactness, though he did not prove this theorem.

[4]Émil Borel (1871–1956), a French mathematician and politician, founded modern measure theory and contributed greatly to the study of divergent series, probability, and game theory.

THEOREM 4.6.8. *Let $f:E \to \Re$ be a continuous function. If $E \subset \Re$ is a compact set, then $f(E)$ is also compact.*

THEOREM 4.6.9

(a) *If E is a nonempty closed and bounded set of real numbers, then $\sup E \in E$ and $\inf E \in E$.*

(b) *A set $E \subset \Re$ is compact if and only if every converging sequence of points in E has a limit in E.*

(c) *A set $E \subset \Re$ is compact if and only if every sequence in E has a subsequence that converges to a point in E.*

(d) *The interval $[0,1]$ has uncountably many points. See Theorem 1.4.5, part (e).*

THEOREM 4.6.10. *(Extreme Value Theorem) A continuous function on a compact set E attains its maximum and minimum values on E.*

Problem 4.6.11. Prove Theorems 4.6.6–4.6.10.

Part 2. Multiplicative, Subadditive, and Additive Functions

Definition 4.6.12. An arithmetical function f is *multiplicative* if f is not identically zero and if $f(mn) = f(m)f(n)$ whenever m and n are relatively prime. A multiplicative function f is *completely multiplicative* if $f(mn) = f(m)f(n)$ for all m and n.

Problem 4.6.13

(a) Prove that the function $f:N \to \Re$ defined by $f(n) = n^r$, r a fixed real number, is completely multiplicative.

(b) Prove that the *divisor function* d, defined by $d(n)$ = the number of the positive divisors of n, is multiplicative, and if $n = (p_1)^{k_1} \cdot (p_2)^{k_2} \cdot \cdots (p_m)^{k_m}$, where $p_1, p_2, \ldots, p_m$ are primes, with $m \in N$, then

$$d(n) = (1+k_1)(1+k_2)\cdots(1+k_m) \equiv \prod_{i=1}^{m}(1+k_i).$$

(c) Prove that the function σ, defined by $\sigma(n)$ = sum of the positive divisors of n, is multiplicative, and if $n = (p_1)^{k_1} \cdot (p_2)^{k_2} \cdot \cdots \cdot (p_m)^{k_m}$, where $p_1, p_2, \ldots, p_m$ are primes, with $m \in N$, then

$$\sigma(n) = \prod_{i=1}^{m} \frac{(p_1)^{k_i+1} - 1}{p_i - 1}.$$

If $\sigma(n) = 2n$, then n is called a *perfect number*.[5]

(d) Prove that the *Möbius*[6] *function* μ, defined for positive integers n by

$$\mu(n) = \begin{cases} 1, \text{ if } n = 1 \\ 0, \text{ if } b^2 \text{ divides } n \text{ for some } b > 1 \\ (-1)^k, \text{ if } n \text{ is the product of } k \text{ distinct primes} \end{cases}$$

is multiplicative.

(e) Prove that $\frac{\mu(n)}{n}$ is multiplicative.

Definition 4.6.14. For each positive integer n, the number of positive integers less than n and relatively prime to n are denoted by $\phi(n)$, *Euler's* ϕ*-function*.

Problem 4.6.15

(a) Show that $\phi(6) = 2$.

(b) Prove that ϕ is multiplicative.

(c) Prove that $\frac{\phi(n)}{n}$ is multiplicative.

Problem 4.6.16

(a) Prove that if f is multiplicative, then $f(1) = 1$.

(b) Prove that if f is multiplicative, then f is completely multiplicative if and only if $f(p^n) = [f(p)]^n$ for all primes p and all $n \in N$.

Problem 4.6.17. Suppose that a function $f: \Re \to \Re$ satisfies $f(x+t) = f(x)f(t)$ for all $x, t \in \Re$.

(a) Give an example of such a function.

(b) Prove that either $f(0) = 0$ or $f(0) = 1$.

(c) Prove that if $f(a) = 0$ for some $a \in \Re$, then $f(x) = 0$ for all $x \in \Re$.

---

[5]Customarily, $\sigma^-(n)$ is the sum of the positive divisors less than n. If $\sigma^-(n) < n$, then n is *deficient*. If $\sigma^-(n) > n$, then n is *abundant*. And if $\sigma^-(n) = n$, then n is *perfect*. There are no prime numbers that are perfect. And every even perfect number must end with a 6 or 28. Also, no one has found any odd perfect numbers, and no one has proved that no such number exists.

[6]August Ferdinand Möbius (1790–1868), a German geometer, topologist, number theorist, statistician, and astronomer, was a student of Gauss. He is known for the "Möbius strip," made by putting a half twist in a paper band before joining the ends. This allows a fly to walk on both sides of the strip without going over the edge.

(d) Prove that if f is continuous at $x = 0$, then f is continuous on $\Re$.

(e) See Exercise 12 of Section 5.1.

Definition 4.6.18. A function $f:\Re \to \Re$ is a *subadditive function* if and only if $f(x+t) \le f(x) + f(t)$ for all $x, t \in \Re$. If the inequality is reversed, then the function is said to be *superadditive*.

Problem 4.6.19. Suppose that a function f is a subadditive.

(a) Prove that if $f(0) = 0$ and if f is continuous at $x = 0$, then f is continuous on all of $\Re$.

(b) Give an example of a subadditive function f that is continuous on all of $\Re$ and $f(0) \ne 0$.

(c) See Exercise 25 in Section 5.3.

Definition 4.6.20. A function $f:\Re \to \Re$ is an *additive function* if and only if $f(x+t) = f(x) + f(t)$ for all $x, t \in \Re$.

Problem 4.6.21. Suppose that f is an additive function.

(a) Prove that $f(nx) = nf(x)$ for all $n \in N$.

(b) Prove that $f(0) = 0$ and $f\left(\frac{1}{n}\right) = \frac{1}{n} f(1)$ for all $n \in N$.

(c) Prove that f is an odd function.

(d) Prove that if $\lim_{x \to 0} f(x) = L$ then $L = 0$.

(e) Prove that if $\lim_{x \to 0} f(x) = L$, then $\lim_{x \to a} f(x)$ is finite for each $a \in \Re$.

(f) Prove that if f is continuous at $x = a$ for any $a \in \Re$, then f is continuous on $\Re$.

(g) Prove that if f is continuous at some point $a \in \Re$ and $f(1) = c$ with c any real constant, then $f(x) = cx$ for all $x \in \Re$. This result is useful when studying Lebesgue[7] measure versus Borel measure on $\Re^n$ in higher levels of real analysis.

(h) Give an example of an additive function.

(i) Prove that f is continuous on $\Re$ if and only if f is bounded on some neighborhood of $x = 0$.

(j) Prove that if f is bounded on some neighborhood of $x = 0$, then f is uniformly continuous on $\Re$.

(k) Show that not every additive function is continuous.

(l) Prove that if f is continuous, then it is homogeneous. See Exercise 34 in Section 5.2.

---

[7]Henri Léon Lebesgue (1875–1941), a French analyst, contributed greatly to the theory of measure (in 1901) and integration (in 1902), as well as to the study of trigonometric series. He also worked in topology and potential theory.

5

Differentiation

5.1 Derivative of a Function

In this chapter we study another class of functions, the differentiable functions. We begin our study with the following definition.

Definition 5.1.1. *(Local)* Suppose that a function $f: D \to \Re$ with $D \subseteq \Re$, a is an accumulation point of D, and $a \in D$. The *derivative of* f *at* $x = a$ is defined by

$$f'(a) = \lim_{x \to a} \frac{f(x) - f(a)}{x - a},$$

provided this limit is finite. If this is the case, we say that f *is differentiable at* $x = a$.

The right-hand side of the limit expression in Definition 5.1.1 is called a *difference quotient*, which can be thought of as the slope of the *secant line* passing through the points $(a, f(a))$ and $(x, f(x))$. This secant line may or may not attain a limit as x approaches a fixed real number a. If the secant line does attain a limit, it will approach a line (see Figure 5.1.1) often called a *tangent line* to the curve f at a point $x = a$ whose slope is given by $f'(a)$ (unless the tangent line is vertical, in which case $f'(a)$ is not finite.[1]) Recall $f'(a)$ must be finite.

---

[1]The concept of a tangent line was first introduced by Fermat around 1630.

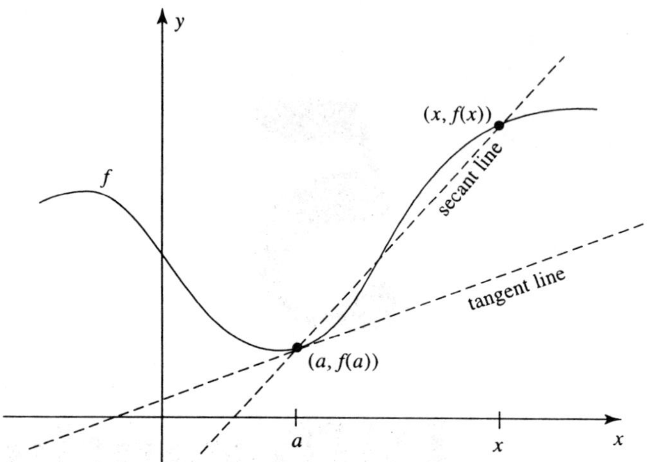

Figure 5.1.1

Also note that x may approach a from both sides or from only one side, depending on the function's domain.

We need to keep in mind, however, that tangent lines provide only a motivation for the definition of a derivative. For example, consider the following function previously discussed in Exercise 2(e) of Section 4.1.

$$f(x) = \begin{cases} x^2 \sin \dfrac{1}{x}, & \text{if } x \neq 0 \\ 0, & \text{if } x = 0. \end{cases}$$

Note that, f is differentiable at $x = 0$ with

$$f'(0) = \lim_{x \to 0} \frac{f(x) - f(0)}{x - 0} = \lim_{x \to 0} \frac{x^2 \sin \frac{1}{x}}{x} = \lim_{x \to 0} x \sin \frac{1}{x} = 0.$$

Since this function f cannot be drawn in the neighborhood of $x = 0$, a converging sequence of secant lines cannot be drawn through $x = 0$ near the origin where the tangent line is horizontal. Next, consider an example where the function has a vertical tangent line, but no derivative, at a particular point.

Example 5.1.2. We will show that the function f, defined by $f(x) = \sqrt[3]{x}$, does not have a derivative at $x = 0$ (i.e., f is not differentiable at $x = 0$). We attempt to evaluate

$$\lim_{x \to 0} \frac{f(x) - f(0)}{x - 0}.$$

This limit is equal to

$$\lim_{x\to 0}\frac{\sqrt[3]{x}}{x}=\lim_{x\to 0}\frac{1}{\sqrt[3]{x^2}}=\lim_{t\to +\infty}\sqrt[3]{t^2}=+\infty.$$

Even though this limit exists, its value is not finite; hence, $f'(0)$ does not exist. Note that the tangent line to this function f at $x=0$ is vertical (see Figure 5.1.2). ■

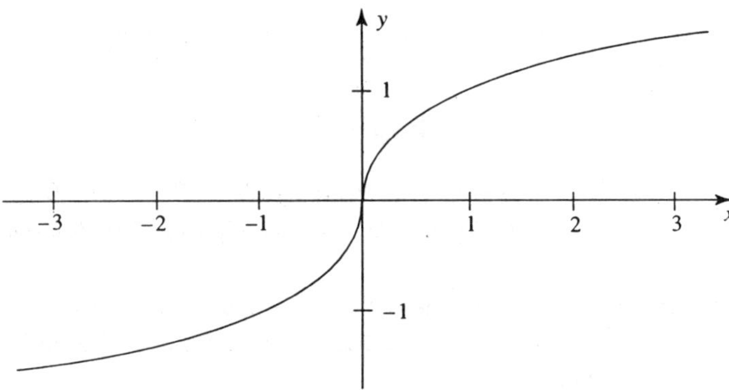

Figure 5.1.2

Compare the previous example with the following one, where the function has neither a tangent line at $x = 0$ nor a derivative.

Example 5.1.3. Show that the function $f(x)=|x|$ is not differentiable at $x=0$.

Proof. To verify this, we attempt to compute

$$\lim_{x\to 0}\frac{f(x)-f(0)}{x-0}.$$

Since the function f is defined differently to the right of zero than to the left, because

$$|x|=\begin{cases} x, & \text{if } x\geq 0\\ -x, & \text{if } x<0,\end{cases}$$

we must evaluate two limits. So we have

$$\lim_{x\to 0^+}\frac{f(x)-f(0)}{x-0}=\lim_{x\to 0^+}\frac{|x|}{x}=\lim_{x\to 0^+}\frac{x}{x}=1, \text{ and}$$
$$\lim_{x\to 0^-}\frac{f(x)-f(0)}{x-0}=\lim_{x\to 0^-}\frac{|x|}{x}=\lim_{x\to 0^-}\frac{-x}{x}=-1.$$

Thus, $\lim_{x\to 0}\frac{f(x)-f(0)}{x-0}$ does not exist. Hence, $f'(0)$ does not exist. Observe in Figure 5.1.3, which illustrates f, that a sharp point exists at the origin. Thus, f has no tangent line at that point. ■

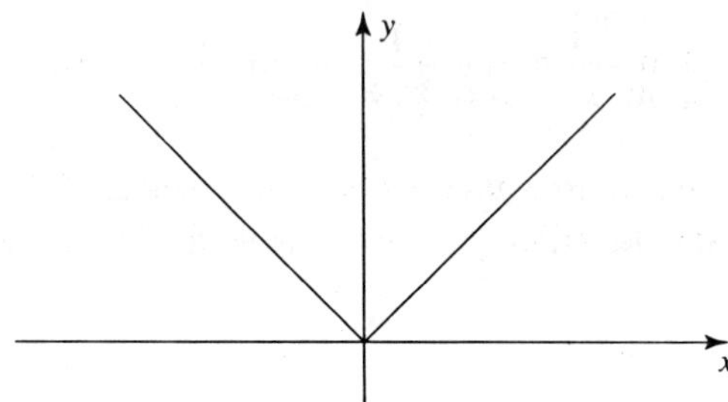

Figure 5.1.3

Definition 5.1.4. *(Global)* Suppose $f: D \to \Re$ with $D \subseteq \Re$. If f is differentiable at every value in its domain D, then we say that f is *differentiable* (on D).

Example 5.1.5. Consider the function f, defined by $f(x) = x^2$. We observe that f is differentiable, because for any real value $x = a$, a limit of the difference quotient exists, which we can write as

$$\lim_{x \to a} \frac{f(x) - f(a)}{x - a} = \lim_{x \to a} \frac{x^2 - a^2}{x - a} = \lim_{x \to a} \frac{(x-a)(x+a)}{x - a} = \lim_{x \to a} (x + a) = 2a.$$

Thus, $f'(a) = 2a$, for any real number a. ■

Let us now observe that if we set $h = x - a$ in Definition 5.1.1 (see Exercise 16 in Section 3.2), a derivative of f at $x = a$ can be written as

$$f'(a) = \lim_{h \to 0} \frac{f(a+h) - f(a)}{h}.$$

We now replace a by x to obtain

$$f'(x) = \lim_{h \to 0} \frac{f(x+h) - f(x)}{h}.$$

This expression defines a new function f', the *derivative of* f, which is derived from f. Here, h can be replaced by Δx, which is a *change* (or *increment*) in x, depending on one's preference. Sometimes $f'(x)$ is denoted by $D_x f$, f_x, $\frac{df}{dx}$, $\frac{dy}{dx}$ where relation involves x and y, $\frac{d}{dx} f(x)$, etc. For instance, if $f(x) = x^2$, we can write $f'(x) = 2x$. This is equivalent to writing $\frac{d}{dx}\left(x^2\right) = 2x$. The notation involving $\frac{d}{dx}$ was introduced by Leibniz[2], whereas the notation involving primes, such as $f'(x)$ or y', was introduced by Lagrange. Newton used $\dot{y}$ to denote y'. *Differentiating,* or *differentiation,* is the process of finding the derivative for a given function.

The concept of differentiability is stronger than that of continuity. We deduce this by considering the next result and recalling Examples 5.1.2 and 5.1.3, whose functions were continuous at $x = 0$ but not differentiable there.

THEOREM 5.1.6. *If a function f is differentiable at a point $x = a$, then f must be continuous at $x = a$.*

Proof. Since $x = a$ is an accumulation point of f, showing that f is continuous at $x = a$ is equivalent to showing that the $\lim_{x \to a} f(x) = f(a)$, which, in turn, is equivalent to showing that the $\lim_{x \to a}[f(x) - f(a)] = 0$. Why? Thus, we write

$$\begin{aligned}\lim_{x \to a}[f(x) - f(a)] &= \lim_{x \to a}\left\{[f(x) - f(a)]\left(\frac{x-a}{x-a}\right)\right\} = \lim_{x \to a}\left[\frac{f(x) - f(a)}{x - a}(x - a)\right] \\ &= \left[\lim_{x \to a}\frac{f(x) - f(a)}{x - a}\right]\left[\lim_{x \to a}(x - a)\right] = f'(a) \cdot 0 = 0.\end{aligned}$$

Note here that $f'(a)$ is a finite number. Why? This completes the proof. ■

Thus, differentiability implies continuity. And from Examples 5.1.2 and 5.1.3, we know that continuity does not imply differentiability. But, if we know a function is differentiable, it must be continuous. Therefore, no continuity implies no differentiability, which is the contrapositive version of Theorem 5.1.6. Consider the next example for an illustration.

Example 5.1.7. Determine if the function $f(x) = \begin{cases} 2x + 1, & \text{if } x \le 0 \\ x^2 + 2x, & \text{if } x > 0 \end{cases}$ is differentiable at $x = 0$.

Answer. Perhaps we want to say that $f'(0) = 2$, but this is incorrect. Why? The function f is not continuous at $x = 0$, so, using the contrapositive version of Theorem 5.1.6, we conclude that $f'(0)$ does not exist. ■

A useful intuitive approach to a differentiable function is that it is continuous and it possesses neither a sharp point nor a vertical tangent line.

---

[2]Gottfried Wilhelm von Leibniz (1646–1716), a great German mathematician, was very concerned with mathematical form and symbolism. The $\frac{dy}{dx}$ and integral notation were his creations. Leibniz developed calculus later than Newton, but independently. He also made great contributions to algebra and discrete mathematics. Leibniz was the founder of the Berlin Academy of Science.

In our next result we express a derivative as the limit of a sequence. Since the idea involved is so similar to the one covered in Theorem 3.2.6, the proof will be omitted.

THEOREM 5.1.8. *Suppose a function $f: D \to \Re$ with $D \subseteq \Re$, a is an accumulation point of D, and $a \in D$. Then the two following conditions are equivalent.*

(a) *f is differentiable at $x = a$.*

(b) *There exists a real number z such that for every sequence $\{x_n\}$ in $D \setminus \{a\}$ converging to a, the sequence $\{y_n\}$, where*

$$y_n = \frac{f(x_n) - f(a)}{x_n - a},$$

converges to z.

Moreover, if condition (b) holds, then $z = f'(a)$.

For better comprehension, view Figure 5.1.4. In view of Theorem 5.1.8, to prove a function is not differentiable at $x = a$, we need to pick two sequences in $D \setminus \{a\}$ that converge to a but whose corresponding difference quotients approach two different values. The following example demonstrates this technique.

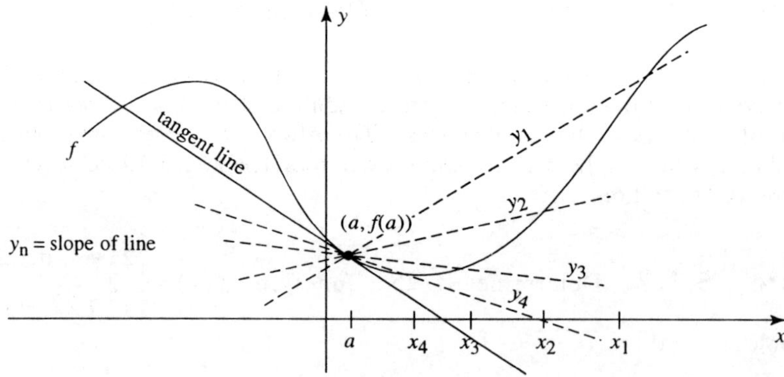

Figure 5.1.4

Example 5.1.9 Recall $f(x) = |x|$ is not differentiable at $x = 0$, because if we define two sequences converging to zero, namely, $\{s_n\}$ and $\{t_n\}$, where $s_n = \frac{1}{n}$ and $t_n = -\frac{1}{n}$, we will have

$$\lim_{n\to\infty} \frac{f(s_n) - f(0)}{s_n - 0} = \lim_{n\to\infty} \frac{f(s_n)}{s_n} = \lim_{n\to\infty} \frac{\frac{1}{n}}{\frac{1}{n}} = 1,$$

and

$$\lim_{n\to\infty}\frac{f(t_n)-f(0)}{t_n-0}=\lim_{n\to\infty}\frac{f(t_n)}{t_n}=\lim_{n\to\infty}\frac{\frac{1}{n}}{-\frac{1}{n}}=-1.$$

Since these values are not equal, $f'(0)$ does not exist. Note, however, that if $x\neq 0$, then $f'(x)=\frac{|x|}{x}$. That is,

$$\frac{d}{dx}|x|=\frac{|x|}{x}=\frac{x}{|x|}=\begin{cases}1, & \text{if } x>0\\ -1, & \text{if } x<0.\end{cases}$$ ■

Remark 5.1.10. Statement (b) in Theorem 5.1.8 can be replaced by

(b)′ There exists a real number z such that for every sequence $\{x_n\}$ in $D\setminus\{a\}$ converging to 0, the sequence $\{y_n\}$, where $y_n=\dfrac{f(a+x_n)-f(a)}{x_n}$ converges to z. ■

We conclude this section with one more example.

Example 5.1.11. Let the functions f and g be defined on an interval (a,b) and

$$g(x)=\lim_{h\to 0}\frac{f(x+h)-f(x-h)}{2h}.$$

(a) If f is differentiable at $x=c$ for $c\in(a,b)$, then prove that $f'(c) = g(c)$.

(b) Find a function f and a point $x=c$ in (a,b) so that $g(c)$ exists but f is not differentiable at $x=c$.

Proof of part (a). We need to show that

$$f'(c)=\lim_{h\to 0}\frac{f(c+h)-f(c-h)}{2h}.$$

We are going to accomplish this using Remark 5.1.10. Let $\{h_n\}$ be any sequence converging to zero such that $c+h_n$ and $c-h_n$ are in the interval (a,b). Observe that the new sequences $\{c\pm h_n\}$ converge to c. Thus, by Remark 5.1.10, sequences $\{s_n\}$ and $\{t_n\}$, where $s_n=\dfrac{f(c+h_n)-f(c)}{h_n}$ and $t_n=\dfrac{f(c-h_n)-f(c)}{-h_n}$, both converge to $f'(c)$. Why? Therefore, we have

$$\begin{aligned}\lim_{n\to\infty}\frac{f(c+h_n)-f(c-h_n)}{2h_n} &= \lim_{n\to\infty}\left[\frac{f(c+h_n)-f(c)}{2h_n}+\frac{f(c-h_n)-f(c)}{-2h_n}\right]\\ &=\frac{1}{2}f'(c)+\frac{1}{2}f'(c)=f'(c).\end{aligned}$$

Hence, by Theorem 3.2.6, we have

$$g(c) = \lim_{h \to 0} \frac{f(c+h) - f(c-h)}{2h} = f'(c).$$

Proof of part (b). Consider $f(x) = |x|$ and $c = 0$. Clearly, $g(0) = 0$, but f is not differentiable at $x = 0$. In fact, any even function that is not differentiable at zero can serve as an example. ■

Note, Example 5.1.11, part (b), is important to consider while studying central differences in numerical analysis. See Part 1 of Section 5.6.

Exercises 5.1

1. **(a)** Rewrite Definition 5.1.1 in an equivalent form using ε and δ.

(b) Use ε and δ to define $f'(a) = \lim_{h \to 0} \frac{f(a+h) - f(a)}{h}$.

2. For each given function f find $f'(x)$, if possible.

(a) $f(x) = c$, c a real constant **(b)** $f(x) = \sqrt{x}$

(c) $f(x) = x^{\frac{3}{2}}$ **(d)** $f(x) = \sqrt[3]{x}$

(e) $f(x) = x^{\frac{2}{3}}$ **(f)** $f(x) = x^{-\frac{1}{2}}$

(g) $f(x) = x^n$, $n \in N$ (See Exercise 16.) **(h)** $f(x) = 2x^2 - x + 1$, $x \geq 0$

3. Determine if each function f is differentiable at the indicated point. If it is, find its derivative at that point; if not explain why not.

(a) $f(x) = \begin{cases} 3x+1, \text{ if } x < 0 \\ x^2 + 3x + 1, \text{ if } x > 0 \end{cases}$ at $x = 0$

(b) $f(x) = \begin{cases} x^2 + 1, \text{ if } x < 1 \\ 2x, \text{ if } x \geq 1 \end{cases}$ at $x = 1$

(c) $f(x) = \begin{cases} x^2 + 2, \text{ if } x \leq 1 \\ 3x, \text{ if } x > 1 \end{cases}$ at $x = 1$

(d) $f(x) = \begin{cases} x^2, \text{ if } x \text{ is rational} \\ 0, \text{ if } x \text{ is irrational} \end{cases}$ at $x = 0$

(e) $f(x) = \begin{cases} x, \text{ if } x \text{ is rational} \\ 0, \text{ if } x \text{ is irrational} \end{cases}$ at $x = 0$

(f) $f(x)=\begin{cases}\sin\frac{1}{x}, & \text{if } x\neq 0\\ 0, & \text{if } x=0\end{cases}$ at $x=0$

(g) $f(x)=\sin x$ at $x=0$ (See Exercise 14.)

(h) $f(x)=\cos x$ at $x=0$

(i) $f(x)=\begin{cases}x^2+x, & \text{if } x=\frac{1}{n},\ n\in N\\ 0, & \text{if } x=0\end{cases}$ at $x=0,1$

(j) $f(x)=\begin{cases}x^2\sin\frac{1}{x}, & \text{if } x \text{ is rational}\\ 0, & \text{if } x \text{ is irrational}\end{cases}$ at $x=0$

4. (a) Discuss differentiability at $x=0$ of functions, defined by

$$f(x)=\begin{cases}x^m\sin\frac{1}{x}, & \text{if } x \text{ is irrational}\\ 0, & \text{if } x \text{ is rational,}\end{cases}$$

with $m = 0, 1, 2,$ or 3.

(b) Discuss differentiability of the function f, defined by

$$f(x)=\begin{cases}x, & \text{if } x \text{ is rational}\\ \sin x, & \text{if } x \text{ is irrational.}\end{cases}$$

5. (a) Find points on the curve $f(x)=\frac{1}{3}x^3+\frac{1}{2}x^2-1$ at which the tangent line is parallel to the line $y=6x$.

(b) Is it true that every cubic, that is, polynomial of degree 3, has at least one point at which the tangent line is horizontal? Explain.

6. Suppose $f:(0,1)\to(0,1)$ is the Dirichlet's function, defined by

$$f(x)=\begin{cases}\frac{1}{q}, & \text{if } x=\frac{p}{q} \text{ in lowest terms with } p,q\in N\\ 0, & \text{if } x \text{ is irrational.}\end{cases}$$

See Example 3.2.10 and Exercise 8 of Section 6.2. Show that f is not differentiable at any point in $(0,1)$.

7. Prove that a function f is differentiable at $x=a$ with $f'(a)=b$, $b\in\Re$, if and only if

$$\lim_{x\to a}\frac{f(x)-f(a)-b(x-a)}{x-a}=0.$$

8. Give an example of a function, if possible, that is defined everywhere and is:

(a) continuous and differentiable at exactly one point.

(b) continuous and differentiable at exactly two points.
(c) continuous at exactly two points and differentiable at exactly one point.
(d) differentiable at exactly two points and continuous at exactly one point.
(e) continuous at countably infinitely many points and differentiable at least at one point but at only finitely many points.
(f) nowhere differentiable.
(g) continuous everywhere but not differentiable anywhere.

9. Suppose the function f is differentiable on $\Re$.

(a) Prove that $\lim_{h\to 0}\dfrac{f(x)-f(x-h)}{h}=f'(x)$.

(b) Find the constant c so that

$$f'(x)=\lim_{h\to 0}\frac{5f(x+h)-f(x)-4f(x-h)}{ch}.$$

10. Suppose the function f is differentiable. Prove that a given limit exists and equals $f'(x)$. Show that the limits might exist with functions f not being differentiable. Observe that all coefficients in the numerator add up to zero. Is that just coincidental? Explain. For more on this topic refer to Part 1 of Section 5.6.

(a) $\lim_{h\to 0}\dfrac{2f(x+h)-f(x)-f(x-h)}{3h}$

(b) $\lim_{h\to 0}\dfrac{3f(x+h)-f(x)-2f(x-h)}{5h}$

(c) $\lim_{h\to 0}\dfrac{f(x+g(h))-f(x)}{g(h)}$, with g a nonzero function of h satisfying $\lim_{x\to 0}g(x)=0$

(d) $\lim_{h\to 0}\dfrac{f(x+2h)+6f(x+h)-7f(x)}{8h}$

(e) $\lim_{h\to 0}\dfrac{-8f(x)+9f(x+h)-f(x+3h)}{6h}$

(f) $\lim_{h\to 0}\dfrac{-f(x+2h)+4f(x+h)-4f(x-h)+f(x-2h)}{4h}$

(g) $\lim_{h\to 0}\dfrac{f(x-ah)-f(x-bh)}{(b-a)h}$, a,b real constants

(h) $\lim_{h\to 0}\dfrac{a^2f(x+bh)-b^2f(x+ah)+\left(b^2-a^2\right)f(x)}{\left(a^2b-ab^2\right)h}$, $a,b\in\Re$ with $0<a<b$

11. If a differentiable function f is even (odd), then prove that f' must be an odd (even) function.

12. Suppose f is differentiable at $x=0$, it is not identically zero, and, for all a and b real, f satisfies $f(a+b)=f(a)f(b)$. Prove that f is differentiable and that $f'(x)=f'(0)f(x)$. See Problem 4.6.17.

13. Assume that all properties of exponents and logarithms are true and prove that $\dfrac{d}{dx}\ln x = \dfrac{1}{x}$.

14. Show that $\dfrac{d}{dx}\sin x = \cos x$ and $\dfrac{d}{dx}\cos x = -\sin x$.

15. If the function $xf(x)$ has a derivative at $x = a \neq 0$, prove that f is differentiable at $x = a$.

16. Prove that $\dfrac{d}{dx}x^n = nx^{n-1}$ with $n \in N$ by following the given steps.

(a) Note that the formula is valid for $x = 0$.

(b) Now assume that $x \neq 0$ and that $f(x) \equiv x^n$. Observe that

$$f'(x) = \lim_{t \to x} \frac{f(t) - f(x)}{t - x} = \lim_{a \to 1} \frac{f(ax) - f(x)}{ax - x},$$

where $t = ax$.

(c) Write

$$\begin{aligned} f'(x) &= \lim_{a \to 1} \frac{(ax)^n - x^n}{ax - x} = x^{n-1} \lim_{a \to 1} \frac{a^n - 1}{a - 1} \\ &= x^{n-1} \lim_{a \to 1}\left(1 + a + a^2 + \cdots + a^{n-1}\right) = x^{n-1}(n) = nx^{n-1}. \end{aligned}$$

See Example 1.3.4.

5.2 Properties of Differentiable Functions

THEOREM 5.2.1. *Suppose functions* $f, g : D \to \Re$ *with* $D \subseteq \Re$ *are differentiable at* $x = a$. *Then* $f \pm g$, fg, $\dfrac{f}{g}$ *if* $g(a) \neq 0$ *are also differentiable at* $x = a$ *and*

(a) $(f \pm g)'(a) = f'(a) \pm g'(a)$.

(b) $(fg)'(a) = f(a)g'(a) + g(a)f'(a)$.

(c) $\left(\dfrac{f}{g}\right)'(a) = \dfrac{g(a)f'(a) - f(a)g'(a)}{[g(a)]^2}$, *provided that* $g(a) \neq 0$.

Parts (b) and (c) are known as the *product rule* and *quotient rule*, respectively.

Proof. Only the proof of part (b) will be given here. Parts (a) and (c) are left to the reader. By performing a standard trick of adding "zero," we write

$$\begin{aligned}\lim_{x\to a}\frac{(fg)(x)-(fg)(a)}{x-a} &= \lim_{x\to a}\frac{f(x)g(x)-f(a)g(x)+f(a)g(x)-f(a)g(a)}{x-a}\\ &= \lim_{x\to a}\left[g(x)\frac{f(x)-f(a)}{x-a}+f(a)\frac{g(x)-g(a)}{x-a}\right]\\ &= \left[\lim_{x\to a}g(x)\right]\left[\lim_{x\to a}\frac{f(x)-f(a)}{x-a}\right]+f(a)\left[\lim_{x\to a}\frac{g(x)-g(a)}{x-a}\right]\\ &= g(a)f'(a)+f(a)g'(a),\end{aligned}$$

because f and g are differentiable, and g is continuous at $x=a$. Thus, since this limit is finite, we have that $(fg)'(a)=g(a)f'(a)+f(a)g'(a)$, ■

COROLLARY 5.2.2. *If functions f_i, $i=1,2,\ldots,n$, are differentiable on $D\subseteq\Re$, then so are functions $f_1+f_2+\cdots+f_n$ and $f_1\cdot f_2\cdot\cdots\cdot f_n$; and*

(a) $(f_1+f_2+\cdots+f_n)'(x)=f_1'(x)+f_2'(x)+\cdots+f_n'(x)$.

(b) $(f_1f_2\cdots f_n)'(x)=f_1'(x)f_2(x)f_3(x)\cdots f_n(x)$ $+f_1(x)f_2'(x)f_3(x)\cdots f_n(x)+\cdots+f_1(x)f_2(x)f_3(x)\cdots f_n'(x)$.

If $f_i=f$ for all $i=1,2,\ldots,n$, then part (b) of Corollary 5.2.2 can be rewritten as

$$(f^n)'(x)=nf^{n-1}(x)f'(x).$$

This, in fact, is true even when n is not a natural number. See Example 5.2.5 and Remark 5.2.7. An even further generalization is given in the following theorem.

THEOREM 5.2.3. *(Chain Rule) Suppose S and T are open intervals in $\Re$ and functions $f:S\to T$ and $g:T\to\Re$. Suppose further that f is differentiable at $x=a\in S$ and g is differentiable at $t=f(a)\in T$. Then the composite function $g\circ f$ is differentiable at $x=a$ and*

$$(g\circ f)'(a)=g'(f(a))f'(a).$$

Proof. The first temptation is to write

$$\begin{aligned}(g\circ f)'(a) &= \lim_{h\to 0}\frac{g(f(a+h))-g(f(a))}{h}\\ &= \lim_{h\to 0}\left[\frac{g(f(a+h))-g(f(a))}{f(a+h)-f(a)}\cdot\frac{f(a+h)-f(a)}{h}\right]\end{aligned}$$

and claim that this value is $g'(f(a))f'(a)$. Unfortunately, $a+h\neq a$ does not imply $f(a+h)\neq f(a)$. To avoid this problem of possible division by zero, we will prove the differentiability of $g\circ f$ at $x=a$ by employing Theorem 5.1.8.

Let $\{x_n\}$ be any sequence in $S\setminus\{a\}$ converging to a. We need to show that the sequence $\{y_n\}$, where

$$y_n = \frac{(g \circ f)(x_n) - (g \circ f)(a)}{x_n - a},$$

converges to $g'(f(a))f'(a)$. Note that since f is differentiable at $x = a$, by Theorem 5.1.8, the sequence $\{z_n\}$, where

$$z_n = \frac{f(x_n) - f(a)}{x_n - a},$$

converges to $f'(a)$. To conclude, consider one more sequence $\{w_n\}$, where

$$w_n = \begin{cases} \dfrac{(g \circ f)(x_n) - (g \circ f)(a)}{f(x_n) - f(a)}, \text{ if } f(x_n) \neq f(a) \\ g'(f(a)), \text{ if } f(x_n) = f(a). \end{cases}$$

The sequence $\{w_n\}$ converges to $g'(f(a))$. Why? In the previous statement we used the fact that f is continuous at $x = a$. The reader should be able to point out where and why this is necessary. Now,

$$\begin{aligned} y_n &= \frac{(g \circ f)(x_n) - (g \circ f)(a)}{x_n - a} \\ &= \begin{cases} \dfrac{(g \circ f)(x_n) - (g \circ f)(a)}{f(x_n) - f(a)} \cdot \dfrac{f(x_n) - f(a)}{x_n - a}, \text{ if } f(x_n) \neq f(a) \\ 0, \text{ if } f(x_n) = f(a) \end{cases} \\ &= \begin{cases} w_n z_n, \text{ if } f(x_n) \neq f(a) \\ 0, \text{ if } f(x_n) = f(a) \end{cases} \\ &= w_n z_n, \text{ in either case.} \end{aligned}$$

Since $\{w_n\}$ converges to $g'(f(a))$ and $\{z_n\}$ converges to $f'(a)$, we conclude that $\{y_n\}$ converges to $g'(f(a))f'(a)$, which completes the proof. ■

It should be noted that in Theorem 5.2.3, S and T cannot be arbitrary subsets of $\Re$. Otherwise, even if a is an accumulation point of S and $f(a)$ is an accumulation point of T, it need not be the case that a is an accumulation point of the domain of $g \circ f$. Thus, we cannot talk about differentiability of $g \circ f$ at a. Try finding an example of such a situation. There are many different proofs of the chain rule in literature. The reader should enhance his or her knowledge by visiting the library.

THEOREM 5.2.4. *(Inverse Function Theorem) Suppose that I is an interval and a function $f: I \to \Re$ is differentiable with $f'(x) \neq 0$, for any $x \in I$. Then* *(a)* *f is an injection,*

(b) *f^{-1} is continuous on $f(I)$,*

(c) *f^{-1} is differentiable on $f(I)$, and*

(d) $(f^{-1})'(y)=\dfrac{1}{f'(x)}$, *with $y=f(x)$. This may be written as* $\dfrac{dx}{dy}=\dfrac{1}{\dfrac{dy}{dx}}$.

Proof. Since $f'(x)\neq 0$ for any $x\in I$, f is strictly monotone and an injection on I. See Exercise 12. Since f is continuous on I, f^{-1} is continuous on $f(I)$. See Theorem 4.3.8. We will employ Theorem 5.1.8 to prove that f^{-1} is differentiable on $f(I)$. To this end, choose any $y_0\in f(I)$ and an arbitrary sequence $\{y_n\}\in f(I)\setminus\{y_0\}$ converging to y_0. Let $x_0=f^{-1}(y_0)$ and $x_n=f^{-1}(y_n)$, for all $n\in N$. Since f^{-1} is injective, $x_n\neq x_0$, for any $n\in N$. In addition, since f^{-1} is continuous on I, it is continuous at x_0. Therefore, $\lim\limits_{n\to\infty} x_n=x_0$. And since f is both differentiable and nonzero, we have

$$\lim_{n\to\infty}\frac{f(x_n)-f(x_0)}{x_n-x_0}=f'(x)\neq 0.$$

Thus,

$$\lim_{n\to\infty}\frac{f^{-1}(y_n)-f^{-1}(y_0)}{y_n-y_0}=\lim_{n\to\infty}\frac{x_n-x_0}{f(x_n)-f(x_0)}=\frac{1}{f'(x_0)}.$$

Hence, f^{-1} is differentiable at any $y_0\in f(I)$. Hence on $f(I)$. And the desired relation is true. ■

THEOREM 5.2.5. *(Power Rule) Assume that $f(x)=x^r$ with any $x>0$ and r representing some rational number. Prove that $\dfrac{d}{dx}x^r=rx^{r-1}$.*

Proof. First we will prove that f is a differentiable function. If $x>0$, $n\in N$, and $g(x)=x^n$, by Exercise 2(g) of Section 5.1, $g'(x)=\dfrac{d}{dx}x^n=nx^{n-1}\neq 0$. Thus by Theorem 5.2.4, the inverse function $g^{-1}(x)=x^{\frac{1}{n}}$ is differentiable. Now consider the function $f(x)=x^r$, where $r=\dfrac{p}{q}$, with $p,q\in Z$ and $q>0$. If $p>0$, then f is differentiable by Corollary 5.2.2. If $p<0$, then f is differentiable by both Theorem 5.2.1, part (c), and Corollary 5.2.2. Now we verify that $\dfrac{d}{dx}x^r=rx^{r-1}$. Let $r=\dfrac{p}{q}$, where $p,q\in Z$, and write $y=x^r=x^{\frac{p}{q}}$. Then $y^q=x^p$. Since $\dfrac{dy}{dx}$ exists, applying the chain rule, we have that

$$q\left(y^{q-1}\right)\frac{dy}{dx} = p\left(x^{p-1}\right).$$

Solving for $\frac{dy}{dx}$, we obtain the desired result. Why? This process involves the so-called *implicit differentiation* because $f(x) = y = x^r$ is written *implicitly* as $y^q = x^p$. ■

Example 5.2.6. Suppose that $k(x) = \sqrt{8x^3 + 1}$. Find $k'(x)$.

Answer. We write $k(x) = (g \circ f)(x)$, where $g(x) = \sqrt{x}$ and $f(x) = 8x^3 + 1$. Thus, we have $g'(x) = \frac{1}{2}x^{-\frac{1}{2}} = \frac{1}{2\sqrt{x}}$ and $f'(x) = 24x^2$. Hence, using the chain rule, we have

$$k'(x) = g'(f(x))f'(x) = \frac{1}{2\sqrt{8x^3 + 1}}\left(24x^2\right) = \frac{12x^2}{\sqrt{8x^3 + 1}}.$$

Observe also that this formula is valid only for $x > -\frac{1}{2}$. ■

Observe that, for example, if $f(x) = x^2$, then a choice exists as to how to compute, say, $f'(3)$. Either a direct computation can be found by determining the limit of a difference quotient with $a = 3$, as given in Section 5.1, or we can first compute $f'(x)$ using definitions from Section 5.1 or the power rule, and then replace x by 3.

Remark 5.2.7. Without proof, assume that all of the familiar differentiation formulas of trigonometric, exponential, and logarithmic functions are true. Some of these formulas are verified in this book. (See Exercises 13 and 14 in Section 5.1) Thus, the function $f(x) = x^r$ discussed in Theorem 5.2.5 can be extended to cover the case when r is any real number. In this situation, using the chain rule enables us to write

$$\frac{d}{dx}x^r = \frac{d}{dx}e^{\ln x^r} = \frac{d}{dx}e^{r\ln x} = e^{r\ln x}\left(\frac{r}{x}\right) = x^r\left(\frac{r}{x}\right) = rx^{r-1}.$$

Often, differentiating $f(x) = x^r$ is accomplished by writing $y = x^r$ and obtaining $\frac{dy}{dx}$ using the so-called *logarithmic differentiation*. First, $y = x^r$ is rewritten as $\ln y = r\ln x$. Then, differentiating both sides, we obtain $\frac{1}{y}\frac{dy}{dx} = \frac{r}{x}$, which yields $\frac{dy}{dx} = rx^{r-1}$.

When differentiating trigonometric functions, the domain must be given in radians. Otherwise, conversion and possibly an application of the chain rule will be required. For example, if $f(x) = \sin x$, where x is given in degrees, then $f'(x) = \frac{\pi}{180}\cos x$. ■

We continue our study of differentiable functions with the next definition.

Definition 5.2.8. Let $f: D \to \Re$. The function f has a *relative (local) maximum* at a point $x = c \in D$ if and only if there exists $\delta > 0$ such that $f(c) \ge f(x)$ for every $x \in (c-\delta, c+\delta) \cap D$.

For a *relative (local) minimum,* we reverse the inequality to $f(c) \le f(x)$. A function f has a *relative (local) extremum* at $x = c$ if it has either a relative maximum or a relative minimum at $x = c$. In either case, the value $f(c)$ is the relative extremum. Thus, intuitively a relative extremum is either the highest or the lowest point the function attains in some small interval containing $x = c$. Note that relative extremum could very well occur at an endpoint of some closed interval. In Figure 5.2.1, we have relative minimum at $x = a$ and $x = d$, relative minimum and relative maximum at every point $(x, f(x))$ with $x \in [b, c]$, and no extremum at $x = e$.

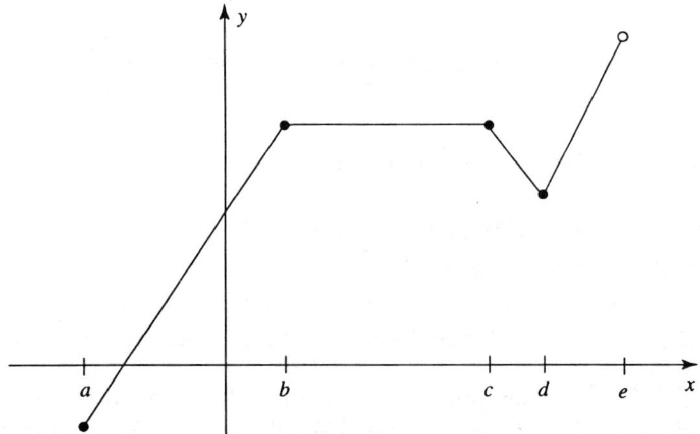

Figure 5.2.1

THEOREM 5.2.9. *Suppose that* $f:(a,b) \to \Re$ *has a relative extremum at* $c \in (a,b)$. *If* f *is differentiable at* $x = c$, *then* $f'(c) = 0$.

Proof. We prove the theorem only for the case where f has a relative maximum at $x = c$. Thus, since f has a relative maximum at $x = c$, there exists $\delta > 0$ such that $f(x) \le f(c)$ for all $|x - c| < \delta$. Thus, for any $h \in (-\delta, \delta)$, we have $f(c+h) - f(c) \le 0$. Therefore, we can write

$$\frac{f(c+h)-f(c)}{h} = \begin{cases} \le 0, \text{ if } h > 0 \\ \ge 0, \text{ if } h < 0, \end{cases}$$

with $h \in (-\delta, \delta)$. Hence, $\lim\limits_{h \to 0^+} \dfrac{f(c+h)-f(c)}{h} \le 0$ and $\lim\limits_{h \to 0^-} \dfrac{f(c+h)-f(c)}{h} \ge 0$. But, since f is differentiable at $x = c$, both of these limits must give the same value. Why? Therefore, $f'(c) = \lim\limits_{h \to 0} \dfrac{f(c+h)-f(c)}{h} = 0$. ■

Find an example of a function that has a relative extremum at $c \in (a,b)$ for which $f'(c) \neq 0$. Does this contradict Theorem 5.2.9? Also, is the converse of Theorem 5.2.9 true? Explain. The points $x = c$ for which $f'(c) = 0$ are called *stationary points*.

Definition 5.2.10. Let $f: D \to \mathfrak{R}$ with $D \subseteq \mathfrak{R}$. A function f is said to have an *absolute (global) maximum* (or simply *maximum*) at a point $x = c \in D$ if and only if $f(c) \geq f(x)$ for all $x \in D$.

For an *absolute minimum* (or simply *minimum*) we reverse the inequality to $f(c) \leq f(x)$. *(Absolute) extremum* means either a maximum or a minimum. Figure 5.2.1 has an absolute minimum at $x = a$ but has no absolute maximum. Review Definition 1.2.15, part (b), and Theorem 4.3.2.

Example 5.2.11. The function $f:[-1,\infty) \to \mathfrak{R}$, defined by $f(x) = 5x^{\frac{2}{3}} - x^{\frac{5}{3}}$ and illustrated in Figure 5.2.2, has a relative and absolute maximum of 6 at $x = -1$, a relative minimum of 0 at $x = 0$, a relative maximum of approximately 4.7622 at $x = 2$, and no absolute minimum. Why? ■

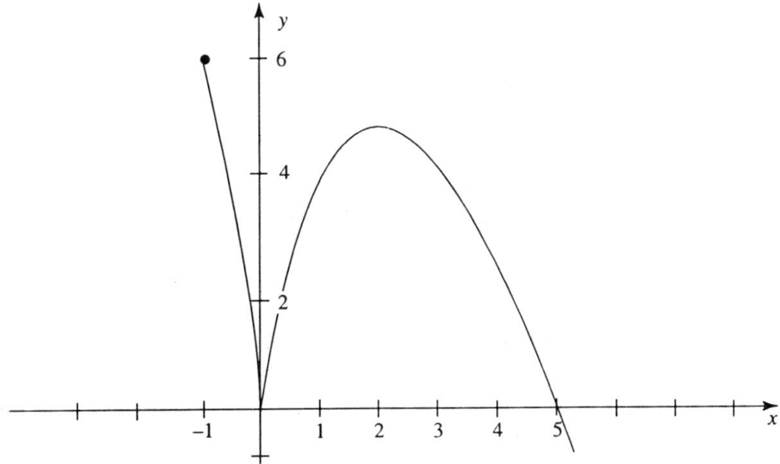

Figure 5.2.2

Notice that if a function f has an extremum at $x = c$, whether absolute or relative, then $f'(c)$ need not be zero. This can be seen in Example 5.2.11 when $x = 0$. See Exercise 23 for further comments. An important observation is that the only possible choices for *extrema* (plural for extremum) are at the points that make the derivative zero and/or the points that make the derivative undefined (provided they are in the domain). These points are called *critical points*. Endpoints of the domain may or may not give rise to an extremum of the function.

THEOREM 5.2.12. *(Rolle's[3] Theorem) Suppose the following three conditions are true for a function f:*

(a) *f is continuous on $[a,b]$;*

(b) *f is differentiable on (a,b); and*

(c) *$f(a)=f(b)$.*

Then there exists some point $c \in (a,b)$ such that $f'(c)=0$.

Remark 5.2.13

(a) Depending on the function, there could be more than one point $c \in (a,b)$ for which the derivative vanishes.

(b) Each condition in Rolle's theorem is essential. See Exercise 7.

(c) Intuitively speaking, if f starts and ends at the same height, when we connect those points in a "smooth" fashion, we see there must be a point between a and b at which the tangent line is horizontal. Examine Figure 5.2.3, where we indicated two such c's and called them c_1 and c_2. ■

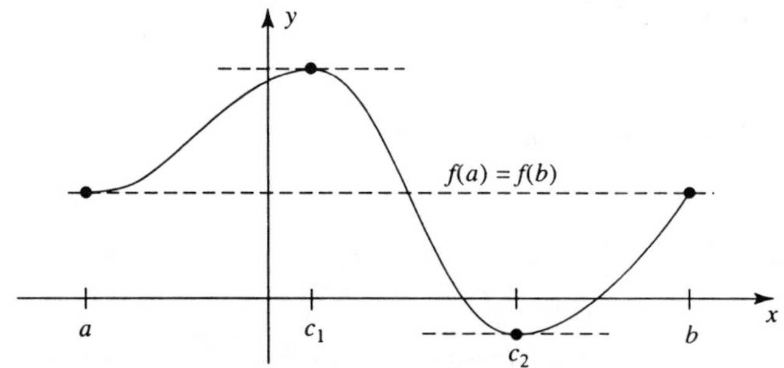

Figure 5.2.3

Proof of Rolle's theorem. First, if f is a constant function, then the result follows from Exercise 2(a) of Section 5.1. Thus, assume f is not constant. Since f is continuous on a compact interval $[a,b]$, it must attain a maximum and a minimum value (see Theorem 4.3.2). We will call them M and m, respectively. Note that $M \neq m$. The remainder of the proof is by contradiction. We will suppose that $f'(x) \neq 0$ for any $x \in (a,b)$, and look for a contradiction. Since f is differentiable on (a,b) and $f'(x) \neq 0$ at any $x \in (a,b)$, then f cannot have any extremum in (a,b). Why? Thus, f has an extremum at a and at b. But $f(a)=f(b)$. Hence $M=m$, which gives a contradiction. The proof is complete. ■

---

[3]Michel Rolle (1652–1719), a French mathematician, is best known for his book on the algebra of equations entitled *Traité d'elgèbre*, published in 1690. In *Traité d'elgèbre*, Rolle firmly established the notation $\sqrt[n]{a}$. In 1846, by Giusto Bellavitis (1803–1880), an Italian mathematician who made significant contributions to algebraic and descriptive geometry, gave Theorem 5.2.12 the name Rolle's theorem as we know it today. Rolle also proved a polynomial version of Rolle's theorem.

THEOREM 5.2.14. *(Mean Value Theorem sometimes known as Lagrange's Mean Value Theorem) Suppose the following conditions are true for a function f:*

(a) f *is continuous on* $[a,b]$;

(b) f *is differentiable on* (a,b).

Then there exists some point $c \in (a,b)$ such that

$$f'(c) = \frac{f(a)-f(b)}{a-b} = \frac{f(b)-f(a)}{b-a}.$$

To understand fully the preceding statement, observe that the last fraction is the slope m of a line joining the endpoints $(a, f(a))$ and $(b, f(b))$. Thus, by the mean value theorem, there must be a point $(c, f(c))$ with $a < c < b$ at which the slope of the tangent line will be equal to m. Examine Figure 5.2.4. Note that Rolle's theorem is a special case of the mean value theorem.

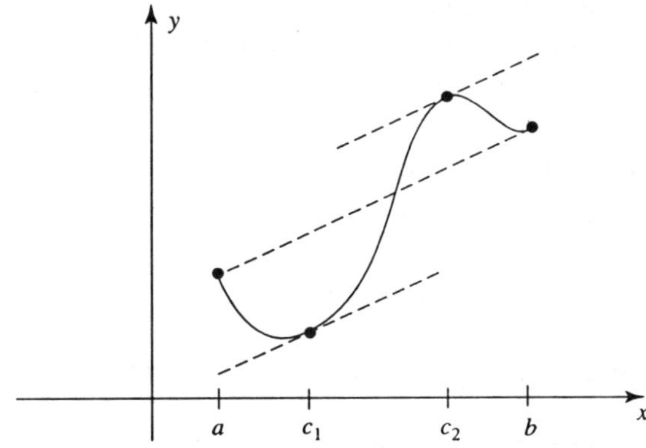

Figure 5.2.4

Remark 5.2.15. Observe also that the conclusion of the mean value theorem could be restated as follows. There exists $\theta \in (0,1)$ such that

$$f(x+h) - f(x) = hf'(x+\theta h),$$

where $x = a$ and $x + h = b$. ■

Proof of the Mean Value Theorem. Consider a new function g defined by

$$g(x) = f(x) - \frac{f(b)-f(a)}{b-a}x.$$

Note that g satisfies all the hypotheses of Rolle's theorem. Why? Hence, by Rolle's theorem, there exists $c \in (a,b)$ such that $g'(c) = 0$. Therefore,

$$f'(c) = \frac{f(b) - f(a)}{b - a},$$

which completes the proof. ■

Observe that the function g in the proof of the mean value theorem is not unique. Another commonly used g is

$$g(x) = f(x) - f(a) - \frac{f(b) - f(a)}{b - a}(x - a).$$

Rolle's theorem and the mean value theorem are important tools in proving results in various areas of mathematics and physics. Theorems are often proven using the mean value theorem, even though they do not mention derivatives. This is pointed out in many exercises and examples and in Section 5.3. The mean value theorem is not true in the context of complex numbers.

Example 5.2.16. Use the mean value theorem to prove:

(a) $\sin x \leq x$ for all $x \geq 0$. See Exercise 51 from Section 1.9.

(b) $ny^{n-1}(x - y) \leq x^n - y^n$ for $0 \leq y \leq x$ and n a natural number.

Proof of part (a). Let $f(x) = x - \sin x$. Then, $f'(x) = 1 - \cos x$. So, by the mean value theorem, there exists $c \in (0, x)$ such that

$$\frac{f(x) - f(0)}{x - 0} = f'(c).$$

Therefore, $f(x) - f(0) = xf'(c)$, which gives $x - \sin x = x(1 - \cos c) \geq 0$.

Proof of part (b). We define the function f by $f(x) = x^n$ with $n \geq 1$. Therefore, by the power rule, $f'(x) = nx^{n-1}$. Suppose $0 \leq y < x$. By the mean value theorem, there exists $c \in (y, x)$ such that

$$\frac{x^n - y^n}{x - y} = nc^{n-1}.$$

Therefore, $x^n - y^n = nc^{n-1}(x - y) \geq ny^{n-1}(x - y)$. Since equality is true when $x = y$, we must have

$$ny^{n-1}(x - y) \leq x^n - y^n$$

for $0 \leq y \leq x$. Explain clearly where the conditions of x and y nonnegative are used. ■

Example 5.2.17. Use the mean value theorem to find an upper and lower bounds of $\sqrt{51}$.

Answer. Apply the mean value theorem to the function $f(x)=\sqrt{x}$ on an interval, say [49,51] to obtain

$$\frac{f(51)-f(49)}{51-49}=\frac{1}{2\sqrt{c}},$$

for some $c\in(49,51)$. Thus, because we have $\sqrt{51}-\sqrt{49}=\frac{1}{\sqrt{c}}$ with $49<c<51<64$ (why 64?), we can write $\frac{1}{8}<\frac{1}{\sqrt{c}}<\frac{1}{7}$, and therefore, $7\frac{1}{8}<\sqrt{51}<7\frac{1}{7}$ (why?), so $\frac{57}{8}<\sqrt{51}<\frac{50}{7}$. ■

THEOREM 5.2.18. *(Cauchy's Mean Value Theorem, sometimes known as the Generalized Mean Value Theorem) If functions f and g are continuous on $[a,b]$ and differentiable on (a,b), then there exists $c\in(a,b)$ such that*

$$f'(c)[g(b)-g(a)]=g'(c)[f(b)-f(a)].$$

Proof. Define a new function k by $k(x)=f(x)[g(b)-g(a)]-g(x)[f(b)-f(a)]$ and then apply Rolle's theorem. Details of this proof are left to the reader. ■

Note that Cauchy's mean value theorem has a similar geometrical interpretation as the mean value theorem. However, in Cauchy's mean value theorem, we consider a "simple continuous curve" which is parametrized (see Section 10.4) by $x=g(t)$ and $y=f(t)$, with $t\in[a,b]$. The slope of the line segment joining the endpoints is given by

$$m=\frac{f(b)-f(a)}{g(b)-g(a)}.$$

Furthermore, the slope of the tangent line to the curve at any point $t=t_0$ is given by $\frac{f'(t_0)}{g'(t_0)}$. Cauchy's mean value theorem guarantees the existence of a value $c\in(a,b)$ for which the slope of the tangent line at c must equal m. Parametric equations are discussed at depth in Section 10.4.

Before we close this section, let us clarify one point that is often incorrectly interpreted.

THEOREM 5.2.19. *Suppose the function f is defined on a neighborhood of $x=a$ with $f'(a)>0$. Then there exists a $\delta>0$ such that $f(a)<f(x_1)$ for some $x_1\in(a,a+\delta)$ and such that $f(x_2)<f(a)$ for some $x_2\in(a-\delta,a)$.*

Remark 5.2.20. Theorem 5.2.19 does *not* say that if $f'(a)>0$, then there exists a neighborhood of a on which f is increasing. The fact that $f'(x)>0$ at only one point $x=a$ is not enough to guarantee that f must be increasing. See Example 5.2.21. For the function f to be increasing on an interval I, $f'(x)$ must be nonnegative for all x in I. See Exercise 10 of this section. Also note that Theorem 5.2.19 can be restated with reverse inequalities. Proof of this theorem is left as an exercise. ■

Example 5.2.21. Consider the function[4] f, defined by

$$f(x)=\begin{cases} x+2x^2\sin\dfrac{1}{x}, & \text{if } x\neq 0 \\ 0, & \text{if } x=0. \end{cases}$$

Verify that

$$f'(x)=\begin{cases} 1+4x\sin\dfrac{1}{x}-2\cos\dfrac{1}{x}, & \text{if } x\neq 0 \\ 1, & \text{if } x=0. \end{cases}$$

Moreover, observe that even though $f'(0)=1>0$, for any $\delta>0$, there exists x_1 and x_2 in the interval $(-\delta,\delta)$ for which $f'(x_1)>0$ and $f'(x_2)<0$. Therefore, f is not monotonic in any neighborhood of $x=0$. See Figure 5.2.5. ■

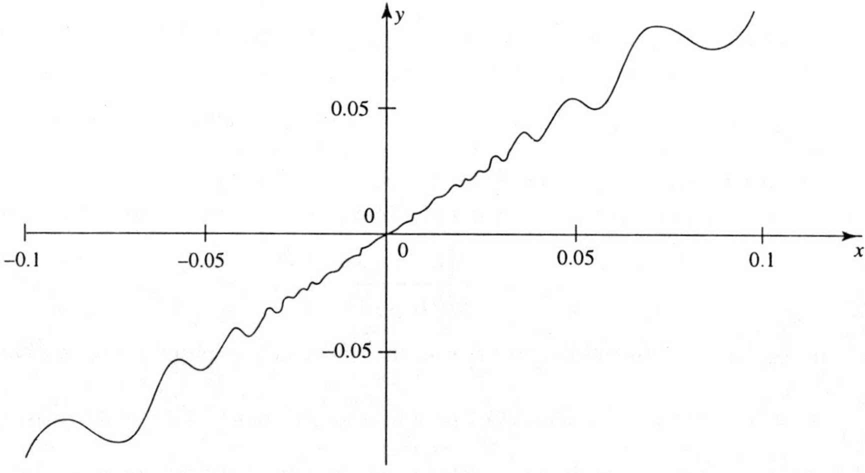

Figure 5.2.5

As we saw in Chapter 4, continuous functions enjoy a very special property, namely, the intermediate value property. The same is true for differentiable functions. In fact, f' must possess an intermediate value property whether f' is continuous or not. See Exercise 41. A contrapositive statement would be, if a function g does not possess the intermediate value property then it does not have an *antiderivative*, that is, there is no function f defined on an interval I such that $f'(x)=g(x)$ with $x\in I$. All continuous functions possess an antiderivative. See Exercise 40.

---

[4]This function was given by Gelbaum and Olmstead in *Counterexamples in Analysis,* published by Holden-Day, 1964.

Exercises 5.2

1. (a) Use the chain rule to evaluate $\frac{d}{dx}\left(2x^3 - x + 3\right)^{-\frac{1}{3}}$.

   (b) Use the chain rule to evaluate $\frac{d}{dx}|x|$ and $\frac{d}{dx}\ln|x|$. See Example 5.1.9 and Exercise 13 from Section 5.1.

   (c) Use logarithmic differentiation to evaluate $\frac{d}{dx}\frac{\sqrt{x^2 - x}}{(3x+1)^{20}\sqrt[3]{2x+1}}$.

   (d) Use the implicit differentiation to find $\frac{dy}{dx}$, when $3xy + x^2y^2 = 1$.

   (e) Write an equation of the tangent line to the curve $xy + y^2 = 3$ at a point $(2,1)$.

   (f) Write an equation of the *normal line*, that is, perpendicular to the tangent line, to the curve $x^2 + xy - y^2 = 1$ at a point $(2,3)$.

   (g) Determine whether or not the function $f : \Re^+ \to \Re$, defined by $f(x) = \frac{1}{x\sqrt[4]{x}}$, is eventually monotone.

   (h) Give an example of functions f and g so that $(fg)' = f'g'$.

   (i) If $f(x) = 2x + \cos x$ and $g(x) = f'(x)$, find $g'(1)$.

   (j) Use implicit differentiation to evaluate $\frac{d}{dx}\arcsin x$ and $\frac{d}{dx}\arctan x$.

   (k) Find $\frac{d}{dx}\arcsin(\sin x)$. See Exercise 10 of Section 1.8.

   (l) If a differentiable function f is even (odd), then prove that f' must be an odd (even) function. See Exercise 11 of Section 5.1.

   (m) Show that $y = x|x|$ satisfies the equation $\frac{dy}{dx} = 2\sqrt{|y|}$.

2. Prove parts (a) and (c) of Theorem 5.2.1.

3. If function f is differentiable and c is a real constant, prove that $\frac{d}{dx}cf(x) = cf'(x)$.

4. Prove Corollary 5.2.2.

5. Prove Theorem 5.2.9 for the case in which f has a relative minimum at $x = c$.

6. Give an example of a function defined on (a,b) that is not continuous on (a,b) but attains both relative and absolute extrema.

7. Give three functions for which exactly two out of three assumptions of Rolle's theorem are satisfied but for which the conclusion does not follow.

8. Consider the function $f:[0,1] \to \Re$, defined by $f(x) = x^3 - x$. What does Rolle's theorem guarantee? Explain.

9. Suppose that the function f is continuous on $[a,b]$, differentiable on (a,b), and $f'(x)=0$ on (a,b). Prove that f must be a constant function on $[a,b]$.

10. **(a)** Suppose that the function f is differentiable on (a,b). Prove that $f'(x)\geq 0$ on (a,b) if and only if f is increasing on (a,b). This exercise, which could serve useful in proving many previously given statements, can be applied to Exercise 11 in Section 2.4.

(b) Is it true that $f'(x)>0$ on (a,b) is equivalent to f being strictly increasing on (a,b)? Explain.

(c) Is it true that if $f(x)=\sqrt[3]{x}$, then $\lim_{x\to 0} f'(x)=+\infty$? Explain.

11. Suppose that functions f and g are continuous on $[a,b]$, differentiable on (a,b), and $f'(x)=g'(x)$ on (a,b). Prove that there exists a real number k such that $f(x)=g(x)+k$ for all $x\in[a,b]$.

12. Suppose that the function f is continuous on $[a,b]$, differentiable on (a,b), and $f'(x)\neq 0$ for any $x\in(a,b)$. Prove that f must be one to one (injective).

13. **(a)** Suppose that the function f is differentiable on (a,b) and f' is bounded on (a,b). Prove that f is uniformly continuous. See Example 4.4.11.

(b) Give an example of a function f that is differentiable, uniformly continuous on (a,b), but f' is not bounded.

14. Suppose that functions f and g are differentiable on (a,b) with $f(c)=g(c)$ for some $c\in(a,b)$. If $f'(x)\leq g'(x)$ for all $x\in[c,b)$, prove that $f(x)\leq g(x)$ for all $x\in[c,b)$. Is the converse true? Explain.

15. *(First Derivative Test)* Suppose that the function f is continuous on $[a,b]$ and differentiable on (a,b), except possibly at $c\in(a,b)$.

(a) Prove that if $f'(x)>0$ for $x\in(a,c)$, and $f'(x)<0$ for $x\in(c,b)$, then f has a relative maximum at $x=c$.

(b) Prove that if $f'(x)<0$ for $x\in(a,c)$, and $f'(x)>0$ for $x\in(c,b)$, then $f(c)$ is a relative minimum of f.

(c) Are the converses of parts (a) and (b) true? Explain.

(d) Find all relative and absolute extrema for the function $f(x)=x^x$, with $x>0$. Also, see Exercise 1(p) in Section 5.4.

(e) Find all extreme values of $f(x)=x^{\frac{2}{3}}(1-x)^{\frac{1}{3}}$. See Exercise 31(c) in Section 5.3.

16. **(a)** Prove that if $a>0$ and b is an arbitrary real number, then the equation $x^3+ax+b=0$ has exactly one real solution.

(b) Suppose that the function f is continuous on $[a,b]$, and differentiable on (a,b). Prove that if $f'(x)\neq 0$ for any x in (a,b), then there exists at most one value $c\in[a,b]$ such that $f(c)=0$. Could we apply this exercise to Exercises 5 and 6 from Section 4.3?

17. Use the mean value theorem to find upper and lower bounds for $\sqrt[3]{43}$.

18. Suppose that the function f is differentiable on (a,b) and there exist $c_1, c_2, \dots, c_n$ such that $f(c_i)=0$ for $i=1,2,\dots,n$. Prove that there exist $n-1$ real numbers d_i, $i=1,2,\dots,n-1$, such that $f'(d_i)=0$. Is it possible for f' to have more roots than f? Explain.

19. Suppose that the function f is continuous on $[a,b]$, differentiable on (a,b), and $f(a)=f(b)=0$. Prove that for each real number α there is some $c\in(a,b)$ such that $f'(c)=\alpha f(c)$.

20. Prove the given inequalities.

(a) $1+rx\le(1+x)^r$ for any real number $r\ge1$ and $x\ge0$. Prove that the equality holds only if $x=0$. See Bernoulli's inequality stated in Exercise 5 from Section 1.3.

(b) $1+x\le e^x$ for $x\in\Re$. Prove that the equality holds only if $x=0$. Use this inequality to reprove Exercise 9 from Section 1.3. See Exercise 12(b) in Section 4.3.

(c) $1-x\le e^{-x}$ for $x\in\Re$. Prove that the equality holds only if $x=0$.

(d) $\dfrac{x}{1+x^2}<\arctan x<x$ for $x>0$.

(e) $|\sin x-\sin y|\le|x-y|$ for $x,y\in\Re$. See Example 5.2.16, part (a). This inequality is useful in proving that $\sin x$ is uniformly continuous. See Exercise 4 from Section 4.4. Furthermore, conclude that $|\sin x+\sin y|\le|x+y|$ for $x,y\in\Re$.

(f) $\dfrac{x}{x+1}\le\ln(x+1)\le x$ for $x\ge0$. Prove that the equality holds only if $x=0$.

(g) $x^n-y^n\le nx^{n-1}(x-y)$ for $n\ge1$, $0\le y\le x$. Compare to part (b) of Example 5.2.16.

(h) $a^tb^{1-t}\le ta+(1-t)b$ for $t\in[0,1]$ with $a>0$ and $b>0$. Prove that the equality is true only if $t=0$, $t=1$, or $a=b>0$. This inequality is used to prove the *Hölder's*[5] *inequality*, which involves sums or integrals depending on its version. See Problem 1.10.22, part (b), and Exercise 14 in Section 6.3. Also, see Exercise 2(b) in Section 10.4.

(i) $\cos x\ge1-\dfrac{x^2}{2}$ for $x\ge0$.

(j) $e(\ln x)\le x$ for all $x>0$. Prove that the equality is true only if $x=e$.

(k) $x^e\le e^x$ for all $x>0$. Prove that the equality is true only if $x=e$.

21. Consider $P(x)=a_nx^n+a_{n-1}x^{n-1}+\cdots+a_1x+a_0$ with $x\in\Re$.

(a) If the coefficients satisfy

---

[5]Ludwig Hölder (1859–1937), a German group theorist, is well known for his work with the summability of series.

$$\frac{a_n}{n+1} + \frac{a_{n-1}}{n} + \cdots + \frac{a_1}{2} + a_0 = 0,$$

then prove that there exists $c \in (0,1)$ such that $P(c) = 0$.

(b) Prove that that between any two roots of P there exists a root of the polynomial Q, where $Q(x) = na_n x^{n-1} + (n-1)a_{n-1}x^{n-2} + \cdots + a_1$.

22. Suppose that functions f and g are defined on a neighborhood of $x = a$ and are differentiable at $x = a$. Use the fact that $fg = \frac{1}{2}(f+g)^2 - \frac{1}{2}f^2 - \frac{1}{2}g^2$ to prove that fg is also differentiable at $x = a$. Compare this proof with the proof of Theorem 5.2.1, part (b).

23. Give an example of a function f that has an absolute maximum at, say, $x = 0$, and f' alternates sign in any neighborhood of zero.

24. **(a)** Complete the proof of Theorem 5.2.18.

(b) Prove Theorem 5.2.18 using the function

$$K(x) = f(x) - f(a) - \frac{f(b) - f(a)}{g(b) - g(a)}[g(x) - g(a)]$$

instead of $k(x)$ as used in the proof of part (a).

25. What is wrong with the following proof of Cauchy's mean value theorem? Since f and g satisfy the hypotheses of the mean value theorem, there exists $c \in (a,b)$ such that

$$\frac{\dfrac{f(b) - f(a)}{b-a}}{\dfrac{g(b) - g(a)}{b-a}} = \frac{f'(c)}{g'(c)}.$$

Now simplify to obtain the desired expression.

26. Verify that if functions f and g are continuous on $[a,b]$, differentiable on (a,b), and $g'(x) \neq 0$ for any $x \in (a,b)$, then

(a) there exists $c \in (a,b)$ such that $\dfrac{f(b) - f(a)}{g(b) - g(a)} = \dfrac{f'(c)}{g'(c)}$.

(b) there exists $k \in (a,b)$ such that $\dfrac{f(k) - f(a)}{g(b) - g(k)} = \dfrac{f'(k)}{g'(k)}$.

Compare this exercise with Theorem 5.2.18.

27. Explain why Cauchy's mean value theorem is indeed a generalization of the mean value theorem.

28. Suppose that the function f satisfies $|f(x) - f(t)| \leq (x-t)^2$ for each $x, t \in \Re$. Prove that f must be a constant function. See Exercise 23 of Section 4.5.

29. Suppose that the function f is differentiable on (a, ∞), where a is any real constant, and suppose that $\lim_{x\to\infty} f'(x)=0$. If the function g is defined by $g(x)=f(x+1)-f(x)$, prove that $\lim_{x\to\infty} g(x)=0$.

30. Suppose that the function f is continuous on $[0,\infty)$, differentiable on $(0,\infty)$, $f(0)=0$, and f' is an increasing function. Define $g(x)=\dfrac{f(x)}{x}$ for $x>0$ and prove that g is increasing for $x>0$.

31. Suppose that the function $f:(a,b)\to\Re$ is differentiable on (a,b) except possibly at $c\in(a,b)$. If $\lim_{x\to c} f'(x)=A$, then prove that f must also be differentiable at $x=c$ and $f'(c)=A$.

32. Prove Theorem 5.2.19.

33. Evaluate the given limits, if possible.

(a) $\lim_{x\to 0}\dfrac{2(x+1)^{\frac{7}{11}}-2}{x}$ **(b)** $\lim_{x\to 0}\dfrac{3x}{e^{2x}-1}$ **(c)** $\lim_{x\to 0}\dfrac{\cos x-1}{x}$

34. A function f with the domain $\Re$ is said to be *homogeneous of degree* n, with n any real number, if and only if $f(tx)=t^n f(x)$ for every $t>0$ such that x and tx are in $\Re$. Prove that if f is differentiable on $\Re$ and homogeneous of degree n, then $xf'(x)=nf(x)$ for all $x\in\Re$. This is known as Euler's equation for homogeneous functions of degree n. See Problem 4.6.21, part (1).

35. Suppose $f:\Re^+\to\Re$ satisfies $\lim_{x\to\infty}(f+f')(x)=0$. Show that $\lim_{x\to\infty} f(x)=0$.

36. Suppose that f and g are differentiable functions on (a,b). Show that between two consecutive roots of f there exists a root of $f'+fg'$.

37. Prove that between two consecutive roots of $f(x)=1-e^x\sin x$ there exists at least one root of $g(x)=1+e^x\cos x$.

38. **(a)** Use Exercise 20(a) to prove that $\lim_{x\to\infty} a^x=+\infty$ if $a>1$.

(b) Verify that $\lim_{x\to-\infty} a^x=0$ if $a>1$.

(c) Verify that $\lim_{x\to\infty} b^x=0$ if $0<b<1$.

(d) Verify that $\lim_{x\to-\infty} b^x=+\infty$ if $0<b<1$.

39. Suppose that $f'(x_0)=0$ in Theorem 5.2.4, that is, the inverse function theorem. Show that $\left(f^{-1}\right)'\left(f(x_0)\right)$ does not exist. Represent this situation graphically.

40. *(Darboux's[6] Intermediate Value Theorem)* Let $f'_+(a) \equiv \lim_{x \to a^+} \frac{f(x)-f(a)}{x-a}$ and $f'_-(a) \equiv \lim_{x \to a^-} \frac{f(x)-f(a)}{x-a}$. Suppose that a function f is differentiable on $[a,b]$ and $f'_+(a) \neq f'_-(b)$. If k is a number between $f'_+(a)$ and $f'_-(b)$, then there exists $c \in (a,b)$ such that $f'(c) = k$.

41. Give an example of a function f that is differentiable on (a,b) but f' is not continuous on (a,b).

42. Find a function f, if possible, for which $f'(x) = \lfloor x \rfloor$, with $x \in [-2,2]$.

43. **(a)** Prove that if a function $f:\Re \to \Re$ is differentiable and $f'(x) = f(x)$ for all x, then $f(x) = ce^x$, where c is some real constant.

(b) Find a function that satisfies $f'(x) = \frac{1}{f(x)}$.

5.3 Higher Order Derivatives

Concavity, n times continuity and uniform differentiability, multiplicity of roots, and Taylor's theorem are topics covered in this section. Knowledge of these topics is useful in differential equations, numerical analysis, and many other areas of mathematics.

Consider a differentiable function f. If the derivative function f' has a derivative at $x = a$, then that number is the *second derivative* of f at $x = a$ and is denoted by $f''(a)$. The *third, fourth, fifth* through the nth *derivatives* are similarly defined. Thus, we can write

$$f''(a) = \lim_{x \to a} \frac{f'(x)-f'(a)}{x-a} = \lim_{h \to 0} \frac{f'(a+h)-f'(a)}{h} = \lim_{n \to \infty} \frac{f'(a+h_n)-f'(a)}{h_n},$$

where $\{h_n\}$ is any sequence converging to 0. However, if we can differentiate a function f over and over on some interval I, then we will obtain the *first, second, third* through the nth *derivative functions*. A number of notations exist for higher order derivative functions. For example, $f''(x)$ can be written as $\frac{d}{dx}f'(x)$. Also, if the relation involves x and y, then $f''(x)$ may be written as $\frac{d^2y}{dx^2}$, $\frac{d^2}{dx^2}(y)$ or $\frac{d}{dx}\left(\frac{dy}{dx}\right)$. By convention, for fourth-order derivatives and up, instead of using primes, we write $f^{(4)}(x)$, $f^{(5)}(x)$, etc.

[6]Jean Gaston Darboux (1842–1917), a French mathematician, excelled in differential geometry and did work with the intermediate value property.

The symbol $f^{(0)}(x)$ stands for a function that is differentiated no times; thus, $f^{(0)}(x) = f(x)$. If we are given a function f which is twice differentiable on an interval I, that is, f'' exists, then to evaluate $f''(a)$ we often find $f''(x)$ and then replace x by a. Thus, we could write

$$f''(a) = \frac{d}{dx} f'(x)\bigg|_{x=a}.$$

This, however, is not always possible. Sometimes, for example, f may have a second derivative only at $x = a$, or f' may be piecewise defined with a change of definition at $x = a$.

Second-order derivatives are useful in studying concavity of functions, acceleration in physics, road planning in engineering, and a great many other areas. For a quick review, consider the following definition.

Definition 5.3.1. A differentiable function f is *concave up* on an interval (a,b) if and only if f' is strictly increasing on (a,b). *Concavity downward* is defined similarly.

Intuitively, this definition says that a differentiable function f is concave up on (a,b) if and only if it lies above its tangent lines. Recall from Section 5.2, Exercise 10, that if the derivative of a function is positive, then that function is strictly increasing. Therefore, if the function f is twice differentiable and $f'' > 0$, then f' must be strictly increasing. Thus, according to Definition 5.3.1, f must be concave up. This informally proves the next theorem. Figure 5.3.1 shows a function f that is twice differentiable, together with several tangent lines.

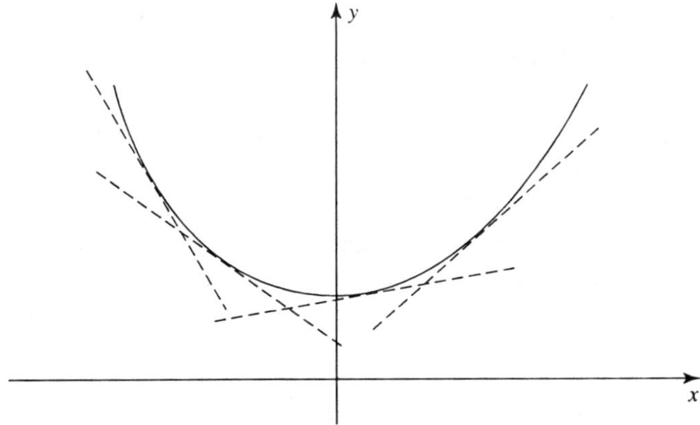

Figure 5.3.1

THEOREM 5.3.2. *If a function f satisfies $f''(x) > 0$ for all $x \in (a,b)$, then f must be concave up on (a,b). Similarly, $f''(x) < 0$ on (a,b) implies that f is concave down on (a,b).*

It should be clear that if $f''(x) > 0$ for all $x \in (a,b)$, then a line segment joining any two points $(x_1, f(x_1))$ and $(x_2, f(x_2))$ with $x_1, x_2 \in (a,b)$ is above the curve f on the interval (x_1, x_2). See Problem 5.6.24, part (m), and the discussion at the end of Section 1.3. Also recall that if for a continuous function f at $x = a$, $f''(x)$ changes sign at $x = a$, then f changes concavity at $x = a$, creating $(a, f(a))$, a *point of inflection.* Note that even if a function changes concavity at $x = a$, where $x = a$ is in the domain of that function, it is incorrect to expect a point of inflection at $x = a$. Why is this the case? Look at the function

$$f(x) = \begin{cases} \dfrac{1}{x}, & \text{if } x \neq 0 \\ 0, & \text{if } x = 0. \end{cases}$$

If g is some continuous function with an inflection point $(a, g(a))$, then $g'(a)$ may or may not be zero. In fact, it may not even exist. See Exercise 2(b). If $(a, g(a))$ is a point of inflection at which the tangent line is horizontal, then $(a, g(a))$ is called a *horizontal point of inflection.* Also, observe that if $h'(x) > 0$ and $h''(x) > 0$ eventually, then $h(x) > 0$ eventually. Why?

Since differentiability implies continuity, we have that if f'' exists on an interval I, then f' must be continuous on I. The converse is not true. For example, the function f, given by

$$f(x) = \begin{cases} x^2 \sin \dfrac{1}{x}, & \text{if } x \neq 0 \\ 0, & \text{if } x = 0, \end{cases}$$

is differentiable everywhere, but f' is not continuous at $x = 0$. Thus, certainly f' will not be differentiable on any interval containing $x = 0$.

Recall that if the nth derivative of a function f, that is, $f^{(n)}(x)$, exists on an interval I, then the function f is *n times differentiable* on I. If we require the nth derivative to be continuous on I, then we say f is *n times continuously differentiable* on I. For example, the function f defined above is one time differentiable, but not one time continuously differentiable. This function f, however, would be one time continuously differentiable if we restricted it to an interval not containing zero. In fact, f would be *infinitely many times continuously differentiable*, that is, derivatives of all orders would exist.

Example 5.3.3. Consider the function f, defined by

$$f(x) = \begin{cases} \exp\left(-\dfrac{1}{x}\right), & \text{if } x > 0 \\ 0, & \text{if } x \leq 0. \end{cases}$$

Show that f is an infinitely differentiable function.

Proof. If $x > 0$, then $f^{(n)}(x)$ exists for any $n \in N$, likewise for $x < 0$. Why? Therefore, we need only to verify that $f^{(n)}$ exists at $x = 0$ for every $n \in N$. If $n = 1$, we consider two limits:

$$\lim_{h \to 0^-} \frac{f(0+h) - f(0)}{h} = 0,$$

and

$$\lim_{h \to 0^+} \frac{f(0+h) - f(0)}{h} = \lim_{h \to 0^+} \frac{f(h)}{h} = \lim_{h \to 0^+} \frac{\exp\left(-\frac{1}{h}\right)}{h} = \lim_{t \to \infty} \frac{t}{e^t} = 0.$$

Why? Therefore, $f'(0) = 0$. One can follow a similar procedure for higher order derivatives. However, to prove that $f^{(n)}(0) = 0$ for each $n \in N$, mathematical induction should be invoked. See Section 1.3. ■

Definition 5.3.4. Suppose the function f is defined on an interval (a,b). Then f is *uniformly differentiable* on (a,b) if and only if f is differentiable on (a,b), and for each $\varepsilon > 0$ there exists $\delta > 0$, such that

$$\left| \frac{f(x) - f(t)}{x - t} - f'(x) \right| < \varepsilon$$

whenever $0 < |x - t| < \delta$, with $x, t \in (a,b)$.

Remark 5.3.5. As pointed out in Exercise 11, uniform differentiability implies continuous differentiability, which in turn implies differentiability, and hence, continuity. Does differentiability imply uniform continuity? ■

The negation of uniform differentiability can be most easily expressed in terms of sequences. That is, a function f is not uniformly differentiable on an interval I if there exists a particular $\varepsilon > 0$ and two sequences $\{x_n\}$ and $\{t_n\}$ in I such that $|x_n - t_n| < \frac{1}{n}$ and

$$\left| \frac{f(x_n) - f(t_n)}{x_n - t_n} - f'(x_n) \right| \geq \varepsilon.$$

See Exercise 11(b). Compare this with the negation of uniform continuity in Remark 4.4.4. To intuitively decide whether or not f is uniformly differentiable, we determine whether the difference quotient, involving points close together, is within ε from the actual derivative at one of the points involved in the difference quotient.

It can be verified that any polynomial and e^x are uniformly differentiable on $[a,b]$. Functions $\sin x$ and $\cos x$ are uniformly differentiable on $\Re$.

A value $x = r$ is said to be a *root* or a *zero* of a function f if and only if $f(r) = 0$. See Definition 1.2.15, part (a). Recall that the root $r = 0$ is of different nature for each of the functions $f_1(x)=x$, $f_2(x)=x^2$, and $f_3(x)=x^3$. Since the function f_1 crosses the x-axis and $f_1'(0)\neq 0$, a "simple" root exists at $r=0$. A "double" root exists at $r=0$ for the function f_2 since it does not cross the x-axis. And the function f_3 crosses the x-axis, changes concavity at $x=0$, and thus gives rise to a "triple" root at $r=0$. Whenever an even root exists for a polynomial, the polynomial just "bounces" off of the x-axis. If a root of a polynomial is odd but not simple, then this root becomes a point of inflection with a horizontal tangent line. The distinction between different varieties of roots is formally given by Definition 5.3.6.

Definition 5.3.6. Consider a function f. Suppose $x=r$ is a solution of the equation $f(x)=0$ (i.e. $f(r)=0$). Then $x=r$ is a *root* (or a *zero*) *of multiplicity* m of f, with $m\in N$, if and only if m is the smallest value for which $f(x)$ can be written as

$$f(x)=(x-r)^m q(x)$$

with $x\neq r$, where $\lim_{x\to r} q(x)\neq 0$. If $m=1$, then the root is called a *simple root.* If $m=2$, then the root is called a *double root.*

The next theorem shows an easier method of determining the multiplicity of a root, provided all appropriate derivatives of the function exist.

THEOREM 5.3.7. *Suppose a function f is m times continuously differentiable. The function f has a root of multiplicity m at $x=r$ if and only if*

$$0=f(r)=f'(r)=\cdots=f^{(m-1)}(r)$$

but $f^{(m)}(r)\neq 0$.

Proof. We are only going to prove the theorem in the case of a simple root, (i.e. $m=1$).

($\Rightarrow$) First, assume f has a simple root at $x=r$. We want to show that $f(r)=0$, and that $f'(r)\neq 0$. Clearly, $f(r)=0$. Also, by definition, we can write $f(x)=(x-r)q(x)$, where $\lim_{x\to r} q(x)\neq 0$. Since f has a continuous first derivative at $x=r$, we can write

$$f'(r)=\lim_{x\to r} f'(x)=\lim_{x\to r}\left[q(x)+(x-r)q'(x)\right]=\lim_{x\to r} q(x)\neq 0.$$

($\Leftarrow$) Suppose $f(r)=0$ and $f'(r)\neq 0$, and show that $x=r$ must be a simple root of f. Since we need to prove that $\dfrac{f(x)}{r-x}=q(x)$, which is $\dfrac{f(x)-f(r)}{x-r}=q(x)$, we are motivated to use the mean value theorem. Thus, we know that there exists a c between x and r such that

$$\frac{f(x)-f(r)}{x-r}=f'(c).$$

Since x varies, c will vary. Therefore, we denote c by $c(x)$ and write

$$f(x) = f(r) + (x - r)f'(c(x)) = (x - r)f'(c(x)).$$

Since f has a continuous first derivative, we have that

$$\lim_{x \to r} f'(c(x)) = f'\left(\lim_{x \to r} c(x)\right) = f'(r) \neq 0.$$

Therefore, if we let $q(x) = (f' \circ c)(x)$, then $f(x) = (x - r)q(x)$, where $\lim_{x \to r} q(x) \neq 0$. Thus, f has a simple root at $x = r$. ■

The final result in this section, known as Taylor's[7] theorem, approximates functions using nth degree polynomials. Naturally some sort of an "error" will be involved. Taylor's theorem, which involves ideas to be covered again in Chapter 8, is another generalization of the mean value theorem and is an important proving tool in the area of numerical analysis.

THEOREM 5.3.8. *(Taylor's Theorem) Suppose for any $n \in N$, a function f has $n+1$ derivatives in a neighborhood of $x = a$. If*

$$\begin{aligned} p_n(x) &= f(a) + f'(a)(x-a) + \frac{f''(a)}{2!}(x-a)^2 + \cdots + \frac{f^{(n)}(a)}{n!}(x-a)^n \\ &= f(a) + \sum_{k=1}^{n} \frac{f^{(k)}(a)}{k!}(x-a)^k, \end{aligned}$$

then Taylor's formula, with the remainder for f, is given by $f(x) = p_n(x) + R_n(x)$, where

$$R_n(x) = \frac{f^{(n+1)}(c)}{(n+1)!}(x-a)^{n+1},$$

for some c between x and a. If $x = a$, then $c = a$.

The functions $p_n(x)$ are called the *nth Taylor polynomial of degree n centered about* $x = a$. The values $\frac{f^{(k)}(a)}{k!}$ are called the Taylor coefficients. $R_n(x)$, given in Theorem 5.3.8, is called *Lagrange's form of the remainder. Cauchy's form of the remainder* is given by

$$R_n(x) = \frac{f^{(n+1)}(c)}{n!}(x-c)^n(x-a),$$

[7]Brook Taylor (1685–1731) was born to a wealthy English family. The theorem for which Taylor is best known did not receive notoriety until Euler applied it in calculus. Still later, full appreciation for Taylor's theorem resulted from the work of Lagrange. Taylor also devoted his writings to magnetism, capillary action, thermometers, and later, to religion and philosophy.

and is proven in Exercise 16 of Section 6.4. A different form of $R_n(x)$ can be found in Exercise 14 of Section 6.4.

Proof of Taylor's theorem. Taylor's formula trivially is true if $x = a$. So we will assume that $x \neq a$ and prove that $R_n(x) = f(x) - p_n(x)$. To accomplish this, we use Rolle's theorem and prove that $R_n(b) = f(b) - p_n(b)$, for any $b \neq a$, where

$$R_n(b) = \frac{f^{(n+1)}(c)}{(n+1)!}(b-a)^{n+1},$$

for some c between a and b. So, we apply Rolle's theorem to a new function F, defined by

$$\begin{aligned} F(x) &= \left[f(b) + f'(x)(b-x) + \frac{f''(x)}{2!}(b-x)^2 + \cdots + \frac{f^{(n)}(x)}{n!}(b-x)^n \right] \\ &\quad + \frac{R_n(b)}{(b-a)^{n+1}}(b-x)^{n+1}. \end{aligned}$$

The reader should verify that $F(a) = f(b) = F(b)$ and that

$$F'(x) = \frac{f^{(n+1)}(x)}{n!}(b-x)^n - \frac{(n+1)R_n(b)}{(b-a)^{n+1}}(b-x)^n.$$

Therefore, by Rolle's theorem, there exists a c between a and b such that $F'(c) = 0$. Hence, we have

$$\frac{f^{(n+1)}(c)}{n!}(b-c)^n = \frac{(n+1)R_n(b)}{(b-a)^{n+1}}(b-c)^n,$$

which reduces to $R_n(b) = \dfrac{f^{(n+1)}(c)}{(n+1)!}(b-a)^{n+1}$. Hence, the proof is complete. ∎

It should be observed that in Taylor's theorem, $p_0(x), p_1(x), p_2(x), \ldots$ and so on become progressively closer and closer to the curve f near $x = a$ as n becomes larger and larger. However, after a certain point these polynomials may move drastically away from the function f. Note that p_1 is the tangent line to f at $x = a$, which generally will not be close to f for very long. In Figure 5.3.2, $f(x) = \sin x$ is graphed together with $p_n(x)$ for $n = 0, 1, \ldots, 7$ on the interval $[0, 7]$. What are the equations of these polynomials?

Taylor's theorem can be used to approximate certain values and/or prove inequalities. The next two examples demonstrate this.

Example 5.3.9. Use Taylor's theorem to approximate $\sin 3°$ to four-decimal-place accuracy, that is, the magnitude of the error is less than 0.5×10^{-4}.

Answer. Let us expand the function $f(x) = \sin x$ about $x = 0$, since 0 is the closest point to $3° = \dfrac{\pi}{60}$ radians, whose sine and cosine values are easily evaluated. Thus, we have

$$\sin x = x - \frac{x^3}{3!} + \frac{x^5}{5!} - \frac{x^7}{7!} + \cdots + R_n(x),$$

where $R_n(x)$ involves $f^{(n+1)}(c)$, which in the case of $\sin x$ is $\pm\sin c$ or $\pm\cos c$, and the term $\dfrac{1}{(n+1)!}x^{n+1}$. We now replace x by $\dfrac{\pi}{60}$ (why not $3°$?) and look for the value of n so that $\left|R_n\left(\dfrac{\pi}{60}\right)\right| < 0.5\times10^{-4}$. Note that if we expand $\sin x$ about a point more distant to 3° than 0, then the value of n in a related Taylor polynomial will be greater.

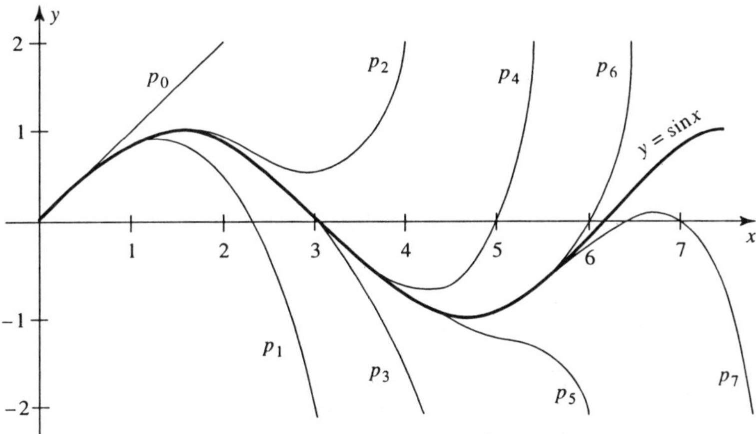

Figure 5.3.2

Thus, since $\left|f^{(n+1)}(c)\right| \le 1$, we can write

$$\left|R_n\left(\frac{\pi}{60}\right)\right| \le \left(\frac{\pi}{60}\right)^{n+1}\frac{1}{(n+1)!}.$$

Since this value is less than 0.5×10^{-4} when $n = 3$, we need only to compute

$$\sin 3° = \sin\frac{\pi}{60} = \frac{\pi}{60} - \frac{\left(\frac{\pi}{60}\right)^3}{3!} \approx 0.0523.$$

Note that $n = 300$ will also make $\left|R_n\left(\dfrac{\pi}{60}\right)\right| < 0.5\times10^{-4}$, but it will require evaluating many more terms in $p_n\left(\dfrac{\pi}{60}\right)$. ■

Example 5.3.10. For $x>0$, use Taylor's theorem to prove that

$$1+\frac{x}{2}-\frac{x^2}{8}\le\sqrt{x+1}\le 1+\frac{x}{2}.$$

Proof. Since Taylor's formula centered about $x=0$ has a remainder given by

$$R_n(x)=\frac{f^{(n+1)}(c)}{(n+1)!}x^{n+1}$$

with $c>0$, upon repeated differentiation of $f(x)=\sqrt{x+1}$ we can conclude that $R_1(x)<0$ and $R_2(x)>0$ for all $x>0$. Since $f(x)=1+\frac{x}{2}+R_1(x)$ and $f(x)=1+\frac{x}{2}-\frac{x^2}{8}+R_2(x)$, the inequality follows. Details are left to the reader. ■

Exercises 5.3

1. Prove Theorem 5.3.2.

2. **(a)** Determine where the function

$$f(x)=\begin{cases}\frac{x^2-1}{x^3}, & \text{if } x\neq 0\\ 0, & \text{if } x=0\end{cases}$$

is concave up and where it is concave down. Then find all points of inflection, if any, and sketch the graph.

(b) Give three functions f, g, and h that are twice differentiable except perhaps at $x=0$, have a point of inflection at the origin, and $f'(0)=0$, $g'(0)$ is finite but not 0, and $h'(0)$ is undefined.

(c) If $f(x)=|x|^3$, compute $f'''(0)$, if possible.

(d) If $x^2+y^2=1$, show that $\frac{d^2y}{dx^2}=-\frac{1}{y^3}$.

(e) Show that $\frac{d^n}{dx^n}\left(x^n\right)=n!$ for any $n\in N$.

3. Give an example of a function that is twice but not three times continuously differentiable.

4. Give an example of a function that is three times differentiable but not three times continuously differentiable.

5. Give an example of a function that is k times differentiable but not $k+1$ times continuously differentiable.

6. Show that *Cauchy's function*

$$f(x)=\begin{cases}\exp\left(-\dfrac{1}{x^2}\right), & \text{if } x>0\\ 0, \text{ if } x\le 0\end{cases}$$

is infinitely differentiable. See Exercise 2(k) of Section 4.1 and Exercise 13 of Section 5.4. Verify that every Taylor polynomial of f about $x=0$ is identically zero. Do the same to the function given in Example 5.3.3. See Example 8.6.2 and Exercise 9 from Section 8.6.

7. Give an example of an infinitely differentiable function f on an interval (a,∞), where a is some real number, such that $\lim_{x\to\infty} f(x)=0$ but $\lim_{x\to\infty} f'(x)\neq 0$.

8. If n is a nonnegative integer and functions f and g have nth order derivatives, prove that

$$\begin{aligned}(fg)^{(n)} &= f^{(n)}g+\binom{n}{1}f^{(n-1)}g'+\binom{n}{2}f^{(n-2)}g''+\cdots+\binom{n}{n}fg^{(n)}\\ &=\sum_{k=0}^{n}\binom{n}{k}f^{(n-k)}g^{(k)},\end{aligned}$$

where $\binom{n}{k}$ denotes the binomial coefficients. This is the *Leibniz rule* for differentiating the product.

9. If $f''(x)=0$, prove that $f(x)=ax+b$ for some real numbers a and b, that is, f is a linear function.

10. If $h(x)=(g\circ f)(x)$, show that under some suitable differentiability assumptions, we will have

$$h''(x)=g''(f(x))[f'(x)]^2+g'(f(x))f''(x).$$

11. **(a)** Prove that if a function f is uniformly differentiable on (a,b), then f' is continuous on (a,b).
(b) Is the converse of part (a) true? Explain.
(c) Is the converse of part (a) true when an interval (a,b) is replaced by $[a,b]$? Explain.

12. Show that $f(x)=\dfrac{1}{x}$ is uniformly differentiable on (a,∞) for any $a>0$. Is f uniformly differentiable on $(0,\infty)$? Explain.

13. Find all roots of the given functions and determine their multiplicity.

(a) $f(x)=x^{\frac{2}{3}}$ **(b)** $f(x)=e^x-x-1$

(c) $f(x)=\begin{cases}x^2, \text{ if } x\ge 0\\ -\sqrt[3]{x}, \text{ if } x<0\end{cases}$ **(d)** $f(x)=\begin{cases}3(x-1)^2, \text{ if } x\ge 1\\ 0, \text{ if } x<1\end{cases}$

14. Prove that if a function f is continuously differentiable and has a root $x = r$ of multiplicity m, $m > 1$, $m \in N$, then the function $g(x) = \dfrac{f(x)}{f'(x)}$ has a simple root at $x = r$.

15. If $f(a) = 0$ and $x = a$ is a point of inflection of a function f, does it mean that $x = a$ must be a triple root of f? Explain.

16. Complete the details in the proof of Theorem 5.3.8.

17. Explain why Taylor's theorem is indeed a generalization of the mean value theorem.

18. Prove Taylor's theorem using Cauchy's mean value theorem.

19. Write Taylor's formula for $f(x) = e^x$ centered about $x = 0$.

20. Find the nth Taylor polynomial for the given functions centered about $x = a$.

(a) $f(x) = \dfrac{1}{1-x}$, $a = 0$ **(b)** $f(x) = \sin x$, $a = 0$

(c) $f(x) = \cos x$, $a = 0$ **(d)** $f(x) = \ln x$, $a = 1$

21. Use Taylor's theorem to approximate the given values to four decimal place accuracy.

(a) $\sin 92°$ **(b)** $\cos 58°$

22. **(a)** Use Taylor's theorem to prove $1 - \dfrac{x^2}{2} \le \cos x$ for all $x \in \Re$. See Exercise 20(i) in Section 5.2.

(b) Prove that $x \le \dfrac{\pi}{2} \sin x$ for $0 \le x \le \dfrac{\pi}{2}$. See Problem 5.6.24, part (m).

(c) Use concavity of the function $f(x) = \ln \dfrac{1}{x}$ to prove the inequality given in Exercise 20(h) of Section 5.2.

23. Show the binomial theorem is a special case of Taylor's theorem.

24. Could Taylor's polynomial $p_n(x)$ be written as $\displaystyle\sum_{k=0}^{n} \frac{f^{(k)}(a)}{k!}(x-a)^k$? Explain.

25. Consider a function f for which $f(0) = 0$ and $f''(x) \le 0$ for all $x > 0$. If $a \ge 0$ and $b \ge 0$, then prove that $f(a+b) \le f(a) + f(b)$. See Definition 4.6.18.

26. Follow the given steps in reproving the binomial theorem. That is, prove that

$$(a+b)^n = \sum_{k=0}^{n} \binom{n}{k} a^k b^{n-k},$$

where $a, b \in \Re$, $n = 0, 1, 2, \ldots$ and $\dbinom{n}{k}$ is a binomial coefficient. See Theorem 1.3.7.

(a) Without using the binomial theorem, establish that if $p(x) = x^n$, then $p'(x) = nx^{n-1}$. See Exercise 2(g) in Section 5.1.

(b) If f is a polynomial of degree n and $x^0 = 1$ even if $x = 0$, then show that we can write

$$f(x) = f(0) + \sum_{k=1}^{n} \frac{f^{(k)}(0)}{k!} x^k = \sum_{k=0}^{n} \frac{f^{(k)}(0)}{k!} x^k.$$

(c) If $g(x) = (x+b)^n$, verify that $g^{(k)}(0) = (k!)\binom{n}{k} b^{n-k}$ for natural numbers $1 \le k \le n$ and $b \in \Re$.

(d) Conclude that $(a+b)^n$ is given as requested.

27. **(a)** *(Jensen's Inequality)* Suppose that a twice differentiable function f is concave downward on an interval (a,b). If $x_i \in (a,b)$ for $i = 1, 2, \ldots, n,$ prove that

$$\frac{f(x_1) + f(x_2) + \cdots + f(x_n)}{n} \le f\left(\frac{x_1 + x_2 + \cdots + x_n}{n}\right),$$

with the equality only when $x_1 = x_2 = \cdots = x_n$. Furthermore, if f is concave upward on (a,b), prove that the above inequality is reversed. See Theorem 1.3.15.

(b) Use part (a) to prove that $\sin x_1 + \sin x_2 \le 2 \sin \frac{x_1 + x_2}{2}$, for $x_1, x_2 \in (0, \pi)$ with the equality only if $x_1 = x_2$. Can you prove this inequality using trigonometric identities?

(c) Use part (a) to prove the *arithmetic-geometric mean inequality*. That is, prove that

$$\sqrt[n]{x_1 \cdot x_2 \cdot \cdots \cdot x_n} \le \frac{x_1 + x_2 + \cdots + x_n}{n},$$

with $x_1, x_2, \ldots, x_n \in \Re^+$. See Exercise 9 in Section 1.3.

(d) Use part (a) to prove that $\left(\frac{x+y}{2}\right)^2 \le \frac{x^2 + y^2}{2}$ for all $x, y \in \Re$.

28. If a is a real number, and a function f satisfies $f'(x) < 0 < f''(x)$ for all $x < a$ and $f'(x) > 0 > f''(x)$ for all $x > a$, prove that $f'(a)$ does not exist.

29. *(Generalized Rolle's Theorem)* Suppose the function f is continuous on $[a,b]$, n times differentiable on (a,b), and $f(x_0) = f(x_1) = \cdots = f(x_n) = 0$ for some $x_0, x_1, \ldots, x_n \in [a,b]$ with $n \in N$. Prove that there exists $c \in (a,b)$ such that $f^{(n)}(c) = 0$. Compare this with Exercise 18 in Section 5.2.

30. *(Second Derivative Test)* Suppose f'' is continuous on an open interval containing a critical point c such that $f'(c) = 0$.

(a) If $f''(c) > 0$, prove that f has a local minimum at $x = c$.

(b) If $f''(c) < 0$, prove that f has a local maximum at $x = c$.

(c) If $f''(c) = 0$, show that f may or may not have an extremum at $x = c$.

31. **(a)** Use the second derivative test to find the extreme values of the function $f(x) = x^4 - \frac{7}{2}x^2 + 3x + \frac{1}{2}$.

(b) Use the first derivative test to find the extreme values of f given in part (a).

(c) Find the extreme values of $f(x) = x^{\frac{2}{3}}(1-x)^{\frac{4}{3}}$. See Exercise 15(e) in Section 5.2.

(d) State some advantages and disadvantages of the second derivative test.

5.4* L'Hôpital's Rule

Chapter 3 covered limits of functions in great depth. In particular, limits of the form $\frac{0}{0}$ were discussed and evaluated. Often the evaluation of those limits meant dividing expressions, factoring, expressing a value of 1 "cleverly," rationalizing numerators or denominators, and using geometry or some other methods. Sometimes, however, these methods do not work. For instance, how would we apply any of these mentioned methods in evaluating $\lim_{x \to 0} \frac{x}{1 - e^x}$? To evaluate more complicated limits and/or simplify the evaluation process of previously studied limits, say $\lim_{x \to 0} \frac{\sin x}{x}$, we introduce a new method called L'Hôpital's[8] rule. These methods, usually encountered in calculus classes, enable us to evaluate limits of functions that are not only in an *indeterminate form* $\frac{0}{0}$ but also $\frac{\infty}{\infty}$, $0 \cdot \infty$, $\infty - \infty$, 1^∞, ∞^0, and 0^0 (provided they exist). Recall that the definition of a derivative creates a limit of the form $\frac{0}{0}$.

THEOREM 5.4.1. *(L'Hôpital's Rule for the $\frac{0}{0}$ form) Suppose functions f and g are continuous on $[a,b]$ and differentiable on (a,b). Assume also that* $\lim_{x \to a^+} f(x) = 0$,

[8]Guillaume Francois Antoine Marquis De L'Hôpital (1661–1704) was born into the French nobility. L'Hôpital showed talent in mathematics quite early; at age 15, he solved a difficult problem on cycloids posed by Pascal. In 1696, L'Hôpital published the first textbook on differential calculus in which L'Hôpital's rule first appeared. Actually, however, L'Hôpital's rule and most of the material in the textbook were due to Johann Bernoulli, L'Hôpital's tutor.

$\lim_{x \to a^+} g(x) = 0$, *and* $g'(x) \neq 0$ *for* x *"near"* a.

$$(a)\quad \textit{If } \lim_{x \to a^+} \frac{f'(x)}{g'(x)} = L, \textit{ then } \lim_{x \to a^+} \frac{f(x)}{g(x)} = L.$$

$$(b)\quad \textit{If } \lim_{x \to a^+} \frac{f'(x)}{g'(x)} = +\infty, \textit{ then } \lim_{x \to a^+} \frac{f(x)}{g(x)} = +\infty.$$

$$(c)\quad \textit{If } \lim_{x \to a^+} \frac{f'(x)}{g'(x)} = -\infty, \textit{ then } \lim_{x \to a^+} \frac{f(x)}{g(x)} = -\infty.$$

Remark 5.4.2. Theorem 5.4.1 would also be valid if $x \to a^+$ were replaced by $x \to b^-$ or $x \to x_0$ for any $x_0 \in (a,b)$. In fact, if the conditions of the theorem were to be valid on suitable intervals, $x \to a^+$ could also be replaced by $x \to +\infty$ or $x \to -\infty$. Furthermore, the implication in Theorem 5.4.1 cannot be reversed. For example, $\lim_{x \to \infty} \frac{x - \cos x}{x} = 1$. Why? Note that the differentiation step from Theorem 5.4.1 would yield $\lim_{x \to \infty} \frac{1 + \sin x}{1}$, which does not exist. Why? To avoid erroneous results, it is essential to check the conditions of Theorem 5.4.1 and the theorems that follow, before applying L'Hôpital's rule. ■

Example 5.4.3. Compute $\lim_{x \to 0} \frac{x}{1 - e^x}$, if possible.

Answer. Since all conditions of Theorem 5.4.1 are satisfied, we differentiate the numerator and the denominator of $\frac{x}{1 - e^x}$ separately, to obtain

$$\lim_{x \to 0} \frac{x}{1 - e^x} \underset{\text{L'H}}{=} \lim_{x \to 0} \frac{1}{-e^x} = -1.$$ ■

Proof of Theorem 5.4.1, part (a). Theorem 3.2.6 and Cauchy's mean value theorem are keys to the proof. Begin with any sequence $\{x_n\}$ such that $a < x_n < b$ and $\lim_{n \to \infty} x_n = a$, and show that $\lim_{n \to \infty} \left(\frac{f}{g}\right)(x_n) = L$, provided that $\lim_{n \to \infty} \frac{f'(t_n)}{g'(t_n)} = L$ for any sequence $\{t_n\}$ tending to a with $a < t_n < b$. For each $n \in N$ fixed at a time, we apply Cauchy's mean value theorem to the functions f and g on the interval $[a, x_n]$. Thus, there exists a $c \in (a, x_n)$, for which

$$\frac{f(x_n) - f(a)}{g(x_n) - g(a)} = \frac{f'(c_n)}{g'(c_n)}.$$

Therefore, $\frac{f(x_n)}{g(x_n)} = \frac{f'(c_n)}{g'(c_n)}$. Why? Hence, since $\lim_{n\to\infty} \frac{f'(t_n)}{g'(t_n)} = L$ for any sequence $\{t_n\}$, where $\lim_{n\to\infty} t_n = a$ and $a < t_n < b$, we can write

$$\lim_{n\to\infty} \frac{f(x_n)}{g(x_n)} = \lim_{n\to\infty} \frac{f'(c_n)}{g'(c_n)} = L$$

for $a < c_n < x_n < b$. Part (a) is proven.

Proof of Theorem 5.4.1, part (b). Assume that $\lim_{x\to a^+} \frac{f'(x)}{g'(x)} = +\infty$. Then the definition of a limit is used to show that $\lim_{x\to a^+} \frac{f(x)}{g(x)} = +\infty$. Let an arbitrary positive real number M be given. We must find $\delta > 0$, so that if $a < x < a+\delta$, then $\frac{f(x)}{g(x)} > M$. In addition, by definition, $\lim_{x\to a^+} \frac{f'(x)}{g'(x)} = +\infty$ means that $\frac{f'(x)}{g'(x)} > M^*$ for any positive real M^*, if $a < x < a+\delta_0$, for some $\delta_0 > 0$. In particular, this is true for $M = M^*$. Now choose $\delta = \delta_0$ and let $x \in (a, a+\delta_0)$. Applying Cauchy's mean value theorem to f and g on $[a, x]$, we obtain a $c \in (a, x)$ such that

$$\frac{f'(c)}{g'(c)} = \frac{f(x) - f(a)}{g(x) - g(a)},$$

which in turn equals $\frac{f(x)}{g(x)}$. Since $\frac{f'(c)}{g'(c)} > M$, we have $\frac{f(x)}{g(x)} > M$. The proof is complete. ■

THEOREM 5.4.4. *Suppose the function f is twice differentiable on (a,b) and $x \in (a,b)$. Then*

$$f''(x) = \lim_{h\to 0} \frac{f(x+h) - 2f(x) + f(x-h)}{h^2}.$$

Proof. Evaluate the limit by applying L'Hôpital's rule and employing Example 5.1.11. Details are left to the reader. ■

Remark 5.4.5. Note that the preceding limit does not define f''. The statement is not an "if and only if" statement. Verify that the function $f(x) = x|x|$ satisfies the preceding limit and that f' is not differentiable. To be more precise, $f''(0)$ does not exist. Why? Refer to Part 1 of Section 5.6 and to numerical analysis for further details on this central difference. ■

THEOREM 5.4.6. *(L'Hôpital's Rule for the $\frac{\infty}{\infty}$ form) Suppose functions f and g are differentiable on (a,b), with $g'(x) \neq 0$ for x "near" a. Assume that $\lim_{x\to a^+} f(x) = +\infty$ and that $\lim_{x\to a^+} g(x) = +\infty$. If $\lim_{x\to a^+} \frac{f'(x)}{g'(x)} = L$, $+\infty$, or $-\infty$, then $\lim_{x\to a^+} \frac{f(x)}{g(x)}$ has that same value.*

Note that as in Remark 5.4.2, we can rewrite the Theorem 5.4.6 (under suitable conditions) to have the same conclusion in cases where

$$\begin{array}{lcl} \lim_{x\to a^+} f(x) = \pm\infty & \text{and} & \lim_{x\to a^+} g(x) = \pm\infty \\ \lim_{x\to a^-} f(x) = \pm\infty & \text{and} & \lim_{x\to a^-} g(x) = \pm\infty \\ \lim_{x\to +\infty} f(x) = \pm\infty & \text{and} & \lim_{x\to +\infty} g(x) = \pm\infty \\ \lim_{x\to -\infty} f(x) = \pm\infty & \text{and} & \lim_{x\to -\infty} g(x) = \pm\infty. \end{array}$$

Clearly, much remains to be proven. The proofs are omitted.

Remark 5.4.7. The indeterminate form $\frac{\infty}{\infty}$ is very useful in determining which expression, numerator or denominator, approaches infinity "more quickly." For example, consider the two functions f and g, where $f(x) = (\ln x)^n$, with any fixed natural number n, and $g(x) = x^r$, with any fixed positive real number r. It can be shown, as in Exercise 1(i), that x^r goes to infinity more quickly than $(\ln x)^n$ for any choices of r and n that fit the preceding restrictions. Therefore, x^r is of higher order. See Section 2.3. Thus, provided that x is sufficiently large, $(\ln x)^n < x^r$, no matter how large n and how small r. This inequality can be proven by showing that $\lim_{x\to\infty} \frac{(\ln x)^n}{x^r} = 0$. Why? Also verify that $\lim_{x\to\infty} \frac{x^n}{e^x} = 0$ for any fixed $n \in N$. This means that, provided that x is sufficiently large, $x^n < e^x$ for any choice of $n \in N$. Using this approach, we can verify that growing exponential functions approach infinity more quickly than do polynomials with positive leading coefficients. This proves important when analyzing economics problems. In finance and banking, the exponential growth is sometimes to our advantage. The limits mentioned here are very useful when we study infinite series in Chapter 8. Also, see Remark 5.4.13. ■

L'Hôpital's rules can also be stated for sequences. One version is stated without proof below.

THEOREM 5.4.8. *Consider the sequences $\{a_n\}$ and $\{b_n\}$, where $\{b_n\}$ increases to $+\infty$. Then*

$$\lim_{n\to\infty} \frac{a_{n+1}-a_n}{b_{n+1}-b_n} = L \textit{ implies that } \lim_{n\to\infty} \frac{a_n}{b_n} = L.$$

Remark 5.4.9. Recall from Theorem 3.1.6 that if f is a function, where $n \in N$, $x \in \Re$, and $\lim_{x\to\infty} f(x)$ exists, then so does $\lim_{n\to\infty} f(n)$, and the values are equal. This implication, however, cannot be reversed. For instance, if $f(x) = \sin \pi x$, then $\lim_{n\to\infty} \sin \pi n = 0$, but $\lim_{x\to\infty} \sin \pi x$ does not exist (see Figure 3.1.2). All this means is that we can employ L'Hôpital's rules to evaluate limits of sequences ($n \to \infty$) by replacing n with x and evaluating a stronger limit. If the stronger limit does not exist, then the existence of the limit of the sequence is unknown. For instance, to evaluate $\lim_{n\to\infty} x_n$, where $x_n = \dfrac{\ln \sin \frac{1}{n}}{\ln \tan \frac{1}{n}}$, we first calculate $\lim_{x\to\infty} \left(\dfrac{\ln \sin \frac{1}{x}}{\ln \tan \frac{1}{x}} \right)$ to obtain the value of 1 for the desired limit of the sequence. Details are left to the reader. Sometimes, of course, we do not wish to use this approach. For example, to evaluate $\lim_{n\to\infty} \dfrac{n^n}{n!}$ (see Exercise 11(b) of Section 2.3), we do not shift to real numbers in hopes of applying L'Hôpital's rule. Why not? ■

With the use of algebraic manipulations and/or logarithmic and exponential functions, we attempt to evaluate limits that have other indeterminate forms: $0 \cdot \infty$, $\infty - \infty$, 1^∞, ∞^0, and 0^0. To avoid tedious and uninspiring statements, the techniques will be demonstrated only by examples. Remember, L'Hôpital's rule for evaluating the limit of $\dfrac{f}{g}$ involves $\dfrac{f'}{g'}$, not $\left(\dfrac{f}{g}\right)'$. Moreover, L'Hôpital's rules apply only to indeterminate forms $\dfrac{0}{0}$ or $\dfrac{\pm\infty}{\pm\infty}$.

Example 5.4.10. Evaluate the $\lim_{x\to\infty} x \ln\left(1 + \dfrac{1}{x}\right)$, which has an indeterminate form of $0 \cdot \infty$.

Answer. We can write that

$$\lim_{x\to\infty} x \ln\left(1+\frac{1}{x}\right) = \lim_{x\to\infty} \frac{\ln\left(1+\frac{1}{x}\right)}{\frac{1}{x}} \underset{\text{L'H}}{=} \lim_{x\to\infty} \frac{\frac{1}{1+1/x}\left(-\frac{1}{x^2}\right)}{-\frac{1}{x^2}} = \lim_{x\to\infty} \frac{1}{1+1/x} = 1. \quad ■$$

After combining and rewriting, the indeterminate form $\infty - \infty$ is handled in a similar fashion, resulting in the form of either $\dfrac{0}{0}$ or $\dfrac{\pm\infty}{\pm\infty}$.

Example 5.4.11. Evaluate the $\lim_{x\to\infty}\left(1+\frac{r}{x}\right)^x$, when $r\in\Re$, if possible. See Exercise 10(a) from Section 3.1.

Answer. Note that the indeterminate form here is 1^∞. This expression needs to be rewritten, if possible, in the form of either $\frac{0}{0}$ or $\frac{\pm\infty}{\pm\infty}$.

Procedure 1. Define y to be

$$y=\lim_{x\to\infty}\left(1+\frac{r}{x}\right)^x.$$

Then, due to the continuity of the logarithmic function (see Remark 4.1.5), we have that

$$\begin{aligned}\ln y &=\ln\left[\lim_{x\to\infty}\left(1+\frac{r}{x}\right)^x\right]=\lim_{x\to\infty}\ln\left(1+\frac{r}{x}\right)^x=\lim_{x\to\infty}x\left(1+\frac{r}{x}\right)\\ &=\lim_{x\to\infty}\frac{\ln(1+r/x)}{1/x}\underset{\text{L'H}}{=}\lim_{x\to\infty}\frac{\left(-\frac{r}{x}\right)\frac{1}{1+r/x}}{-1/x^2}=\lim_{x\to\infty}\frac{r}{1+r/x}=r.\end{aligned}$$

Thus, we have that $y=e^r$. See Part 1 of Section 2.8 and Exercise 10 of Section 3.1.

Procedure 2. Here, we write that

$$\lim_{x\to\infty}\left(1+\frac{r}{x}\right)^x=\lim_{x\to\infty}\exp\left[\ln\left(1+\frac{r}{x}\right)^x\right]=\exp\left[\lim_{x\to\infty}\ln\left(1+\frac{r}{x}\right)^x\right]=\ldots=\exp(r)=e^r.\ \blacksquare$$

Remark 5.4.12. Note that if $r=1$, then $\lim_{x\to\infty}\left(1+\frac{r}{x}\right)^x$ would define the irrational number e. Likewise, if x were replaced by $\frac{1}{x}$, then $\lim_{t\to 0}(1+t)^{\frac{1}{t}}=e$. See Exercise 17 of Section 3.3. Indeterminate forms of ∞^0 and 0^0 are handled in a similar fashion. Keep in mind that even though $\infty+\infty$, $-\infty-\infty$, $\infty\cdot\infty$, 0^∞, $0^{-\infty}$, ∞^1, and ∞^∞ may seem indeterminate, they are not. Their values are given by ∞, $-\infty$, ∞, 0, ∞, ∞, and ∞, respectively, provided that the functions are appropriately defined. ■

Remark 5.4.13. The following very common limits can be proven using certain inequalities, or L'Hôpital's rule. Also see Remark 5.4.7.

(a) $\lim_{x\to 0^+}x^p\ln x=0$, with p any positive constant.

(b) $\lim_{x\to\infty}\frac{\ln x}{x^p}=0$, with p any positive constant.

(c) $\lim_{x\to\infty}\frac{x^p}{e^x}=0$, with p any constant. (Also see Example 2.3.8.)

(d) $\lim\limits_{x\to\infty}\dfrac{a^x}{b^x}=\begin{cases}0, \text{ if } a<b\\ \infty, \text{ if } a>b,\end{cases}$ where a and b are positive constants. (See Exercise 38 in Section 5.2.)

Exercises 5.4

1. Evaluate the given limits, if possible.

(a) $\lim\limits_{x\to1}\dfrac{1-x+\ln x}{x^3-3x+2}$ **(b)** $\lim\limits_{x\to\infty}\dfrac{x^{-\frac{4}{3}}}{\sin\frac{1}{x}}$

(c) $\lim\limits_{x\to0^-}\dfrac{\tan x}{x^2}$ **(d)** $\lim\limits_{x\to0^+} x\ln x$

(e) $\lim\limits_{x\to\infty} x\sin\dfrac{k}{x}$, with $k\in\Re$ (See Remark 2.2.4 and Example 3.3.8.)

(f) $\lim\limits_{x\to\infty}\left(\sqrt{x^4+5x^2-1}-x^2\right)$ **(g)** $\lim\limits_{x\to\infty}(\sinh x-x)$

(h) $\lim\limits_{x\to0^+}\dfrac{\ln x}{\csc x}$ **(i)** $\lim\limits_{x\to\infty}\dfrac{(\ln x)^n}{x^r}$, $n\in N$, $r\in(0,\infty)$

(j) $\lim\limits_{x\to\infty}\dfrac{x^n}{e^x}$ where $n\in N$ **(k)** $\lim\limits_{x\to\infty}\left(\dfrac{x}{x+1}\right)^x$

(l) $\lim\limits_{x\to0}(1-\sin x)^{\frac{1}{x}}$ **(m)** $\lim\limits_{x\to\infty}\left(\sin\dfrac{1}{x}\right)^x$

(n) $\lim\limits_{x\to0^-}\sqrt[x]{x}$ **(o)** $\lim\limits_{n\to\infty} n\left(\sqrt[n]{2}-1\right)$, $n\in N$

(p) $\lim\limits_{x\to0^+} x^x$ (See Exercise 15(d) from Section 5.2.)

(q) $\lim\limits_{x\to0^+}(2x)^{\frac{1}{x}}$ **(r)** $\lim\limits_{n\to\infty}\left(1+\dfrac{k}{n^2}\right)^n$, $k\in\Re$

(s) $\lim\limits_{x\to0^+}\arctan(x\ln x)$ **(t)** $\lim\limits_{x\to0}\dfrac{\sqrt{-x^2-2x}}{x}$

2. Prove Theorem 5.4.1, part (c).

3. Use Theorem 5.4.1 to prove that if functions f and g are differentiable with $g'(x)\neq0$ on (M,∞) for some $M\in\Re$, and if functions f and g satisfy $\lim\limits_{x\to\infty} f(x)=\lim\limits_{x\to\infty} g(x)=0$, then

$$\lim_{x\to\infty}\frac{f(x)}{g(x)}=\lim_{x\to\infty}\frac{f'(x)}{g'(x)},$$

provided that the second limit exists.

4. Complete the proof of Theorem 5.4.4. Is the converse true, that is, if the given limit is satisfied at $x = x_0$, then is f twice differentiable at $x = x_0$? Explain.

5. **(a)** Suppose the function f is twice differentiable. Prove that $f''(x) \geq 0$ for all x if and only if

$$f\left(\frac{a+b}{2}\right) \leq \frac{f(a)+f(b)}{2},$$

for all $a, b \in \Re$. Compare this result with Exercises 25 and 27 of Section 5.3. Also, see Definition 5.6.23.

(b) If $a, b \in (0, \pi)$ and $x = \frac{a+b}{2}$, prove that $\left(\frac{\sin x}{x}\right)^2 \geq \frac{\sin a}{a} \cdot \frac{\sin b}{b}$. Compare this result with Exercise 27(b) in Section 5.3.

6. Suppose the function f is twice differentiable. Prove that each of the following limits represents $f''(x)$.

(a) $\lim_{h \to 0} \frac{f(x+3h) - 3f(x+h) + 2f(x)}{3h^2}$

(b) $\lim_{h \to 0} \frac{2f(x+3h) - 3f(x+2h) + f(x)}{3h^2}$

(c) $\lim_{h \to 0} \frac{-f(x+3h) + 4f(x+2h) - 5f(x+h) + 2f(x)}{h^2}$

7. Consider $f : \Re^+ \to \Re$ defined by $f(x) = \sqrt[x]{x}$.

(a) Compute $\lim_{x \to 0^+} f(x)$. See Exercise 4(e) of Section 3.3 and Exercise 1(n) in this section.

(b) Compute $\lim_{x \to \infty} f(x)$.

(c) Determine where f is increasing and where f is decreasing.

(d) Find all relative extrema. Find $\max f$, $\min f$, $\sup f$, and $\inf f$, if possible.

(e) Find all vertical and horizontal asymptotes.

(f) Sketch the graph of f.

8. Evaluate $\lim_{x \to 0} (x \cot x)$ by writing

(a) $x \cot x$ as $(\cos x)\left(\frac{x}{\sin x}\right)$.

(b) the limit in an indeterminate form $\frac{0}{0}$.

(c) the limit in an indeterminate form $\frac{\infty}{\infty}$.

9. If possible, give examples of indeterminate form ∞^0 and 0^0 that do not yield a value of 1.

10. Explain why form 0^∞ yields a value of 0.

11. **(a)** Attempt to evaluate $\lim_{x\to\infty} \frac{\sqrt{x^2+1}}{x}$ using L'Hôpital's rule.

(b) Evaluate $\lim_{x\to\infty} \frac{\sqrt{x^2+1}}{x}$. **(c)** Evaluate $\lim_{x\to-\infty} \frac{\sqrt{x^2+1}}{x}$.

12. If a and b are any two positive real numbers, show that the geometric mean, $\sqrt{ab}$, is given by

$$\sqrt{ab} = \lim_{x\to\infty}\left(\frac{\sqrt[x]{a}+\sqrt[x]{b}}{2}\right)^x.$$

See Part 3 of Section 1.10.

13. **(a)** Prove $\lim_{x\to 0} \frac{\exp\left(-1/x^2\right)}{x^n} = 0$ for all $n \in N$.

(b) If f is Cauchy's function

$$f(x) = \begin{cases} \exp\left(-\frac{1}{x^2}\right), & \text{if } x > 0 \\ 0, \text{ if } x \le 0, \end{cases}$$

use the result of part (a) to prove $\lim_{x\to 0} f^{(k)}(x) = 0$ for $k = 0,1,2,\ldots$. See Exercise 6 in Section 5.3 and Exercise 9 in Section 8.6.

14. Consider functions f and g, defined by

$$f(x) = \begin{cases} x+2, \text{ if } x \ne 0 \\ 0, \text{ if } x = 0 \end{cases} \text{ and } g(x) = \begin{cases} x^2+2x+1, \text{ if } x \ne 0 \\ 0, \text{ if } x = 0. \end{cases}$$

Find $\lim_{x\to 0} \frac{f'(x)}{g'(x)}$ and $\lim_{x\to 0} \frac{f(x)}{g(x)}$. Does this contradict L'Hôpital's rule?

15. Consider the *sinc function* f defined by

$$f(x) = \text{sinc } x \equiv \begin{cases} \frac{\sin x}{x}, \text{ if } x \ne 0 \\ 1, \text{ if } x = 0, \end{cases}$$

which was studied in Exercise 18(b) of Section 2.3, Remark 3.2.11, and Exercise 2(c) of Section 4.1. Look ahead to Exercises 19-22 in Section 6.5, as well as Exercises 1(e) and 6(b) in Section 8.6.

(a) Is f continuous at $x = 0$? Explain.

(b) Is f differentiable at $x = 0$? If so, find $f'(0)$.

(c) How many roots does f have? What is the multiplicity of each root? Explain.

(d) What is $\sup f$? What is $\max f$? How many relative extrema are there? If the relative extremum occurs at $x = c$, show that $|f(c)| = \dfrac{1}{\sqrt{1+c^2}}$.

(e) Prove that

$$\frac{2}{\pi} = \sqrt{\frac{1}{2}} \cdot \sqrt{\frac{1}{2} + \frac{1}{2}\sqrt{\frac{1}{2}}} \cdot \sqrt{\frac{1}{2} + \frac{1}{2}\sqrt{\frac{1}{2} + \frac{1}{2}\sqrt{\frac{1}{2}}}} \cdot \cdots .$$

This analytical procedure of approximating π using "continued roots" was first given by Viète[9] in 1593.

(f) Evaluate the infinite product

$$\frac{1}{2}\sqrt{\frac{1}{2} + \frac{1}{2} \cdot \frac{1}{2}}\sqrt{\frac{1}{2} + \frac{1}{2}\sqrt{\frac{1}{2} + \frac{1}{2} \cdot \frac{1}{2}}}\sqrt{\frac{1}{2} + \frac{1}{2}\sqrt{\frac{1}{2} + \frac{1}{2}\sqrt{\frac{1}{2} + \frac{1}{2} \cdot \frac{1}{2}}}} \cdots .$$

(g) If x is a measure of an angle in degrees instead of radians, calculate $\lim_{x \to 0} \dfrac{\sin x}{x}$ and a derivative of $\sin x$. See Remark 5.2.7.

16. If $f'(c)$ exists and if $\lim_{x \to c} f'(x)$ is finite with c in an interval (a, b), prove that f' is continuous at $x = c$. Is the result still true if $\lim_{x \to c} f'(x)$ is not finite? Explain.

17. Consider the sequence $\{a_n\}$ with $a_n = \sqrt[n]{\dfrac{3n+2}{n+1}}$.

(a) Is $\{a_n\}$ eventually monotone? Explain.

(b) Does $\{a_n\}$ converge? If so, find the limit.

18. Will functions $y_1 = 1.0000001^x$ and $y_2 = x^{1000000}$ intersect? If so, how many times? And if so, any idea where?

19. Suppose that f is a function that is twice continuously differentiable and $f(0) = 0$. If $g(x) = \dfrac{f(x)}{x}$ for $x \neq 0$ and $g(0) = f'(0)$, prove that g is continuously differentiable. Compare this result to Exercise 30 in Section 5.2.

---

[9]François Viète (1540–1603) was a French algebraist, arithmetist, cryptanalyst, and geometer. He used letters to represent both constants and variables. Viète was given credit for finding a trigonometric solution of a general cubic equation in one variable.

5.5* Review

Label each statement as true or false. If a statement is true, prove it. If not,

(**i**) give an example of why it is false, and

(**ii**) if possible, correct it to make it true, and then prove it.

1. If $f(x)=(4x-1)^x$, then $f'(x)=x(4x-1)^{x-1}(4)$.

2. The function $f(x)=x^3+x-\sqrt{2}$ has exactly one real root.

3. If a function f is differentiable and oscillatory, then f' must also be oscillatory.

4. If the function f is differentiable at $x=a$, then $f'(a)$ is unique.

5. If the function f is defined by $f(x)=\begin{cases}2x+1, & \text{if } x\le 0\\ x^2+2x, & \text{if } x>0,\end{cases}$ then $f'(0)=2$.

6. If a function f^2 is differentiable, then f must be differentiable.

7. If the function f is defined by

$$f(x)=\begin{cases}x^2\sin\dfrac{1}{x^2}, & \text{if } x\ne 0\\ 0, & \text{if } x=0,\end{cases}$$

then f is differentiable and f' is unbounded on $[-1,1]$.

8. If the function f is defined by

$$f(x)=\begin{cases}x^4\left[\exp\left(-\dfrac{x^2}{4}\right)\right]\sin\dfrac{8}{x^3}, & \text{if } x\ne 0\\ 0, & \text{if } x=0,\end{cases}$$

then f' exists and is bounded on $[-1,1]$, but f' does not attain its absolute extrema on $[-1,1]$.

9. A function f differentiable at $x=a$ implies that there exists $\varepsilon>0$ such that f is differentiable on $(a-\varepsilon,a+\varepsilon)$.

10. If a function f is continuous on $[a,b]$, differentiable on (a,b), and $f(a)=f(b)=0$, then for each real number α there is some $c\in(a,b)$ such that $f'(c)=\alpha[f(c)]^2$. Compare this problem with Exercise 19 of Section 5.2.

11. It is possible to have a function that is differentiable everywhere but whose derivative is not continuous everywhere.

12. If f' is bounded, then f is bounded.

13. It is possible to have one point that is both a relative maximum and an inflection point for a function.

14. If we have an inflection point at $x=a$, then $f'(a)$ must be zero.

15. If f is a twice differentiable function in a neighborhood of $x = a$ at which point f has a double root, then f attains a relative extremum at $x = a$.

16. If $\lim_{x\to\infty} f'(x) = 1$ then $\lim_{x\to\infty} \frac{f(x)}{x} = 1$.

17. If a function f is twice differentiable, then

$$f''(x) = \lim_{h\to 0} \frac{f(x+h) - [f(x) + hf'(x)]}{h^2}.$$

18. A limit involving the forms $\frac{0}{0}$ or ∞^0 must yield the value of 1.

19. The $\lim_{x\to\infty}\left(x - \sqrt{x^2 + x}\right) = 0$.

20. All four of the following limits give the same result:

$$\lim_{x\to\infty}\left(1+\frac{1}{x}\right)^{2x};\ \lim_{x\to\infty}\left(1+\frac{1}{2x}\right)^{x};\ \lim_{x\to\infty}\left(1+\frac{2}{x}\right)^{x};\ \lim_{x\to\infty}\left[\frac{1}{x}+\frac{1}{2}\exp\left(\frac{2}{x}\right)\right]^{x}$$

21. The function $f(x) = x^3 + 4x^2 - x - 2$ is one to one on an interval $[-1,0]$.

22. $\sqrt{1+x} < 1 + \frac{1}{2}x$ for $x > 0$.

23. Consider a family of functions f_n with $n = 0, 1, 2, \ldots$, where

$$f_n(x) = \begin{cases} x^n \sin\frac{1}{x}, & \text{if } x \neq 0 \\ 0, & \text{if } x = 0. \end{cases}$$

Then

(a) f_0 is continuous and differentiable everywhere except at $x = 0$.

(b) f_0 is not continuous nor differentiable at $x = 0$.

(c) f_1 is continuous everywhere and differentiable everywhere except at $x = 0$.

(d) f_2 is continuous and differentiable everywhere, but f_2' is not continuous at zero.

(e) f_3 is continuous and differentiable everywhere, and f_3' is continuous everywhere but not differentiable at $x = 0$.

24. If L is the tangent line to a function f at a point $x = a$, then

(a) L does not intersect f at any point.

(b) L intersects f at at most one point.

(c) L cannot cross f at $x = a$.

(d) L cannot cross f at any other point than $x = a$.

25. Suppose that a function f is twice differentiable on an interval $[a,b]$, a sequence $\{x_n\}$ of mutually distinct points of $[a,b]$ converges to t, and $f(x_n) = 0$ for all $n \in N$. Then $f(t) = f'(t) = f''(t) = 0$.

26. If a, b, and c are positive real numbers, then

$$\lim_{h\to 0}\left(\frac{a^h+b^h+c^h}{3}\right)^{\frac{1}{h}}=\sqrt[3]{abc}.$$

27. $$\lim_{x\to 0^+} x^{x^{x^{\cdot^{\cdot^{\cdot^x}}}}}=\begin{cases}0, & \text{if the number of } x\text{'s is odd}\\ 1, & \text{if the number of } x\text{'s is even.}\end{cases}$$

28. If $f(x)=ax^3+bx^2+cx+d$, $a\neq 0$, has a local minimum, then f has a local maximum and the extrema are located at $x=\dfrac{-b\pm\sqrt{b^2-3ac}}{3a}$.

29. L'Hôpital's rule, when it applies, always makes evaluation of a limit easier and faster.

30. If f is differentiable, then $f'(x)=\lim\limits_{h\to 0}\dfrac{-4f(x+3h)+8f(x+2h)-5f(x)}{6h}$.

31. If $\lim\limits_{n\to\infty} n\left[f\left(a+\dfrac{1}{n}\right)-f(a)\right]$ is finite, then f must be differentiable at $x=a$.

32. If functions f and g are differentiable and $f(x)g(x)=1$, then $\dfrac{f'(x)}{f(x)}+\dfrac{g'(x)}{g(x)}=0$.

33. A function f exists that does not satisfy all conditions of Rolle's theorem, yet its conclusion holds.

34. If a function f is continuous and one to one, then f must be differentiable.

35. If a function f is differentiable and $f'(a)=0$ at some point $x=a$, then f cannot be one to one.

36. If a function f is increasing on an interval (a,b), then f has to have a nonnegative derivative at every point in (a,b).

37. If a function f is differentiable and uniformly continuous on an interval (a,b), then f' must be bounded on (a,b).

38. If functions f and g are differentiable and $g'(x)\neq 0$ for any $x\in\Re$, and $\lim\limits_{x\to a}\dfrac{f'(x)}{g'(x)}$ does not exist, where a is a fixed real number or $\pm\infty$, then $\lim\limits_{x\to a}\dfrac{f(x)}{g(x)}$ does not exist.

39. If $f:(a,b)\to\Re$ has a relative extremum at $c\in(a,b)$, then either $f'(c)=0$ or $f'(c)$ does not exist.

40. If f is twice differentiable on (a,b), with $f'(x)>0$ for all $x\in(a,b)$ except that $f'(c)=0$ with $a<c<b$, then f has an inflection point at $x=c$.

41. If f is twice continuously differentiable on (a,b), with $f'(x)>0$ for all $x\in(a,b)$ except that $f'(c)=0$ with $a<c<b$, then f has an inflection point at $x=c$.

42. If a function f is continuously differentiable on an interval I, then f must be uniformly continuous on I.

43. The function $f(x)=e^{-x}-\frac{1}{2}x^2+x-1$ has a root of multiplicity 3 at $x=0$.

44. If $0<a<b$, then $\frac{x+a-4b}{b}<\left(\frac{4a}{x-a}\right)\ln\left(\frac{2a}{x+a}\right)<\frac{x-3a}{b}$, for all $x\in(a,b)$.

45. If $f'(x)>g'(x)$ for all $x\in(a,b)$, then $f(x)>g(x)$ for all $x\in(a,b)$.

46. Suppose that the function f is continuous on $[0,\infty)$ and $f(0)=0$. If $f'(x)>0$ for all $x>0$, then $f(x)>0$ for all $x>0$.

47. Suppose that a function f is differentiable on $(0,1]$ with $|f'(x)|<1$. If $\{a_n\}$ is a sequence with $a_n=f\left(\frac{1}{n}\right)$, $n\in N$, then $\lim_{n\to\infty}a_n$ must be finite.

48. If $x=a$ is a triple root of a three times differentiable function f, then f must have an inflection point at $x=a$.

49. For a natural number n we have $n^{\sqrt{n+1}}<(n+1)^{\sqrt{n}}$, if $n<7$ and $n^{\sqrt{n+1}}>(n+1)^{\sqrt{n}}$, if $n\geq 7$.

50. The function $r(x)=|x|$ is not a rational function. See Exercise 8 of Section 3.4 and Exercise 3 of Section 4.5.

51. If f is twice differentiable then $f''(x)=\lim_{h\to 0}\frac{f(x+h)-2f(x+2h)+f(x+3h)}{3h^2}$.

52. It is possible for f' to have more roots than f has.

5.6* Projects

Part 1. Approximation of Derivatives

If the function f is twice differentiable, then we will have

$$f(x)=f(a)+f'(a)(x-a)+\frac{f''(c)}{2}(x-a)^2,$$

for some c between a and x. Why? Thus, we can rewrite this as

$$f(x+h)=f(x)+hf'(x)+\frac{h^2}{2}f''(c),$$

with $h\in\Re$. Why? Solving for $f'(x)$, we have that

$$f'(x)=\frac{f(x+h)-f(x)}{h}+\frac{h}{2}f''(c).$$

If h approaches 0, the "error" term $\frac{h}{2}f''(c)$ must also approach 0. Why? The preceding difference quotient can be rewritten as $\frac{\Delta f(x)}{h}+\mathcal{O}(h)$ since the highest power of h in the error term involved is h. Expression $\Delta f(x)=f(x+h)-f(x)$ is called the *first forward difference of* f. Write $f(x-h)$ in a similar way and obtain

$$f'(x)=\frac{f(x)-f(x-h)}{h}+\frac{h}{2}f''(c)=\frac{\nabla f(x)}{h}+\mathcal{O}(h),$$

where $\nabla f(x)=f(x)-f(x-h)$ is called the *first backward difference of* f. See Section 5.1.

Problem 5.6.1

(a) Find a constant c so that

$$f'(x)=\frac{3f(x+h)-f(x)-2f(x-h)}{ch}+\mathcal{O}(h).$$

(b) Write $f(x+h)$ and $f(x+2h)$ using Taylor's theorem. Then, verify that under suitable assumptions

$$f''(x)=\frac{f(x+2h)-2f(x+h)+f(x)}{h^2}+\mathcal{O}(h).$$

Observe that if we were to define $\Delta^2 f(x)=\Delta(\Delta f(x))$, then we could rewrite the preceding as

$$f''(x)=\frac{\Delta^2 f(x)}{h^2}+\mathcal{O}(h).$$

We can do the same with backward differences. In general, we can define that $\Delta^n f(x)=\Delta\left(\Delta^{n-1}f(x)\right)$ and $\nabla^n f(x)=\nabla\left(\nabla^{n-1}f(x)\right)$. Thus, we would obtain that

$$\frac{d^n f}{dx^n}=\frac{\Delta^n f(x)}{h^n}+\mathcal{O}(h) \text{ and } \frac{d^n f}{dx^n}=\frac{\nabla^n f(x)}{h^n}+\mathcal{O}(h).$$

Problem 5.6.2. Show that the "operators" Δ and ∇ are linear, that is, show that

$$\Delta(af(x)+bg(x))=a\Delta f(x)+b\Delta g(x) \text{ and}$$
$$\nabla(af(x)+bg(x))=a\nabla f(x)+b\nabla g(x),$$

where f and g are functions and $a,b\in\Re$ are constants. See Section 11.3.

Problem 5.6.3. Expand $\Delta^5 f(x)$ and $\nabla^5 f(x)$.

Problem 5.6.4. Why do the coefficients in Problem 5.6.3 add up to zero? See Exercise 10 of Section 5.1. Do you see any pattern? Use the pattern to expand $\Delta^9 f(x)$ and $\nabla^9 f(x)$.

Sometimes combinations of formulas might give better results. For instance, a "biased" representation can be used to write that

$$f'(x) = \frac{\frac{1}{3}\nabla f(x) + \frac{2}{3}\Delta f(x)}{h} + \mathcal{O}(h).$$

Why? Similarly, Exercise 10(b) in Section 5.1 states that

$$f'(x) \approx \frac{\frac{2}{5}\nabla f(x) + \frac{3}{5}\Delta f(x)}{h}.$$

The value will not, however, be much more accurate unless we derive formulas that involve $\mathcal{O}(h^2)$. This can be done for f' by combining $f(x+h)$ and $f(x-h)$ to yield

$$f'(x) = \frac{f(x+h) - f(x-h)}{2h} + \mathcal{O}(h^2).$$

See Example 5.1.11.

The *central difference* is given by $\delta f(x) \equiv f\left(x + \frac{1}{2}h\right) - f\left(x - \frac{1}{2}h\right)$, which does not incorporate the value of $f(x)$ itself. Also, $\delta f\left(x + \frac{1}{2}h\right) = f(x+h) - f(x)$ and $\delta\left(\delta^n f\right) = \delta^{n+1} f$. (See Exercises 12(c) and 12(d) in Section 11.3.) This gives

$$\frac{d^n f}{dx^n} = \frac{\Delta^n f\left(x - \frac{n}{2}h\right) + \nabla^n f\left(x + \frac{n}{2}h\right)}{2h^n} + \mathcal{O}(h^2), \quad \text{if } n \text{ is even, and}$$

$$\frac{d^n f}{dx^n} = \frac{\Delta^n f\left(x - \frac{n-1}{2}h\right) + \nabla^n f\left(x + \frac{n-1}{2}h\right)}{2h^n} + \mathcal{O}(h^2), \quad \text{if } n \text{ is odd.}$$

See Theorem 5.4.4. For other formulas and more details refer to the area of numerical analysis.

Problem 5.6.5. Verify that under suitable conditions we can write

(a) that $f'(x) = \dfrac{-f(x+2h) + 4f(x+h) - 3f(x)}{2h} + \mathcal{O}(h^2)$, and

$$f'(x) = \frac{f(x-2h) - 4f(x-h) + 3f(x)}{2h} + \mathcal{O}(h^2), \text{ with an average given by}$$

$$f'(x) = \frac{-f(x+2h) + 4f(x+h) - 4f(x-h) + f(x-2h)}{4h} + \mathcal{O}(h^2).$$

(b) $$f'(x)=\frac{2f(x+3h)-9f(x+2h)+18f(x+h)-11f(x)}{6h}+\mathcal{O}(h^3).$$

Would a circumstance ever arise in which the last of the representations in part (a) for $f'(x)$ be preferred? Explain.

Part 2. Lipschitz Condition

The Lipschitz condition, which is widely used in mathematics and was briefly introduced in Section 4.4, is the topic covered in this section.

Definition 5.6.6. We say that a function $f:D\to\mathfrak{R}$, with $D\subseteq\mathfrak{R}$, satisfies a *Lipschitz condition of order* α on D if and only if there exists a real number $L>0$, called a *Lipschitz constant*, and a real constant $\alpha>0$, such that

$$|f(x)-f(t)|\le L|x-t|^{\alpha}$$

for all $x,t\in D$. If $\alpha=1$, then we say that f satisfies the *Lipschitz condition*, or f is a *Lipschitz function*, or simply f is *Lipschitzian.*

The next several theorems and remarks and problems point out the relationship of the Lipschitz condition to other properties already discussed in this book.

THEOREM 5.6.7. *Suppose that a function f, defined on an interval I, satisfies the Lipschitz condition of order α. Then*

(a) *if $\alpha>0$, f is uniformly continuous on I (see Theorem 4.4.10); and*

(b) *if $\alpha>1$, f is differentiable on I. In fact, f must be constant on I.*

Remark 5.6.8. The converse of part (a) in the theorem above does not hold. It can be shown that $f(x)=\sqrt[3]{x}$, on $I=$ neighborhood of zero, is uniformly continuous but not a Lipschitz function. See Exercises 1(d) and 7(c) of Section 4.4. Furthermore, a Lipschitzian function need not be differentiable (see part (b) of Problem 5.6.11). Thus, part (b) of Theorem 5.6.7 fails if $\alpha=1$. Also, worth noting is that if f is differentiable it is also a Lipschitz function provided that the tangent lines are bounded away from the vertical line. Formally stated, this is the following theorem. ■

THEOREM 5.6.9. *Suppose a function f is differentiable on an interval I. Then f' is bounded if and only if f is Lipschitzian.*

Proof. ($\Rightarrow$) Suppose $|f'(x)|\le L$ for some real constant L with $x\in I$. Choose any x_1, $x_2\in I$ with $x_1<x_2$. By the mean value theorem, there exists a $c\in(x_1,x_2)$ such that

$$f'(c) = \frac{f(x_1) - f(x_2)}{x_1 - x_2}.$$

Taking absolute values, we can rewrite this as

$$|f(x_1) - f(x_2)| = |f'(c)||x_1 - x_2| \le L|x_1 - x_2|.$$

Hence, f satisfies the Lipschitz condition.

($\Leftarrow$) The function f, differentiable on I, implies that for any $x_0 \in I$, we have that $\lim_{x \to x_0} \frac{f(x) - f(x_0)}{x - x_0} = f'(x_0)$. The function f, Lipschitzian on I, implies that $|f(x_1) - f(x_2)| \le L|x_1 - x_2|$ for any $x_1, x_2 \in I$. This can be written as $\left|\frac{f(x) - f(x_0)}{x - x_0}\right| \le L$ if we pick $x_1 = x$, $x_2 = x_0$, and $x \ne x_0$. Thus $|f'(x_0)| \le L$. Why? Hence f' is bounded on I. The proof is complete. ■

Compare Theorem 5.6.9 to Example 4.4.11.

Problem 5.6.10. Determine whether or not the given functions are Lipschitzian. Explain.

(a) $f(x) = \frac{1}{x}$, with $x \in (0,1)$

(b) $f(x) = x^2 \sin\frac{1}{x}$, with $x \in (0,1]$ (Compare with Exercise 7(d)–(f), in Section 4.4.)

(c) $f(x) = x^{\frac{3}{2}} \sin\frac{1}{x}$, with $x \in (0,1)$

(d) $f(x) = x^2 \sin\left[\exp\left(\frac{1}{x}\right)\right]$, with $x \in (0,1)$

(e) $f : \Re \to \Re$, where f is a polynomial of degree larger than 1. See Exercise 36 of Section 4.5.

Problem 5.6.11. Give an example of a function that is

(a) continuous but neither differentiable nor Lipschitzian.
(b) Lipschitzian but not differentiable.
(c) differentiable but not Lipschitzian.

Problem 5.6.12. Prove Theorem 5.6.7.

Problem 5.6.13. Suppose that the function f is differentiable on $[a,b]$. Which of the following conditions is more restrictive?

(a) f is a Lipschitz function with Lipschitz constant $L \in (0,1)$, that is, f is a contraction.

(b) $|f'(x)| < 1$ for all $x \in [a,b]$.

Problem 5.6.14. There exists a function f that is both continuously differentiable and uniformly continuous but not Lipschitzian.

Part 3. Functions of Bounded Variation

Definition 5.6.15. For a compact interval $[a,b]$, a collection of points $P = \{x_0, x_1, \ldots, x_n\}$ satisfying $a = x_0 < x_1 < \cdots < x_n = b$ is called a *partition* of $[a,b]$.

Definition 5.6.16. Suppose that a function f is defined on $[a,b]$ and $P = \{x_0, x_1, \ldots, x_n\}$ is a partition of $[a,b]$. The number

$$V_f(a,b) \equiv \sup\left\{ \sum_{k=1}^{n} \left| f(x_k) - f(x_{k-1}) \right| \,\middle|\, P \text{ is a partition of } [a,b] \text{ with } n \in N \right\}$$

is called the *(total) variation of f* on the interval $[a,b]$. If $V_f(a,b) < M$ for some constant M, then f is said to be of *bounded variation*.

Example 5.6.17. The function $f(x) = \sin x$, with $x \in [a,b]$, is of bounded variation.

Proof. Since, by Exercise 20(e) from Section 5.2, $|\sin x - \sin t| \le |x - t|$ for any $x, t \in \Re$, no matter what the partition of $[a,b]$ is, we can write

$$\sum_{k=1}^{n} \left| f(x_k) - f(x_{k-1}) \right| \le \sum_{k=1}^{n} |x_k - x_{k-1}| = b - a. \qquad \blacksquare$$

Problem 5.6.18. If the functions f, g, and h are defined by

$$f(x) = \begin{cases} \sin\dfrac{1}{x}, \text{ if } 0 < x \le 1 \\ 0, \text{ if } x = 0, \end{cases} \qquad g(x) = \begin{cases} x\sin\dfrac{1}{x}, \text{ if } 0 < x \le 1 \\ 0, \text{ if } x = 0, \end{cases}$$

$$\text{and } h(x) = \begin{cases} x^2 \sin\dfrac{1}{x}, \text{ if } 0 < x \le 1 \\ 0, \text{ if } x = 0, \end{cases}$$

then prove that h is of bounded variation and that f and g are not. See Example 10.7.5.

Problem 5.6.19

(a) Prove that if $f:[a,b]\to\mathfrak{R}$ is monotone, then f is of bounded variation. Is $f(x)=\sqrt[3]{x}$ of bounded variation?

(b) Prove that if a function f is continuous on $[a,b]$, and has a bounded derivative on (a,b), then f is of bounded variation.

(c) Show that

$$f(x)=\begin{cases} x^2\cos\dfrac{1}{x^2}, \text{ if } 0<x\le 1 \\ 0, \text{ if } x=0 \end{cases}$$

is differentiable on $[0,1]$, but not of bounded variation.

(d) Prove that a function $f:[a,b]\to\mathfrak{R}$ is of bounded variation if and only if f can be written as the difference of two increasing functions.

(e) Prove that if a function f is of bounded variation on both $[a,b]$ and $[b,c]$, then f is of bounded variation on $[a,c]$ and $V_f(a,c)=V_f(a,b)+V_f(b,c)$.

(f) Prove that if functions f and g are of bounded variation on $[a,b]$, then so are functions $f\pm g$, fg, and $\dfrac{f}{g}$ provided g is bounded away from zero.

(g) Prove that if a function $f:[a,b]\to\mathfrak{R}$ and $V_f(a,b)<M$ with M a real constant, then $|f(x)|\le|f(a)|+M$ for all $x\in[a,b]$.

(h) Give an example of a function that is of bounded variation and has an unbounded derivative. See parts (a) and (b) above.

(i) Prove that if a function f is Lipschitzian, then f is of bounded variation, but not conversely. Conclude that the function g in Problem 5.6.18 is not Lipschitzian. See Exercise 7(e) in Section 4.4.

(j) Give an example of a function that satisfies Lipschitz condition of order $\alpha<1$ but is not of bounded variation.

(k) Prove that if a function f is of bounded variation, then only countably many points of discontinuity exist and they are all simple.

(l) Prove that if $f:[a,b]\to\mathfrak{R}$ is of bounded variation, then it is bounded.

Part 4. Absolutely Continuous Functions

We will briefly introduce the concept of absolute continuity of a function and some of its consequences. Absolutely continuous functions occur in Lebesgue's theory of differentiation and integration.

Definition 5.6.20. A function $f:[a,b]\to\mathfrak{R}$ is *absolutely continuous (on $[a,b]$)* if and only if for any given $\varepsilon>0$ there exists $\delta>0$ such that

$$\sum_{k=1}^{n} \left| f(b_k) - f(a_k) \right| < \varepsilon$$

for every n nonoverlapping intervals (a_k, b_k) of $[a,b]$, $n \in N$ with $\sum_{k=1}^{n}(b_k - a_k) < \delta$.

Absolute continuity is more restrictive than uniform continuity. If in Definition 5.6.20, $n = 1$, then absolute continuity is identical to uniform.

Problem 5.6.21

(a) Prove that if a function f has a bounded derivative on $[a,b]$, then f is absolutely continuous.

(b) Prove that if a function f is absolutely continuous, then it is continuous and of bounded variation but not conversely.

(c) Prove that if functions f and g are absolutely continuous, then so are functions $f \pm g$, fg, and $\frac{f}{g}$ provided g is bounded away from zero.

(d) Prove that an absolutely continuous and differentiable function is Lipschitzian if and only if it has a bounded derivative.

(e) Give an example of an absolutely continuous function that is not differentiable.

(f) Prove that if a function is Lipschitzian, then it is absolutely continuous.

(g) Prove that $f(x) = \sqrt{x}$, with $x \in [0,1]$, is absolutely continuous, but not Lipschitzian.

(h) Prove that

$$f(x) = \begin{cases} x^2 \cos\dfrac{1}{x^2}, & \text{if } 0 < x \le 1 \\ 0, & \text{if } x = 0 \end{cases}$$

is differentiable, and thus continuous, but not absolutely continuous. See Problem 5.6.19, part (c), and the discussion following Example 6.4.1.

(i) Give an example of a function that is of bounded variation, but not absolutely continuous.

(j) Give an example of a function f which is absolutely continuous but $\sqrt{f}$ is not.

(k) If $f:[a,b] \to [0,\infty)$ is monotone and absolutely continuous, prove that $\sqrt{f}$ is absolutely continuous.

Part 5. Convex Functions

Definition 5.6.22. A function $f: I \to \Re$, with I an interval, is *convex (on I)* if and only if

$$f((1-t)a+tb) \le (1-t)f(a)+tf(b),$$

whenever $a,b \in I$ and $t \in [0,1]$. If the inequality $\le$ is replaced by $<$, then f is said to be *strictly convex*. In addition, if a function $-g$ is convex, then g is said to be *concave*. Similarly, g is *strictly concave*, if $-g$ is strictly convex.

See Exercise 2(b) in Section 10.4.

Definition 5.6.23. A function $f: I \to \Re$, with I an open interval, is *midpoint convex (on I)* if and only if

$$f\left(\frac{x+t}{2}\right) \le \frac{1}{2}[f(x)+f(t)]$$

for all $x,t \in I$.

See Definition 1.3.14 and Exercise 5 of Section 5.4.

Problem 5.6.24

(a) Give a geometrical interpretation of a convex function.

(b) Give an example of a convex function that is not twice differentiable.

(c) If a function f is convex on I, show that $f(x) \le \max\{f(a), f(b)\}$ with $a,b \in I$ and x between a and b.

(d) Show that any linear function is both convex and concave. In fact, those are the only functions that are both convex and concave.

(e) If a function f is convex on I and $a<b<c$ are points in I, then prove that

$$\frac{f(b)-f(a)}{b-a} \le \frac{f(c)-f(a)}{c-a} \le \frac{f(c)-f(b)}{c-b}.$$

(f) If a function f is convex on (a,b), $f'_+(x) = \lim_{h \to 0^+} \dfrac{f(x+h)-f(x)}{h}$, and $f'_-(x) = \lim_{h \to 0^-} \dfrac{f(x+h)-f(x)}{h}$, then prove that $f'_+(x)$ and $f'_-(x)$ are finite, and that $f'_-(x) \le f'_+(x)$ at each $x \in (a,b)$.

(g) If a function f is convex on (a,b), prove that f is continuous on (a,b).

(h) Give an example of a function that is convex on $[a,b]$, but continuous at neither $x=a$ nor $x=b$.

(i) Prove that if f and g are convex functions, then so is $f+g$.

(j) Prove that if a function f is convex on (a,b), then f is bounded below on (a,b).

(k) Prove that if a function f is convex on (a,b), then f is differentiable on (a,b), except, perhaps, at countably many points of (a,b).

(l) If a function f is convex on (a,b) and m is any real number satisfying $m \in [f'_-(c), f'_+(c)]$ for some $c \in (a,b)$, then prove that $f(x) \geq m(x-c) + f(c)$ for all $x \in (a,b)$. The line $y = m(x-c) + f(c)$ is called a *supporting line*.

(m) If a function f is twice differentiable on (a,b), then prove that f is convex if and only if $f''(x) \geq 0$ for all $x \in (a,b)$. Show that

$$|x| \leq \frac{\pi}{2}|\sin x| \text{ for } 0 \leq |x| \leq \frac{\pi}{2}.$$

See Exercise 22(b) in Section 5.3.

(n) If a function f is twice differentiable on (a,b), then prove that f is strictly convex if and only if f is concave upward. See Theorem 5.3.2.

(o) Suppose that $a > 0$ and $b \geq 0$ are real constants and

$$f(x) = (ax+b)^r.$$

Prove that

(i) f is convex on $[0,\infty)$, if $r \in [1,\infty)$.

(ii) f is concave on $[0,\infty)$, if $r \in (0,1]$.

(iii) f is strictly convex on $[0,\infty)$, if $r \in (1,\infty)$.

(iv) f is strictly concave on $[0,\infty)$, if $r \in (0,1)$.

(p) Prove that if a function f is continuous and convex on $[a,b]$, then f is absolutely continuous on $[a,b]$.

(q) Prove that if a function f is continuous and midpoint convex on (a,b), then f is convex on (a,b).

(r) Prove that if $g(x) = \ln(f(x))$ is a convex function on (a,b), that is, f is *logarithmically convex on* (a,b), then f is convex on (a,b).

6

Integration

6.1 Riemann Integral

There are numerous applications for an "integral." Here, resorting to the area under the curve for motivation, we concentrate on the theory of integrals rather than on their usage. And of the many different integrals there are to choose from, we will study the one that is most familiar, namely the Riemann[1] integral. Cauchy introduced this concept of an integral and Riemann refined it. In more advanced texts, integration due to Darboux, Stieltjes[2], Lebesgue, and others is studied. Unlike other concepts of analysis, the integral cannot be as readily defined. First, we need to familiarize ourselves with other underlying ideas, such as a "partition," which was introduced in Section 5.6.

Definition 6.1.1. A *partition* P of a closed interval $[a,b]$ is a finite collection of points $\{x_k\}_{k=0}^{n}$ that satisfies $a = x_0 < x_1 < x_2 < \ldots < x_n = b$. The *norm (mesh) of a partition* P is defined by

---

[1]Georg Friedrich Bernhard Riemann (1826–1866), a German mathematician, contributed greatly to number theory and complex analysis. Modern relativity theory is based on Riemannian geometry. Riemann also did work in potential theory, topology, and mathematical physics. He was Gauss' protégé.

[2]Jean Thomas Stieltjes (1856–1894), a famous French analyst and number theorist, defined the Stieltjes integral. He also studied divergent and conditionally convergent series and spherical harmonics.

$$|P| = \max\{\Delta x_k \mid k = 1, 2, \ldots, n\}$$

with $\Delta x_k = x_k - x_{k-1}$. Each interval $[x_{k-1}, x_k]$ is called a *subinterval* of $[a, b]$. In addition, Q is called a *refinement* of P if Q is a partition of $[a, b]$ and contains all the points of P along with a finite number of additional points from the interval $[a, b]$, that is, $P \subseteq Q$.

Remark 6.1.2. Suppose a function $f:[a,b] \to \Re$ is bounded and $P = \{x_0, x_1, x_2, \ldots, x_n\}$ is a partition of $[a, b]$, with Δx_k denoting $x_k - x_{k-1}$. Also, for each $k = 1, 2, \ldots, n$, we define

$$M_k(f) = \sup\{f(x) \mid x \in [x_{k-1}, x_k]\} \text{ and } m_k(f) = \inf\{f(x) \mid x \in [x_{k-1}, x_k]\}$$

When only one function is involved, the notation can be shortened to M_k and m_k, respectively. The *upper* and *lower sums*, sometimes referred to as *upper* and *lower Darboux sums*, are, respectively, defined by

$$U(P, f) = \sum_{k=1}^{n} M_k \Delta x_k \text{ and } L(P, f) = \sum_{k=1}^{n} m_k \Delta x_k .$$

A *Riemann sum* is defined by

$$S(P, f) = \sum_{k=1}^{n} f(c_k) \Delta x_k$$

with any $c_k \in [x_{k-1}, x_k]$.

If a function f is positive, then the $U(P, f)$, $L(P, f)$, and $S(P, f)$ represent the shaded areas, respectively, in Figure 6.1.1. Note that although $U(P, f)$ and $L(P, f)$ are unique, there are infinitely many different Riemann sums corresponding to any fixed partition of a given interval. ■

Next, we will state a preliminary result.

LEMMA 6.1.3. *Suppose that a function $f:[a,b] \to \Re$ is bounded and $P = \{x_0, x_1, x_2, \ldots, x_n\}$ is a partition of $[a, b]$. Then there exist two real numbers m and M such that*

$$m(b-a) \le L(P, f) \le S(P, f) \le U(P, f) \le M(b-a)$$

for any $c_k \in [x_{k-1}, x_k]$.

A proof of Lemma 6.1.3 is left to the reader. Note that if f is a continuous function on $[a, b]$, then the real numbers m and M, as labeled in this lemma, may actually represent absolute minimum and maximum values. See the extreme value theorem,

Theorem 4.3.2. If f is not continuous on $[a,b]$, then f need not possess a minimum or maximum value.

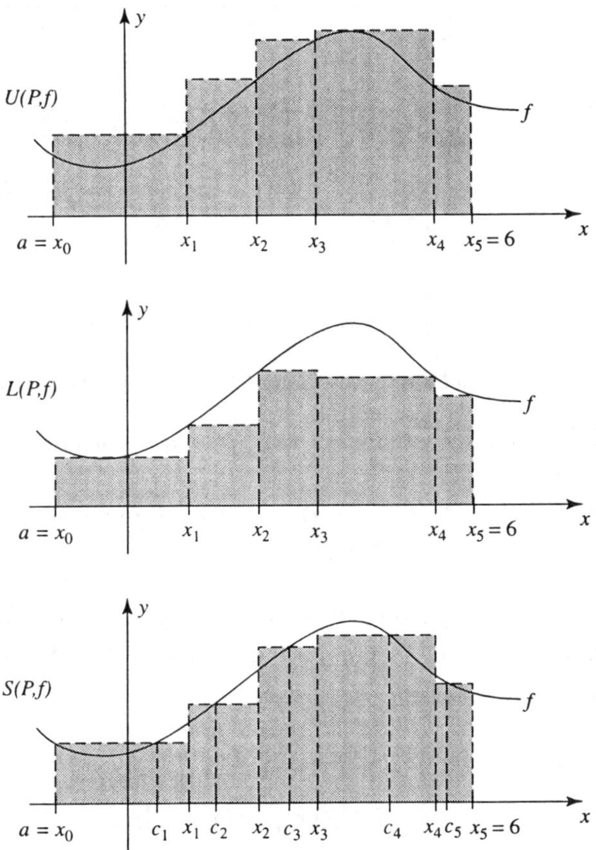

Figure 6.1.1

Example 6.1.4. Consider the function $f(x)=x^2$ with $x \in [0,2]$ and a particular partition P given by

$$P = \left\{0, \frac{2}{n}, \frac{4}{n}, \ldots, 2 - \frac{2}{n} = \frac{2(n-1)}{n}, 2\right\}.$$

Since f is increasing on $[0,2]$, $f(x_k)$ is the maximum value for f and $f(x_{k-1})$ is the minimum value for f on each subinterval $[x_{k-1}, x_k]$. Thus, since $\Delta x_k = \dfrac{2}{n}$ on each subinterval, we have

$$\begin{aligned}U(P,f) &= \sum_{k=1}^{n} M_k \Delta x_k = \sum_{k=1}^{n} f(x_k)\frac{2}{n} = \frac{2}{n}\sum_{k=1}^{n} f\left(\frac{2k}{n}\right) = \frac{2}{n}\sum_{k=1}^{n}\frac{4k^2}{n^2} \\ &= \frac{2}{n}\cdot\frac{4}{n^2}\sum_{k=1}^{n}k^2 = \frac{8}{n^3}\cdot\frac{n(n+1)(2n+1)}{6} = \frac{4(n+1)(2n+1)}{3n^2}, \text{ and}\end{aligned}$$

$$\begin{aligned}L(P,f) &= \sum_{k=1}^{n} m_k \Delta x_k = \sum_{k=1}^{n} f(x_{k-1})\frac{2}{n} = \frac{2}{n}\sum_{k=1}^{n} f\left(\frac{2(k-1)}{n}\right) = \frac{2}{n}\sum_{k=1}^{n}\frac{4(k-1)^2}{n^2} \\ &= \frac{8}{n^3}\sum_{k=1}^{n}(k-1)^2 = \frac{8}{n^3}\sum_{i=1}^{n-1}i^2 = \frac{8}{n^3}\frac{(n-1)n(2(n-1)+1)}{6} = \frac{4(n-1)(2n-1)}{3n^2}.\end{aligned}$$

We could have also expanded $\sum_{k=1}^{n}(k-1)^2$ and used the necessary formulas from Section 1.3. Observe that the upper and lower sums depended on n, the size of the partition. ■

LEMMA 6.1.5. *Suppose a function $f:[a,b]\to\Re$ is bounded, and let P and Q be any two partitions of $[a,b]$.*

(a) If $P\subseteq Q$, then $L(P,f)\le L(Q,f)$ and $U(Q,f)\le U(P,f)$.

(b) *Always,* $L(P,f)\le U(Q,f)$.

Proof of part (a). If $P=Q$, then there is nothing to prove. Suppose that $P=\{x_0,x_1,x_2,\dots,x_n\}$ and that Q contains one additional point of $[a,b]$, say, $c\in[x_{i-1},x_i]$, where $1\le i\le n$. Let

$$\begin{aligned}m_i &= \inf\{f(x)\,|\,x\in[x_{i-1},x_i]\}, \\ r_1 &= \inf\{f(x)\,|\,x\in[x_{i-1},c]\}, \text{ and} \\ r_2 &= \inf\{f(x)\,|\,x\in[c,x_i]\}.\end{aligned}$$

Then, $m_i=\min\{r_1,r_2\}$. So we can write

$$\begin{aligned}L(P,f) &= \sum_{k=1}^{n} m_k\Delta x_k = \sum_{k=1}^{i-1} m_k(x_k-x_{k-1}) + m_i(x_i-x_{i-1}) + \sum_{k=i+1}^{n} m_k(x_k-x_{k-1}) \\ &\le \sum_{k=1}^{i-1} m_k(x_k-x_{k-1}) + r_1(c-x_{i-1}) + r_2(x_i-c) + \sum_{k=i+1}^{n} m_k(x_k-x_{k-1}) \\ &= L(Q,f).\end{aligned}$$

See Figure 6.1.2 for a visualization of this situation.

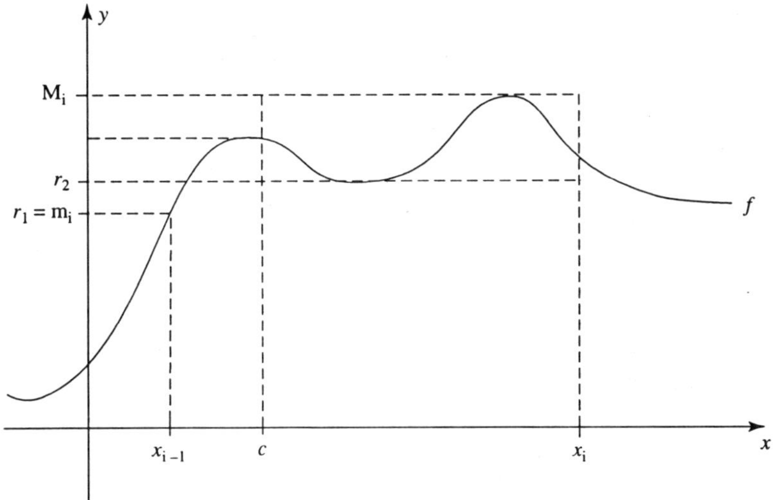

Figure 6.1.2

A similar set of steps can be followed to prove that $U(P,f) \geq U(Q,f)$. Finally, if Q contained n points that were not in P, the preceding argument would be repeated n times.

Proof of part (b). By Lemma 6.1.3, $L(P,f) \leq U(P,f)$ for any partition P of $[a,b]$. Since in our situation, P and Q are arbitrary partitions of $[a,b]$, so is their refinement $P \cup Q$. Thus, by part (a), we can write

$$L(P,f) \leq L(P \cup Q, f) \leq U(P \cup Q, f) \leq U(Q,f).$$

The proof is complete. ■

As seen in Lemma 6.1.5 and Example 6.1.4, lower sums increase as the partition gets "finer," that is, as $|P|$ decreases, whereas upper sums decrease. Since lower and upper sums are bounded, their respective supremum and infimum must exist. The symbols

$$\underline{\int_a^b} f, \qquad \underline{\int_a^b} f\,dx, \quad \text{and} \quad \underline{\int_a^b} f(x)\,dx,$$

all represent the supremum of the lower sums, and are referred to as *lower integrals.* Similarly, we define the *upper integral.* Little dashes used in these symbols are referred to as *underbar* for lower integral and as *overbar* for upper integral. Thus, we write

$$\underline{\int_a^b} f = \sup\{L(P,f) \mid P \text{ a partition of } [a,b]\} = \sup_P L(P,f), \text{ and}$$

$$\overline{\int_a^b} f = \inf\{U(P,f) \mid P \text{ a partition of } [a,b]\} = \inf_P U(P,f).$$

Combining this with Lemma 6.1.5 yields

$$L(P,f) \le \underline{\int_a^b} f \le \overline{\int_a^b} f \le U(P,f),$$

where P is any partition of $[a,b]$ and $f:[a,b] \to \Re$ is a bounded function. The variable x used in the preceding is a *dummy variable*. That is, we can write

$$\underline{\int_a^b} f(x)\,dx = \underline{\int_a^b} f(t)\,dt \quad \text{and} \quad \overline{\int_a^b} f(x)\,dx = \overline{\int_a^b} f(t)\,dt.$$

Definition 6.1.6. A bounded function $f:[a,b] \to \Re$ is *Riemann integrable* on $[a,b]$ if and only if

$$\underline{\int_a^b} f(x)\,dx = \overline{\int_a^b} f(x)\,dx = A,$$

a real number. Whenever this is the case, $f \in R[a,b]$, meaning f is in the set of all Riemann integrable functions on $[a,b]$. The *Riemann integral* of f on $[a,b]$, whose value is given by A, is a real number denoted by $\int_a^b f$, $\int_a^b f\,dx$, or $\int_a^b f(x)\,dx$. The symbol $\int$ is called an *integral sign* and $f(x)\,dx$ is called an *integrand*. In elementary calculus, we referred to the Riemann integral as a *definite integral. Integrating* or *integration* is the process of evaluating the Riemann integral. The expression represents $\int_a^b 1\,dx = b - a$.

Example 6.1.7. Use upper and lower sums computed in Example 6.1.4 to evaluate $\int_0^2 x^2\,dx$.

Answer. Since $\lim\limits_{n\to\infty} U(P,f)=\dfrac{8}{3}$, we must have that $\overline{\int_0^2} x^2dx \le \dfrac{8}{3}$. Why? Similarly, $\underline{\int_0^2} x^2dx \ge \dfrac{8}{3}$. Hence, $\underline{\int_0^2} x^2dx = \overline{\int_0^2} x^2dx$, which yields that $f \in R[0,2]$ and $\int_0^2 x^2dx = \dfrac{8}{3}$. ■

Later, quicker methods for evaluating Riemann integrals will be used. Next, consider a bounded, nonnegative function f that is not integrable.

Example 6.1.8. If $f:[-2,3]\to\Re$ is defined by $f(x)=\begin{cases}0, & \text{if } x \text{ is rational}\\ 4, & \text{if } x \text{ is irrational,}\end{cases}$ then, f is not Riemann integrable, that is, $\int_{-2}^{3} f(x)\,dx$ does not exist.

Proof. Let $P=\{x_0,x_1,x_2,\ldots,x_n\}$ be any partition of $[-2,3]$. Since each subinterval $[x_{k-1},x_k]$ contains both rational and irrational values, we have that $M_k=4$ and $m_k=0$, for all $k=1,2,\ldots,n$. Therefore,

$$U(P,f)=\sum_{k=1}^{n} M_k\Delta x_k = 4\sum_{k=1}^{n}\Delta x_k = 4[3-(-2)]=20, \text{ and}$$
$$L(P,f)=\sum_{k=1}^{n} m_k\Delta x_k = 0\sum_{k=1}^{n}\Delta x_k = 0[3-(-2)]=0.$$

Thus, since lower and upper integrals are not equal, $f \notin R[-2,3]$. ■

Exercises 6.1

1. Prove Lemma 6.1.3.
2. From Lemma 6.1.5, prove that $U(Q,f)\le U(P,f)$.
3. Prove that a constant function $f(x)=c$, $c\in\Re$, is Riemann integrable on any interval $[a,b]$ and $\int_a^b f(x)\,dx = c(b-a)$.
4. If $f(x)\le g(x)\le h(x)$ for all $x\in[a,b]$, and f and h are Riemann integrable on $[a,b]$, then so is g. True or false? Explain.
5. If a function $f:[-2,3]\to\Re$, defined by

$$f(x) = \begin{cases} 2|x|+1, & \text{if } x \text{ is rational} \\ 0, & \text{if } x \text{ is irrational,} \end{cases}$$

prove that f is not Riemann integrable.

6. If a function $f:[a,b] \to \Re$ is bounded and nonnegative, prove that $\underline{\int_a^b} f \ge 0$.

7. Find bounded functions $f,g:[a,b] \to \Re$ such that $\overline{\int_a^b} (f+g) < \overline{\int_a^b} f + \overline{\int_a^b} g$.

8. Prove that if functions $f,g:[a,b] \to \Re$ are bounded, then $\overline{\int_a^b} (f+g) \le \overline{\int_a^b} f + \overline{\int_a^b} g$.

9. Use upper and lower sums to calculate $\int_0^2 \left(x^3 - x^2 + 3x + 2\right) dx$.

10. Give an example of a function $f:[0,2] \to \Re$ that is not Riemann integrable but

(a) $|f| \in R[0,2]$. **(b)** $f^2 \in R[0,2]$.

11. If $f:[a,b] \to \Re$ is Riemann integrable and $f(x) \ge 0$ for all $x \in [a,b]$, how does the value of $\int_a^b f$ compare to the area of the region bounded by the curve $f(x)$, x-axis, and the vertical lines $x=a$ and $x=b$? How about if $f(x) \le 0$ on $[a,b]$?

12. Evaluate $\int_0^2 (x+[x])\,dx$ and $\int_0^1 \sqrt{1-(x-1)^2}\,dx$, if possible.

13. Relating integrals to areas, explain why

$$\int_0^{\frac{\pi}{2}} \sin x\,dx = \frac{\pi}{2} - \int_0^1 \arcsin x\,dx.$$

6.2 Integrable Functions

As with the definition of differentiation, the definition of a Riemann integrable function is not easy to use. In this section, a number of theorems are presented that will help us determine with less difficulty whether a given function is Riemann integrable. These results, however, do not indicate how the value of a Riemann integral, if it exists, is to be found.

THEOREM 6.2.1. *Let a function $f:[a,b]\to\Re$ be bounded. Then f is Riemann integrable on $[a,b]$ if and only if for any arbitrary $\varepsilon>0$ there exists a partition P such that $U(P,f)-L(P,f)<\varepsilon$.*

Proof. ($\Rightarrow$) Suppose that $f\in R[a,b]$. Then

$$\overline{\int_a^b} f = \int_a^b f = \underline{\int_a^b} f = \sup_P L(P,f) = A,$$

with A a real constant. Let an arbitrary $\varepsilon>0$ be given. Since $A-\frac{\varepsilon}{2}$ is not an upper bound for the set of all lower sums, there must be a partition P_1 of $[a,b]$ such that $A-\frac{\varepsilon}{2}<L(P_1,f)$. Similarly, there exists a partition P_2 of $[a,b]$ such that $A+\frac{\varepsilon}{2}>U(P_2,f)$. If $P=P_1\cup P_2$, then we can write

$$A-\frac{\varepsilon}{2}<L(P_1,f)\le L(P,f)\le U(P,f)\le U(P_2,f)<A+\frac{\varepsilon}{2}.$$

Thus, $L(P,f)$ and $U(P,f)$ are within ε of each other. Hence, we found a partition P of $[a,b]$ such that $U(P,f)-L(P,f)<\varepsilon$.

($\Leftarrow$) Suppose a function $f:[a,b]\to\Re$ is bounded and an arbitrary $\varepsilon>0$ is given. Also, suppose that there exists a partition P of $[a,b]$ such that $U(P,f)-L(P,f)<\varepsilon$. To show that $f\in R[a,b]$, we need to show that

$$0\le \overline{\int_a^b} f - \underline{\int_a^b} f < \varepsilon, \text{ which, in turn, will imply that } \overline{\int_a^b} f = \underline{\int_a^b} f.$$

To this end, since $U(P,f)<L(P,f)+\varepsilon$, we have that

$$L(P,f)\le \underline{\int_a^b} f \le \overline{\int_a^b} f \le U(P,f)<L(P,f)+\varepsilon.$$

Subtracting $L(P,f)$ from the left- and right-hand sides yields the desired result. ■

We can now use Theorem 6.2.1 to prove some very useful results.

THEOREM 6.2.2. *If a function $f:[a,b]\to\Re$ is monotone, then f is Riemann integrable.*

Proof. Suppose that f is increasing on $[a,b]$. The case in which f is decreasing is left to Exercise 1 at the end of this section. Since f is increasing on a compact interval $[a,b]$, it must be bounded on $[a,b]$. Now, let any $\varepsilon>0$ be given. Then there exists a positive real number K such that

$$f(b)-f(a)<\boxed{\frac{\varepsilon}{K}}.$$

We need to find a partition P of $[a,b]$ so that $U(P,f)-L(P,f)<\varepsilon$. To this end, choose any $P=\{x_0,x_1,\ldots,x_n\}$ such that $|P|<K$. Since f is increasing, we have $M_k=f(x_k)$ and $m_k=f(x_{k-1})$, that is, M_k and m_k are functional values of the right and left endpoints, respectively, of each subinterval $[x_{k-1},x_k]$. Therefore, we can write

$$\begin{aligned}U(P,f)-L(P,f) &= \sum_{k=1}^{n}\left[f(x_k)-f(x_{k-1})\right](x_k-x_{k-1})\\ &< K\sum_{k=1}^{n}\left[f(x_k)-f(x_{k-1})\right]=K[f(b)-f(a)]<K\cdot\frac{\varepsilon}{K}=\varepsilon.\end{aligned}$$

Hence, by Theorem 6.2.1, $f\in R[a,b]$. ■

Remark 6.2.3. By Theorem 6.2.2, the function $f:[-1,1]\to\Re$ defined by $f(x)=\frac{|x|}{x}$, if $x\neq 0$, and $f(0)=0$, is integrable. However, f has no antiderivative. See the end of Section 5.2. ■

THEOREM 6.2.4. *If a function $f:[a,b]\to\Re$ is continuous, then f is Riemann integrable.*

Proof. Let $\varepsilon>0$ be given. Since f is continuous on a compact interval $[a,b]$, f is uniformly continuous on $[a,b]$. Therefore, there exists $\delta>0$ such that

$$|f(x)-f(t)|<\boxed{\frac{\varepsilon}{b-a}},$$

whenever $|x-t|<\delta$ and $x,t\in[a,b]$. Now, choose a partition $P=\{x_0,x_1,\ldots,x_n\}$ of $[a,b]$ with $|P|<\delta$. Observe that by the extreme value theorem there exist t_k and s_k in $[x_{k-1},x_k]$ such that $M_k=f(t_k)$ and $m_k=f(s_k)$ for all $k=1,2,\ldots,n$. Since $\Delta x_k=x_k-x_{k-1}<\delta$, we have $|t_k-s_k|<\delta$, and so $0\le f(t_k)-f(s_k)<\frac{\varepsilon}{b-a}$. Thus, we can write

$$\begin{aligned}U(P,f)-L(P,f) &= \sum_{k=1}^{n}\left[f(t_k)-f(s_k)\right](x_k-x_{k-1})\\ &< \frac{\varepsilon}{b-a}\sum_{k=1}^{n}(x_k-x_{k-1})=\frac{\varepsilon}{b-a}\cdot(b-a)=\varepsilon.\end{aligned}$$

Hence, by Theorem 6.2.1, $f\in R[a,b]$. ■

Example 6.2.5. Since the function $f(x) = \text{sinc}\, x$ is continuous on any compact interval $[a,b]$, by Theorem 6.2.4, it is Riemann integrable. See Exercise 15 in Section 5.4. ■

Remark 6.2.6. None of the results thus far gave any clue to the value of the Riemann integral of a given function. However, if we can prove that a given function f is integrable and then find a value that falls between the upper and lower integrals, the Riemann integral will be equal to that value. This will be illustrated using the function $f(x) = x^3$, with $x \in [0,1]$. Recall that the Riemann sum $S(P,f)$ falls between $L(P,f)$ and $U(P,f)$. See Lemma 6.1.3. Thus, for any partition $P = \{x_0, x_1, \dots, x_n\}$ of $[0,1]$, we conveniently choose a new function $g(x) = \frac{1}{4}x^4$. Then, applying the mean value theorem to g on each subinterval $[x_{k-1}, x_k]$, we obtain $c_k \in [x_{k-1}, x_k]$ such that

$$\frac{g(x_k) - g(x_{k-1})}{x_k - x_{k-1}} = g'(c_k).$$

Now, $g(x_k) - g(x_{k-1}) = g'(c_k)(x_k - x_{k-1}) = f(c_k)(x_k - x_{k-1})$. Therefore,

$$\sum_{k=1}^{n} f(c_k)(x_k - x_{k-1}) = \sum_{k=1}^{n} [g(x_k) - g(x_{k-1})] = g(1) - g(0) = \frac{1}{4},$$

which falls between $L(P,f)$ and $U(P,f)$. Hence, $\underline{\int_0^1} f \leq \frac{1}{4} \leq \overline{\int_0^1} f$. So, $\int_0^1 x^3 dx = \frac{1}{4}$. ■

Recall that in calculus we evaluated the Riemann integral of a function f on $[a,b]$ by subdividing the interval $[a,b]$ into n equal pieces and then computing $\lim_{k \to \infty} f(x_k)(x_k - x_{k-1})$. Something similar was done in Example 6.1.7. The next result, whose proof is left as Exercise 4, verifies this procedure.

THEOREM 6.2.7. *Suppose a function $f:[a,b] \to \Re$ is bounded, and suppose that there exists a sequence $\{P_n\}$ of partitions of $[a,b]$ such that*

$$\lim_{n \to \infty} [U(P_n, f) - L(P_n, f)] = 0.$$

Then $f \in R[a,b]$ and $\int_a^b f = \lim_{n \to \infty} U(P_n, f) = \lim_{n \to \infty} L(P_n, f)$.

Remark 6.2.8. Theorem 6.2.7 can readily be rewritten to involve Riemann sums. Thus, suppose a function $f:[a,b] \to \Re$ is bounded. Then $f \in R[a,b]$ if and only if each sequence $\{P_n\}$ of partitions of $[a,b]$ whose norm converges to 0, the sequence $\{S(P_n, f)\}$ converges, for every choice of $S(P_n, f)$ associated with a partition P_n. In fact, when the last mentioned limit actually exists, the limiting value is the value of the integral. See Exercise 6. ■

Exercises 6.2

1. Prove Theorem 6.2.2 in the case where the function f is decreasing.

2. By Theorem 6.2.2, the function $f(x) = \lfloor x \rfloor$ with $x \in [0,4]$ is Riemann integrable. Thus, for any $\varepsilon > 0$ given, there exists a partition P of $[0,4]$ such that $U(P,f) - L(P,f) < \varepsilon$. Find such a partition.

3. If a function f is Riemann integrable on $[a,b]$, prove that the value of the Riemann integral is unique.

4. Prove Theorem 6.2.7.

5. Let a function $f:[a,b] \to \Re$ be bounded. Prove that f is Riemann integrable on $[a,b]$ if and only if for any arbitrary $\varepsilon > 0$, there exists $\delta > 0$ such that $U(P,f) - L(P,f) < \varepsilon$ whenever P is a partition of $[a,b]$ with $|P| < \delta$.

6. Let a function $f:[a,b] \to \Re$ be bounded. Prove that f is Riemann integrable on $[a,b]$ if and only if there exists a real number A, such that for any arbitrary $\varepsilon > 0$ there exists $\delta > 0$ such that $|S(P,f) - A| < \varepsilon$ for every choice of $S(P,f)$ associated with a partition P of $[a,b]$, where $|P| < \delta$. A is the value of $\int_a^b f$.

7. Suppose that a function $f:[a,b] \to \Re$ is continuous and nonnegative. Prove that if $\int_a^b f = 0$, then $f(x) = 0$, for all $x \in [a,b]$.

8. Consider Dirichlet's function $f:[0,1] \to \Re$, defined by

$$f(x) = \begin{cases} \dfrac{1}{q}, \text{ if } x = \dfrac{p}{q} \text{ in lowest terms with } p, q \in N \\ 0, \text{ otherwise.} \end{cases}$$

Prove that $\int_0^1 f(x)dx = 0$. Does this contradict Exercise 7? Also, see Example 3.2.10 and Exercise 6 of Section 5.1.

9. Use Remark 6.2.8 to evaluate $\int_0^2 \left(x^3 - x^2 + 3x + 2\right)dx$. See Exercise 9 in Section 6.1.

10. Let a function $f:[0,1] \to \Re$ be bounded, and let $\{a_n\}$ be a sequence defined by

$$a_n = \frac{1}{n}\sum_{k=1}^{n} f\left(\frac{k}{n}\right).$$

(a) If f is Riemann integrable, prove that $\{a_n\}$ converges to $\int_0^1 f(x)\,dx$. See Exercise 24 in Section 7.1.

(b) Give an example of a function f for which the sequence $\{a_n\}$ converges, but f is not Riemann integrable.

(c) If for a function f, the sequence $\{a_n\}$ diverges, is it true that f is not Riemann integrable? Explain.

11. Give an example of a function $f:[0,1] \to \Re$ that is

(a) bounded but $f \notin R[0,1]$.

(b) $f \in R[0,1]$ but not monotone.

(c) $f \in R[0,1]$ but neither continuous nor monotone.

12. If a function $f:[a,b] \to \Re$ is Riemann integrable on $[a,b]$ and $[c,d] \subseteq [a,b]$, then prove that $f \in R[c,d]$.

13. If a function $f:[a,b] \to \Re$ is bounded and $f(x) = 0$ except at finitely many values in $[a,b]$, prove that $\int_a^b f = 0$.

14. Suppose $f:[0,1] \to \Re$ is defined by

$$f(x) = \begin{cases} \dfrac{1}{n}, & \text{if } x \in \left(\dfrac{1}{n+1}, \dfrac{1}{n}\right] \\ 0, & \text{if } x = 0. \end{cases}$$

Verify that $f \in R[0,1]$. In fact it can be verified that $\int_0^1 f(x)\,dx = \dfrac{\pi^2}{6} - 1$.

6.3 Properties of the Riemann Integral

We are now ready to turn our attention to some basic algebraic properties of the Riemann integral. The first theorem is the so-called "linear property."

THEOREM 6.3.1

(a) *If functions $f,g:[a,b]\to\Re$ are Riemann integrable, then $f+g\in R[a,b]$, and*

$$\int_a^b (f+g)=\int_a^b f+\int_a^b g.$$

(b) *If $f:[a,b]\to\Re$ is Riemann integrable and c is a real constant, then $cf\in R[a,b]$, and*

$$\int_a^b cf = c\int_a^b f.$$

Proof of part (a). Let $\varepsilon>0$ be given. We will show that there exists a partition P such that $U(P,f+g)-L(P,f+g)<\varepsilon$. Since $f,g\in R[a,b]$, by Theorem 6.2.1 there exist partitions Q and R of $[a,b]$ such that

$$U(Q,f)-L(Q,f)<\boxed{\frac{\varepsilon}{2}}, \text{ and } U(R,g)-L(R,g)<\boxed{\frac{\varepsilon}{2}}.$$

Now let P be the refinement $Q\cup R$ and denote it by $P=\{x_0,x_1,\ldots,x_n\}$. Therefore, P is a partition of $[a,b]$, and from Lemma 6.1.5 we have that

$$U(P,f)-L(P,f)\le U(Q,f)-L(Q,f)<\frac{\varepsilon}{2}, \text{ and}$$
$$U(P,g)-L(P,g)\le U(R,g)-L(R,g)<\frac{\varepsilon}{2},$$

with the sums given by $[U(P,f)+U(P,g)]-[L(P,f)+L(P,g)]<\varepsilon$. Now, let us examine the function $f+g$. Note that if $x\in[x_{k-1},x_k]$, then we have $(f+g)(x)=f(x)+g(x)\le M_k(f)+M_k(g)$, where

$$M_k(f)\equiv\sup\{f(x)\,|\,x\in[x_{k-1},x_k]\} \text{ and } M_k(g)\equiv\sup\{g(x)\,|\,x\in[x_{k-1},x_k]\}.$$

Thus, since $f+g$ is bounded on each subinterval $[x_{k-1},x_k]$, the

$$\sup\{(f+g)(x)\,|\,x\in[x_{k-1},x_k]\}\equiv M_k(f+g)\le M_k(f)+M_k(g).$$

See Exercise 55 in Section 1.9. Therefore,

$$\sum_{k=1}^n M_k(f+g)\Delta x_k\le\sum_{k=1}^n[M_k(f)+M_k(g)]\Delta x_k=\sum_{k=1}^n M_k(f)\Delta x_k+\sum_{k=1}^n M_k(g)\Delta x_k.$$

Hence, $U(P,f+g)\le U(P,f)+U(P,g)$.

Using a similar argument with lower sums, we can obtain $L(P,f)+L(P,g)\le L(P,f+g)$. With subtraction yielding

$$U(P,f+g)-L(P,f+g)\le [U(P,f)+U(P,g)]-[L(P,f)+L(P,g)]<\varepsilon .$$

So, by Theorem 6.2.1, $f+g\in R[a,b]$.

Next, we need to show that $\int_a^b (f+g)=\int_a^b f+\int_a^b g$. To this end, observe that

$$(*)\qquad L(P,f)+L(P,g)\le L(P,f+g)\le \int_a^b (f+g)\le U(P,f+g)\le U(P,f)+U(P,g).$$

And since $f,g\in R[a,b]$, we also have that

$$L(P,f)+L(P,g)\le \int_a^b f+\int_a^b g\le U(P,f)+U(P,g),$$

which is equivalent to

$$-U(P,f)-U(P,g)\le -\left[\int_a^b f+\int_a^b g\right]\le -L(P,f)-L(P,g)$$

Adding this to line $(*)$ we obtain

$$\begin{aligned}&-[U(P,f)-L(P,f)]-[U(P,g)-L(P,g)]\\ &\qquad\le \int_a^b (f+g)-\left[\int_a^b f+\int_a^b g\right]\le [U(P,f)-L(P,f)]+[U(P,g)-L(P,g)].\end{aligned}$$

Therefore,

$$\left|\int_a^b (f+g)-\left[\int_a^b f+\int_a^b g\right]\right|\le [U(P,f)-L(P,f)]+[U(P,g)-L(P,g)]\to 0$$

as $|P|\to 0$. Hence, the desired conclusion follows. ■

Proof of part (b) is left as an exercise.

THEOREM 6.3.2. *If functions* $f, g:[a,b] \to \Re$ *are Riemann integrable and* $f(x) \le g(x)$, *for all* $x \in [a,b]$, *then* $\int_a^b f \le \int_a^b g$.

Some alterations of this result are possible, but not the converse. See Exercise 1. Also note that Theorem 6.3.2 follows from Exercise 6 in Section 6.1. Why?

THEOREM 6.3.3. *Suppose a function* $f:[a,b] \to \Re$ *is bounded, and suppose* $c \in (a,b)$. *Then,* $f \in R[a,b]$ *if and only if* $f \in R[a,c]$ *and* $f \in R[c,b]$. *Moreover, if* f *is Riemann integrable on* $[a,b]$, *then* $\int_a^b f = \int_a^c f + \int_c^b f$.

Proof. ($\Rightarrow$) This direction follows from Exercise 12 of Section 6.2.

($\Leftarrow$) Suppose $f \in R[a,c]$, and $f \in R[c,b]$. And show that $f \in R[a,b]$. Let $P_1 = \{x_0, x_1, \dots, x_i\}$ be a partition of $[a,c]$, and $P_2 = \{x_i, x_{i+1}, \dots, x_n\}$ be a partition of $[c,b]$, where $i < n$ and $i, n \in N$. Then $P = \{x_0, x_1, \dots, x_i, x_{i+1}, \dots, x_n\}$ is a partition of $[a,b]$. As before, $m_k(f) \equiv \inf\{f(x) \mid x \in [x_{k-1}, x_k]\}$ and $\Delta x_k = x_k - x_{k-1}$, for all $k = 1, 2, \dots, n$. Also,

$$L(P,f) = \sum_{k=1}^{n} m_k \Delta x_k = \sum_{k=1}^{i} m_k \Delta x_k + \sum_{k=i+1}^{n} m_k \Delta x_k = L(P_1,f) + L(P_2,f).$$

But, $L(P,f) \le \underline{\int_a^b} f$. Therefore, $L(P_1,f) + L(P_2,f) \le \underline{\int_a^b} f$. And since the lower integral is an upper bound for all the partitions P_1 of $[a,c]$ and P_2 of $[c,b]$, we have that the

$$\sup\{L(P_1,f) \mid P_1 \text{ is a partition of } [a,c]\}$$
$$+\sup\{L(P_2,f) \mid P_2 \text{ is a partition of } [c,b]\} \le \underline{\int_a^b} f.$$

Therefore, $\underline{\int_a^c} f + \underline{\int_c^b} f \le \underline{\int_a^b} f$. Now, since $f \in R[a,c]$ and $f \in R[c,b]$, we have that $\int_a^c f + \int_c^b f \le \underline{\int_a^b} f$. And using a similar argument with upper sums, we can obtain $\overline{\int_a^b} f \le \int_a^c f + \int_c^b f$. Thus, since $\int_a^b f \le \overline{\int_a^b} f$, combining our statements we obtain that

$$\int_a^c f + \int_c^b f \le \underline{\int_a^b} f \le \overline{\int_a^b} f \le \int_a^c f + \int_c^b f.$$

Therefore, since expressions on opposite ends of the inequality are the same, equality must hold. Thus, $\underline{\int_a^b} f = \overline{\int_a^b} f = \int_a^b f$. Hence, $f \in R[a,b]$, and $\int_a^b f = \int_a^c f + \int_c^b f$. ■

Can you reprove Theorem 6.3.3 by implementing part (a) of Theorem 6.3.1?

THEOREM 6.3.4. *Suppose a function $f:[a,b] \to B$ is a surjection and $f \in R[a,b]$. Furthermore, if B is a closed and bounded set and a function $g: B \to \Re$ is continuous, then $g \circ f \in R[a,b]$.*

Refer to library sources for a proof of this theorem.

COROLLARY 6.3.5. *If a function $f:[a,b] \to \Re$ is Riemann integrable, then so is the nth power function f^n, with $n \in N$.*

Proof. If we let $g(x) = x^n$ with $n \in N$, then by Theorem 6.3.4, $(g \circ f)(x) = f^n(x)$ is Riemann integrable on $[a,b]$. ■

Remark 6.3.6. As observed in Corollary 6.3.5, Theorem 6.3.4 can be very useful. See exercises at the end of this section for further practice using Theorem 6.3.4. Do not make the mistake of concluding that a composition of Riemann integrable functions is Riemann integrable. See Exercise 6. All conditions for Theorem 6.3.4 must be satisfied before drawing a conclusion. ■

Definition 6.3.7. If a function $f:[a,b] \to \Re$ is Riemann integrable, then we can define $\int_b^a f$ by $\int_b^a f = -\int_a^b f$.

With Definition 6.3.7 on hand, Theorem 6.3.3 is valid for $c > b$ or $c < a$ as well, provided f is integrable on the necessary intervals. Two existence results for the Riemann integral follow. The first of these theorems has a number of different versions; we have chosen a more common version.

THEOREM 6.3.8. *(First Mean Value Theorem for Integrals, sometimes known as Bonnet's[3] Mean Value Theorem) Suppose a function $f:[a,b] \to \Re$ is continuous and*

---

[3]Pierre Ossian Bonnet (1819–1892), a French mathematician well known in analysis and differential geometry, corrected and expanded Cauchy's method of solving partial differential equations of the first order having a number of variables.

a function $g:[a,b]\to\mathfrak{R}$ *is Riemann integrable and nonnegative. Then there exists* $c\in[a,b]$ *such that* $\int_a^b fg = f(c)\int_a^b g$.

Proof. Since f is continuous on $[a,b]$, by the extreme value theorem, Theorem 4.3.2, there exists m and M, real constants, such that $m\le f(x)\le M$ for all $x\in[a,b]$. And since $g(x)\ge 0$ for all $x\in[a,b]$, we have $mg(x)\le f(x)g(x)\le Mg(x)$. Assuming that product of integrable functions is integrable (see Exercise 4), we have that

$$m\int_a^b g \le \int_a^b fg \le M\int_a^b g.$$

Now if $\int_a^b g = 0$, then $\int_a^b fg = 0$ and the desired equality is true. If $\int_a^b g\ne 0$, then we can divide by it to obtain $m\le \int_a^b fg \Big/ \int_a^b g\le M$. And by Bolzano's intermediate value theorem, Theorem 4.3.3, there exists $c\in[a,b]$ such that $f(c)=\int_a^b fg \Big/ \int_a^b g$, which proves the desired result. ■

What happens when the assumptions on g are changed to $g(x)\le 0$ for all $x\in[a,b]$? The familiar mean value theorem for integrals encountered in most elementary calculus textbooks is actually a corollary to the preceding result. Its proof is left as an exercise.

COROLLARY 6.3.9. *If a function* $f:[a,b]\to\mathfrak{R}$ *is continuous, then there exists* $c\in[a,b]$ *such that* $\int_a^b f = f(c)(b-a)$. *The number* $f(c)$ *is called the average value of* f *on* $[a,b]$.

Geometrically, the average value AV for a nonnegative continuous function f is the height of a rectangle whose base is $b-a$ and whose area is equal to the area of the region between $f(x)$ and the x-axis between lines $x=a$ and $x=b$. (See Figure 6.3.1.) The average value is a generalization of Definition 1.10.14, part (c). Why? The idea of Corollary 6.3.9 is demonstrated in Figure 6.3.1.

THEOREM 6.3.10. *(Second Mean Value Theorem for Integrals) If a function* $f:[a,b]\to\mathfrak{R}$ *is monotone, then there exists* $c\in[a,b]$ *such that* $\int_a^b f = f(a)(c-a)+f(b)(b-c)$.

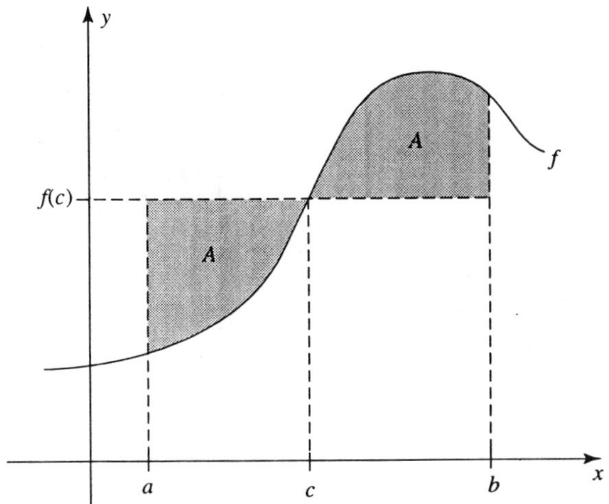

Figure 6.3.1

Proof. Since f is monotone, by Theorem 6.2.2 it is Riemann integrable. Let the function $g:[a,b]\to\Re$ be defined by $g(x)=f(a)(x-a)+f(b)(b-x)$. Since g is a linear function, it is continuous and thus Riemann integrable. In addition, $g(a)=f(b)(b-a)$ and $g(b)=f(a)(b-a)$. Again, since f is monotone, the values of $f(k)(b-a)$ for any $k\in[a,b]$ and $\int_a^b f$ are between $g(a)$ and $g(b)$. Since g is continuous, by Bolzano's intermediate value theorem there exists a particular $k\in[a,b]$, say $k=c$, such that $g(c)=\int_a^b f$. Hence, $\int_a^b f=g(c)=f(a)(c-a)+f(b)(b-c)$, which is what we wished to prove. ∎

Theorem 6.3.10 is revealed in studying the increasing function f in Figure 6.3.2. Before we close the section, consider Exercise 10 at the end of this section. A bounded function $f:[a,b]\to\Re$ with finitely many discontinuities is Riemann integrable. Note also that Dirichlet's function in Exercise 8 of Section 6.2 and the function in Exercise 8 of Section 6.7 both have infinitely many discontinuities on $[0,1]$, yet they are also Riemann integrable. However, the function f as given in Example 6.1.8 is not. The reason is that this last-mentioned function has just too many points of discontinuity. This topic of how many discontinuities is too many is addressed in full when the concept of a "measure" is discussed in real analysis texts.

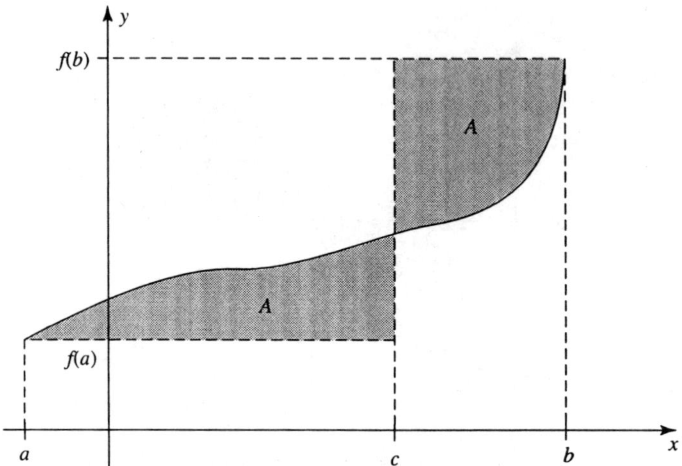

Figure 6.3.2

Exercises 6.3

1. Suppose that f and g are Riemann integrable on $[a,b]$ and are such that $\int_a^b f \le \int_a^b g$.

(a) Is it true that $f(x) \le g(x)$ for all $x \in [a,b]$? Explain.
(b) Is it true that there exists $c \in [a,b]$ such that $f(c) \le g(c)$? Explain.

2. If a function $f:[a,b] \to \Re$ is Riemann integrable, and if there exist real constants m and M such that $m \le f(x) \le M$ for all $x \in [a,b]$, prove that

$$m(b-a) \le \int_a^b f \le M(b-a).$$

3. **(a)** Prove part (b) of Theorem 6.3.1.
(b) Prove Theorem 6.3.2 by using Exercise 6 of Section 6.1.

4. **(a)** If functions $f, g:[a,b] \to \Re$ are Riemann integrable, then prove that $fg \in R[a,b]$. Note that in general $\int_a^b f \cdot \int_a^b g \ne \int_a^b fg$.
(b) Give an example of two functions $f, g:[a,b] \to \Re$ that are not Riemann integrable, but fg is.

(c) Give an example of two functions $f,g:[a,b]\to\mathfrak{R}$, where $f\in R[a,b]$, $g\notin R[a,b]$, but fg is Riemann integrable.

(d) Prove or find a counterexample to the following statements in which functions $f,g:[a,b]\to\mathfrak{R}$ are such that both f and fg are Riemann integrable.

(i) If f is continuous on $[a,b]$, then $g\in R[a,b]$.

(ii) If f is monotone on $[a,b]$, then $g\in R[a,b]$.

(iii) If f is strictly monotone on $[a,b]$, then $g\in R[a,b]$.

5. (a) If a function $f:[a,b]\to\mathfrak{R}$ is Riemann integrable, prove that $|f|\in R[a,b]$ and that $\left|\int_a^b f\right|\le\int_a^b|f|$.

(b) Give an example of a function f that is not Riemann integrable on $[a,b]$ but $|f|$ is.

6. Give an example of two Riemann integrable functions whose composition is not Riemann integrable.

7. (a) Suppose a function $f:[a,b]\to\mathfrak{R}$ is positive and continuous on $[a,b]$. See Exercise 11 of Section 4.3. Prove that $\frac{1}{f}$ is Riemann integrable on $[a,b]$.

(b) Give an example of a function $f:[a,b]\to\mathfrak{R}$ that is Riemann integrable and positive on $[a,b]$, but $\frac{1}{f}\notin R[a,b]$.

8. If functions $f,g:[a,b]\to\mathfrak{R}$ are Riemann integrable, prove that $f\vee g$ and $f\wedge g$ are also. See Exercises 4(c) and 4(d) in Section 4.1.

9. If a function $f:[a,b]\to\mathfrak{R}$ is Riemann integrable and a function $g:[a,b]\to\mathfrak{R}$ is obtained by altering values of f at finite number of points, prove that g is Riemann integrable and that $\int_a^b f=\int_a^b g$. See Exercise 13 in Section 6.2. Would the above conclusion hold if g was undefined at a finite number of points in $[a,b]$?

10. (a) If a function $f:[a,b]\to\mathfrak{R}$ is bounded and has at most finitely many discontinuities on $[a,b]$, prove that $f\in R[a,b]$.

(b) If a function $f:D\to\mathfrak{R}$ and if there exists an accumulation point a of D, prove that $\int_a^a f=0$. Observe that if a is not an accumulation point of D, then it makes no sense to write such an expression.

11. Determine which of the functions $f_n:[0,1]\to\mathfrak{R}$ with $n=0,1$, and 2, defined by

$$f_n(x)=\begin{cases} x^n \sin\frac{1}{x}, \text{ if } 0<x\le 1 \\ 0, \text{ if } x=0 \end{cases}$$

are Riemann integrable on $[0,1]$. Explain.

12. If $a_n=\sum_{k=1}^{n}\frac{1}{k}-\int_1^n \frac{1}{x}dx$, prove that the sequence $\{a_n\}$ converges.[4]

13. Prove the *Cauchy–Schwarz inequality*, which states that if functions $f,g:[a,b]\to\Re$ are Riemann integrable, then

$$\left(\int_a^b fg\right)^2 \le \left(\int_a^b f^2\right)\left(\int_a^b g^2\right).$$

When does the equality hold? See Exercise 10(a) of Section 1.7.

14. Generalizing the Cauchy–Schwarz inequality, *Hölder's inequality* states that if functions $f,g:[a,b]\to\Re$ are Riemann integrable and if there exist real numbers $p,q\in(1,\infty)$ satisfying $\frac{1}{p}+\frac{1}{q}=1$, then

$$\int_a^b |fg| \le \left(\int_a^b |f|^p\right)^{\frac{1}{p}}\left(\int_a^b |g|^q\right)^{\frac{1}{q}}.$$

Prove this inequality. See Exercise 20(h) in Section 5.2.

15. Prove *Minkowski inequality*, which states that if functions $f,g:[a,b]\to\Re$ are Riemann integrable, then

$$\left(\int_a^b |f+g|^p\right)^{\frac{1}{p}} \le \left(\int_a^b |f|^p\right)^{\frac{1}{p}}+\left(\int_a^b |g|^p\right)^{\frac{1}{p}}$$

for the case when $p=2$. Actually, the result is true for any real $p\ge 1$. Can you prove that? When does the equality hold? See Exercise 10(b) from Section 1.7.

---

[4]The sequence $a_n=\sum_{k=1}^{n}\frac{1}{k}-\int_1^n \frac{1}{x}dx$ converges to a value known as *Euler's constant* and is customarily denoted by γ. Some mathematicians also refer to γ as *Mascheroni's constant* in honor of Lorenzo Mascheroni (1750–1800), the Italian geometer and analyst. The approximate value for γ is 0.57721566. It is not known whether γ is rational or irrational. a_n can also be written as $a_n=1+\frac{1}{2}+\frac{1}{3}+\cdots+\frac{1}{n}-\ln n$. Also, see Exercises 27 and 28 in Section 6.6 and Example 6.8.2.

16. **(a)** Prove Corollary 6.3.9 by showing that it is a consequence of Theorem 6.3.8.

(b) Prove Corollary 6.3.9 without using Theorem 6.3.8.

17. Find a function $f:[a,b]\to\Re$ that is continuous but for which there is no value $c\in[a,b]$ such that

$$\int_a^b f = f(a)(c-a)+f(b)(b-c).$$

18. Find the average value of $f(x)=x^2+1$ on $[0,2]$.

6.4 Integration in Relation to Differentiation

Section 6.3 centered around determining whether or not a given function was Riemann integrable. We will now state and prove results that will often be used in actually finding the value of the Riemann integral, provided it exists. Before presenting the first theorem, however, a specific example rules out the absurd, yet often heard, statement that integration is "opposite" to differentiation.

Example 6.4.1. Give an example of a function that is differentiable but whose derivative is not Riemann integrable.

Answer. Choose, say $f:[-1,1]\to\Re$ defined by

$$f(x)=\begin{cases} x^2\sin\dfrac{1}{x^2}, & \text{if } x\neq 0\\ 0, & \text{if } x=0.\end{cases}$$

Then,

$$f'(x)=\begin{cases} 2x\sin\dfrac{1}{x^2}-\dfrac{2}{x}\cos\dfrac{1}{x^2}, & \text{if } x\neq 0\\ 0, & \text{if } x=0.\end{cases}$$

Why? Next, observe that f' is not Riemann integrable on $[-1,1]$ since it is unbounded. Hence, f differentiable does not imply that f' is Riemann integrable. ■

For those who studied Section 5.6, recall from Problem 5.6.21, part (b), that if a function f is absolutely continuous on $[a,b]$, then f is continuous and of bounded variation on $[a,b]$. Now, by Problem 5.6.18, part (d), f can be written as a difference of two increasing functions p and q. Since p is increasing on $[a,b]$, it can be shown that

p' is integrable. Similarly for q. The moral here is that if f is absolutely continuous, then f' is Riemann integrable. By Problem 5.6.21, part (a), we know that if a function has a bounded derivative it is absolutely continuous. Therefore, the function given in Example 6.4.1 is not absolutely continuous and its derivative is unbounded. See Problem 5.6.21, part (h).

As pointed out in Remark 6.2.3, not every Riemann integrable function f has an antiderivative. However, if it does, and we can readily find it, then the evaluation of the value of the Riemann integral of f can be done quickly. Observe that not every antiderivative of a function can be written in *closed form*.[5] For example, $f(x) = \exp\left(x^2\right)$ has no antiderivatives that can be written in closed form. In Chapter 8, power series will be used to represent these antiderivatives. See Exercise 2 in Section 8.6. Furthermore, recall in Exercise 11 of Section 5.2 that all antiderivatives for a given function differ by an additive constant when they exist. One procedure for how the Riemann integral can be obtained is given in the next theorem.

THEOREM 6.4.2. *(Fundamental Theorem of Calculus) Suppose a function $f:[a,b]\to\Re$ is Riemann integrable on $[a,b]$, and suppose a function $g:[a,b]\to\Re$ is an antiderivative for f. Then $\int_a^b f = g(b)-g(a)$.*

It is customary to write

$$\int_a^b f(x)\,dx = g(x)\Big|_a^b = g(b)-g(a).$$

Proof. Let $P_n = \{x_0, x_1, x_2, \dots, x_n\}$ be any partition of $[a,b]$. By the mean value theorem applied to g on the general subinterval $[x_{k-1}, x_k]$, there exists $c_k \in (x_{k-1}, x_k)$ such that

$$g(x_k)-g(x_{k-1}) = g'(c_k)(x_k - x_{k-1}) = f(c_k)(x_k - x_{k-1}).$$

Therefore,

$$\sum_{k=1}^{n} f(c_k)(x_k - x_{k-1}) = \sum_{k=1}^{n}\left[g(x_k)-g(x_{k-1})\right] = g(x_n)-g(x_0) = g(b)-g(a).$$

If $m_k = \inf\{f(x) \mid x\in[x_{k-1},x_k]\}$ and $M_k = \sup\{f(x) \mid x\in[x_{k-1},x_k]\}$, then $m_k \le f(c_k) \le M_k$. Thus

[5]*Closed form* means that the expression may be expressed as a finite algebraic combination of elementary functions.

$$\sum_{k=1}^{n} m_k\left(x_k - x_{k-1}\right) \le \sum_{k=1}^{n} f(c_k)\left(x_k - x_{k-1}\right) \le \sum_{k=1}^{n} M_k\left(x_k - x_{k-1}\right).$$

Therefore, $L(P,f) \le g(b) - g(a) \le U(P,f)$. Now, since $f \in R[a,b]$, the value of $\int_a^b f$ must equal $g(b) - g(a)$. ■

Note that Theorem 6.4.2 would fail if g', that is, f, was not Riemann integrable. Recall Example 6.4.1. Another popular method in calculus for evaluating Riemann integrals is the use of "integration by parts." The validity of that procedure is proved next.

THEOREM 6.4.3. *(Integration by Parts) Suppose functions $f, g:[a,b] \to \Re$ are differentiable and $f', g' \in R[a,b]$, then*

$$\int_a^b (fg')(x)\,dx = f(b)g(b) - f(a)g(a) - \int_a^b (f'g)(x)\,dx.$$

Proof. Observe that both f and g are continuous and hence Riemann integrable. By Exercise 4(a) of Section 6.3, the products fg' and $f'g$ are both Riemann integrable. Consequently, from Theorem 6.3.1, since $(fg)' = fg' + f'g$, $(fg)' \in R[a,b]$ and

$$\int_a^b (fg)'(x)\,dx = \int_a^b (fg')(x)\,dx + \int_a^b (f'g)(x)\,dx.$$

But, by the fundamental theorem of calculus we have that

$$\int_a^b (fg)'(x)\,dx = f(x)g(x)\Big|_a^b = f(b)g(b) - f(a)g(a).$$

Hence, the result follows. ■

In the preceding result, the value of the Riemann integral is found by actually evaluating two, hopefully easier, Riemann integrals.

THEOREM 6.4.4. *(Indefinite Riemann Integral Theorem) Suppose a function $f:[a,b] \to \Re$ is Riemann integrable. Furthermore, we define a function $F:[a,b] \to \Re$ by $F(x) = \int_a^x f(t)\,dt$. Then*

(a) *F is continuous on $[a,b]$, and thus uniformly continuous.*

(b) *if f is continuous at $x = c \in [a,b]$, then F is differentiable at $x = c$ and $F'(c) = f(c)$.*

Proof of part (a). Since $f \in R[a,b]$, f is bounded on $[a,b]$. Thus, there exists a positive real constant M such that $|f(x)| \le M$ for all $x \in [a,b]$. Let $\varepsilon > 0$ be given. To show that F is uniformly continuous on $[a,b]$, we need to find $\delta > 0$ so that if $|x-t| < \delta$ for any $t, x \in [a,b]$, then $|F(x) - F(t)| < \varepsilon$. To this end, choose $\delta = \dfrac{\varepsilon}{M}$. Then, if $a \le t \le x \le b$ and $|x-t| < \delta$, we can write

$$\begin{aligned}|F(x)-F(t)| &= \left|\int_a^x f(u)\,du - \int_a^t f(u)\,du\right| = \left|\int_t^x f(u)\,du\right| \quad \text{(Why?)}\\ &\le \int_t^x |f(u)|\,du \le \int_t^x M\,du = M(x-t) < M\cdot\frac{\varepsilon}{M} = \varepsilon.\end{aligned}$$

Hence, F is uniformly continuous on $[a,b]$, and thus continuous on $[a,b]$.

Proof of part (b). Suppose f is continuous at $x = c \in [a,b]$, and let $\varepsilon > 0$ be given. Then, there exists $\delta > 0$ such that

$$|f(x) - f(c)| < \boxed{\varepsilon} \quad \text{for all } x \in [a,b] \text{ with } |x-c| < \delta.$$

We will show that

$$\lim_{x\to c} \frac{F(x)-F(c)}{x-c} = f(c).$$

So if $0 < |x-c| < \delta$, we can write

$$\begin{aligned}\left|\frac{F(x)-F(c)}{x-c} - f(c)\right| &= \left|\frac{1}{x-c}\int_c^x f(u)\,du - f(c)\right| = \left|\frac{1}{x-c}\left(\int_c^x f(u)\,du - \int_c^x f(c)\,du\right)\right|\\ &\le \frac{1}{|x-c|}\left|\int_c^x |f(u)-f(c)|\,du\right| < \frac{1}{|x-c|}\left|\int_c^x \varepsilon\,du\right| = \frac{1}{|x-c|}\cdot\varepsilon|x-c| = \varepsilon. \text{ Why?}\end{aligned}$$

Hence, F is differentiable at $x = c$ and $F'(c) = f(c)$. ■

Remark 6.4.5. Here are several observations pertaining to Theorem 6.4.4.

(a) Due to Exercise 12 of Section 6.2, the function F is well defined for all $x \in [a,b]$. In fact, $F(a) = 0$ and $F(b) = \int_a^b f$.

(b) The process of Riemann integration creates a "nicer" function than the original one.

(c) Part (b) of Theorem 6.4.4 shows that a continuous function f must have an antiderivative. The family of all antiderivatives for a continuous function

f that differ by an additive constant is often denoted by an *indefinite integral* $\int f$, $\int f\,dx$ or $\int f(x)\,dx$. Thus,

$$\frac{d}{dx}\int f(x)\,dx = f(x) \text{ and } \int f'(x)\,dx = f(x) + C,$$

with C a real constant. Worth noting is that integration by parts that involves an indefinite integral yields the formula

$$\int (fg')(x)\,dx = f(x)g(x) - \int (f'g)(x)\,dx,$$

with C a real constant.

(d) A corollary to Theorem 6.4.4, known as the *Second Fundamental Theorem of Calculus*, says that if $f:[a,b] \to \mathfrak{R}$ is continuous, and $F:[a,b] \to \mathfrak{R}$ is defined by $F(x) = \int_a^x f(t)\,dt$, then $F'(x) = f(x)$ for all $x \in [a,b]$.

(e) Using Theorem 6.4.4 gives a slicker proof for a weaker version of the fundamental theorem of calculus, given as Exercise 3.

(f) If f is continuous and positive, then F gives the area under the curve as shaded in Figure 6.4.1. ■

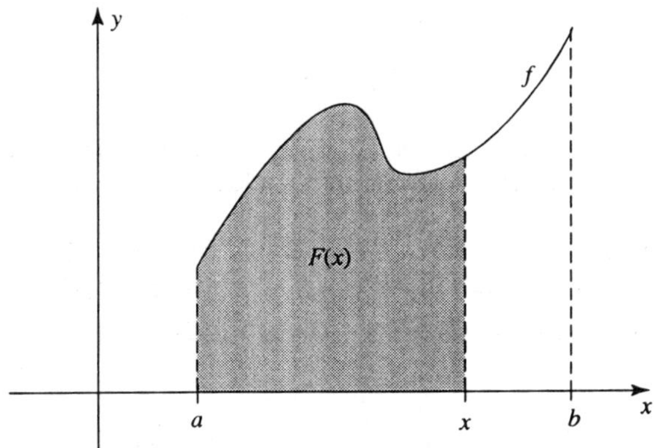

Figure 6.4.1

If in part (c) of Remark 6.4.5 we let $u = f(x)$ and $v = g(x)$, then the integration by parts formula can be written in a familiar way as

$$\int u\,dv = uv - \int v\,du.$$

For definite integrals this can be written as

$$\int_{v_1}^{v_2} u\,dv = (u_2 v_2 - u_1 v_1) - \int_{u_1}^{u_2} v\,du.$$

This formula, in turn, is used to illustrate the areas of 4 regions A_1, A_2, A_3, and A_4, where $A_1 = \int_{v_1}^{v_2} u\,dv$, $A_2 = \int_{u_1}^{u_2} v\,du$, $A_3 = u_2 v_2 =$ area of a large rectangle, and $A_4 = u_1 v_1 =$ area of a small rectangle. See Figure 6.4.2.

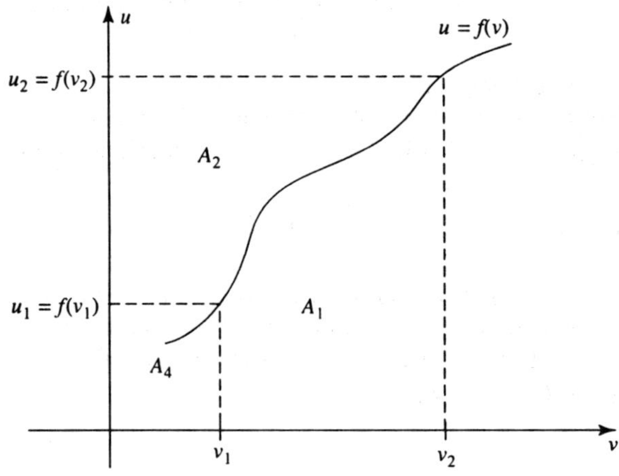

Figure 6.4.2

Another popular method used in calculus for evaluating Riemann integrals is recognized as a change of variables, commonly called a *substitution*.

THEOREM 6.4.6. *(Change of Variable Theorem) Suppose there exists a differentiable function* $g:[c,d] \to [a,b]$, *and* $g' \in R[c,d]$. *Also, suppose a function* $f:[a,b] \to \Re$ *is continuous with* $a = g(c)$ *and* $b = g(d)$. *Then,*

$$\int_a^b f(x)\,dx = \int_c^d (f \circ g)(x) g'(x)\,dx.$$

Proof. Define functions F and h by

$$F(x) = \int_a^x f(t)\,dt \text{ with } x \in [a,b] \text{ and } h(x) = (F \circ g)(x) \text{ with } x \in [c,d].$$

Since f is continuous, F is differentiable and $F' = f$. Furthermore, since g is differentiable, so is h. Therefore, by the chain rule, Theorem 5.2.3,

$$h'(x) = (F' \circ g)(x)g'(x) = (f \circ g)(x)g'(x)$$

for all $x \in [c,d]$. Since h' is the product of two Riemann integrable functions, by Exercise 4(a) of Section 6.3, $h' \in R[c,d]$. Hence,

$$\int_c^d (f \circ g)(x)g'(x)\,dx = \int_c^d h'(x)\,dx = h(d) - h(c) \quad \text{(By Theorem 6.4.2.)}$$

$$= F(g(d)) - F(g(c)) = F(b) - F(a) = \int_a^b f(x)\,dx. \qquad \blacksquare$$

The following is a consequence of Theorem 6.4.4.

THEOREM 6.4.7. *Suppose a function $g:[c,d] \to [a,b]$ (not necessarily a surjection) is differentiable, and a function $f:[a,b] \to \mathfrak{R}$ is continuous. If*

$$F(x) = \int_a^{g(x)} f(t)\,dt$$

with $x \in [c,d]$, then F is differentiable and $F'(x) = (f \circ g)(x)g'(x)$.

Proof. Define a function $G:[a,b] \to \mathfrak{R}$ by $G(x) = \int_a^x f(t)\,dt$. Then, $F(x) = (G \circ g)(x)$, with $x \in [c,d]$. By the indefinite integral theorem we have that $G'(x) = f(x)$. Also, since G and g are both differentiable, by the chain rule, we have

$$F'(x) = (G' \circ g)(x)g'(x) = (f \circ g)(x)g'(x),$$

which is the desired result. $\blacksquare$

Exercises 6.4

1. **(a)** Suppose a function $f:[a,b] \to \mathfrak{R}$ is differentiable and $f' \in R[a,b]$. Prove that $\int_a^b f'(x)\,dx = f(b) - f(a)$.

(b) Verify that

$$\int_a^b x^r\,dx = \left.\frac{x^{r+1}}{r+1}\right|_a^b$$

if r is a real number different from -1 and $[a,b]$ is any interval.

2. If a function $f:[a,b]\to\Re$ is continuous, find an antiderivative for f.

3. Suppose a function $f:[a,b]\to\Re$ is continuously differentiable on $[a,b]$. Then use Theorem 6.4.4 to prove that $f'\in R[a,b]$ and that

$$\int_a^b f'(x)\,dx = f(b)-f(a).$$

Why is this a weaker result than Theorem 6.4.2?

4. Give an example of a function f for which the weaker version of the fundamental theorem of calculus, as given in Exercise 3, does not apply, but for which Theorem 6.4.2 does.

5. Evaluate the given expressions, if possible.

(a) $\displaystyle\int_0^{\frac{1}{2}} \frac{x^3}{\sqrt{1-x^2}}\,dx$ (Use Theorem 6.4.6, if possible.)

(b) $\displaystyle\int_0^{\frac{1}{2}} \frac{x^3}{\sqrt{1-x^2}}\,dx$ (Use Theorem 6.4.3, if possible.)

(c) $\displaystyle\int \sin(\ln x)\,dx$

(d) $\displaystyle\int_0^1 \arctan x\,dx$

(e) $\displaystyle\int_1^2 \frac{\sin(\ln x)}{x}\,dx$

(f) $\displaystyle\int_0^1 \exp\sqrt{x}\,dx$

(g) $\displaystyle\int_0^1 \frac{\exp\sqrt{x}}{\sqrt{x}}\,dx$

(h) $\displaystyle\int \frac{dx}{4x^2+4x+2}$

(i) $\displaystyle\int e^x\left(\frac{1}{x}-\frac{1}{x^2}\right)dx$

(j) $\displaystyle\int x^{-4}\exp\left(-\frac{1}{x}\right)dx$

(k) $\displaystyle\int_1^2 \ln x\,dx$

(l) $\displaystyle\int_{-3}^2 x^{-2}\,dx$

(m) $\displaystyle\int \frac{\sin\sqrt{x}}{\sqrt{x}}\,dx$

(n) $\displaystyle\int_0^1 x\arctan x\,dx$

(o) $\displaystyle\int \frac{dx}{\left(x^2+4\right)^2}$

(p) $\displaystyle\int \frac{3x^2+5x+2}{x^3+2x^2+2x}\,dx$

6. (a) If a function $f:[a,b]\to\Re$ is continuous and $F(x)=\int_x^b f(t)\,dt$ with $x\in[a,b]$, prove that F is differentiable and $F'(x)=-f(x)$.

(b) If f is continuous on $\Re$, and g and h are differentiable on $\Re$, show that

$$\frac{d}{dx}\int_{g(x)}^{h(x)} f(t)\,dt = f\big(h(x)\big)h'(x)-f\big(g(x)\big)g'(x).$$

7. If $F(x)$ is as given, find $F'(x)$ when possible.

(a) $F(x)=\int_1^x \sqrt{2+t^3}\,dt$ (b) $F(x)=\int_{-x}^1 \sqrt{t^2+1}\,dt$

(c) $F(x)=\int_{2x}^{x^2} \arctan\left(t^2\right)dt$ (d) $F(x)=\int_{\sqrt{x}}^{x^3} \sqrt{t}\sin t\,dt$

8. Suppose a function $f:[a,b]\to\Re$ is Riemann integrable. If a function $F:[a,b]\to\Re$ is defined by $F(x)=\int_a^x f(t)\,dt$, prove that F is Lipschitzian. Compare this result with Theorem 6.4.4.

9. Suppose a function $f:[0,b]\to\Re$, $b>0$, is continuous and $f(x)\neq 0$, for all $x\in(0,b)$. Moreover, suppose that $[f(x)]^2=2\int_0^x f(t)\,dt$ for all $x\in[0,b]$. Prove that $f(x)=x$ for all $x\in[0,b]$.

10. Evaluate the given limits when possible. See Exercise 10 in Section 6.2.

(a) $\lim\limits_{n\to\infty}\frac{1}{n^{16}}\sum_{k=1}^n k^{15}$ (b) $\lim\limits_{n\to\infty}\frac{1}{n^{15}}\sum_{k=1}^n k^{15}$

(c) $\lim\limits_{n\to\infty}\frac{1}{n^{17}}\sum_{k=1}^n k^{15}$ (d) $\lim\limits_{n\to\infty} n^{-\frac{3}{2}}\sum_{k=1}^n \sqrt{k}$

(e) Conclude from part (d) that when n is large we can write

$$\sqrt{1}+\sqrt{2}+\sqrt{3}+\cdots+\sqrt{n}\approx\frac{2}{3}n^{\frac{3}{2}}.$$

(f) Show that $\lim\limits_{n\to\infty}\sum_{k=1}^n \frac{n}{k^2+n^2}=\frac{\pi}{4}$.

11. **(a)** If an odd function $f:[-a,a]\to\Re$, $a\in\Re$, is Riemann integrable, prove that $\int_{-a}^{a} f\,dx = 0$.

(b) If an even function $f:[-a,a]\to\Re$, $a\in\Re$, is Riemann integrable, prove that $\int_{-a}^{a} f\,dx = 2\int_{0}^{a} f\,dx$.

12. Determine whether or not the given statements are true.

(a) If f is continuous on $\Re$, then $\frac{d}{dx}\int_0^x f(t)\,dt = \int_0^x \left[\frac{d}{dt}f(t)\right]dt$.

(b) If $f'(x)=\sin x-\cos x$, then $f(x)=\int_0^x (\sin t-\cos t)\,dt$.

13. If a function $f:\Re\to\Re$ is periodic on $\Re$ with period p, prove that for any real constant a we have the following:

(a) $\int_0^a f = \int_p^{a+p} f$ **(b)** $\int_0^p f = \int_a^{a+p} f$

14. Prove that the remainder term in Taylor's theorem, Theorem 5.3.8, can be replaced by

$$R_n(x) = \frac{1}{n!}\int_a^x f^{(n+1)}(t)(x-t)^n\,dt.$$

15. Use Exercise 14 and the first mean value theorem for integrals, Theorem 6.3.8, to prove Taylor's theorem, Theorem 5.3.8, in the case $f^{(n+1)}$ is continuous.

16. Use Exercise 14 and Corollary 6.3.9 to prove *Cauchy's form of the remainder*

$$R_n(x) = \frac{f^{(n+1)}(c)}{n!}(x-c)^n(x-a)$$

in Taylor's theorem for the case when $f^{(n+1)}$ is continuous.

17. Find a function f that satisfies $f'(x)=2xf(x)$ for all x and $f(0)=1$. See Exercise 13 in Section 8.4.

18. Prove that $\ln x \le 2\left(\sqrt{x}-1\right)$ for all $x\ge 1$.

19. Determine the values of x for which $G(x)=\int_0^x (1+[t])\,dt$ is differentiable and find its derivative.

20. Archimedes discovered that the area under a parabolic arch is always $\frac{2}{3}$ times the base times the height, as shown in Figure 6.4.3.

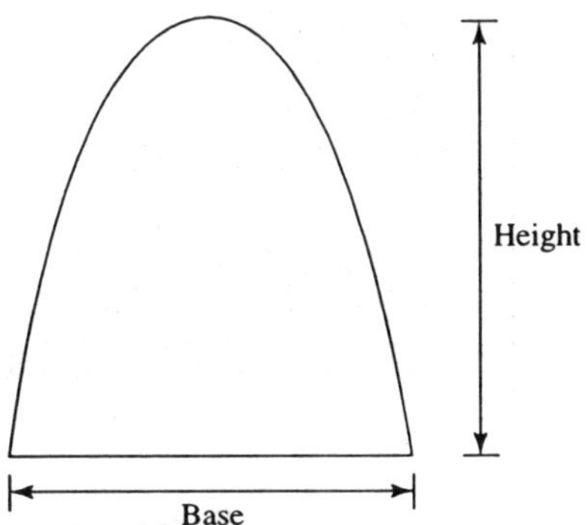

Figure 6.4.3

(a) Prove that this is indeed a true discovery.

(b) Illustrate the validity of part (a) on the parabola $y=-x^2+4x$, with $x\in[0,4]$.

21. (a) Show that a continuous function f is symmetric with respect to the point (a,b) if and only if $f(a-x)+f(a+x)=2f(a)$. See Exercise 14 in Section 1.2.

(b) Show that if f is continuous on $[a,b]$, and f is symmetric with respect to the midpoint $\frac{a+b}{2}$ of $[a,b]$, then

$$f(x)+f(a+b-x)=2f\left(\frac{a+b}{2}\right)=f(a)+f(b)$$

for all $x\in[a,b]$.

(c) Show that if f is continuous on $[a,b]$, and is symmetric with respect to the midpoint $\frac{a+b}{2}$, then

$$\int_a^b f(x)\,dx=(b-a)f\left(\frac{a+b}{2}\right).$$

See Figure 6.4.4.

(d) Use part (c) to evaluate $\int_1^3 \frac{dx}{4+2^x}$.

(e) Use part (c) to evaluate $\int_0^{2\pi} \frac{dx}{1+e^{\sin x}}$.

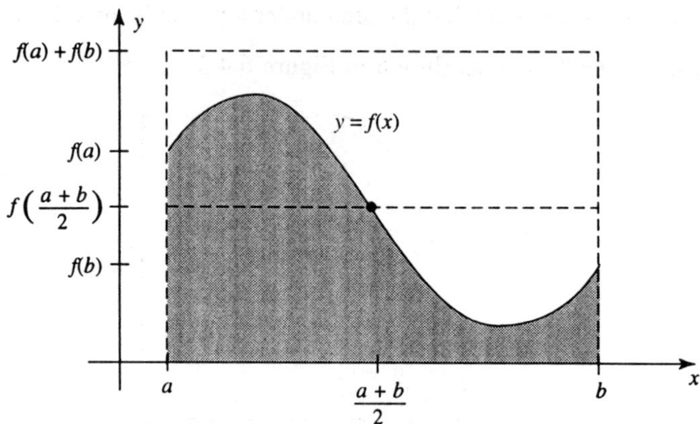

Figure 6.4.4

22. Suppose f is defined on $[0,1]$, and is such that $0 < f'(x) \le 1$ and $f(0) = 0$. Prove that

$$\left[\int_0^1 f(x)\,dx\right]^2 \ge \int_0^1 [f(x)]^3\,dx.$$

23. **(a)** *(Generalized Second Mean Value Theorem for Integrals)* Let $f:[a,b] \to \Re$ be a function that is continuous and increasing. And let $g:[a,b] \to \Re$ be nonnegative and Riemann integrable. Prove that there exists $c \in [a,b]$ such that

$$\int_a^b fg = f(a)\int_a^c g + f(b)\int_c^b g.$$

(b) Explain why the result in part (a) is a generalization of Theorem 6.3.10.

6.5 Improper Integral

In light of Definition 6.1.6, Riemann integrals, often called *definite integrals*, are defined for bounded functions on compact intervals. Definite integrals may be extended to include unbounded functions and/or intervals that are not compact. The resulting integrals are referred to as *improper*.

Definition 6.5.1. Suppose a function $f:[a,\infty)\to\Re$ is Riemann integrable on $[a,b]$ for every $b\ge a\in\Re$. If $\lim_{b\to\infty}\int_a^b f$ is a finite number A, then the Riemann integral of f is *improper* on $[a,\infty)$, is denoted by $\int_a^\infty f$, and *converges* to A. If the preceding limit is not finite, then $\int_a^\infty f$ *diverges.*

In a similar fashion, if $\lim_{a\to-\infty}\int_a^b f$ is finite, then we define it to be equal to $\int_{-\infty}^b f$. These two improper Riemann integrals satisfy a *linear property* (see Exercise 1). Observe also that for the function f in Definition 6.5.1, $\int_a^\infty f$ converges if and only if $\lim_{b\to\infty}\int_b^\infty f=0$. Why? Also, $\int_a^\infty f$ converging to A, is equivalent to writing $\int_a^\infty f=A$.

Example 6.5.2. Evaluate $\int_1^\infty xe^{-x}\,dx$, if possible.

Answer. Since $f(x)=xe^{-x}$ is continuous on $[1,\infty)$, it is Riemann integrable on $[1,b]$ for all $b>1$. Therefore, we can write

$$\int_1^\infty xe^{-x}\,dx=\lim_{b\to\infty}\int_1^b xe^{-x}\,dx=\lim_{b\to\infty}\left(-\frac{b}{e^b}+\frac{1}{e}+\int_1^b e^{-x}\,dx\right)=\frac{2}{e}. \qquad\blacksquare$$

Remark 6.5.3. Note that Theorem 6.3.2 can be extended to the improper Riemann integral defined at the beginning of the section. Now, suppose functions $f,g:[a,\infty)\to\Re$ are Riemann integrable on $[a,b]$ for every $b\ge a\in\Re$, and $0\le f(x)\le g(x)$ for all $x\ge a$.

(a) If $\int_a^\infty g$ converges, then $\int_a^\infty f$ converges.

(b) If $\int_a^\infty f$ diverges, then $\int_a^\infty g$ diverges.

This *comparison test for improper integrals* is especially useful in determining whether or not the given improper Riemann integral converges without necessarily desiring to know its value. For example, we can verify that $\int_{\frac{3}{2}}^{\infty} \frac{dx}{\sqrt{x^4+2}}$ converges simply by noting that

$0 \le \frac{1}{\sqrt{x^4+2}} \le \frac{1}{\sqrt{x^4}} = \frac{1}{x^2}$ for all $x \ge \frac{3}{2}$, and that $\int_{\frac{3}{2}}^{\infty} \frac{1}{x^2}\,dx$ converges. Why? ■

But, what if we were to determine the convergence or divergence of $\int_{\frac{3}{2}}^{\infty} \frac{dx}{\sqrt{x^4-2}}$ instead? Since $\frac{1}{\sqrt{x^4-2}}$ and $\frac{1}{\sqrt{x^4+2}}$ tend to 0 at about the same rate, we would probably like to say that $\int_{\frac{3}{2}}^{\infty} \frac{dx}{\sqrt{x^4-2}}$ converges since $\int_{\frac{3}{2}}^{\infty} \frac{dx}{\sqrt{x^4+2}}$ converges. Unfortunately, we cannot compare $\frac{1}{\sqrt{x^4-2}}$ to $\frac{1}{x^2}$ as was done in Remark 6.5.3. Thus, either we find a different function to compare $\frac{1}{\sqrt{x^4-2}}$ with, or use the same function $\frac{1}{x^2}$ but a different test. We choose the latter. Suppose functions f and g are of the same order, that is $\lim_{x\to\infty} \frac{f(x)}{g(x)} = M \ne 0$, then $\int_a^{\infty} f$ and $\int_a^{\infty} g$ both converge or both diverge. Now, can we use the function $\frac{1}{x^2}$ to show that $\int_{\frac{3}{2}}^{\infty} \frac{dx}{\sqrt{x^4-2}}$ converges? This test, presented in Exercise 17 known as a *limit comparison test for improper integrals*, can be further generalized. Comparison and limit comparison tests will again be presented for infinite series in Section 7.2.

Another type of improper integral occurs when a function becomes infinite over the interval of integration. We stereotype these improper integrals into four categories, each defined in the next four definitions.

Definition 6.5.4. Suppose an unbounded function $f:[a,b) \to \Re$ is Riemann integrable on $[a,c]$ for every $c \in [a,b)$. We define an improper Riemann integral $\int_a^b f$ by $\lim_{c\to b^-} \int_a^c f$ provided this limit is a finite number A. Then we say that $\int_a^b f$ *converges* to

A and write $\int_a^b f = A$. If the limit is not finite, then $\int_a^b f$ *diverges.*

Example 6.5.5. Evaluate $\int_0^8 (8-x)^{-\frac{4}{3}}\,dx$, if possible.

Answer. Since $f(x) = (8-x)^{-\frac{4}{3}}$ is continuous on $[0,8)$, we write

$$\int_0^8 (8-x)^{-\frac{4}{3}}\,dx = \lim_{c\to 8^-} \int_0^c (8-x)^{-\frac{4}{3}}\,dx = \lim_{c\to 8^-}\left(\frac{3}{\sqrt[3]{8-c}} - \frac{3}{2}\right) = +\infty.$$

Why? But since this limit is not finite, the given expression diverges. ■

Definition 6.5.6. Suppose an unbounded function $f:(a,b]\to\Re$ is Riemann integrable on $[c,b]$ for every $c\in(a,b]$, then $\int_a^b f = \lim_{c\to a^+}\int_c^b f$ provided this limit is finite.

Remark 6.5.7. Although in the previous two definitions the given limits are true for the bounded functions, the Riemann integrals for those functions would not be considered improper. Generally, these definitions are used when the functions under consideration are unbounded at endpoints of their interval of integration. The function $f(x) = \frac{1}{\sqrt{x}}$ on $(0,1]$ is an example of an improper Riemann integrable function, whereas $g(x) = \frac{1}{x^2}$ on $(0,1]$ is not. Why? See Exercise 3(b). ■

Now, what happens if a function f becomes unbounded within its interval of integration, that is, other than at the endpoints? If the interval is $[a,\infty)$ or $(-\infty,b]$ and a continuous function f grows without bound as x gets large, then trivially, an improper Riemann integral for f cannot exist. Try to visualize the growing area under this curve. Now, if f is not continuous, then the improper Riemann integral may very well exist. See Exercise 15(b). What if a function is unbounded in a neighborhood of some real number c? We define such a situation as follows.

Definition 6.5.8. Suppose an unbounded function $f:[a,c)\cup(c,b]\to\Re$ is Riemann integrable on $[a,t]\cup[k,b]$ for all $t\in[a,c)$ and all $k\in(c,b]$. Then,

$$\int_a^b f = \int_a^c f + \int_c^b f = \lim_{t\to c^-}\int_a^t f + \lim_{k\to c^+}\int_k^b f$$

provided both limits are finite.

If one of the preceding limits is not finite, then $\int_a^b f$ diverges. In addition, the values t and k in the definition must be kept independent of each other. See Exercise 7. Often, improper integrals appear to be definite. Evaluating an improper Riemann integral using methods designed for definite integrals can lead to erroneous results. For example, $\int_{-3}^{2} \frac{1}{x^2}\,dx$, as given in Exercise 5(l) of Section 6.4, is an improper integral that diverges. However, if we incorrectly evaluate it using the formula from Exercise 1(b) of Section 6.4 we will obtain $-\frac{2}{3}$, which could not possibly represent an area under the positive function $\frac{1}{x^2}$.

Definition 6.5.9. If both $\int_{-\infty}^{a} f$ and $\int_{a}^{\infty} f$ converge for some real constant a, then $\int_{-\infty}^{\infty} f$ converges and its value is given by $\int_{-\infty}^{\infty} f = \int_{-\infty}^{a} f + \int_{a}^{\infty} f$.

See Exercise 18 in Section 6.7.

Definition 6.5.10. The *Cauchy Principal Value* (CPV) of $\int_{-\infty}^{\infty} f$ is defined by $\lim_{a\to\infty} \int_{-a}^{a} f$ provided it is finite.

Example 6.5.11. Discuss convergence of $\int_{1}^{\infty} \frac{dx}{\sqrt{|(x-1)(x-2)(x-5)|}}$.

Answer. Observe that the function $f:(1,\infty)\to\Re$, defined by

$$f(x)=\frac{1}{\sqrt{|(x-1)(x-2)(x-5)|}}$$

is continuous except at 2 and 5. Thus, to evaluate this improper Riemann integral we must split it at values between 1, 2, 5, and $+\infty$. Therefore, we write

$$\int_{1}^{\infty} f = \int_{1}^{1.5} f + \int_{1.5}^{2} f + \int_{2}^{3} f + \int_{3}^{5} f + \int_{5}^{8} f + \int_{8}^{\infty} f.$$

The integral on the left-hand side of the equality converges if and only if each integral on the right-hand side is finite. This is, in fact, the case here. Compare this with Exercise 8 in Section 7.2. ■

Definition 6.5.12. Suppose that $-\infty < a < b \le \infty$ and $f:[a,b) \to \Re$ is such that $\int_a^b f$ exists.

(**a**) If $\int_a^b |f|$ converges, then we say that $\int_a^b f$ *converges absolutely* and that f is an *absolutely improper Riemann integrable* function.

(**b**) If $\int_a^b |f|$ diverges, then we say that $\int_a^b f$ *converges conditionally* and that f is a *conditionally improper Riemann integrable* function.

For every nonnegative improper Riemann integrable function $f:[a,b) \to \Re$ with $-\infty < a < b \le \infty$, $\int_a^b f$ converges absolutely since $|f| = f$.

THEOREM 6.5.13. *Suppose a function $f:[a,b) \to \Re$ with $-\infty < a < b \le \infty$ and $\int_a^b f$ is absolutely convergent. Then $\int_a^b f$ converges and $\left|\int_a^b f\right| \le \int_a^b |f|$.*

Proof. Suppose $\int_a^b f$ is absolutely convergent. Define a function $g:[a,b) \to \Re$ by $g(x) = |f(x)| - f(x)$. Then $0 \le g(x) \le 2|f(x)|$ for all $x \in [a,b)$. And since f is an absolutely improper Riemann integrable function, by the comparison test from Remark 6.5.3, part (a), $\int_a^b g$ converges. Now since $f = |f| - g$, by Exercise 1, f is an improper Riemann integrable function. The desired inequality is obtained from the fact that $\left|\int_a^c f\right| \le \int_a^c |f|$ for any $c \in [a,b)$. ■

Example 6.5.14. Prove that for the function $f:[1,\infty) \to \Re$, defined by $f(x) = \frac{\sin x}{x^p}$, with $p > 1$, $\int_1^\infty f$ converges absolutely.

Proof. Observe that $\left|\dfrac{\sin x}{x^p}\right| \le \dfrac{1}{x^p}$ and $\int_1^\infty x^{-p}\,dx$ converges for $p > 1$. See Exercise 3.

Thus, by the comparison test, $\int_1^\infty \left|\dfrac{\sin x}{x^p}\right| dx$ converges for $p > 1$. Hence, by Definition 6.5.12, part (a), f is absolutely improper Riemann integrable. ■

Remark 6.5.15. The converse of Theorem 6.5.13 is not true. That is, if $\int_a^b f$ converges, then $\int_a^b |f|$ need not converge (see Exercise 19). ■

Tests for convergence of improper Riemann integrals make up the final two results of this section.

THEOREM 6.5.16. *(Abel's[6] Test) Suppose that $-\infty < a < b \le \infty$, and that a function $f:[a,b) \to \mathfrak{R}$ is bounded and monotonic. If a function $g:[a,b) \to \mathfrak{R}$ is an improper Riemann integrable, then so is the product fg. That is, $\int_a^b fg$ is convergent.*

THEOREM 6.5.17. *(Dirichlet's Test) Suppose that $-\infty < a < b \le \infty$, and that a function $f:[a,b) \to \mathfrak{R}$ is differentiable and satisfies $\lim_{x \to b^-} f(x) = 0$ with $\int_a^b f'$ converging absolutely. Furthermore, if a function $g:[a,b) \to \mathfrak{R}$ is continuous and the antiderivative $G(x) = \int_a^x g(t)\,dt$ is bounded for all $x \in [a,b)$, then $\int_a^b fg$ converges.*

Example 6.5.18. Verify that $\int_1^\infty x^p \sin x\,dx$ diverges for $p > 0$.

Proof. Suppose that the given improper Riemann integral converges for some $p > 0$ and consider two functions $f(x) = x^{-p}$ and $g(x) = x^p \sin x$. Also, due to the contradictory assumptions, the function $G(x) = \int_1^x t^p \sin t\,dt$ is bounded. By Dirichlet's test, it follows

[6]Niels Henrik Abel (1802–1829), a Norwegian algebraist and analyst, made fundamental contributions to theories of transcendental and elliptic functions, infinite series, and groups.

that $\int_1^\infty fg\,dx = \int_1^\infty \sin x\,dx$ converges. This is, however, a contradiction. Why? ■

Remark 6.5.19. Arguing that a function f does not approach 0 is not enough to verify that a given improper Riemann integral of f diverges. See Exercise 15(a). ■

All of the results given in this section are true for each type of an improper integral.

Exercises 6.5

1. Prove that if functions $f_1, f_2 : [a,\infty) \to \Re$ are improper Riemann integrable, with c_1 and c_2 real constants, then $c_1 f_1 + c_2 f_2$ is improper Riemann integrable and

$$\int_a^\infty (c_1 f_1 + c_2 f_2)\,dx = c_1 \int_a^\infty f_1\,dx + c_2 \int_a^\infty f_2\,dx.$$

2. Give an example that would negate the sometimes true statement that for functions $f_1, f_2 : [a,\infty) \to \Re$, with c_1 and c_2 real constants, we have

$$\int_a^\infty (c_1 f_1 + c_2 f_2)\,dx = c_1 \int_a^\infty f_1\,dx + c_2 \int_a^\infty f_2\,dx.$$

Does this contradict Exercise 1? Explain.

3. **(a)** Find a real number p which would make the function $f(x) = \frac{1}{x^p}$ improper Riemann integrable on $[1,\infty)$.

(b) Find a real number p which would make the function $f(x) = \frac{1}{x^p}$ improper Riemann integrable on $(0,1]$.

(c) How large is the area of the region between $f(x) = \frac{1}{x}$ and the x-axis on the interval $[1,\infty)$? Explain.

(d) *(Gabriel's horn)* How large is the volume of a solid obtained by revolving $f(x) = \frac{1}{x}$ about the x-axis on the interval $[1,\infty)$? One may need to review methods covered in elementary calculus texts.

4. Suppose an increasing function $f : [a,\infty) \to \Re$ is Riemann integrable on $[a,b]$ for all $b \ge a \in \Re$, and $f(x) \ge 0$ for all $x \ge a$. Prove that f is improper Riemann integrable if and only if $f(x) = 0$ for all $x \ge a$.

5. Verify that $\int_a^b f = \lim_{c \to b^-} \int_a^c f$ as given in Definition 6.5.4, can be rewritten as $\int_a^b f = \lim_{h \to 0^+} \int_a^{b-h} f$.

6. Using integration, prove that the function $f:(0,\infty) \to \Re$ defined by $f(x) = \dfrac{1}{x^2}$ is not symmetrical with respect to the line $y = x$. See Exercise 6 in Section 1.8.

7. **(a)** Explain why Definition 6.5.8 did not define $\int_a^b f$ by $\lim_{\varepsilon \to 0^+} \left(\int_a^{c-\varepsilon} f + \int_{c+\varepsilon}^b f \right)$, where this expression is defined to be the *Cauchy principal value* of $\int_a^b f$.

(b) Find the Cauchy principal value (CPV) of $\int_{-1}^1 \dfrac{1}{|x|} dx$, if it exists.

(c) Find CPV of $\int_{-1}^2 \dfrac{4}{x-1} dx$, if it exists.

8. If $\int_{-\infty}^{\infty} f$ converges to A, prove that CPV of $\int_{-\infty}^{\infty} f = A$. Is the converse true?

9. Evaluate the given expressions, provided they are finite.

(a) $\int_0^1 \ln x \, dx$ **(b)** $\int_0^{\infty} \dfrac{dx}{\sqrt{x}(x+1)}$ **(c)** $\int_0^1 x \ln x \, dx$

(d) $\int_0^{\infty} \dfrac{1}{x} \ln \dfrac{1}{x} dx$ **(e)** $\int_{-\infty}^{\infty} \dfrac{dx}{x^2+1}$ **(f)** $\int_0^{\frac{\pi}{2}} \left(\sec^2 x - \sec x \tan x\right) dx$

(g) $\int_0^{\infty} x^n e^{-x} \, dx$ for $n = 0, 1, 2, 3$ (Compare the values of the integral with $n!$.)

(h) $\int_0^{\frac{\pi}{2}} \ln(\sin x) \, dx$ (See Exercise 2(d) in Section 12.2.)

(i) $\int_0^{\infty} \dfrac{\ln x}{x^2+1} dx$

10. Determine whether or not the given expressions converge or diverge.

(a) $\int_0^\infty \exp(-x^2)\,dx$ (See Remark 6.6.7 and Exercise 2(a) in Section 8.6.)

(b) $\int_0^{\frac{\pi}{2}} \frac{dx}{x\sin x}$ **(c)** $\int_1^\infty \frac{dx}{\sqrt{1+x^2}}$ **(d)** $\int_0^1 \frac{dx}{x(x+1)}$

(e) $\int_1^\infty \frac{dx}{x^2(x+1)}$ **(f)** $\int_2^\infty \frac{dx}{x^3-4}$ **(g)** $\int_1^\infty \frac{dx}{3^x-2}$

(h) $\int_0^1 (\ln x)^2\,dx$ **(i)** $\int_0^\infty \frac{1}{\sqrt{x+x^3}}\,dx$ **(j)** $\int_0^\infty \sin(e^x)\,dx$

11. **(a)** Suppose $f:[a,\infty)\to\Re$ is nonnegative and improper Riemann integrable. Prove that if there exists $M>0$ such that $\int_a^b f \le M$ for every $b \ge a$, then $\int_a^\infty f$ is convergent.

(b) Prove part (a) of Remark 6.5.3.

12. Prove that $\int_0^\infty x^2\exp(-x^2)\,dx = \frac{1}{2}\int_0^\infty \exp(-x^2)\,dx$. See Exercise 10(a).

13. If a function $f:\Re\to\Re$ is even and Riemann integrable on any interval $[a,b]$, then prove that $\int_{-\infty}^\infty f = 2\int_0^\infty f$. If f is an odd function, can we say that $\int_{-\infty}^\infty f = 0$? Explain.

14. Evaluate $\lim_{x\to\infty} \frac{1}{x}\int_1^x \sqrt{4+\exp(-t)}\,dt$, if this limit exists.

15. **(a)** Give an example of a bounded function $f:[0,\infty)\to\Re$ that is improper Riemann integrable on $[0,\infty)$, but $\lim_{x\to\infty} f(x)\neq 0$.

(b) Give an example of an unbounded function $f:[0,\infty)\to\Re$ that is improper Riemann integrable on $[0,\infty)$.

16. Find the area of the region bounded by curves $f(x)=\ln x$, $y=0$, $x=0$, and $x=1$, if possible.

17. Prove the following result. Suppose functions $f,g:[a,\infty)\to\Re$ are such that $f(x)\geq 0$, $g(x)>0$, $f,g\in R[a,b]$ for every $b\geq a$, and $\lim_{x\to\infty}\frac{f(x)}{g(x)}=M$.

(**a**) If $0<M<\infty$, then both $\int_a^\infty f$ and $\int_a^\infty g$ either converge or diverge.

(**b**) If $M=0$ and $\int_a^\infty g$ converges, then $\int_a^\infty f$ converges.

(**c**) If $M=\infty$ and $\int_a^\infty g$ diverges to $+\infty$, then $\int_a^\infty f$ diverges to $+\infty$.

18. Give an example of functions $f,g:[a,\infty)\to\Re$ that are positive, $\lim_{x\to\infty}\frac{f(x)}{g(x)}=+\infty$, and where g is an improper Riemann integrable but f is not. Does this contradict problem 17? Explain.

19. Verify that the function $f:[0,\infty)\to\Re$ defined by $f(x)=\operatorname{sinc} x$ is conditionally improper Riemann integrable. Compare this exercise with Example 6.5.14.

20. Determine whether the given functions are absolutely improper Riemann integrable, conditionally improper Riemann integrable, or not improper Riemann integrable.

(**a**) $f(x)=\dfrac{1}{\sqrt{x+1}}$, with $x\geq 0$ (**b**) $f(x)=\dfrac{\cos x}{x\sqrt{x}}$, with $x\geq 2$

(**c**) $f(x)=\dfrac{1}{x}\sin\dfrac{1}{x}$, with $x\in(0,1]$ (**d**) $f(x)=\dfrac{\sin x}{\sqrt{x}}$, with $x\in(0,1]$

21. (**a**) Discuss convergence of $\int_1^\infty \frac{\sin x}{x^p}\,dx$ for various values of p. (See Examples 6.5.14 and 6.5.18 and Exercise 19.)

(**b**) Use Dirichlet's test to show that $\int_1^\infty \frac{\sin x}{xe^x}\,dx$ converges.

22. Verify that the function $f(x)=\dfrac{\sin^2 x}{x^2}$ is absolutely improper Riemann integrable on $[0,\infty)$ and that $\int_0^\infty f=\int_0^\infty \operatorname{sinc} x\,dx$. Notice that the first improper Riemann integral converges absolutely, whereas the second does not. See Exercise 19.[7]

23. Prove Dirichlet's test.

24. Use substitution in such a way that the resulting integral will be a definite integral. In parts (b) and (c), evaluate the resulting definite integral. This process will be useful when computing improper integrals numerically.

(a) $\displaystyle\int_2^\infty \frac{1}{1+x^4}\,dx$ **(b)** $\displaystyle\int_0^\infty \frac{1}{\left(1+x^2\right)^2}\,dx$ **(c)** $\displaystyle\int_0^\infty \frac{x^2}{1+x^6}\,dx$

---

[7]It is customary to define the *sine integral* by $\mathrm{Si}(x)=\displaystyle\int_0^x \frac{\sin t}{t}\,dt$. It can be proved generally using contour integrals in complex analysis that $\displaystyle\lim_{x\to\infty} \mathrm{Si}(x)=\int_0^\infty \mathrm{sinc}\,x\,dx=\frac{\pi}{2}$. Also, see Exercise 25 of Section 6.6, and Section 9.7 in the *Instructor's Supplement*. The *cosine integral* is defined by $\mathrm{Ci}(x)=-\displaystyle\int_x^\infty \frac{\cos t}{t}\,dt$. Both sine and cosine integrals are even functions. Their graphs are given in Figure 6.5.1.

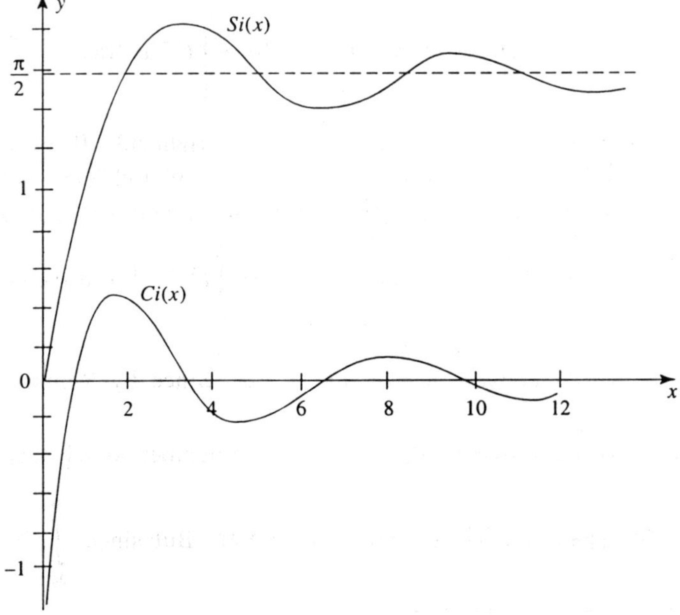

Figure 6.5.1

6.6* Special Functions

Many functions important in both applied mathematics and mathematical physics are defined in terms of improper integrals. Here, in giving just a few of these functions, we hope to stimulate both the interest and desire to pursue further studies in the area of special functions. Material from this section indirectly involves functions in two variables that will be studied more in depth in later chapters. Some integrals will be of the form $\int_a^b f(x,t)\,dt$. In order to evaluate these integrals we simply fix x and treat it as if it was temporarily a constant. x is called a *parameter*.

Definition 6.6.1. The function $\Gamma:\mathfrak{R}^+ \to \mathfrak{R}$ defined by $\Gamma(x)=\int_0^\infty t^{x-1}e^{-t}dt$ is called the *gamma function*.

The above definition of the gamma function was first given by Euler, though Weierstrass defined the gamma function by an infinite product[8] to be introduced in Part 3 of Section 7.6.

Remark 6.6.2. The gamma function is defined by an improper Riemann integral that converges for all $x>0$. We verify this by breaking up the integral at 1, and writing

$$\int_0^\infty t^{x-1}e^{-t}dt = \int_0^1 t^{x-1}e^{-t}dt + \int_1^\infty t^{x-1}e^{-t}dt.$$

Consider the convergence of the first integral on the right side. If $x>1$, then $t^{x-1}e^{-t}$ is bounded for all $t\in[0,1]$, and the integral is thus not improper. Due to continuity, $t^{x-1}e^{-t}$ is Riemann integrable on $[0,1]$. If $0<x\le 1$ and $t>0$, then $t^{x-1}e^{-t}<t^{x-1}$. Since $\int_0^1 t^{x-1}\,dt$ converges, by the comparison test $\int_0^1 t^{x-1}e^{-t}dt$ also converges.

Now consider the convergence of $\int_0^1 t^{x-1}e^{-t}dt$. Since by Remark 5.4.13, part (c), $\lim_{t\to\infty} t^{x+1}e^{-t}=0$, we know that there exists a real number $M>1$ such that $t^{x+1}e^{-t}\le 1$ for all $t\ge M$. Therefore, $t^{x-1}e^{-t}\le t^{-2}$ for $t\ge M$. But since $\int_M^\infty t^{-2}\,dt$ converges, by

[8]This approach to the gamma function can be found in Karl Stromberg's book *An Introduction to Classical Real Analysis* published in 1981 by Wadsworth.

Exercise 3 of Section 6.5 and the comparison test, $\int_M^\infty t^{x-1}e^{-t}dt$ also converges. And since $t^{x-1}e^{-t}$ is Riemann integrable for $t \in [1, M]$, we can conclude that

$$\int_1^\infty t^{x-1}e^{-t}dt = \int_1^M t^{x-1}e^{-t}dt + \int_M^\infty t^{x-1}e^{-t}dt$$

converges. Hence, the gamma function Γ is indeed well defined. ■

We have actually evaluated $\Gamma(n)$ with $n = 0, 1, 2,$ and 3 in Exercise 9(g) of Section 6.5. In Theorem 6.6.4, we will in fact verify that $\Gamma(n) = (n-1)!$ for all $n \in N$, which will in turn prove that $n! = \int_0^\infty t^n e^{-t}\,dt$, and that Γ is an extension of the factorial, sometimes called the *generalized factorial function*. We will elaborate on this later.

LEMMA 6.6.3. *If* $x > 0$, *then* $\Gamma(x+1) = x\Gamma(x)$.

THEOREM 6.6.4. *If* $n \in N$, *then* $\Gamma(n) = (n-1)!$.

Proofs of Lemma 6.6.3 and Theorem 6.6.4 are left as exercises.

Remark 6.6.5. It is customary to extend the domain of the gamma function to include all negative real numbers excluding the negative integers and 0. We do this by defining Γ recursively as $\Gamma(x+1) = x\Gamma(x)$ with $x \in \Re^- \setminus \{-1, -2, \ldots\}$. This resulting new function is also called the gamma function and is well-defined. To see this, we need only verify that Γ is well defined for all negative real numbers excluding 0 and the negative integers. To this end, consider some arbitrary real number $x < 0$. Then, there exists a positive integer k such that $-k < x < 0$. Now we write

$$\begin{aligned}\Gamma(x+k) &= (x+k-1)\Gamma(x+k-1) = (x+k-1)(x+k-2)\Gamma(x+k-2) \\ &= (x+k-1)(x+k-2)\cdots(x)\Gamma(x).\end{aligned}$$

Thus, $\Gamma(x) = \dfrac{\Gamma(x+k)}{(x+k-1)(x+k-2)\cdots(x)}$. And since by Definition 6.6.1, the right-hand side is well defined, so is the left-hand side, provided $x \neq 0, -1, -2, \ldots$. The graph of Γ is given in Figure 6.6.1. ■

Remark 6.6.6. Using the gamma function, the factorial symbol can be extended. For example, $\left(\frac{3}{2}\right)!$ is defined to be $\Gamma\left(\frac{5}{2}\right)$, and $\left(-\frac{3}{2}\right)! = \Gamma\left(-\frac{1}{2}\right) = -2\Gamma\left(\frac{1}{2}\right)$. ■

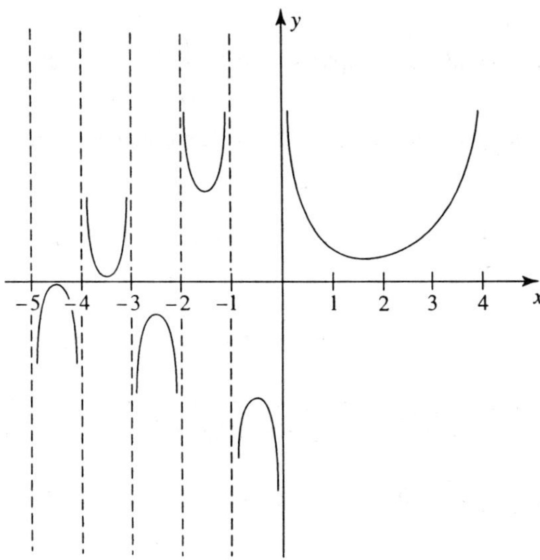

Figure 6.6.1

Remark 6.6.7. Some of the more important properties dealing with the gamma function Γ follow.

(a) $\Gamma\left(\frac{1}{2}\right)=2\int_0^\infty \exp\left(-x^2\right)dx=\sqrt{\pi}$. See Exercises 10(a) and 12 in Section 6.5.

(b) $\Gamma(x)=2\int_0^\infty t^{2x-1}e^{-t^2}\,dt$, for $x>0$. **(c)** $\Gamma(x)=\int_0^1\left(\ln\frac{1}{t}\right)^{x-1}dt,\ x>0$.

(d) $\Gamma\left(n+\frac{1}{2}\right)=\frac{(2n)!\sqrt{\pi}}{4^n n!}$, for $n=0,1,2,\ldots$.

(e) If α is a positive constant, then $\alpha^{-x}\Gamma(x)=\int_0^\infty t^{x-1}e^{-\alpha t}\,dt$, for $x>0$.

(f) If α is a positive constant, then $\Gamma\left(\frac{1}{\alpha}\right)=\alpha\int_0^\infty \exp\left(-t^\alpha\right)dt$.

(g) $\frac{\Gamma(x)\Gamma(y)}{2\Gamma(x+y)}=\int_0^{\frac{\pi}{2}}\left(\cos^{2x-1}u\right)\left(\sin^{2y-1}u\right)du=\frac{1}{2}\int_0^1 t^{x-1}(1-t)^{y-1}\,dt,\ x,y>0$.

(h) $\int_0^{\frac{\pi}{2}}\cos^k x\,dx=\int_0^{\frac{\pi}{2}}\sin^k x\,dx=\left[\sqrt{\pi}\Gamma\left(\frac{k+1}{2}\right)\right]\Big/\left[2\Gamma\left(\frac{k}{2}+1\right)\right]$, for $k>-1$. In case $k=n\in N$, we can write

$$\int_0^{\frac{\pi}{2}} \sin^n x\,dx = \begin{cases} \dfrac{(n-1)!!}{n!!}, & \text{if } n \geq 3 \text{ is odd} \\ \dfrac{(n-1)!!}{n!!}\left(\dfrac{\pi}{2}\right), & \text{if } n \text{ is even.} \end{cases}$$

See Exercise 19 from Section 2.4.

(i) Γ has derivatives of all orders and $\Gamma^{(n)}(x) = \int_0^{\infty} t^{x-1} e^{-t} (\ln t)^n\,dt$ for $x > 0$ and $n \in N$. Thus, Γ is concave upward for $x > 0$.

(j) Γ is logarithmically convex on $(0, \infty)$, which can be proved using Hölder's inequality. See Problem 5.6.24, part (r), and Exercise 14 from Section 6.3, which can be revised for the case of improper integrals.

(k) For large positive values α, $\Gamma(\alpha+1)$ can be approximated by *Stirling's*[9] *formula*

$$\Gamma(\alpha+1) \approx \sqrt{2\pi\alpha}\left(\frac{\alpha}{e}\right)^{\alpha}$$

which can also be rewritten as $\lim\limits_{x\to\infty} \dfrac{e^x \Gamma(x+1)}{x^x \sqrt{2\pi x}} = 1$. A weak form of Stirling's formula is $n^n e^{1-n} < n! < n^n e^{1-n} \sqrt{n}$, with $n \in N$. See Problem 6.8.3.

(l) *Wallis'*[10] *formula,*

$$\left(\frac{2\cdot 2}{1\cdot 3}\right)\left(\frac{4\cdot 4}{3\cdot 5}\right)\left(\frac{6\cdot 6}{5\cdot 7}\right)\left(\frac{8\cdot 8}{7\cdot 9}\right)\cdot \cdots = \frac{\pi}{2},$$

which can also be written as

$$\lim_{n\to\infty}\left[\frac{2\cdot 4\cdot \cdots \cdot 2n}{1\cdot 3\cdot \cdots \cdot (2n-1)}\right]^2 \left(\frac{1}{2n+1}\right) = \frac{\pi}{2},$$

can be used to prove that $\Gamma\left(\frac{1}{2}\right) = \sqrt{\pi}$. See Problem 6.8.1 and Remark 7.6.17, part (a). ■

---

[9]James Stirling (1692–1770), a famous Scottish mathematician, studied infinite series, interpolation, and quadrature. At one point Stirling was a manager of a mining company, later publishing a paper on the ventilation of mine shafts. Stirling is also known for special numbers in probability named after him. Stirling numbers of the second kind are closely related to Bell numbers, named after Eric Temple Bell (1883–1960), a native of Scotland who moved to the United States in 1903. The formula above, although attributed to Stirling, was actually discovered by Abraham De Moivre.

[10]John Wallis (1616–1703), an English mathematician and physicist, introduced the symbol of infinity, developed the study of negative and fractional exponents, and prepared the way for calculus.

Example 6.6.8. Use the fact that $\Gamma\left(\frac{6}{5}\right) \approx 0.92$ in evaluating $\int_0^\infty x^5 \exp\left(-x^5\right) dx$.

Answer. Using Definition 6.6.1, we write

$$0.92 \approx \Gamma\left(\frac{6}{5}\right) = \int_0^\infty t^{\frac{1}{5}} e^{-t}\, dt = \int_0^\infty x e^{-x^5}\left(5x^4 dx\right) = 5\int_0^\infty x^5 \exp\left(-x^5\right) dx.$$

Therefore, the value of the desired improper Riemann integral is approximately given by $\frac{0.92}{5} = 0.184.$ ■

According to Remark 6.7.7, part (i), Γ has continuous derivatives of all orders on $\Re^+$. Therefore, the *digamma function*, also known as the *psi function* $\psi : \Re^+ \to \Re$, can be defined by

$$\psi(x) = \frac{d}{dx} \ln \Gamma(x) = \frac{\Gamma'(x)}{\Gamma(x)}.$$

Note that since $\Gamma(x+1) = x\Gamma(x)$, $\ln \Gamma(x+1) = \ln x + \ln \Gamma(x)$. Thus, differentiating both sides, we obtain $\psi(x+1) = \frac{1}{x} + \psi(x)$. In the case of $x = n \in N$, the preceding expression can be written as

$$\begin{aligned}
\psi(n+1) &= \psi(n) + \frac{1}{n} = \psi(n-1) + \frac{1}{n-1} + \frac{1}{n} = \cdots = \psi(1) + 1 + \frac{1}{2} + \frac{1}{3} + \cdots + \frac{1}{n} \\
&= \frac{\Gamma'(1)}{\Gamma(1)} + 1 + \frac{1}{2} + \frac{1}{3} + \cdots + \frac{1}{n} = \Gamma'(1) + 1 + \frac{1}{2} + \frac{1}{3} + \cdots + \frac{1}{n} \\
&= \int_0^\infty e^{-t}(\ln t)\, dt + 1 + \frac{1}{2} + \frac{1}{3} + \cdots + \frac{1}{n} = -\gamma + \sum_{k=1}^{n} \frac{1}{k}.
\end{aligned}$$

It can be verified that the preceding constant γ is the Euler's constant. See Exercise 12 of Section 6.3 and Exercise 28 of this section.

Definition 6.6.9. If $x, y > 0$, then $\frac{\Gamma(x)\Gamma(y)}{\Gamma(x+y)}$ defines a new function called the *beta function* B, which is in the two variables x and y.

Since we are not attempting to study functions in more than one variable at this point, only one exercise is devoted to the beta function. Another special function defined as an integral is the *error function*

$$\operatorname{erf} x = \frac{2}{\sqrt{\pi}} \int_0^x \exp\left(-t^2\right) dt \text{ where } x > 0.$$

The error function is used when studying normal distributions in statistics, discussing certain heat conduction problems in applied mathematics, and in a number of other areas.

Remark 6.6.10. Since $\lim_{x\to\infty} \operatorname{erf} x = \frac{2}{\sqrt{\pi}} \int_0^\infty \exp\left(-t^2\right) dt$, substituting $u = t^2$ and using Remark 6.6.7, part (a), we have $\lim_{x\to\infty} \operatorname{erf} x = \frac{2}{\sqrt{\pi}} \int_0^\infty \frac{1}{2} e^{-u} u^{-\frac{1}{2}}\, du = \frac{1}{\sqrt{\pi}} \Gamma\left(\frac{1}{2}\right) = 1.$ ■

Remark 6.6.11. The *complementary error function* $\operatorname{erfc} x = \frac{2}{\sqrt{\pi}} \int_x^\infty \exp\left(-t^2\right) dt$, with $x \ge 0$. And since $\operatorname{erfc} x = \frac{2}{\sqrt{\pi}} \int_0^\infty \exp\left(-t^2\right) dt - \frac{2}{\sqrt{\pi}} \int_0^x \exp\left(-t^2\right) dt$, by Remark 6.6.7, part (a), we can conclude that $\operatorname{erfc} x = 1 - \operatorname{erf} x$. ■

Remark 6.6.12. In statistics, the *cumulative distribution function* Φ *for the standardized normal (Gaussian) distribution* is given by

$$\Phi(x) = \frac{1}{\sqrt{2\pi}} \int_{-\infty}^{x} \exp\left(\frac{-t^2}{2}\right) dt$$

Certainly, $\Phi(x)$ is related to $\operatorname{erf} x$. Also, $\Phi(0) = \frac{1}{2}$ and $\lim_{x\to\infty} \Phi(x) = 1$. Why? ■

Several other special functions are the *Bessel*[11] *function of the first kind of order* p defined by

$$J_p(x) = \lim_{n\to\infty} \sum_{k=0}^{n} \frac{(-1)^k}{k!\,\Gamma(k+p+1)} \left(\frac{x}{2}\right)^{2k+p} \quad \text{where } x > 0;$$

the *Bessel function of the second kind of order* p, at times referred to as the *Neumann*[12]

---

[11]Friedrich Wilhelm Bessel (1784–1846), a German astronomer and mathematician, contributed greatly to the area of special functions and differential equations. He was the first person to measure the distance of a fixed star from the sun. Equations that bear his name often come up in nuclear and plasma physics.

[12]Karl Gottfried Neumann (1832–1925), was a celebrated German analyst and potential theorist born in Königsberg, Prussia, now Kaliningrad, Russia. The function $Y_p(x)$ is sometimes written as $N_p(x)$. It is also true that if $x > 0$, we have $N_p(x) = \frac{1}{\sin p\pi}\left[(\cos p\pi) J_p(x) - J_{-p}(x)\right]$ if p is not an integer, and $N_n(x) = \lim_{p\to n} \frac{1}{\sin p\pi}\left[(\cos p\pi) J_p(x) - J_{-p}(x)\right]$ if p is a nonnegative integer. Karl was the son of Franz Neumann. See footnote to Problem 9.5.12.

function or the *Weber*[13] *function*, defined by

$$Y_p(x) = \frac{2}{\pi}\left\{\left(\ln\frac{x}{2}+\gamma\right)J_p(x) - \frac{1}{2}\sum_{k=0}^{p-1}\frac{(p-k-1)!}{k!}\left(\frac{x}{2}\right)^{2k-p}\right.$$
$$\left. + \frac{1}{2}\lim_{n\to\infty}\sum_{k=0}^{n}\left[(-1)^{k+1}\left(\sum_{i=1}^{k}\frac{1}{i}+\sum_{i=1}^{k+p}\frac{1}{i}\right)\left(\frac{1}{k!(k+p)!}\left(\frac{x}{2}\right)^{2k+p}\right)\right]\right\}$$

where $x \geq 0$ and γ is Euler's constant; *Legendre*[14] *polynomials*, defined by

$$P_n(x) = \frac{1}{2^n n!}\frac{d^n}{dx^n}\left(x^2-1\right)^n, \quad n \in N \cup \{0\}$$
$$= \sum_{k=0}^{I}\frac{(-1)^k(2n-2k)!x^{n-2k}}{2^n k!(n-k)!(n-2k)!}$$

where $I = \frac{n}{2}$ or $I = \frac{n-1}{2}$, whichever is an integer (see Part 2 of Section 9.5); *Hermite polynomials*, defined by

$$H_n(x) = \frac{2^{n+1}}{\sqrt{\pi}}\exp\left(x^2\right)\int_0^{\infty}t^n\left[\exp\left(-t^2\right)\right]\cos\left(2xt-\frac{n\pi}{2}\right)dt$$
$$= \frac{\sqrt{\pi}}{2}(-1)^n\exp\left(x^2\right)\frac{d^{n+1}}{dx^{n+1}}(\operatorname{erf} x)$$
$$= n!\sum_{k=0}^{\left[\frac{n}{2}\right]}(-1)^k\frac{(2x)^{n-2k}}{k!(n-2k)!}, \text{ with } n \in N \cup \{0\};$$

Laguerre[15] *polynomials*, defined by

$$L_n(x) = \frac{e^x}{n!}\frac{d^n}{dx^n}\left(x^n e^{-x}\right), \text{ with } n \in N \cup \{0\};$$

Chebyshev[16] *polynomials of the first kind*, defined by

---

[13]Heinrich Weber (1842–1913) was a German mathematician.

[14]Adrien Marie Legendre (1752–1833), a French analyst and number theorist, greatly contributed to the area of differential equations and special functions theory.

[15]Edmond Nicolas Laguerre (1834–1886), a French analyst and geometer, is considered one of the founders of modern geometry. In addition, he contributed to analytical and infinitesimal geometry as well as to the theory of algebraic equations and linear systems.

[16]Pafnuti Lvovich Chebyshev (1821–1894), a famous Russian mathematician, made contributions in algebra, analysis, geometry, probability, and number theory. He also proved Bertrand's postulate. See the footnote in Theorem 7.4.5. Letter T is used to represent his polynomials due to another transliteration of "Chebyshev" from Russian, namely Tchebycheff.

$$T_n(x) = \cos(n \arccos x) = \frac{1}{2}\left(x + \sqrt{x^2 - 1}\right)^n + \frac{1}{2}\left(x - \sqrt{x^2 - 1}\right)^n$$

with $x \in [-1,1]$ and $n \in N \cup \{0\}$; and *Chebyshev polynomials of the second kind*, defined by

$$U_n(x) = \frac{\sin[(n+1)\arccos x]}{\sqrt{1-x^2}}$$

with $n \in N \cup \{0\}$ and $x \in [-1,1]$.

Finally, let us briefly look at Laplace[17] transforms. In general, a *transform* is a device that changes objects from one form to another. The resulting entity is hopefully easier to work with. The *Fourier*[18] *transform*, which is the earliest transform, dating back to around 1811, involves a complex number i and is especially used by electrical engineers in analysis of signals. Other common transforms are the *Hankel*[19] *transform*, the *Mellin*[20] *transform*, and the *Stieltjes transform*. A number of areas in mathematics rely heavily on the uses of transforms.

Definition 6.6.13. The *Laplace transform* of a function $f:[0,\infty) \to \Re$ is defined by $F(x) = \int_0^\infty e^{-xt} f(t)\,dt$ provided this improper integral is finite. Other common notations used for $F(x)$ are $\mathcal{L}\{f\}$ or $\mathcal{L}\{f(t)\}$.

Definition 6.6.14. A function f is said to be of *exponential order* if there exists a real constant a and positive constants t_0 and M such that $|f(t)| < Me^{at}$ for all $t \ge t_0$ at which $f(t)$ is defined. That is, $\lim_{t\to\infty} f(t)e^{-at}$ is not infinite. If f is of exponential order corresponding to some definite constant a, then we say that f is of exponential order e^{at}.

Certainly, any bounded function is of exponential order. In addition, if a given function f is of exponential order e^{at}, with $a > 0$, then f is also of exponential order e^{bt} for any $b > a$. Why?

---

[17]Pierre Simon Marquis de Laplace (1749–1827), a French-born genius, is best known for his monumental work in celestial mechanics, differential equations, and probability in mathematics.

[18]Jean Baptiste Joseph Baron de Fourier (1768–1830), a French mathematician and friend of Cauchy and Napoleon, studied heat conduction in physics and contributed to mathematical analysis. He is most famous for widely applicable series, which bears his name.

[19]Hermann Hankel (1839–1873), a famous German analyst and geometer, is known for the complex integral representation of the gamma function.

[20]Robert Hjalmar Mellin (1854–1933), a Finnish mathematician, is known in analysis and mathematical physics.

THEOREM 6.6.15. *Suppose a function* $f:[0,\infty)\to\Re$ *is continuous and of exponential order* e^{at}, $a\in\Re^+$. *Then,* $\mathcal{L}\{f(t)\}$ *exists for* $x>a$.

Proof of Theorem 6.6.15 is left as an exercise. Note that the hypotheses in Theorem 6.6.15 are sufficient but not necessary. That is, there are functions f that do not satisfy all the conditions of Theorem 6.6.15 for which $\mathcal{L}\{f\}$ exists. One such function is $f(t)=\frac{1}{\sqrt{t}}$, which is not continuous on $[0,\infty)$. See Exercise 26(c).

Exercises 6.6

1. Prove Lemma 6.6.3.

2. Prove Theorem 6.6.4.

3. **(a)** Prove that $\lim\limits_{x\to0^+}\Gamma(x)=+\infty$.

(b) Prove that $\lim\limits_{x\to0^-}\Gamma(x)=-\infty$.

(c) Conclude that Γ has an odd vertical asymptote at $x=0$, and thus has an infinite discontinuity at $x=0$.

4. **(a)** Prove Remark 6.6.7, parts (b), (c), and (h).

(b) Use Remark 6.6.7, part (g), to prove that

$$\frac{\Gamma(x)\Gamma(x)}{2\Gamma(2x)}=2^{1-2x}\int_0^{\frac{\pi}{2}}\sin^{2x-1}u\,du.$$

(c) Use part (b) and Remark 6.6.7, part (h), to prove that

$$\Gamma\left(x+\frac{1}{2}\right)=\frac{\sqrt{\pi}\,\Gamma(2x)}{2^{2x-1}\Gamma(x)},\quad x>0.$$

(d) Use part (c) to prove Remark 6.6.7, part (d).

5. **(a)** Prove that $n!!=2^{\frac{n}{2}}\sqrt{\frac{2}{\pi}}\,\Gamma\left(1+\frac{n}{2}\right)$, if $n=1,3,5,\ldots$. See Exercise 19(b) of Section 2.4.

(b) Prove that $(2n+1)!!=\frac{2^{n+1}}{\sqrt{\pi}}\Gamma\left(n+\frac{3}{2}\right)$, if $n\in N\cup\{0\}$.

(c) Compute $\left(-\frac{5}{2}\right)!$.

6. **(a)** Use the fact that $\Gamma\left(\frac{5}{3}\right)\approx0.89$ to calculate $\int_0^\infty x^4\exp\left(-x^3\right)dx$.

(b) Prove that $\int_0^{\infty} \sqrt{x} \exp(-x^3)\, dx = \frac{\sqrt{\pi}}{3}$.

(c) Use the fact that $\Gamma(x)\Gamma(1-x) = \frac{\pi}{\sin \pi x}$, if x is not an integer, to evaluate $\int_0^{\frac{\pi}{2}} \sqrt{\tan x}\, dx$. See Exercise 8 for further comments.

(d) Prove[21] that $\int_0^{\frac{\pi}{2}} \sqrt{\sin 2x}\, dx = \sqrt{\frac{2}{\pi}} \left[\Gamma\left(\frac{3}{4}\right)\right]^2$.

(e) Compute $\int_0^1 \sqrt[3]{\ln x}\, dx$.

7. Prove that the gamma function $\Gamma: \Re^+ \to \Re$ attains an absolute minimum at a value $c \in (1,2)$.

8. **(a)** Provided x is not an integer, we can prove, using infinite products covered in Section 7.6, that $\Gamma(x)\Gamma(1-x) = \pi \csc \pi x$. In fact, if $x = \frac{1}{2}$, then as in Remark 6.6.7, part (a), we can conclude that $\Gamma\left(\frac{1}{2}\right) = \sqrt{\pi}$. Use the first statement to prove that when x is not an integer, $\psi(1-x) - \psi(x) = \pi \cot \pi x$.

(b) Use Exercise 4(c) to prove that $\psi(x) + \psi\left(x + \frac{1}{2}\right) + 2\ln 2 = 2\psi(2x)$.

(c) Verify that $\psi\left(\frac{1}{2}\right) = -\gamma - 2\ln 2$.

9. Prove that the beta function B with $x, y > 0$ can be written as

(a) $B(x,y) = \int_0^1 t^{x-1}(1-t)^{y-1}\, dt$.

(b) $B(x,y) = 2\int_0^{\frac{\pi}{2}} (\sin\theta)^{2x-1}(\cos\theta)^{2y-1}\, d\theta$.

---

[21]Actually, since $\Gamma\left(\frac{1}{2} - x\right) = \frac{\pi \sec \pi x}{\Gamma\left(\frac{1}{2} + x\right)}$, we can write $\Gamma\left(\frac{3}{4}\right) = \frac{\sqrt{2}\pi}{\Gamma\left(\frac{1}{4}\right)}$. Also, $\Gamma\left(\frac{3}{4}\right) = \sqrt[4]{\pi}\sqrt{u}$, where u is called the *ubiquitous constant*, denoting the common mean of 1 and $\frac{1}{\sqrt{2}}$.

(c) $B(x,y)=\int_0^\infty \frac{u^{x-1}}{(1+u)^{x+y}}\,du.$ (d) $B(x,y)=B(y,x).$

(e) $B(x,y)=\frac{x+y}{x}B(x+1,y).$ (f) $B(x,1)=\frac{1}{x}.$

(g) Evaluate $\int_0^\infty \frac{x^3}{(1+x)^5}\,dx.$

10. (a) Prove that $\operatorname{erf} x$ is an odd function.

(b) Evaluate $\int \exp\left(-x^2\right)(\operatorname{erf} x)\,dx.$

11. If $x>0$, prove that

(a) $\Phi(x)=\frac{1}{\sqrt{2\pi}}\int_0^x \exp\left(-\frac{t^2}{2}\right)dt+\frac{1}{2}.$

(b) $\Phi(x)=\frac{1}{2}\operatorname{erf}\left(\frac{x}{\sqrt{2}}\right)+\frac{1}{2}.$

(c) $\operatorname{erf} x=2\Phi\left(\sqrt{2}x\right)-1.$

12. Prove that if k is a real constant and $x>0$, then

(a) $\frac{d}{dx}\left[x^pJ_p(kx)\right]=kx^pJ_{p-1}(kx).$

(b) $\frac{d}{dx}\left[x^{-p}J_p(kx)\right]=-kx^{-p}J_{p+1}(kx).$

(c) $\frac{d}{dx}\left[J_p(kx)\right]=kJ_{p-1}(kx)-\frac{p}{x}J_p(kx).$

(d) $\frac{d}{dx}\left[J_p(kx)\right]=-kJ_{p+1}(kx)+\frac{p}{x}J_p(kx).$

(e) $\frac{d}{dx}\left[J_p(kx)\right]=\frac{k}{2}\left[J_{p-1}(kx)-J_{p+1}(kx)\right].$

(f) $J_p(kx)=\frac{kx}{2p}\left[J_{p-1}(kx)+J_{p+1}(kx)\right].$

(g) $x\frac{d}{dx}\left[J_p(x)\right]-pJ_p(x)=-xJ_{p+1}(x).$

13. (a) Prove that $J_n(-x)=(-1)^nJ_n(x)$ for $n=0,1,2,\ldots$.

(b) Define $\frac{1}{\Gamma(n)}=0$ when n is a nonpositive integer, and prove that $J_{-m}(x)=(-1)^mJ_m(x)$, when m is a positive integer.

(c) Prove that between any two successive positive real roots of J_0 there exists a real root of J_1.

(d) Prove that between any two successive positive real roots of J_1 there exists a real root of J_0.

(e) Evaluate $\int x^3 J_0(x)\,dx$.

14. If $y = J_p(x)$ satisfies the *Bessel's equation of order* p

$$x^2 \frac{d^2y}{dx^2} + x\frac{dy}{dx} + \left(x^2 - p^2\right)y = 0.$$

(a) Show that $z = xJ_1(x)$ satisfies the equation $x\dfrac{d^2z}{dx^2} - \dfrac{dz}{dx} + xz = 0$.

(b) Show that $z = J_0\left(2\sqrt{x}\right)$ satisfies the equation $x\dfrac{d^2z}{dx^2} + \dfrac{dz}{dx} + z = 0$.

(c) Show that $z = J_p(kx)$ satisfies the equation

$$x^2 \frac{d^2z}{dx^2} + x\frac{dz}{dx} + \left(k^2x^2 - p^2\right)z = 0.$$

(d) Find an equation where one of the solutions is given by $J_1\left(2\sqrt{x}\right)$.

(e) Prove that $y(x) = \dfrac{u(x)}{\sqrt{x}}$, transforms the Bessel's equation of order p to

$$\frac{d^2u}{dx^2} + \left[1 + \left(\frac{1}{4} - p^2\right)\frac{1}{x^2}\right]u = 0.$$

(f) Show that $y(x) = \sqrt{\dfrac{2}{\pi x}}\sin x + \sqrt{\dfrac{2}{\pi x}}\cos x$ satisfies the Bessel's equation of order $\dfrac{1}{2}$. In fact, it can be verified that $\sqrt{\dfrac{2}{\pi x}}\sin x = J_{\frac{1}{2}}(x)$ and $\sqrt{\dfrac{2}{\pi x}}\cos x$ $= J_{-\frac{1}{2}}(x) = -Y_{\frac{1}{2}}(x)$. See Exercise 12 in Section 8.6. From this result, it can be shown that *spherical Bessel functions* $J_{\frac{n}{2}}(x)$, with n an odd integer, can be written in closed form.

15. **(a)** Use the fact that $H_0(x) = 1$,

$$\frac{d}{dx}H_n(x) = 2nH_{n-1}(x), \text{ and } H_{n+1}(x) = 2xH_n(x) - 2nH_{n-1}(x)$$

for all $n \in N$, to prove that $y = H_n(x)$ satisfies the *Hermite equation*

$$\frac{d^2y}{dx^2} - 2x\frac{dy}{dx} + 2ny = 0.$$

(b) Show that $y = \exp\left(-\frac{x^2}{2}\right) H_n(x)$ satisfies *Weber's equation*

$$\frac{d^2y}{dx^2} = \left(x^2 - 2n - 1\right)y.$$

(c) Write polynomials that represent $L_n(x)$ for $n = 0,1,2,3$.

16. (a) Write polynomials that represent $T_n(x)$ for $n = 0,1,2,3,4$.

(b) Show that $y = T_n(x)$ satisfies the *Chebyshev equation*

$$\left(1-x^2\right)\frac{d^2y}{dx^2} - x\frac{dy}{dx} + n^2 y = 0$$

for all $n = 0,1,2,\ldots$.

(c) Prove that $T_n(x) = 2xT_{n-1}(x) - T_{n-2}(x)$ for all $n = 2,3,\ldots$.

(d) Prove that

$$\int_{-1}^{1} T_k(x) T_m(x) \frac{dx}{\sqrt{1-x^2}} = \begin{cases} 0, & \text{if } m \neq k \\ \frac{\pi}{2}, & \text{if } m = k > 0 \\ \pi, & \text{if } m = k = 0. \end{cases}$$

(e) Prove that $T_m\left(T_n(x)\right) = T_{mn}(x)$ for all $m,n > 0$.

17. Using Definition 6.6.13, evaluate the given expressions, if possible.

(a) $\mathcal{L}\{1\}$ (b) $\mathcal{L}\{t\}$

(c) $\mathcal{L}\left\{e^{at}\right\}$ with $a \in \Re$ (d) $\mathcal{L}\{\sin kt\}$, $k \in \Re$

(e) $\mathcal{L}\{\cos kt\}$, $k \in \Re$ (f) $\mathcal{L}\left\{\frac{1}{\sqrt{t}}\right\}$

(g) $\mathcal{L}\left\{\frac{1}{t^2}\right\}$ (h) $\mathcal{L}\{\sinh t\}$

18. (a) Prove that Laplace transforms satisfy the linear property. That is, prove that if a and b are real constants, and if for functions $f,g:\Re^+ \to \Re$ both $\mathcal{L}\{f\}$ and $\mathcal{L}\{g\}$ exist, then $\mathcal{L}\{af(t) + bg(t)\} = a\mathcal{L}\{f(t)\} + b\mathcal{L}\{g(t)\}$.

(b) Use part (a) to evaluate $\mathcal{L}\left\{\sin^2 t\right\}$.

19. Determine whether or not the given function is of exponential order.

(a) $f(t) = \frac{1}{t}$ (b) $f(t) = \exp\left(t^2\right)$ (c) $f(t) = t^r$, $r \in \Re^+$

20. (a) Prove Theorem 6.6.15.

(b) Prove that if the hypotheses of Theorem 6.6.15 are satisfied, then $\mathcal{L}\{|f|\}$ exists for $x > a$.

(c) Prove that if the hypotheses of Theorem 6.6.15 are satisfied, then $\int_0^\infty \exp(-xt)f(t)\,dt$ converges absolutely for $x > a$.

(d) Prove that if the hypotheses of Theorem 6.6.15 are satisfied, then $\lim_{x\to\infty} \mathcal{L}\{f(t)\} = \lim_{x\to\infty} F(x) = 0$.

(e) Conclude that there is no function $f:[0,\infty)\to\Re$ that is continuous and of exponential order for which $\mathcal{L}\{f(t)\} = \sin x$.

21. (a) Suppose f has a derivative that is piecewise continuous on $[0,b]$ for every $b\in\Re^+$, and suppose f is of exponential order e^{at} for $a\in\Re$. Prove that

$$\mathcal{L}\{f'(t)\} = x\,\mathcal{L}\{f(t)\} - f(0),\ \text{for } x > a.$$

(b) Use part (a) and Exercise 17(e) to compute $\mathcal{L}\{\sin kt\}$ for $k\in\Re$. See Exercise 17(d).

(c) Prove that under conditions similar to those in part (a), we have

$$\mathcal{L}\{f''(t)\} = x^2\,\mathcal{L}\{f(t)\} - xf(0) - f'(0).$$

Use this formula to compute $\mathcal{L}\{\sin kt\}$ and $\mathcal{L}\{t^2\}$.

(d) Let $F(x) = \mathcal{L}\{f(t)\}$, where f is piecewise continuous function on $[0,b]$ for every $b\in\Re^+$, and f is of exponential order e^{at} for $a\in\Re$. If interchanging the order of integration and differentiation is permitted (see Theorem 12.2.6), prove that for $x > a$, we have

$$\mathcal{L}\{t^n f(t)\} = (-1)^n \frac{d^n F}{dx^n}.$$

Use this formula to compute $\mathcal{L}\{t^2 \sin kt\}$.

(e) Suppose the function f is piecewise continuous function on $[0,b]$ for every $b\in\Re^+$, and f is of exponential order e^{at} for $a\in\Re$. Moreover, suppose $\lim_{t\to 0^+} \frac{f(t)}{t}$ is finite. Using iterated integration, it can be proved that $\mathcal{L}\left\{\frac{f(t)}{t}\right\} = \int_x^\infty F(z)\,dz$ for $x > a$, where $F(z) = \mathcal{L}\{f(t)\}$. Use this formula to compute $\mathcal{L}\left\{\frac{\sinh t}{t}\right\}$ and $\mathcal{L}\left\{\frac{\sin t}{t}\right\}$. See Exercise 11 in Section 8.6 and Problem 6.8.6, part (g).

22. If a function f is continuous on $[0,\infty)$, p-periodic, and of exponential order e^{at}, with $a \in \Re$, prove that $\mathcal{L}\{f(t)\} = \dfrac{1}{1-e^{-xp}} \displaystyle\int_0^p e^{-xt} f(t)\,dt$.

23. **(a)** If for a function f, $F(x) = \mathcal{L}\{f(t)\}$ when $x > a$, prove that $\mathcal{L}\left\{e^{bt} f(t)\right\} = F(x-b)$ when $x > a+b$.

(b) Let f be such that $f(t) = 0$ for $t < 0$. Define g by $g(t) = f(t-a)$, $a \in \Re$. Prove that $\mathcal{L}\{g(t)\} = e^{-at}\,\mathcal{L}\{f(t)\}$.

24. **(a)** Suppose a function $f:[0,\infty) \to \Re$ is both continuous and of exponential order e^{at}, for $a \in \Re$. Prove that $\mathcal{L}\left\{\displaystyle\int_0^t f(u)\,du\right\} = \dfrac{1}{x}\mathcal{L}\{f(t)\}$.

(b) The *Convolution Theorem* says that if two functions $f, g:[0,\infty) \to \Re$ are both continuous and of exponential order, then

$$\mathcal{L}\left\{\int_0^t f(u)g(t-u)\,du\right\} = \mathcal{L}\{f(t)\} \cdot \mathcal{L}\{g(t)\}.$$

The preceding integral is customarily denoted by $f(t) * g(t)$ and is called the *convolution of the functions f and g*. Use this statement to evaluate $\mathcal{L}\left\{\displaystyle\int_0^t e^u \sin(t-u)\,du\right\}$. Observe that part (a) is just a special case of the convolution theorem.

(c) Prove that $0 * f = f * 0 = 0$, where 0 denotes the function that is identically 0 for all t.

25. Prove that $\displaystyle\int_0^\infty \operatorname{sinc} x\,dx = \frac{\pi}{2}$ by following the given steps.

(a) Write $\mathcal{L}\{f(t)\} = F(x)$ as an integral.

(b) Suppose that in part (a) we can integrate both sides with respect to x from 0 to ∞ and interchange the order of integration. Write down what you obtain.

(c) Replace $f(t)$ by $\sin at$, where $a > 0$. What do you get?

(d) See the footnote in Exercise 22 from Section 6.5.

26. **(a)** Prove that $\mathcal{L}\left\{t^r\right\} = \dfrac{\Gamma(r+1)}{x^{r+1}}$ for every real constant $r > -1$.

(b) Find $\mathcal{L}\left\{\sqrt{t}\right\}$. **(c)** Find $\mathcal{L}\left\{\dfrac{1}{\sqrt{t}}\right\}$. See Exercise 17(f).

(d) Prove that $\mathcal{L}\left\{t^n\right\} = \frac{n}{x}\mathcal{L}\left\{t^{n-1}\right\}$ for all $n \in N$.

27. Prove that Euler's constant γ can be written as $\gamma = \int_0^1 \frac{1-e^{-t}}{t}\,dt - \int_1^\infty \frac{1}{te^t}\,dt$. See Exercise 12 of Section 6.3 and Example 6.8.12.

28. Prove that Euler's constant γ can be written as $\gamma = -\int_0^\infty e^{-t}(\ln t)\,dt = -\Gamma'(1)$.

6.7* Review

Label each statement as true or false. If a statement is true, prove it. If not,
(i) give an example of why it is false, and
(ii) if possible, correct it to make it true, and then prove it.

1. An improper Riemann integral can diverge without being infinite.
2. The function $f:[0,1] \to \Re$ defined by

$$f(x) = \begin{cases} n, & \text{if } x = \frac{1}{n} \text{ with } n \in N \\ 0, & \text{otherwise} \end{cases}$$

is Riemann integrable.
3. If a function f is bounded on $[a,b]$, then f is Riemann integrable on $[a,b]$.
4. If $f \in R[a,b]$, $f(x) \geq 0$ for all $x \in [a,b]$, and $f(x) \not\equiv 0$, then $\int_a^b f(x)\,dx > 0$.
5. If functions $f, g:[a,b) \to \Re$ are improper Riemann integrable, and $\int_a^b f^2$ and $\int_a^b g^2$ converge for $-\infty < a < b \leq +\infty$, then $\int_a^b fg$ converges absolutely.
6. If $n = 0, 1, 2, \ldots$, then $n! = \int_0^1 (-\ln x)^n\,dx$.
7. If the improper integral $\int_a^b f$ converges, then so does $\int_a^b f^2$.
8. The function $f:[0,1] \to \Re$ defined by

$$f(x) = \begin{cases} 1, \text{ if } x = \dfrac{1}{n} \text{ with } n \in N \\ 0, \text{ otherwise,} \end{cases}$$

is Riemann integrable. Compare this with Example 6.1.8 and Exercise 8 of Section 6.2.

9. If both $\mathcal{L}\{f(t)\}$ and $\mathcal{L}\{g(t)\}$ exist, then $\mathcal{L}\{(fg)(t)\}$ must also exist and

$$\mathcal{L}\{f(t)\} \cdot \mathcal{L}\{g(t)\} = \mathcal{L}\{(fg)(t)\}.$$

10. If $x > 0$, then $\mathcal{L}\{\text{erf}\,(t)\} = \dfrac{1}{x}\exp\left(\dfrac{1}{4}x^2\right)\text{erfc}\left(\dfrac{x}{2}\right)$.

11. $\displaystyle\int_0^1 \frac{x^4}{\sqrt{1-x^2}}\,dx = \frac{\pi}{4}$.

12. $B(x,x) = \dfrac{\sqrt{\pi}\Gamma(x)}{2^{2x-1}\Gamma\left(x+\dfrac{1}{2}\right)}$.

13. The function $f:[0,\infty) \to \mathfrak{R}$, defined by

$$f(x) = \begin{cases} n, \text{ if } n - \dfrac{1}{n^2} \le x \le n \text{ with } n \in N \\ 0, \text{ otherwise,} \end{cases}$$

is improper Riemann integrable.

14. If $f:[a,b] \to \mathfrak{R}$ is continuous and $f(x) = g(x) - h(x)$ for all $x \in [a,b]$, then $\displaystyle\int_a^b f = \int_a^b g - \int_a^b h.$

15. $f(t)*g(t) = g(t)*f(t)$.

16. If a function $f:[a,b] \to \mathfrak{R}$ is of bounded variation, then $f \in R[a,b]$.

17. If $0 \le g(x) \le G(x)$ for all $x \in [a,\infty)$ and $\displaystyle\int_a^\infty G$ converges, then $\displaystyle\int_a^\infty g$ converges.

18. If $\displaystyle\int_{-\infty}^\infty f$ is convergent and $a, b \in \mathfrak{R}$, then $\displaystyle\int_{-\infty}^a f + \int_a^\infty f = \int_{-\infty}^b f + \int_b^\infty f$.

19. Suppose there exist functions $f, g:[a,b] \to \mathfrak{R}$ with f Riemann integrable and $f(x) = g(x)$ for all $x \in [a,b]$ except for countably infinite number of points in $[a,b]$, then $g \in R[a,b]$. See Exercise 10(a) of Section 6.3.

20. If a function is of exponential order, then $\mathcal{L}\{f(t)\}$ exists.

21. If g is a convex function on $\mathfrak{R}$ and a function $f:[0,1] \to \mathfrak{R}$ is Riemann integrable, then $\int_0^1 (g \circ f) \geq g\left(\int_0^1 f\right)$. This inequality is known as *Jensen's inequality* where the left-hand side represents the weighted average for the function g and the left-hand side represents a composition. See Definition 1.10.24 and Exercise 27 of Section 5.3.

22. If a function $f:[0,1] \to \Re$ is Riemann integrable, then

$$\int_0^1 \exp(f) \geq \exp\left(\int_0^1 f\right).$$

23. The gamma function is logarithmically convex. See Problem 5.6.24, part (r).

24. If a function $f:[a,b] \to \Re$ is Riemann integrable and $f(x) \geq 0$ for all $x \in [a,b]$, then $g = \sqrt{f} \in R[a,b]$.

25. If functions $f, g:[a,b] \to \Re$ are such that $f \in R[a,b]$ and $g \notin R[a,b]$, then $fg \notin R[a,b]$.

26. If functions $f, g:[a,b] \to \Re$ are such that $f \in R[a,b]$ and $g \notin R[a,b]$, then $fg \in R[a,b]$.

27. If a function $f \in R[a,b]$, then $\frac{d}{dx}\int_a^x f(t)\,dt = f(x)$ with $x \in [a,b]$.

28. If functions f and g are defined on $[a,b]$ but are not Riemann integrable on $[a,b]$, then $f + g \notin R[a,b]$.

29. If $f \in R[a,b]$, then $f^2 \in R[a,b]$.

30. Every Riemann integrable function has an antiderivative.

31. Every uniformly continuous function on $[a,b]$ is Riemann integrable on $[a,b]$.

32. If a function $f:[0,2] \to \Re$ is increasing, then $\underline{\int}_0^2 f = \overline{\int}_0^2 f$.

33. There exists a bounded function $f:[a,b] \to \Re$ such that $\underline{\int}_a^b f \geq \overline{\int}_a^b f$.

34. If we can find a real number A such that $\underline{\int}_a^b f \leq A \leq \overline{\int}_a^b f$, then $f \in R[a,b]$.

35. There exists a nonnegative function $f:[a,b] \to \Re$ for which $\int_a^b f = 0$ but $f(x) \not\equiv 0$.

36. $\int_0^\infty \exp\left(-x^3\right)dx = \frac{1}{3}\Gamma\left(\frac{1}{3}\right)$.

37. The functions $G_1(x) \equiv \int_a^x f(t)\,dt$ and $G_2(x) \equiv \int_b^x f(t)\,dt$ have the same derivative, and G_1 and G_2 differ by a constant $|f(b) - f(a)|$.

38. If $f:[0,1] \to \Re$ is a continuous function that satisfies

$$\int_0^x f(t)\,dt = \int_x^1 f(t)\,dt$$

for all $x \in [0,1]$, then $f(x) = 0$ for all $x \in [0,1]$.

39. $\int_{-\infty}^{\infty} \frac{1}{\sqrt{2\pi}} \exp\left(-\frac{x}{2}\right)dx = 1.$

40. If a function $f:[a,b] \to \Re$ is absolutely continuous, then $V_f(a,b) = \int_a^b |f'|$. See Part 3 of Section 5.8.

41. $\Gamma\left(\frac{2n+1}{2}\right) = \frac{1 \cdot 3 \cdot \cdots \cdot (2n-1)}{2^n}\sqrt{\pi}$ for $n \in N$. See Remark 6.6.7, part (d).

42. If Legendre polynomials $y = P_n(x)$ satisfy the *Legendre equation*

$$\left(1 - x^2\right)\frac{d^2y}{dx^2} - 2x\frac{dy}{dx} + n(n+1)y = 0,$$

then this equation can be transformed to

$$(\sin\theta)\frac{d^2y}{d\theta^2} + (\cos\theta)\frac{dy}{d\theta} + n(n+1)(\sin\theta)y = 0.$$

43. $\mathrm{Si}(x) = \frac{1}{\pi}\int_0^{\pi x} \mathrm{sinc}\, t\, dt.$

44. If p is not a nonnegative integer, then Y_p is a solution to Bessel's equation of order p.

45. The equation $\frac{d^2y}{dx^2} + cx^n y = 0$, with $x > 0$ and c and n positive real constants, can be transformed to Bessel's equation

$$t^2\frac{d^2z}{dt^2}+t\frac{dz}{dt}+\left[t^2-\frac{1}{(n+2)^2}\right]z=0,$$

with $t>0$, using substitutions $y=\sqrt{x}z$ and $t=\frac{2\sqrt{c}}{n+2}x^{\frac{n}{2}+1}$.

46. $f*1=f$, where 1 denotes the function that is identically 1 for all t.

47. $\int_0^\infty \sin\left(x^2\right)dx$ converges.[22] See Exercise 15(a) from Section 6.5.

48. If $a_n=\frac{n!e^n}{n^{n+\frac{1}{2}}}$, then the sequence $\{a_n\}$ converges to $\sqrt{2\pi}$. See part (k) of Remark 6.6.7.

49. If f and g are continuous and 1-periodic on $\Re$, then

$$\lim_{n\to\infty}\int_0^1 f(x)g(nx)\,dx=\left[\int_0^1 f(x)\,dx\right]\left[\int_0^1 g(x)\,dx\right].$$

---

[22]It is customary to define $S(x)=\int_0^x \sin\left(t^2\right)dt$, known as a *Fresnel sine integral. Fresnel cosine integral* is defined by $C(x)=\int_0^x \cos\left(t^2\right)dt$. It can be verified that $\lim_{x\to\infty}S(x)=\lim_{x\to\infty}C(x)=\frac{1}{2}\sqrt{\frac{\pi}{2}}$. Both $S(x)$ and $C(x)$ are odd functions whose Maclaurin series are given in Exercise 2(c) of Section 8.6 and Exercise 40 of Section 8.7, respectively. Graphs of $S(x)$ and $C(x)$ are given in Figure 6.7.1.

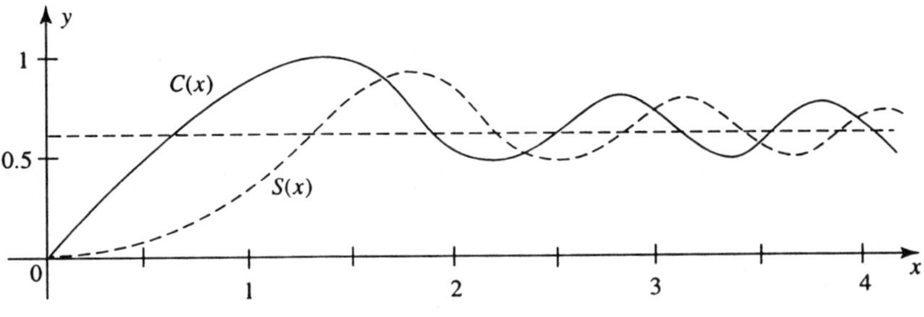

Figure 6.7.1

Fresnel integrals are named after Augustin Jean Fresnel (1788–1827), a French mathematician and physicist, mostly known for his work in optics. Fresnel is one of the founders of the wave theory of light. He also developed the use of compound lenses instead of mirrors in lighthouses.

6.8* Projects

Part 1. Wallis' Formula

Problem 6.8.1. Use the following outline to prove *Wallis' formula*. See Remark 6.6.7, part (l).

(a) Verify that if $0 < x < \frac{\pi}{2}$ and $n \in N$, then $\sin^{2n+1} x < \sin^{2n} x < \sin^{2n-1} x$.

(b) Integrate the inequalities in part (a) from 0 to $\frac{\pi}{2}$. Then use Remark 6.6.7, part (h), to obtain

$$\frac{(2)(4)(6)\cdot\;\cdots\;\cdot(2n)}{(1)(3)(5)\cdot\;\cdots\;\cdot(2n+1)} < \frac{(1)(3)(5)\cdot\;\cdots\;\cdot(2n-1)}{(2)(4)(6)\cdot\;\cdots\;\cdot(2n)}\frac{\pi}{2} < \frac{(2)(4)(6)\cdot\;\cdots\;\cdot(2n-2)}{(1)(3)(5)\cdot\;\cdots\;\cdot(2n-1)}.$$

(c) Draw the final conclusion.

Part 2. Euler's Summation Formula

Suppose a continuously differentiable function $f:[0,\infty)\to\Re$ is decreasing and positive. For convenience, $f(n)$ is denoted by f_n, with $n \in N$. In attempting to evaluate the nonnegative expression $f_0 + f_1 + \cdots + f_n - \int_0^n f$, observe that

$$\sum_{k=1}^{n}\int_{k-1}^{k} kf' = \sum_{k=1}^{n} k\left(f_k - f_{k-1}\right) = -\left(f_0 + f_1 + \cdots + f_n\right) + (n+1)f_n.$$

Note that if $x \in (k-1,k)$, then $k = \lfloor x \rfloor + 1$. Hence,

$$f_0 + f_1 + \cdots + f_n = (n+1)f_n - \int_0^n \left(\lfloor x \rfloor + 1\right)f'(x)\,dx.$$

Integration by parts yields $(n+1)f_n = nf_n + f_n = \int_0^n xf'(x)\,dx + \int_0^n f(x)\,dx + f_n$. Why?

Therefore,

$$\begin{aligned}\sum_{k=0}^{n} f_k - \int_0^n f(x)\,dx &= f_n + \int_0^n \left(x - \lfloor x \rfloor - 1\right)f'(x)\,dx \\ &= \frac{1}{2}\left(f_0 + f_n\right) + \int_0^n \left(x - \lfloor x \rfloor - \frac{1}{2}\right)f'(x)\,dx,\end{aligned}$$

which is *Euler's Summation Formula*. This formula relates the integral of a function to the sum of its functional values. Obviously, this formula can be used to approximate integrals by sums and vice versa.

Example 6.8.2. Applying Euler's summation formula to $f(x)=\dfrac{1}{1+x}$, with n replaced by $n-1$, we obtain

$$1+\frac{1}{2}+\frac{1}{3}+\cdots+\frac{1}{n}-\ln n=\frac{1}{2}+\frac{1}{2n}-\int_0^{n-1}\frac{g(x)}{(x+1)^2}\,dx,$$

where $g(x)=x-\lfloor x\rfloor-\dfrac{1}{2}$. Since $|g(x)|\le\dfrac{1}{2}$ and $g(x+1)=g(x)$, the preceding integral converges and is of the form $\displaystyle\int_1^n\frac{g(x)}{x^2}\,dx$. Why? It follows that

$$\lim_{n\to\infty}\left(1+\frac{1}{2}+\frac{1}{3}+\cdots+\frac{1}{n}-\ln n\right)=\gamma \text{ with } \gamma=\frac{1}{2}-\int_1^\infty\frac{g(x)}{x^2}\,dx.$$

Recall that γ is Euler's constant. See Exercises 27 and 28 of Section 6.6. ■

Problem 6.8.3. Use the following outline to prove Stirling's formula

$$\lim_{n\to\infty}\frac{n!}{\sqrt{2\pi n}}\left(\frac{e}{n}\right)^n=1.$$

See Remark 6.6.7, part (k).

(a) Use Euler's summation formula with $f(x)=\ln(1+x)$ to obtain

$$\begin{aligned}\ln 1+\ln 2+\cdots+\ln n &=\int_1^n\ln x\,dx+\frac{1}{2}\ln n+\int_1^n\frac{g(x)}{x}\,dx\\ &=\left(n+\frac{1}{2}\right)\ln n-(n-1)+\int_1^n\frac{g(x)}{x}\,dx,\end{aligned}$$

where $g(x)=x-[x]-\dfrac{1}{2}$.

(b) Show that the improper integral $\displaystyle\int_1^\infty\frac{g(x)}{x}\,dx$ is convergent.

(c) Conclude that $\ln(n!)=\left(n+\dfrac{1}{2}\right)\ln n-n+a_n$, where

$$a_n = 1 + \int_1^n \frac{g(x)}{x} dx \to A \text{ as } n \to \infty.$$

(d) Find A by writing both

$$\begin{aligned} 2\ln(2\cdot 4\cdot \cdots \cdot 2n) &= 2\ln\left[2^n(1\cdot 2\cdot \cdots \cdot n)\right] \\ &= (2n+1)\ln 2n - 2n - \ln 2 + 2a_n, \end{aligned}$$

and

$$\ln(2n+1)! = \left(2n+\frac{3}{2}\right)\ln(2n+1) - (2n+1) + a_{2n+1}.$$

Notice that subtraction of the second equality from the first yields

$$\ln\frac{(2)(4)\cdot \cdots \cdot(2n)}{(1)(3)\cdot \cdots \cdot(2n-1)\sqrt{2n+1}} = (2n+1)\ln\left(1-\frac{1}{2n+1}\right) + 1 - \ln 2 + 2a_n - a_{2n+1}.$$

Now, using both Wallis' formula and letting $n \to \infty$ we obtain $\ln\sqrt{\frac{\pi}{2}} = -1 + 1 - \ln 2 + 2A - A$ which gives the value of A.

(e) Write $\ln(n!) = \left(n+\frac{1}{2}\right)\ln n - n + \ln\sqrt{2\pi} - b_n$, where $b_n = \int_n^\infty \frac{g(x)}{x} dx \to 0$ as $n \to \infty$.

(f) Draw a final conclusion.

Part 3. Inverse Laplace Transforms

Definition 6.8.4. If $\mathcal{L}\{f(t)\} = F(x)$, then $f(t)$ is called an *inverse Laplace transform* of $F(x)$. See Definition 6.6.13.

Problem 6.8.5. If k is any real constant and $f(t) = \begin{cases} 1, & \text{if } 0 < t < 2 \\ k, & \text{if } t = 2 \\ 1, & \text{if } t > 2, \end{cases}$ show that

$\mathcal{L}\{f(t)\} = \frac{1}{x}$, if $x > 0$.

From Problem 6.8.5, it is clear that no matter what k is, $\mathcal{L}\{f(t)\}$ is the same. This creates great difficulty in deciding on an inverse transform. However, since only one continuous inverse transform exists for each individual function, barring a Laplace transform, it is that one transform we normally pick and label $\mathcal{L}^{-1}\{F(x)\}$. Thus, "$\mathcal{L}\{f(t)\} = F(x)$ if and only if $\mathcal{L}^{-1}\{F(x)\} = f(t)$" is not a true statement.

Problem 6.8.6

(a) *(Lerch's[23] Theorem)* If f and g are continuous on $[0,\infty)$ for which $\mathcal{L}\{f\}=\mathcal{L}\{g\}$, prove that $f \equiv g$. This will prove that a continuous function is uniquely determined by its Laplace transform.

(b) Prove that $\mathcal{L}^{-1}$ satisfies the linear property.

(c) Evaluate $\mathcal{L}^{-1}\left\{\dfrac{n!}{x^{n+1}}\right\}$ for any $n \in N$.

(d) Evaluate $\mathcal{L}^{-1}\left\{\dfrac{4}{x^2+6x+13}\right\}$. See Exercise 17 of Section 6.6.

(e) Evaluate $\mathcal{L}^{-1}\left\{\dfrac{1}{x(x^2+1)}\right\}$.

(f) Evaluate $\mathcal{L}^{-1}\left\{\dfrac{1}{x(x^2+1)}\right\}$ by using the convolution theorem given in Exercise 24(b) of Section 6.6, with $f(t)=\mathcal{L}^{-1}\left\{\dfrac{1}{x}\right\}$ and $g(t)= \mathcal{L}^{-1}\left\{\dfrac{1}{x^2+1}\right\}$.

(g) Use Exercise 21(e) from Section 6.6 to compute $\mathcal{L}^{-1}\left\{\arctan\dfrac{1}{x}\right\}$. Now find $\mathcal{L}\left\{\dfrac{\sin t}{t}\right\}$. See Exercise 11 of Section 8.6.

(h) Find a function f satisfying $f(0)=1$ and $f'(t)-3f(t)=e^{2t}$ by first computing $\mathcal{L}\{f(t)\}$ and then its inverse.

(i) Repeat part (h) for the case in which $f'(t)=1-\int\limits_0^t e^{-2x}f(t-x)\,dx$. This is known as a *Volterra[24] integro-differential equation* and is used to study population growth based on hereditary influences.

(j) Use Laplace transforms to find the function $y(t)$ such that

$$\frac{d^2y}{dt^2}+y=t,\ y(0)=1,\ \text{and } y'(0)=0.$$

---

[23]Mathias Lerch (1860–1922) was a mathematician born in what is today the Czech Republic. Lerch, who studied under Weierstrass and Kronecker in Berlin, is predominantly known for introducing new methods and ideas in the areas of analysis and number theory, on which he wrote 238 papers.

[24]Vito Volterra (1860–1940), a renowned Italian mathematician and physicist who studied integral equations, developed the theory of integro-differential equations and contributed greatly to functional analysis. The term "integral equation" was first introduced by Du Bois-Reymond.

7

Infinite Series

7.1 Convergence

Suppose $\{a_n\}$ is a sequence of real numbers as defined in Chapter 2. In this chapter we will study the convergence or divergence of the sum of its terms; that is, we will determine whether an expression $a_1 + a_2 + \cdots + a_n + \cdots$ converges or diverges. In other words, we will determine whether or not the preceding expression approaches a finite real value. An infinite sum like this has no meaning to us unless we define it as a limit of a finite sum.

Definition 7.1.1. Suppose p is an integer, often a nonnegative one, and $\{a_n\}_{n=p}^{\infty}$ is a sequence of real numbers. The expression $a_p + a_{p+1} + a_{p+2} + \cdots$ is called an *infinite series* and is denoted by $\sum_{k=p}^{\infty} a_k$ or $\sum_{p}^{\infty} a_k$. When $p=1$, the infinite series $a_1 + a_2 + \cdots + a_n + \cdots$ is denoted by $\sum_{k=1}^{\infty} a_k$ or $\sum a_k$. The symbol $\sum$ is called a *sigma*, and the values $a_1, a_2, \ldots, a_n, \ldots$ are called *terms* of the series with a_n the *nth term* of the series.

For any sequence $\{a_n\}_{n=p}^{\infty}$, we can define a related sequence $\{S_n\}$, where

$$\begin{aligned} S_1 &= a_p \\ S_2 &= a_p + a_{p+1} \\ S_3 &= a_p + a_{p+1} + a_{p+2} \\ &\vdots \\ S_n &= a_p + a_{p+1} + a_{p+2} + \cdots + a_{p+n-1} = \sum_{k=1}^{n} a_{p+k-1}. \end{aligned}$$

The sequence $\{S_n\}$ is called the *sequence of partial sums* of the series $\sum_{k=p}^{\infty} a_k$. See Exercise 15 of Section 2.2. Naturally, if $p = 1$, the subscripts match up nicely. Subscripts are *dummy variables*. Thus, we could express $\sum_{k=p}^{\infty} a_k$ by, say, $\sum_{n=p}^{\infty} a_n$.

Definition 7.1.2. Suppose $\{a_n\}_{n=p}^{\infty}$ is a sequence of real numbers. We say that the series $\sum_{k=p}^{\infty} a_k$ *converges* to some real value S if and only if the sequence of partial sums $\{S_n\}$ converges to S. If $\sum_{k=p}^{\infty} a_k$ converges to S, then S is called the *sum* of the series and $\sum_{k=p}^{\infty} a_k = S$.

Remark 7.1.3. Suppose $\{a_n\}_{n=p}^{\infty}$ is a sequence of real numbers. Definition 7.1.2 suggests that the series $\sum_{k=p}^{\infty} a_k$ can be written as $\lim_{n\to\infty} \sum_{k=p}^{n} a_k$ and that it converges if and only if this limit is finite. In addition, since the sequence of partial sums $\{S_n\}$ determines the behavior of series $\sum_{k=p}^{\infty} a_k$, we say that $\sum_{k=p}^{\infty} a_k$ *diverges to* $+\infty$ (or $-\infty$) if and only if $\lim_{n\to\infty} S_n$ diverges to $+\infty$ (or $-\infty$). ■

Example 7.1.4. Consider the sequence $\{a_n\}$ with $a_n = \dfrac{1}{2^n}$, $n \in N$. Since the sequence of partial sums $\{S_n\}$ with $S_n = \dfrac{1}{2} + \dfrac{1}{2^2} + \cdots + \dfrac{1}{2^n} = 1 - \dfrac{1}{2^n}$ converges, the series $\sum a_k$ converges. See Exercise 19(a) of Section 2.1, Exercise 17 of Section 2.3, and Remark 7.1.15, part (c). In fact, since $\lim_{n\to\infty} S_n = 1$, we can write $\sum_{k=1}^{\infty}\frac{1}{2^k} = 1$. See Example 7.1.14. ■

Example 7.1.5. Prove that $\sum_{k=1}^{\infty}\frac{k}{k+1}$ diverges to $+\infty$. See Exercise 15 in Section 4.1.

Proof. Since $\frac{n}{n+1} \geq \frac{1}{2}$ for all $n \in N$ (why?), we write $S_n = \frac{1}{2} + \frac{2}{3} + \cdots + \frac{n}{n+1} > n \cdot \frac{1}{2}$. And since $\lim_{n\to\infty}\frac{n}{2} = +\infty$, by the comparison test, Theorem 2.3.2, the sequence $\{S_n\}$ diverges to $+\infty$. And the result follows. ■

Example 7.1.6. The series $\sum_{k=5}^{\infty}(-1)^k$ diverges because the limit of partial sums does not exist. Why? ■

Example 7.1.7. Determine whether the $\sum a_k$ with $a_k = \frac{1}{k(k+1)}$ converges or diverges.

Answer. Since $a_k = \frac{1}{k(k+1)} = \frac{1}{k} - \frac{1}{k+1}$, obtained by the partial fraction decomposition, we write

$$S_n = a_1 + a_2 + \cdots + a_n = \left(1 - \frac{1}{2}\right) + \left(\frac{1}{2} - \frac{1}{3}\right) + \cdots + \left(\frac{1}{n} - \frac{1}{n+1}\right) = 1 - \frac{1}{n+1},$$

due to the *telescoping* nature of the series. Thus, since $\lim_{n\to\infty} S_n = 1$, the series converges to 1. See Exercise 2(r) of Section 1.3 and Exercise 16 of Section 2.2. ■

Remark 7.1.8. If $\sum_{k=1}^{\infty} a_k$ and $\sum_{k=1}^{\infty} b_k$ differ only in their first finite number of terms, that is, $a_k = b_k$ eventually; then they either both converge or diverge. Certainly, if they converge, their sums will most likely be different. Also, see Corollary 7.1.20. ■

THEOREM 7.1.9. *If $\sum a_k$ converges, then $\lim_{n\to\infty} a_n = 0$. That is, if $\sum a_k$ converges, then the sequence $\{a_n\}$ converges to zero.*

Proof. Suppose $\sum a_k$ converges and $\{S_n\}$ is its sequence of partial sums. Then $a_n = S_n - S_{n-1}$. And since $\sum a_k$ converges, the sequence $\{S_n\}$ also converges to, say, S. Thus, $\lim_{n\to\infty} S_{n-1} = S$ as well. See Remark 2.1.8, part (c). Now, since both limits are finite, we can write

$$\lim_{n\to\infty} a_n = \lim_{n\to\infty}\left(S_n - S_{n-1}\right) = \lim_{n\to\infty} S_n - \lim_{n\to\infty} S_{n-1} = S - S = 0.$$

Hence, $\{a_n\}$ converges to zero. ■

The converse of Theorem 7.1.9 is false. See Example 7.1.12. The usefulness of Theorem 7.1.9 might be of some concern since it is the convergence or divergence of a series that one usually needs to determine and not the limit of its terms. For that very reason we customarily use the contrapositive version of Theorem 7.1.9, often referred to as the divergence test (or *nth term test*).

Theorem 7.1.10. *(Divergence Test) If a sequence $\{a_n\}$ does not converge to zero, then the series $\sum a_k$ diverges.*

Example 7.1.11. Determine whether $\sum a_k$ with $a_k = \dfrac{k!}{\frac{3k!}{2}+1}$ converges or diverges.

Answer. Since $\lim_{k\to\infty} \dfrac{k!}{\frac{3k!}{2}+1} = \dfrac{2}{3}$ (why?), which is not zero, by the divergence test this series diverges. ■

Example 7.1.12. Verify by an example that if a sequence $\{a_n\}$ converges to zero, then $\sum a_k$ may or may not converge.

Proof. We need to give an example of each situation. First, pick any converging series, say $\sum 0$. Next consider $\sum \ln\left(\dfrac{k+1}{k}\right)$, which diverges but whose terms $a_k = \ln\left(\dfrac{k+1}{k}\right)$ tend to zero. To verify this, observe that

$$S_n = \ln 2 + \ln\frac{3}{2} + \cdots + \ln\frac{n+1}{n} = \ln\left(2\cdot\frac{3}{2}\cdot\ \cdots\ \cdot\frac{n+1}{n}\right) = \ln(n+1).$$

And since $\lim_{n\to\infty} S_n = \lim_{n\to\infty} \ln(n+1) = +\infty$, the series diverges to $+\infty$. However, note that

$$\begin{aligned}\lim_{n\to\infty} a_n &= \lim_{n\to\infty} \ln\left(\frac{n+1}{n}\right) = \ln\left(\lim_{n\to\infty}\frac{n+1}{n}\right). \text{ (Why?)}\\ &= \ln 1 = 0.\end{aligned}$$

■

Example 7.1.13. Prove that the *harmonic series*

$$\sum_{k=1}^{\infty}\frac{1}{k}=1+\frac{1}{2}+\frac{1}{3}+\frac{1}{4}+\cdots$$

diverges. See Exercise 7(d) of Section 2.5 and Exercise 3 of Section 2.6.

Proof. Consider the sequence of partial sums $\{S_n\}$, where $S_n=\sum_{k=1}^{n}\frac{1}{k}$. Then we have

$$S_2=1+\frac{1}{2}$$

$$S_4=1+\frac{1}{2}+\frac{1}{3}+\frac{1}{4}>1+\frac{1}{2}+\frac{1}{4}+\frac{1}{4}=1+\frac{2}{2}$$

$$S_8=1+\frac{1}{2}+\cdots+\frac{1}{8}>1+\frac{1}{2}+\left(\frac{1}{4}+\frac{1}{4}\right)+\left(\frac{1}{8}+\frac{1}{8}+\frac{1}{8}+\frac{1}{8}\right)=1+\frac{3}{2}$$

$$\begin{aligned}S_{16}&=1+\frac{1}{2}+\cdots+\frac{1}{16}\\&>1+\frac{1}{2}+\left(\frac{1}{4}+\frac{1}{4}\right)+\left(\frac{1}{8}+\cdots+\frac{1}{8}\right)+\left(\frac{1}{16}+\cdots+\frac{1}{16}\right)=1+\frac{4}{2},\end{aligned}$$

and so on. In general, we can prove by mathematical induction that $S_{2^n}>1+\frac{n}{2}$. Thus, by the comparison test, the sequence $\{S_{2^n}\}$ diverges to $+\infty$. Therefore, since $\{S_n\}$ has a diverging subsequence $\{S_{2^n}\}$, by Theorem 2.6.5, $\{S_n\}$ diverges. Hence, so does the harmonic series. ■

The harmonic series would be another example for Example 7.1.12.

THEOREM 7.1.14. *If a is a nonzero real number, then the geometric series $\sum_{k=0}^{\infty}ar^k$, although often written as $\sum_{k=1}^{\infty}ar^{k-1}$, with $r\in\Re$, has the property that*

$$\sum_{k=1}^{\infty}ar^{k-1}=\begin{cases}\dfrac{a}{1-r}, & \text{if } |r|<1\\ \text{diverges}, & \text{if } |r|\geq 1.\end{cases}$$

See Exercise 10(b) of Section 2.4.

Proof. If $|r|\geq 1$, then $\lim_{n\to\infty}ar^n\neq 0$, and the series diverges by the divergence test. Thus, suppose that $|r|<1$. If $S_n=a+ar+ar^2+\cdots+ar^{n-1}$ is the nth term of the sequence of partial sums of the given series, often referred to as a *geometric progression*, then $rS_n = ar + ar^2 + ar^3 + \cdots + ar^n$. Subtraction yields $S_n - rS_n = a - ar^n$, which produces $S_n = \frac{a(1-r^n)}{1-r}$. See Example 1.3.4 and Example 1.5.5. We now need to calculate the limiting value of the sequence $\{S_n\}$. Since $|r|<1$, by Theorem 2.1.13, we have $\lim_{n\to\infty} r^n = 0$. Thus,

$$\lim_{n\to\infty} S_n = \lim_{n\to\infty} \frac{a(1-r^n)}{1-r} = \frac{a}{1-r}.$$

And the series converges to $\frac{a}{1-r}$. See Exercise 18 of Section 2.1. ■

Remark 7.1.15

(a) If $|r| \geq 1$, then $\sum_{k=1}^{\infty} ar^{k-1}$ converges only if $a = 0$. If this is the case, clearly $\sum_{k=1}^{\infty} ar^{k-1}$, the series of zeros, converges to zero.

(b) A converging geometric series has an easy formula for its sum. There are few types of series for which the actual sum can be computed. See Remark 7.4.17.

(c) The formula $\frac{1}{2} + \frac{1}{2^2} + \cdots + \frac{1}{2^n} = 1 - \frac{1}{2^n}$ used in Example 7.1.4 may not only be proven by induction but also by writing

$$\frac{1}{2} + \frac{1}{2^2} + \cdots + \frac{1}{2^n} = \frac{\frac{1}{2}\left[1 - \left(\frac{1}{2}\right)^n\right]}{1 - \frac{1}{2}} = \frac{\frac{1}{2} - \left(\frac{1}{2}\right)^{n+1}}{\frac{1}{2}} = 1 - \frac{1}{2^n}.$$

(d) The formula $\frac{1}{1-x} = \sum_{k=0}^{\infty} x^k$, with $|x|<1$, can also be obtained through "long division." ■

Example 7.1.16. Use Theorem 7.1.14 to prove that $\sum_{k=1}^{\infty} \frac{1}{2^k} = 1$. See Example 7.1.4.

Proof. Since $\sum_{k=1}^{\infty} \frac{1}{2^k}$ is a geometric series with $a = \frac{1}{2}$ and $r = \frac{1}{2}$, by Theorem 7.1.14, this series converges to $\frac{a}{1-r} = \frac{\frac{1}{2}}{1-\frac{1}{2}} = 1.$ ■

Example 7.1.17. Write the repeating decimal $1.4\overline{32}$ as a quotient of two integers. See Exercise 19(b) from Section 2.1.

Procedure 1. Write

$$\begin{aligned} 1.4\overline{32} &= 1.4 + 0.032 + 0.00032 + 0.0000032 + \cdots \\ &= 1.4 + 0.032 + (0.032)(0.01) + (0.032)(0.01)^2 + \cdots \\ &= 1.4 + \frac{0.032}{1-0.01} = \frac{1.418}{0.99} = \frac{709}{495}. \end{aligned}$$

Procedure 2. Write $x = 1.4323232...$ and multiply both sides by 100 to get $100x = 143.23232....$ Subtraction yields $99x = 141.8$, which gives $x = \frac{709}{495}$. Note that by writing $x = 1.4\overline{32}$, we are assuming that the series behind the scenes actually converges. We simply found what this series converges to provided that it indeed converges. The proof that this series converges still remains. ■

THEOREM 7.1.18. *If $\sum a_k$ and $\sum b_k$ are infinite series that converge to the real numbers A and B, respectively, then the series $\sum ca_k$, with c any real constant, and $\sum (a_k \pm b_k)$ both converge, and*

(a) $\sum ca_k = cA$, *and*

(b) $\sum (a_k \pm b_k) = A \pm B$.

Proof of Theorem 7.1.18 is straightforward and is thus left as an exercise. As a consequence of Theorem 7.1.18, converging series satisfy the linear property.

THEOREM 7.1.19. *(Cauchy Criterion for Series) An infinite series $\sum a_k$ converges if and only if for each $\varepsilon > 0$ there exists a natural number n^* such that if $n \geq m \geq n^*$ with $m, n \in N$, then*

$$|a_m + a_{m+1} + \cdots + a_n| = \left| \sum_{k=m}^{n} a_k \right| < \varepsilon.$$

Equivalently, Theorem 7.1.19 says that $\sum a_k$ converges if and only if for each $\varepsilon > 0$ there exists $n^* \in N$ such that if $n \geq n^*$ and $i \geq 0$ with $i \in N$, then

$$|a_n + a_{n+1} + \cdots + a_{n+i}| < \varepsilon.$$

Proof of Theorem 7.1.19. Outline: $\sum a_k$ converges $\Leftrightarrow$ $\{S_n\}$ converges $\Leftrightarrow$ $\{S_n\}$ is Cauchy $\Leftrightarrow$ $|S_n - S_{m-1}| < \varepsilon$ for all $n \geq m \geq n^*$ with some $n^* \in N$ $\Leftrightarrow$ $|a_m + a_{m+1} + \cdots + a_n| < \varepsilon$ for all $n \geq m \geq n^*$. ■

Observe that if $\sum a_k$ converges, then by Theorem 7.1.19, there will exist an $n^* \in N$ large enough so that for any $n \geq n^*$, $|a_n| < \varepsilon$. Why? The contrapositive of this theorem is equivalent to the divergence test. In addition, the Cauchy criterion for series does not tell us what the sum of a converging series is, but rather tells us that convergence is determined by the "tail" end of a series. The first finite number of terms has no effect on whether or not a series converges. This is formulated as follows.

COROLLARY 7.1.20. *Suppose $\{a_n\}$ is a sequence of real numbers. If $\sum_{k=1}^{\infty} a_k$ converges, then $\sum_{k=p}^{\infty} a_k$ converges for any $p \in N$. And conversely, if $\sum_{k=p}^{\infty} a_k$ converges for some $p \in N$, then $\sum_{k=1}^{\infty} a_k$ converges, provided that all the terms are defined.*

Exercises 7.1

1. Determine whether the given series converges or diverges. Explain yourself. In the case of a converging geometric or telescoping series find the sum.

(a) $\sum_{k=2}^{\infty} \frac{(-1)^k}{2^{k-1}}$

(b) $\sum \frac{3^{k-1} - 2^k}{6^k}$

(c) $\sum \frac{1}{k^2 + 3k + 2}$

(d) $\sum \frac{1}{4k^2 - 1}$

(e) $\sum \frac{k}{2k+1}$

(f) $\sum \left(\sqrt{k+1} - \sqrt{k}\right)$ (See Exercise 11(f) from Section 2.2.)

(g) $\sum \ln \frac{1}{k}$

(h) $\sum_{k=2}^{\infty} \ln \frac{k^2 - 1}{k^2}$

(i) $\sum \ln \frac{k}{k+1}$ (Compare with Example 7.1.12.)

(j) $\sum k \sin \frac{1}{k}$

(k) $\sum_{k=2}^{\infty} \left(\frac{1}{k}\right)^{\frac{1}{\ln k}}$ (See Exercise 19 in Section 3.3.)

(l) $\sum \left(\frac{k-2}{k}\right)^k$

2. Use Exercise 2(w) from Section 1.3 to prove that $\sum \frac{1}{4k^2 - 1} = \frac{1}{2}$. See Exercise 1(d).

3. Use Theorem 7.1.14 to write the given expressions as a quotient of two integers.

(a) $1.\overline{9}$ (See Exercise 19(b) from Section 2.1.)

(b) $3+1+\frac{1}{2}+\frac{1}{3}+\frac{1}{9}+\frac{1}{27}+\cdots$ (c) $3.2\overline{15}$

4. Determine whether or not the given statements are true. Explain.
(a) If both $\sum a_k$ and $\sum b_k$ diverge, then $\sum(a_k+b_k)$ diverges.
(b) If both $\sum a_k$ and $\sum b_k$ diverge, then $\sum(a_k+b_k)$ converges.
(c) If $\sum a_k$ converges and $\sum b_k$ diverges, then $\sum(a_k+b_k)$ diverges.
(d) If $\sum a_k$ converges and $\sum b_k$ diverges, then $\sum(a_k+b_k)$ converges.
(e) Series $\sum_{k=2}^{\infty}\left(\frac{1}{k}-\frac{1}{2^k}\right)$ converges. (f) Series $\sum_{k=1}^{\infty}\frac{1-3^k}{2^k}$ converges to 4.

5. Find all the values of x for which the given series converges.
(a) $\sum_{k=0}^{\infty}3(x-2)^k$ (b) $\sum_{k=3}^{\infty}\frac{x^{k-1}}{5^k}$
(c) $\sum_{k=1}^{\infty}(\arctan x)^k$ (d) $\sum_{k=2}^{\infty}(\ln x)^k$

6. (a) Give an example of two series $\sum a_k$ and $\sum b_k$ that differ in the first five terms yet converge to the same value.
(b) Give an example of two series $\sum a_k$ and $\sum b_k$ that differ in infinitely many terms yet converge to the same value.
(c) Give an example of two series $\sum a_k$ and $\sum b_k$ that differ in infinitely many but not all terms and yet converge to the same value.

7. Prove Theorem 7.1.18.

8. Suppose $\sum a_k$ is a series of positive terms that converges. Prove that $\sum\frac{1}{a_k}$ is divergent. Is the converse true? See Exercise 1 in Section 7.5.

9. (a) Suppose that $\{a_n\}$ is a sequence of nonnegative real numbers. Prove that $\sum a_k$ is convergent if and only if its sequence of partial sums is bounded.
(b) Give an example of a series that diverges and whose sequence of partial sums is bounded.

10. Suppose that $\{a_n\}$ is a sequence of real numbers and $b_n=a_n-a_{n+1}$ for all $n\in N$. See Exercise 20 in Section 7.5.
(a) Prove that $\sum b_k$ converges if and only if the sequence $\{a_n\}$ converges.
(b) In the case that $\sum b_k$ converges, find its sum.

11. Give an example of two series $\sum a_k$ and $\sum b_k$ that converge to real numbers A and B, respectively, but the series $\sum a_k b_k$ converges to a value different from AB. See Exercises 16 below, 11(a) in Section 7.2, and 7 in Section 7.5.

12. Give an example of two series $\sum a_k$ and $\sum b_k$ that converge to real numbers A and B, respectively, with $b_n \neq 0$ for any $n \in N$, but the series $\sum \frac{a_k}{b_k}$ converges to a value different from $\frac{A}{B}$.

13. Give an example of two converging series $\sum a_k$ and $\sum b_k$ with $b_n \neq 0$ for any $n \in N$, but the series $\sum \frac{a_k}{b_k}$ diverges.

14. If $\{a_n\}$ and $\{b_n\}$ are two sequences of positive real numbers such that

$$b_n = \frac{a_1 + a_2 + \cdots + a_n}{n},$$

prove that $\sum b_k$ diverges to $+\infty$. See Exercise 36 of Section 2.7.

15. **(a)** If $f(x) = \sum_{k=0}^{\infty} x^k = 1 + x + x^2 + \cdots$, give the domain of the function f and graph it.

(b) Do the same as in part (a) for $g(x) = \sum_{k=1}^{\infty} x^k$ and for $h(x) = 1 - x - x^2 - \cdots$. (See Exercise 3(a) of Section 1.3.)

(c) Use the result of part (a) to rewrite $\frac{x^2 - 2}{x - 1}$ as a series.

(d) Use long division on $\frac{x^2 - 2}{x - 1}$ by writing the expression in the numerator and in the denominator in ascending order. Compare this result with part (c).

16. If a series $\sum |a_k|$ converges and a sequence $\{b_n\}$ is bounded, prove that $\sum a_k b_k$ converges.

17. Prove that the series $\sum a_k = 1 - 1 + \frac{1}{2} - \frac{1}{2} + \frac{1}{3} - \frac{1}{3} + \cdots$ converges to 0.

18. **(a)** Show that $\sum \frac{1}{k^2}$ converges.

(b) Use part (a) to show that $\sum \frac{1}{k!}$ converges. See Exercise 8 in Section 7.5.

(c) Show that $\sum \frac{k}{(k+1)!} = 1$.

19. **(a)** Use Exercise 2(u) of Section 1.3 to prove that $\sum \frac{1}{k^2} \leq \frac{7}{4}$. See Exercise 7(b) of Section 2.4 and Remark 7.2.2, part (b).

(b) Use Exercise 2(v) of Section 1.3 to prove that $\sum \frac{1}{k^3} \le \frac{3}{2}$.

(c) Use Exercise 2(s) of Section 1.3 to prove that $\sum \frac{1}{\sqrt{k}}$ diverges.

20. If f is Riemann integrable on $[a,b]$ and $a_n = \frac{b-a}{n} \sum_{k=1}^{n} f\left(a + k\frac{b-a}{n}\right)$, then the sequence $\{a_n\}$ converges to $\int_a^b f$. True or false? See Exercise 10 in Section 6.2 and Exercise 10 in Section 6.4.

21. **(a)** Prove that if the series $\sum |a_k|$ converges, then so does $\sum a_k$.

(b) Use part (a) to verify that $\sum \frac{(-1)^k}{k^2}$ converges.

22. Observe that $1 + x + x^2 + x^3 + \cdots = \frac{1}{1-x}$. Replace x by $\frac{1}{x}$ and subtract 1 from both sides to obtain $\frac{1}{x} + \frac{1}{x^2} + \frac{1}{x^3} + \cdots = \frac{1}{x-1}$. Add these two series to obtain

$$\cdots + \frac{1}{x^3} + \frac{1}{x^2} + \frac{1}{x} + 1 + x + x^2 + x^3 + \cdots = 0.$$

Why is this equation preposterous? What is wrong with the preceding "proof"?

23. Prove that the functions $f_n : [0,1] \to \mathfrak{R}$, defined by $f_n(x) = x^{n-1}$ with $n \in N$, partition the square S, where $S = \{(a,b) \mid 0 \le a \le 1 \text{ and } 0 \le b \le 1\}$ into infinitely many regions whose area adds up to 1.

7.2 Tests for Convergence

Theorem 7.2.1. *(Integral Test) Suppose that a function* $f:[1,\infty) \to \mathfrak{R}$ *is continuous, nonnegative, and decreasing and* $f(k) = a_k$ *for all* $k \in N$. *Then* $\sum_{k=1}^{\infty} a_k$ *converges if and only if* $\int_1^{\infty} f(x)\,dx$ *converges.*

Proof. This theorem will be proven by analyzing the area under the curve f (see Figure 7.2.1). Since f is decreasing, $f(k+1) \le f(x) \le f(k)$ for all $x \in [k, k+1]$. And since f is continuous, f is Riemann integrable. Therefore, by Theorem 6.3.2,

$$\int_k^{k+1} f(k+1)\,dx \le \int_k^{k+1} f(x)\,dx \le \int_k^{k+1} f(k)\,dx,$$

which is equivalent to

$$a_{k+1} = f(k+1) \le \int_k^{k+1} f(x)\,dx \le f(k) = a_k.$$

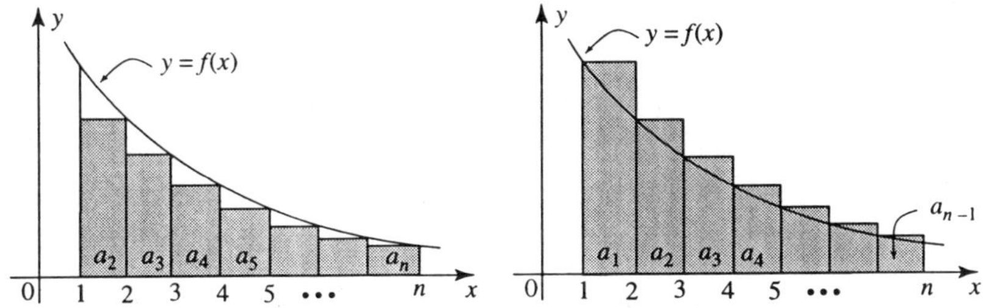

Figure 7.2.1

Hence,

$$a_2 + a_3 + \cdots + a_{k+1} \le \int_1^{k+1} f(x)\,dx \le a_1 + a_2 + \cdots + a_k.$$

Thus, $S_{k+1} - a_1 \le \int_1^{k+1} f(x)\,dx \le S_k$. If $\sum a_k$ converges, then $\{S_n\}$ is bounded, by Exercise 9(a) of Section 7.1. Therefore, $\int_1^{k+1} f(x)\,dx$ is bounded for all $k \in N$. But since $\int_1^{k+1} f(x)\,dx$ increases as k increases, $\int_1^{\infty} f(x)\,dx$ converges. Next, suppose that $\int_1^{\infty} f(x)\,dx$ converges. Then, since $S_{k+1} - a_1 \le \int_1^{k+1} f(x)\,dx \le \int_1^{\infty} f(x)\,dx$, $S_{k+1} - a_1$ is bounded. Thus, S_{k+1} is bounded and certainly increasing, which in turn, implies that the sequence $\{S_n\}$ is bounded and increasing. Hence, $\sum a_k$ converges. ■

Remark 7.2.2

(a) In Theorem 7.2.1, the domain of the function f can be changed to $[p,\infty)$, where p is any natural number not necessarily equal to one. Then $\sum_{k=p}^{\infty} f(k)$ converges if and only if $\int_p^{\infty} f$ converges.

(b) The integral test can be easily misinterpreted. If a series and integral both converge, they need not converge to the same value. The integral test only tells us that

$$\int_p^{\infty} f \leq \sum_{k=p}^{\infty} f(k) \leq f(p) + \int_p^{\infty} f.$$

For example, $\int_1^{\infty} \frac{1}{x^2}\,dx = 1$ and, therefore, $\sum \frac{1}{k^2}$ converges. See Exercise 18(a) in Section 7.1. However, in Problem 12.10.8, we will see that $\sum \frac{1}{k^2}$ converges not to 1 but to $\frac{\pi^2}{6}$. Also, see Exercise 7(b) of Section 2.4 and Example 9.2.4.

(c) Each condition in Theorem 7.2.1 is necessary for the result to be true. See Exercise 15 at the end of this section. ■

Example 7.2.3. *(p-series)* Prove that $\sum \frac{1}{k^p}$ converges if $p > 1$, and diverges if $p \leq 1$. A converging p-series is called *hyperharmonic*. See Example 7.1.13 and Exercise 3 of Section 6.5.

Proof. If $p \leq 0$, by the divergence test, the p-series diverges. Now, assume that $p > 0$. Since all of the conditions in the integral test are satisfied by the function $f(x) = \frac{1}{x^p}$ (why?), we can evaluate the improper integral to obtain

$$\int_1^{\infty} \frac{1}{x^p}\,dx = \begin{cases} \dfrac{1}{p-1}, & \text{if } p > 1 \\ \infty, & \text{if } 0 < p \leq 1. \end{cases}$$

Therefore, the result follows by the integral test. ■

The next two theorems should sound familiar. See the comparison tests covered in Section 6.5.

THEOREM 7.2.4. *(Comparison Test) Consider the infinite series* $\sum a_k$ *and* $\sum b_k$ *with* $b_k \geq 0$ *for all* $k \in N$.

(*a*) *If* $\sum b_k$ *converges and* $|a_k| \leq Mb_k$ *eventually with* $M \in \Re^+$, *then* $\sum |a_k|$ *converges, and by Exercise 21 of Section 7.1,* $\sum a_k$ *converges.*

(*b*) *If* $\sum b_k$ *diverges and* $|a_k| \geq mb_k$ *eventually with* $m \in \Re^+$, *then* $\sum |a_k|$ *diverges.*

Proof of part (a). Suppose that $\sum b_k$ converges. And suppose that there exists $n^* \in N$ such that $|a_k| \leq Mb_k$ for all $k \geq n^*$. Since $\sum b_k$ converges, $\sum Mb_k$ converges. Therefore,

$$\sum_{k=n^*}^{n} |a_k| \leq M \sum_{k=n^*}^{n} b_k \equiv S.$$

Thus, the sequence of partial sums of $\sum |a_k|$ is increasing and bounded above and, hence, convergent. Thus, the proof of part (a) is complete.

Proof of part (b) is left as an exercise. ■

The comparison test often proves beneficial in determining convergence or divergence of a series. The difficulty with using this test is deciding in which direction to turn, that is, do we "feel" that the series at hand converges or diverges, having a repertoire of converging and diverging series at our fingertips, and having the capability of comparing series with each other term by term. Thus, we first "guess" whether the series at hand converges or diverges by examining the dominant feature of the series. If it converges, we need to compare it with a converging series that, term by term, is larger. Otherwise, we need to compare the series at hand with a diverging series that, term by term, is smaller. Illustrations are given in the next two examples.

Example 7.2.5. Test the given series for convergence or divergence.

(**a**) $\sum \frac{5}{3^k + 2}$ (**b**) $\sum \frac{1}{2 + \sqrt{k}}$ (**c**) $\sum_{k=2}^{\infty} \frac{1}{k+5}$

Answer to part (a). $\sum \frac{5}{3^k + 2} = 5 \sum \frac{1}{3^k + 2}$. Observe that the denominator is very much like 3^k for large values of k. Therefore, the dominant feature is $\frac{1}{3^k}$, which suggests convergence. Thus, we can attempt to compare the given series with the converging geometric series $\sum \frac{1}{3^k}$. Since $\frac{1}{3^k + 2} \leq \frac{1}{3^k}$ for all $k \in N$, by the comparison test, $\sum \frac{1}{3^k + 2}$ converges. Hence, $\sum \frac{5}{3^k + 2}$ converges.

Answer to part (b). Here, we might wish to compare $\frac{1}{2+\sqrt{k}}$ to $\frac{1}{\sqrt{k}}$, since $\sum \frac{1}{\sqrt{k}}$ diverges and has terms very similar to $\sum \frac{1}{2+\sqrt{k}}$. Unfortunately, $\frac{1}{2+\sqrt{k}} \le \frac{1}{\sqrt{k}}$ for any $k \in N$. This inequality goes the wrong way for the comparison test. Now, either we lose faith and attempt to compare the original series to one that converges or we look for a different diverging series to it compare with. Choosing the latter, observe that $\frac{1}{2+\sqrt{k}} \ge \frac{1}{k}$, for all $k \ge 5$. Therefore, by the comparison test the original series diverges.

Answer to part (c). Consider the following options in verifying the divergence of series $\sum_{k=2}^{\infty} \frac{1}{k+5}$. A p-series with $p < 1$ cannot be used. Why? However, since $k+5 < 2k$ for $k > 5$, we have $\sum_{6}^{\infty} \frac{1}{k+5} > 2\sum_{6}^{\infty} \frac{1}{k}$. Since $\sum_{6}^{\infty} \frac{1}{k}$ diverges, so does $2\sum_{6}^{\infty} \frac{1}{k}$. Thus, by the comparison test, $\sum_{6}^{\infty} \frac{1}{k+5}$ diverges, and hence, by Corollary 7.1.20, so does $\sum_{2}^{\infty} \frac{1}{k+5}$. Another option in verifying divergence, though lengthy and thus omitted, is the integral test. Perhaps the easiest method of showing divergence is to relabel $k+5=i$ to obtain $\sum_{k=2}^{\infty} \frac{1}{k+5} = \sum_{i=7}^{\infty} \frac{1}{i} = \sum_{k=7}^{\infty} \frac{1}{k}$. Divergence follows readily. ■

Example 7.2.6. Determine whether the series $\sum \left(\frac{k+47}{3k}\right)^k$ converges or diverges.

Answer. Since $\lim_{k\to\infty} \frac{k+47}{3k} = \frac{1}{3}$, our first inclination is that this series closely resembles the converging geometric series $\sum \left(\frac{1}{3}\right)^k$. However, the sequence $\{a_n\}$, with $a_n = \frac{n+47}{3n}$, is decreasing toward $\frac{1}{3}$. Why? Thus, $\frac{n+47}{3n} \ge \frac{1}{3}$ for any $n \in N$, which gives an inequality going the wrong way. On the other hand, there does exist a $p \in \left(\frac{1}{3}, 1\right)$, say, $p = \frac{2}{3}$ such that $\frac{n+47}{3n} \le p$ eventually. Why? Thus, since $\sum p^k$ converges, our original series converges also. ■

The comparison test is not always easy to apply. In Example 7.2.5, part (a), we were fortunate to have a positive sign in the term $\frac{1}{3^k+2}$. If $\frac{1}{3^k+2}$ were changed to $\frac{1}{3^k-2}$, the comparison would be more difficult, even though for large values of k the terms are very similar. We should not believe that both series converge or diverge simply because the terms of two series seem very similar. Consider the two series

$$\sum \frac{1}{k^{1.00001}} \text{ and } \sum \frac{1}{k^{0.9999}}.$$

Although terms of these series seem much alike, the first series converges, whereas the second series diverges to plus infinity. The next test, which is a generalization of the comparison test, is often easier to use.

THEOREM 7.2.7. *(Limit Comparison Test) Suppose that a positive termed series $\sum a_k$ is to be tested for convergence or divergence.*

(a) If there exists a converging series $\sum b_k$ with $b_k > 0$, such that $\lim_{n\to\infty} \frac{a_n}{b_n}$ is finite, then $\sum a_k$ converges.

(b) If there exists a diverging series $\sum b_k$ with $b_k \geq 0$, such that $\lim_{n\to\infty} \frac{b_n}{a_n}$ is finite, then $\sum a_k$ diverges.

Remark 7.2.8. A special case of Theorem 7.2.7 can be phrased as follows. Suppose that there exist two positive-termed series $\sum a_k$ and $\sum b_k$ such that $\lim_{n\to\infty} \frac{a_n}{b_n} = L > 0$. Then either both series converge or both diverge. In this situation, we are not concerned about whether the terms from a converging or diverging series are in the numerator or denominator when evaluating the limit. The only time for this concern is when $L = 0$ or $L = +\infty$. If for two series $\sum a_k$ and $\sum b_k$, with $a_k, b_k \geq 0$ for all $k \in N$, $\lim_{n\to\infty} \frac{a_n}{b_n} = 0$, then divergence of $\sum b_k$ does not imply divergence of $\sum a_k$. Choose $a_n = \frac{1}{n^2}$ and $b_n = \frac{1}{n}$ to see why the previous sentence is correct. ■

Proof of Theorem 7.2.7 part (a). Suppose that there exists a converging series $\sum b_k$ with $b_k > 0$, such that $\lim_{n\to\infty} \frac{a_n}{b_n} = L \geq 0$, and L is finite.

Case 1. Suppose that $L > 0$. Then there exist real constants M and $n^* \in N$ such that $\frac{L}{2} < \frac{a_n}{b_n} < M$ for all $n \geq n^*$. Therefore, for all $n \geq n^*$, we have $\frac{1}{2} L b_n < a_n < M b_n$. If $\sum b_k$ converges, then so does $\sum M b_k$, and hence, $\sum a_k$ converges, proving the first part of our theorem. If $\sum a_k$ converges, then $\sum \frac{1}{2} L b_k$ converges, and thus $\sum b_k$ converges. Proof of Remark 7.2.8 is now complete.

Case 2. Suppose L = 0. Then since $\lim_{n\to\infty} \frac{a_n}{b_n} = 0$, there exist real numbers M and

$n^* \in N$ such that $\frac{a_n}{b_n} \le M$ for all $n \ge n^*$. Thus, for all $n \ge n^*$ we have $0 < a_n \le Mb_n$. Therefore, if $\sum b_k$ converges, then $\sum Mb_k$ converges, and thus, $\sum a_k$ converges, completing the proof of Theorem 7.2.7, part (a).

Proof of Theorem 7.2.7 part (b). Suppose that $\lim_{n\to\infty} \frac{b_n}{a_n} = K \ge 0$, K finite. As in the proof of part (a), consider two cases. The case in which $K > 0$ is identical to case 1 in the proof of part (a), letting $L = \frac{1}{K}$. Thus, consider only the case in which $K = 0$. However, there exist real numbers M and $n^* \in N$ such that $\frac{b_n}{a_n} \le M$ for all $n \ge n^*$. Thus, $0 \le b_n \le Ma_n$. Therefore, since $\sum b_k$ diverges, by the comparison test, $\sum Ma_k$ diverges. Hence, $\sum a_k$ diverges. The proof is complete. ■

Example 7.2.9. Test the series $\sum \frac{2k}{3k^3 - 4k + 5}$ for convergence or divergence.

Answer. The leading terms in both numerator and denominator, ignoring constant multiples, give rise to the term $\frac{1}{n^2}$. Thus, since $\sum \frac{1}{k^2}$ converges, we will attempt to verify that the original series converges. Applying the limit comparison test, we write

$$\lim_{n\to\infty} \frac{\frac{2n}{3n^3 - 4n + 5}}{\frac{1}{n^2}} = \frac{2}{3}.$$

Thus, by Theorem 7.2.7, the original series converges. ■

THEOREM 7.2.10. *(Cauchy's Condensation Test) Suppose that* $\{a_n\}$ *is a sequence of nonnegative real numbers such that* $a_n \ge a_{n+1}$ *eventually. Then* $\sum_{k=1}^{\infty} a_k$ *converges if and only if* $\sum_{k=0}^{\infty} 2^k a_{2^k}$ *converges.*

Proof. Without loss of generality, assume that $a_n \ge a_{n+1}$ for all $n \in N$. Furthermore, let S_n and T_m denote the nth and mth partial sums $\sum_{k=1}^{n} a_k$ and $\sum_{k=0}^{m} 2^k a_{2^k}$, respectively. For $n \le 2^m$, since $a_n \ge a_{n+1}$ for all $n \in N$, we write

$$\begin{aligned} S_n &= a_1 + a_2 + \cdots + a_n \\ &\le a_1 + (a_2 + a_3) + (a_4 + a_5 + a_6 + a_7) + \cdots + \left(a_{2^m} + \cdots + a_{2^{m+1}-1}\right) \\ &\le a_1 + 2a_2 + 4a_4 + \cdots + 2^m a_{2^m} = T_m. \end{aligned}$$

If $n \ge 2^m$, we write

$$\begin{aligned} S_n &= a_1 + a_2 + \cdots + a_n \\ &\ge a_1 + a_2 + (a_3 + a_4) + \cdots + \left(a_{2^{m-1}+1} + \cdots + a_{2^m}\right) \\ &\ge \frac{1}{2} a_1 + a_2 + 2a_4 + \cdots + 2^{m-1} a_{2^m} = \frac{1}{2} T_m. \end{aligned}$$

Thus, $T_m \le 2S_n$. In either case, the sequences of partial sums of $\sum_{k=1}^{\infty} a_k$ and $\sum_{k=0}^{\infty} 2^k a_{2^k}$ are both bounded or both unbounded. This proves the theorem. Why? ■

Example 7.2.11. Use Cauchy's condensation test to prove that $\sum \frac{1}{k^p}$ converges if $p > 1$ and diverges if $0 < p \le 1$. See Example 7.2.3.

Proof. Since $p > 0$ Cauchy's condensation test is applicable. Thus, $\sum \frac{1}{k^p}$ converges if and only if $\sum 2^k \frac{1}{\left(2^k\right)^p}$ converges. But, $\sum 2^k \frac{1}{\left(2^k\right)^p} = \sum \left(\frac{1}{2^{p-1}}\right)^k$ is a geometric series that converges if and only if $p - 1 > 0$, i.e. $p > 1$. Hence, $\sum \frac{1}{k^p}$ converges if and only if $p > 1$. ■

Exercises 7.2

1. Test the given series for convergence or divergence.

(a)	$\sum_{k=2}^{\infty} \frac{1}{\ln k}$	**(b)**	$\sum \frac{1}{\sqrt{k} + 1.1}$	**(c)**	$\sum \ln\left(1 + \frac{1}{2^k}\right)$
(d)	$\sum \frac{(2k)!!}{(2k+1)!!}$	**(e)**	$\sum \frac{1}{k!}$	**(f)**	$\sum \frac{k+1}{k^2}$
(g)	$\sum_{k=2}^{\infty} \frac{1}{(\ln k)^k}$	**(h)**	$\sum \frac{\ln k}{k^2}$	**(i)**	$\sum \frac{1}{k\sqrt[k]{k}}$

(j) $\sum \frac{\sin k}{k^2}$ **(k)** $\sum \frac{\ln k}{k}$ **(l)** $\sum_{k=2}^{\infty} \frac{1}{(\ln k)^2}$

(m) $\sum \frac{1}{k^k}$ **(n)** $\sum \frac{\sqrt{k+1}-\sqrt{k}}{\sqrt{k}}$ **(o)** $\sum \frac{1}{\left(1+\frac{1}{k}\right)^{k \ln k}}$

(p) $\sum \frac{k+\ln k}{k^2+1}$ **(q)** $\sum \tan \frac{1}{k}$ **(r)** $\sum \left(\sqrt[k]{k}-1\right)$

(s) $\sum_{k=3}^{\infty} \frac{\sqrt{k}}{\sqrt{k^3}+1}$ **(t)** $\sum \frac{k}{e^k}$ **(u)** $\sum e^k \sin\left(2^{-k}\right)$

(v) $\sum 2^k \sin\left(e^{-k}\right)$ **(w)** $\sum \frac{1}{(2k+1)!!}$ **(x)** $\sum_{k=2}^{\infty} \frac{1}{(\ln k)^{\ln k}}$

(y) $\sum \sin \frac{1}{k}$ **(z)** $\sum \sin \frac{1}{k^2}$

2. Find a positive real value p that yields convergence of the *logarithmic p-series* $\sum_{k=2}^{\infty} \frac{1}{k(\ln k)^p}$ using

(a) the integral test. **(b)** Cauchy's condensation test.

3. **(a)** Attempt to use the limit comparison test with the series $\sum \frac{1}{k!}$ to determine convergence of the series $\sum \frac{k^2}{k!}$. See Exercise 1(e) above.

(b) Attempt to use the limit comparison test with the series $\sum \frac{1}{k!}$ to determine convergence of the series $\sum \frac{(k+2)^2}{(k+2)!}$.

(c) Does the outcome in part (b) shed any light on the convergence or divergence of $\sum \frac{k^2}{k!}$? How does the series $\sum \frac{k^2}{k!}$ differ from the series $\sum \frac{(k+2)^2}{(k+2)!}$? Explain.

4. Prove part (b) of Theorem 7.2.4.

5. If $\sum a_k$, with $a_k \geq 0$, converges, prove that $\sum (a_k)^2$ converges.

6. If $\sum a_k$, with $a_k \geq 0$, converges, prove that $\sum \sqrt{a_k a_{k+1}}$ converges.

7. **(a)** Suppose $\{a_n\}$ is a decreasing sequence of positive real numbers such that $\sum a_k$ converges. Prove that $\lim_{n\to\infty} na_n$.

(b) Prove that part (a) is false if the sequence is not decreasing.

(c) Prove that there exists a decreasing sequence of positive real numbers $\{a_n\}$ for which the sequence $\{na_n\}$ converges to zero but $\sum a_k$ diverges.

8. Prove that $\int_1^{\infty} \frac{1}{\sqrt{x(x-0.3)(x+2)}}dx$ converges.

9. For which values of p is the given series convergent? Explain.

(a) $\sum \frac{\ln k}{k^p}$ (See Exercise 1(h) and 1(k).)

(b) $\sum_{k=2}^{\infty} \frac{1}{k^p \ln k}$ (c) $\sum_{k=2}^{\infty} \frac{1}{k(\ln k)(\ln \ln k)^p}$

10. Suppose that $\{p_n\}$ is a sequence of prime numbers 2, 3, 5, 7, 11, 13, ... It can be shown that $\lim_{n\to\infty} \frac{n \ln n}{p_n} = 1$. Use this fact to prove that $\sum \frac{1}{p_k}$ diverges to $+\infty$.

11. Suppose that $\{a_n\}$ and $\{b_n\}$ are sequences of positive real numbers. Determine whether or not the given statements are true. Explain.

(a) $\sum a_k$ and $\sum b_k$ converge if and only if $\sum a_k b_k$ converges. See Part 1 of Section 7.6, Exercises 11 and 16 of Section 7.1, and Exercise 7 of Section 7.4.

(b) $\sum a_k$ and $\sum b_k$ converge if and only if $\sum \sqrt{(a_k)^2 + (b_k)^2}$ converges.

12. The *zeta function* z is defined by $z(x) = \sum_{k=1}^{\infty} \frac{1}{k^x}$. What is the domain of z? See part (e) of Remark 7.6.17, and Exercise 17 of Section 8.4.

13. If both $\sum (a_k)^2$ and $\sum (b_k)^2$ converge, then verify that $\sum a_k b_k$ and $\sum (a_k + b_k)^2$ also converge and that

$$\left(\sum a_k b_k\right)^2 \le \left[\sum (a_k)^2\right]\left[\sum (b_k)^2\right]$$ *(Cauchy–Schwarz inequality)*, and

$$\left[\sum (a_k + b_k)^2\right]^{\frac{1}{2}} \le \left[\sum (a_k)^2\right]^{\frac{1}{2}} + \left[\sum (b_k)^2\right]^{\frac{1}{2}}$$ *(Minkowski inequality)*.

See Exercise 10 of Section 1.7, and Exercises 13 and 15 of Section 6.3.

14. If $a_n \ge 0$ for all $n \in N$ and $\sum a_k$ converges, prove that $\sum \frac{\sqrt{a_k}}{k}$ converges.

15. Give an example, if possible, of a function $f:[1,\infty) \to \Re$ for which

(a) $\sum_{k=1}^{\infty} f(k)$ converges but $\int_1^{\infty} f(x)\,dx$ diverges.

(b) $\int_1^{\infty} f(x)\,dx$ converges but $\sum_{k=1}^{\infty} f(k)$ diverges.

See Remark 7.2.2, part (c).

16. If $f:(0,1] \to \Re$ is defined by $f(x) = 2nx$ for $\frac{1}{n+1} < x \le \frac{1}{n}$ and $n \in N$ assuming that $\sum \frac{1}{k^2} = \frac{\pi^2}{6}$, show that $\int_0^1 f = \frac{\pi^2}{6}$.

7.3 Ratio and Root Tests

THEOREM 7.3.1. *(Fundamental Ratio Test sometimes known as Ratio Comparison Test) Suppose that $\{a_n\}$ and $\{b_n\}$ are sequences of real numbers and $n^* \in N$ such that $a_n, b_n > 0$ and $\frac{a_{n+1}}{a_n} \le \frac{b_{n+1}}{b_n}$ for all $n \ge n^*$. If $\sum b_k$ converges, then $\sum a_k$ converges.*

Proof. Suppose that for all $n \ge n^* \in N$, all of the conditions of the theorem are true. Then $\frac{a_{n+1}}{b_{n+1}} \le \frac{a_n}{b_n}$ for all $n \ge n^*$. Therefore, the sequence $\{c_n\}$, with $c_n = \frac{a_n}{b_n}$, is decreasing for all $n \ge n^*$. Thus, $\frac{a_n}{b_n} \le \frac{a_{n^*}}{b_{n^*}} \equiv M$, which implies that $a_n \le Mb_n$, where $M \in \Re$ is an upper bound for the sequence $\{c_n\}_{n=n^*}^{\infty}$. Therefore, since $\sum b_k$ converges, $\sum Mb_k$ converges also, and hence, $\sum a_k$ converges by the comparison test. ■

Remark 7.3.2. Theorem 7.3.1 can be rewritten to prove divergence of the series $\sum a_k$. See Exercise 1 at the end of this section. ■

THEOREM 7.3.3. *(Ratio Test sometimes known as d'Alembert's[1] Ratio Test or Generalized Ratio Test) Suppose that $\{a_n\}$ is a sequence of real numbers and $n^* \in N$. If $a_n > 0$ for all $n \ge n^*$ and*

---

[1]Jean Le Rond d'Alembert (1717–1783), who proved the ratio test in 1768, was a world renowned French mathematician, philosopher, and physicist. He pioneered the use of differential equations in physics.

(*a*) *if* $\frac{a_{n+1}}{a_n} \le \alpha$ *for all* $n \ge n^*$, *with* α *a constant satisfying* $\alpha \in [0,1)$, *then* $\sum a_k$ *converges.*

(*b*) *if* $\frac{a_{n+1}}{a_n} \ge 1$ *for all* $n \ge n^*$, *then* $\sum a_k$ *diverges.*

Proof of part (a). Since $\alpha = \frac{\alpha^{n+1}}{\alpha^n}$, we have $\frac{a_{n+1}}{a_n} \le \alpha = \frac{\alpha^{n+1}}{\alpha^n}$. Therefore, since $\sum \alpha^k$ is a converging geometric series, by the fundamental ratio test, $\sum a_k$ converges.

Proof of part (b). If $\frac{a_{n+1}}{a_n} \ge 1$ for all $n \ge n^* \in N$, then $a_{n+1} \ge a_n$ for all $n \ge n^*$. Thus, $\{a_n\}$ is increasing, which implies that $a_n \ge a_{n^*} > 0$ for all $n \ge n^*$. Therefore, $\lim_{n\to\infty} a_n \ne 0$. Hence, by the divergence test, $\sum a_k$ diverges, and the proof is complete. ■

Remark 7.3.4

(a) Note that the term $\frac{a_{n+1}}{a_n}$ in part (a) of the ratio test must be bounded away from 1. If $a_n > 0$ and $\frac{a_{n+1}}{a_n} < 1$, for all $n \ge n^*$, then the series $\sum a_k$ may or may not converge. This can be demonstrated by using the p-series, which only converges for certain values of p, diverges otherwise, and yet satisfies the preceding inequality for all real p. See exercise 2 at the end of this section for details.

(b) An analysis of the ratio of consecutive terms of a sequence was also studied in Theorem 2.3.7. ■

The familiar ratio test studied in calculus is a consequence of Theorem 7.3.3 and is given next.

COROLLARY 7.3.5. *(Cauchy's Ratio Test) Suppose that* $\{a_n\}$ *is a sequence of real numbers and* $n^* \in N$. *If* $a_n > 0$ *for all* $n \ge n^*$ *and*

(*a*) $\lim_{n\to\infty} \frac{a_{n+1}}{a_n} < 1$, *then* $\sum a_k$ *converges.*

(*b*) $\lim_{n\to\infty} \frac{a_{n+1}}{a_n} > 1$, *then* $\sum a_k$ *diverges.*

Remark 7.3.6

(a) Observe that Corollary 7.3.5 says nothing about convergence or divergence in cases where the limit is 1 or the limit does not exist. The reason is that either convergence or divergence can happen. See Exercise 2(d).

(b) Proof of Corollary 7.3.5 uses Theorem 7.3.3 and is left as an exercise.
(c) Theorem 7.3.3 is referred to as a *stronger*, or *sharper*, result than Corollary 7.3.5. That is, there are series for which convergence or divergence can be determined using Theorem 7.3.3 but not Corollary 7.3.5. In addition, when Theorem 7.3.3 does not apply, then neither does Corollary 7.3.5. See Example 7.3.7.
(d) Just as the comparison tests from Section 7.2 are, the limit version of the ratio test, that is, Cauchy's ratio test, is easier to apply than the term-by-term comparisons used in the generalized ratio test.
(e) In both ratio tests, the convergence or divergence of a series is determined without using any other series or function.
(f) Ratio tests are commonly employed when terms of the series at hand involve products, quotients, and fixed powers. ■

Example 7.3.7. Give an example of a sequence for which the ratio test is a stronger result than Cauchy's ratio test. See Remark 7.3.6, part (c).

Answer. Consider the sequence $\{a_n\}$, where

$$a_n = \begin{cases} \dfrac{1}{2^{\frac{n+1}{2}} 3^{\frac{n-1}{2}}}, & \text{if } n \text{ is odd} \\ \dfrac{1}{2^{\frac{n}{2}} 3^{\frac{n}{2}}}, & \text{if } n \text{ is even.} \end{cases}$$

Since for this sequence,

$$\frac{a_{n+1}}{a_n} = \begin{cases} \dfrac{1}{3}, & \text{if } n \text{ is odd} \\ \dfrac{1}{2}, & \text{if } n \text{ is even,} \end{cases} \qquad \text{(why?)}$$

by the ratio test, $\sum a_k$ converges. However, $\lim\limits_{n\to\infty} \dfrac{a_{n+1}}{a_n}$ does not exist. Thus, Cauchy's ratio test is inconclusive. ■

Now we turn to another popular test for series.

THEOREM 7.3.8. *(Root Test) Suppose that $\{a_n\}$ is a sequence of real numbers and $n^* \in N$. If $a_n \ge 0$, for all $n \ge n^*$, and*

(a) *if $\sqrt[n]{a_n} \le \alpha$ for all $n \ge n^*$ with α a constant satisfying $\alpha \in [0,1)$, then $\sum a_k$ converges.*

(b) *if $\sqrt[n]{a_n} \ge 1$, for all $n \ge n^*$, then $\sum a_k$ diverges.*

COROLLARY 7.3.9. *(Cauchy's Root Test) Suppose that $\{a_n\}$ is a sequence of real numbers and $n^* \in N$. If $a_n \geq 0$, for all $n \geq n^*$, and*

(a) $\lim_{n\to\infty} \sqrt[n]{a_n} < 1$, *then* $\sum a_k$ *converges.*

(b) $\lim_{n\to\infty} \sqrt[n]{a_n} > 1$, *then* $\sum a_k$ *diverges.*

Proof of Theorem 7.3.8 and Corollary 7.3.9 are left as exercises. Corollary 7.3.9 sheds no light on convergence for the case when $\lim_{n\to\infty} \sqrt[n]{a_n} = 1$.

Example 7.3.10. Give an example of a sequence for which the root test is sharper than Cauchy's root test.

Proof. Consider the sequence $\{a_n\}$ consisting of the terms $\frac{1}{2}, \frac{1}{3}, \frac{1}{2^2}, \frac{1}{3^2}, \frac{1}{2^3}, \frac{1}{3^3}, \ldots$. That is, if $n \in N$, we have

$$a_n = \begin{cases} \dfrac{1}{2^{\frac{n+1}{2}}}, & \text{if } n \text{ is odd} \\ \dfrac{1}{3^{\frac{n}{2}}}, & \text{if } n \text{ is even.} \end{cases} \quad \text{Then,} \quad \lim_{n\to\infty} \sqrt[n]{a_n} = \begin{cases} \dfrac{1}{\sqrt{2}}, & \text{if } n \text{ is odd} \\ \dfrac{1}{\sqrt{3}}, & \text{if } n \text{ is even.} \end{cases}$$

Hence, $\sum a_k$ converges by Theorem 7.3.8. Corollary 7.3.9 is inconclusive. Why? Moreover, is it correct to write $\sum a_k$ as $\sum \frac{1}{2^k} + \sum \frac{1}{3^k}$? Why? ■

Example 7.3.11. Give an example of a sequence for which the root test is stronger than the ratio test.

Proof. Consider the sequence given in Example 7.3.10, and observe that

$$\frac{a_{n+1}}{a_n} = \begin{cases} \left(\dfrac{2}{3}\right)^{\frac{n+1}{2}}, & \text{if } n \text{ is odd} \\ \dfrac{1}{2}\left(\dfrac{3}{2}\right)^{\frac{n}{2}}, & \text{if } n \text{ is even.} \end{cases}$$

Why? Both ratio tests are inconclusive. Why? However, as verified in Example 7.3.10, $\sum a_k$ converges by the root test. ■

Remark 7.3.12. If $\{a_n\}$ is a sequence of positive real numbers and $\sum a_k$ converges by Cauchy's ratio test, then $\sum a_k$ converges by Cauchy's root test. That is,

Cauchy's root test is stronger than Cauchy's ratio test. More accurately, if $\{a_n\}$ is a sequence of positive real numbers and a sequence $\{b_n\}$ with $b_n = \frac{a_{n+1}}{a_n}$ converges to a real value L, then the sequence $\{c_n\}$ with $c_n = \sqrt[n]{a_n}$ converges to L. Proof of this statement is left as an exercise for more ambitious students. ■

Exercises 7.3

1. Prove that if $a_n, b_n > 0$, $\frac{a_{n+1}}{a_n} \geq \frac{b_{n+1}}{b_n}$ for all $n \geq n^* \in N$, and $\sum b_k$ diverges, then $\sum a_k$ diverges.

2. Consider the sequence $\{a_n\}$, where $a_n = \frac{1}{n^p}$, p a positive real constant. See Example 7.2.3.

(a) Show that $\frac{a_{n+1}}{a_n} < 1$, for all $n \in N$.

(b) Determine for which values of p the series $\sum a_k$ converges and for which values it diverges.

(c) Compute $\lim_{n \to \infty} \frac{a_{n+1}}{a_n}$.

(d) What can be concluded from parts (a) to (c)?

3. Prove Corollary 7.3.5.

4. Determine whether the given series converges or diverges. Also, show that Cauchy's ratio test is inconclusive.

(a) $\sum_{k=2}^{\infty} \frac{1}{k \ln k}$ (See Exercises 2 and 9(b) in Section 7.2.)

(b) $\sum_{k=1}^{\infty} \frac{k!}{(k+2)!}$ (See Exercise 1(c) in Section 7.1.)

5. **(a)** If possible, give an example of a sequence $\{a_n\}$ such that $\sum a_k$ converges, and the sequence $\{c_n\}$, with $c_n = \frac{a_{n+1}}{a_n}$, for all $n \in N$, diverges.

(b) If possible, give an example of a sequence $\{b_n\}$ such that $\sum b_k$ diverges, and the sequence $\{c_n\}$, with $c_n = \frac{b_{n+1}}{b_n}$, for all $n \in N$, diverges.

6. If possible, give an example of a series $\sum a_k$ for which the ratio test is conclusive but $\lim_{n\to\infty} \frac{a_{n+1}}{a_n} = 1$.

7. Whenever possible, use the tests covered in this section to determine convergence or divergence of each given series.

(a) $\sum_{k=2}^{\infty} \frac{k^2 2^{k+1}}{3^k}$ **(b)** $\sum \frac{(k!)^2}{(2k)!}$

(c) $\sum e^k \sin\left(2^{-k}\right)$ (See Exercise 1(u) of Section 7.2.)

(d) $\sum 2^k \sin\left(e^{-k}\right)$ (See Exercise 1(v) of Section 7.2.)

(e) $\sum \frac{3^{2k+1}}{k^{2k}}$

(f) $\sum \left(\sqrt[k]{k} - 1\right)^k$ (Compare with Exercise 1(r) of Section 7.2.)

(g) $\sum \frac{k!}{k^k}$ (See Exercise 11(b) in Section 2.3. Also, compare to Exercise 1(m) in Section 7.2.)

(h) $\sum_{k=2}^{\infty} \frac{k}{(\ln k)^k}$ (Compare to Exercise 1(g) in Section 7.2.)

(i) $\sum \left(1 + \frac{1}{k}\right)^k$

8. Verify that the sequence $\{a_n\}$ with $a_n = \frac{2^n n!}{n^n}$ converges to zero. See Exercise 3(d) of Section 8.5.

9. If $f(x) = \sum_{k=1}^{\infty} kx^k$, find the domain of the function f. See Exercise 18(a) of this section and Exercise 13(a) in Section 8.5.

10. Determine whether the series $\sum a_k$ converges or diverges, where sequence $\{a_n\}$ is defined recursively as

(a) $a_1 = 1$, $a_{n+1} = 2a_n$ for all $n \in N$. See Exercise 7(c) in Section 1.3.

(b) $a_1 = 2$, $a_{n+1} = \frac{1}{n} a_n$ for all $n \in N$.

(c) $a_1 = 1$, $a_{n+1} = \frac{n+1}{n} a_n$ for all $n \in N$. See Exercise 7(b) in Section 1.3.

11. Prove Theorem 7.3.8.

12. Prove Corollary 7.3.9.

13. Using both the comparison test and the root test, show that the series

$$\sum a_k = \frac{1}{3} + \frac{1}{2^2} + \frac{1}{3^3} + \frac{1}{2^4} + \frac{1}{3^5} + \cdots$$

converges. Also, explain why the ratio test is inconclusive.

14. Consider the series $\sum a_k$ where $a_{2n} = \frac{1}{3^n}$ and $a_{2n-1} = \frac{1}{3^{n+1}}$ with $n \in N$. Show that Cauchy's root test is conclusive. Explain why the ratio test is inconclusive.

15. (a) Give an example of a series $\sum a_k$ for which both of Cauchy's ratio and root tests are inconclusive, but either $\lim_{n\to\infty} \frac{a_{n+1}}{a_n} = 1$ or $\lim_{n\to\infty} \sqrt[n]{a_n} = 1$.

(b) Give an example of a sequence $\{a_n\}$ for which $\lim_{n\to\infty} \frac{a_{n+1}}{a_n} = \lim_{n\to\infty} \sqrt[n]{a_n} = 1$.

16. If a series $\sum a_k$ converges by either Cauchy's ratio or Cauchy's root test and p is any natural number, then $\sum k^p a_k$ converges by that same test. True or false?

17. Prove the statement given in Remark 7.3.12.

18. (a) Use infinite series to prove that if $|x| < 1$, then $\lim_{n\to\infty} nx^n = 0$. See Exercise 17 of Section 2.1.

(b) Consider the sequence $\{a_n\}$ with $a_n = n^2 x(1-x)^n$ and where x is any real value satisfying $x \in [0,1]$. Prove that $\{a_n\}$ converges to 0. See Exercise 1(j) of Section 8.1.

19. Let c be a positive real constant and define a *continued root* of c by the sequence $\{a_n\}$, where $a_1 = c$ and $a_{n+1} = c^{\frac{1}{a_n}} = \sqrt[a_n]{c}$, for all $n \in N$.

(a) If $c \in (0,1)$, prove that $\{a_n\}$ converges.

(b) Evaluate $\sqrt[\cdot^{\cdot^{\sqrt[c]{c}}}]{\sqrt[c]{c}}$ for $c \in (0,1)$. Compare with Exercise 15 in Section 2.4 and Exercise 16 in Section 4.1.

(c) If $c \in (0,1)$, prove that the series $\sum a_k$ converges.

20. If J_p is a Bessel function as defined in Section 6.6, show that J_p is defined at least for $x \in (0,\infty)$, and $J_p(0) = 0$ for $p > 0$. What is $J_0(0)$ equal to?

7.4 Absolute and Conditional Convergence

Most previously studied tests have been for series $\sum a_k$ in which a_k was eventually positive. In this section, we discuss convergence of series in which the positive and negative terms alternate, as well as series without any particular pattern of positive and negative terms. If $\{a_n\}$ is a sequence of nonnegative terms, then both $(-1)^{n+1}a_k$ and $(-1)^n a_k$ will change every other sign of $\{a_n\}$ to a minus creating a new, so-called *alternating sequence*. An alternating series is defined in a similar fashion. Observe also that $\cos n\pi = (-1)^n$ (see Exercise 2(i) from Section 1.3) and $\sin\dfrac{(2n-1)\pi}{2} = (-1)^{n+1}$ for all $n \in N$.

Definition 7.4.1. If $\{a_n\}$ is a sequence of nonnegative real numbers, then series $\sum(-1)^{k+1}a_k$ and $\sum(-1)^k a_k$ are called *alternating series.*

An alternating series is a special case of infinite products. See Part 3 of Section 7.6.

THEOREM 7.4.2. *(Leibniz's Alternating Series Test) If $\{a_n\}$ is a sequence of real numbers satisfying both*

(a) $0 \le a_{n+1} \le a_n$ *eventually, and*

(b) $\lim_{n\to\infty} a_n = 0$,

then $\sum(-1)^{k+1}a_k$ *converges.*

Proof. Without loss of generality, suppose that $0 \le a_{n+1} \le a_n$ for all $n \in N$. If $\{S_n\}$ is the sequence of partial sums of the infinite series $\sum(-1)^{k+1}a_k$, then

$$S_2 \le S_4 \le S_6 \le \cdots \le S_5 \le S_3 \le S_1.$$

Since $\{S_{2n}\}$ is increasing and bounded above by S_1, it converges to, say, S. Also, since $\{S_{2n-1}\}$ is decreasing and bounded below by S_2, so it converges to, say, T. But $S_{2n+1} - S_{2n} = a_{2n+1}$ gives us

$$\lim_{n\to\infty} S_{2n+1} - \lim_{n\to\infty} S_{2n} = \lim_{n\to\infty} a_{2n+1},$$

yielding $T - S = 0$. Hence, $S = T$, implying that $\lim_{n\to\infty} S_n = S$. Thus, the alternating series converges. ■

Remark 7.4.3

(a) In Theorem 7.4.2 conditions (a) and (b) are independent of each other. See Exercise 1.

(b) Alternating series exist that do not satisfy both conditions (a) and (b) in Theorem 7.4.2, but still converge. See Exercise 4.

(c) The series $\sum(-1)^k$ is an example of an alternating series that diverges.

(d) To show that the sequence $\{a_n\}$ in Theorem 7.4.2 is decreasing, it is sometimes possible to show that $a'(x) \le 0$ eventually, where $x \in \Re$ and $a(n) = a_n$. This method may not work every time, even if $a(x)$ is differentiable. See Exercise 4(b). ■

Example 7.4.4. Verify that the *alternating harmonic series* $\sum \frac{(-1)^{k+1}}{k}$ converges.

Proof. Since the sequence $\{a_n\}$, with $a_n = \frac{1}{n}$ decreases and converges to zero, by Theorem 7.4.2, our series converges. In fact, Example 8.5.16 will verify that this series converges to $\ln 2$. ■

Example 7.4.5. Determine whether or not the series $\sum (-1)^{k+1} \frac{\ln k}{k}$ converges.

Answer. Observe that

$$\lim_{n\to\infty} \frac{\ln n}{n} = \lim_{x\to\infty} \frac{\ln x}{x} \underset{\text{L'H}}{=} \lim_{x\to\infty} \frac{\frac{1}{x}}{1} = 0.$$

The sequence $\{a_n\}$ with $a_n = \frac{\ln n}{n}$ eventually decreases. To show that $\{a_n\}$ decreases for $n \ge 3$, consider the function $f(x) = \frac{\ln x}{x}$. We will show that f is decreasing for $x \ge 3$. To this end, note that $f'(x) = \frac{1 - \ln x}{x^2}$, and that $1 - \ln x \le 0$ for $x \ge e$. Thus, f decreases when $x \ge e$. Consequently, $\{a_n\}$ decreases when $n \ge e$, namely, $n \ge 3$. Hence, by Leibniz's alternating series test, the original series converges. ■

Although the alternating harmonic series $\sum \frac{(-1)^{k+1}}{k}$ converges, the sum of its absolute values, that is, the harmonic series diverges. This is not the case for many other series. For example, the series $\sum \frac{(-1)^{k+1}}{k^2}$ as well as its absolute value, that is, $\sum \left| \frac{(-1)^{k+1}}{k^2} \right| = \sum \frac{1}{k^2}$, converges. See Exercise 21(b) of Section 7.1.

Definition 7.4.6. An infinite series $\sum a_k$ is *absolutely convergent* if the series $\sum |a_k|$ converges.

If a series is absolutely convergent, then we say that it *converges absolutely*. The series $\sum \frac{(-1)^{k+1}}{k^2}$ is an example of such a series, as are converging series with nonnegative terms. In addition, $\sum a_k$, with $a_k = \frac{\sin k}{k^2}$, which is neither alternating nor eventually of one sign, converges absolutely, since $\left| \frac{\sin n}{n^2} \right| \le \frac{1}{n^2}$. Due to Theorem 7.4.7, most tests from previous sections can be rewritten inserting absolute values in appropriate places to yield absolutely convergent tests.

THEOREM 7.4.7. *If $\sum a_k$ converges absolutely, then $\sum a_k$ converges.*

Thus, if $\sum |a_k|$ converges, then $\sum a_k$ converges. The converse is false. Why? For contrast, however, observe that if $\{|a_n|\}$ converges, then $\{a_n\}$ need not converge. The converse, though, is true. Proof of Theorem 7.4.7 lies in Exercise 21(a) of Section 7.1. Furthermore, if $\sum a_k$ diverges, then $\sum |a_k|$ must also diverge.

Definition 7.4.8. A series $\sum a_k$ is *conditionally convergent* if $\sum a_k$ converges and $\sum |a_k|$ diverges.

For example, an alternating harmonic series is a conditionally converging series. The importance of distinguishing between absolute and conditional convergence will become clear in the discussion of rearrangements of series at the end of this section.

Next, consider series with minus signs "sprinkled" randomly throughout nonnegative terms. An example of this generalization of an alternating series is the series $\sum \frac{\sin k}{k^2}$ discussed earlier. The following new notation should prove helpful.

Definition 7.4.9. Suppose that $\{a_n\}$ is any sequence of real numbers. For all $n \in N$, define

$$a_n^+ = \begin{cases} a_n, & \text{if } a_n \geq 0 \\ 0, & \text{if } a_n < 0, \end{cases} \quad \text{and} \quad a_n^- = \begin{cases} a_n, & \text{if } a_n \leq 0 \\ 0, & \text{if } a_n > 0. \end{cases}$$

Example 7.4.10. Suppose that the sequence $\{a_n\}$ consists of the values $2, 5, -\frac{1}{2}, 0, -2, 3, 1, \frac{1}{2}, -4, \ldots$. Then the sequence $\{a_n^+\}$ consists of the values $2, 5, 0, 0, 0, 3, 1, \frac{1}{2}, 0, \ldots$, and the sequence $\{a_n^-\}$ consists of the values $0, 0, -\frac{1}{2}, 0, -2, 0, 0, 0, -4, \ldots$. ■

Remark 7.4.11

(a) The sequence $a_n = a_n^+ + a_n^-$ for all $n \in N$.

(b) If $S_n^+ = \sum_{k=1}^{n} a_k^+$ and $S_n^- = \sum_{k=1}^{n} a_k^-$, then

$$S_n = \sum_{k=1}^{n} a_k = S_n^+ + S_n^-. \text{ Also define } S_n{}^* = \sum_{k=1}^{n} |a_k| = S_n^+ - S_n^-.$$

(c) Sequences $\{S_n^+\}$ and $\{S_n^-\}$ both converge if and only if $\{S_n{}^*\}$ converges. That is, if $\sum a_k^+$ and $\sum a_k^-$ both converge, then so does $\sum |a_k|$, that is,

$\sum a_k$ converges absolutely. Furthermore, if either $\sum a_k^+$ or $\sum a_k^-$ but not both diverges, then both $\sum |a_k|$ and $\sum a_k$ diverge. Moreover, if both $\sum a_k^+$ and $\sum a_k^-$ diverge, then $\sum |a_k|$ diverges, but $\sum a_k$ need not diverge, in which case $\sum a_k$ would converge conditionally. ■

Example 7.4.12. Find two infinite series $\sum a_k$ and $\sum b_k$, where $\sum a_k^+$, $\sum a_k^-$, $\sum b_k^+$, and $\sum b_k^-$ all diverge, but $\sum a_k$ converges, while $\sum b_k$ diverges.

Answer. Suppose that $a_n = \dfrac{(-1)^{n+1}}{n}$. Observe that $\sum a_k$ converges, since $\sum a_k^+ = 1 + 0 + \dfrac{1}{3} + 0 + \dfrac{1}{5} + \cdots = \sum \dfrac{1}{2k-1} = +\infty$, and $\sum a_k^- = 0 - \dfrac{1}{2} + 0 - \dfrac{1}{4} + 0 - \dfrac{1}{6} + \cdots = -\sum \dfrac{1}{2k} = -\infty$. Next, suppose that $b_n = (-1)^n$. Clearly $\sum b_k$, $\sum b_k^+$, and $\sum b_k^-$ all diverge. ■

Now let us turn our attention to grouping and rearranging series. Inserting pairs of parentheses to obtain a new series is called *grouping*.

THEOREM 7.4.13. *If $\sum a_k$ converges to S and $\sum b_k$ is a regrouping of $\sum a_k$, then $\sum b_k$ also converges to S.*

Thus, inserting parentheses is permitted, but removing them is perhaps not. Proof of Theorem 7.4.13 is left as an exercise.

Definition 7.4.14. Suppose that $f : N \to N$ is some bijection. Then $\sum a_{f(k)}$ is called a *rearrangement* of the series $\sum a_k$.

Intuitively, this idea is very simple. We start with $\sum a_k$ and change the order of its terms. The resulting series is a rearrangement of $\sum a_k$.

THEOREM 7.4.15. *If $\sum a_k$ converges absolutely to S and $\sum b_k$ is a rearrangement of $\sum a_k$, then $\sum b_k$ converges absolutely to S.*

The key word in Theorem 7.4.15 is "absolutely." If $\sum a_k$ converges but does not converge absolutely, that is, $\sum a_k$ converges conditionally, then Theorem 7.4.15 fails to be true.

THEOREM 7.4.16. *If $\sum a_k$ converges conditionally, then it can be rearranged either to converge to any real number or to diverge to $+\infty$ or $-\infty$. It can even be rearranged to oscillate between any two chosen real numbers.*

Proof. Suppose that $\sum a_k^+ = +\infty$, $\sum a_k^- = -\infty$, and $\sum a_k = \sum a_k^+ + \sum a_k^-$. Why is this true? Let A be any real number. We will rearrange $\sum a_k$ so that its rearrangement converges to A. To this end, let n_1 be the smallest natural number such that

$$a_1^+ + a_2^+ + \cdots + a_{n_1}^+ > A.$$

Next, choose the smallest natural number n_2 so that

$$a_1^+ + a_2^+ + \cdots + a_{n_1}^+ + a_1^- + a_2^- + \cdots + a_{n_2}^- < A.$$

Now, choose the smallest $n_3 \in N$ so that

$$a_1^+ + a_2^+ + \cdots + a_{n_1}^+ + a_1^- + a_2^- + \cdots + a_{n_2}^- + a_{n_1+1}^+ + a_{n_1+2}^+ + \cdots + a_{n_3}^+ > A.$$

And so the process continues. Thus, enough positive terms are being picked up to make their sum just larger than A. And enough negative terms are being added to make the partial sum of the resulting series less than A. Again enough positive terms are being added to make the partial sum greater than A. Continuing this process, all terms of the series are exhausted and the resulting series not only converges to A but is a rearrangement of $\sum a_k$. Proofs of the remaining parts of Theorem 7.4.16 are left to the reader. ■

Remark 7.4.17. As pointed out in Remark 7.1.15, part (b), the sum S of a converging series $\sum a_k$ can rarely be found and often one has to be satisfied with an approximation of S. However, ways do exist to find the exact value of S. Our study of geometric and telescoping series in Section 7.1 gives examples. Also, computing a general term for the sequence of partial sums $\{S_n\}$ is sometimes possible. Then, $\lim_{n\to\infty} S_n$ would yield the sum of a series. Furthermore, if the sum of one or more series is known, then often using algebra on series causes sums of other series to become apparent. See Exercise 17. Other methods used to find sums of series are Maclaurin,[2] Taylor, and other power series covered in Chapter 8. Fourier series covered in Chapter 9 together with differentiation and Riemann integration also enable us to find sums for a vast number of infinite series. Other more sophisticated methods will follow from further studies of mathematics. ■

---

[2]Colin Maclaurin (1698–1746) was a prominent Scottish mathematician and physicist. His father died when he was six months old and mother when he was nine years old. At the age of 17, he received a Master's degree from Glasgow University and in 1719 he met Isaac Newton. Maclaurin used his geometric ideas to defend Newton's analytical methods, which were bitterly attacked by other major mathematicians. In addition, Maclaurin introduced the method of generating conics and was the first to give a correct theory for distinguishing maxima and minima. He also demonstrated that a homogeneous rotating fluid mass revolves in elliptic fashion.

Exercises 7.4

1. **(a)** Find a sequence $\{a_n\}$ that satisfies condition (a) but not condition (b) of Theorem 7.4.2.

(b) Find a sequence $\{a_n\}$ that satisfies condition (b) but not condition (a) of Theorem 7.4.2.

(c) Is there a sequence $\{a_n\}$ for which $\sum(-1)^{k+1}a_k$ converges and that satisfies condition (a) but not condition (b) of Theorem 7.4.2? Explain.

2. Determine convergence or divergence of the given series.

(a) $\sum \frac{k}{(-2)^{k-1}}$ **(b)** $\sum(-1)^{k+1}\left(\frac{3k+2}{4k^2-3}\right)$

(c) $\sum \frac{(-e)^k}{k^4}$ **(d)** $\sum(-1)^{k-1}\left(\frac{k}{4k^2-3}\right)$

(e) $\sum \frac{(-1)^{k+1}}{k\sqrt[k]{k}}$ (See Exercise 1(g) of Section 5.2.)

3. Determine for which $p \in N$ the series $\sum(-1)^k \frac{(\ln k)^p}{k}$ converges. See Exercise 9.

4. **(a)** Give an example of a converging alternating series $\sum(-1)^{k+1}a_k$ for which the sequence $\{a_n\}$ is not decreasing. See Exercise 1(c) of this section.

(b) Give an example of a differentiable function $a(x)$ with $x \geq 1$, that is not decreasing, but the sequence $\{a_n\}$ is decreasing, where $a_n = a(n)$. See part (d) of Remark 7.4.3.

5. Give an example of a series $\sum a_k$ that consists of nonzero terms, $\lim_{n\to\infty}\left|\frac{a_{n+1}}{a_n}\right| = 1$, and

(a) $\sum a_k$ converges absolutely. **(b)** $\sum a_k$ converges conditionally.

(c) $\sum a_k$ diverges.

6. Determine whether the given series converges absolutely, converges conditionally, or diverges.

(a) $\sum(-1)^k \frac{\arctan k}{k^2}$ **(b)** $\sum(-1)^k k \tan\frac{1}{k}$

(c) $\sum \frac{(-1)^k}{k \ln k}$ **(d)** $\sum \frac{(-1)^k}{\sqrt{k(k+1)}}$

(e) $\sum(-1)^k \ln\left(1+\frac{1}{k}\right)$ (See Example 7.1.12.)

7. Give an example of two series $\sum a_k$ and $\sum b_k$ that converge but $\sum a_k b_k$ diverges. See Exercises 11 and 16 of Section 7.1.

8. Find all values of x for which the series $\sum \frac{(-1)^k (x-3)^k}{k+1}$

(a) converges absolutely. **(b)** converges conditionally.
(c) converges. **(d)** diverges.

9. If p and q are two positive real constants, prove that the series $\sum (-1)^k \frac{(\ln k)^p}{k^q}$ converges. See Exercise 9(a) in Section 7.2.

10. Consider the series $\sum a_k = 1 - 1 + \frac{1}{2} - \frac{1}{2} + \frac{1}{3} - \frac{1}{3} + \cdots$. Prove that $\sum a_k$ converges conditionally and find its sum. See Exercise 17 of Section 7.1.

11. Suppose the sequence $\{a_n\}$ is defined by

$$a_n = \begin{cases} -\frac{1}{n^2}, & \text{if } n \text{ is odd} \\ \frac{1}{n}, & \text{if } n \text{ is even.} \end{cases}$$

Show that the alternating series $\sum a_k$ diverges even though its terms tend to zero. Compare to Exercise 4(c) in Section 7.1.

12. **(a)** Prove Theorem 7.4.13.
(b) Demonstrate the validity of Theorem 7.4.13 on the series

$$\sum a_k = 1 - \frac{1}{2} + \frac{1}{3} - \frac{1}{4} + \frac{1}{5} - \frac{1}{6} + \cdots,$$

and on the grouped series

$$\sum b_k = \left(1 - \frac{1}{2}\right) + \left(\frac{1}{3} - \frac{1}{4}\right) + \left(\frac{1}{5} - \frac{1}{6}\right) + \cdots.$$

13. Give an example of a diverging series that, when grouped appropriately, converges.

14. What are your thoughts about the following argument?

$$\begin{aligned} 0 &= 0 + 0 + 0 + \cdots \\ &= (1-1) + (1-1) + (1-1) + \cdots \\ &= 1 - 1 + 1 - 1 + 1 - 1 + \cdots \\ &= 1 + (-1+1) + (-1+1) + (-1+1) + \cdots \\ &= 1 + 0 + 0 + 0 + \cdots \\ &= 1 \end{aligned}$$

15. Prove that $\sum a_k = 1 + \frac{1}{2} - \frac{1}{3} + \frac{1}{4} + \frac{1}{5} - \frac{1}{6} + \cdots$ converges conditionally. Also, what can one say about $\sum a_k^+$ and $\sum a_k^-$? See Example 7.4.12.

16. Determine whether the given statements are true or false. Explain clearly.

(a) If $\sum a_k$ and $\sum b_k$ converge, then $\sum (a_k + b_k)$ converges. (See Exercise 4 in Section 7.1.)

(b) A conditionally convergent series can be rearranged to form an absolutely convergent series.

(c) If $\sum |a_k|$ converges, then $\sum |a_k|$ converges absolutely.

(d) If $\sum |a_k|$ diverges, then $\sum |a_k|$ converges conditionally.

(e) If $a_n \geq 0$ for all $n \in N$ and $\sum a_k$ converges, then $\sum \sin(a_k)$ converges.

(f) $\displaystyle\sum_{k=2}^{\infty} (-1)^k \frac{\ln k}{k^p}$ converges conditionally if $0 < p \leq 1$.

(g) $\displaystyle\sum_{k=2}^{\infty} (-1)^k \frac{\ln k}{k^p}$ converges absolutely if $p > 1$.

17. Assume that $\displaystyle\sum \frac{1}{k^2} = \frac{\pi^2}{6}$. See part (b) of Remark 7.2.2 and Example 9.2.4. Prove that

(a) $\displaystyle\sum \frac{1}{(2k-1)^2} = \frac{\pi^2}{8}$.

(b) $\displaystyle\frac{\pi^2}{24} = \frac{1}{2^2} + \frac{1}{4^2} + \frac{1}{6^2} + \cdots$.

(c) $\displaystyle\frac{\pi^2}{12} = 1 - \frac{1}{2^2} + \frac{1}{3^2} - \frac{1}{4^2} + \cdots$.

18. What are your thoughts about the following argument?

$$\begin{aligned}
\ln 2 &= 1 - \frac{1}{2} + \frac{1}{3} - \frac{1}{4} + \frac{1}{5} - \frac{1}{6} + \cdots && \text{(See Example 7.4.4.)} \\
&= 1 - \frac{1}{2} - \frac{1}{4} + \frac{1}{3} - \frac{1}{6} - \frac{1}{8} + \frac{1}{5} - \frac{1}{10} - \frac{1}{12} + \cdots && \text{(Rearrange)} \\
&= \left(1 - \frac{1}{2}\right) - \frac{1}{4} + \left(\frac{1}{3} - \frac{1}{6}\right) - \frac{1}{8} + \left(\frac{1}{5} - \frac{1}{10}\right) - \frac{1}{12} + \cdots && \text{(Group)} \\
&= \frac{1}{2} - \frac{1}{4} + \frac{1}{6} - \frac{1}{8} + \frac{1}{10} - \frac{1}{12} + \cdots && \text{(Simplify)} \\
&= \frac{1}{2}\left(1 - \frac{1}{2} + \frac{1}{3} - \frac{1}{4} + \frac{1}{5} - \frac{1}{6} + \cdots\right) && \text{(Factor)} \\
&= \frac{1}{2}\ln 2. && \text{(Relabel)}
\end{aligned}$$

19. Complete the proof of Theorem 7.4.16.

20. **(a)** Prove that if a sequence $\{a_n\}$ converges, then $\displaystyle\lim_{n\to\infty} (a_{n+1} - a_n) = 0$, but not conversely. See Exercise 8 of Section 2.5.

(b) Prove that a sequence $\{a_n\}$ converges if and only if $\sum (a_{k+1} - a_k)$ converges. See Exercise 10 of Section 7.1.

(c) A sequence $\{a_n\}$ is of *bounded variation* if and only if $\sum |a_{k+1} - a_k|$ is convergent. Prove that if a sequence $\{a_n\}$ is of bounded variation, then $\sum a_k$ is convergent, but not conversely.

(d) Prove that if $\{a_n\}$ is a converging monotone sequence, then it is of bounded variation.

(e) Determine whether or not the given sequence $\{a_n\}$, where $a_n = \dfrac{(-1)^n}{\sqrt{n}}$ is of bounded variation. Do the same when $a_n = \dfrac{(-1)^n}{n^2}$.

(f) Suppose that $\{a_n\}$ is a sequence such that $|a_{n+1} - a_n| \le b_n$, where $\sum b_k$ is some convergent series. Prove that $\{a_n\}$ converges.

7.5* Review

Label each statement as true or false. If a statement is true, prove it. If not,

(i) give an example of why it is false, and

(ii) if possible, correct it to make it true, and then prove it.

1. If $\sum a_k$ is a series of positive terms that diverges, then $\sum \dfrac{1}{a_k}$ converges.
2. If $\{a_n\}$ converges, then $\sum a_k$ converges.
3. If $\sum a_k$ diverges and c is a real constant, then $\sum ca_k$ also diverges.
4. If $\sum (\alpha a_k + \beta b_k)$ converges, where α and β are real constants, then both $\sum a_k$ and $\sum b_k$ converge.
5. If $\sum a_k$ converges, then $\sum (a_k)^2$ converges.
6. If $\sum (a_k)^2$ converges, then $\sum a_k$ converges.
7. If $\left(\sum a_k\right)^2$ converges, then $\sum a_k$ converges.
8. The series $\displaystyle\sum_{k=0}^{\infty} \frac{1}{k!}$ converges to e.
9. If $\sum a_k$ converges absolutely, then $\sum (a_k)^2$ converges.
10. The series $\sum \dfrac{(k!)^2}{(2k)!}$ diverges.

11. $\sum \frac{1}{9k^2 - 3k - 2} = \frac{1}{3}$.

12. The series $\sum \sin \frac{\pi}{k}$ diverges.

13. The series $\sum x^{\ln(\ln k)}$ diverges, for all $x > 0$.

14. The series $\sum \frac{k!}{(2k-1)!!}$ converges.

15. The series $\sum \frac{(-1)^k}{\sqrt[p]{k}}$ converges if $p \geq 3$ with $p \in N$.

16. If $\sum a_k$ diverges, then $\sum |a_k|$ diverges.

17. The sequence $\{a_n\}$, with $a_n = n^x x^n$ and $0 < x < 1$, converges to zero.

18. If $\sum a_k$ and $\sum b_k$ are series of positive real numbers such that $\lim_{n\to\infty} \frac{a_n}{b_n} = 1$, then $\sum a_k$ converges if and only if $\sum b_k$ converges.

19. If $\sum (a_k)^2$ and $\sum (b_k)^2$ both converge, then $\sum a_k b_k$ converges absolutely.

20. The sequence $\{a_n\}$ with $a_n = \frac{\sqrt[n]{n!}}{n}$ converges to e.

21. If a series $\sum a_k$ converges, then $\sum \frac{a_k}{k}$ converges.

22. There exists a continuous nonnegative function $f:[1,\infty) \to \Re$ for which $\sum f(k)$ converges but $\int_1^\infty f$ diverges.

23. If $a_n > 0$, for all $n \in N$, and $\sum a_k$ diverges, then $\sum \frac{a_k}{a_k + 1}$ diverges. See Exercise 27 of Section 2.7 and Exercise 15 of Section 3.3.

24. If $a_n \neq 1$, for any $n \in N$, and $\sum a_k$ converges absolutely, then $\sum \frac{a_k}{a_k + 1}$ converges absolutely.

25. If both $\sum (a_k)^2$ and $\sum (b_k)^2$ converge, then $\sum (a_k + b_k)^2$ converges.

26. The series $\sum \frac{\sqrt{k+1} - \sqrt{k}}{k}$ converges. See Exercise 1(n) in Section 7.2.

27. The series $\sum \frac{1}{k\sqrt{k}}$ converges.

28. If $\lim_{n\to\infty} \frac{a_n}{b_n} = +\infty$ and $\sum b_k$ diverges, then $\sum a_k$ diverges.

29. If $a_n = -1, -\frac{1}{2}, \frac{1}{3}, -\frac{1}{4}, -\frac{1}{5}, \frac{1}{6}, -\frac{1}{7}, -\frac{1}{8}, \frac{1}{9}, \ldots$, then $\sum a_k^+ = \sum \frac{1}{3k}$.

30. If $a_n = -1, -\frac{1}{2}, \frac{1}{3}, -\frac{1}{4}, -\frac{1}{5}, \frac{1}{6}, -\frac{1}{7}, -\frac{1}{8}, \frac{1}{9}, \ldots$, then $\sum_{k=1}^{n} a_k^+ = \sum_{k=1}^{n} \frac{1}{3k}$.

31. If $\sum a_k$ diverges, then $\sum a_k^+$ diverges.

32. If $a_n > 0$, for all $n \in N$, and $\lim_{n\to\infty} \frac{a_{n+1}}{a_n}$ does not exist, then $\sum a_k$ diverges.

33. If the sequence of partial sums for a series $\sum a_k$ is bounded above, then $\sum a_k$ must converge.

34. The series $\sum \frac{k^2 - 2k + 8}{k^4 + \sqrt{5-k}}$ converges.

35. If a sequence $\{a_n\}$ is of bounded variation, then $\{a_n\}$ converges.

36. If $\{a_n\}$ diverges, then $\sum a_k$ diverges.

37. $\sum_{k=0}^{\infty} \frac{r^k}{k!}$ converges absolutely for any $r \in \Re$.

38. If $\sum a_k$ converges and $a_n \geq 0$, for all $n \in N$, then $\sum \sqrt{a_k}$ converges.

39. If $\sum a_k$ converges absolutely, $a_n \neq 0$ for any $n \in N$, then $\sum \frac{1}{|a_k|}$ diverges.

40. $\sum (a_k + b_k) = \sum a_k + \sum b_k$.

41. The series $\sum \frac{1}{\sqrt{k} + \sqrt{k+1}}$ converges. See Exercise 1(f) in Section 7.1.

42. If the absolutely convergent series $\sum a_k$ converges to A, then $\sum |a_k|$ must converge to $|A|$.

43. $\sum \frac{1}{k^{1+\varepsilon}}$ converges for each $\varepsilon > 0$. See Exercise 1(i) in Section 7.2.

7.6* Projects

Part 1. Summation by Parts

LEMMA 7.6.1. *If $\{u_n\}_{n=1}^{\infty}$ and $\{v_n\}_{n=0}^{\infty}$ are sequences of real numbers, then*

$$\sum_{k=1}^{n} v_k (u_{k+1} - u_k) + \sum_{k=1}^{n} u_k (v_k - v_{k-1}) = u_{n+1} v_n - u_1 v_0.$$

This can be verified by simplifying the left-hand side. Furthermore, if we let $u_{k+1} - u_k = \Delta u_k$ and $v_{k+1} - v_k = \Delta v_k$, then the preceding expression becomes

$$\sum_{k=1}^{n} v_k \Delta u_k + \sum_{k=1}^{n} u_k \Delta v_{k-1} = u_{n+1}v_n - u_1 v_0,$$

called the *summation by parts formula.* Does this remind you of integration by parts? Summation by parts can be used in a variety of places, not just with infinite series and integration. For example, when computing the sums of powers of integers, choose $u_k = v_k = k$ to get a familiar result, or if computing the sum of a geometric series, choose $u_k = a$ and $v_k = r^k$, where r is a real number different from 1. In the latter, Lemma 7.6.1 gives us that $\sum_{k=1}^{n} ar^{k-1} = \frac{a\left(1-r^n\right)}{1-r}$. Other applications of Lemma 7.6.1 involve differentiating powers of x, reducing the order of the powers of sine or cosine, and writing proofs of many familiar results pertaining to Fibonacci sequence. Next, replacing u_n by b_n and v_n by S_n, where S_n is the sequence of partial sums of the series $\sum a_k$, Lemma 7.6.1 leads us to

$$\sum_{k=1}^{n} a_k b_k = S_n b_{n+1} - \sum_{k=1}^{n} S_k \left(b_{k+1} - b_k\right).$$

This can be rewritten as

$$(*) \qquad \sum_{k=n+1}^{m} a_k b_k = \sum_{k=n+1}^{m} S_k \left(b_k - b_{k+1}\right) + S_m b_{m+1} - S_n b_{n+1}$$

for $m > n$. Show why $(*)$ is true.

THEOREM 7.6.2. *(Dirichlet's Test) If a series $\sum a_k$ has bounded partial sums, and a sequence $\{b_n\}$ is decreasing and converging to 0, then $\sum a_k b_k$ converges.*

In proving Theorem 7.6.2, use equation $(*)$ and Cauchy's criterion. Dirichlet's test, which can also be written in terms of a Riemann integral, can be used to prove Liebniz's alternating series test and to determine convergence of more stubborn series.

Problem 7.6.3. Use Dirichlet's test and a suitable trigonometric identity for $\sum_{k=1}^{n} \sin k$ to prove that $\sum_{k=1}^{\infty} \frac{\sin k}{k}$ converges.

In fact, using complex values or Fourier series, it can be proven that

$$\sum_{k=1}^{\infty} \frac{\sin kx}{k} = \begin{cases} \dfrac{\pi - x}{2}, & \text{if } x \in (0, 2\pi) \\ 0, & \text{if } x = 0, 2\pi. \end{cases}$$

See Section 8.4, and Exercises 1(d) and 16 of Section 9.3. So, if $x = 1$, we have

$\sum_{k=1}^{\infty} \frac{\sin k}{k} = \frac{\pi - 1}{2}$. And, if $x = \frac{\pi}{2}$, we have that $\frac{\pi}{4} = 1 - \frac{1}{3} + \frac{1}{5} - \frac{1}{7} + \cdots$. See Exercise 11(b) in Section 8.5. Similarly, using complex values, it can easily be proven that for all $0 < x < 2\pi$, $\sum_{k=1}^{\infty} \frac{\cos kx}{k} = -\ln\left(2 \sin \frac{x}{2}\right)$.

Problem 7.6.4. *(Abel's Test)* If $\sum a_k$ converges and a sequence $\{b_n\}$ is bounded and monotonic, prove that series $\sum a_k b_k$ converges. See Exercise 7 of Section 7.4.

Problem 7.6.5. If $\sum a_k$ converges, and the sequence $\{b_n\}$ is of bounded variation prove that $\sum a_k b_k$ converges. See Exercise 20 of Section 7.4.

Problem 7.6.6. Suppose that $\sum a_k$ has bounded partial sums, and a sequence $\{b_n\}$ tends to 0 and is of bounded variation. Prove that $\sum a_k b_k$ converges.

Part 2. Multiplication of Series

If the two power series $\sum_{k=0}^{\infty} a_k x^k$ and $\sum_{k=0}^{\infty} b_k x^k$, discussed in Section 8.5, were multiplied without regard to rigor, and then if like terms were collected, we would obtain

$$a_0 b_0 + (a_0 b_1 + a_1 b_0)x + (a_0 b_2 + a_1 b_1 + a_2 b_0)x^2 + \cdots = \sum_{k=0}^{\infty} c_k x^k,$$

where $c_k = \sum_{i=0}^{k} a_i b_{k-i}$ for $n = 0, 1, 2, \ldots$. See Exercise 41 in Section 8.7. Two series of numbers may be multiplied in the same way.

Definition 7.6.7. The *Cauchy product* of the series $\sum_{k=0}^{\infty} a_k$ and $\sum_{k=0}^{\infty} b_k$ is the series $\sum_{k=0}^{\infty} c_k$, where $c_k = \sum_{i=0}^{k} a_i b_{k-i}$.

Problem 7.6.8. If $\sum_{k=0}^{\infty} a_k$ and $\sum_{k=0}^{\infty} b_k$ converge absolutely to A and B, respectively, prove that $\sum_{k=0}^{\infty} c_k$ converges absolutely to AB, where $c_k = \sum_{i=0}^{k} a_i b_{k-i}$.

THEOREM 7.6.9. *(Mertens'[3] Theorem) If $\sum_{k=0}^{\infty} a_k$ converges to A and $\sum_{k=0}^{\infty} b_k$ converges to B, and $\sum_{k=0}^{\infty} |b_k|$ converges (not necessarily to $|B|$), then $\sum_{k=0}^{\infty} c_k$ converges to AB, where $c_k = \sum_{i=0}^{k} a_i b_{k-i}$.*

If absolute convergence is removed from both of the series $\sum_{k=0}^{\infty} a_k$ and $\sum_{k=0}^{\infty} b_k$, then the Cauchy product need not converge. An example of such a case is when $a_n = b_n = (-1)^{n+1}\dfrac{1}{\sqrt{n+1}}$. Furthermore, $\sum_{k=0}^{\infty} |b_k|$ does not have to converge to $|B|$. See Exercise 42 in Section 7.5.

THEOREM 7.6.10. *(Abel's Theorem) Suppose that $\sum_{k=0}^{\infty} a_k$ and $\sum_{k=0}^{\infty} b_k$ converge to A and B, respectively. If $c_k = \sum_{i=0}^{k} a_i b_{k-i}$ and $\sum_{k=0}^{\infty} c_k$ converges, then $\sum_{k=0}^{\infty} c_k = AB$.*

Note that in this result given by Abel in 1826, we do not assume absolute convergence of either of the two given series, but, we do assume convergence of their Cauchy product.

Example 7.6.11. Determine the sum of the given series

$$\frac{1}{1}+\frac{1}{2}+\frac{1}{3}+\frac{1}{4}+\frac{1}{6}+\frac{1}{8}+\frac{1}{9}+\frac{1}{12}+\frac{1}{16}+\frac{1}{18}+\frac{1}{24}+\frac{1}{27}+\cdots.$$

Answer. Think of $\sum \frac{1}{2^k}$, $\sum \frac{1}{3^k}$, and Theorem 7.4.15. ■

Part 3. Infinite Products

Definition 7.6.12. If $\{a_n\}$ is a sequence of real numbers then $\prod_{k=1}^{1} a_k \equiv a_1$ and $\prod_{k=1}^{n} a_k \equiv \left(\prod_{k=1}^{n-1} a_k\right)(a_n) \equiv P_n =$ *sequence of partial products* of $\prod_{k=1}^{\infty} a_k$. The notation

[3]Franz Mertens (1840–1927), whose above stated theorem was delivered in 1875, studied in Berlin and later taught in Cracow and Vienna. Mertens contributed to the areas of geometry, number theory, and algebra.

$\prod_{k=1}^{\infty} a_k$, also denoted by $\prod a_k$, is called an *infinite product*. An infinite product $\prod_{k=1}^{\infty} a_k$ *converges* to a real number P if and only if $\lim_{n\to\infty} P_n = P$. If the sequence $\{P_n\}$ diverges then $\prod a_k$ *diverges*. Moreover, if $\lim_{n\to\infty} P_n = 0$, then $\prod a_k$ *diverges to 0.* Note that

$n! = \prod_{k=1}^{n} k$.

THEOREM 7.6.13. *(Cauchy Criterion for Infinite Products) An infinite product $\prod a_k$ converges if and only if for any $\varepsilon > 0$ there exists $n^* \in N$ such that $\left|\frac{P_m}{P_n} - 1\right| < \varepsilon$ whenever $m, n \geq n^*$.*

Problem 7.6.14. Prove that if $\prod a_k$ converges, then $\lim_{n\to\infty} a_n = 1$. What can you say about a_n if $\prod (1 + a_k)$ converges?

Problem 7.6.15. Verify that

(a) If $a_n = (-1)^n$, then $P_n = (-1)^n$. Thus, $\prod a_k$ diverges.

(b) If $a_n = \dfrac{n+1}{n}$, then $P_n = n+1$. Thus, $\prod a_k$ diverges.

(c) If $a_n = \dfrac{n}{n+1}$, then $P_n = \dfrac{1}{n+1}$. Thus, $\prod a_k$ diverges to 0.

(d) $\prod \dfrac{k^2 - 1}{k^2}$ converges to 0 and has only one 0 factor.

(e) $\prod \dfrac{k-1}{k}$ diverges to 0 and has only one 0 factor.

(f) $\prod \left[1 + (-1)^k\right]$ diverges to 0.

(g) If $a_{2n-1} = \dfrac{n+1}{n}$ and $a_{2n} = \dfrac{n}{n+1}$ then $\prod a_k = 1$.

(h) The converse of Problem 7.6.14 is false.

Problem 7.6.16. Prove the following statements.

(a) If $a_n > 0$ for all $n \in N$, then $\prod a_k$ converges if and only if $\sum \ln a_k$ converges.

(b) If $a_n \geq 0$, then $\prod (1 + a_k)$, $\prod (1 - a_k)$, and $\sum a_k$ either all converge or all diverge.

Remark 7.6.17. A few observations:

(a) From Wallis' formula, Remark 6.7.7, part (l), we can obtain that $\frac{\pi}{2}=\prod\frac{4k^2}{4k^2-1}$, and $\sqrt{\pi}=\lim_{n\to\infty}\frac{1}{\sqrt{n}}\prod_{k=1}^{n}\frac{2k}{2k-1}$.

(b) $\operatorname{sinc} x=\prod\cos\frac{x}{2^k}$.

(c) $\frac{2}{\pi}=\prod\cos\frac{\pi}{2^{k+1}}$.

(d) $\frac{2}{\pi}=\prod a_k$, where $a_1=\sqrt{\frac{1}{2}}$ and $a_{n+1}=\sqrt{\frac{1+a_n}{2}}$. See Exercise 15(e) of Section 5.4.

(e) In Exercise 12 of Section 7.2, the zeta function $z:(1,\infty)\to\Re$, defined by $z(x)=\sum\frac{1}{k^x}$ and then further discussed in Exercise 17 of Section 8.4, can also be written as $z(x)=\prod\frac{1}{1-(p_k)^{-x}}$, where p_k denotes the kth prime number. This is called *Euler's product for the zeta function*. In general, if $\{f_n\}$ is a sequence of completely multiplicative functions (see Definition 4.6.12), with $f_1(x)\equiv 1$ and such that $\sum f_n(x)$ converges absolutely, then $\sum_{k=1}^{\infty}f_k(x)=\prod_{k=1}^{\infty}\frac{1}{1-f_{p_k}(x)}$, where p_k is the kth prime number. One can also write that $z(x)\Gamma(x)=\int_0^{\infty}\frac{t^{x-1}}{e^t-1}dt$. ■

Part 4. Cantor Set

We will define the Cantor set by defining its complement. From the interval $[0,1]$, pick out the open middle thirds, that is, the interval $\left(a_1^{(1)},b_1^{(1)}\right)\equiv\left(\frac{1}{3},\frac{2}{3}\right)$. From the remaining intervals pick out open middle thirds, that is, the intervals $\left(a_1^{(2)},b_1^{(2)}\right)\equiv\left(\frac{1}{9},\frac{2}{9}\right)$ and $\left(a_2^{(2)},b_2^{(2)}\right)\equiv\left(\frac{7}{9},\frac{8}{9}\right)$. From the remaining intervals pick out open middle thirds again, that is, the intervals $\left(a_1^{(3)},b_1^{(3)}\right)\equiv\left(\frac{1}{27},\frac{2}{27}\right)$, $\left(a_2^{(3)},b_2^{(3)}\right)\equiv\left(\frac{7}{27},\frac{8}{27}\right)$, $\left(a_3^{(3)},b_3^{(3)}\right)\equiv\left(\frac{19}{27},\frac{20}{27}\right)$, and $\left(a_4^{(3)},b_4^{(3)}\right)\equiv\left(\frac{25}{27},\frac{26}{27}\right)$. And so on. The complement of the union of all these intervals $\left(a_k^{(n)},b_k^{(n)}\right)$ is the Cantor set C. Thus, if C_1 is the complement of $\left(a_1^{(1)}, b_1^{(1)}\right)$, then $C_1 = \left[0, \frac{1}{3}\right] \cup \left[\frac{2}{3}, 1\right]$. If C_2 is the complement of $\left(a_1^{(2)}, b_1^{(2)}\right)$ and $\left(a_2^{(2)}, b_2^{(2)}\right)$, then $C_2 = \left[0, \frac{1}{9}\right] \cup \left[\frac{2}{9}, \frac{1}{3}\right] \cup \left[\frac{2}{3}, \frac{7}{9}\right] \cup \left[\frac{8}{9}, 1\right]$. In general, C_{n+1} is the subset of C_n obtained by removing open middle thirds if the 2^n number of parts, that is, components of C_n. Therefore, the *Cantor set* C is given by $\bigcap_{n=0}^{\infty} C_n$. The Cantor set is sometimes called a *ternary set*.

Problem 7.6.18. Show that the Cantor set C is a closed subset of $\Re$.

Problem 7.6.19. Show that C is compact.

Problem 7.6.20. Show that the total length of the removed intervals is 1.

Together with knowledge of the concept of a "measure" we could conclude that the Cantor set C has measure 0. With a little more work, it can be shown that C is an uncountable and nowhere dense subset of $\Re$. The reason the Cantor set C is also called ternary set lies amidst its elements. Let us consider an element $x \in (0,1)$. Then x can be written in decimal notation as $0.x_1x_2x_3 \ldots$ where $x_n \in \{0, 1, 2, \ldots, 9\}$ for each $n \in N$. Thus, in decimal notation $x = \sum_{k=1}^{\infty} x_k \left(\frac{1}{10^k}\right)$. This means that the decimal $0.2587 = 2\left(\frac{1}{10}\right) + 5\left(\frac{1}{10^2}\right) + 8\left(\frac{1}{10^3}\right) + 7\left(\frac{1}{10^4}\right) = \sum_{k=1}^{\infty} x_k \left(\frac{1}{10^k}\right)$, where $x_1 = 2$, $x_2 = 5$, $x_3 = 8$, $x_4 = 7$, and all other x_ns are 0. Repeating decimals, such as 0.13434..., written as $0.1\overline{34}$, may also be written as a series. See Example 7.1.17. The values of $x \in (0,1)$ may be written in a ternary expansion as opposed to decimal expansion using powers of $\frac{1}{3}$ instead of $\frac{1}{10}$. For example, we can write $\frac{1}{4}$ as $\frac{1}{4} = 0\left(\frac{1}{3}\right) + 2\left(\frac{1}{3^2}\right) + 0\left(\frac{1}{3^3}\right) + 2\left(\frac{1}{3^4}\right) + 0\left(\frac{1}{3^5}\right) + 2\left(\frac{1}{3^6}\right) + \cdots = \frac{2}{3^2} + \frac{2}{3^4} + \frac{2}{3^6} + \cdots$. This is a converging geometric series with a sum of $\frac{1}{4}$. Thus, in ternary expansion we may write $\frac{1}{4} = 0.\overline{02}$, and we say that $\frac{1}{4}$ is represented by $.\overline{02}$ in base 3.

Problem 7.6.21

(a) What real number does ternary $.\overline{002}$ represent?

(b) Show that the real number $\frac{1}{3}$ can be represented in a ternary expansion as $.1\overline{0}$ and $.0\overline{2}$.

(c) Show the real number $\frac{1}{9}$ can be written as both $.01\overline{0}$ and $.00\overline{2}$ in base 3.

Remark 7.6.22. From the above discussion the following observations arise.

(a) A ternary expansion of $x \in (0,1)$ is unique except when x has a ternary expansion ending in a sequence of 2s. For then, x has a ternary expansion ending in a sequence of 0s, and vice versa.

(b) The numbers that require a 1 in the first position of their ternary expansions lie in the open interval $\left(\frac{1}{3}, \frac{2}{3}\right)$. Thus, these numbers are not in the Cantor set C.

(c) The numbers that require 1 in the second position of their ternary expansions lie in the union of the intervals $\left(\frac{1}{9}, \frac{2}{9}\right)$ and $\left(\frac{7}{9}, \frac{8}{9}\right)$.

(d) Continuing in this manner, we see that if $x \in (0,1)$ is in the Cantor set C, its ternary expansion must consist only of 0s and 2s. ■

Problem 7.6.23. The real number $\frac{1}{3}$ may be written as $.1\bar{0}$ in base 3, but $\frac{1}{3}$ is an element of the Cantor set C. Does this contradict part (c) of Remark 7.6.22?

Problem 7.6.24. Determine whether or not $\frac{1}{13}$, $\frac{11}{12}$, and $\frac{25}{27}$ are in the Cantor set C. Give the ternary expansion for each.

8

Sequences and Series of Functions

8.1 Pointwise Convergence

We are now ready to generalize real-numbered terms of sequences and series to terms consisting of real-valued functions. That is, consider a *sequence of functions* $\{f_n\}$, where the terms $f_1, f_2, \ldots$ are functions with a common domain $D \subseteq \Re$. If a particular value $x_0 \in D$ is chosen, then the sequence $\{f_n(x_0)\}$ is a sequence of real numbers to which all previously discussed theory applies. We were teased for the first time by this concept when the sequence $\{a_n\}$, with $a_n = r^n$, for $r \in (-1,1)$, was considered in Theorem 2.1.13. This could very well have been written as $f_n(x) = x^n$, with $x \in (-1,1)$ (see Exercise 14 of Section 3.3). In Theorem 2.1.13, our focus was on the limit of such a sequence. To find the limit for the preceding sequence of functions, x is fixed and the limit of the real constants is sought. The particular value $x = \frac{1}{2}$ was chosen in Figure 8.1.1 to demonstrate this idea. Notice that the y-coordinates for $x = \frac{1}{2}$ tend to 0. In fact, for any other $x \in (-1,1)$ the y-coordinates, i.e. values $f_n(x)$, also tend to 0. Thus, the "limiting function" is $f(x) = 0$ for all $x \in (-1,1)$.

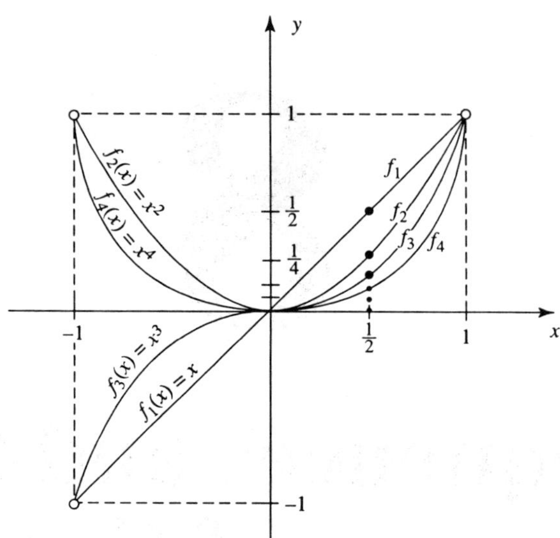

Figure 8.1.1

Definition 8.1.1. A sequence of functions $\{f_n\}$, where for each $n \in N$, $f_n : D \to \Re$ with $D \subseteq \Re$, *converges (pointwise)* on D to a function f if and only if for each $x_0 \in D$ the sequence of real numbers $\{f_n(x_0)\}$ converges to the real number $f(x_0)$.

Remark 8.1.2

(a) The function f in Definition 8.1.1 to which the sequence of functions converges is called the *pointwise limit function*, the *limit*, or the *limiting function* whose domain is also D. If a sequence of functions $\{f_n\}$ converges pointwise on D, then it is *convergent* on D.

(b) If f is the limit function for a sequence of functions $\{f_n\}$, then $\lim_{n\to\infty} f_n(x) = f(x)$ for all $x \in D$.

(c) By Definition 2.1.2, $\lim_{n\to\infty} f_n(x) = f(x)$ for all $x \in D$ if and only if for each $\varepsilon > 0$ there exists a positive integer n^* such that $|f_n(x_0) - f(x_0)| < \varepsilon$ whenever $n \geq n^*$ and x_0 is any value in D. This limit is called the pointwise limit since the integer n^* often depends on the value of ε and the location of x_0.

(d) If a sequence of functions $\{f_n\}$ converges pointwise, then the pointwise limit is unique. There exist sequences of functions $\{f_n\}$ that have no pointwise limit. See Exercise 2(a).

(e) The pointwise limit for a sequence of functions $\{f_n\}$, if it exists, very likely has strikingly different properties than the functions f_n do. The examples and exercises of this section will demonstrate this.

(f) Pointwise convergence for a sequence of functions $\{f_n\}$ can be tested visually using a *vertical line test*, very similar to the one used in Chapter 1 when we tested whether a given relation was a function. Here, we can also draw a vertical line $x = x_0$ for any fixed x_0 in the domain D for the sequence of functions f_n. But the distance between the points $(x_0, f_n(x_0))$ and $(x_0, f(x_0))$ must tend to zero as n tends to $+\infty$ for the function f to be the pointwise limit for the sequence of functions $\{f_n\}$.

(g) As in Chapter 2, the subscripts n of f_n are natural numbers unless otherwise indicated. ■

Example 8.1.3. Consider the sequence of functions $\{f_n\}$, where $f_n:[0,1]\to\Re$ are defined by $f_n(x) = x^n$. Prove that this sequence of functions converges to the function

$$f(x) = \begin{cases} 0, \text{ if } 0 \le x < 1 \\ 1, \text{ if } x = 1. \end{cases}$$

See Theorem 2.1.13 and Exercise 14 of Section 3.3.

Proof. If $x = 0$, then the sequence $\{f_n\}$ reduces simply to a sequence of zeros, which obviously converges to 0. If $x = 1$, then $f_n(x) = 1$ for all $n \in N$ and $\{f_n\}$ converges to 1. Now, assume that $x \in (0,1)$, and prove that $\{f_n\}$ must converge to the zero function, that is, show that $\lim_{n\to\infty} x^n = 0$ for all $x \in (0,1)$. To this end, let $\varepsilon > 0$ be given and choose any $x_0 \in (0,1)$. We need to find $n^* \in N$ so that $|f_n(x_0) - f(x_0)| < \varepsilon$ provided $n \ge n^*$. Thus, choose $n^* > \boxed{\dfrac{\ln\varepsilon}{\ln x_0}}$. Then, if $n \ge n^*$, we have

$$|f_n(x_0) - f(x_0)| = |(x_0)^n - 0| = (x_0)^n < \varepsilon.$$

The last inequality holds true since, when we solve for n, we obtain $n > \dfrac{\ln\varepsilon}{\ln x_0}$. But $n > \dfrac{\ln\varepsilon}{\ln x_0}$ is true if $n \ge n^*$ and n^*, in turn, satisfies $n^* > \dfrac{\ln\varepsilon}{\ln x_0}$. Note that this is how n^* was chosen and why it was placed in a box. Review the discussion at the beginning of Section 2.2. ■

Recall that if each of the functions f_n are continuous on D, then $\lim_{x\to a} f_n(x) = f_n(a)$ for any $a \in D$. In Example 8.1.3 each f_n is continuous. However, the limiting function f is discontinuous at $x = 1 \in D$. Thus, $\lim_{x\to 1} f(x) \ne f(1)$. That is,

$$\lim_{x\to 1}\left[\lim_{n\to\infty} f_n(x)\right] \neq \lim_{n\to\infty}\left[\lim_{x\to 1} f_n(x)\right].$$

Hence, two limits cannot necessarily be interchanged. Also, observe that in Example 8.1.3 it is impossible to find n^* that depends only on ε and works for all $x \in [0,1]$. This will be discussed further in Section 8.2.

Example 8.1.4. Find the pointwise limit of the sequence $\{f_n\}$, where $f_n : [0,1] \to \Re$ with $f_n(x) = \dfrac{x}{n}$, and then prove the result.

Answer. To find the limiting function we first fix the value of x by replacing it with $x_0 \in [0,1]$, and then compute $\lim_{n\to\infty} f_n(x_0)$, which yields

$$\lim_{n\to\infty} \frac{x_0}{n} = x_0 \lim_{n\to\infty} \frac{1}{n} = x_0 \cdot 0 = 0.$$

Therefore, the limiting function is $f(x) = 0$, for all $x \in [0,1]$. To prove this, choose any $\varepsilon > 0$ and fix x by calling it $x_0 \in [0,1]$. Then,

$$\left|f_n(x_0) - f(x_0)\right| = \left|\frac{x_0}{n} - 0\right| = \frac{x_0}{n} < \varepsilon, \text{ if } n > \frac{x_0}{\varepsilon}.$$

Hence, if $n^* > \dfrac{x_0}{\varepsilon}$ and $n \geq n^*$, we have $\left|f_n(x_0) - 0\right| < \varepsilon$. Thus, $\lim_{n\to\infty} f_n(x) = 0$ for all $x \in [0,1]$. ■

Remark 8.1.5

(a) Example 8.1.4 is illustrated in Figure 8.1.2. The heavily marked curve is the limiting function $f(x) = 0$. Observe that all of the functions f_n as well as f are continuous, differentiable, integrable, and bounded.

(b) An important observation in Example 8.1.4 is that n^* depends on ε and x_0, but does not have to, because, since $x_0 \in [0,1]$, we can write

$$\left|f_n(x_0) - f(x_0)\right| = \frac{x_0}{n} \leq \frac{1}{n} < \varepsilon$$

if $n > \dfrac{1}{\varepsilon}$. Therefore, we can choose $n^* > \dfrac{1}{\varepsilon}$. Then, $n \geq n^*$ will yield the desired inequality. Compare this with Example 8.1.3, where n^* had to depend on both ε and x_0. The importance of all this hard work and elaboration should become apparent in Sections 8.2 and 8.3.

(c) Finding the limiting function is not always easy. Often, different choices of x_0 in the domain D for a sequence of functions result in a different limiting value. For a more complicated sequence of functions more sophisticated methods may be needed to compute the limit. See Exercise 1 from this section and Example 8.2.7, part (a).

(d) The functions in Figure 8.1.2 were obtained by fixing n and then graphing the corresponding function f_n on the domain $[0,1]$. That is, for $n=1$, the function $y=x$ on domain $[0,1]$ was graphed to give f_1; for $n=2$, the function $y=\frac{x}{2}$ on domain $[0,1]$ was graphed to give f_2; and so on. When working with sequence of functions, try graphing a few f_ns to obtain an intuitive feel for what the sequence resembles. A graphing calculator or computer is very helpful. ■

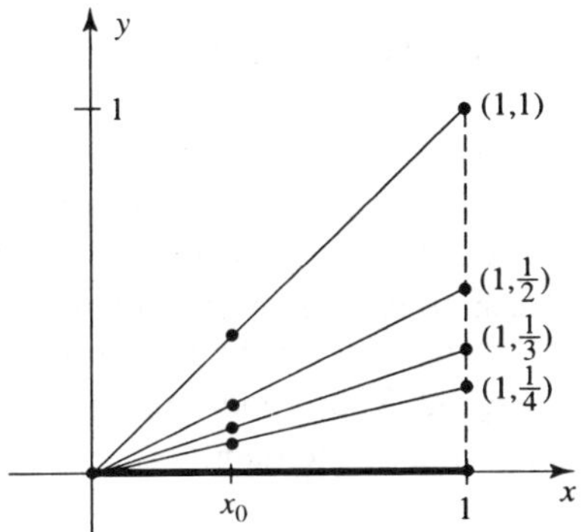

Figure 8.1.2

Example 8.1.6. Consider the sequence $\{f_n\}$, where $f_n:(0,1)\to\Re$, with $f_n(x)=nx^n$. Find the limiting function f and show that $\lim_{n\to\infty}\int_0^1 f_n(x)\,dx \neq \int_0^1 f(x)\,dx$.

Answer. Recall that, by Exercise 17 of Section 2.1 and Exercise 18(a) of Section 7.3, the limiting function f is zero. Thus, $\int_0^1 f(x)\,dx = 0$. Now,

$$\lim_{n\to\infty}\int_0^1 f_n(x)\,dx = \lim_{n\to\infty}\int_0^1 nx^n\,dx = \lim_{n\to\infty} n\int_0^1 x^n\,dx = \lim_{n\to\infty}\frac{n}{n+1} = 1 \neq 0.$$ ■

Remark 8.1.7. Example 8.1.6 proves that, in general, we cannot interchange a limit and an integral. That is, often

$$\lim_{n\to\infty}\int_a^b f_n(x)\,dx \neq \int_a^b\left[\lim_{n\to\infty} f_n(x)\right]dx.$$

In Section 8.3 we will give conditions for when the two preceding expressions are equal. Likewise, $\lim_{n\to\infty} (f_n)'(x)$ need not always equal $\frac{d}{dx}\left[\lim_{n\to\infty} f_n(x)\right]$. This follows from not being able to interchange the order of two limits, since a derivative is a limit of the difference quotient. But this idea is not simple. See Exercises 2(j) and 5(b) of this section and Exercise 7 in Section 8.2. ■

Definition 8.1.8. (Negation of Definition 8.1.1) Suppose that $f_n : D \to \Re$ for all $n \in N$. Then the sequence $\{f_n\}$ *does not pointwise converge*, that is, *diverges* on D if and only if there is $x_0 \in D$ such that the sequence $\{f_n(x_0)\}$ diverges.

Remark 8.1.9. Divergence of a sequence $\{f_n\}$ can be due to unboundedness or oscillation, just as is the case with real numbers. And since we can define functions $f_n : D \to \Re$ by constant values, that is, f_ns are horizontal lines, examples of diverging sequences from Chapter 2 can be used here as examples of diverging sequences as well. Observe that if the domain is changed for the functions f_n, then the resulting sequence might very well be convergent. Thus, one cannot talk about the convergence of a sequence $\{f_n\}$ without specifying the domain. ■

Remark 8.1.10. Be very careful to not confuse $f_n(x)$ with $f(x_n)$. The symbol $f(x_n)$ stands for one function f that is to be evaluated at values x_n along the horizontal axis producing a sequence of real numbers. The symbol $f_n(x)$ stands for n functions graphed on one coordinate plane with some common domain D resulting in possibly one limit of real numbers for each value of x. ■

Before moving on to uniform convergence in Section 8.2, observe a less popular form of convergence whose importance can be surprising.

Definition 8.1.11. A sequence $\{f_n\}$, with $f_n : [a,b] \to \Re$, *converges in the mean* to a function f if and only if $\lim_{n\to\infty}\left(\int_a^b [f_n(x) - f(x)]^2\,dx\right)^{\frac{1}{2}} = 0$.

Remark 8.1.12

(a) Definition 8.1.11 says that $\{f_n\}$ converges in the mean to a function f if and only if the sequence $\{d_n\}$, with $d_n = \left(\int_a^b [f_n(x) - f(x)]^2\,dx\right)^{\frac{1}{2}}$ converges to 0. The sequence $\{d_n\}$ consists of only real numbers. Using Definition 9.1.1, d_n could be written as $\|f_n - f\|$. The function f need not be unique.

(b) A sequence of functions $\{f_n\}$ on $[a,b]$ exists that converges in the mean to one function while converging pointwise to another function.

(c) A sequence of functions $\{f_n\}$ on $[a,b]$ exists that converges in the mean, but diverges pointwise. See Exercise 4 of this section.

(d) A sequence of functions $\{f_n\}$ on $[a,b]$ exists that converges pointwise, but not in the mean.

(e) If a sequence $\{f_n\}$ converges both pointwise to f and in the mean to g on $[a,b]$, and if f and g are continuous on $[a,b]$, then $f \equiv g$.

(f) If each f_n is Riemann integrable on $[a,b]$ and $\{f_n\}$ converges uniformly on $[a,b]$, then $\{f_n\}$ converges in the mean on $[a,b]$. ■

Definition 8.1.13. Let $\{f_n\}$ be a sequence of functions, where $f_n : D \to \Re$ with $D \subseteq \Re$.

(a) A sequence $\{f_n\}$ is *pointwise bounded* on D if and only if there exists a function $M : D \to \Re$ such that $|f_n(x)| \le M(x)$, for all $x \in D$ and $n \in N$.

(b) A sequence $\{f_n\}$ is *uniformly bounded* on D if and only if there exists a real constant K such that $|f_n(x)| \le K$, for all $x \in D$ and $n \in N$.

Exercises 8.1

1. Find the pointwise limit for the sequence $\{f_n\}$, where

(a) $f_n : [0,1] \to \Re$ with $f_n(x) = \dfrac{x^n}{n}$.

(b) $f_n : [-1,1] \to \Re$ with $f_n(x) = \dfrac{nx}{1+n^2x^2}$.

(c) $f_n : [0,\infty) \to \Re$ with $f_n(x) = \dfrac{x^n}{1+x^{2n}}$.

(d) $f_n : \Re^+ \to \Re$ with $f_n(x) = \dfrac{1}{n}e^{-n^2x^2}$.

(e) $f_n : [0,1] \to \Re$ with $f_n(x) = nxe^{-nx^2}$.

(f) $f_n : [0,1] \to \Re$ with $f_n(x) = \dfrac{\sin nx}{\sqrt{n}}$.

(g) $f_n : [0,\pi] \to \Re$ with $f_n(x) = (\sin x)^n$.

(h) $f_n : [0,\infty) \to \Re$ with $f_n(x) = \dfrac{x}{n}\exp\left(-\dfrac{x}{n}\right)$.

(i) $f_n:[0,\infty)\to\Re$ with $f_n(x)=\dfrac{x^n}{1+x^n}$. (See Exercise 27 of Section 2.7, Exercise 15 of Section 3.3, Exercise 2(p) of Section 4.2, and Exercises 23 and 24 of Section 7.5.)

(j) $f_n:[0,1]\to\Re$ with $f_n(x)=n^2x(1-x)^n$. (See Exercise 18(b) of Section 7.3.)

(k) $f_n:[0,1]\to\Re$ with $f_n(x)=nx^n(1-x)$. (See Exercise 1 in Section 8.7.)

(l) $f_n:\Re\to\Re$ with $f_n(x)=\begin{cases}1, & \text{if } x\in[-n,n]\\ 0, & \text{otherwise.}\end{cases}$ (This function is called the *characteristic function* of the interval $[-n,n]$.)

(m) $f_n:\Re\to\Re$ with $f_n(x)=\dfrac{1}{nx^2+1}$.

(n) $f_n:\Re\to\Re$ with $f_n(x)=\dfrac{1}{n\left(x^2+1\right)}$.

(o) $f_n:\Re\to\Re$ with $f_n(x)=\dfrac{1}{x^2+n}$. (The curves f_1 in parts (m), (n), and (o) are called the *Witch of Agnesi*. See Exercises 14 and 15 in Section 10.4.)

2. Give an example of a sequence $\{f_n\}$ where

(a) $\{f_n\}$ has no limit.

(b) each f_n is not Riemann integrable, but the limiting function is.

(c) each f_n is Riemann integrable, but the limiting function is not.

(d) each f_n is bounded, but the limiting function is not.

(e) each f_n is unbounded, but the limiting function is bounded.

(f) each f_n is bounded, but the limiting function is improper Riemann integrable and unbounded.

(g) each f_n is discontinuous, but the limiting function is differentiable.

(h) each f_n is continuous and not differentiable, but the limiting function is differentiable.

(i) the sequence $\{|f_n|\}$ converges pointwise, but $\{f_n\}$ does not.

(j) $\lim\limits_{n\to\infty}(f_n)'(x)\neq\dfrac{d}{dx}\left[\lim\limits_{n\to\infty}f_n(x)\right]$.

(k) $\lim\limits_{n\to\infty}\displaystyle\int_a^b f_n(x)\,dx\neq\int_a^b\left[\lim_{n\to\infty}f_n(x)\right]dx$, and f_n are different from those in Example 8.1.6.

(l) each f_n is uniformly continuous, but the limiting function is not.

(m) each f_n is not a constant function, and $\{f_n\}$ diverges.

3. Suppose that $f_n, g_n: D\to\Re$, $\{f_n\}$ converges pointwise to f, and $\{g_n\}$ converges pointwise to g. Prove that $\{f_n\pm g_n\}$ converge pointwise to $f\pm g$.

4. Consider the sequence $\{f_n\}$ with $f_n:[0,1]\to\Re$, defined by $f_{2^n+k}(x)=1$ for $\frac{k}{2^n}\le x\le\frac{k+1}{2^n}$, with $k=0,1,2,\ \ldots\ ,2^n-1$, and 0 otherwise. Prove that $\{f_n\}$ converges in the mean to $f(x)=0$, but does not converge pointwise.

5. Give an example of a sequence $\{f_n\}$ with $f_n:D\to\Re$ and for which if $x=a\in D$, then

   (a) $\lim\limits_{x\to a}\left[\lim\limits_{n\to\infty} f_n(x)\right]\neq\lim\limits_{n\to\infty}\left[\lim\limits_{x\to a} f_n(x)\right]$.

   (b) $\lim\limits_{x\to a}\left[\lim\limits_{n\to\infty}\frac{f_n(x)-f_n(a)}{x-a}\right]\neq\lim\limits_{n\to\infty}\left[\lim\limits_{x\to a}\frac{f_n(x)-f_n(a)}{x-a}\right]$.

6. (a) Show that the sequence $\{f_n\}$, where $f_n:\Re^+\to\Re$ with $f_n(x)=x-\frac{1}{n}$, is pointwise bounded but not uniformly bounded.

   (b) Prove that if a sequence $\{f_n\}$ is bounded uniformly, then it is also bounded pointwise.

7. Suppose that $\{f_n\}$ is the sequence of functions $f_n:\Re^+\to\Re$, defined by $f_n(x)=n^2xe^{-nx}$. Show that the pointwise limit function is zero, but that the function containing all of the maximum values for the functions f_n is given by the unbounded function $g(x)=\left(\frac{1}{e}\right)\left(\frac{1}{x}\right)$.

8.2 Uniform Convergence

Definition 8.2.1. A sequence of functions $\{f_n\}$, where for each $n\in N$ we have $f_n:D\to\Re$ with $D\subseteq\Re$, *converges uniformly* to a function f if and only if for each $\varepsilon>0$ there exists $n^*\in N$ such that $|f_n(x)-f(x)|<\varepsilon$, for all $x\in D$ and $n\ge n^*$. Such a function f is called the *uniform limit* of $\{f_n\}$.

Remark 8.2.2

(a) Notice the difference between Definition 8.2.1 and Remark 8.1.2, part (c). In Remark 8.1.2, part (c), the value of n^* depended not only on ε but also on the value of x, since n^* was computed individually for each choice of x in the domain. In Definition 8.2.1, only one n^* is chosen to work for all x in the domain. The uniform limit is a global property that generalizes the pointwise limit. That is, if n^* works for all of the values of x, then it must work for each particular one.

(b) The pointwise limit must exist for the uniform limit to exist. And if the uniform limit exists, then it must be equal to the pointwise limit and thus be unique. Therefore, to prove that a sequence $\{f_n\}$ converges uniformly using Definition 8.2.1, we must first find the pointwise limit function f, choose an arbitrary $\varepsilon > 0$, and then find $n^* \in N$ so that the condition $|f_n(x) - f(x)| < \varepsilon$ is true for all $n \geq n^*$. The easiest way to find such an n^* is to bound the expression $|f_n(x) - f(x)|$ by a term that goes to 0 as n tends to $+\infty$ without involving x. See Remark 8.1.5, part (b), Example 8.2.3, and others that follow.

(c) In Section 8.3 we will see that the uniform limit often preserves properties of the functions f_n.

(d) The intuitive concept behind the uniform limit is simple. Once the pointwise limit function f is known, f is the uniform limit of a sequence $\{f_n\}$ if eventually, that is, for all $n \geq n^*$, the graphs of all the functions f_n stay within a small two-dimensional "cylinder" or "tube," that is, the strip bounded by the graphs of $f - \varepsilon$ and $f + \varepsilon$. (See Figure 8.2.1.)

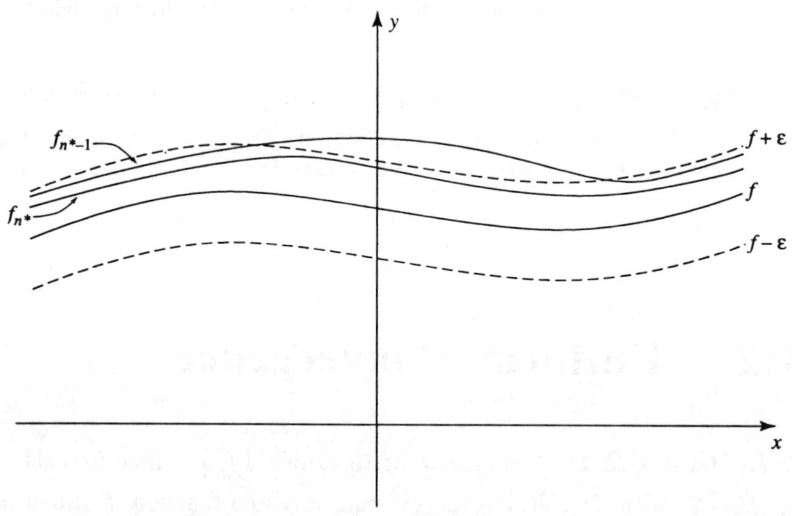

Figure 8.2.1

(e) The horizontal line $y = \varepsilon$ is drawn to see if the curve $|f_n(x) - f(x)|$ is below this line eventually, that is, for sufficiently large values of n. If so, then the sequence $\{f_n\}$ converges uniformly to f. ■

Example 8.2.3. Prove that the sequence $\{f_n\}$ with $f_n : [0,1] \to \Re$ defined by $f_n(x) = \dfrac{x}{n}$ converges uniformly.

Proof. In view of Example 8.1.4 and Remark 8.1.5, part (b), we will prove that the pointwise limit, that is, the zero function, is the uniform limit. To this end, let $\varepsilon > 0$ be given. Then

$$|f_n(x) - f(x)| = \left|\frac{x}{n} - 0\right| = \frac{x}{n} \le \frac{1}{n}.$$

But, $\frac{1}{n} < \varepsilon$ if $n > \frac{1}{\varepsilon}$. Thus, choose any $n^* > \frac{1}{\varepsilon}$. Then, for all $n \ge n^*$ and any $x \in [0,1]$, we have $|f_n(x) - f(x)| < \varepsilon$. Note that n^* depends only on the size of ε. ■

Example 8.2.4. Determine whether the pointwise limit of the sequence $\{f_n\}$, where $f_n(x) = x^n$, is uniform on the indicated interval.

(a) $[0,1]$ **(b)** $[0,1)$ **(c)** $[0,k]$ with $k \in (0,1)$

Answer. According to Example 8.1.3, the pointwise limit function is

$$f(x) = \begin{cases} 0, \text{ if } 0 \le x < 1 \\ 1, \text{ if } x = 1. \end{cases}$$

We will determine whether the three choices for the domain of the functions f_n converge uniformly to this function f.

Answer to part (a). In Figure 8.2.2, the limiting function is highlighted and the curves $f - \varepsilon$ and $f + \varepsilon$ are dotted in. At $x = 1$, a problem exists since the "cylinder" has a width of one point. Of course if $\varepsilon > 3$, then all of the functions f_n are within 3 of the function f. In Figure 8.2.2, $\varepsilon = \frac{1}{4}$ and as can easily be seen, every function f_n must leave the bottom "cylinder" to get to 1. Therefore, $\{f_n\}$ does not converge uniformly to f. This is only an intuitive argument. A formal proof is given in Example 8.2.6, part (a), and should be quite easy after Remark 8.2.5.

Answer to part (b). Here, the point $(1,1)$ is missing from the graph. But again, every function f_n has to leave the bottom "cylinder" to reach the missing point $(1,1)$. Thus, eventually each f_n will cross the line $y = \varepsilon$ for any $\varepsilon \in (0,1)$. Therefore, again, convergence is not uniform. A formal proof of this problem is identical to the one in part (a). Basically, in parts (a) and (b) there is not one $n^* \in N$ that works for all x in these two domains so that the inequality $|f_n(x) - f(x)| < \varepsilon$ will hold true.

Answer to part (c). Since the domain is bounded away from $x = 1$, a gap is left between the points k and 1 so that the functions f_n with a large enough subscript do not escape the "cylinder" until after the point k. Thus, for any $\varepsilon > 0$, we can write

$$|f_n(x) - f(x)| = x^n \le k^n.$$

But, $k^n < \varepsilon$ if $n > \frac{\ln \varepsilon}{\ln k}$. Observe that $\ln k$ is a negative number. Therefore, if $n^* > \frac{\ln \varepsilon}{\ln k}$, then for all $n \geq n^*$ and any $x \in [0,k]$, we have $|f_n(x) - f(x)| < \varepsilon$. Hence, $\{f_n\}$ converges uniformly to $f(x) = 0$ on $[0,k]$ with $k \in (0,1)$. ■

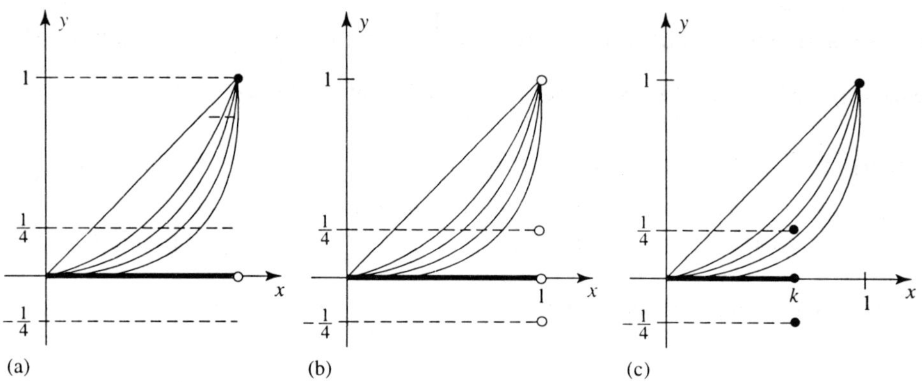

Figure 8.2.2

Remark 8.2.5. A sequence of functions $\{f_n\}$, where $f_n : D \to \Re$ with $D \subseteq \Re$, does not converge uniformly

(a) if the sequence $\{f_n\}$ has no pointwise limit,

(b) if for every function $f : D \to \Re$ there exists $\varepsilon > 0$ such that given any $n^* \in N$ there exists $x \in D$ and $n > n^*$ such that $|f_n(x) - f(x)| \geq \varepsilon$,

(c) if f is the pointwise limit and there exists $\varepsilon > 0$ and a sequence $\{x_n\}$ in D such that $|f_n(x_n) - f(x_n)| \geq \varepsilon$, eventually,

(d) if the pointwise limit f is not continuous, provided f_ns are continuous on D (see Section 8.3),

(e) if the maximum "error" between f_n and the pointwise limit function f does not tend to 0 as n tends to $+\infty$ (in fact, $\{f_n\}$ does not converge uniformly to its pointwise limit f if and only if the sequence $\{M_n\}$ with $M_n = \sup_{x \in D} |f_n(x) - f(x)|$ does not converge to 0, see Exercise 6), or

(f) if the sequence $\{f_n\}$ does not satisfy the Cauchy criterion. See Theorem 8.2.8. ■

Example 8.2.6. Prove that the sequence $\{f_n\}$, where $f_n(x) = x^n$, does not converge uniformly on the indicated interval.

(a) $[0,1]$ **(b)** $[0,1)$

Proof of part (a). Since the limiting function f on $[0,1]$ is not continuous, by Remark 8.2.5, part (d), the convergence is not uniform. Also observe that by part (b), f is not a uniform limit on the smaller domain $[0,1)$, and thus cannot be a uniform limit on the larger domain.

Proof of part (b). Since the functions increase rapidly near $x = 1$, a problem is suspected in that region. Thus, choose a sequence $\{x_n\}$ in the interval $[0,1)$ that converges to 1. Suppose $x_n = \sqrt[n]{\frac{1}{2}}$. Then, if $\varepsilon = \frac{1}{4}$, say,

$$\left|f_n(x_n) - f(x_n)\right| = \frac{1}{2} \ge \frac{1}{4} = \varepsilon.$$

Therefore, by Remark 8.2.5, part (c), the sequence $\{f_n\}$ does not converge uniformly on $[0,1)$. ■

Example 8.2.7. Determine whether or not the given sequence $\{f_n\}$ with $f_n : [0,\infty) \to \Re$, converges uniformly.

(a) $f_n(x) = x^n e^{-nx}$ **(b)** $f_n(x) = \dfrac{nx}{1 + n^2 x^2}$

Answer to part (a). First, the pointwise limit is sought using the first derivative test, Exercise 15 of Section 5.2, rather than the usual argument of fixing x and then computing the limit. Thus, the absolute maximum of each f_n is attained at $x = 1$ with a maximum value of e^{-n}. Hence,

$$\left|f_n(x)\right| \le e^{-n}, \text{ for all } x \ge 0.$$

Therefore, the pointwise limit function is $f(x) = 0$. See Remark 8.2.5, part (e). Now let $\varepsilon > 0$ be given. Then,

$$\left|f_n(x) - f(x)\right| = x^n e^{-nx} \le e^{-n}$$

for all $x \ge 0$. Solving $e^{-n} < \varepsilon$ for n, we obtain $n > \ln\frac{1}{\varepsilon}$. Therefore, if $n^* > \ln\frac{1}{\varepsilon}$, then for all $n \ge n^*$, we have $\left|f_n(x) - f(x)\right| < \varepsilon$. Hence, convergence is uniform to 0. See Figure 8.2.3.

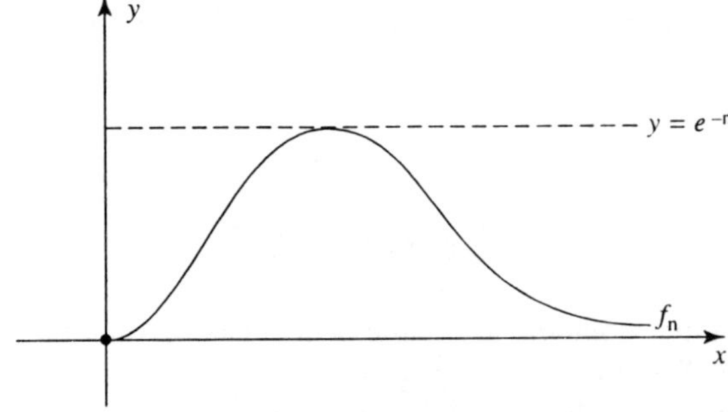

Figure 8.2.3

Answer to part (b). By Exercise 1(b) of Section 8.1, the sequence $\{f_n\}$ converges pointwise to $f(x)=0$. However, the functions f_n have humps of height $\frac{1}{2}$ at $x_n = \frac{1}{n}$. See Figure 8.2.4.

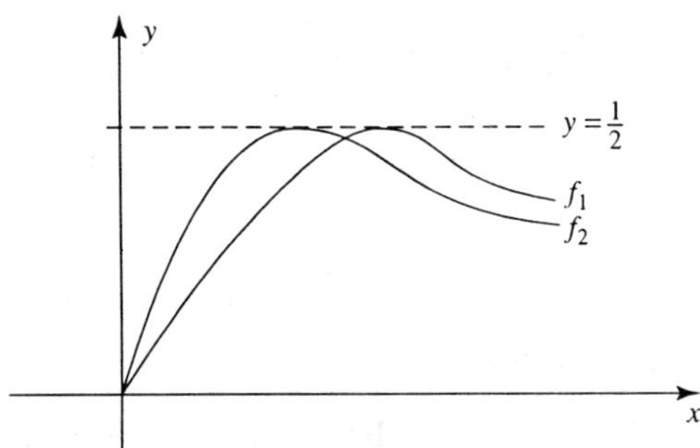

Figure 8.2.4

Thus, to disprove uniform convergence to $f(x)=0$, choose $\varepsilon = \frac{1}{4}$ and $x_n = \frac{1}{n}$. Then,

$$\left|f_n(x_n) - f(x_n)\right| = \left|f_n\left(\frac{1}{n}\right) - 0\right| = \frac{1}{2} \geq \frac{1}{4}.$$

And since $f(x)=0$ is the only possibility for the uniform limit but does not satisfy Definition 8.2.1, by Remark 8.2.5, part (c), the sequence $\{f_n\}$ does not converge uniformly. ■

THEOREM 8.2.8. *(Cauchy Criterion for Sequences) A sequence of functions $\{f_n\}$ converges uniformly on D if and only if for each $\varepsilon > 0$ there exists $n^* \in N$ such that for all $x \in D$, $|f_n(x) - f_m(x)| < \varepsilon$ whenever $n, m \geq n^*$.*

As in Section 2.5, the advantage of the Cauchy criterion for determining whether or not a sequence $\{f_n\}$ converges uniformly is that the pointwise limit need not be computed.

Exercises 8.2

1. **(a)** Determine whether or not the sequences of functions from Exercise 1 in Section 8.1 converge uniformly.

(b) Determine whether or not the sequence $\{f_n\}$, with $f_n : (-1,1) \to \Re$, defined by $f_n(x) = \dfrac{1-x^n}{1-x}$, converges uniformly. See the geometric series test from Section 7.1.

(c) Determine whether or not the domain D on which the sequence $\{f_n\}$, with $f_n : D \to \Re$, defined by $f_n(x) = e^{-nx}$, converges uniformly.

(d) Determine whether or not the sequence $\{f_n\}$, defined by $f_n(x) = n^2 x^2 e^{-nx}$, is uniformly convergent on $D = [0, \infty)$. How about on $D = \Re^+$ or $D = [1, \infty)$? Explain.

2. If sequences $\{f_n\}$ and $\{g_n\}$ converge uniformly on D to functions f and g, respectively, prove the sequences $\{f_n \pm g_n\}$ converge uniformly to $f \pm g$ on D.

3. If $\{f_n\}$ is a sequence of bounded functions that converges uniformly to f on D, prove that

(a) f is bounded. Note that this need not be true if convergence is not uniform. See Exercise 2(d) of Section 8.1.

(b) $\{f_n\}$ is uniformly bounded.

4. Give examples of sequences $\{f_n\}$ and $\{g_n\}$ that converge uniformly, but the sequence $\{f_n g_n\}$ does not.

5. If $\{f_n\}$ and $\{g_n\}$ are sequences of bounded functions that converge uniformly on D to functions f and g, respectively, prove that the sequence $\{f_n g_n\}$ converges uniformly to fg on D.

6. Prove that a sequence $\{f_n\}$, with the pointwise limit f on D, converges uniformly to f on D if and only if the sequence $\{M_n\}$, with

$$M_n = \sup_{x \in D} |f_n(x) - f(x)|,$$

converges to zero. See Remark 8.2.5, part (e). Use this result to determine whether the sequences $\{f_n\}$ and $\{g_n\}$, where $f_n(x) = x^n$ and $g_n(x) = \dfrac{x^n}{x^n + 1}$ for $x \in [0, 1]$, converge uniformly. See Example 8.1.3 and Exercise 1(i) in Section 8.1.

7. Give an example of a sequence $\{f_n\}$ that converges uniformly to f on D, but the sequence $\{f_n'\}$ does not converge uniformly to f' on D or any other function. See Exercise 2(j) of Section 8.1.

8. Give an example, if possible, of a sequence $\{f_n\}$ of discontinuous functions on D, that converges uniformly to a

(a) continuous function on D. **(b)** discontinuous function on D.

9. Prove Theorem 8.2.8.

8.3 Properties of Uniform Convergence

THEOREM 8.3.1. *If $\{f_n\}$ is a sequence of continuous functions that converges uniformly to a function f on $D \subseteq \Re$, then f is continuous on D.*

Proof. Let $\varepsilon > 0$ be given. Since convergence is uniform there exists $n^* \in N$ such that for all $x \in D$,

$$|f_n(x) - f(x)| < \boxed{\frac{\varepsilon}{3}}, \text{ provided } n \geq n^*.$$

We will prove that f is continuous on D by proving that f is continuous at an arbitrary point $x_0 \in D$. Thus, we need to find $\delta > 0$ such that if $|x - x_0| < \delta$ and $x \in D$, then $|f(x) - f(x_0)| < \varepsilon$. Since the functions f_n are continuous on D, they are continuous at $x = x_0$. Therefore, we will choose a particular f_n, say, f_{n^*}. And since f_{n^*} is continuous at x_0, there exists $\delta_1 > 0$ such that

$$|f_{n^*}(x) - f_{n^*}(x_0)| < \boxed{\frac{\varepsilon}{3}}, \text{ if } |x - x_0| < \delta_1 \text{ and } x \in D.$$

Therefore, with some addition and subtraction, for all $x \in D$, we have

$$\begin{aligned} |f(x) - f(x_0)| &\leq |f(x) - f_{n^*}(x)| + |f_{n^*}(x) - f_{n^*}(x_0)| + |f_{n^*}(x_0) - f(x_0)| \\ &< \frac{\varepsilon}{2} + \frac{\varepsilon}{2} + \frac{\varepsilon}{2} = \varepsilon \text{ if } |x - x_0| < \delta_1. \end{aligned}$$

Now choose $\delta = \delta_1$ to complete the proof. ∎

Remark 8.3.2

(a) The contrapositive of Theorem 8.3.1 was used in Remark 8.2.5, part (d).

(b) The converse of Theorem 8.3.1 is false. That is, a sequence of continuous functions $\{f_n\}$ converging pointwise to a continuous function f on D, does not imply that the convergence is uniform. See Example 8.2.4, part (b) and Exercise 7.

(c) If $\{f_n\}$ is a sequence of continuous functions that converges uniformly to a function f on D, then the symbols " $\lim\limits_{n\to\infty}$ " and " $\lim\limits_{x\to a}$ " for any $a \in D$ can be interchanged. See Exercise 5(a) in Section 8.1.

(d) If $\{f_n\}$ is a sequence of functions that converges uniformly to a continuous function f on D, then nothing can be said about the continuity of the functions f_n. See Example 8.2.4, part (c), and Exercise 8 of Section 8.2. ∎

THEOREM 8.3.3. *If* $\{f_n\}$ *is a sequence of continuous functions that converges uniformly to a function* f *on* $[a,b]$, *then*

$$\lim_{n\to\infty}\int_a^b f_n(x)\,dx = \int_a^b\left[\lim_{n\to\infty} f_n(x)\right]dx.$$

Proof. By Theorem 8.3.1, f is continuous on $[a,b]$. Thus, together with results from Chapter 6, the functions f, f_n, and $f_n - f$ are all Riemann integrable on $[a,b]$. To prove the result, let a constant L represent the right-hand side of the equality. Then, for any given $\varepsilon > 0$, we must find $n^* \in N$ such that

$$\left|\int_a^b f_n(x)\,dx - L\right| < \varepsilon \quad \text{if } n \ge n^*.$$

Since the sequence $\{f_n\}$ converges uniformly to f, there exists $n_1 \in N$ such that for all $x \in [a,b]$,

$$|f_n(x) - f(x)| < \boxed{\frac{\varepsilon}{b-a}}, \text{ if } n \ge n_1.$$

Thus, for all $x \in [a,b]$,

$$\left|\int_a^b f_n(x)\,dx - \int_a^b f(x)\,dx\right| = \left|\int_a^b [f_n(x) - f(x)]\,dx\right| \le \int_a^b |f_n(x) - f(x)|\,dx < \int_a^b \frac{\varepsilon}{b-a}\,dx = \varepsilon,$$

if $n \ge n_1$. Hence, choose $n^* = n_1$ and the proof is complete. ■

In Theorem 8.3.3, the continuity condition on the functions f_n can be relaxed without altering the conclusion. See Exercise 2 in this section. The following theorem is a generalization of both Theorem 8.3.3 and Exercise 2 in this section. The result, which is a special case of a theorem on Lebesgue integrals, is interesting since it does not refer to uniform convergence.

THEOREM 8.3.4. *(Arzelà's[1] Theorem) If* $\{f_n\}$ *is a uniformly bounded sequence of Riemann integrable functions that converge pointwise to a Riemann integrable function* f *on* $[a,b]$, *then*

$$\int_a^b f(x)\,dx = \lim_{n\to\infty}\int_a^b f_n(x)\,dx.$$

---

[1]Cesare Arzelà (1847–1912), an Italian analyst, taught at Bologna and is recognized for contributions in sequences of functions.

THEOREM 8.3.5. *If a sequence of continuously differentiable functions* $\{f_n\}$ *converges pointwise to a function* f *on an interval* I, *and the sequence* $\{f_n'\}$ *converges uniformly to a function* g *on* I, *then* f *is continuously differentiable and* $f' = g$ *on* I. *That is,* $\{f_n'\}$ *converges uniformly to* f' *on* I.

Proof. Let $x = a$ be any fixed value in I. Then, by Exercise 3 of Section 6.4,

$$\int_a^x f_n'(t)\,dt = f_n(x) - f_n(a)$$

for any $x \in I$. Now, taking the limit of both sides and writing the right-hand side first, we have

$$\begin{aligned} f(x) - f(a) &= \lim_{n\to\infty}\left[f_n(x) - f_n(a)\right] = \lim_{n\to\infty}\int_a^x f_n'(t)\,dt \\ &= \int_a^x \left[\lim_{n\to\infty} f_n'(t)\right] dt \qquad \text{(By Theorem 8.3.3)} \\ &= \int_a^x g(t)\,dt \qquad \text{(By hypothesis)} \end{aligned}$$

Next, differentiating both sides of the resulting equality, namely $f(x) = \int_a^x g(t)\,dt + f(a)$, and applying Theorem 6.4.4 we obtain

$$f'(x) = \frac{d}{dx}\int_a^x g(t)\,dt = g(x).$$

Where did we use the fact that the sequence $\{f_n'\}$ converges uniformly to g on I? (See Exercise 3.) ■

In view of Exercise 7 of Section 8.2, the statement "$\{f_n(x)\}$ converging to $f(x)$ implies that $\{f_n'(x)\}$ converges uniformly to $f'(x)$" is false. In order for $\{f_n'(x)\}$ to converge uniformly to $f'(x)$ we need the additional assumption that the sequence $\{f_n'(x)\}$ converges uniformly. Then we can show, as we did above, that its limit is indeed $f'(x)$. However, the assumption that $\{f_n(x)\}$ converges to $f(x)$ on the entire interval I may be relaxed to convergence simply at one point of I. We state this result next.

THEOREM 8.3.6. *Suppose that $\{f_n\}$ is a sequence of functions that are differentiable on an interval I such that the sequence $\{f_n(x_0)\}$ converges at some point $x = x_0 \in I$. If the sequence $\{f_n'\}$ converges uniformly to a function g on I, then $\{f_n\}$ must converge uniformly to f on I, where $f'(x) \equiv g(x)$.*

See Exercise 9. Also, note that if in Theorem 8.3.6 the functions f_n are assumed to be continuously differentiable on $[a,b]$, a proof becomes considerably easier.

Exercises 8.3

1. Suppose that the sequence $\{f_n\}$ with $f_n:[2,5] \to \Re$ is defined by $f_n(x) = \dfrac{x^n}{1+x^{2n}}$. Find $\lim_{n\to\infty} \int_2^5 f_n(x)\,dx$.

2. If $\{f_n\}$ is a sequence of Riemann integrable functions that converges uniformly to a function f on $[a,b]$, prove that f is Riemann integrable on $[a,b]$ and $\int_a^b f(x)\,dx = \lim_{n\to\infty} \int_a^b f_n(x)\,dx$.

3. In the proof of Theorem 8.3.5, where did we use the fact that the sequence $\{f_n'\}$ converges uniformly to g on I? Explain in detail.

4. *(Dini's[2] Theorem)* Let $\{f_n\}$ be a monotonic sequence of continuous functions on a compact domain D which converges pointwise to a continuous function f on D. Show that $\{f_n\}$ converges uniformly to f on D.

5. Four conditions of Dini's theorem are as follows:
- **(a)** functions f_n are continuous on D.
- **(b)** function f is continuous on D.
- **(c)** $f_n(x) \le f_{n+1}(x)$ for each $x \in D$ and all $n \in N$.
- **(d)** D is compact.

Show that if any one of these four conditions is not satisfied, then Dini's theorem need not be true, that is, $\{f_n\}$ need not converge uniformly.

6. Suppose $\{f_n\}$ is a sequence of continuous functions that converges to f on D.

---

[2]Ulisse Dini (1845–1918), an Italian mathematician, contributed greatly to the area of analysis, and did work in infinitesimal geometry and differential equations. A square in Pisa is named in his honor.

Prove that if $\{x_n\}$ is any sequence converging to a point $c \in D$, then the sequence of real numbers $\{f_n(x_n)\}$ converges to $f(c)$, that is, $\lim_{n\to\infty} f_n(x_n) = f(c)$. Is the converse true? Explain. See Exercise 10 in Section 8.7.

7. If possible, find a sequence $\{f_n\}$ of noncontinuous functions that converges uniformly to a noncontinuous function f. Explain.

8. If $\{f_n(x)\}$ is defined and bounded on $[0,\infty)$ and converges to 0, then $\lim_{n\to\infty} \int_0^\infty f_n(x)\,dx = 0$. True or false? Explain.

9. **(a)** Prove Theorem 8.3.6 in the case the functions f_n are assumed to be continuously differentiable on $[a,b]$.

(b) Prove Theorem 8.3.6 as stated.

8.4 Pointwise and Uniform Convergence of Series

Definition 8.4.1. Suppose that $\{f_n\}$ is a sequence of functions, where $f_n : D \to \Re$ with $D \subseteq \Re$. A sequence of functions $\{F_n\}$, defined by $F_n(x) = \sum_{k=1}^{n} f_k(x)$ is called the *sequence of partial sums* of the *infinite series of functions* $\sum_{k=1}^{\infty} f_k \equiv \sum f_k$. An infinite series of functions *converges* (*pointwise*) to a *pointwise sum* $F : D \to \Re$ if and only if the corresponding sequence of partial sums converges to the function F on D. In addition, the convergence of a series of functions is *uniform* to a *uniform sum* F if and only if the sequence of partial sums converges uniformly to F on D. A function F is *represented* by an infinite series if the series converges to F. The infinite series $\sum f_k(x)$ is *uniformly bounded* on D if and only if there exists a real constant K such that $\sum_{k=1}^{n} |f_k(x)| \le K$, for all $x \in D$ and $n \in N$.

An infinite series of functions is defined parallel to an infinite series of real numbers. If a series of functions converges, whether uniformly or only pointwise, then the resulting expression is called the *limit* and is a function with domain D. Recall Theorem 7.1.14 rewritten next as Remark 8.4.2. Also, using part (b) of Definition 8.1.13 we have a parallel definition of uniform boundedness of the infinite series. For the sequence of partial sums, say $\{\overline{F}_n(x)\}$ with $\overline{F}_n(x) \equiv \sum_{k=1}^{n} |f_k(x)|$, is uniformly bounded if and only if $\sum f_k(x)$ is uniformly bounded.

Remark 8.4.2. Recall that by Theorem 7.1.14, the geometric series $\sum_{k=1}^{\infty} x^{k-1}$ converges to the pointwise limit $\frac{1}{1-x}$ if and only if $x \in (-1,1)$. Here we also assume that $x^0 = 1$ even if $x = 0$. Also, from Definition 8.4.1, the sequence of partial sums $\{F_n\}$, with $F_n(x) = \sum_{k=1}^{n} x^{k-1}$, can be written as $F_n(x) = \frac{1-x^n}{1-x}$, which converges (pointwise) to $\frac{1}{1-x}$ provided $x \in (-1,1)$. See Exercise 1(b) in Section 8.2. ■

THEOREM 8.4.3. *Suppose* $\sum_{k=1}^{\infty} f_k$ *is a series of continuous functions. If* $\sum_{k=1}^{\infty} f_k$ *converges uniformly to* F *on* D*, then* F *is continuous on* D.

Proof. Since each f_k is continuous on D, each partial sum $\sum_{k=1}^{n} f_k$ is continuous on D. Since the infinite series $\sum_{k=1}^{\infty} f_k$ converges uniformly to F, by Definition 8.4.1, the corresponding sequence of partial sums also converges uniformly. Hence, by Theorem 8.3.1, F is continuous on D. ■

Remark 8.4.4

- **(a)** If a function F is the pointwise limit for a series of functions, it need not be the uniform limit. However, if F is the uniform limit for a series of functions, then it is also the pointwise limit.
- **(b)** A series of functions has a unique limit, be it pointwise or uniform.
- **(c)** An ε-definition of pointwise and uniform convergence for an infinite series of functions can be written. This approach is rarely used.
- **(d)** According to Theorem 8.4.3, an infinite series $\sum f_k$ does not converge uniformly to a function F on D if the functions f_k are continuous on D but F is not. See Example 8.4.5, part (a).
- **(e)** If a series $\sum f_k$ of nonnegative and continuous functions converges to a continuous function F on a compact domain D, then the convergence is uniform on D. This is often referred to as *Dini's theorem.*
- **(f)** Exercise 6 of Section 8.2 can be applied to a sequence of partial sums to determine uniform convergence. See Example 8.4.5, part (b). ■

Example 8.4.5. Determine whether or not the series of functions $\sum f_k$ converges uniformly for each given f_n.

(a) $f_n:[0,1]\to\Re$, with $f_n(x)=\dfrac{x}{(x+1)^{n-1}}$

(b) $f_n:[-a,a]\to\Re$, with $a\in(0,1)$ and $f_n(x)=x^{n-1}$ (See Remark 8.4.2.)

Answer to part (a). If $x=0$, then all of the functions f_n are zero functions. Thus, $F(0)=\sum f_k(0)=\sum 0=0$. If $x\in(0,1]$, then $\dfrac{1}{2}\le\dfrac{1}{x+1}<1$, and we have the converging geometric series from Theorem 7.1.14. Hence, with $a=x$ and $r=\dfrac{1}{x+1}$, we have

$$F(x)=\sum\frac{x}{(x+1)^{k-1}}=x\sum\left(\frac{1}{x+1}\right)^{k-1}=\frac{x}{1-\dfrac{1}{x+1}}=x+1.$$

And, thus,

$$F(x)=\begin{cases}0, & \text{if } x=0\\ x+1, & \text{if } 0<x\le 1.\end{cases}$$

Now, by Theorem 8.4.3, Remark 8.4.4, part (d), and the fact that F is not continuous, this series converges pointwise to F but not uniformly.

Answer to part (b). As in Remark 8.4.2, the sequence of partial sums $\{F_n\}$ for the $\sum f_k$ is given by $F_n(x)=\dfrac{1-x^n}{1-x}$. To show that the uniform limit for this series of functions is $F(x)=0$ for all $x\in[-a,a]$, we will employ Exercise 6 from Section 8.2. To this end, we write

$$M_n=\sup_{x\in[-a,a]}\left|F_n(x)-F(x)\right|=\sup_{|x|\le a<1}\frac{|x|^n}{1-x}=\frac{a^n}{1-a}.$$

But, $\lim\limits_{n\to\infty}M_n=\left(\dfrac{1}{1-a}\right)\left(\lim\limits_{n\to\infty}a^n\right)=0$. Thus, the sequence of partial sums $\{F_n\}$ converges uniformly to F. Hence, the given infinite series converges uniformly to F. ■

THEOREM 8.4.6. *(Cauchy Criterion for Series) Suppose $\{f_n\}$ is a sequence of functions defined on D. The series $\sum f_k$ converges uniformly on D if and only if given any $\varepsilon>0$ there exists $n^*\in N$ such that*

$$\left|f_{n+1}(x)+\cdots+f_m(x)\right|=\left|\sum_{k=n+1}^{m} f_k(x)\right|<\varepsilon$$

for all $m>n\geq n^*$ *and every* $x\in D$.

Remark 8.4.7

(a) Theorem 8.4.6 says that if $\sum f_k$ converges uniformly on D, then, in particular by letting $m=n+1$, we must have $\left|f_{n+1}(x)\right|<\varepsilon$, for all $n\geq n^*$ and every $x\in D$.

(b) If we are able to find $\varepsilon>0$ and a sequence $\{x_n\}$ in D such that $\left|f_n(x_n)\right|\geq\varepsilon$ for all n, then $\sum f_k$ does not converge uniformly on D. See Example 8.4.8. Does this remind you of the divergence test for a series of real numbers?

(c) In using the Cauchy criterion for series, the pointwise limit is not needed to test the series for uniform convergence.

(d) From Theorem 8.4.6 it follows that if $\lim\limits_{n,m\to\infty}\left|\sum_{k=n+1}^{m} f_k(x)\right|=0$, then $\sum f_k$ converges uniformly. Why? ■

Example 8.4.8. Determine whether or not the series $\sum f_k$, where $f_n:(0,1)\to\Re$, with $f_n(x)=\dfrac{n}{n+1}x^n$ converges uniformly.

Answer. Since $\sum x^k$ converges for $x\in(0,1)$ and $0<\dfrac{n}{n+1}x^n<x^n$, by the comparison test the pointwise limit exists. However, if we choose $\varepsilon=\dfrac{1}{3}$ and $x_n=\sqrt[n]{\dfrac{n+1}{n+2}}\in(0,1)$, then $\left|f_n(x_n)\right|=\dfrac{n}{n+2}\geq\dfrac{1}{3}$, for all $n\in N$. Hence, by Remark 8.4.7, part (b), the given series does not converge uniformly. ■

Definition 8.4.9. A series of functions $\sum f_k$ is

(a) *absolutely convergent* on D if $\sum\left|f_k\right|$ converges pointwise on D.

(b) *absolutely uniformly (normally) convergent* on D if $\sum\left|f_k\right|$ converges uniformly on D.

Remark 8.4.10

(a) Absolute convergence of $\sum f_k$ on D implies convergence of $\sum f_k$ on D. And, if $\sum f_k$ is absolutely uniformly convergent on D, then $\sum f_k$ converges uniformly. The converse is false. See Exercise 6.

(b) Do not confuse absolute convergence with uniform convergence. A sequence exists that converges uniformly on D but not absolutely, and hence, not absolutely uniformly. Again see Exercise 6. Furthermore, a sequence exists that converges absolutely on D but not uniformly. See Exercise 15(c).

(c) Often, a quick way of proving absolute uniform convergence, without even knowing the pointwise limit, is given in Theorem 8.4.11. It relates convergence of a series of real numbers to a series of functions. ■

THEOREM 8.4.11. *(Weierstrass M-Test) Suppose that $\{f_n\}$ is a sequence of functions defined on D. And suppose that $\{M_n\}$ is a sequence of nonnegative real numbers such that $|f_n(x)| \le M_n$, for all $x \in D$ and each $n \in N$. If the series $\sum M_k$ converges, then $\sum |f_k|$ converges uniformly on D. And hence, $\sum f_k$ converges uniformly and absolutely on D.*

Proof of Theorem 8.4.11 is straightforward and, thus, is left as an exercise. Compare Theorem 8.4.11 to Cauchy criterion, Theorem 8.4.6.

Example 8.4.12. Since $\left|\frac{\sin nx}{n^p}\right| \le \frac{1}{n^p}$, for all $x \in \Re$, and $\sum \frac{1}{k^p}$ converges, by Weierstrass M-test the series $\sum \left|\frac{\sin kx}{k^p}\right|$, with $p > 1$ a real constant, converges absolutely uniformly on $\Re$. Hence, this series of functions converges absolutely and uniformly on $\Re$. ■

Since series exist that converge uniformly but not absolutely, uniform convergence alone will never follow from the Weierstrass M-test. Thus, infinite series exist that converge uniformly, but for which the Weierstrass M-test does not apply. (See Exercise 6.) The same discussion applies to series that converge absolutely but not uniformly. In addition, series exist that converge absolutely and uniformly for which the Weierstrass M-test does not apply. The converse of Theorem 8.4.11 is false. See Exercise 12 in Section 8.7. The following two theorems, which are used to test for uniform convergence, are more difficult to apply than the Weierstrass M-test. These tests extend ones given in Section 7.6. The coverage of the remainder of this section is optional.

THEOREM 8.4.13. *(Abel's Test for Uniform Convergence) Suppose that there exists a series of functions $\sum f_k$ that is uniformly convergent on D, and a sequence of functions $\{g_n\}$ that is uniformly bounded with $\{g_n(x)\}$ eventually monotonic for each $x \in D$. Then, $\sum f_k g_k$ converges uniformly on D.*

THEOREM 8.4.14. *(Dirichlet's Test for Uniform Convergence) Suppose that there exist two sequences of functions $\{f_n\}$ and $\{g_n\}$ such that $\sum f_k$ is uniformly bounded on D and $\{g_n\}$ is uniformly convergent on D. In addition, for each $x \in D$ the sequence $\{g_n(x)\}$ is eventually monotonic and tends to 0. Then $\sum f_k g_k$ converges uniformly on D.*

In first attempting a proof of Theorem 8.4.13 one might be tempted to write

$$\left|\sum_{k=n+1}^{m} f_k g_k\right| \le \sum_{k=n+1}^{m} |f_k g_k| = \sum_{k=n+1}^{m} |f_k||g_k| \le K \sum_{k=n+1}^{m} |f_k|,$$

for all $x \in D$, since $\{g_n\}$ is uniformly bounded by, say, K. But, $\sum |f_k|$ need not converge. Use of both a summation by parts formula from Section 7.7 and the Cauchy criterion for series, Theorem 8.4.6, is recommended. See Exercises 19 and 20 for proofs of Theorems 8.4.13 and 8.4.14. Note that since both Abel's and Dirichlet's tests are stated for functions, they clearly apply to constant functions, that is, to $\sum a_k f_k(x)$ and $\sum a_k b_k$, where a_n, b_n are constants. In other words, any series $\sum u_k$ may be expressed as $\sum f_k g_k$, but Abel's or Dirichlet's test will not apply unless we are able to factor u_k into $f_k g_k$ in such a way that all the conditions of the desired theorem are satisfied.

Consider for example $\sum_{k=1}^{\infty} \frac{\sin kx}{k^p}$, where $0 < p \le 1$. (See Example 8.4.12.) Although the Weierstrass M-test does not apply, (why?), Dirichlet's test can be used to prove uniform convergence at certain values of x. Since from Exercise 24(a) we know that

$$\sum_{k=1}^{n} \sin kx = \frac{\cos \frac{1}{2}x - \cos\left(n+\frac{1}{2}\right)x}{2\sin\frac{x}{2}}, \quad x \ne 2m\pi, \text{ with } m \text{ an integer},$$

then for $F_n(x) \equiv \sum_{k=1}^{n} \sin kx$, we have that $|F_n(x)| \le \frac{1}{\sin\frac{\delta}{2}}$ for $x \in [\delta, 2\pi - \delta]$ and $0 < \delta < \pi$. Therefore, $\{F_n(x)\}$ is uniformly bounded on $[\delta, 2\pi-\delta]$. If $g_n(x) = \frac{1}{n^p}$, a constant function, it is uniformly convergent for $0 < p \le 1$. So, by Dirichlet's test, the series $\sum_{k=1}^{\infty} \frac{\sin kx}{k^p}$ with $0 < p \le 1$ converges uniformly for $x \in [\delta, 2\pi - \delta]$, where $0 < \delta < \pi$. In fact, the series converges uniformly for $x \in \bigcup_{\substack{m=-\infty \\ m\in Z}}^{\infty} [2m\pi + \delta, 2(m+1)\pi - \delta]$, where $0 < \delta < \pi$. See Problem 7.6.3 and Exercise 1(d) of Section 9.3.

Due to uniform convergence, a few nice results are acquired. We will also prove that sometimes the symbols $\sum_{k=1}^{\infty}$ and $\int_a^b$, as well as $\sum_{k=1}^{\infty}$ and $\frac{d}{dx}$, can be interchanged, allowing us to differentiate or integrate a given series term by term.

THEOREM 8.4.15. *Suppose that $\sum f_k$ is a series of Riemann integrable functions that converges uniformly to a function F on $[a, b]$. Then F is Riemann integrable and*

$$\int_a^b F(x)\,dx = \sum_{k=1}^{\infty} \int_a^b f_k(x)\,dx.$$

Example 8.4.16. Recall that if $|x|<1$, then $\dfrac{1}{1-x} = 1 + x + x^2 + x^3 + \cdots$ and, thus, $\sum_{k=0}^{\infty} x^k$ converges pointwise. However, if I is any compact interval inside $(-1,1)$, then $\sum_{k=0}^{\infty} x^k$ converges uniformly on I. In particular, if $a \in (0,1)$, then $\sum_{k=0}^{\infty} x^k$ converges uniformly on $[-a,a]$. See Example 8.4.5, part (b). Now, by Theorem 8.4.15, this series is integrable from 0 to t for $t \in [-a,a]$. And we integrate by integrating each term of the series from 0 to t. Thus,

$$\int_0^t \frac{1}{1-x}\,dx = \int_0^t \left(1 + x + x^2 + x^3 + \cdots\right) dx = \int_0^t 1\,dx + \int_0^t x\,dx + \int_0^t x^2\,dx + \cdots.$$

Therefore,

$$-\ln(1-t) = t + \frac{t^2}{2} + \frac{t^3}{3} + \cdots$$

giving us the familiar series

$$-\ln(1-x) = \ln\frac{1}{1-x} = \sum_{k=1}^{\infty} \frac{x^k}{k},$$

which holds for any $x \in [-a,a]$. And since by Cauchy's ratio test this series converges for each $x \in (-1,1)$ and the above a is an arbitrary number in $(0,1)$, the preceding equality is true for any $x \in (-1,1)$. Now, choose $x = \dfrac{1}{2}$. Then,

$$\ln 2 = \frac{1}{2} + \frac{1}{8} + \frac{1}{24} + \frac{1}{64} + \cdots = \sum \frac{1}{k2^k}.$$

Thus, a series representation is found for the function $\ln(1-x)$ and for the value of $\ln 2$. See Example 8.5.16. ■

THEOREM 8.4.17. *Suppose that $\{f_n\}$ is a sequence of continuously differentiable functions on an interval I such that $\sum f_k$ converges pointwise to a function F on I, and $\sum f_k'$ converges uniformly on I. Then F is differentiable, and $F'(x) = \sum f_k'(x)$ for all $x \in I$.*

Theorem 8.4.17 can be improved as follows.

THEOREM 8.4.18. *Suppose that $\{f_n\}$ is a sequence of functions differentiable on an interval I and such that $\sum f_k(a)$ converges for some point $x = a \in I$. If the series $\sum f'_k$ converges uniformly to G on I, then $\sum f_k(x)$ must converge uniformly to F on I, where $F'(x) \equiv G(x)$.*

Remark 8.4.19

(a) Consider the series $\sum_{k=1}^{\infty} \frac{(-1)^k}{k} \sin \frac{x}{k}$ which converges at $x = 0$. Since when differentiated term by term we obtain the uniformly convergent series $\sum_{k=1}^{\infty} \frac{(-1)^k}{k^2} \cos \frac{x}{k}$ (recall Weierstrass M-test), and since all other conditions of Theorem 8.4.18 are satisfied, the given series converges uniformly on $\Re$ to $f(x)$, and $\sum_{k=1}^{\infty} \frac{(-1)^k}{k^2} \cos \frac{x}{k} = f'(x)$ on $\Re$.

(b) It should be noted that term-by-term differentiation of a series of functions may either create or destroy convergence. For example, consider the series $\sum \sin \frac{x}{k}$ which diverges for each $x \in \Re$. Why? When differentiated term by term, we obtain an absolutely convergent series $\sum \frac{1}{k} \cos \frac{x}{k}$ on $\Re$ which converges uniformly on any compact subset of $\Re$. Why? See the discussion at the beginning of Section 9.4. Now consider the series $\sum \frac{1}{k^2} \sin[(k!)x]$, which converges for every real value of x. Term-by-term differentiation, however, produces a divergent series for each value of x. Less severe examples where convergence is lost at only one or two values of x do exist. Consider the power series given in Exercise 12(a) of Section 8.5. ■

Exercises 8.4

1. Determine whether or not the given series converges uniformly on the indicated interval. Explain.

   (a) $\sum \frac{x^k}{k^2}$ with $x \in [0,1]$ **(b)** $\sum x^k e^{-kx}$ with $x \in [0, \infty)$

   (c) $\sum k^r e^{-kx}$ with $x \in [a, \infty)$, where $a > 0$ and $r \in \Re$ a constant

   (d) $\sum \frac{1}{x^k + 1}$ with $x \in (0,1]$ **(e)** $\sum \frac{1}{x^k + 1}$ with $x \in (1, \infty)$

(f) $\sum \frac{1}{1+k^2x^2}$ with $x \in (0,1]$ (g) $\sum \frac{1}{1+k^2x^2}$ with $x \in (1,\infty)$

(h) $\sum_{k=0}^{\infty} \exp(-kx)$ with $x \in \Re^+$ (i) $\sum_{k=0}^{\infty} \exp(-kx)$ with $x \in [a,\infty)$ and $a > 0$

(j) $\sum x^k$ with $x \in [-a,a]$, where $|a| < 1$. See Exercise 12(c) in Section 8.5.

2. Find the pointwise limit for the given series and show that it is not uniform.

(a) $\sum_{k=1}^{\infty} \frac{x^2}{\left(x^2+1\right)^k}$ on $\Re$ (b) $\sum_{k=1}^{\infty} x(1-x)^k$ on $[0,1]$

3. Suppose that $\sum f_k$ and $\sum g_k$ converge uniformly to functions F and G on D, respectively. Prove that $\sum \left(f_k \pm g_k\right)$ converge uniformly to $F \pm G$ on D.

4. Prove Theorem 8.4.6.

5. If $f_n(x) \geq f_{n+1}(x) \geq 0$ for all $x \in D$ and $\{f_n\}$ converges uniformly to 0 on D, prove that $\sum (-1)^k f_k$ converges uniformly on D.

6. Prove that $\sum \frac{(-1)^k}{k+x^2}$ is uniformly convergent on $[0,\infty)$, but not absolutely convergent on $[0,\infty)$.

7. Prove Theorem 8.4.11.

8. If a series of real numbers $\sum a_k$ converges absolutely, prove that $\sum a_k \sin kx$ converges uniformly on $\Re$.

9. Find a set on which the series $\sum kx^k(1-x)^k$ converges absolutely and uniformly. Explain. Compare this with Exercises 1(j) and 1(k) of Section 8.1. Also, see Exercise 44 in Section 8.7.

10. Prove Theorem 8.4.15.

11. Use $\frac{1}{1-x} = 1 + x + x^2 + x^3 + \cdots$ if $|x| < 1$ to find a series representation for

(a) $f(x) = \frac{1}{1-2x}$. (b) $f(x) = \frac{1}{x}$.

(c) $f(x) = \frac{a}{x-b}$ with a and b positive real constants.

(d) $f(x) = \frac{5x-1}{x^2-x-2}$. (e) $f(x) = \frac{2}{(1-2x)^2}$.

12. (a) Write a series representation for $f(x) = \frac{1}{1-x}$.

(b) Using part (a), write the series representation for $g(x) = f(-x)$.

(c) Using part (b), write the series representation for $h(x) = g\left(x^2\right)$.

(d) Use Theorem 8.4.15 to show that

$$\arctan x = x - \frac{x^3}{3} + \frac{x^5}{5} - \frac{x^7}{7} + \cdots = \sum_{k=0}^{\infty} (-1)^k \frac{x^{2k+1}}{2k+1}$$

if $|x| < 1$. This series is sometimes referred to as *Gregory's*[3] *series*. See Exercise 11(a) of Section 8.5.

(e) Use Theorem 8.4.15 and part (b) to show that

$$\ln(x+1) = x - \frac{x^2}{2} + \frac{x^3}{3} - \frac{x^4}{4} + \cdots = \sum_{k=0}^{\infty} (-1)^k \frac{x^{k+1}}{k+1}$$

if $|x| < 1$. See Exercise 14(b) in Section 8.5.

(f) Use Theorem 8.4.17 to show that

$$\frac{1}{(1-x)^2} = 1 + 2x + 3x^2 + \cdots = \sum_{k=1}^{\infty} kx^{k-1}, \text{ if } |x| < 1.$$

(g) Use long division to obtain series representation in parts (a)–(d) and (f). See part (d) of Exercise 15 in Section 7.1.

13. If $f(x) = \sum_{k=0}^{\infty} \frac{x^{2k}}{k!}$ for all $x \in \Re$, show that $f'(x) = 2xf(x)$ for all x. Do you recognize the function f? See Exercise 17 of Section 6.4.

14. Prove Theorem 8.4.17.

15. (a) Prove that $\sum_{k=0}^{\infty} \frac{x^k}{k!}$ converges uniformly on $[-a, a]$, where a is any fixed real constant.

(b) Prove that $\sum_{k=0}^{\infty} \frac{x^k}{k!}$ converges absolutely on $\Re$. See Exercise 37 of Section 7.5.

(c) Prove that $\sum_{k=0}^{\infty} \frac{x^k}{k!}$ converges pointwise, but not uniformly on $\Re$.

(d) Use Theorem 8.4.17 to show that if $f(x) = \sum_{k=0}^{\infty} \frac{x^k}{k!}$, then $f'(x) = f(x)$ for all $x \in \Re$.

---

[3]James Gregory (1638–1675), a Scottish mathematician, was the first to distinguish between convergent and divergent series. He came up with a series used to approximate π, and was involved with calculating the distance from the earth to the sun.

(e) Use Exercise 8 of Section 7.5 to show that $e^x = \sum_{k=0}^{\infty} \frac{x^k}{k!}$. See Example 8.6.5.

(f) Show that $\sum_{k=2}^{\infty} \frac{(-1)^k}{k!} = \frac{1}{e}$, and $\sum_{k=0}^{\infty} \frac{(k+1)^2}{k!} = 5e$.

16. If $\sum a_k$ converges absolutely, then prove that $\int_0^1 \left(\sum_{k=1}^{\infty} a_k x^k \right) dx = \sum_{k=1}^{\infty} \frac{a_k}{k+1}$.

17. Prove that the zeta function given by the series $z(x) = \sum_{k=1}^{\infty} \frac{1}{k^x}$ converges uniformly on $[a, \infty)$, with $a > 1$. Then show that z is differentiable and that

$$z'(x) = -\sum_{k=1}^{\infty} \frac{\ln k}{k^x}$$

for $x > 1$. See Exercise 12 of Section 7.2.

18. Evaluate $\sum \frac{1}{k3^k}$.

19. Prove Theorem 8.4.13.

20. Prove Theorem 8.4.14.

21. If $\sum_{k=0}^{\infty} a_k$ converges, prove that $\sum_{k=0}^{\infty} a_k x^k$ converges uniformly to, say, F, on the interval $[0,1]$. Verify that $\lim_{x \to 1^-} \sum_{k=0}^{\infty} a_k x^k = \sum_{k=0}^{\infty} a_k$. See Exercise 10 in Section 8.5.

22. *(Dedekind's Test for Uniform Convergence)* Suppose that there exist two sequences of functions $\{f_n\}$ and $\{g_n\}$ such that $\sum f_k$ is uniformly bounded on D, $\{g_n\}$ converges uniformly to 0 on D, and $\sum |g_{k+1}(x) - g_k(x)|$ converges uniformly to 0 on D. Prove that $\sum f_k g_k$ converges uniformly on D. See Exercise 20(c) in Section 7.4.

23. *(Du Bois-Reymond[4] Test for Uniform Convergence)* Suppose that there exist two sequences of functions $\{f_n\}$ and $\{g_n\}$ such that $\sum f_k$ is uniformly bounded on D and $\sum |g_{k+1}(x) - g_k(x)|$ and $\{g_n\}$ are uniformly bounded on D. Prove that $\sum f_k g_k$ converges uniformly on D.

---

[4]Paul David Gustav Du Bois-Reymond (1831-1889), born in Germany, studied both medicine and mathematics. Du Bois-Reymond contributed mostly to real analysis and differential equations.

24. **(a)** If $F_n(x)=\sum_{k=1}^{n}\sin kx$, prove that $F_n(x)=\dfrac{\cos\frac{1}{2}x-\cos\left(n+\frac{1}{2}\right)x}{2\sin\frac{1}{2}x}$.

(b) If $G_n(x)=\sum_{k=1}^{n}\cos kx$, prove that $G_n(x)=\dfrac{\sin\left(n+\frac{1}{2}\right)x-\sin\frac{1}{2}x}{2\sin\frac{1}{2}x}$. (See Exercise 8 in Section 9.2.)

25. Suppose that $f(x)=\sum\dfrac{(-1)^k}{k^2e^{kx}}$ for $x\geq 0$, show that $f'(x)=\sum\dfrac{(-1)^{k+1}}{ke^{kx}}$ for $x\geq 0$.

26. Show that $\sum\dfrac{\cos kx}{k}$ converges uniformly on any closed interval $[a,b]\subset(0,2\pi)$.

8.5 Power Series

A power series is an important type of an infinite series of functions. The following discussion should be a review, although many of the proofs may be new. Polynomials are a special type of power series since they have a finite number of terms.

Definition 8.5.1. If $\{a_n\}$ is a sequence of real numbers with $n=0,1,2,\ldots$, then the series of functions of the form

$$a_0+a_1(x-a)+a_2(x-a)^2+a_3(x-a)^3+\cdots$$

is called a *power series centered at* $x=a$, with a any fixed real number.

Remark 8.5.2

(a) Sometimes the series of functions in Definition 8.5.1 is referred to as a *power series in* $x-a$. The sigma notation

$$a_0+\sum_{k=1}^{\infty}a_k(x-a)^k$$

may be used to represent this series. However, it is customary to abbreviate the notation further by writing

$$a_0+\sum_{k=1}^{\infty}a_k(x-a)^k=\sum_{k=0}^{\infty}a_k(x-a)^k.$$

Note that in this situation, the term $(x-a)^0$ is assumed to equal 1 even if $x=a$. (See Exercise 24 of Section 5.3.)

(b) A power series converges (pointwise) for at least one value of x, namely $x=a$. The pointwise sum at $x=a$ is given by a_0.

(c) Our first encounter with a power series in this text was the geometric series given in Section 7.1. With a geometric series, if $|x|<1$ then $\frac{1}{1-x}=\sum_{k=0}^{\infty} x^k$. This series *represents* the function $f(x)=\frac{1}{1-x}$ and is a *power series representation (expansion)* for $\frac{1}{1-x}$. We further dealt with power series in Exercise 9 from Section 7.3, Exercise 8 of Section 7.4, and on a number of occasions in Section 8.4. ■

Example 8.5.3. Find all of the values of x for which the given power series converges.

(a) $\sum_{k=1}^{\infty} 3(x-2)^k$ (Compare to Exercise 5(a) of Section 7.1.)

(b) $\sum_{k=0}^{\infty} \frac{(x-1)^k}{2^k(k+1)}$

(c) $\sum_{k=0}^{\infty} (k!)(x+3)^k$

(d) $\sum_{k=0}^{\infty} \frac{x^k}{k!}$ (See Exercise 15(c) of Section 8.4.)

Answer to part (a). This geometric series starting at $k=1$ converges if and only if $|x-2|<1$, that is, $1<x<3$.

Answer to part (b). Let $a_n = \frac{(x-1)^n}{2^n(n+1)}$. Since

$$\lim_{n\to\infty}\left|\frac{a_{n+1}}{a_n}\right| = \frac{|x-1|}{2} \quad \text{(why?)}$$

by Cauchy's ratio test $\sum_{k=0}^{\infty} \frac{(x-1)^k}{2^k(k+1)}$ converges absolutely if $\frac{|x-1|}{2}<1$, that is, $-1<x<3$, and diverges if $\frac{|x-1|}{2}>1$. Now if $x=-1$, then $\sum_{k=0}^{\infty} \frac{(x-1)^k}{2^k(k+1)}$ reduces to the conditionally converging alternating harmonic series $\sum_{k=0}^{\infty} \frac{(-1)^k}{k+1}$. But if $x=3$, $\sum_{k=0}^{\infty} \frac{(x-1)^k}{2^k(k+1)}$ reduces to the diverging harmonic series $\sum_{k=0}^{\infty} \frac{1}{k+1}$. Hence, $\sum_{k=0}^{\infty} \frac{(x-1)^k}{2^k(k+1)}$ converges if and only if $x\in[-1,3)$. Note that an alternate approach could have been Cauchy's root test.

Answer to part (c). Let $a_n=(n!)(x+3)^n$. Then

$$\lim_{n\to\infty}\left|\frac{a_{n+1}}{a_n}\right| = |x+3| \lim_{n\to\infty}(n+1) = +\infty, \text{ if } x \neq -3,$$

and $+\infty$ is never <1. Hence, by Cauchy's ratio test, $\sum_{k=0}^{\infty}(k!)(x+3)^k$ converges only at its center $x=-3$.

Answer to part (d). Let $a_n = \dfrac{x^n}{n!}$. Then $\lim_{n\to\infty}\left|\dfrac{a_{n+1}}{a_n}\right| = |x| \lim_{n\to\infty}\dfrac{1}{n+1} = 0$ for all $x \in \Re$.

Hence, $\sum_{k=0}^{\infty}\dfrac{x^k}{k!}$ converges for all real x. ■

Remark 8.5.4. The preceding examples demonstrate the following three possibilities for convergence of a power series:

(a) convergence only at its center $x=a$ (see Example 8.5.3, part (c)),

(b) convergence for all values an equal positive finite distance to the right and left of the center $x=a$ (endpoints of an interval should be tested separately for convergence, see Example 8.5.3, parts (a) and (b)), and

(c) convergence for all real values, that is, on an interval with an infinite distance to the right and left of the center $x=a$ (see Example 8.5.3, part (d)).

It needs to be proved that no other possibility exists. ■

THEOREM 8.5.5. *If* $\sum_{k=0}^{\infty}a_k(x-a)^k$ *converges when* $x = x_0 + a$ *with* x_0 *and* a *real numbers, that is,* $\sum_{k=0}^{\infty}a_k(x_0)^k$ *converges, then* $\sum_{k=0}^{\infty}a_k(x-a)^k$ *converges absolutely for all* x*, satisfying* $|x-a|<|x_0|$.

Proof. Suppose that $\sum_{k=0}^{\infty}a_k(x-a)^k$ converges when $x = x_0 + a$. Then the series of real numbers $\sum_{k=0}^{\infty}a_k(x_0)^k$ converges. If $x_0 = 0$, there is nothing to prove. Therefore, suppose that $x_0 \neq 0$. Note that in either case, the terms of this series must tend to 0. Thus, $\lim_{n\to\infty} a_n(x_0)^n = 0$. Hence, terms of this series are bounded by a positive real number, say, M. Thus,

$$\left|a_n(x-a)^n\right| = \left|a_n(x_0)^n\right|\left|\frac{x-a}{x_0}\right|^n \le M\left|\frac{x-a}{x_0}\right|^n.$$

And since $|x-a|<|x_0|$, $M\sum_{k=0}^{\infty}\left|\frac{x-a}{x_0}\right|^k$ is a converging geometric series. Hence, by the comparison test the original power series converges absolutely. ■

COROLLARY 8.5.6. *If* $\sum_{k=0}^{\infty} a_k(x-a)^k$ *diverges when* $x=x_1+a$, *with* x_1 *and* a *real numbers, then* $\sum_{k=0}^{\infty} a_k(x-a)^k$ *diverges for all* x, *satisfying* $|x-a|>|x_1|$.

The proof of Corollary 8.5.6 is left as an exercise. Thus, from the previous theorem and corollary, if a power series converges at a value x_0 units away from its center $x=a$, then it converges absolutely for all of the values closer to a than x_0. And if a power series diverges at some value x_1 units away from its center $x=a$, then it diverges for all of the values even farther away from its center than x_1.

Definition 8.5.7. The *interval of convergence* for a power series $\sum_{k=0}^{\infty} a_k(x-a)^k$ is the set I of all real values x for which this power series converges. If R is a positive real number and

$$(a-R,a+R)\subseteq I\subseteq[a-R,a+R],$$

then R is called the *radius of convergence* for this power series.

Refer to Remark 8.5.4. When a power series converges only at its center, by convention, we say that the series has radius of convergence $R=0$, and when a power series converges for all real numbers $R=\infty$. Thus, Example 8.5.3, part (a), has a radius of convergence $R=1$, part (b) has a radius of convergence $R=2$, part (c), has a radius of convergence $R=0$, and part (d) has a radius of convergence $R=\infty$.

THEOREM 8.5.8. *Suppose* R *is the radius of convergence for the power series* $\sum_{k=0}^{\infty} a_k(x-a)^k$, *satisfying* $0\le R\le\infty$. *Then* $\sum_{k=0}^{\infty} a_k(x-a)^k$

(*a*) *converges absolutely if* $|x-a|<R$.

(*b*) *converges uniformly if* $|x-a|\le r$ *for any* $r\in(0,R)$.

(*c*) *diverges if* $|x-a|>R$.

Due to Remark 8.5.4, Theorem 8.5.5, Corollary 8.5.6, and parts (a) and (c) of Theorem 8.5.8 are not surprising. Proof of part (b) is left to the reader. The interval in part (b) need not be symmetrical. The power series $\sum_{k=0}^{\infty} a_k(x-a)^k$ converges uniformly on $[a-R+\varepsilon_1, a+R-\varepsilon_2]$ for any choice of positive real numbers ε_1 and ε_2.

THEOREM 8.5.9. *Let R be the radius of convergence for $\sum_{k=0}^{\infty} a_k(x-a)^k$. And suppose $\lim_{n\to\infty}\left|\frac{a_{n+1}}{a_n}\right| = L$. Then*

(a) *if L is a nonzero finite real number, $R = \frac{1}{L}$.*

(b) *if $L = 0$, $R = \infty$.*

(c) *if $L = \infty$, $R = 0$.*

Proof of this theorem is left as an exercise.

Remark 8.5.10. From Theorem 8.5.9 we see that only the coefficients a_k of the terms $(x-a)^k$ determine the radius of convergence for the series $\sum_{k=0}^{\infty} a_k(x-a)^k$. Although a very quick way of finding the radius of convergence, Theorem 8.5.9 is quite restrictive since the desired limit must attain one of the indicated values. For example, consider the power series $\sum_{k=1}^{\infty} a_k x^k$, where

$$a_n = \begin{cases} 1, & \text{if } n \text{ is odd} \\ \frac{1}{n}, & \text{if } n \text{ is even.} \end{cases}$$

Theorem 8.5.9 does not apply, although the radius of convergence is 1. (See Exercise 7.) Often, d'Alembert's ratio test or the root test is more desirable for finding the radius of convergence. And better yet is *Cauchy–Hadamard*[5] *theorem*, which states that each power series has to possess a radius of convergence and goes on to finding it using "limit superior," which is discussed in Section 8.8. ■

THEOREM 8.5.11. *If a function f is represented by the power series $\sum_{k=0}^{\infty} a_k(x-a)^k$, with a radius of convergence $R \in (0,\infty)$, then f is continuous on the interval $(a-R, a+R)$.*

The proof of Theorem 8.5.11 is left as an exercise. The following theorem under certain conditions extends Theorem 8.5.8, part (b), and Theorem 8.5.11 to larger intervals.

---

[5]Jacques Salomon Hadamard (1865–1963), a French mathematician, was known for his work on infinitesimal calculus, proof of the prime number theorem, and initiation of the theory of functional analysis. Mathematics was Hadamard's worst subject until grade 7, when a good mathematics teacher turned him toward this subject.

THEOREM 8.5.12. *(Abel's Theorem) Suppose $R \in (0,\infty)$ is the radius of convergence for the power series $\sum_{k=0}^{\infty} a_k (x-a)^k$.*

(a) *If $\sum_{k=0}^{\infty} a_k R^k$ converges, then $\sum_{k=0}^{\infty} a_k (x-a)^k$ converges uniformly on $[a, a+R]$.*

(b) *If $\sum_{k=0}^{\infty} a_k (-R)^k$ converges, then $\sum_{k=0}^{\infty} a_k (x-a)^k$ converges uniformly on $[a-R, a]$.*

Combining Theorems 8.5.8 and 8.5.12, $\sum_{k=0}^{\infty} a_k (x-a)^k$ converges uniformly on $\left[a-R+\varepsilon_1, a+R-\varepsilon_2\right]$ for any $\varepsilon_1, \varepsilon_2 \geq 0$, depending on the convergence of series $\sum_{k=0}^{\infty} a_k (\pm R)^k$, because if $\sum_{k=0}^{\infty} a_k R^k$ converges and $\sum_{k=0}^{\infty} a_k (-R)^k$ diverges, then $\varepsilon_1 > 0$ and $\varepsilon_2 \geq 0$. On the other hand, if $\sum_{k=0}^{\infty} a_k R^k$ diverges and $\sum_{k=0}^{\infty} a_k (-R)^k$ converges, then $\varepsilon_2 > 0$ and $\varepsilon_1 \geq 0$. Also, if both series converge, then both ε_1 and ε_2 are greater than or equal to 0. Proof of Abel's theorem is left as a more challenging exercise. Observe that using Abel's theorem, the conclusion of Theorem 8.5.11 can be extended to give continuity of a power series on an entire interval of convergence, whether or not it includes endpoints $x = a-R$ and $x = a+R$. We state this formally as the following corollary.

COROLLARY 8.5.13. *Suppose that a function f is represented by the power series $\sum_{k=0}^{\infty} a_k (x-a)^k$, which has a radius of convergence $R \in (0,\infty)$. And suppose $\sum_{k=0}^{\infty} a_k (x-a)^k$ converges at $a+R$. Then f and its power series are continuous at $a+R$. Similarly, if $\sum_{k=0}^{\infty} a_k (x-a)^k$ converges at $a-R$, then f and its power series are continuous at $a-R$.*

Remark 8.5.14

(a) Note that the converse of Corollary 8.5.13 is not true. For example, consider the series $\sum_{k=0}^{\infty} (-1)^k x^k$ which converges to $f(x) = \dfrac{1}{1+x}$ on $(-1,1)$. However, $\lim_{x \to 1^-} f(x) = \dfrac{1}{2}$, that is, f, is continuous on $(-1,1]$, but the power series at $x=1$, i.e. $\sum_{k=0}^{\infty}(-1)^k$ diverges. By placing additional conditions, called *Tauberian*[6] *hypothesis*, on the a_ns, one can obtain a converse to the revised theorem. There is a large number of such results known as *Tauberian Theorems*. The simplest of these theorems is given at the end of this section as Theorem 8.5.19.

(b) From Example 8.5.3, part (b), $\sum_{k=0}^{\infty}\frac{(x-1)^k}{2^k(k+1)}$ converges absolutely on $(-1,3)$ and conditionally at $x=-1$, and thus converges on $[-1,3)$. Due to Abel's theorem, for any $\varepsilon\in(0,4)$, this power series converges uniformly on $[-1,3-\varepsilon]$. Hence, $\lim_{x\to a} f(x)=f(a)$ for all $a\in[-1,3)$, that is

$$\lim_{x\to a}\sum_{k=0}^{\infty}\frac{(x-1)^k}{2^k(k+1)}=\sum_{k=0}^{\infty}\frac{(a-1)^k}{2^k(k+1)}=\sum_{k=0}^{\infty}\lim_{x\to a}\frac{(x-1)^k}{2^k(k+1)}.$$

However, by Theorem 8.5.11 and Corollary 8.5.13, the function the series represents is continuous on the entire interval $[-1,3)$. ■

Remark 8.5.15. Due to Theorem 8.4.15, a power series $\sum_{k=0}^{\infty}a_k(x-a)^k$, with radius of convergence R, can be integrated term by term for all x on an interval $[a-r,a+r]$ for any $r\in(0,R)$. See Example 8.5.16. Thus, if b is any fixed value in $(a-R,a+R)$, then there exists $r_0 \in (0,R)$ such that $b \in \left[a-r_0,a+r_0\right] \subset (a-R,a+R)$. Hence,

$$\int_0^b\left(\sum_{k=0}^{\infty}a_k(x-a)^k\right)dx=\sum_{k=0}^{\infty}\left(\int_0^b a_k(x-a)^k\,dx\right)$$

for all $x\in(a-R,a+R)$. See Exercise 16 of Section 8.4. ■

Example 8.5.16. Recall that $\frac{1}{1-x}=\sum_{k=0}^{\infty}x^k$, if $|x|<1$. Upon term-by-term integration, we have that $-\ln(1-x)=\sum_{k=1}^{\infty}\frac{x^k}{k}$, if $|x|<1$. Both of the preceding power series have a radius of convergence $R=1$. And, since the second series converges when $x=-1$, due to Corollary 8.5.13,

[6]Alfred Tauber (1866–1942), a mathematician from Slovakia, is credited for research on function theory, potential theory, and differential equations. Tauber died in a Nazi concentration camp in Theresienstadt, Germany.

$$-\ln(1-x)=\sum_{k=1}^{\infty}\frac{x^k}{k}, \text{ if } -1\le x<1.$$

So letting $x=-1$, we obtain $\ln 2=\sum_{k=1}^{\infty}\frac{(-1)^{k+1}}{k}$. Compare this expression for $\ln 2$ with the one given in Example 8.4.16. Also see Example 7.5.4. ■

We now turn to differentiation of the series $\sum_{k=0}^{\infty} a_k(x-a)^k$, which represents some function f on $(a-R,a+R)$ with R again as the radius of convergence. In part (h) of Problem 8.8.4 we will prove that if a power series $\sum_{k=0}^{\infty} a_k x^k$ has a radius of convergence R, then the power series $\sum_{k=0}^{\infty}\frac{a_k}{k+1}x^{k+1}$ and $\sum_{k=1}^{\infty} ka_k x^{k-1}$ also have radius of convergence R. Therefore, in differentiating the series $\sum_{k=0}^{\infty} a_k(x-a)^k$ term by term, we obtain

$$\sum_{k=0}^{\infty} ka_k(x-a)^{k-1}=\sum_{k=1}^{\infty} ka_k(x-a)^{k-1},$$

which also has a radius of convergence R and converges uniformly on $[a-r,a+r]$ for any $r\in(0,R)$. Therefore, by Theorem 8.4.17, f is differentiable with its derivative given by $f'(x)=\sum_{k=1}^{\infty} ka_k(x-a)^{k-1}$ for any $x\in(a-R,a+R)$. Using a similar argument,

$$f''(x)=\sum_{k=2}^{\infty} k(k-1)a_k(x-a)^{k-2}$$

for any $x\in(a-R,a+R)$, and so on. This leads us to the conclusion that, if a function f can be represented by the power series $\sum_{k=0}^{\infty} a_k(x-a)^k$ on an interval $(a-R,a+R)$, then f has derivatives of all orders. The converse, however, is not true. Functions do exist that have derivatives of all orders but no power series representation. See Example 8.6.2 and Exercise 9 in Section 8.6.

Moving in a somewhat different direction, consider again

$$\frac{1}{1-x}=\sum_{k=0}^{\infty} x^k=1+x+x^2+\cdots, \text{ with } |x|<1.$$

Antiderivatives can be found term by term to obtain

$$-\ln(1-x) = C + x + \frac{x^2}{2} + \frac{x^3}{3} + \cdots, \text{ with } |x| < 1.$$

The constant of integration C is then computed by choosing any convenient $x \in (-1,1)$, say, $x = 0$. Then, $C = 0$ and we obtain the familiar formula given in Example 8.5.16.

As seen with the number $\ln 2$, real numbers often have more than one series representation. Functions, as we will see, have at most one power series.

THEOREM 8.5.17. *Suppose that one function f is represented by two power series $\sum_{k=0}^{\infty} a_k(x-a)^k$ and $\sum_{k=0}^{\infty} b_k(x-a)^k$, which converge on some interval $(a-K, a+K)$ for some $K > 0$. That is, suppose that*

$$\sum_{k=0}^{\infty} a_k(x-a)^k = \sum_{k=0}^{\infty} b_k(x-a)^k$$

on some interval $(a-K, a+K)$ with $K \in \Re^+$. Then $a_n = b_n$, for all $n = 0,1,2, \ldots$.

Proof. Let $f(x) = \sum_{k=0}^{\infty} a_k(x-a)^k$ and $g(x) = \sum_{k=0}^{\infty} b_k(x-a)^k$. Then $f(a) = a_0$ and $g(a) = b_0$. But $f(x) = g(x)$, for all x in the domain $(a-K, a+K)$. In particular, $f(a) = g(a)$, yielding $a_0 = b_0$. Next, $a_1 = f'(a)$ and $b_1 = g'(a)$. Why? Thus, here also, $a_1 = b_1$. Continuing this process leads to the desired result. ■

Note that by mathematical induction, we can prove that $a_n = \dfrac{f^{(n)}(a)}{n!}$ for all $n = 0, 1, 2, \ldots$. This expression could have been used in the preceding proof. (See Theorem 8.6.1.)

Remark 8.5.18. If R is the radius of convergence for a power series, then R is also the radius of convergence for its term-by-term derivative and antiderivative. However, differentiation may destroy convergence at the endpoints. (See Exercise 12.) Also, if the series is not a power series, term-by-term differentiation might not work. For example, $\sum_{k=1}^{\infty} \frac{\sin(k!x)}{k^2}$ converges for every value of x, but if we differentiate this series term by term we will obtain a series that diverges for each value of x. Why? (See Remark 8.4.19.) ■

THEOREM 8.5.19. *(Tauber's First Theorem of 1897) If the power series $\sum_{k=0}^{\infty} a_k x^k = f(x)$ for $|x| < 1$, $\lim_{n\to\infty} na_n = 0$, and $\lim_{x\to 1^-} f(x) = A$, then $\sum_{k=0}^{\infty} a_k = A$.*

Exercises 8.5

1. Suppose $\left|a_n(x-a)^n\right| \le \left|b_n(x-a)^n\right|$ for all $n = 0, 1, 2, \ldots$ and $\sum_{k=0}^{\infty}\left|b_k(x-a)^k\right|$ converges for all $x \in I$.

   (a) Is it true that $\sum_{k=0}^{\infty}\left|a_k(x-a)^k\right|$ converges for all $x \in I$? Explain.

   (b) If $\sum_{k=0}^{\infty} b_k(x-a)^k$ has radius of convergence R, is it true that $\sum_{k=0}^{\infty} a_k(x-a)^k$ has radius of convergence R? Explain.

2. (a) Give an example of a power series centered at $x = a$ that converges on $[a-R, a+R]$ and diverges otherwise.

   (b) Give an example of a power series centered at $x = a$ that converges on $(a-R, a+R]$ and diverges otherwise.

   (c) Give an example of a power series that converges absolutely on $[a-R, a+R]$ and diverges otherwise.

   (d) Write the polynomial $p(x) = x^2 - 3x + 5$ as a power series centered at $x = 1$.

3. Find both the radius and interval of convergence for the given series. Also, determine the values of x for which the series converges absolutely and those for which the series converges conditionally.

   (a) $\sum_{k=1}^{\infty} k(x-2)^k$

   (b) $\sum_{k=1}^{\infty} \frac{1}{k}\left(\frac{x}{2}\right)^k$

   (c) $\sum_{k=0}^{\infty} k\left(-\frac{1}{3}\right)^k (x-2)^k$

   (d) $\sum_{k=1}^{\infty} \frac{2^k k!}{k^k} x^k$ (See Exercise 8 of Section 7.3.)

   (e) $\sum_{k=0}^{\infty} \frac{k}{3^{2k-1}}(x-1)^{2k}$

   (f) $\sum_{k=1}^{\infty} \frac{(x-1)^k}{k4^k}$

   (g) $\sum_{k=2}^{\infty} \frac{x^k}{(\ln k)^k}$

   (h) $\sum_{k=1}^{\infty} (-1)^k \frac{1 \cdot 3 \cdot 5 \cdot \cdots \cdot (2k-1)}{3 \cdot 6 \cdot 9 \cdot \cdots \cdot (3k)} x^k$

4. Prove Corollary 8.5.6.

5. Prove Theorem 8.5.8, part (b).

6. Prove Theorem 8.5.9.

7. Prove that $\sum_{k=1}^{\infty} a_k x^k$, with

$$a_n = \begin{cases} 1, & \text{if } n \text{ is odd} \\ \dfrac{1}{n}, & \text{if } n \text{ is even} \end{cases}$$

has a radius of convergence $R = 1$, and that $\lim_{n\to\infty}\left|\frac{a_{n+1}}{a_n}\right|$ does not exist.

8. Prove Theorem 8.5.11.

9. If R is the radius of convergence for $\sum_{k=0}^{\infty} a_k x^k$, then for what values does the series $\sum_{k=0}^{\infty} a_k x^{-k}$ converge? Explain.

10. Suppose that the series $\sum a_k$ of real numbers converges conditionally. Prove that the power series $\sum_{k=1}^{\infty} a_k x^k$ has a radius of convergence $R = 1$. Compare to Exercise 21 in Section 8.4.

11. **(a)** Show that

$$\sum_{k=0}^{\infty}(-1)^k \frac{x^{2k+1}}{2k+1} \quad \text{for } |x| \le 1$$

is a power series representation for $\arctan x$. Compare this representation with the one given in Exercise 12(d) of Section 8.4. Also, recall that the domain of $\arctan x$ is $\Re$.

(b) What do we acquire when $x = 1$ in part (a)?

(c) Use part (a) to show that

$$\frac{\pi}{6} = \sum_{k=0}^{\infty}\left(-\frac{1}{3}\right)^k \frac{1}{\sqrt{3}(2k+1)}.$$

12. **(a)** Show that $f(x) = \sum_{k=2}^{\infty} \frac{x^k}{k(k-1)}$ has a radius of convergence $R = 1$ and an interval of convergence $[-1, 1]$. Find all x for which this series converges uniformly.

(b) Show that $f'(x) = \sum_{k=1}^{\infty} \frac{x^k}{k}$ has a radius of convergence $R = 1$ and an interval of convergence $[-1, 1)$. Find all x for which this series converges uniformly.

(c) Show that $f''(x) = \sum_{k=0}^{\infty} x^k$ has a radius of convergence $R = 1$ and an interval of convergence $(-1, 1)$. Graph $f''(x)$. Does this series converge uniformly on $(-1, 1)$? See Exercise 15 of Section 7.1, Exercise 1(b) of Section 8.2, and Exercise 1(j) of Section 8.4.

13. **(a)** Find the function whose power series representation is given by $\sum_{k=1}^{\infty} kx^k$, for $|x|<1$. See Exercise 9 of Section 7.3.

(b) Find the sum of $\sum_{k=1}^{\infty} \frac{k}{2^k}$.

(c) Find the function whose power series representation is given by $\sum_{k=2}^{\infty} k(k-1)x^k$ for $|x|<1$. Compare with Exercise 12(a) of this section.

(d) Find the sum of $\sum_{k=2}^{\infty} \frac{k^2-k}{2^k}$. **(e)** Find the sum of $\sum_{k=1}^{\infty} \frac{k^2}{2^k}$.

14. Find a power series representation for the function

(a) $\ln x$ and $\frac{1}{x}$ centered at $x=a$, with a any positive real number.

(b) $\ln(x+1)$ centered at $x=0$. See Exercise 12(e) of Section 8.4. Does the interval of convergence include $x=1$, thus enabling us to calculate $\ln 2$? Explain.

(c) $\frac{x}{x^2+1}$ centered at $x=0$. Does the interval of convergence include $x=1$? See Exercise 12(c) of Section 8.4.

(d) $2\tanh^{-1} x \equiv \ln\frac{1+x}{1-x}$ centered at $x=0$. See Example 3.5.20, part (e).

15. Suppose that J_p is the Bessel function of order p, defined in Remark 6.6.12.

(a) What is the domain of J_0? See Exercise 20 in Section 7.3.

(b) What is the domain of J_1?

(c) Show that if $f(x)=J_0(x)$, then $xf''(x)+f'(x)+xf(x)=0$, that is, $J_0(x)$ solves the differential equation $xy''+y'+xy=0$

(d) Show that if $y=J_1(x)$, then $x^2y''+xy'+(x^2-1)y=0$.

16. Use the definition of J_p as given in Section 6.6 to prove that

$$xJ_n'(x)=nJ_n(x)-xJ_{n+1}(x), \text{ with } n\in N.$$

See Exercise 12(g) of Section 6.6.

17. Prove Abel's theorem, Theorem 8.5.12.

18. **(a)** Use the fact that $\frac{1}{1-x}=\sum_{k=0}^{\infty} x^k$ for $|x|<1$ to find the power series representation for $\frac{x}{1-x}$ with $0<x<1$.

(b) Show that $\sum_{k=0}^{\infty} \frac{x^{2^k}}{1-x^{2^{k+1}}} = \frac{x}{1-x}$ with $0<x<1$.

(c) Show that $-\sum_{k=1}^{\infty} x^{-k} = \frac{1}{1-x}$. For what values of x is this equality valid?

(d) In view of parts (a)–(c), is Theorem 8.5.17 contradicted? Explain.

8.6 Taylor Series

If a function f is represented by a power series centered at $x = a$, with a positive radius of convergence R, then

$$f(x) = \sum_{k=0}^{\infty} a_k (x-a)^k$$

with $x \in (a-R, a+R)$. As pointed out in Section 8.5, such a function f has derivatives of all orders, which are computed by differentiating the function's power series representation term by term. Each resulting series will have the same radius of convergence. Therefore,

$$f'(x) = \sum_{k=1}^{\infty} k a_k (x-a)^{k-1}, \quad f''(x) = \sum_{k=2}^{\infty} k(k-1) a_k (x-a)^{k-2},$$

and, in general, it can be proved that

$$f^{(n)}(x) = \sum_{k=n}^{\infty} k(k-1) \cdots (k-n+1) a_k (x-a)^{k-n}.$$

Evaluating each one of these at its center, which is always in the interval of convergence, we obtain $f(a) = a_0, f'(a) = a_1, f''(a) = 2a_2, \ldots$. And in general, $f^{(n)}(a) = (n!)a_n$, telling us that the coefficients in the power series are $a_n = \frac{f^{(n)}(a)}{n!}$. Placing this coefficient representation into the power series, we have

$$f(x) = \sum_{k=0}^{\infty} \frac{f^{(k)}(a)}{k!}(x-a)^k = f(a) + f'(a)(x-a) + \frac{f''(a)}{2!}(x-a)^2 + \cdots$$

with a radius of convergence R. This resulting series representation for f is called a *Taylor series centered at* $x = a$. When $a = 0$, this series is known as the *Maclaurin series* for the function f. We have just proven Theorem 8.6.1.

THEOREM 8.6.1. *If a function f can be written as*

$$f(x)=\sum_{k=0}^{\infty}a_k(x-a)^k$$

where x is in some open interval about $x=a$, then $a_n=\dfrac{f^{(n)}(a)}{n!}$ for all $n=0,1,2,\ldots$, and

$$f(x)=\sum_{k=0}^{\infty}\frac{f^{(k)}(a)}{k!}(x-a)^k .$$

Recall Taylor's theorem, Theorem 5.3.8, from which the "tail" end of Taylor's series was considered as the remainder

$$R_n(x)=\sum_{k=n+1}^{\infty}\frac{f^{(k)}(a)}{k!}(x-a)^k ,$$

where $R_n(x)$ may be given as a finite term. Both Lagrange and Cauchy forms of the remainder $R_n(x)$ were given in Theorem 5.3.8. And in Exercise 14 of Section 6.4 an integral form was given. Other forms are due to Young,[7] Schlomlich[8] –Roche, etc.

Example 8.6.2. If possible, find a power series representation centered at $x=0$ for the function $f:\Re\to\Re$, defined by

$$f(x)=\begin{cases}\exp\left(-\dfrac{1}{x}\right), \text{ if } x>0\\ 0, \text{ if } x\le 0.\end{cases}$$

Answer. We need to find a Maclaurin series with a positive radius of convergence R whose values are equal to the functional values of f, at least on the interval $I=(-R,R)$. To this end, by Example 5.3.3, f has derivatives of all orders and $f^{(n)}(0)=0$, for each $n=0,1,2,\ldots$. Thus, the coefficients a_n in the series are given by $a_n=\dfrac{f^{(n)}(0)}{n!}=0$, but this gives rise to a power series representing the zero function, which matches with our functional values of f only when $x\le 0$ not for $x>0$. Hence, this function f, which is infinitely differentiable at $x=0$, cannot be represented by a Maclaurin series. ∎

When will an infinitely differentiable function have a power series expansion?

---

[7]William Henry Young (1863–1942), an English mathematician, developed topics in the theory of integration and contributed to the theory of Fourier series, orthogonal series, and functions of several variables.

[8]Oskar Xaver Schlömlich (1823–1901), a German analyst, is recognized for solving differential equations using infinite series and for work with Bessel functions.

THEOREM 8.6.3. *An infinitely differentiable function f in an open interval I of $x = a$ has a Taylor series expansion centered at $x = a$ if and only if $\lim_{n\to\infty} R_n(x) = 0$, for every $x \in I$.*

Proof. Considering Taylor's theorem, Theorem 5.3.8, recall that the nth Taylor polynomial p_n is the nth term of the sequence of partial sums for a Taylor series centered at $x = a$. Moreover, $p_n(x) = f(x) - R_n(x)$. Therefore, the sequence of partial sums converges to f if and only if $\lim_{n\to\infty} p_n(x) = f(x)$, but this is equivalent to saying that $\lim_{n\to\infty} R_n(x) = 0$, for every $x \in I$. ■

Example 8.6.4. If possible, find the Maclaurin series for $f(x) = \sin x$. (See Example 5.3.9.)

Answer. Note first that $f^{(2n)}(x) = (-1)^n \sin x$ and $f^{(2n+1)}(x) = (-1)^n \cos x$ for $n = 0, 1, 2, \ldots$ and any $x \in \Re$. Why? Thus, at $x = a = 0$, the coefficients for the Taylor series are given by $a_{2n} = 0$ and $a_{2n+1} = \dfrac{(-1)^n}{(2n+1)!}$, for $n = 0, 1, 2, \ldots$, which give rise to the Maclaurin series

$$\sum_{k=0}^{\infty} \frac{(-1)^k}{(2k+1)!} x^{2k+1}.$$

Hence, if $f(x) = \sin x$ has a power series representation about $x = 0$, then it is given by the preceding series. To determine whether such a representation exists, we will compute $\lim_{n\to\infty} R_n(x)$, where $R_n(x)$ is Lagrange's form of the remainder. Since $\left|f^{(n+1)}(x)\right| = |\cos x|$ or $\left|f^{(n+1)}(x)\right| = |\sin x|$, for all n, $\left|f^{(n+1)}(x)\right| \le 1$, for any real x. Therefore, for some real c,

$$0 \le |R_n(x)| = \left|\frac{f^{(n+1)}(c)}{(n+1)!}\right| |x|^{n+1} \le \frac{|x|^{n+1}}{(n+1)!}.$$

However, from Example 8.5.3, part (d), and Theorem 7.1.9, $\lim_{n\to\infty} \dfrac{|x|^{n+1}}{(n+1)!} = 0$. Hence, the previously given series is indeed the power series representation for $f(x) = \sin x$, on all of $\Re$. Why all of $\Re$? ■

Example 8.6.5. Use Theorem 8.6.3 to find a Maclaurin series expansion for $f(x) = e^x$. See Exercise 15 of Section 8.4.

Answer. Since $f^{(n)}(x) = e^x$ for all $n = 0, 1, 2, \ldots$ and $x \in \Re$, the coefficients for the

Maclaurin series are given by $a_n = \frac{1}{n!}$. Therefore, a representation is given by $\sum_{k=0}^{\infty} \frac{1}{k!} x^k$ $= \sum_{k=0}^{\infty} \frac{x^k}{k!}$ wherever $\lim_{n\to\infty} R_n(x) = 0$. Lagrange's form of the remainder is

$$R_n(x) = \frac{e^c}{(n+1)!} x^{n+1}$$

for some c between 0 and x. Since e^c is a real number not depending on n and $\lim_{n\to\infty} \frac{|x|^{n+1}}{(n+1)!} = 0$, we have that $\lim_{n\to\infty} R_n(x) = 0$, for any real x. Hence, $e^x = \sum_{k=0}^{\infty} \frac{x^k}{k!}$ for all $x \in R$. ■

Often, the Taylor series is used to compute derivatives and antiderivatives of functions, evaluating them at certain values to find some nonobvious limits. And in integration, for example, the function $f(x) = \exp\left(-x^2\right)$ has no antiderivative in a "closed" and simple form. Thus, the Taylor series is invoked to compute $\int_0^x \exp\left(-t^2\right) dt$. (See Exercise 2.) The Taylor series can also be used to verify formulas and identities such as $\frac{d}{dx} e^x = e^x$, $\frac{d}{dx} \sin x = \cos x$, $\sin^2 x + \cos^2 x = 1$, and $\sin(x+y) = \sin x \cos y +$ $\cos x \sin y$, or inequalities such as $1 + x \le e^x$, for all $x > 0$. (See Exercises 3 and 7.) Furthermore, power series expansions can be found for functions $\tan x$, $\arcsin x$, and many more. These expansions enable us to find sums of certain series of real numbers. Furthermore, on small intervals, the Taylor series is perfect for finding approximations for functions and their derivatives. (See Section 5.6.) The Taylor series plays a major role in numerical analysis and mathematical physics. And since the concept of the Taylor series representation is so important, a special term for the functions represented by them is used.

Definition 8.6.6. A function f is *analytic* on a neighborhood I of $x = a$ if and only if there exists a power series expansion centered at $x = a$ that converges to f on I.

Exercises 8.6

1. If possible, write a Taylor series expansion for the given functions centered at the given value $x = a$. That is, determine whether or not the given functions are analytic on some open interval about $x = a$.

(a) $f(x) = e^x$ at $x = a \in \Re$ **(b)** $f(x) = \cos x$ at $x = \dfrac{\pi}{4}$

(c) $f(x) = \dfrac{x}{1-x^2}$ at $x = 0$ **(d)** $f(x) = \cos^2 x$ at $x = 0$

(e) $f(x) = \operatorname{sinc} x$ at $x = 0$ (See Exercise 15 from Section 5.4.)

(f) $f(x) = \begin{cases} \dfrac{e^x - 1}{x}, & \text{if } x \neq 0 \\ 1, & \text{if } x = 0 \end{cases}$ at $x = 0$ (The function $\dfrac{1}{f}$ is of special interest since it gives rise to *Bernoulli's numbers*.[9])

(g) $f(x) = \exp\left(-x^2\right)$ at $x = 0$

(h) $f(x) = \ln x$ at $x = 1$ (See Exercise 14(a) in Section 8.5.)

(i) $f(x) = |x|$ at $x = 0$

(j) $f(x) = \cosh x$ at $x = 0$[10] (The function $\dfrac{1}{f}$ is of special interest since it gives rise to *Euler's numbers*.)

2. Use the Maclaurin series to compute

(a) $\displaystyle\int_0^x \exp\left(-t^2\right) dt$. (See Exercise 1(g) of this section and Exercise 10(a) of Section 6.5.)

(b) the antiderivative of $f(x) = \exp\left(-x^2\right)$.

(c) $\displaystyle\int_0^x \sin\left(t^2\right) dt$. (See the footnote to Exercise 47 in Section 6.7.)

3. Use the Taylor series to show that

(a) $\left(e^x\right)^{-1} = e^{-x}$. **(b)** $2 \sin x \cos x = \sin 2x$.

---

[9]Jacob Bernoulli (1654–1705), also known as Jacques, was a celebrated Swiss mathematician, astronomer, and Johann Bernoulli's older brother. Jacob discovered the *isochrone*, the curve along which a particle travels with uniform vertical velocity, studied the catenary, the curve of a suspended string (see the footnote for Exercise 1(k)), was one of the first to use polar coordinates, and in 1690 was the first to use the term integral. The Bernoulli numbers, the Bernoulli differential equation, and the Bernoulli distribution in probability theory are all named after Jacob Bernoulli.

[10]A function of the type $y = a \cosh \dfrac{x}{a}$, the curve of a suspended string, is called a *catenary*, the name given by Huygens in 1690. The hyperbolic function $y = \cosh x$ is the average of the curves e^x and e^{-x}. The catenary curve was first studied by Leibniz, Huygens, Gregory, Johann Bernoulli, and Jacob Bernoulli. (See Euler's numbers in Part 3 of Section 8.10 of *Instructor's Supplement*.) Christiaan Huygens (1629–1695), a mathematician from the Netherlands, is best known for his pendulum clock, which greatly increased the accuracy of time measurement.

4. Use Taylor series to find values of given expressions.

(a) $\sum_{k=1}^{\infty}\frac{k}{(k+1)!}$ (See Exercise 1(f).) **(b)** $\sum_{k=1}^{\infty}\frac{1}{2k^2-k}$

(c) $\frac{1}{2}-\frac{1}{2^3}+\frac{1}{2^5}-\frac{1}{2^7}+\cdots$

(d) $\tanh^{-1}\left(\frac{1}{3}\right)$ (See Exercise 14(d) of Section 8.5.)

5. **(a)** Suppose that $f(x)=\arctan x$. Find $f^{(40)}(0)$ and $f^{(5)}(0)$.

(b) Suppose that $f(x)=\sum_{k=0}^{\infty}a_k x^k$, for all $x\in(-R,R)$. If f is an even function, prove that $a_{2n+1}=0$ for all $n\in N$. If f is odd, prove that $a_{2n}=0$ for all $n=0,1,2,\ldots$.

(c) If $f(x)=\sum_{k=0}^{\infty}\frac{2^{-k}}{k+1}(x-1)^k$, find the Taylor series for f centered at $x=1$.

6. Use the Taylor series to compute the given limits.

(a) $\lim_{x\to 0}\frac{e^x-1}{x}$ (See Exercise 1(f).)

(b) $\lim_{x\to 0}\operatorname{sinc} x$ (See Exercise 15 of Section 5.4.)

(c) $\lim_{x\to\infty} x\sin\frac{1}{x}$ (See Example 3.3.8.)

(d) $\lim_{x\to 1}\frac{\ln x}{x-1}$

(e) $\lim_{x\to 0}\left(\frac{1}{\sin x}-\frac{1}{x}\right)$

7. Use the Taylor series to prove the given inequalities. (See Example 5.3.10.)

(a) $1+x\le e^x$ for all $x\ge 0$ (See Exercise 20(b) of Section 5.2.)

(b) $x^n<(n!)e^x$ for all $x\ge 0$ (See Exercise 19(c) of Section 6.6.)

(c) $1-x\le e^{-x}$ for all $x\ge 0$ (See Exercise 20(c) of Section 5.2.)

(d) $\frac{x}{1+x^2}<\arctan x<x$ for all $x\in[0,1)$ (See Exercise 20(d) of Section 5.2.)

(e) $\ln(x+1)\le x$ for all $x\in[0,1]$ (See Exercise 20(f) of Section 5.2.)

8. Use the Maclaurin series to approximate e to four-decimal-place accuracy. (See Example 5.3.9.)

9. Consider Cauchy's function, as defined in Exercise 6 of Section 5.3.

(a) Show that f is infinitely differentiable for all $x\in\Re$.

(b) Show that $f^{(n)}(0)=0$ for all $n=0,1,2,\ldots$.

(c) Compute the Taylor series $F(x)=\sum_{k=0}^{\infty}\frac{f^{(k)}(0)}{k!}x^k$. Then find its radius of convergence.

(d) Conclude that $f(x)=F(x)$ only when $x=0$.

(e) Explain why Theorem 8.6.3 is not contradicted.

10. If an *imaginary (complex) number* i is defined by $i=\sqrt{-1}$, and if the Maclaurin series expansions for $\sin x$ and $\cos x$ are true for the case in which $x=it$, with t any real number, show that *Euler's formula* given by

$$e^{it}=\cos t+i\sin t$$

is true for all real t. Also, use Euler's formula to prove *De Moivre's formula.* That is, for $n\in Z$, prove that

$$(\cos t+i\sin t)^n=\cos nt+i\sin nt.$$

(Recall De Moivre's footnote in Section 1.10.)

11. Using Exercise 1(e) of this section, show that if $x>1$, then $\mathcal{L}\,\{\operatorname{sinc} t\}=\arctan\frac{1}{x}$. See Exercise 21(e) in Section 6.6 and Problem 6.8.6, part (g).

12. (a) Use Exercise 41 of Section 6.7 or Exercise 4(c) of Section 6.6 to prove that

$$J_{\frac{1}{2}}(x)=\sqrt{\frac{2}{\pi x}}\sin x,$$

where $J_p(x)$ is defined in Section 6.6. See Exercise 14(f) of Section 6.6.

(b) Use Exercise 12(a) of Section 6.6 to prove that

$$J_{-\frac{1}{2}}(x)=\sqrt{\frac{2}{\pi x}}\cos x.$$

(c) Use Exercise 12(f) of Section 6.6 to show that

$$J_{\frac{3}{2}}(x)=\sqrt{\frac{2}{\pi x}}\left(\frac{\sin x}{x}-\cos x\right).$$

See Exercise 14(f) of Section 6.6.

13. (a) In Remark 7.1.15, part (d), we used the long division to obtain the power series. Use the same idea with power series representations for $\sin x$ and $\cos x$ to obtain a power series representation for $\tan x$. Find the interval of convergence of the resulting series.

(b) Evaluate $\lim_{x\to 0}\left(\frac{1}{\sin x}-\frac{1}{x}\right)$ by employing power series division. See Exercise 6(e).

(c) Use computations from part (b) to verify that if $|x|$ is small, then $\frac{1}{\sin x} - \frac{1}{x} \approx \frac{x}{6}$, which in turn gives $\csc x \approx \frac{1}{x} + \frac{x}{6}$.

8.7* Review

Label each statement as true or false. If a statement is true, prove it. If not,
(i) give an example of why it is false, and
(ii) if possible, correct it to make it true, and then prove it.

1. Suppose $\{f_n\}$ is a sequence of functions with $f_n : [0,1] \to \Re$ and $f_n(x) = n^k x^n (1-x)$, k any real constant. Then $\{f_n\}$ converges pointwise to the zero function. See Exercise 1(k) of Section 8.1.

2. If a sequence of functions $\{f_n\}$ converges uniformly on domains D_1 and D_2, then it converges uniformly on $D_1 \cup D_2$.

3. If a sequence of functions $\{f_n\}$ converges uniformly to a continuous function f on D, then f is uniformly continuous.

4. Functions $\{f_n\}$ have to be uniformly continuous on D to converge uniformly on D.

5. There exists a sequence $\{f_n\}$ of everywhere discontinuous functions converging uniformly to an everywhere continuous function f.

6. There exists a sequence of functions $\{f_n\}$ that converges uniformly to 0 on $[0,\infty)$, but $\int_0^\infty f_n(x)\,dx$ does not converge to 0.

7. There exists a sequence of functions $\{f_n\}$ that converges uniformly to 0 on $[0,\infty)$, but $\int_0^\infty f_n(x)\,dx$ diverges to $+\infty$.

8. If $\sum a_k$ is absolutely convergent, then $\sum_{k=0}^{\infty} a_k x^k$ is absolutely convergent on $[-1,1]$.

9. If the interval of convergence for $\sum_{k=0}^{\infty} a_k x^k$ is $(-R,R]$, then this series converges conditionally at $x = R$.

10. If $\{x_n\}$ is any sequence converging to some point in a domain D, and if $\{f_n\}$ is a sequence of functions with $f_n : D \to \Re$ and $\{f_n(x_n)\}$ converging to 0, then $\{f_n\}$ converges uniformly on D. (See Exercise 6 of Section 8.3.)

11. If $\sum f_k$ is a series of continuous functions converging to a continuous function on D, then the convergence is uniform.

12. If $\sum f_k$ is absolutely uniformly convergent on some domain D, then $\sum M_k$ with $M_n = \sup_{x \in D} f_n(x)$ converges.

13. If $\sum f_k$ is uniformly convergent on D, then $\{f_n\}$ converges uniformly to 0 on D.

14. If a sequence of functions $\{f_n\}$ converges uniformly to 0 on D, then $\sum f_k$ converges uniformly on D.

15. If a sequence $\{a_n\}_{n=0}^{\infty}$ is bounded and $\sum_{k=0}^{\infty} a_k$ diverges, then the power series $\sum_{k=0}^{\infty} a_k x^k$ has a radius of convergence $R = 1$.

16. The radius of convergence for the power series $\sum_{k=0}^{\infty} (\sin k) x^k$ is $R = 1$.

17. $\sum \left(\frac{\ln x}{x} \right)^k$ converges uniformly on $[1, \infty)$.

18. $\sum \frac{kx^2}{k^3 + x^3}$ converges uniformly on $[0, a]$ with $a \in \Re^+$.

19. $\sum (x \ln x)^k$ converges uniformly on $(0,1]$.

20. If a function f has bounded derivatives of all orders on an interval $(-r, r)$, then f is analytic on $(-r, r)$.

21. The function $f(x) = 1 + x + x^2 + \cdots$ satisfies the equation $f'(x) = [f(x)]^2$ for all $x \in (-1, 1)$.

22. If the series $\sum_{k=0}^{\infty} a_k x^k$ diverges when $x = \frac{1}{3}$, then it also diverges when $x = -\frac{2}{3}$.

23. If a function $f : \Re \to \Re$ is uniformly continuous and if for each $n \in N$ the functions f_n are defined by $f_n(x) = f\left(x + \frac{1}{n} \right)$, then the sequence $\{f_n\}$ converges uniformly to f on $\Re$.

24. The sequence $\{f_n\}$ with $f_n : [0,1] \to \Re$ and $f_n(x) = \frac{1}{1 + x^{2n}}$ converges pointwise to $f(x) = 1$.

25. The sequence $\{f_n\}$ converges uniformly on a domain D only if the functions f_n are bounded on D.

26. If $f(x) = \lim\limits_{n\to\infty} \dfrac{1}{1+n^2 x}$, then $f(x) = \begin{cases} 0, & \text{if } x \neq 0 \\ 1, & \text{if } x = 0. \end{cases}$

27. The sequence of functions $\{f_n\}$ with $f_n : \Re^+ \to \Re$ and $f_n(x) = \dfrac{nx}{1+nx}$, converges uniformly to 1.

28. If the power series $\sum\limits_{k=0}^{\infty} a_k (x-3)^k$ converges when $x = 7$, then its radius of convergence $R \geq 7$.

29. If the power series $\sum\limits_{k=0}^{\infty} a_k (x-3)^k$ diverges when $x = 7$, then its radius of convergence $R \leq 4$.

30. Suppose that a sequence of functions $\{f_n\}$ converges pointwise to a function f on $[a,b]$. If

$$\lim_{n\to\infty} \int_a^b f_n(x)\,dx \neq \int_a^b f\,dx,$$

then $\{f_n\}$ does not converge uniformly to f on $[a,b]$.

31. Suppose that a sequence of functions $\{f_n\}$ converges pointwise to f on $[a,b]$. If

$$\lim_{n\to\infty} \int_a^b f_n(x)\,dx = \int_a^b f\,dx$$

then $\{f_n\}$ converges uniformly to f on $[a,b]$.

32. If $\{f_n\}$ is a sequence of uniformly continuous functions converging uniformly to a function f on a domain D, then f is uniformly continuous on D.

33. If $|x| > 1$, then $\sum\limits_{k=1}^{\infty} k^{-x^2}$ converges pointwise.

34. If $\sum f_k(x)$ converges uniformly on a domain D and $g : D \to \Re$ is a bounded function, then $\sum g(x) f_k(x)$ converges uniformly on D.

35. If $f_n : (0,1) \to \Re$ are defined by $f_n(x) = \dfrac{x^n}{1+x^n}$, then the sequence $\{f_n\}$ converges uniformly. (See Exercise 1(i) in Section 8.1.)

36. If the sequence $\{f_n\}$ is uniformly bounded on D, then $\{f_n\}$ converges uniformly on D.

37. If the sequence $\{f_n\}$ converges uniformly on D, then $\{f_n\}$ is uniformly bounded on D.

38. Every pointwise convergent sequence of functions contains a uniformly convergent subsequence.

39. If $\{f_n\}$ is a sequence of continuous and decreasing functions that converges pointwise, then this convergence is uniform.

40. The Maclaurin series for the Fresnel cosine integral $C(x)$ is given by $x - \frac{x^5}{5 \cdot 2!} + \frac{x^9}{9 \cdot 4!} - \frac{x^{13}}{13 \cdot 6!} + \cdots$, with the interval of convergence being $(-\infty, \infty)$. (See the footnote to Exercise 47 in Section 6.7.)

41. If $\sum_{k=0}^{\infty} a_k x^k$ and $\sum_{k=0}^{\infty} b_k x^k$ have as their radius of convergence R_1 and R_2, respectively, then $\sum_{k=0}^{\infty} c_k x^k = \left(\sum_{k=0}^{\infty} a_k x^k\right)\left(\sum_{k=0}^{\infty} b_k x^k\right)$, for $c_k = \sum_{i=0}^{k} a_i b_{k-i}$ and $|x| < \min\{R_1, R_2\}$. (See Part 2 of Section 7.6.)

42. Series $\sum_{k=1}^{\infty} \frac{x^k}{k^2 R^k}$, $\sum_{k=1}^{\infty} \frac{x^k}{kR^k}$, $\sum_{k=1}^{\infty} \frac{(-x)^k}{kR^k}$, and $\sum_{k=0}^{\infty} \frac{x^k}{R^k}$ have interval of convergence $[-R, R]$, $[-R, R)$, $(-R, R]$, and $(-R, R)$, respectively.

43. Since $f(x) = \frac{1}{1+x}$ is continuous for all $x \neq 1$ and $\frac{1}{1+x} = \sum_{k=0}^{\infty} (-1)^k x^k$ on $(-1, 1)$, we can conclude that $\lim_{x \to 1^-} \sum_{k=0}^{\infty} (-1)^k x^k = \sum_{k=0}^{\infty} \lim_{x \to 1^-} (-1)^k x^k$.

44. $\sum_{k=0}^{\infty} (-1)^k (1-x) x^k$ converges uniformly on $[0, 1]$, but $\sum_{k=0}^{\infty} (1-x) x^k$ converges only pointwise on $[0, 1]$.

45. If the sequence $\{f_n\}$ converges uniformly to f on D, then $\{(f_n)^2\}$ converges uniformly to f^2 on D.

8.8* Projects

Part 1. Limit Superior

Often convergence of a sequence is an unnecessarily restrictive assumption. Many results are obtained by using weaker assumptions such as those of limit superior or limit inferior. These concepts were introduced in Section 2.6. A few examples and properties follow.

Problem 8.8.1. Prove that the given statements are true concerning the two sequences of real numbers $\{a_n\}$ and $\{b_n\}$.

(a) If $a_n = (-1)^n$, then $\liminf_{n\to\infty} a_n = -1$ and $\limsup_{n\to\infty} a_n = 1$.

(b) If $a_n = (-1)^n n$, then $\liminf_{n\to\infty} a_n = -\infty$ and $\limsup_{n\to\infty} a_n = +\infty$.

(c) If $a_n = n^2 \sin^2\left(\frac{n\pi}{2}\right)$, then $\liminf_{n\to\infty} a_n = 0$ and $\limsup_{n\to\infty} a_n = +\infty$.

(d) $\liminf_{n\to\infty} a_n = +\infty$ only if $\lim_{n\to\infty} a_n = +\infty$, i.e. only if $\{a_n\}$ diverges to $+\infty$.

(e) A sequence $\{a_n\}$ converges to A if and only if $\liminf_{n\to\infty} a_n = \limsup_{n\to\infty} a_n = A$.

(f) If $\liminf_{n\to\infty} a_n < \limsup_{n\to\infty} a_n$, then $\{a_n\}$ oscillates.

(g) If $\{a_n\}$ is unbounded above, then $\limsup_{n\to\infty} a_n = +\infty$.

(h) If $\{a_n\}$ is unbounded below, then $\liminf_{n\to\infty} a_n = -\infty$.

(i) $\liminf_{n\to\infty} a_n \le \limsup_{n\to\infty} a_n$.

(j) $\limsup_{n\to\infty}(a_n + b_n) \le \limsup_{n\to\infty} a_n + \limsup_{n\to\infty} b_n$. See Exercise 55 of Section 1.8.

(k) $\liminf_{n\to\infty}(a_n + b_n) \ge \liminf_{n\to\infty} a_n + \liminf_{n\to\infty} b_n$.

(l) If $a_n \le b_n$ eventually, then $\limsup_{n\to\infty} a_n \le \limsup_{n\to\infty} b_n$ and $\liminf_{n\to\infty} a_n \le \liminf_{n\to\infty} b_n$.

(m) If a_n and $b_n \ge 0$ and $\lim_{n\to\infty} a_n = L > 0$, then

$$\limsup_{n\to\infty}(a_n b_n) = \left(\limsup_{n\to\infty} a_n\right)\left(\limsup_{n\to\infty} b_n\right).$$

(n) If a_n and $b_n > 0$ and if both $\limsup_{n\to\infty} a_n$ and $\limsup_{n\to\infty} b_n$ are either finite or infinite, then $\limsup_{n\to\infty}(a_n b_n) \le \left(\limsup_{n\to\infty} a_n\right)\left(\limsup_{n\to\infty} b_n\right)$.

(o) If $a_n > 0$, then $\liminf_{n\to\infty} \frac{a_{n+1}}{a_n} \le \liminf_{n\to\infty} \sqrt[n]{a_n} \le \limsup_{n\to\infty} \sqrt[n]{a_n} \le \limsup_{n\to\infty} \frac{a_{n+1}}{a_n}$.

(p) If $c_n = \frac{1}{n}(a_1 + a_2 + \cdots + a_n)$, then $\liminf_{n\to\infty} a_n \le \liminf_{n\to\infty} c_n \le \limsup_{n\to\infty} c_n \le \limsup_{n\to\infty} a_n$.

(q) *(Ratio Test)* If $\sum a_k$ is a series of nonzero terms, then $\limsup_{n\to\infty}\left|\frac{a_{n+1}}{a_n}\right| < 1$ implies that $\sum a_k$ converges absolutely. In addition, $\liminf_{n\to\infty}\left|\frac{a_{n+1}}{a_n}\right| > 1$ implies that $\sum a_k$ diverges.

(r) *(Root Test)* If $\alpha = \limsup_{n\to\infty} \sqrt[n]{|a_n|}$, then $\alpha < 1$ implies that $\sum a_k$ converges absolutely, and if $\alpha > 1$, then $\sum a_k$ diverges.

THEOREM 8.8.2. *(Cauchy–Hadamard Theorem) Suppose that* $\sum_{k=0}^{\infty} a_k x^k$ *is any power series,* $\alpha = \limsup_{n\to\infty} \sqrt[n]{|a_n|}$, *and* R *is the radius of convergence.*

(a) *If* $\alpha = +\infty$, *then* $R = 0$.

(b) *If* $\alpha = 0$, *then* $R = \infty$.

(c) *If* $\alpha \in (0, \infty)$, *then* $R = \dfrac{1}{\alpha}$.

Proof of this result is straightforward using the root test from Problem 8.8.1, part (r).

THEOREM 8.8.3. *(Abel's Limit Theorem) Suppose that* $r > 0$ *and* $\sum_{k=0}^{\infty} a_k r^k$ *converges, then* $\sum_{k=0}^{\infty} a_k x^k$ *converges absolutely for* $|x| < r$, *and* $\lim_{x\to r^-} \sum_{k=0}^{\infty} a_k x^k = \sum_{k=0}^{\infty} a_k r^k$.

Compare this result with Theorem 8.5.12.

Problem 8.8.4. Prove the following results using straightforward methods from Theorem 8.8.2 and material from Sections 8.4 and 8.5.

(a) If the power series $\sum_{k=1}^{\infty} a_k x^k$ has infinitely many coefficients a_n that are nonzero integers and R is the radius of convergence, then $R \le 1$.

(b) If sequence $\{a_n\}$ consists of the terms $2, 3, 2^2, 3^2, 2^3, 3^3, \ldots$, then $\sum_{k=1}^{\infty} a_k x^k$ has a radius of convergence $R = \dfrac{1}{\sqrt{3}}$. See part (a).

(c) If a sequence $\{a_n\}$ consists of the terms $\dfrac{1}{2}, \dfrac{1}{3}, \dfrac{1}{2^2}, \dfrac{1}{3^2}, \dfrac{1}{2^3}, \dfrac{1}{3^3}, \ldots$, then $\sum_{k=1}^{\infty} a_k x^k$ has a radius of convergence $R = \sqrt{2}$. See Example 7.3.10.

(d) If $\lim_{n\to\infty} a_n = 0$ and $|x| < 1$, then $\sum_{k=1}^{\infty} a_k x^k$ converges.

(e) If a sequence $\{a_n\}$ is bounded, $\sum a_k$ diverges and $|x| < 1$, then $\sum_{k=1}^{\infty} a_k x^k$ converges. See Exercise 7 of Section 8.5.

(f) If $\sum a_k x^k$ has a radius of convergence R, then $\sum (a_k)^2 x^k$ has a radius of convergence R^2 and $\sum_{k=1}^{\infty} a_k x^{2k}$ has a radius of convergence $\sqrt{R}$. See Exercises 13 and 15 of Section 8.4.

(g) Show that $1 - \frac{1}{4} + \frac{1}{7} - \frac{1}{10} + \cdots = \frac{1}{3}\left(\ln 2 + \frac{\pi}{\sqrt{3}}\right)$.

(h) Prove that if $\sum_{k=0}^{\infty} a_k x^k$ has a radius of convergence R, then both $\sum_{k=0}^{\infty} \frac{a_k}{k+1} x^{k+1}$ and $\sum_{k=1}^{\infty} k a_k x^{k-1}$ do also.

Part 2. Irrationality of e

THEOREM 8.8.5. *The number e is irrational.*

Proof. Suppose to the contrary that $e = \frac{p}{q}$, where p and q are integers with $q > 1$. Thus, $\frac{p}{q} = \sum_{k=1}^{\infty} \frac{1}{k!}$. Why? Hence, $p(q-1)! = (q!)\sum_{k=1}^{q} \frac{1}{k!} + (q!)\sum_{k=q+1}^{\infty} \frac{1}{k!}$. Why? Therefore,

$$0 < p(q-1)! - (q!)\sum_{k=1}^{q} \frac{1}{k!} = \frac{1}{q+1} + \frac{1}{(q+1)(q+2)} + \frac{1}{(q+1)(q+2)(q+3)} + \cdots \le \frac{1}{q} < 1.$$

Why? But both $p(q-1)!$ and $(q!)\sum_{k=1}^{q} \frac{1}{k!}$ are integers. Thus, their difference is an integer and is, in fact, between 0 and 1. Why? Hence, we have a contradiction, and the proof is complete. ■

In 1873, Hermite proved that e is transcendental. The proof goes beyond the scope of this text.

Part 3. An Everywhere Continuous but Nowhere Differentiable Function

It is believed that Bolzano knew of functions that are continuous everywhere but nowhere differentiable. But not until 40 years later, in 1872, did Weierstrass announce the findings. We will follow Weierstrass' construction of such a function. See Exercise 8(g), Section 5.1.

THEOREM 8.8.6. *A function exists that is continuous for all* $x \in \Re$, *but not differentiable at any real* x.

Construction. Verify the following steps in detail.

Step 1. Let g be the periodic extension of the function $y = |x|$ with $x \in [-1,1]$ to all of $\Re$. Then $g(x+2) = g(x)$, for all x.

Step 2. For any $s, t \in \Re$, we have $|g(s) - g(t)| \le |s - t|$.

Step 3. For each $n = 0, 1, 2, \ldots$, define functions g_n by $g_n(x) = \left(\frac{3}{4}\right)^n g\left(4^n x\right)$. Graph g_0, g_1 and g_2.

Step 4. Define the function $f: \Re \to \Re$ by $f(x) = \sum_{k=0}^{\infty} g_k(x) = \sum_{k=0}^{\infty} \left(\frac{3}{4}\right)^k g\left(4^k x\right)$. We will show that f is continuous for all $x \in \Re$. To this end, observe that $0 \le g(x) \le 1$. Thus, $0 \le g_n(x) \le \left(\frac{3}{4}\right)^n$ for all $x \in \Re$. Now, using Weierstrass M-test, $\sum_{k=0}^{\infty} g_k(x)$ converges uniformly. Hence, f is continuous on all of $\Re$.

Step 5. We will now prove that f is not differentiable at any $x \in \Re$. We choose an arbitrary x and let m be any natural number. Then, we define a sequence $\{y_m\}$ by $y_m = \dfrac{f(x+h_m) - f(x)}{h_m}$ where $\{h_m\}$, with $h_m \ne x$, converges to 0. We will show that $\{y_m\}$ diverges, and invoke Remark 5.1.10 to obtain the desired conclusion. To this end, for $m = 1, 2, 3, \ldots$ let $h_m = \pm \frac{1}{2}\left(4^{-m}\right)$, where the sign is chosen so that there is no integer between $4^m x$ and $4^m(x + h_m)$. Thus, $f(x+h_m) - f(x) = \sum_{k=0}^{\infty} \left(\frac{3}{4}\right)^k \left[g\left(4^k x + 4^k h_m\right) - g\left(4^k x\right)\right]$.

Step 6. If $n > m$, then $g\left(4^n x + 4^n h_m\right) - g\left(4^n x\right) = 0$. But if $n \le m$, then $\left|g\left(4^n x + 4^n h_m\right) - g\left(4^n x\right)\right| \le 4^n |h_m|$.

Step 7. Hence,

$$|y_m| = \left| \sum_{k=0}^{m} \left(\frac{3}{4}\right)^k \frac{g\left(4^k x + 4^k h_m\right) - g\left(4^k x\right)}{h_m} \right| \ge 3^m - \sum_{k=0}^{m-1} 3^k = \frac{1}{2}\left(3^m + 1\right).$$

And since this expression tends to $+\infty$, f is not differentiable on $\Re$. The construction is complete. ■

Part 4. Equicontinuity

By the Bolzano–Weierstrass theorem for sequences, every bounded sequence of real numbers has a converging subsequence. Is this true for a sequence of functions?

Definition 8.8.7. A sequence of functions $\{f_n\}$, where $f_n: D \to \Re$ with $D \subseteq \Re$, is said to be *equicontinuous* on D if and only if for any $\varepsilon > 0$ there exists $\delta > 0$, such that $|f_n(x) - f_n(t)| < \varepsilon$ whenever $|x - t| < \delta$, for all $x, t \in D$ and any $n \in N$.

In Definition 8.8.7, every f_n is uniformly continuous on D. Moreover, δ is independent of the choices of x, t, and n.

Problem 8.8.8. Find a sequence of functions $\{f_n\}$, where

(a) $\{f_n\}$ is uniformly bounded but not equicontinuous.

(b) the functions f_n are continuous, $\{f_n\}$ converges uniformly to f on an open interval I, and $\lim_{n\to\infty} f_n(x_n) = f(x)$, for any sequence $\{x_n\}$ in D, but $\{f_n\}$ is not equicontinuous on D.

(c) $\{f_n\}$ is equicontinuous on $[a, b]$, but has no uniformly convergent subsequence.

Problem 8.8.9. Prove the given statements.

(a) If $\{f_n\}$ is equicontinuous on a compact set D and converges pointwise to a function f, then $\{f_n\}$ converges uniformly to f on D.

(b) If $\{f_n\}$ is a sequence of continuous functions that converges uniformly on a compact set D, then $\{f_n\}$ is uniformly bounded and equicontinuous on D.

(c) If $\{f_n\}$ is equicontinuous and converges pointwise to a function f on D, then f is uniformly continuous on D.

(d) *(Arzelà–Ascoli[11] Theorem)* If a sequence $\{f_n\}$ is equicontinuous and pointwise bounded on a compact set D, then $\{f_n\}$ is uniformly bounded and contains a uniformly convergent subsequence.

---

[11]Giulio Ascoli (1843–1896), an Italian analyst, taught at Milan and is recognized for contributions in equicontinuity and Fourier series.

9

Fourier Series

9.1* Introduction

In Chapter 8 we associated a power series with a function that has derivatives of all orders at some fixed point. Now we wish to relate some *trigonometric* series, that is, series involving sines and/or cosines with their corresponding piecewise continuous function. Fourier analysis[1] gives an essential background for a number of areas in mathematics, physics, and other sciences. For the sake of terminology we first state the definition of the norm and a few properties involving both sine and cosine functions.

Definition 9.1.1. The *norm of the continuous function f with respect to the weight function w*, where $w(x) > 0$ for all $x \in [a, b]$, is defined by

$$\left(\int_a^b w(x)[f(x)]^2 \, dx \right)^{\frac{1}{2}},$$

[1]The ideas discussed here from Fourier analysis were contained in a paper, which Joseph Fourier submitted for publication to the Academy of Sciences in Paris in 1807. The paper was at first rejected for its lack of rigor. In the 18th century Daniel Bernoulli and Euler had already worked with trigonometric series known today as Fourier series. However, Fourier placed these series on a firmer foundation, thus receiving credit as such.

and denoted by $\|f\|$. See Definition 1.10.24. In the case $w(x) \equiv 1$, the *norm* of f is

$$\|f\| = \left(\int_a^b [f(x)]^2 \, dx \right)^{\frac{1}{2}}.$$

See Definition 8.1.11, part (a) of Remark 8.1.12, and Problem 10.9.5.

Remark 9.1.2. The sequences of functions $\{f_n\}_{n=0}^{\infty}$ and $\{g_n\}_{n=0}^{\infty}$, where $f_n(x) = \cos\frac{n\pi x}{L}$ and $g_n(x) = \sin\frac{n\pi x}{L}$, have interesting properties.

(a) $2L$ is the period for all functions f_n and g_n as well as all their *linear combinations*, that is, all finite sums of multiples of f_n and g_n by a real number. See Exercise 4(c) in Section 4.2.

(b) $\displaystyle\int_{-L}^{L} \cos\frac{n\pi x}{L}\, dx = 0$ and $\displaystyle\int_{-L}^{L} \sin\frac{n\pi x}{L}\, dx = 0$

(c) For any m and n positive integers, $\displaystyle\int_{-L}^{L} \left(\cos\frac{m\pi x}{L}\right)\left(\sin\frac{n\pi x}{L}\right) dx = 0.$

(d) If $m \neq n$, then $\displaystyle\int_{-L}^{L} f_m(x) f_n(x)\, dx = 0$. That is, if $m \neq n$ we have $\displaystyle\int_{-L}^{L} \left(\cos\frac{m\pi x}{L}\right)\left(\cos\frac{n\pi x}{L}\right) dx = 0$. Due to this fact, the sequence $\{f_n\}_{n=0}^{\infty}$ is a sequence of *orthogonal* functions on $[-L, L]$. See Exercise 9.

(e) If $m = n \neq 0$, then $\displaystyle\int_{-L}^{L} \left(\cos\frac{m\pi x}{L}\right)\left(\cos\frac{n\pi x}{L}\right) dx = \int_{-L}^{L} \left(\cos\frac{n\pi x}{L}\right)^2 dx = L$. That is, $\left\|\cos\frac{n\pi x}{L}\right\| = \sqrt{L}$ if $n \neq 0$.

(f) If $m \neq n$, then functions g_n are orthogonal to g_m on $[-L, L]$. That is, $\displaystyle\int_{-L}^{L} \left(\sin\frac{m\pi x}{L}\right)\left(\sin\frac{n\pi x}{L}\right) dx = 0.$

(g) If $m = n \neq 0$, then $\displaystyle\int_{-L}^{L} \left(\sin\frac{m\pi x}{L}\right)\left(\sin\frac{n\pi x}{L}\right) dx = \int_{-L}^{L} \left(\sin\frac{n\pi x}{L}\right)^2 dx = L$. That is, $\left\|\sin\frac{n\pi x}{L}\right\| = \sqrt{L}$ if $n \neq 0$. ■

Results in parts (b)–(g) remain unaltered even when the limits of integration, $-L$ and L, are changed to a and $a+2L$, respectively, for any $a \in \Re$. See Exercise 3.

THEOREM 9.1.3. *If the series*

$$A+\sum_{k=1}^{\infty}\left(a_k \cos\frac{k\pi x}{L}+b_k \sin\frac{k\pi x}{L}\right)$$

converges uniformly to an integrable function f on the interval $(-L, L)$, then we have

(a) $A=\dfrac{a_0}{2}$, *where* $a_0=\dfrac{1}{L}\displaystyle\int_{-L}^{L} f(x)\,dx$,

(b) $a_n=\dfrac{1}{L}\displaystyle\int_{-L}^{L} f(x)\cos\frac{n\pi x}{L}\,dx$ *for* $n \in N$, *and*

(c) $b_n=\dfrac{1}{L}\displaystyle\int_{-L}^{L} f(x)\sin\frac{n\pi x}{L}\,dx$ *for* $n \in N$.

Proof of part (a). Integration with the help of Remark 9.1.2 part (b) gives

$$\begin{aligned}\int_{-L}^{L} f(x)\,dx &= \int_{-L}^{L} A\,dx+\int_{-L}^{L}\sum_{k=1}^{\infty}\left(a_k \cos\frac{k\pi x}{L}+b_k \sin\frac{k\pi x}{L}\right)dx \\ &= A\int_{-L}^{L} dx+\sum_{k=1}^{\infty}\left(a_k\int_{-L}^{L}\cos\frac{k\pi x}{L}\,dx+b_k\int_{-L}^{L}\sin\frac{k\pi x}{L}\,dx\right) \\ &= 2LA+\sum_{k=1}^{\infty}\left(a_k\cdot 0+b_k\cdot 0\right)=2LA.\end{aligned}$$

Observe that due to uniform convergence, the interchanging of the summation and integration symbols is valid. Therefore, $A=\dfrac{1}{2L}\displaystyle\int_{-L}^{L} f(x)\,dx$. Thus, if $a_0=\dfrac{1}{L}\displaystyle\int_{-L}^{L} f(x)\,dx$, then $A=\dfrac{a_0}{2}$.

Proof of part (b). Multiply the expression

$$f(x)=A+\sum_{k=1}^{\infty}\left(a_k \cos\frac{k\pi x}{L}+b_k \sin\frac{k\pi x}{L}\right)$$

by $\cos\dfrac{m\pi x}{L}$ with $m \in N$, then integrate from $-L$ to L, and use Remark 9.1.2 to obtain

$$\int_{-L}^{L} f(x)\cos\frac{m\pi x}{L}\,dx =$$
$$= A\int_{-L}^{L}\cos\frac{m\pi x}{L}\,dx + \sum_{k=1}^{\infty}\left(a_k\int_{-L}^{L}\cos\frac{m\pi x}{L}\cos\frac{k\pi x}{L}\,dx + b_k\int_{-L}^{L}\cos\frac{m\pi x}{L}\sin\frac{k\pi x}{L}\,dx\right)$$
$$= a_m L.$$

The desired result follows. ■

Proof of part (c) is left as an exercise.

Definition 9.1.4. If f is Riemann integrable on the interval $(-L, L)$, then the *Fourier coefficients* of f on $(-L, L)$ are the real numbers

$$a_0 = \frac{1}{L}\int_{-L}^{L} f(x)\,dx,$$

$$a_n = \frac{1}{L}\int_{-L}^{L} f(x)\cos\frac{n\pi x}{L}\,dx \text{ for } n \in N, \text{ and}$$

$$b_n = \frac{1}{L}\int_{-L}^{L} f(x)\sin\frac{n\pi x}{L}\,dx \text{ for } n \in N.$$

The *Fourier series* for the function f on $(-L, L)$ is the trigonometric series

$$\frac{a_0}{2} + \sum_{k=1}^{\infty}\left(a_k\cos\frac{k\pi x}{L} + b_k\sin\frac{k\pi x}{L}\right).$$

Remark 9.1.5

(a) In Definition 9.1.4 nothing is said about convergence of the Fourier series. Discussion of convergence will take place later. Definition 9.1.4 only states that the Fourier series for a function f is given by the specified trigonometric series. This is formally written as

$$f(x) \sim \frac{a_0}{2} + \sum_{k=1}^{\infty}\left(a_k\cos\frac{k\pi x}{L} + b_k\sin\frac{k\pi x}{L}\right).$$

The tilde symbol[2] " ~ " cannot be replaced by the equals symbol " = " unless we know that the series converges to $f(x)$ for all x in the domain of f. See Example 9.1.6. The symbol " ~ " is used to remind us as to which function is used in the computation of Fourier coefficients.

---

[2]The use of the tilde notation, since 1903, is due to the mathematician named Hurwitz.

(b) Even if the Fourier series for f does converge, it need not have $f(x)$ for its sum. Two different functions f_1 and f_2 may have identical Fourier coefficients. A simple example is given in Exercise 6.

(c) There may very well be several different trigonometric series corresponding to one function. This will be discussed in depth in Section 9.3.

(d) Convergence of the Fourier series for f need not be uniform. For example, consider the function given in Example 9.1.6 below. Since f is not continuous, by Theorem 8.4.3, convergence to f cannot be uniform.

(e) The value of f at the endpoints $x=-L$ and $x=L$, or the lack of it, does not alter the definition of the Fourier coefficients of f. Hence, the Fourier series for f remains the same. It is its convergence that may be affected. Thus, in Definition 9.1.4 we could have referred to the Fourier series for the function f on $[-L,L]$, $[-L,L)$, or $(-L,L]$.

(f) Choice for the Fourier coefficients given in Definition 9.1.4 was motivated by Theorem 9.1.3. Also, the formula for a_0, in Definition 9.1.4, can be obtained from the formula for a_n by allowing n to be 0. This, however, might not be the case after the integral for a_n is evaluated. See Exercise 5.

(g) Notice that the first term in the Fourier series for f is

$$\frac{a_0}{2}=\frac{1}{2L}\int_{-L}^{L} f(x)\,dx,$$

the average value of $f(x)$ over the length of the period $2L$.

(h) Again, the Fourier coefficients remain unaltered even when the limits of integration $-L$ and L are changed to a and $a+2L$, respectively, for any $a\in\mathfrak{R}$. ■

Example 9.1.6. Compute the Fourier series representation for the function $f:(-\pi,\pi]\to\mathfrak{R}$ defined by $f(x)=\begin{cases}0, \text{ if } -\pi<x<0\\ 1, \text{ if } 0\le x\le\pi.\end{cases}$

Answer. Here, $a_0=\frac{1}{\pi}\int_{-\pi}^{\pi} f(x)dx=\frac{1}{\pi}\int_0^{\pi} dx=1$. If $n\ge 1$,

$$a_n=\frac{1}{\pi}\int_{-\pi}^{\pi} f(x)\cos nx\,dx=\frac{1}{\pi}\int_0^{\pi}\cos nx\,dx=0, \text{ and}$$

$$b_n=\frac{1}{\pi}\int_{-\pi}^{\pi} f(x)\sin nx\,dx=\frac{1}{\pi}\int_0^{\pi}\sin nx\,dx=\frac{1-\cos n\pi}{n\pi}.$$

Therefore, the Fourier series for f is given by

$$\frac{a_0}{2}+\sum_{k=1}^{\infty}\left(a_k\cos\frac{k\pi x}{L}+b_k\sin\frac{k\pi x}{L}\right)$$
$$=\frac{1}{2}+\sum_{k=1}^{\infty}\left(\frac{1-\cos k\pi}{k\pi}\sin kx\right)=\frac{1}{2}+\sum_{k=1}^{\infty}\frac{1}{k\pi}\sin 2kx$$
$$=\frac{1}{2}+\frac{1}{\pi}\sin 2x+\frac{1}{2\pi}\sin 4x+\frac{1}{3\pi}\sin 6x+\cdots.$$

Observe that the value of this series at $x=0$ is $\frac{1}{2}$, although $f(0)=1$. Clearly, this trigonometric series is not equal to $f(x)$, at least not for all values of x. ■

Now that we have a little experience, let us observe the basic differences between Taylor and Fourier series. Taylor coefficients are evaluated at one point, whereas Fourier coefficients are evaluated over an entire interval. Taylor series use derivatives; Fourier series use integrals. Very "smooth" functions may have both series representations, although more general functions may have Fourier series and not Taylor series. Since evaluation of Fourier coefficients depends on the ability to integrate, a review of Chapter 6 and perhaps an acquaintance with the computer would be wise. Often, repeated integration by parts is necessary. See Remark 6.4.5. When performing calculations by hand, applying Kronecker[3] formula often saves time. This formula, which comes from the integration by parts, says that if p is a polynomial of degree n, and if f is a continuous function, then

$$\int p(x)f(x)dx=p(x)F_1(x)-p'(x)F_2(x)+p''(x)F_3(x)-\cdots+(-1)^n p^{(n)}(x)F_{n+1}(x)+C,$$

where C is the constant of integration, F_1 an antiderivative of f, F_2 an antiderivative of F_1, and so on. Differentiate the right hand side of this formula to verify its validity.

Example 9.1.7. Compute the Fourier series for the function $f(x)=x^3$ on $[-\pi,\pi)$.

Answer. Since f is odd, functions $f(x)\cos\frac{n\pi x}{L}$ are odd. Why? Thus, by Exercise 11(a) of Section 6.4 we have that $a_n=0$ for all $n=0,1,2,\ldots$. And, using the Kronecker integration formula, for $n\in N$

$$b_n=\frac{1}{L}\int_{-L}^{L}f(x)\sin\frac{n\pi x}{L}dx=\frac{1}{\pi}\int_{-\pi}^{\pi}x^3\sin nx\,dx=\frac{2}{\pi}\int_{0}^{\pi}x^3\sin nx\,dx$$

---

[3]Leopold Kronecker (1823–1891), a German algebraist and algebraic number theorist, greatly relied on intuition. The function δ_{ij}, defined as 0 if $i\neq j$ and 1 if $i=j$, is named after him and is called *Kronecker's delta*. See Definition 9.5.6.

$$= \frac{2}{\pi}\left[\left(x^3\right)\left(-\frac{\cos nx}{n}\right) - \left(3x^2\right)\left(-\frac{\sin nx}{n^2}\right) + (6x)\left(\frac{\cos nx}{n^3}\right) - (6)\left(\frac{\sin nx}{n^4}\right)\right]\Bigg|_0^{\pi}$$

$$= 2(-1)^{n+1}\frac{n^2\pi^2 - 6}{n^3}.$$

Hence, $f(x) = x^3 \sim 2\sum_{k=1}^{\infty}(-1)^{k+1}\left(\frac{k^2\pi^2 - 6}{k^3}\right)\sin kx$ for $x \in [-\pi, \pi)$. ■

Exercises 9.1

1. When $n = 0$, evaluate $\left\|\cos\frac{n\pi x}{L}\right\|$ on $[-L, L]$.
2. Verify all parts of Remark 9.1.2.
3. For any piecewise continuous function f that is p-periodic, verify that

$$\int_a^{a+p} f(x)\,dx = \int_0^p f(x)\,dx = \int_{-\frac{p}{2}}^{\frac{p}{2}} f(x)\,dx.$$

4. (a) Prove part (c) of Theorem 9.1.3.
   (b) Is the function f in Theorem 9.1.3 periodic? If so, what is its period?
5. For the function f defined as

$$f(x) = \begin{cases} 0, & \text{if } -\pi < x < 0 \\ \pi - x, & \text{if } 0 < x < \pi, \end{cases}$$

compute a_n as given in Definition 9.1.4, and show that it is undefined when $n = 0$. Find a_0 using the formula given in Definition 9.1.4. Also, find the Fourier series for f.

6. Consider the functions $f_1, f_2, f_3 : [-\pi, \pi] \to \Re$, defined by

$$f_1(x) = \begin{cases} x, & \text{if } x \neq 0 \\ 1, & \text{if } x = 0, \end{cases} \quad f_2(x) = \begin{cases} x, & \text{if } x \neq 0 \\ 2, & \text{if } x = 0, \end{cases} \quad f_3(x) = x.$$

Write the Fourier series for each of these functions and evaluate them at $x = 0$. How do the Fourier series compare? How do their values at $x = 0$ differ from $f_1(0)$, $f_2(0)$, and $f_3(0)$, respectively? What about when $x = \pi$?

7. Compute the Fourier series for each given function f.

(a) $f(x)=\begin{cases}-1, \text{ if } -\pi<x<0\\ 0, \text{ if } x=0\\ 1, \text{ if } 0<x<\pi\end{cases}$ **(b)** $f(x)=\begin{cases}0, \text{ if } -\pi<x<0\\ x, \text{ if } 0\le x<\pi\end{cases}$

(c) $f(x)=5$, if $x\in[-4,4]$ **(d)** $f(x)=x^2$ on $(-\pi,\pi]$

(e) $f(x)=x$ on $(-\pi,\pi)$ (See Exercise 5.)

(f) $f(x)=\sin\frac{x}{2}$, if $x\in[-2\pi,2\pi]$ **(g)** $f(x)=\sin\frac{x}{2}$, if $x\in[-\pi,\pi]$

(h) $f(x)=|\sin x|$ if $-\pi<x<\pi$ **(i)** $f(x)=|x|$ if $-1\le x<1$

(j) $f(x)=\begin{cases}2, \text{ if } -\pi<x<0\\ -3, \text{ if } 0<x<\pi\end{cases}$ **(k)** $f(x)=\cos^2 x$ on $(-\pi,\pi)$

8. Use a computer to graph functions $F_n(x)$, with $n=1,2,\ldots,10$, where

$$F_n(x)=\frac{a_0}{2}+\sum_{k=1}^{n}\left(a_k\cos\frac{k\pi x}{L}+b_k\sin\frac{k\pi x}{L}\right)$$

for each Fourier series in Exercise 7. See Definition 8.4.1.

9. The sequence of functions $\{f_n(x)\}$ is *orthogonal with respect to the weight function* $w(x)$ on the interval $[a,b]$ if and only if $\int_a^b f_m(x)f_n(x)w(x)dx=0$ whenever $m\neq n$. See Remark 9.1.2, part (d), and Part 1 of Section 9.5.

(a) Hermite polynomials $H_n(x)$, defined in Section 6.6, are orthogonal on $(-\infty,\infty)$ with respect to the weight function $w(x)=\exp\left(-x^2\right)$. Verify this fact for H_0, H_1, and H_2. See Exercise 15 in Section 6.6.

(b) Chebyshev polynomials $T_n(x)$, defined in Section 6.6, are orthogonal on $[-1,1]$ with respect to the weight function $w(x)=\frac{1}{\sqrt{1-x^2}}$. Verify this fact for T_0, T_1, and T_2. See Exercise 16, part (d), in Section 6.6.

(c) Laguerre polynomials $L_n(x)$, defined in Section 6.6, are orthogonal on $[0,\infty)$ with respect to the weight function $w(x)=e^{-x}$. Verify this fact for L_0, L_1, and L_2.

9.2* Convergence

Definition 9.2.1. The function f is said to be *piecewise smooth* on the interval $[a,b]$ if both f and f' are piecewise continuous on $[a,b]$. See Definition 4.2.12.

The following is a *Fourier Theorem*, sometimes called a *Theorem of Dirichlet*.

THEOREM 9.2.2. *Suppose that the function f is piecewise smooth on the interval $[-L, L]$ and $f(x+2L) = f(x)$, then the Fourier series for f converges to $\frac{1}{2}\left[f(x^+) + f(x^-)\right]$ at each value of $x \in \Re$.*

Remark 9.2.3

(a) The value $\frac{1}{2}\left[f(x^+) + f(x^-)\right]$ is the arithmetic mean, that is, the average of the right- and left-hand limits of the function f at x.

(b) If f is continuous at $x = a$, then the Fourier series for f at the point $x = a$ converges to $f(a)$. If f is $2L$-periodic, and if f' is continuous on $\Re$, then the Fourier series converges to f. And we can write

$$f(x) = \frac{a_0}{2} + \sum_{k=1}^{\infty}\left(a_k \cos\frac{k\pi x}{L} + b_k \sin\frac{k\pi x}{L}\right), \quad x \in \Re.$$

(c) Theorem 9.2.2 is stated only for periodic functions with period $2L$. However, it also applies to functions defined on the open interval $(-L, L)$, since those functions may be periodically extended. Review Definition 4.2.10. The value of f at the endpoints is irrelevant. Why?

(d) Piecewise continuity of f ensures the existence of Fourier coefficients. Why?

(e) Hypotheses of Theorem 9.2.2 are sufficient but not necessary. There are unbounded functions with converging Fourier series, like $f(x) = x^{-\frac{1}{3}}$. Also, the function $f(x) = \sqrt{|x|}$ on $[-L, L]$ is not piecewise smooth but yet has a convergent Fourier series.

(f) Theorem 9.2.2 does not give the most general sufficient conditions. However, relaxing these conditions down to continuity alone is not enough to guarantee convergence of Fourier series, as was demonstrated in 1873 by Du Bois-Reymond. At present there are no known necessary and sufficient conditions for convergence of Fourier series.

(g) Theorem 9.2.2 only addresses pointwise convergence.

(h) In this section Theorem 9.2.2 will be proven under stronger assumptions; that is, f is $2L$-periodic and f'' is continuous on $\Re$. See the proof of Theorem 9.2.5. For a proof of Theorem 9.2.2 with its stated conditions, see Exercise 13. ■

Example 9.2.4. Use the Fourier series of $f(x) = x^2$ on $[-\pi, \pi]$ to show that

(a) $\sum_{k=1}^{\infty}\frac{1}{k^2} = \frac{\pi^2}{6}$. (See Exercise 2 in this section, Exercise 17 in Section 7.4, and Problem 12.10.8.)

(b) $\sum_{k=1}^{\infty}\frac{(-1)^{k+1}}{k^2}=\frac{\pi^2}{12}$. (See Exercise 4 in this section and Exercise 17(c) in Section 7.5.)

(c) $\sum_{k=1}^{\infty}\frac{1}{(2k-1)^2}=\frac{\pi^2}{8}$. (See Exercise 1 in Section 9.4 and Exercise 17(a) in Section 7.4.)

Answer. Observe that f and f' are continuous on $[-\pi,\pi]$ and that the periodic extension of f is continuous on $\Re$. Thus, the Fourier series for f agrees with the functional values of f at every $x\in\Re$. To compute the Fourier coefficients the Definition 9.1.4 is invoked. Thus, $a_0=\frac{2}{3}\pi^2$ and $a_n=\frac{4(-1)^n}{n^2}$ for $n\in N$. To compute b_ns, Definition 9.1.4 again may be used, or we may observe that since x^2 is an even function, $x^2\sin nx$ is odd, and thus, the Riemann integral over the symmetric interval $[-\pi,\pi]$ must be 0 for all $n\in N$. Therefore, the Fourier series for f on $[-\pi,\pi]$ is given by

$$\frac{\pi^2}{3}+4\sum_{k=1}^{\infty}\frac{(-1)^k}{k^2}\cos kx.$$

Due to Theorem 9.2.2, we can write that

$$x^2=\frac{\pi^2}{3}+4\sum_{k=1}^{\infty}\frac{(-1)^k}{k^2}\cos kx$$

for all $x\in\Re$. Thus, if $x=\pi$ we obtain the answer to part (a). And if $x=0$, we obtain the answer to part (b). By combining parts (a) and (b) we obtain the result in part (c). ■

As in Example 9.2.4, coming up with a function whose Fourier series at some point gives the appropriate value is not obvious. Also, remember, there are many ways to represent a given real number by an infinite series.

Observe that in order for the function f in Theorem 9.2.2 to be $2L$-periodic, the values $f(-L)$ and $f(L)$ may not be different. Thus, in starting with a smooth function f defined only on $[-L,L]$ with $f(-L)\neq f(L)$, the Fourier series for f will agree with $f(x)$ for all $x\neq 2Lk$, with k an integer. At values $x=2Lk$, in view of Theorem 9.2.2, the Fourier series will converge to $\frac{1}{2}\left[f(L^-)+f(-L^+)\right]$. See Example 9.1.6.

Again, if the function f on $[-L,L]$ is not continuous or if $f(-L)\neq f(L)$, then convergence of the Fourier series for f on $[-L,L]$ cannot possibly be uniform. Why? But, if these conditions are satisfied in addition to those from Theorem 9.2.2, then convergence of the Fourier series is uniform. We formally write this as the next theorem.

THEOREM 9.2.5. *Suppose that the function f is continuous on $\Re$ and $f(x+2L)=f(x)$ for all $x\in\Re$. Moreover, suppose that f' is piecewise continuous on $[-L,L]$. Then the Fourier series for f converges uniformly and absolutely to f on $\Re$.*

Proof. Here, a stronger version of Theorem 9.2.5 will be proven. That is, assume that f is $2L$-periodic and that f'' is continuous on $\Re$.

Before we start, observe that f'' is bounded on $\Re$, and thus, $|f''(x)|\le M$ for some $M\in\Re$. Also, $\left|\cos\frac{n\pi x}{L}\right|\le 1$ and $\left|\sin\frac{n\pi x}{L}\right|\le 1$. Using integration by parts twice, we can write that

$$a_n=\frac{1}{L}\int_{-L}^{L} f(x)\cos\frac{n\pi x}{L}\,dx=-\frac{L}{n^2\pi^2}\int_{-L}^{L} f''(x)\cos\frac{n\pi x}{L}\,dx.$$

Why? Thus,

$$\begin{aligned}|a_n| &=\frac{L}{n^2\pi^2}\left|\int_{-L}^{L} f''(x)\cos\frac{n\pi x}{L}\,dx\right|\le\frac{L}{n^2\pi^2}\int_{-L}^{L}|f''(x)|\left|\cos\frac{n\pi x}{L}\right|dx\\ &\le\frac{L}{n^2\pi^2}\int_{-L}^{L} M\,dx=\frac{L}{n^2\pi^2}M(2L)=\frac{2ML^2}{n^2\pi^2}\equiv M_n\end{aligned}$$

for all $n\in N$. Similarly, $|b_n|\le M_n$. And, therefore, we have that

$$\begin{aligned}&\left|\frac{a_0}{2}+\sum_{k=1}^{\infty} a_k\cos\frac{k\pi x}{L}+b_k\sin\frac{k\pi x}{L}\right|\\ &\qquad\le\frac{|a_0|}{2}+\sum_{k=1}^{\infty}\left(|a_k|\left|\cos\frac{k\pi x}{L}\right|+|b_k|\left|\sin\frac{k\pi x}{L}\right|\right)\\ &\qquad\le\frac{|a_0|}{2}+\sum_{k=1}^{\infty}\left(|a_k|+|b_k|\right)\le\frac{1}{2}|a_0|+\sum_{k=1}^{\infty}2M_k.\end{aligned}$$

Since this series converges, (why?), by Weierstrass M-test, that is, Theorem 8.4.11, the Fourier series for the function f converges both uniformly and absolutely on $\Re$. ■

Theorem 9.2.5, like Theorem 9.2.2, is stated for continuous functions that are periodic with period $2L$. However, Theorem 9.2.5 also applies to continuous functions on $[-L,L]$ for which $f(L)=f(-L)$. Why? Also, note that the Fourier series given in Example 9.1.6 satisfies both that $\left|\frac{1}{k\pi}\sin 2kx\right|\le\frac{1}{k\pi}$ and that $\sum_{k=1}^{\infty}\frac{1}{k\pi}$ diverges. This does not mean that the Fourier series does not converge uniformly. In fact, it converges uniformly on any closed interval that does not include $2n\pi$, with $n\in N$. Why?

Note that not every trigonometric series is a Fourier series for some function. One can verify that the series $\sum_{k=2}^{\infty}\frac{\cos kx}{\ln k}$ is a Fourier series. See Exercise 14. The series $\sum_{k=2}^{\infty}\frac{\sin kx}{\ln k}$ is not a Fourier series, although it converges for all $x \in \Re$. See Exercise 5 in Section 9.4.

Example 9.2.6. Show that the series

$$\frac{1}{3}+\frac{4}{\pi^2}\sum_{k=1}^{\infty}\frac{(-1)^k}{k^2}\cos k\pi x$$

for $x \in [-1,1]$ is a Fourier series.

Answer. Since $\left|\frac{(-1)^k}{k^2}\cos k\pi x\right| \le \frac{1}{k^2}$, the given series converges uniformly to, say, $S(x)$.

Why? Is it a Fourier series? That is, are the coefficients $\frac{1}{3}$ for $n=0$ and $\frac{4}{\pi^2}\frac{(-1)^n}{n^2}$ for $n \in N$ indeed Fourier coefficients of $S(x)$? We will check this next. From Definition 9.1.4, we have

$$\begin{aligned} a_0 &= \frac{1}{1}\int_{-1}^{1} S(x)dx = \int_{-1}^{1}\left[\frac{1}{3}+\frac{4}{\pi^2}\sum_{k=1}^{\infty}\frac{(-1)^k}{k^2}\cos k\pi x\right]dx \\ &= \int_{-1}^{1}\frac{1}{3}dx+\frac{4}{\pi^2}\sum_{k=1}^{\infty}\left[\frac{(-1)^k}{k^2}\int_{-1}^{1}\cos k\pi\, dx\right] = \frac{1}{3}\cdot 2+0=\frac{2}{3}. \end{aligned}$$

$$\begin{aligned} a_n &= \int_{-1}^{1}\left[\frac{1}{3}+\frac{4}{\pi^2}\sum_{k=1}^{\infty}\frac{(-1)^k}{k^2}\cos k\pi x\right]\cos n\pi x\, dx \\ &= \frac{1}{3}\int_{-1}^{1}\cos n\pi x\, dx+\frac{4}{\pi^2}\sum_{k=1}^{\infty}\left[\frac{(-1)^k}{k^2}\int_{-1}^{1}\cos k\pi x\cos n\pi x\, dx\right] \quad \text{(Why?)} \\ &= \frac{4}{\pi^2}\frac{(-1)^n}{n^2},\quad n \in N. \text{ Why?} \end{aligned}$$

$$\begin{aligned} b_n &= \int_{-1}^{1}\left[\frac{1}{3}+\frac{4}{\pi^2}\sum_{k=1}^{\infty}\frac{(-1)^k}{k^2}\cos k\pi x\right]\sin n\pi x\, dx \\ &= \frac{1}{3}\int_{-1}^{1}\sin n\pi x\, dx+\frac{4}{\pi^2}\sum_{k=1}^{\infty}\left[\frac{(-1)^k}{k^2}\int_{-1}^{1}\cos k\pi x\sin n\pi x\, dx\right] \\ &= 0+0=0,\quad n \in N. \end{aligned}$$

Hence, by Definition 9.1.4, $\frac{1}{3}+\frac{4}{\pi^2}\sum_{k=1}^{\infty}\frac{(-1)^k}{k^2}\cos k\pi x$ is in fact a Fourier series for S on $[-1,1]$. It turns out that $S(x)=x^2$. ■

Exercises 9.2

1. Verify that the Fourier series for the function $f(x)=\begin{cases}0, \text{ if } -\pi<x<0\\ x, \text{ if } 0\le x<\pi\end{cases}$ on $[-\pi,\pi]$ converges for every value of $x\in\Re$. What value does this Fourier series converge to at $x=0$? How about at $x=\pi$?

2. Use the Fourier series for $f(x)=\frac{\pi^2-3x^2}{12}$ on $[-\pi,\pi]$ to prove that

$$\sum_{k=1}^{\infty}\frac{1}{k^2}=\frac{\pi^2}{6}.$$

See Example 9.2.4, part (a).

3. **(a)** Use the Fourier series for $f(x)=x$ on $[-\pi,\pi]$ to show that

$$\frac{\pi}{4}=\sum_{k=1}^{\infty}\frac{(-1)^{k+1}}{2k-1}.$$

(b) Use the Fourier series for the *square wave function*

$$f(x)=\begin{cases}-1, \text{ if } -\pi<x<0\\ 1, \text{ if } 0<x<\pi\end{cases}$$

with period 2π to show $\frac{\pi}{4}=\sum_{k=1}^{\infty}\frac{(-1)^{k+1}}{2k-1}$. See Exercise 7(a) in Section 9.1.

4. Use the Fourier series for $f(x)=4-x^2$ on $[-2,2]$ to show that

$$\sum_{k=1}^{\infty}\frac{(-1)^{k+1}}{k^2}=\frac{\pi^2}{12}.$$

See part (b) of Example 9.2.4.

5. Without computing the series, determine whether the Fourier series for $f(x)=x^2+x$ on $(-1,1)$ converges pointwise on $(-1,1)$. Does it converge uniformly?

6. Consider the series $\sum \dfrac{\sin[(2k-1)x]}{2k-1}$.

(a) Find the sum of this series on $[-\pi, \pi]$.

(b) Does this series converge uniformly on $[0, \pi]$? Explain.

(c) Does this series converge uniformly on any closed interval $[a,b] \subset (0,\pi)$? Explain.

7. If the Fourier series for the function f given in Definition 9.1.4 converges to S at $x = L$, show that it must also converge to S at $x = -L$.

8. Prove that

$$\frac{1}{2} + \sum_{k=1}^{n} \cos kx = \frac{\sin\left(n+\frac{1}{2}\right)x}{2\sin\left(\frac{1}{2}x\right)}$$

for any $x \in \Re$. See Exercise 24(b) in Section 8.4.

9. If f is piecewise continuous on $(-\pi, \pi)$, and partial sums of a Fourier series are given by

$$F_n(x) = \frac{a_0}{2} + \sum_{k=1}^{n} \left(a_k \cos kx + b_k \sin kx\right)$$

with Fourier coefficients a_0, a_k, and b_k, show that

(a) $F_n(x) = \dfrac{1}{2\pi} \displaystyle\int_{-\pi}^{\pi} f(x+t) \frac{\sin\left(n+\frac{1}{2}\right)t}{\sin\left(\frac{1}{2}t\right)} dt$, for $x \in (-\pi, \pi)$, and

(b) $F_n(x) = \dfrac{1}{2\pi} \displaystyle\int_{0}^{\pi} [f(x+t) + f(x-t)] \frac{\sin\left(n+\frac{1}{2}\right)t}{\sin\left(\frac{1}{2}t\right)} dt$, for $x \in (-\pi, \pi)$.

10. Prove that

(a) $\dfrac{1}{\pi} \displaystyle\int_{0}^{\pi} \frac{\sin\left(n+\frac{1}{2}\right)t}{\sin\left(\frac{1}{2}t\right)} dt = 1.$ **(b)** $\dfrac{1}{\pi} \displaystyle\int_{-\pi}^{0} \frac{\sin\left(n+\frac{1}{2}\right)t}{\sin\left(\frac{1}{2}t\right)} dt = 1.$

11. If $F(x)$ is continuous on $[-\pi, \pi]$ and $F_n(x)$ is as given in Exercise 9(b), prove that

(a) for $x \in (-\pi, \pi)$

$$F_n(x) - F(x) = \frac{1}{2\pi} \int_{0}^{\pi} [f(x+t) + f(x-t) - 2F(x)] \frac{\sin\left(n+\frac{1}{2}\right)t}{\sin\left(\frac{1}{2}t\right)} dt.$$

(b) the Fourier series for f on $[-\pi, \pi]$ converges to $F(x)$ if and only if

$$\lim_{n\to\infty} \int_{0}^{\pi} [f(x+t) + f(x-t) - 2F(x)] \frac{\sin\left(n+\frac{1}{2}\right)t}{\sin\left(\frac{1}{2}t\right)} dt = 0.$$

12. If the Fourier series for the piecewise continuous function f converges uniformly to $f(x)$ on $(-L, L)$, prove *Parseval's identity*,[4] that is, that

$$\frac{1}{L}\int_{-L}^{L}[f(x)]^2\,dx = \frac{(a_0)^2}{2} + \sum_{k=1}^{\infty}\left[(a_k)^2 + (b_k)^2\right],$$

provided the integral exists.

13. Follow the given steps in order to prove a variation of Theorem 9.2.2.

(a) If f is piecewise continuous on $(-\pi, \pi)$, prove that

$$\lim_{n\to\infty}\int_{-\pi}^{\pi} f(x)\left[\sin\left(n+\tfrac{1}{2}\right)x\right]dx = 0.$$

(b) Use Exercises 9(a) and 10 of Section 9.2 to prove that

$$F_n(x) - \frac{f(x^+)+f(x^-)}{2}$$
$$= \frac{1}{\pi}\int_{-\pi}^{0}\frac{f(t+x)-f(x^-)}{2\sin\frac{1}{2}t}\left[\sin\left(n+\tfrac{1}{2}\right)t\right]dt + \frac{1}{\pi}\int_{0}^{\pi}\frac{f(t+x)-f(x^+)}{2\sin\frac{1}{2}t}\left[\sin\left(n+\tfrac{1}{2}\right)t\right]dt.$$

(c) If f and f' are piecewise continuous on $(-\pi, \pi)$, prove that

$$\lim_{n\to\infty} F_n(x) = \frac{1}{2}\left[f(x^+)+f(x^-)\right]..$$

14. Show that $\sum_{k=2}^{\infty}\frac{\cos kx}{\ln k}$ and $\sum_{k=2}^{\infty}\frac{\sin kx}{k^3-3}$ are Fourier series.

9.3* Fourier Cosine and Fourier Sine Series

As was witnessed earlier, given that the function f is either odd or even can drastically reduce computations in finding Fourier coefficients. Recall that the product of an odd and an even function is odd, and that the product of two odd functions is even. For an odd function f on the symmetric interval $[-L, L]$ or $(-L, L)$, we have

[4]Marc Antoine Parseval Des Chênes (1755–1836), a French mathematician, wrote poems criticizing the Napoleonic government and was forced to flee France.

$$\int_{-L}^{L} f(x)\cos\frac{n\pi x}{L}\,dx = 0, \text{ and}$$

$$\int_{-L}^{L} f(x)\sin\frac{n\pi x}{L}\,dx = 2\int_{0}^{L} f(x)\sin\frac{n\pi x}{L}\,dx.$$

See Exercise 11 in Section 6.4. Note that the above integrals remain the same on $(-L, L)$ whether or not the values $f(-L)$ and/or $f(L)$ are defined. Therefore, an odd function on $(-L, L)$ with the arbitrary values $f(-L)$ and $f(L)$, has only sine terms in its Fourier series, that is, $a_n = 0$ for all $n = 0,1,2, \ldots$. Similarly, an even function on $(-L, L)$, with the arbitrary values $f(-L)$ and $f(L)$, has no sine terms in its Fourier series, that is, $b_n = 0$ for all $n = 1,2,3, \ldots$. This brings us to the next result.

THEOREM 9.3.1. *Suppose the function f is Riemann integrable on $[-L, L]$.*

(a) *If f is an even function, then the Fourier series for f on $(-L, L)$ is*

$$\frac{a_0}{2} + \sum_{k=1}^{\infty} a_k \cos\frac{k\pi x}{L},$$

where $a_0 = \frac{2}{L}\int_0^L f(x)\,dx$ *and* $a_n = \frac{2}{L}\int_0^L f(x)\cos\frac{n\pi x}{L}\,dx$ *for* $n \in N$. *The given trigonometric series is often referred to as the Fourier cosine series for f on $(0, L)$.*

(b) *If f is an odd function, then the Fourier series for f on $(-L, L)$ is*

$$\sum_{k=1}^{\infty} b_k \sin\frac{k\pi x}{L},$$

where $b_n = \frac{2}{L}\int_0^L f(x)\sin\frac{n\pi x}{L}\,dx$ *for* $n \in N$. *The given trigonometric series is often referred to as the Fourier sine series for f on $(0, L)$.*

In the above definitions of the Fourier cosine and Fourier sine series for f, the interval $(0, L)$ can be changed to $[0, L)$, $(0, L]$, or $[0, L]$. Both the Fourier cosine and Fourier sine series converge for piecewise smooth functions f on $(0, L)$, since Theorem 9.2.2 can be applied to both even and odd extensions of f to the interval $(-L, L) \setminus \{0\}$. See part (c) of Remark 9.2.3. Thus, the Fourier cosine series for f at $x = 0$ converges to $f(0^+)$, and at $x = L$ to $f(L^-)$. The Fourier sine series for f converges to 0 at both $x = 0$ and $x = L$. Why? ■

Example 9.3.2. Compute the Fourier sine and Fourier cosine series for the function $f(x)=2x$ on $[0,1)$.

Answer. Since we are to find the Fourier sine series for f on $[0,1)$, we are actually looking for the Fourier series for an extension g of f to the interval $(-1,1)$, where g is an odd function that agrees with f on $[0,1)$. Commonly g is called an *odd extension* of f to the interval $(-1,1)$. Thus, the Fourier sine series for f on $[0,1)$ is identical to the Fourier series for its odd extension on $(-1,1)$. See Figure 9.3.1. Therefore, for $n \in N$ integration by parts reveals that

$$b_n = \frac{2}{1}\int_0^1 2x(\sin n\pi x)\,dx = -\frac{4}{n\pi}\cos n\pi = (-1)^{n+1}\frac{4}{n\pi}.$$

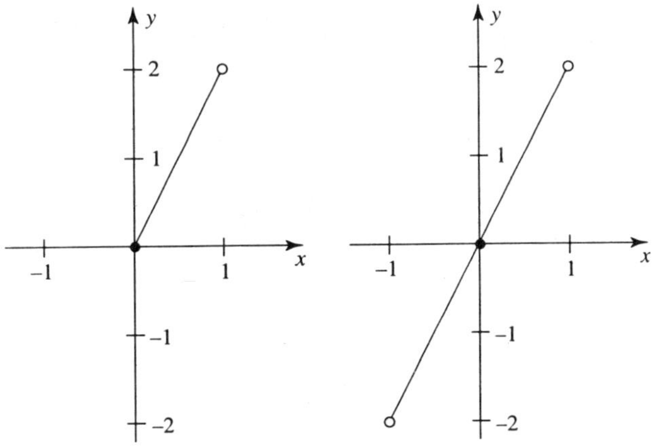

f on [0,1)

odd extension of f to (−1,1)

Figure 9.3.1

Recall that $\cos n\pi = (-1)^n$ according to Exercise 2(i) in Section 1.3. Hence, the Fourier sine series for f is

$$\sum_{k=1}^{\infty} b_k \sin k\pi x = \sum_{k=1}^{\infty}(-1)^{k+1}\frac{4}{k\pi}\sin k\pi x.$$

In fact, due to Theorem 9.2.2, we can say that $2x = \sum_{k=1}^{\infty}(-1)^{k+1}\frac{4}{k\pi}\sin k\pi x$ whenever x is not an odd integer. For contrast, the Fourier cosine series for $f(x)=2x$ on $[0,1)$ is identical to the Fourier series for the *even extension* of f to the interval $(-1,1)$. See Figure 9.3.2. Thus, $a_0 = \frac{2}{1}\int_0^1 2x\,dx = 2$. And for $n \in N$,

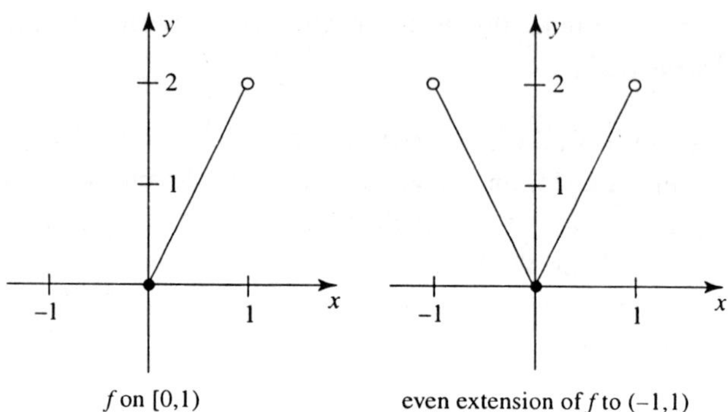

f on $[0,1)$ even extension of f to $(-1,1)$

Figure 9.3.2

$$a_n = \frac{2}{1}\int_0^1 2x(\cos n\pi x)\,dx = \begin{cases} -\dfrac{8}{n^2\pi^2}, \text{ if } n \text{ is odd} \\ 0, \text{ if } n \text{ is even.} \end{cases}$$

Therefore, the Fourier cosine series for $f(x) = 2x$ on $[0,1)$ is

$$\frac{a_0}{2} + \sum_{k=1}^{\infty} a_k \cos k\pi x = 1 + \sum_{\substack{k=1 \\ \text{odd}}}^{\infty} \left(-\frac{8}{k^2\pi^2}\right)\cos k\pi x = 1 - \frac{8}{\pi^2}\sum_{k=1}^{\infty} \frac{\cos(2k-1)\pi x}{(2k-1)^2}.$$

Note that we can write

$$2x = 1 - \frac{8}{\pi^2}\sum_{k=1}^{\infty} \frac{\cos(2k-1)\pi x}{(2k-1)^2}$$

for all $x \in \Re$. Why? Also, observe that both the Fourier sine and Fourier cosine series are equal to the function $f(x) = 2x$ on $[0,1)$, but that the Fourier cosine series converges on $(0,1)$ more rapidly. Why? ■

In Example 9.3.2, we found two trigonometric series for $f(x) = 2x$ that converge to f on $[0,1)$. Are there any other ones? What if we consider the periodic extension of f to the interval $[-1,1)$. See Figure 9.3.3. Its Fourier coefficients are given by

$$\begin{aligned} a_0 &= 2, \\ a_n &= 0, \text{ for } n \in N, \text{ and} \\ b_n &= -\frac{2}{n\pi}\left[1 + (-1)^n\right], \text{ for } n \in N. \end{aligned}$$

Verify these computations. Thus, the Fourier series for the periodic extension of f is

$$\frac{a_0}{2} + \sum_{k=1}^{\infty}\left(a_k \cos\frac{k\pi x}{L} + b_k \sin\frac{k\pi x}{L}\right) = 2 - \sum_{k=1}^{\infty} \frac{2}{k\pi}\sin 2k\pi x.$$

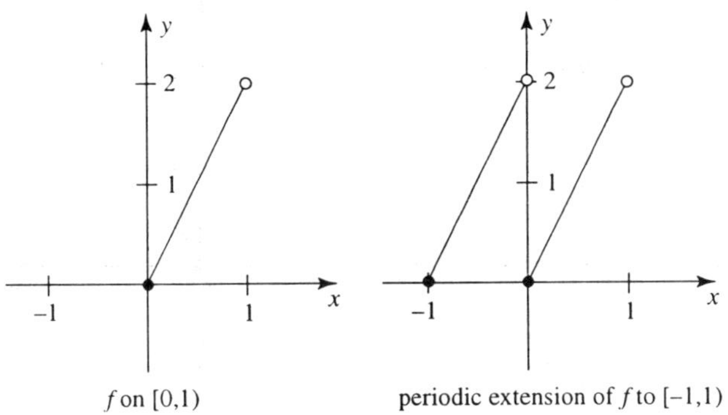

Figure 9.3.3

Here, too, is another trigonometric series that converges to $f(x) = 2x$ on $[0,1)$. As we see, there are a number of trigonometric series converging to $f(x) = 2x$ on $[0,1)$. To be more creative yet, consider the function g defined by

$$g(x) = \begin{cases} 2x, \text{ if } 0 \leq x \leq 1 \\ 0, \text{ if } -1 \leq x < 0. \end{cases}$$

See Figure 9.3.4.

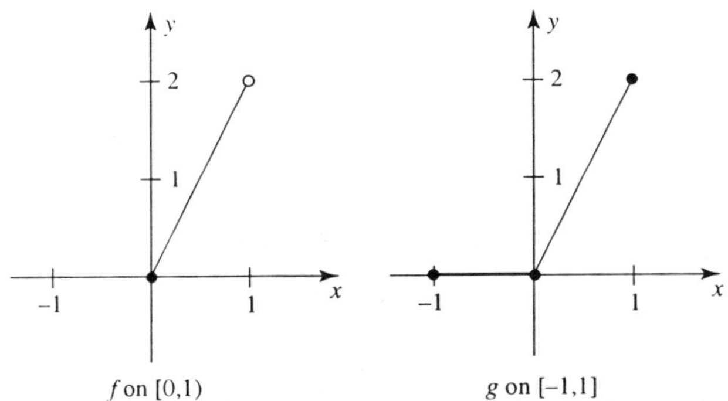

Figure 9.3.4

The Fourier series for g will give rise to yet another trigonometric series equal to $f(x) = 2x$ on $[0,1)$. For a better visualization, the nth partial sums of all these different trigonometric series for $f(x) = 2x$ on $[0,1)$ discussed above should be graphed for values of $n = 12$ or more. See Exercise 8 in Section 9.1. In Example 9.3.2, the function, due to its domain, was equal to its Fourier sine series. If in that example the domain was $[0,1]$ instead of $[0,1)$, then the function and its Fourier sine series would not be equal to each other at all values of x. See Figure 9.3.5.

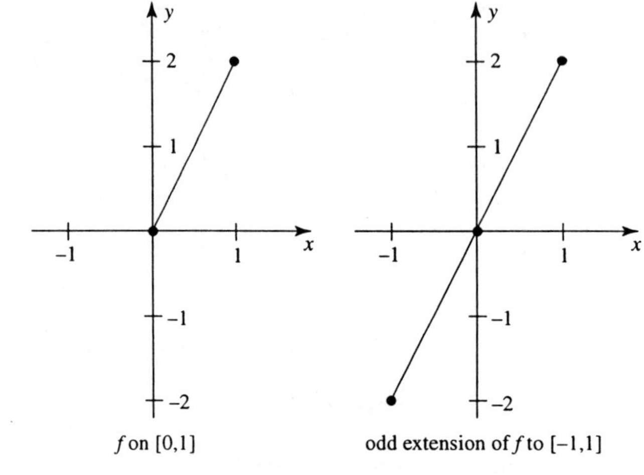

f on [0,1] odd extension of f to [−1,1]

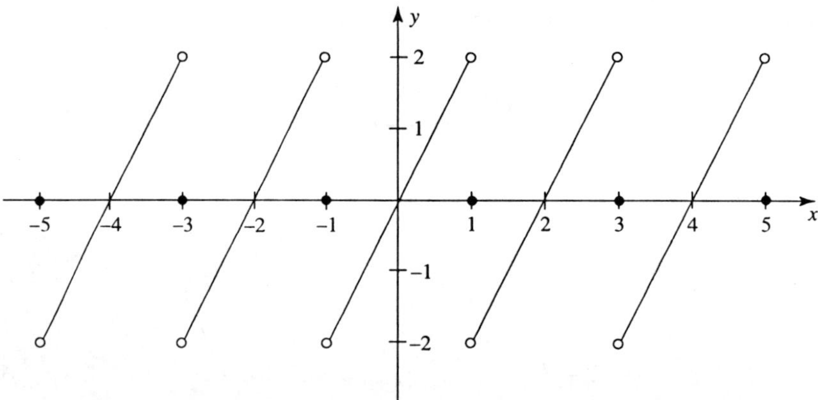

Fourier series for f on [−5,5]

Figure 9.3.5

Example 9.3.3. Compute the Fourier series for the function $f(x) = 2x - 2$ on the interval $[1,3]$.

Answer. The desired Fourier series is identical to the Fourier series for $g(x) = 2x + 2$ on the interval $[-1,1]$. See Figure 9.3.6. In fact, the Fourier series for $g(x) = 2x + 2$ on $[-1,1]$ is also the Fourier series for the function $h(x) = 2x + 2$ on $(-1,1)$, with $h(x+2) = h(x)$. The function h is a periodic extension of $g(x) = 2x + 2$, with $x \in (-1,1)$ and with period equal to 2. See the graph of h in Figure 9.3.7. Computation of the desired Fourier series is left as Exercise 3. ■

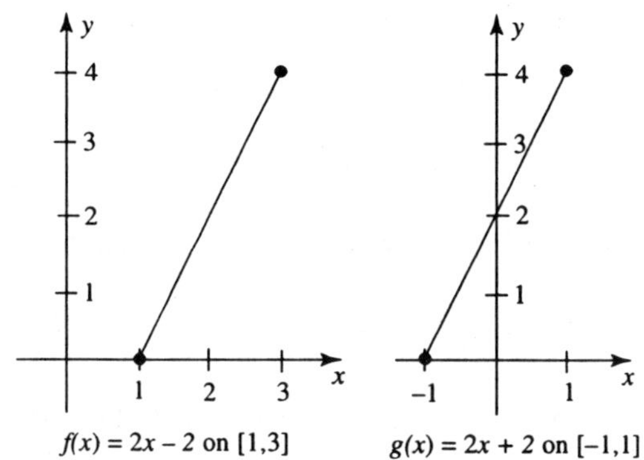

Figure 9.3.6

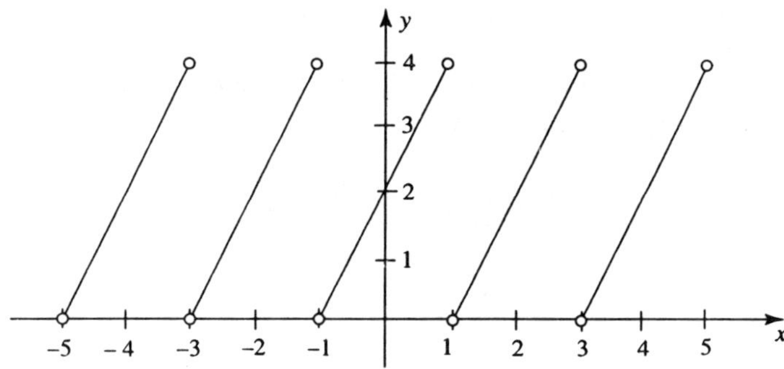

Figure 9.3.7

In view of Euler's formula (review Exercise 10 in Section 8.6), the Fourier series for f on $(-L, L)$ may be written in *complex exponential* form as

$$\begin{aligned}
&\frac{a_0}{2} + \sum_{k=1}^{\infty}\left(a_k \cos\frac{k\pi x}{L} + b_k \sin\frac{k\pi x}{L}\right) \\
&= \frac{a_0}{2} + \sum_{k=1}^{\infty}\frac{a_k}{2}\left[\exp\left(\frac{ik\pi x}{L}\right) + \exp\left(-\frac{ik\pi x}{L}\right)\right] + \sum_{k=1}^{\infty}\frac{b_k}{2i}\left[\exp\left(\frac{ik\pi x}{L}\right) - \exp\left(-\frac{ik\pi x}{L}\right)\right] \\
&= \frac{a_0}{2} + \sum_{k=1}^{\infty}\left[\frac{a_k - ib_k}{2}\exp\left(\frac{ik\pi x}{L}\right) + \frac{a_k + ib_k}{2}\exp\left(-\frac{ik\pi x}{L}\right)\right] \\
&= \sum_{k=-\infty}^{\infty} c_k \exp\left(\frac{ik\pi x}{L}\right),
\end{aligned}$$

where $c_0 = \dfrac{a_0}{2}$, $c_n = \dfrac{a_n - ib_n}{2}$, and $c_{-n} = \dfrac{a_n + ib_n}{2}$, for $n \in N$.

Before closing, we note that the Fourier series for the function f on $(-L, L)$ may be written in yet another form, namely, the phase angle form. That is, we need to find values c_n and d_n so that

$$a_n \cos\frac{n\pi x}{L} + b_n \sin\frac{n\pi x}{L} = c_n \cos\left(\frac{n\pi x}{L} + d_n\right).$$

Expanding the right-hand side of the above expression, we obtain

$$c_n \cos\left(\frac{n\pi x}{L} + d_n\right) = c_n\left(\cos\frac{n\pi x}{L}\right)(\cos d_n) - c_n\left(\sin\frac{n\pi x}{L}\right)(\sin d_n).$$

Therefore, $c_n \cos d_n = a_n$, and $-c_n \sin d_n = b_n$. Squaring both equations and then adding, we obtain $c_n = \sqrt{(a_n)^2 + (b_n)^2}$. Lastly, dividing the above equations, we obtain $\tan d_n = -\frac{b_n}{a_n}$, which yields $d_n = \arctan\left(-\frac{b_n}{a_n}\right)$. Hence, if c_n and d_n are as above, then the Fourier series for f on $(-L, L)$ can be written in the *phase angle* form, also called the *harmonic* form or *shifted cosine* form, as

$$\frac{a_0}{2} + \sum_{k=1}^{\infty} c_k \cos\left(\frac{k\pi x}{L} + d_k\right).$$

The phase angle form can be written in the *shifted sine* form as well.

Exercises 9.3

1. Compute the Fourier sine and the Fourier cosine series for each given function f, and determine for which values of x the series agrees with the given function.

(a) $f(x) = 1$ on $[0, \pi]$

(b) $f(x) = x$ on $[0, 2]$ (See Exercise 7(e) in Section 9.1.)

(c) $f(x) = \sin x$ on $(0, \pi]$ (See Exercise 7(h) in Section 9.1.)

(d) $f(x) = \frac{\pi - x}{2}$ on $(0, 2\pi)$ (See Problem 7.6.3.)

(e) $f(x) = x(1 - x)$ on $[0, 1)$

(f) $f(x) = \sin x$ on $[0, 2\pi)$

2. Write four different trigonometric series that equal $f(x) = 2 - x$ on the interval $(0, 2)$.

3. Compute the Fourier series for $f(x) = 2x - 2$ on $[1, 3]$. See Example 9.3.3.

4. Determine the differences between the Fourier series for $f(x) = \begin{cases} -1, \text{ if } -2 \le x < 0 \\ 1, \text{ if } 0 \le x \le 2 \end{cases}$ on $[-2, 2]$ and the Fourier sine series for $g(x) = 1$ on $[0, 2]$.

5. (a) Verify that on $\Re$, the Fourier series for $f(x)=|\sin x|$ with $x\in[-\pi,\pi]$ converges to the periodic extension of $f(x)$. See Exercise 7(h) in Section 9.1.

(b) Conclude that $|\sin x| = \frac{2}{\pi} - \frac{4}{\pi}\sum_{k=1}^{\infty}\frac{\cos 2kx}{4k^2-1}$. Then, using this equality, verify that $\sum_{k=1}^{\infty}\frac{1}{4k^2-1} = \frac{1}{2}$ and $\sum_{k=1}^{\infty}\frac{(-1)^k}{4k^2-1} = \frac{1}{2} - \frac{\pi}{4}$. See Exercise 1(d) in Section 7.1.

(c) Verify that the Fourier cosine series for $f(x)=\sin x$ with $x\in[0,\pi]$ converges to the periodic extension of f on $\Re$. See Exercise 1(c) above.

6. Specify the limiting function of the Fourier sine series for f on $[-1,1]$, where $f(x)=x$ with $x\in(0,1]$.

7. (a) Write the Fourier sine series for $f(x)=1$ on $[0,\pi]$.

(b) Can you use part (a) to find an infinite series representation for $\frac{\pi}{4}$? If so, find it. If not, explain why not.

8. Use the Fourier cosine series for $f(x)=x^4$ on $[0,\pi]$ to show that $\frac{\pi^4}{90} = \sum_{k=1}^{\infty}\frac{1}{k^4}$.

9. Use the Fourier cosine series for $\sin x$ on $[0,\pi]$ to find the Fourier series for the function f on $[-\pi,\pi]$ where $f(x)=\begin{cases}0, \text{ if } -\pi\le x<0\\ \sin x, \text{ if } 0\le x\le\pi.\end{cases}$ See Exercise 7(h) in Section 9.1.

10. Prove that the Fourier cosine series for $f(x)=x$ on $[0,\pi]$ converges uniformly on $\Re$.

11. Find the Fourier series that agrees with $f(x)=x^2-2$ for all $x\in[2,3]$.

12. Write the shifted cosine form of the Fourier series for the function f, where $f(x)=x^2$ on $(0,3]$ and $f(x+3)=f(x)$ for all $x\in\Re$.

13. (a) Prove that the Fourier coefficients of the function $f+g$ are the sum of the Fourier coefficients of f and g.

(b) Prove that the Fourier coefficients of the function cf, with c a real constant, are c times the Fourier coefficients of f.

(c) Use parts (a) and (b) to find the Fourier series for $f(x)=3x-2x^2$ on $[-\pi,\pi]$. See Exercises 7(d) and 7(e) in Section 9.1.

(d) Find the Fourier series for $f(x)=2+\sin x+2\sin\frac{x}{2}$ on $[-\pi,\pi]$.

14. Prove that if functions f and g differ only by an additive constant c, then the Fourier coefficients of f and the Fourier coefficients of g differ only in the value

a_0. Also, prove that the difference is precisely $2c$.

15. Without computing the series, determine whether the Fourier sine series and the Fourier cosine series for $f(x)=x^2+x$ on $(-1,0)$ converge pointwise on $(-1,0)$. Do they converge uniformly?

16. Prove that $\frac{\pi}{2}-\sum_{k=1}^{\infty}\frac{\sin kx}{k}$ converges uniformly to $\frac{x}{2}$ on any closed interval $[a,b]$ $\subset(0,2\pi)$. Moreover, show that $\frac{\pi}{2}-\sum_{k=1}^{\infty}\frac{\sin kx}{k}$ converges pointwise to

$$F(x)\equiv\begin{cases}\frac{x}{2}, \text{ if } x\in(0,2\pi)\\ \frac{\pi}{2}, \text{ if } x=0,2\pi.\end{cases}$$

See Problem 7.6.3.

9.4* Differentiation and Integration

There are ways to simplify the process of finding Fourier series for functions. Among them is the differentiation and integration of already calculated series. We will study these methods next. But first, recall that the Fourier series for the function $f(x)=x$ on $(-\pi,\pi)$ is given by $2\sum_{k=1}^{\infty}(-1)^{k+1}\frac{\sin kx}{k}$. See Problem 7.6.3, Section 8.4, and Exercise 7(e) from Section 9.1. Due to Theorem 9.2.2, we can write $x=2\sum_{k=1}^{\infty}(-1)^{k+1}\frac{\sin kx}{k}$ for $x\in(-\pi,\pi)$. Term-by-term differentiation of this series would lead to $1=2\sum_{k=1}^{\infty}(-1)^{k+1}\cos kx$. The absurdity of this equality lies in the fact that the terms of the series do not tend to 0. In addition, since $f(-\pi)\neq f(\pi)$ for $f(x)=x$ on $[-\pi,\pi]$, term-by-term differentiation of the Fourier series for f is impossible.

THEOREM 9.4.1. *Suppose that the function f is continuous and periodic on $\Re$ with period $2L$, that is, $f(x+2L)=f(x)$ for all $x\in\Re$. Moreover, suppose that f' and f'' are piecewise continuous on $[-L,L]$. Then, the Fourier series for f' can be obtained from term-by-term differentiation of the Fourier series for f.*

Therefore, according to Theorem 9.4.1, if f is continuous on $[-L,L]$, $f(-L)=f(L)$, and f' is piecewise smooth, then for

$$f(x)=\frac{a_0}{2}+\sum_{k=1}^{\infty}\left(a_k\cos\frac{k\pi x}{L}+b_k\sin\frac{k\pi x}{L}\right),$$

we have

$$f'(x)\sim\sum_{k=1}^{\infty}\frac{k\pi}{L}\left(-a_k\sin\frac{k\pi x}{L}+b_k\cos\frac{k\pi x}{L}\right).$$

In fact, since the hypotheses of Theorem 9.4.1 imply that f' is piecewise smooth, it follows from Theorem 9.2.2 that the above tilde can be replaced by an equal sign, except at the points where f' is discontinuous. Observe this in the following example.

Example 9.4.2. Suppose that the Fourier series for $f(x)=|x|$ on $[-\pi,\pi]$ is

$$\frac{\pi}{2}-\frac{4}{\pi}\sum_{k=1}^{\infty}\frac{\cos(2k-1)x}{(2k-1)^2}.$$

See Exercise 7(i) from Section 9.1. Can Theorem 9.4.1 be used to compute the Fourier series for $g(x)=\dfrac{|x|}{x}$ on $[-\pi,\pi]$? How about to compute the Fourier series for $h(x)=\dfrac{|x|}{x}$ on $(-\pi,\pi]$, and $h(x+2\pi)=h(x)$ for all $x\in\Re$?

Answer. Since f is continuous on $[-\pi,\pi]$, $f(-\pi)=f(\pi)$. And since f' is piecewise smooth on $[-\pi,\pi]$, Theorem 9.4.1 applies. From Theorem 9.2.2,

$$f(x)=|x|=\frac{\pi}{2}-\frac{4}{\pi}\sum_{k=1}^{\infty}\frac{\cos(2k-1)x}{(2k-1)^2}.$$

Therefore, a legal term-by-term differentiation yields

$$g(x)=\frac{|x|}{x}\sim\frac{4}{\pi}\sum_{k=1}^{\infty}\frac{\sin(2k-1)x}{2k-1},$$

which is the Fourier series for g on $[-\pi,\pi]$. Note that g is undefined at $x=0$ and $g(-\pi)\neq g(\pi)$. Thus, since continuity exists at all values in the domain except at 0, π, and $-\pi$, "$\sim$" may be replaced by "$=$" in the equation when $x\neq-\pi$, 0, and π. Note that the Fourier series for h is identical to that of g, and $h(x)=\dfrac{4}{\pi}\displaystyle\sum_{k=1}^{\infty}\frac{\sin(2k-1)x}{2k-1}$ for all $x\neq n\pi$ where n is an integer. ■

As we have seen in Section 8.4, term-by-term integration of Fourier series can be performed under weak conditions. Moreover, from Theorem 6.4.4, since integration is in some sense a smoothing process, convergence of a Fourier series is enhanced. We present this as the next result.

THEOREM 9.4.3. *Suppose that the function f is piecewise continuous on $[-L, L]$ with the Fourier series*

$$f(x) \sim \frac{a_0}{2} + \sum_{k=1}^{\infty}\left(a_k \cos\frac{k\pi x}{L} + b_k \sin\frac{k\pi x}{L}\right).$$

Then, for every $x \in [-L, L]$,

$$\int_{-L}^{x} f(t)\,dt = \int_{-L}^{x} \frac{a_0}{2}\,dt + \sum_{k=1}^{\infty} \int_{-L}^{x}\left(a_k \cos\frac{k\pi t}{L} + b_k \sin\frac{k\pi t}{L}\right)dt.$$

Therefore, if conditions of Theorem 9.4.3 are satisfied, we have

$$\int_{-L}^{x} f(t)\,dt = \frac{1}{2}a_0(x+L) + \sum_{k=1}^{\infty}\frac{L}{k\pi}\left[a_k \sin\frac{k\pi x}{L} - b_k\left(\cos\frac{k\pi x}{L} - \cos k\pi\right)\right].$$

Why? Note that term-by-term integration holds even for functions f whose Fourier series do not converge (pointwise) to f. Observe further that integration does not have to start at $-L$, but rather at any value in $[-L, L)$. Thus, we also have that

$$\int_{0}^{x} f(t)\,dt = \frac{1}{2}a_0 x + \sum_{k=1}^{\infty}\frac{1}{k}\left[a_k \sin kx + b_k(1-\cos kx)\right].$$

Example 9.4.4. Recall that the Fourier series for $f(x) = x$ on $[-\pi, \pi]$ is given by $x \sim 2\sum_{k=1}^{\infty}(-1)^{k+1}\frac{\sin kx}{k}$. Integrating this series term by term, we obtain the equality

$$\int_{-\pi}^{x} t\,dt = 2\sum_{k=1}^{\infty}(-1)^{k+1}\frac{1}{k}\int_{-\pi}^{x}\sin kt\,dt$$

for $x \in [-\pi, \pi]$. Therefore, $\frac{x^2}{2} - \frac{\pi^2}{2} = 2\sum_{k=1}^{\infty}(-1)^k\frac{1}{k^2}(\cos kx - \cos k\pi)$. And $x^2 = 4\sum_{k=1}^{\infty}\frac{(-1)^k}{k^2}\cos kx + C$, where $C = \pi^2 - 4\sum_{k=1}^{\infty}\frac{1}{k^2}$. Since from Example 9.2.4, part (a), $\sum_{k=1}^{\infty}\frac{1}{k^2} = \frac{\pi^2}{6}$, we have $C = \pi^2 - 4\left(\frac{\pi^2}{6}\right) = \frac{\pi^2}{3}$. Now, replacing C by $\frac{\pi^2}{3}$ in the series, we obtain the Fourier series for x^2, which we had in Example 9.2.4. The above definite integral could be changed to the indefinite integral. To compute the constant of integration C, either find a_0 since $C = \frac{a_0}{2}$ or evaluate the resulting series at a particular value of x. This last method, however, does not always work successfully. ■

Exercises 9.4

1. **(a)** Use the Fourier series for $f(x) = |x|$ on $[-\pi, \pi]$ to show that $\frac{\pi^2}{8} = \sum_{k=1}^{\infty} \frac{1}{(2k-1)^2}$. See Exercise 17(a) in Section 7.4.

(b) Use the Fourier series for $f(x) = \begin{cases} \pi, \text{if } -\pi < x < 0 \\ x, \text{if } 0 < x < \pi \end{cases}$ on $[-\pi, \pi]$ to show that $\frac{\pi^2}{8} = \sum_{k=1}^{\infty} \frac{1}{(2k-1)^2}$.

(c) Use the Fourier series for $f(x) = 2x$ on $[-\pi, \pi]$ to show that $\frac{\pi^2}{8} = \sum_{k=1}^{\infty} \frac{1}{(2k-1)^2}$.

2. Use theorems in this section and the fact that $x = 2\sum_{k=1}^{\infty} (-1)^{k+1} \frac{\sin kx}{k}$ for $x \in (-\pi, \pi)$ to find a trigonometric series for $f(x) = 3x^2 - 2$ on $(-\pi, \pi)$.

3. Use the Fourier series for $f(x) = x$ on $[-\pi, \pi]$ in writing the Fourier series for $f(x) = x^3$ on $[-\pi, \pi]$.

4. Use the Fourier sine series for $f(x) = x^2$ on $[0,1]$ to prove that

(a) $\sum_{k=1}^{\infty} \frac{(-1)^{k+1}}{(2k-1)^3} = \frac{\pi^3}{32}$. **(b)** $\sum_{k=1}^{\infty} \frac{1}{(2k-1)^4} = \frac{\pi^4}{96}$.

5. Prove that $\sum_{k=2}^{\infty} \frac{\sin kx}{\ln k}$ is not a Fourier series.

9.5* Projects

Part 1. Series of Orthogonal Functions

The purpose of this section is to represent functions by series that are generalizations of

Fourier series, that is, series that involve terms other then sines and/or cosines.[5] First, review definitions given in part (d) of Remark 9.1.2 and Exercise 9 of Section 9.1. Also, see Definition 10.9.6.

Definition 9.5.1. The sequence of functions $\{f_n(x)\}$ is *orthogonal with respect to the weight function* w on the interval $[a,b]$ if and only if $\int_a^b f_m(x)f_n(x)w(x)dx = 0$ whenever $m \neq n$. If $w(x) \equiv 1$, then $\{f_n(x)\}$ is said to be *orthogonal* on $[a,b]$.

Orthogonal means perpendicular in a sophisticated way. A better understanding of this concept will occur after studying the dot product in Section 10.3 and Part 1 of Section 10.9.

Problem 9.5.2. Verify the following:

(a) $\sin x$ and $\cos x$ are orthogonal on $[-\pi, \pi]$.

(b) $\sin x$ and $\sin 2x$ are orthogonal on $[-\pi, \pi]$.

(c) $\sin \frac{n\pi x}{L}$ with $n \in N$ are orthogonal on $[0, L]$.

(d) $\sin(n \ln x)$ with $n \in N$ are orthogonal with respect to the weight function $w(x) = \frac{1}{x}$ on $\left[1, e^{\pi}\right]$.

(e) If f_1 and f_2 are orthogonal with respect to the weight function $w(x) = x$ on the interval $[1,5]$, then the functions $\sqrt{x}f_1$ and $\sqrt{x}f_2$ are orthogonal on $[1,5]$.

(f) Hermite polynomials $H_n(x)$, with $n = 0,1,2, \ldots$, are orthogonal with respect to the weight function $w(x) = \exp\left(-x^2\right)$ on $(-\infty, \infty)$. See Exercise 9(a) in Section 9.1.

(g) Chebyshev polynomials $T_n(x) = \cos(n \arccos x)$, with $n = 0,1,2, \ldots$, are orthogonal with respect to the weight function $w(x) = \frac{1}{\sqrt{1-x^2}}$ on $[-1,1]$.

See Exercise 9(b) in Section 9.1.

THEOREM 9.5.3. *(Sturm[6]–Liouville Theorem) Consider the equation*

$$\frac{d}{dx}\left[p(x)\frac{dy}{dx}\right] + [q(x) + \lambda w(x)]y = 0,$$

---

[5]Josef Hoëné de Wroński (1778–1853), a Polish mathematician, introduced series expansions of functions where coefficients in the series are determinants, also referred to as Wronskians.

[6]Charles Jacques François Sturm (1803–1855), born in Switzerland and best known in both fluid mechanics and differential equations, was the first to accurately determine the speed of sound in water.

where $p(x) > 0$ and has a continuous derivative on the interval $[a,b]$, q and w are continuous, and $w(x) > 0$ on $[a,b]$. If $\lambda_1, \lambda_2, \lambda_3, \ldots$ are distinct values of λ for which the above equation has nontrivial solutions $y_1, y_2, y_3, \ldots$, (i.e. the functions $y_1, y_2, y_3, \ldots$ that satisfy the equation but are not identically zero,) and which also satisfy the boundary conditions

$$A_1 y(a) + A_2 y'(a) = 0 \text{ with } A_1 \text{ and } A_2 \text{ not both zero, and}$$
$$B_1 y(b) + B_2 y'(b) = 0 \text{ with } B_1 \text{ and } B_2 \text{ not both zero,}$$

then the sequence of functions $\{y_n(x)\}$ is orthogonal with respect to the weight function w on the interval $[a,b]$.

Proof of this theorem can be found in many introductory texts to differential equations. We now move on to another important concept.

Definition 9.5.4. The function f is said to be *normalized with respect to the weight function* w on $[a,b]$ if and only if $\int_a^b [f(x)]^2 w(x)\,dx = 1$. If $w(x) \equiv 1$, then f is said to be *normalized* on $[a,b]$.

Using the notation from Definition 9.1.1, we can say that f is normalized with respect to the weight function w on $[a,b]$, if and only if $\|f\| = 1$.

Problem 9.5.5. Show that $f(x) = \sqrt{\frac{2}{\pi}} \sin x$ is normalized with respect to $w(x) = 1$ on $[0, \pi]$.

Definition 9.5.6. The sequence of functions $\{f_n(x)\}$ is *orthonormal with respect to the weight function* w on the interval $[a,b]$ if and only if

$$\int_a^b f_m(x) f_n(x) w(x)\,dx = \begin{cases} 0, \text{if } m \neq n \\ 1, \text{if } m = n \end{cases} \equiv \delta_{mn},$$

the *Kronecker's delta*. The functions $f_n(x)$ are said to be *orthonormal with respect to the weight function* w on $[a,b]$.

Thus, an orthonormal sequence of functions is a sequence of orthogonal functions that are all normalized.

Problem 9.5.7. Show that if $f_n(x) = \sqrt{\frac{2}{\pi}} \sin nx$ on $[0, \pi]$, then the functions $f_n(x)$ are orthonormal on $[0, \pi]$.

Problem 9.5.8

(a) Suppose that $\{f_n(x)\}$ is orthogonal with respect to w on $[a,b]$. Show that $\{k_n f_n(x)\}$ is orthogonal with respect to w on $[a,b]$ for $k_n \in \Re$ and $n \in N$.

(b) Suppose that $\{f_n(x)\}$ is a sequence of nonzero orthogonal functions with respect to the weight function $w(x) > 0$ on $[a,b]$. Show that if $n \in N$ and

$$k_n = \frac{1}{\sqrt{\int_a^b [f_n(x)]^2 w(x)\,dx}},$$

that is, $k_n = \frac{1}{\|f\|}$, then the sequence $\{k_n f_n(x)\}$ is orthonormal with respect to w on $[a,b]$.

With the orthonormal sequence $\{f_n(x)\}$ on $[a,b]$ at hand, can a function f be expressed in an infinite series using these functions? If so, what would the coefficients be? To answer these questions, we will assume that $f(x) = \sum_{k=1}^{\infty} c_k f_k(x)$ for all $x \in [a,b]$, and attempt to find the coefficients c_k. To this end, multiply both sides of the expression above by $f_n(x)w(x)$, where w is the weight function, and integrate from a to b to obtain

$$\begin{aligned}\int_a^b f(x)f_n(x)w(x)\,dx &= \int_a^b \left(\sum_{k=1}^{\infty} c_k f_k(x) f_n(x) w(x) \right) dx \\ &= \sum_{k=1}^{\infty} \left(c_k \int_a^b f_k(x) f_n(x) w(x)\,dx \right) = c_k,\end{aligned}$$

where the second equality holds only if convergence of the series is uniform. The last equality holds since the sequence $\{f_n(x)\}$ is orthonormal with respect to w on $[a,b]$.

Definition 9.5.9. If the sequence $\{f_n(x)\}$ is orthonormal with respect to the weight function w on the interval $[a,b]$, and the function f is such that $(f)(f_n)(w)$ is Riemann integrable on $[a,b]$, then the series $\sum_{k=1}^{\infty} c_k f_k(x)$, where

$$c_n = \int_a^b f(x) f_n(x) w(x)\, dx,$$

is called the *generalized Fourier series* for f, and c_ns are the *generalized Fourier coefficients*.

As in Section 9.2, convergence of this series needs to be carefully discussed.

Problem 9.5.10. Show that Definition 9.1.4 is a special case of Definition 9.5.9 by considering the sequence $\{f_n(x)\}$, where for $x \in [-L, L]$, we have

$$f_1(x) = 1,$$
$$f_{2n}(x) = \cos\frac{n\pi x}{L}, \ n \in N, \text{ and}$$
$$f_{2n+1}(x) = \sin\frac{n\pi x}{L}, \ n \in N,$$

Part 2. Legendre Polynomials

Legendre polynomials, which in Section 6.6 were defined by

$$P_n(x) = \sum_{k=0}^{I} \frac{(-1)^k (2n-2k)!\, x^{n-2k}}{2^n k!(n-k)!(n-2k)!},$$

where $I = \frac{n}{2}$ or $I = \frac{n-1}{2}$, whichever is an integer, possess several basic properties that we will endeavor to establish here. Problems below can be worked out in an order different from that which is presented.

Problem 9.5.11

(a) Compute P_0, P_1, P_2, and P_3. Also, note that $P_n(x)$ is of degree n for $n = 0, 1, 2, \ldots$.

(b) Show that both P_{2n} with $n = 0, 1, 2, \ldots$ is an even function and P_{2n-1} with $n = 0, 1, 2, \ldots$ is an odd function, that is, $P_n(-x) = (-1)^n P_n(x)$ for $n = 0, 1, 2, \ldots$.

(c) Expand $f(x) = 2x^3 - x^2 - 3$ in terms of Legendre polynomials.

Problem 9.5.12. Show that, for a fixed n with $n = 0, 1, 2, \ldots$, $y = P_n(x)$ is a solution[7] of Legendre's equation

$$\left(1-x^2\right)\frac{d^2y}{dx^2}-2x\frac{dy}{dx}+n(n+1)y=0.$$

Problem 9.5.13. Show that Legendre's equation can be written in *self-adjoint* form as

$$\frac{d}{dx}\left[\left(1-x^2\right)\frac{dy}{dx}\right]+n(n+1)y=0,$$

and in trigonometric form as

$$(\sin\theta)\frac{d^2y}{d\theta^2}+(\cos\theta)\frac{dy}{d\theta}+n(n+1)(\sin\theta)y=0.$$

See Exercise 42 in Section 6.7.

Problem 9.5.14. Show that the sequence $\{P_n\}_{n=0}^{\infty}$ is a sequence of orthogonal functions on $[-1,1]$, that is, show that $\int_{-1}^{1} P_m(x)P_n(x)\,dx=0$ if $m\neq n$ with $m,n=0,1,2,\ldots$.

Problem 9.5.15. Show that $\int_{-1}^{1} P_n(x)q(x)\,dx=0$ for any polynomial $q(x)$ of degree at most $n-1$, with $n=0,1,2,\ldots$.

Problem 9.5.16. Show that if $Q_n(x)$ is a polynomial of degree n, and

$$\int_{-1}^{1} Q_n(x)P_k(x)\,dx=0$$

for each $k=0,1,2,\ldots,n-1$, then $Q_n(x)=cP_n(x)$ for some real constant c.

---

[7] A different solution to Legendre's equation is given by $Q_n(x)=\frac{1}{2}\int_{-1}^{1}\frac{P_n(t)}{x-t}dt$, which can also be expressed in terms of the hypergeometric function F as
$Q_n(x)=\frac{\sqrt{\pi}\,\Gamma(n+1)}{(2x)^{n+1}\Gamma\left(n+\frac{3}{2}\right)}F\left(\frac{1}{2}n+\frac{1}{2},\frac{1}{n}+1;n+\frac{3}{2};x^{-2}\right)$. $Q_n(x)$ is called a *Neumann formula for Legendre functions of the second kind*, named after Franz Ernst Neumann (1798–1895), a German mathematical physicist and crystallographer.

Problem 9.5.17. Prove *Rodrigues'*[8] *formula* $P_n(x)=\dfrac{1}{2^n n!}\dfrac{d^n}{dx^n}\left(x^2-1\right)^n$ for $n=0,1,2,\ldots$.

Problem 9.5.18. Prove that $P_n(x)$ with $n=0,1,2,\ldots$ are generated by $\dfrac{1}{\sqrt{1-2tx+t^2}}$. That is, show that if $|t|<1$ and $|x|<1$, then

$$\frac{1}{\sqrt{1-2tx+t^2}}=\sum_{k=0}^{\infty}P_k(x)t^k.$$

Problem 9.5.19. Use the generating function for $P_n(x)$ to prove that

$$\int_{-1}^{1}\left[P_n(x)\right]^2 dx=\frac{2}{2n+1}$$

with $n=0,1,2,\ldots$. That is, show that $\|P_n(x)\|=\sqrt{\dfrac{2}{2n+1}}$ for $n=0,1,2,\ldots$.

Problem 9.5.20. Find k_n, with $n=0,1,2,\ldots$, so that the sequence $\{k_nP_n(x)\}$ is orthonormal on $[-1,1]$.

Problem 9.5.21. Using the orthonormal functions from Problem 9.5.20, find the first few terms in the generalized Fourier series for $f(x)=\sin x$ on $[-1,1]$.

Problem 9.5.22. Prove the recurrence formula

$$(n+1)P_{n+1}(x)=(2n+1)xP_n(x)-nP_{n-1}(x)$$

for $n\in N$. Recall that $P_0(x)=1$.

Problem 9.5.23. Prove that for $n=0,1,2,\ldots$ $P_n(1)=1$ and $P_n(-1)=(-1)^n$.

Problem 9.5.24. Use Rodrigues' formula to prove that $P_n(x)$ is indeed equal to the sum

$$P_n(x)=\sum_{k=0}^{I}\frac{(-1)^k(2n-2k)!x^{n-2k}}{2^n k!(n-k)!(n-2k)!},$$

where $I=\dfrac{n}{2}$ or $I=\dfrac{n-1}{2}$, whichever is an integer, that is, $I=\left[\dfrac{n}{2}\right]$.

---

[8]Olinde Rodrigues (1794–1851) was a French economist and reformer. In his early years, Rodrigues had mathematical training, and later he developed formulas for certain special functions.

10

Vector Calculus

10.1* Cartesian Coordinates in $\Re^3$

The Cartesian coordinate system on a plane $\Re^2$ was discussed at the end of Section 1.1. Recall that the ordered pair (a,b) is used to represent a point on the two coordinate axes called x-axis and y-axis. In the space $\Re^3$ the *ordered triple* (a,b,c) is used on the coordinate axes made up of three straight lines called the x-axis, the y-axis and the z-axis, all of which intersect each other at right angles. View Figure 10.1.1 to see the positive direction of the axes and the location of the point (a,b,c), where the values $a,b,$ and c were chosen to be positive. The three planes, namely the xy-plane (which is the plane that contains both the x-axis and y-axis), xz-plane, and yz-plane are called *coordinate planes*. These planes divide space into eight parts called *octants*. In Figure 10.1.1 the point (a,b,c) is located in the *first octant*. The *Cartesian product* $\Re\times\Re\times\Re$ is the set of all ordered triples (a,b,c) in space and is denoted by $\Re^3 = \{(a,b,c) \mid a,b,c\in\Re\}$. The space $\Re^3$ is also referred to as a *three-dimensional rectangular coordinate system,* or as a *three-space*. Graphs of equations involving the three variables $x,y,$ and z in $\Re^3$ represent *surfaces*. Clarity of context under discussion is important. For example, $x=1$ is a point in $\Re^1$, a line in $\Re^2$, and a plane in $\Re^3$. To make a clear distinction, $x = 1$ in $\Re^2$ should be written as $\{(x, y) | x = 1\}$, and in $\Re^3$ as $\{(x, y, z) | x = 1\}$.

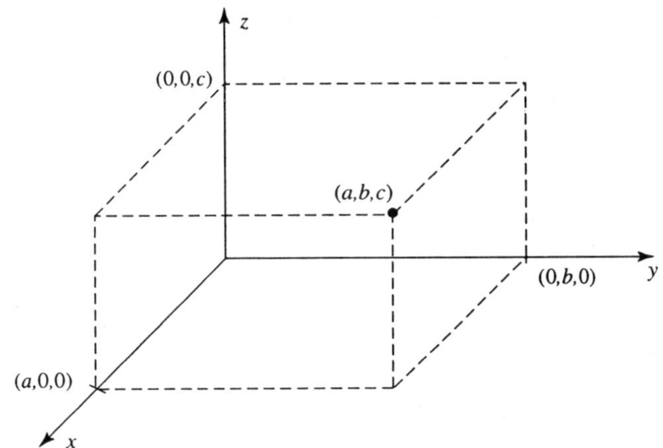

Figure 10.1.1

THEOREM 10.1.1. *(Euclidean Distance Formula) If* $P_1(x_1, y_1, z_1)$ *and* $P_2(x_2, y_2, z_2)$ *are two points in* $\Re^3$, *then the distance* $d(P_1, P_2)$ *between them is given by*

$$d(P_1, P_2) = \sqrt{(x_2 - x_1)^2 + (y_2 - y_1)^2 + (z_2 - z_1)^2}.$$

A *sphere* is the set of all points a specified distance (called the *radius*) from a fixed point (called the *center*). Thus, if the center is at (x_0, y_0, z_0) and the radius is r, $r > 0$, then an equation of the sphere is

$$(x - x_0)^2 + (y - y_0)^2 + (z - z_0)^2 = r^2.$$

This equation is an extension of the definition of a circle given in Section 1.2. If $r = 0$, the sphere is *degenerate* and consists of only one point.

Exercises 10.1

1. Prove the distance formula, Theorem 10.1.1.

2. Determine whether the triangle with vertices $(3, 3, 3)$, $(2, 1, -1)$, and $(5, 4, 2)$ is a right triangle.

3. If $P_1(x_1, y_1, z_1)$ and $P_2(x_2, y_2, z_2)$ are endpoints of a line segment (see part (a) of Definition 1.2.17), show that the midpoint $M(m_1, m_2, m_3)$ is when

$$m_1 = \frac{x_1 + x_2}{2},\ m_2 = \frac{y_1 + y_2}{2},\ m_3 = \frac{z_1 + z_2}{2}.$$

That is, show that the coordinates of the midpoint are averages of the corresponding coordinates of the endpoints.

4. Find all points equidistant from the point $(1, 1, -2)$ to the point $(0, 2, 5)$.

5. Find an equation of the sphere with center $(1, -2, 3)$ and that
(a) passes through the point $(-1, -3, 2)$.
(b) touches in one point the xz-plane.

6. Find an equation of the sphere with a diameter consisting of the endpoints $(4, 1, -2)$ and $(-3, 3, 0)$.

7. Find the center and radius of the sphere $x^2 + y^2 + z^2 - x + 2y + 3z = 1$.

8. What does the inequality $x^2 + y^2 + z^2 - x + 2y + 3z < 1$ represent geometrically in $\Re^3$?

10.2* Vectors in $\Re^3$

If P and Q are two distinct points in $\Re^3$, then there is exactly one line through both P and Q. The part of the line containing P, Q, and all the points between P and Q is called the *line segment* $\overline{PQ}$. If direction is important, that is if we wish to draw the line segment $\overline{PQ}$ starting at P and finishing at Q, then we have a *directed line segment* from P to Q denoted by $\overrightarrow{PQ}$. The point P is the *initial point.* A point Q is the *terminal point.* If P and Q are the same point, then $\overrightarrow{PQ}$ is *degenerate* and consists of only one point. The length of $\overrightarrow{PQ}$ is called the *magnitude* of $\overrightarrow{PQ}$, and the arrow used to represent $\overrightarrow{PQ}$ points in the direction from P to Q. Clearly, if $P \neq Q$, then $\overrightarrow{QP}$ is the directed line segment having the same magnitude as $\overrightarrow{PQ}$ though pointing in the opposite direction. Thus, $\overrightarrow{QP} = -\overrightarrow{PQ}$. Nondegenerate directed line segments are *equivalent* if they have the same magnitude and the same direction. A *vector*[1] is a directed line segment. It is

[1]The term *vector* comes from the Latin word *vectus*, meaning "to carry." Thus, if one is to carry something from the point P to the point Q, the trip is represented by a directed line segment, that is, a vector.

customary to represent vectors by $\mathbf{v}$, $\vec{v}$, or $\overrightarrow{PQ}$ where P is the initial point and Q the terminal point of the vector $\vec{v}$. Also, If P is the point (x_1, y_1, z_1) and Q is the point (x_2, y_2, z_2), then the vector $\overrightarrow{PQ}$ is represented by $\langle x_2 - x_1, y_2 - y_1, z_2 - z_1 \rangle$. Do not confuse this notation with the notation for finite continued fractions. Observe that $\langle x_2 - x_1, y_2 - y_1, z_2 - z_1 \rangle$ also represents the equivalent line segment from the origin to the point $(x_2 - x_1, y_2 - y_1, z_2 - z_1)$. The *zero vector,* that is $\mathbf{0}$, equivalently $\vec{0}$, represents a degenerate vector. Thus, $\vec{0} = \langle 0,0 \rangle$ in $\Re^2$ and $\vec{0} = \langle 0,0,0 \rangle$ in $\Re^3$. Other special vectors in $\Re^3$ are $\mathbf{i} \equiv \vec{i} = \langle 1,0,0 \rangle$, $\mathbf{j} \equiv \vec{j} = \langle 0,1,0 \rangle$, and $\mathbf{k} \equiv \vec{k} = \langle 0,0,1 \rangle$.[2] In $\Re^2$ we write $\vec{i} = \langle 1,0 \rangle$ and $\vec{j} = \langle 0,1 \rangle$. A *unit vector* is a vector whose magnitude is 1. And when dealing with vectors, real numbers are referred to as *scalars*. If $\vec{u} = \langle x_1, y_1, z_1 \rangle$, then the real numbers x_1, y_1, and z_1 are called *components* of the vector $\vec{u}$.

Definition 10.2.1. If $\vec{u} = \langle x_1, y_1, z_1 \rangle$, $\vec{v} = \langle x_2, y_2, z_2 \rangle$, and c is a real number, then

(a) the *magnitude* of $\vec{u}$ is $\|\vec{u}\|$ and is defined by

$$\|\vec{u}\| = \sqrt{(x_1)^2 + (y_1)^2 + (z_1)^2}\,.$$

(b) the *vector addition* $\vec{u} + \vec{v}$ is defined by

$$\vec{u} + \vec{v} = \langle x_1, y_1, z_1 \rangle + \langle x_2, y_2, z_2 \rangle = \langle x_1 + x_2, y_1 + y_2, z_1 + z_2 \rangle.$$

(c) the *scalar multiplication* $c\vec{u}$ is defined by

$$c\vec{u} = c\langle x_1, y_1, z_1 \rangle = \langle cx_1, cy_1, cz_1 \rangle.$$

If $\vec{u} = \langle x_1, y_1, z_1 \rangle$, P_1 is the point (x_1, y_1, z_1), and 0 is the origin, then the distance $d(0,P) = \|\vec{u}\|$. By $-\vec{v}$ we mean $(-1)\vec{v}$. Thus, the *difference* $\vec{u} - \vec{v}$ is the sum $\vec{u} + (-1)\vec{v}$. See Figure 10.2.1 for a geometrical representation of vector addition and vector subtraction. Also, division of a vector by a scalar is defined as a multiplication by the scalar's reciprocal. Division of two vectors is not defined. Multiplication of two vectors is defined in the next section. The following are some vector properties that will require proving (see Exercise 1).

---

[2]The notation $\vec{i}, \vec{j}, \vec{k}$ was introduced by William Rowan Hamilton (1805–1865), Ireland's leading mathematician of his time. Hamilton, who learned several languages by the age of 10 and became a professor of astronomy at 22, was very active in the areas of optics and matrix theory, and paved the way for the development of the vector theory.

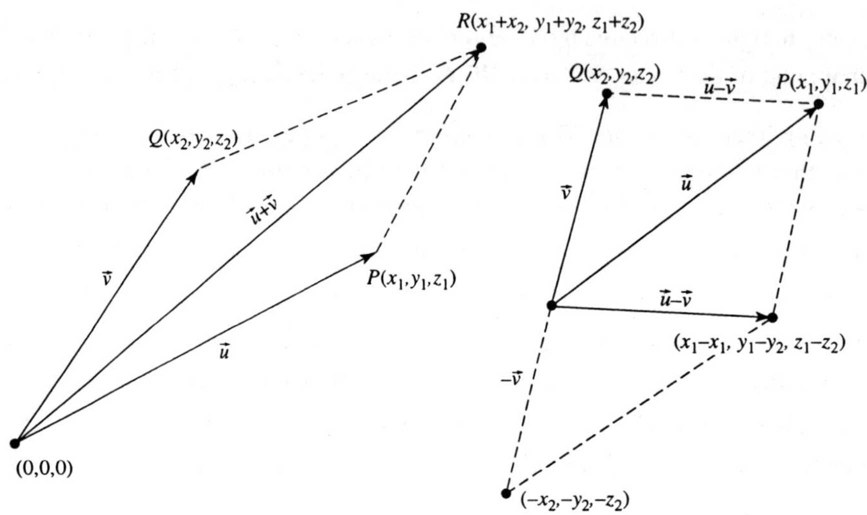

Figure 10.2.1

THEOREM 10.2.2. *If $\vec{u}$, $\vec{v}$, and $\vec{w}$ are vectors in $\Re^3$, and c and d are scalars, then*

(a)	$\vec{u}+\vec{v}=\vec{v}+\vec{u}$.	*(b)*	$\vec{u}+(\vec{v}+\vec{w})=(\vec{u}+\vec{v})+\vec{w}$.
(c)	$\vec{u}+\vec{0}=\vec{u}$.	*(d)*	$\vec{u}+(-\vec{u})=\vec{0}$.
(e)	$1\vec{u}=\vec{u}$.	*(f)*	$c(\vec{u}+\vec{v})=c\vec{u}+c\vec{v}$.
(g)	$(c+d)\vec{u}=c\vec{u}+d\vec{u}$.	*(h)*	$(cd)\vec{u}=c(d\vec{u})$.
(i)	$0\vec{u}=\vec{0}$.	*(j)*	$c\vec{0}=\vec{0}$.

(k) $c\vec{u}=\vec{0}$ *implies that* $c=0$ *or* $\vec{u}=\vec{0}$.

(l) $\|\vec{u}\|\geq 0$ *with* $\|\vec{u}\|=0$ *if and only if* $\vec{u}=\vec{0}$.

(m) $\|c\vec{u}\|=|c|\|\vec{u}\|$.

(n) $\|\vec{u}+\vec{v}\|\leq\|\vec{u}\|+\|\vec{v}\|$. (See part (e) of Theorem 1.7.2, Exercise 10(b) in Section 1.7, and Theorem 10.3.6.)

(o) $\vec{u}=\langle x_1, y_1, z_1\rangle$ *if and only if* $\vec{u}=x_1\mathbf{i}+y_1\mathbf{j}+z_1\mathbf{k}$.

Definition 10.2.3. If $\vec{v}_1, \vec{v}_2, \ldots, \vec{v}_n$ are n vectors, $n\in N$, and $c_1, c_2, \ldots, c_n$ are n scalars, then the vector $\vec{u}=c_1\vec{v}_1+c_2\vec{v}_2+\cdots+c_n\vec{v}_n$ is a *linear combination* of the vectors $\vec{v}_1, \vec{v}_2, \ldots, \vec{v}_n$.

Definition 10.2.4. Let $\vec{u}$ represent $\overrightarrow{OP}$ and $\vec{v}$ represent $\overrightarrow{OQ}$, where $\vec{v}$ is not a scalar multiple of $\vec{u}$. Then the *angle θ between $\vec{u}$ and $\vec{v}$* is the angle POQ. If $\theta=\dfrac{\pi}{2}$, then the vectors $\vec{u}$ and $\vec{v}$ are *orthogonal*, that is, *perpendicular*. If vectors $\vec{u}$ and $\vec{v}$ are scalar multiples, that is, if $\vec{v}=c\vec{u}$ for some scalar c, then the vectors $\vec{u}$ and $\vec{v}$ are *parallel*. If $c>0$, then $\theta=0$ and the vectors have the *same direction*. If $c<0$, then $\theta=\pi$, and the vectors have *opposite directions*.

From Definition 10.2.4 it follows that $\vec{0}$ is parallel and perpendicular to every vector. Why? Also, the angle θ between the two vectors $\vec{u}$ and $\vec{v}$ is such that $\theta \in [0, \pi]$. At times, a vector having the same direction as $\vec{u}$, but whose length is 1, will need to be found. To find such a *unit vector*, $\vec{u}$ is multiplied by the scalar $\frac{1}{\|\vec{u}\|}$. If this is done, we say that $\vec{u}$ was *normalized* and $\frac{1}{\|\vec{u}\|}\vec{u}$ is a unit vector having the same direction as $\vec{u}$. See Exercise 7 for verification. Also, $\frac{\vec{u}}{\|\vec{u}\|}$ is referred to as the *direction* of $\vec{u}$. Since $\|\vec{u}\| \cdot \frac{\vec{u}}{\|\vec{u}\|} = \vec{u}$, every vector can be written as the product of its length and direction.

Though all vectors discussed in this section were in $\Re^3$, the results are also true for vectors in $\Re^2$. When doing proofs, breaking down a vector into its components should be left as a last resort since this procedure often produces a lot of messy algebra. In the next section we will have more vector-related tools to use in proving.

Exercises 10.2

1. Prove Theorem 10.2.2.

2. If $\vec{u}$ and $\vec{v}$ are parallel to $\vec{w}$, show that every linear combination $c\vec{u} + d\vec{v}$, with $c, d \in \Re$, is also parallel to $\vec{w}$.

3. If $\vec{u}$ and $\vec{v}$ are vectors such that $\vec{u} + \vec{v} = \vec{u}$, show that $\vec{v} = \vec{0}$, and thus that the additive identity is unique.

4. If $\vec{u}$ and $\vec{v}$ are vectors such that $\vec{u} + \vec{v} = \vec{0}$, then show that $\vec{v} = -\vec{u}$, and thus that the additive inverse is unique.

5. Show that $-(-\vec{u}) = \vec{u}$.

6. If $\vec{u} = x_1\mathbf{i} + y_1\mathbf{j} + z_1\mathbf{k}$ and $\vec{v} = x_2\mathbf{i} + y_2\mathbf{j} + z_2\mathbf{k}$, show that $\vec{u} = \vec{v}$ if and only if $x_1 = x_2$, $y_1 = y_2$, and $z_1 = z_2$. That is, show that $\vec{u}$ and $\vec{v}$ are equal if and only if their corresponding components are equal.

7. If $\vec{u}$ is not degenerate, show that the vector $\vec{v} \equiv \frac{1}{\|\vec{u}\|}\vec{u}$ is a unit vector with the same direction as $\vec{u}$.

8. Write the vector $\vec{u} = 2\mathbf{i} + 3\mathbf{j} - 4\mathbf{k}$ as a

(a) product of the length and the direction.

(b) a linear combination of the vectors $\mathbf{i}, \mathbf{j}$, and $\mathbf{k}$.

(c) a linear combination of the vectors $\langle 1,1,0 \rangle$, $\langle -1,1,-2 \rangle$, and $\langle 0,1,-2 \rangle$.

10.3* Dot Product and Cross Product

Suppose that $\vec{u} = \langle x_1, y_1, z_1 \rangle$ and $\vec{v} = \langle x_2, y_2, z_2 \rangle$ are two nonzero vectors. From Definition 10.2.4, $\vec{u}$ and $\vec{v}$ are perpendicular if the angle between them is $\frac{\pi}{2}$. But, how is the angle between two nonzero vectors determined in just looking at the components of the vectors? From the Pythagorean theorem, $\vec{u}$ and $\vec{v}$ are perpendicular if and only if $\|\vec{u}\|^2 + \|\vec{v}\|^2 = \|\vec{u} - \vec{v}\|^2$. Why? In terms of components, this equality is equivalent to

$$\left[(x_1)^2 + (y_1)^2 + (z_1)^2\right] + \left[(x_2)^2 + (y_2)^2 + (z_2)^2\right] = (x_1 - x_2)^2 + (y_1 - y_2)^2 + (z_1 - z_2)^2,$$

which simplifies to $x_1x_2 + y_1y_2 + z_1z_2 = 0$. The expression on the left of the equal sign is called the dot product of $\vec{u}$ and $\vec{v}$.

Definition 10.3.1. If $\vec{u} = \langle x_1, y_1, z_1 \rangle$ and $\vec{v} = \langle x_2, y_2, z_2 \rangle$ are two vectors, then the *dot product*[3] $\vec{u} \cdot \vec{v}$ is defined as $\vec{u} \cdot \vec{v} = x_1x_2 + y_1y_2 + z_1z_2$.

Though the dot product acts on two vectors, it produces a scalar. The following properties of the dot product follow almost trivially. Their verification is left to Exercise 1.

THEOREM 10.3.2. *If $\vec{u}$, $\vec{v}$, and $\vec{w}$ are three vectors, and c and d are scalars, then*

(a) $\vec{u} \cdot \vec{v} = \vec{v} \cdot \vec{u}$.
(b) $\vec{u} \cdot (\vec{v} + \vec{w}) = \vec{u} \cdot \vec{v} + \vec{u} \cdot \vec{w}$.
(c) $(\vec{u} + \vec{v}) \cdot \vec{w} = \vec{u} \cdot \vec{w} + \vec{v} \cdot \vec{w}$.
(d) $\vec{u} \cdot \vec{u} = \|\vec{u}\|^2$.
(e) $\vec{0} \cdot \vec{u} = 0$.
(f) $c\vec{u} \cdot d\vec{v} = cd(\vec{u} \cdot \vec{v})$.
(g) $\vec{u} \cdot \vec{u} = 0$ *if and only if* $\vec{u} = \vec{0}$.

THEOREM 10.3.3. *If θ is the angle between the nonzero vectors $\vec{u}$ and $\vec{v}$, then $\vec{u} \cdot \vec{v} = \|\vec{u}\|\|\vec{v}\|\cos\theta$.*

Proof. Case 1. Suppose that $\vec{v} \neq c\vec{u}$ for any scalar c. Also, suppose $\vec{u} = x_1\mathbf{i} + y_1\mathbf{j} + z_1\mathbf{k}$ and $\vec{v} = x_2\mathbf{i} + y_2\mathbf{j} + z_2\mathbf{k}$. Then, the points $(0,0,0)$, $P(x_1, y_1, z_1)$, and $Q(x_2, y_2, z_2)$ form

---

[3]The dot product is a *Euclidean inner product* for $\Re^3$. The general inner product of two vectors $\vec{u}$ and $\vec{v}$ is defined in Part 1 of Section 10.10 and denoted by $(\vec{u}, \vec{v})$.

a triangle. Applying the law of cosines[4] to the angle θ at the origin we obtain

$$\left\|\overrightarrow{PQ}\right\|^2 = \|\vec{u}\|^2 + \|\vec{v}\|^2 - 2\|\vec{u}\|\|\vec{v}\|\cos\theta.$$

Therefore,

$$(x_2 - x_1)^2 + (y_2 - y_1)^2 + (z_2 - z_1)^2 =$$
$$(x_1)^2 + (y_1)^2 + (z_1)^2 + (x_2)^2 + (y_2)^2 + (z_2)^2 - 2\|\vec{u}\|\|\vec{v}\|\cos\theta,$$

which reduces to $x_1x_2 + y_1y_2 + z_1z_2 = \|\vec{u}\|\|\vec{v}\|\cos\theta$. And therefore, $\vec{u}\cdot\vec{v} = \|\vec{u}\|\|\vec{v}\|\cos\theta$, which we were to prove.

Case 2. Suppose $\vec{v} = c\vec{u}$ for some scalar c. Note that $\vec{u}$ and $\vec{v}$ are parallel. Moreover, if $c > 0$, we have $\theta = 0$. And if $c < 0$, we have $\theta = \pi$. First, assume that $c > 0$. Since $\cos\theta = \cos 0 = 1$, using Theorem 10.3.2 we have the following equalities:

$$c\|\vec{u}\|^2 = |c|\|\vec{u}\|^2 = |c|\|\vec{u}\|^2\cos\theta$$
$$c(\vec{u}\cdot\vec{u}) = |c|\|\vec{u}\|\|\vec{u}\|\cos\theta$$
$$\vec{u}\cdot c\vec{u} = \|\vec{u}\||c|\|\vec{u}\|\cos\theta$$
$$\vec{u}\cdot c\vec{u} = \|\vec{u}\|\|c\vec{u}\|\cos\theta$$
$$\vec{u}\cdot\vec{v} = \|\vec{u}\|\|\vec{v}\|\cos\theta.$$

If $c < 0$, we have $\theta = \pi$, and thus, $\cos\theta = -1$. Therefore,

$$c\|\vec{u}\|^2 = -c\|\vec{u}\|^2(-1) = -c\|\vec{u}\|^2\cos\theta = |c|\|\vec{u}\|^2\cos\theta,$$

and the conclusion follows as was done for $c > 0$. ■

COROLLARY 10.3.4. *If θ is the angle between the nonzero vectors $\vec{u}$ and $\vec{v}$, then* $\cos\theta = \dfrac{\vec{u}\cdot\vec{v}}{\|\vec{u}\|\|\vec{v}\|}$.

Thus, Theorem 10.3.3 is used to calculate the dot product, whereas Corollary 10.3.4 is used to calculate the angle between the vectors. Clearly, the vectors are orthogonal with

---

[4]**THEOREM** *(Law of Cosines) For a triangle with sides a, b, and c, and with opposite angles α, β, and γ, respectively, we have*

$$c^2 = a^2 + b^2 - 2ab\cos\gamma,\ b^2 = a^2 + c^2 - 2ac\cos\beta,\ \textit{and}\ a^2 = b^2 + c^2 - 2bc\cos\alpha.$$

(The first formula reduces to the Pythagorean theorem in the case $\gamma = \frac{\pi}{2}$.)

$\theta = \frac{\pi}{2}$ if and only if $\vec{u} \cdot \vec{v} = 0$. The dot product allows us to prove a number of results with a degree of ease, as we will see next.

THEOREM 10.3.5. *(Cauchy–Schwarz Inequality) If $\vec{u}$ and $\vec{v}$ are two vectors, then $|\vec{u} \cdot \vec{v}| \le \|\vec{u}\|\|\vec{v}\|$.*

See Exercise 10(a) in Section 1.7.

Proof. If either $\vec{u}$ or $\vec{v}$ is $\vec{0}$, equality holds. Suppose neither $\vec{u}$ nor $\vec{v}$ is $\vec{0}$, and suppose θ is the angle between them. Then, since $|\cos\theta| \le 1$, we have

$$|\vec{u} \cdot \vec{v}| = \|\vec{u}\|\|\vec{v}\||\cos\theta| \le \|\vec{u}\|\|\vec{v}\|.$$ ■

THEOREM 10.3.6. *(Triangle Inequality) If $\vec{u}$ and $\vec{v}$ are two vectors, then $\|\vec{u} + \vec{v}\| \le \|\vec{u}\| + \|\vec{v}\|$.*

See Exercise 10(b) in Section 1.7, part (e) of Theorem 1.7.2, and part (n) of Theorem 10.2.2.

Proof. Using previously presented properties of the dot product, we can write that

$$\begin{aligned} \|\vec{u} + \vec{v}\|^2 &= (\vec{u} + \vec{v}) \cdot (\vec{u} + \vec{v}) \\ &= (\vec{u} \cdot \vec{u}) + (\vec{v} \cdot \vec{u}) + (\vec{u} \cdot \vec{v}) + (\vec{v} \cdot \vec{v}) \\ &\le \|\vec{u}\|^2 + 2\|\vec{u}\|\|\vec{v}\| + \|\vec{v}\|^2 \\ &= (\|\vec{u}\| + \|\vec{v}\|)^2, \end{aligned}$$

which leads to the desired conclusion. ■

Recall from part (o) of Theorem 10.2.2 that every nonzero vector $\vec{u}$ can be written as a sum of the vectors parallel to the vectors $\mathbf{i}$, $\mathbf{j}$, and $\mathbf{k}$. Now let us write the vector $\vec{u}$ as a sum of the two orthogonal vectors $\vec{v}$ and $\vec{w}$, where $\vec{v}$ is parallel to some nonzero vector of our choice. Let us begin with two nonzero vectors $\vec{u}$ and $\vec{v}$. We are going to project $\vec{u}$ vertically onto $\vec{v}$ and search for the resulting vector. Three possibilities arise.

Case 1. Suppose the angle θ between $\vec{u}$ and $\vec{v}$ is acute, that is, $0 < \theta < \frac{\pi}{2}$. We are looking for the vector $\overrightarrow{OP}$. See Figure 10.3.1. Since the triangle OPQ is a right triangle, we have $\frac{\|\overrightarrow{OP}\|}{\|\vec{u}\|} = \cos\theta$. Thus, the length of $\overrightarrow{OP}$ is $\|\overrightarrow{OP}\| = \|\vec{u}\|\cos\theta$. Now, the direction of $\overrightarrow{OP}$ is $\dfrac{\vec{v}}{\|\vec{v}\|}$. According to Section 10.2, every vector can be written as a product of its length and direction. Thus, the vector $\overrightarrow{OP} = \left(\|\vec{u}\|\cos\theta\right)\dfrac{\vec{v}}{\|\vec{v}\|}$.

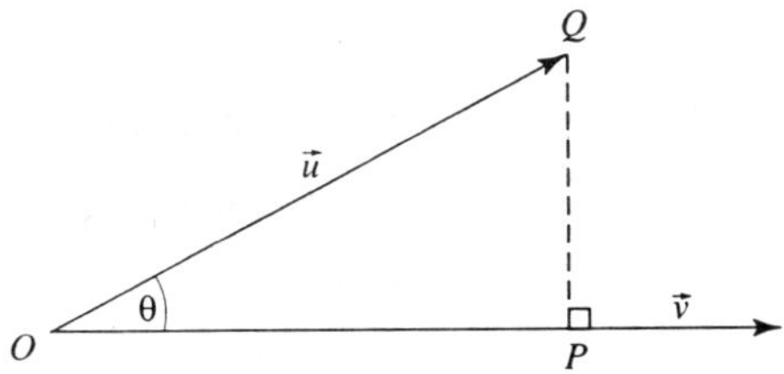

Figure 10.3.1

Case 2. Suppose the angle θ between $\vec{u}$ and $\vec{v}$ is obtuse, that is, $\dfrac{\pi}{2} < \theta < \pi$. Again, we are looking for the vector $\overrightarrow{OP}$. See Figure 10.3.2.

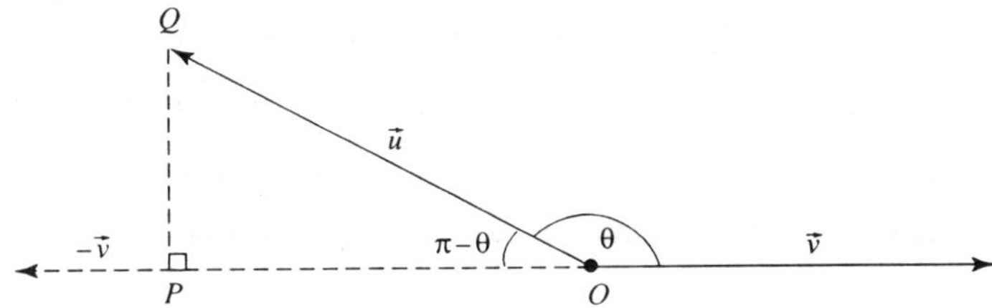

Figure 10.3.2

Since the triangle OPQ is a right triangle, we have that

$$\frac{\left\|\overrightarrow{OP}\right\|}{\|\vec{u}\|} = \cos(\pi - \theta) = \cos\pi\cos\theta + \sin\pi\sin\theta = -\cos\theta .$$

Thus, the length of $\overrightarrow{OP}$ is $\left\|\overrightarrow{OP}\right\| = -\|\vec{u}\|\cos\theta$, and the direction of $\overrightarrow{OP}$ is $-\dfrac{\vec{v}}{\|\vec{v}\|}$. Therefore, $\overrightarrow{OP} = \left(-\|\vec{u}\|\cos\theta\right)\left(-\dfrac{\vec{v}}{\|\vec{v}\|}\right) = \left(\|\vec{u}\|\cos\theta\right)\dfrac{\vec{v}}{\|\vec{v}\|}$.

So, if $\theta \in \left(0, \dfrac{\pi}{2}\right) \cup \left(\dfrac{\pi}{2}, \pi\right)$, then using Theorem 10.3.3, the vector $\overrightarrow{OP}$ can be written as

$$\vec{OP} = (\|\vec{u}\|\cos\theta)\frac{\vec{v}}{\|\vec{v}\|} = \left(\frac{\vec{v}}{\|\vec{v}\|}\cdot\vec{u}\right)\frac{\vec{v}}{\|\vec{v}\|} = \frac{1}{\|\vec{v}\|^2}(\vec{v}\cdot\vec{u})\vec{v} = \frac{1}{\vec{v}\cdot\vec{v}}(\vec{v}\cdot\vec{u})\vec{v} = \frac{\vec{v}\cdot\vec{u}}{\vec{v}\cdot\vec{v}}\vec{v}.$$

In the case $\theta = 0$ or $\theta = \pi$, $\vec{OP} = \vec{u}$, and if $\theta = \frac{\pi}{2}$, $\vec{OP} = \vec{0}$.

Definition 10.3.7. The vector $\vec{OP}$ discussed above is called an *orthogonal projection*, or just a *projection* of $\vec{u}$ onto $\vec{v}$ and denoted $\text{proj}_{\vec{v}}\,\vec{u}$. Thus, $\text{proj}_{\vec{v}}\,\vec{u} = \frac{\vec{v}\cdot\vec{u}}{\vec{v}\cdot\vec{v}}\vec{v}$.

See Exercise 12. Now of interest is finding a vector $\vec{w}$ which is both orthogonal to the $\text{proj}_{\vec{v}}\,\vec{u}$ and such that when added to the $\text{proj}_{\vec{v}}\,\vec{u}$ produces the vector $\vec{u}$. Few choices exist. The vector $\vec{w} \equiv \vec{u} - \text{proj}_{\vec{v}}\,\vec{u}$ is orthogonal to the $\text{proj}_{\vec{v}}\,\vec{u}$, as Exercise 13 shows. Hence, the vector $\vec{u}$ may be written as the sum of two vectors, one that is parallel to some vector $\vec{v}$, namely $\text{proj}_{\vec{v}}\,\vec{u}$, and the other, namely $\vec{w}$, that is orthogonal to it. (See Exercise 30 in Section 10.8.)

Observe that the dot product and all results to this point hold for vectors in $\Re^2$ as well as in $\Re^3$. Ordered triples are simply replaced by ordered pairs. However, vectors $\langle x_1, y_1, 0\rangle$ and $\langle x_1, y_1\rangle$ are not equal. The notation $\vec{u} = x_1\mathbf{i} + y_1\mathbf{j} + z_1\mathbf{k}$ is more effective than $\vec{u} = \langle x_1, y_1, z_1\rangle$ since $\vec{u} = x_1\mathbf{i} + y_1\mathbf{j}$ represents $\langle x_1, y_1\rangle$ in the context of $\Re^2$, while $\vec{u} = x_1\mathbf{i} + y_1\mathbf{j} = x_1\mathbf{i} + y_1\mathbf{j} + 0\mathbf{k}$ represents $\langle x_1, y_1, 0\rangle$ in the context of $\Re^3$. The vector $\vec{u} = \langle x_1, y_1, z_1\rangle$ remains in $\Re^3$ no matter what the components are.

Definition 10.3.8. If $\vec{u} = \langle x_1, y_1, z_1\rangle$ and $\vec{v} = \langle x_2, y_2, z_2\rangle$ are two vectors, then the *cross product*, also called the *vector product*, $\vec{u}\times\vec{v}$, is defined as

$$\vec{u}\times\vec{v} = (y_1z_2 - y_2z_1)\mathbf{i} - (x_1z_2 - x_2z_1)\mathbf{j} + (x_1y_2 - x_2y_1)\mathbf{k}.$$

Remark 10.3.9

(a) Using determinants, $\vec{u}\times\vec{v}$ can be written as

$$\vec{u}\times\vec{v} = \begin{vmatrix} y_1 & z_1 \\ y_2 & z_2 \end{vmatrix}\mathbf{i} - \begin{vmatrix} x_1 & z_1 \\ x_2 & z_2 \end{vmatrix}\mathbf{j} + \begin{vmatrix} x_1 & y_1 \\ x_2 & y_2 \end{vmatrix}\mathbf{k}.$$

(b) Using symbolic expansion of a 3 by 3 determinant about the first row, we can also write that

$$\vec{u}\times\vec{v} = \begin{vmatrix} \mathbf{i} & \mathbf{j} & \mathbf{k} \\ x_1 & y_1 & z_1 \\ x_2 & y_2 & z_2 \end{vmatrix}.$$

Keep in mind, however, that this 3 by 3 "determinant" is symbolic due to its vector entries.

(c) Since $\vec{u} \times \vec{v}$ is a vector, the cross product is also called a vector product.

(d) The cross product is valid only for vectors involving three components, that is vectors in $\mathfrak{R}^3$.

(e) The definition of the cross product was motivated from the search for a vector perpendicular to two arbitrary vectors $\vec{u}$ and $\vec{v}$ in $\mathfrak{R}^3$. Indeed, $\vec{u} \times \vec{v}$ is orthogonal to both $\vec{u}$ and $\vec{v}$. See Exercise 16. Also note that $\vec{u} \times \vec{u} = \vec{0}$. Why?

(f) The *right hand rule* is used to determine the direction of $\vec{u} \times \vec{v}$. That is, suppose that $\vec{u}$ and $\vec{v}$ have the same initial point. If the index finger of the right hand is $\vec{u}$ and the middle finger is $\vec{v}$, then the extended thumb of the right hand points in the direction of $\vec{u} \times \vec{v}$. ■

The magnitude of $\vec{u} \times \vec{v}$ is given in the next theorem.

THEOREM 10.3.10. *If θ is the angle between the nonzero vectors $\vec{u}$ and $\vec{v}$, then $\|\vec{u} \times \vec{v}\| = \|\vec{u}\|\|\vec{v}\| \sin\theta$.*

Proof. Using both $\vec{u} \cdot \vec{v} = \|\vec{u}\|\|\vec{v}\|\cos\theta$ and $\sin^2\theta = 1 - \cos^2\theta$, show $\|\vec{u} \times \vec{v}\|^2 = (\|\vec{u}\|\|\vec{v}\|)^2 \sin^2\theta$. Details are left as Exercise 17. ■

COROLLARY 10.3.11. *The nonzero vectors $\vec{u}$ and $\vec{v}$ are parallel if and only if $\vec{u} \times \vec{v} = \vec{0}$.*

Proof. $\vec{u}$ and $\vec{v}$ parallel imply that the angle θ between $\vec{u}$ and $\vec{v}$ is either 0 or π. Thus, by Theorem 10.3.10, $\|\vec{u} \times \vec{v}\| = 0$. Thus, $\vec{u} \times \vec{v} = \vec{0}$. Why? Proof of the converse follows similarly. See Exercise 18. ■

Number of properties of the dot and the cross product follow with proofs left to Exercise 19.

THEOREM 10.3.12. *If $\vec{u}$, $\vec{v}$, and $\vec{w}$ are vectors in $\mathfrak{R}^3$, and c and d are scalars, then*

(a) $\vec{u} \times \vec{v} = -(\vec{v} \times \vec{u}) = -\vec{v} \times \vec{u} = \vec{v} \times (-\vec{u})$.

(b) $(c\vec{u}) \times (d\vec{v}) = (cd)(\vec{u} \times \vec{v})$.

(c) $\vec{u} \times (\vec{v} + \vec{w}) = (\vec{u} \times \vec{v}) + (\vec{u} \times \vec{w})$.

(d) $(\vec{u} + \vec{v}) \times \vec{w} = (\vec{u} \times \vec{w}) + (\vec{v} \times \vec{w})$.

(e) $(\vec{u} \times \vec{v}) \cdot \vec{w} = (\vec{v} \times \vec{w}) \cdot \vec{u} = (\vec{w} \times \vec{u}) \cdot \vec{v}$.

(f) *And if $\vec{u} = \langle x_1, y_1, z_1 \rangle$, $\vec{v} = \langle x_2, y_2, z_2 \rangle$, and $\vec{w} = \langle x_3, y_3, z_3 \rangle$, then $\vec{u} \cdot (\vec{v} \times \vec{w})$ can be expressed as a determinant. In fact,*

$$\vec{u}\cdot(\vec{v}\times\vec{w})=\begin{vmatrix} x_1 & y_1 & z_1 \\ x_2 & y_2 & z_2 \\ x_3 & y_3 & z_3 \end{vmatrix}.$$

In closing, an application of the cross product to geometry follows. Consider two nonzero vectors $\vec{u}$ and $\vec{v}$, with the angle θ between them such that $0<\theta<\pi$. Then $\vec{u}$ and $\vec{v}$ can be used to form a parallelogram. See Figure 10.3.3.

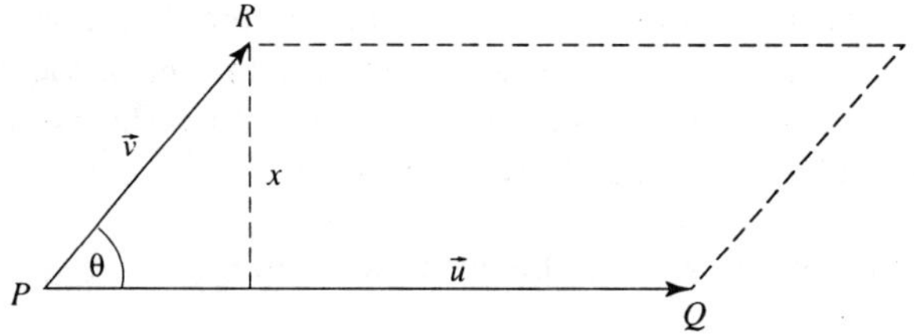

Figure 10.3.3

Observe from Figure 10.3.3 that $\sin\theta=\dfrac{x}{\|\vec{v}\|}$. Thus, $x=\|\vec{v}\|\sin\theta$. And so the area of the parallelogram is given by the $\|\vec{u}\|$ times the height x. Thus the area A is $A=\|\vec{u}\|x=\|\vec{u}\|\|\vec{v}\|\sin\theta=\|\vec{u}\times\vec{v}\|$.

Exercises 10.3

1. Prove Theorem 10.3.2.

2. If $\vec{u}$ and $\vec{v}$ are nonzero vectors, find
   (a) the largest value for $\vec{u}\cdot\vec{v}$. **(b)** the smallest value for $\vec{u}\cdot\vec{v}$.

3. Prove that **i**, **j**, and **k** are mutually orthogonal.

4. If $\vec{u}$ and $\vec{v}$ are vectors, and c and d are scalars, show that

$$\|c\vec{u}+d\vec{v}\|^2=c^2\|\vec{u}\|^2+2cd(\vec{u}\cdot\vec{v})+d^2\|\vec{v}\|^2.$$

5. If $\vec{u}\neq\vec{0}$ and $\vec{u}\cdot\vec{v}=\vec{u}\cdot\vec{w}$, is it true that $\vec{v}=\vec{w}$? Explain.

6. If $\vec{u}\cdot\vec{v}=0$ for every vector $\vec{v}$, what can be said about the vector $\vec{u}$?

7. **(a)** Verify that the squares of the diagonals of a parallelogram must be equal to the sum of the squares of the four sides.
   (b) Verify that the diagonals of a *rhombus*, that is a parallelogram whose sides have equal length, are perpendicular.

8. Verify that a triangle inscribed in a semicircle is a right triangle.

9. If $\vec{u}$ and $\vec{v}$ are vectors, prove that

(a) $\|\vec{u}\| - \|\vec{v}\| \le \|\vec{u} - \vec{v}\|$. See Exercise 6(a) in Section 1.7.

(b) $\left|\|\vec{u}\| - \|\vec{v}\|\right| \le \|\vec{u} - \vec{v}\|$. See Exercise 6(b) in Section 1.7.

(c) $\|\vec{u}\| - \|\vec{v}\| \le \|\vec{u} + \vec{v}\|$.

(d) $\|\vec{u} - \vec{v}\| \le \|\vec{u}\| + \|\vec{v}\|$. (e) $\vec{u} \cdot \vec{v} \le \|\vec{u}\|\|\vec{v}\|$

10. (a) When is $|\vec{u} \cdot \vec{v}| = \|\vec{u}\|\|\vec{v}\|$? (b) When is $\|\vec{u} + \vec{v}\| = \|\vec{u}\| + \|\vec{v}\|$?

11. (a) Show that $\|\vec{u} + \vec{v}\|^2 + \|\vec{u} - \vec{v}\|^2 = 2\left(\|\vec{u}\|^2 + \|\vec{v}\|^2\right)$.

(b) Show that $\vec{u} \cdot \vec{v} = \frac{1}{4}\left(\|\vec{u} + \vec{v}\|^2 - \|\vec{u} - \vec{v}\|^2\right)$.

(c) If $\vec{u} = \langle x, y \rangle$, prove that $|x| \le \|\vec{u}\| \le |x| + |y|$, that is, that

$$\sqrt{x^2} \le \sqrt{x^2 + y^2} \le \sqrt{x^2} + \sqrt{y^2}.$$

12. (a) Show that $\operatorname{proj}_{\vec{v}} \vec{u} = \frac{\vec{v} \cdot \vec{u}}{\vec{v} \cdot \vec{v}} \vec{v}$ for the cases $\theta = 0$, $\frac{\pi}{2}$, and π.

(b) Show that $\|\operatorname{proj}_{\vec{v}} \vec{u}\| = \left|\vec{u} \cdot \frac{\vec{v}}{\|\vec{v}\|}\right|$.

13. Assume that $\vec{u}$ and $\vec{v}$ are vectors. Show that the vector $\vec{v}$, and thus $\operatorname{proj}_{\vec{v}} \vec{u}$, is orthogonal to the vector $\vec{u} - \operatorname{proj}_{\vec{v}} \vec{u}$.

14. Express the vector $\vec{u} = 2\mathbf{i} - 3\mathbf{j} + \mathbf{k}$ as the sum of both a vector parallel to $\vec{v} = -\mathbf{i} - 2\mathbf{k}$ and a vector orthogonal to $\vec{v}$.

15. If $\vec{u} = 2\mathbf{i} + \mathbf{j} + 3\mathbf{k}$ and $\vec{v} = 4\mathbf{i} - \mathbf{j} + 2\mathbf{k}$, find $\vec{u} \times \vec{v}$ and $\vec{u} \times \vec{u}$.

16. Show that the vector $\vec{u} \times \vec{v}$ is orthogonal to both $\vec{u}$ and $\vec{v}$.

17. Prove Theorem 10.3.10.

18. Give a detailed proof of Corollary 10.3.11.

19. Prove Theorem 10.3.12.

20. If $\vec{u} = \langle 2,1,3 \rangle$, $\vec{v} = \langle 4,-1,2 \rangle$, and $\vec{w} = \langle -1,0,1 \rangle$, find

(a) $\vec{u} \times \vec{w}$. (b) $\vec{u} \cdot (\vec{w} \times \vec{u})$. (c) $(\vec{u} \times \vec{v}) \cdot \vec{w}$.

21. (a) Verify that $(\vec{u} + \vec{v}) \times (\vec{u} - \vec{v}) = -2(\vec{u} \times \vec{v})$.

(b) Verify that $\vec{u} \cdot \vec{v} \times \vec{w} = \vec{u} \times \vec{v} \cdot \vec{w}$.

(c) Is it true that $\vec{u} \cdot (\vec{v} \times \vec{w}) = (\vec{u} \cdot \vec{v}) \times \vec{w}$? Explain.

22. If $\vec{u} \times \mathbf{i} = \vec{0}$ and $\vec{u} \times \mathbf{k} = \vec{0}$, what can one say about $\vec{u}$?

23. Prove that for vectors $\vec{u}$, $\vec{v}$, $\vec{w}$, and $\vec{z}$, that

(a) $\vec{u} \times (\vec{v} \times \vec{w}) = (\vec{u} \cdot \vec{w})\vec{v} - (\vec{u} \cdot \vec{v})\vec{w}$.

(b) $(\vec{u} \times \vec{v}) \times \vec{w} = (\vec{u} \cdot \vec{w})\vec{v} - (\vec{v} \cdot \vec{w})\vec{u}$.

(c) $\|\vec{u}\times\vec{v}\|^2 + (\vec{u}\cdot\vec{v})^2 = \|\vec{u}\|^2\|\vec{v}\|^2$, known as *Lagrange's First Identity*.

(d) $(\vec{u}\times\vec{v})\cdot(\vec{w}\times\vec{z}) = (\vec{u}\cdot\vec{w})(\vec{v}\cdot\vec{z}) - (\vec{u}\cdot\vec{z})(\vec{v}\cdot\vec{w})$, known as *Lagrange's Second Identity*.

(e) $(\vec{u}\times\vec{v})\times(\vec{w}\times\vec{z}) = (\vec{u}\times\vec{v}\cdot\vec{z})\vec{w} - (\vec{u}\times\vec{v}\cdot\vec{w})\vec{z}$.

(f) $(\vec{u}\times\vec{v})\cdot(\vec{v}\times\vec{w})\times(\vec{w}\times\vec{u}) = (\vec{u}\cdot\vec{v}\times\vec{w})^2$.

24. Use the cross product to find the area of a triangle determined by the three points $P(2,3)$, $Q(1,-1)$, and $R(-4,0)$.

25. (a) Verify that $\vec{u}\times(\vec{v}\times\vec{w})$ is a linear combination of the vectors $\vec{v}$ and $\vec{w}$.

(b) Verify that $(\vec{u}\times\vec{v})\times\vec{w}$ is a linear combination of the vectors $\vec{u}$ and $\vec{v}$.

26. If $\vec{u}$ and $\vec{v}$ are not parallel, show that the law of cosines, as given in the footnote, is equivalent to Theorem 10.3.3.

27. (a) If $\vec{u}\neq\vec{0}$ and $\vec{u}\times\vec{v} = \vec{u}\times\vec{w}$, is it true that $\vec{v} = \vec{w}$? Explain. See Exercise 5.

(b) If $\vec{u}\neq\vec{0}$, $\vec{u}\cdot\vec{v} = \vec{u}\cdot\vec{w}$, and $\vec{u}\times\vec{v} = \vec{u}\times\vec{w}$, is it true that $\vec{v} = \vec{w}$? Explain.

10.4 Parametric Equations

Consider the line segment $\overline{PQ}$ in the xy-plane where P is the point $(6,5)$ and Q is the point $(10,-3)$. An equation of the line containing this line segment is $y=-2x+17$. Why? If a particle travels along the line segment $\overline{PQ}$ and t represents time, then when the time t is known, the position of the particle on $\overline{PQ}$ becomes of interest. Rewriting $y=-2x+17$ between the points P and Q in a *parametrized* form allows the position of the particle on $\overline{PQ}$ to be known. As an example, consider

$$x(t)=t, \text{ and } y(t)=-2t+17, \text{ for } t\in[6,10].$$

Then, if $t=6$ the particle is at P, and if $t=10$ the particle is at Q. For $t\in(6,10)$, the particle travels the straight line between P and Q. Why? The *direction* or the *orientation* of this parametrized line segment is said to be from P to Q. Now, if the time t is to be in the interval $[0,4]$, a different *parametrization* is written as

$$x(t)=t+6, \text{ and } y(t)=-2(t+6)+17=-2t+5, \text{ for } t\in[0,4].$$

And for a particle to travel the line segment $\overline{PQ}$ twice as fast, the parametrization

$$x(t)=2t+6, \text{ and } y(t)=-2(2t+6)+17=-4t+5, \text{ for } t\in[0,2]$$

is chosen. If the particle is to travel from Q to P, the parametrization

$$x(t) = 10 - \frac{1}{2}t, \text{ and } y(t) = t - 3, \text{ for } t \in [0,8]$$

is chosen. Note that using this last parametrization causes the particle to travel at half the speed the very first parametrization created, that is, $\overline{PQ}$ is *traced* at half the speed. Let us look at one more parametrization for $\overline{PQ}$, where

$$x(t) = 8 - 2\cos t, \text{ and } y(t) = 1 + 4\cos t, \text{ for } t \in [0, 2\pi].$$

In this parametrization, at both $t = 0$ and $t = 2\pi$ the particle is at the point P, and at $t = \pi$ the particle is at the point Q. Exercise 23 of Section 10.6 will allow us to verify that the particle starts at P from a stand-still, accelerates to reach its maximum speed from P to Q at the midpoint $(8,1)$, and then slows down until the point Q where it stops, turns around, and returns to P, again reaching its greatest speed at the midpoint.

Definition 10.4.1. Let $x = f(t)$, $y = g(t)$, $z = h(t)$, with functions f, g, and h having common domain D, where D is usually an interval. The collection of points $(x, y, z) = (f(t), g(t), h(t))$ is called a *curve* or *image* C in $\Re^3$. The equations $x = f(t)$, $y = g(t)$, and $z = h(t)$ with $t \in D$ are called *parametric equations* for this curve C, and the variable t is called a *parameter*. Moreover, the curve C is said to be *parametrized* by, or given *parametrically* by, $x = f(t)$, $y = g(t)$, and $z = h(t)$ with $t \in D$. These functions give a *parametrization* for the curve C which is often denoted by $\gamma(t) = (f(t), g(t), h(t))$ for $t \in D$.

In the case $h(t) \equiv 0$, the curve $\gamma(t) = (f(t), g(t))$ is in $\Re^2$. In this section we have seen that many parametrizations produce the same curve.

Example 10.4.2. A directed line segment from the point $P_0(x_0, y_0)$ to the point $P_1(x_1, y_1)$ can be parametrized by

$$x(t) = x_0 + t(x_1 - x_0), \text{ and } y(t) = y_0 + t(y_1 - y_0), \text{ for } t \in [0,1].$$

An elaboration of this parametrization is in Exercise 2. ■

Example 10.4.3. Consider the curves parametrized by $\gamma_1(t) = (2t + 6, -4t + 5)$ for $t \in [0,2]$ and $\gamma_2(t) = (3 - 5\cos \pi t, 1 + 5\sin \pi t)$ for $t \geq 0$.

(a) What curves do these parametrizations represent?
(b) Do these curves intersect?
(c) If t represents time and a particle travels on each curve, will the particles ever collide?

Answer to part (a). The parametrization γ_1 represents a line segment seen earlier, namely the equation $y = -2x + 17$ between the points $P(6,5)$ and $Q(10,-3)$. Since

$x = 3 - 5\cos \pi t$ implies that $\cos \pi t = \dfrac{3-x}{5}$, and $y = 1 + 5\sin \pi t$ implies that $\sin \pi t = \dfrac{y-1}{5}$, we have that $1 = \cos^2 \pi t + \sin^2 \pi t = \dfrac{(x-3)^2}{25} + \dfrac{(y-1)^2}{25}$, which yields the equation of the circle $(x-3)^2 + (y-1)^2 = 25$ for the curve represented by γ_2.

Answer to part (b). The line segment intersects the circle at points $(6,5)$ and $(8,1)$. Why?

Answer to part (c). The particle on the line segment is at the point of intersection $(6,5)$ with the circle when $t = 0$. But the particle which travels the circle at $t = 0$ has $x(0) = 3 - 5 = -2 \neq 6$. Thus, the particles do not intersect at $(6,5)$. However, the particle on the line segment is at the point of intersection $(8,1)$ when $t = 1$. And the particle that travels the circle at $t = 1$ has $x(1) = 3 - 5\cos \pi = 8$, as well. Thus, the particles collide at $(8,1)$, the only other possible point of intersection. ■

Definition 10.4.4. Assume $[a,b]$ is an interval where $-\infty \leq a < b \leq \infty$. A curve C is *continuous* if it can be parametrized by $\gamma(t) = (f(t), g(t), h(t))$, or $(f(t), g(t))$ in case $h(t) \equiv 0$, with $t \in [a,b]$ with functions f, g, and h continuous on $[a,b]$. The curve C is *smooth* if there is a parametrization $x = f(t)$, $y = g(t)$, and $z = h(t)$ where $t \in [a,b]$, and such that f', g', and h' are continuous on $[a,b]$, and

$$[f'(t)]^2 + [g'(t)]^2 + [h'(t)]^2 > 0$$

for all $t \in (a,b)$. Such parametrization is called *smooth*. The curve C is *piecewise smooth* if the interval $[a,b]$ can be partitioned into subintervals on which C is smooth.

It is customary to assume that if γ is a parametrization of a smooth curve C, then γ is a smooth parametrization. Also, if γ is smooth, then $f,g,h:[a,b] \to \mathfrak{R}$ are continuous and $f',g',h':(a,b) \to \mathfrak{R}$ are continuous and bounded. Why? Due to the notation of $\gamma(t) = (f(t), g(t), h(t))$ with $t \in [a,b]$, γ may be thought of as the function $\gamma:[a,b] \to \mathfrak{R}^3$ where γ's image is the curve C in space. Another way to think of γ is as a *vector-valued function* $\vec{r}(t) = f(t)\mathbf{i} + g(t)\mathbf{j} + h(t)\mathbf{k}$, where $(f(t), g(t), h(t))$ is the endpoint of the vector $\langle f(t), g(t), h(t) \rangle$. Thus, the collection of these endpoints produces the curve C. If $h(t) \equiv 0$, then γ maps into $\mathfrak{R}^2$. Some theory of vector-valued functions will be studied in Section 10.6.

Smooth in Definition 10.4.4 is meant to imply that the particle moving along the parametrized curve C suffers no abrupt changes of direction. The condition

$$\left(\frac{dx}{dt}\right)^2 + \left(\frac{dy}{dt}\right)^2 + \left(\frac{dz}{dt}\right)^2 > 0$$

with $t \in (a,b)$ implies that $f'(t)$, $g'(t)$, and $h'(t)$ are never simultaneously zero, except perhaps at the endpoints of the interval $[a,b]$. Thus the particle neither stops nor reverses direction. Without this condition most anything could happen, as will be pointed out in Exercise 4.

Now of interest are tangent lines to the parametrized curve C in the xy-plane without the elimination of a parameter when computing a slope by differentiation. In view of Exercise 4, the curve C will be assumed to be smooth on the interval $[a,b]$, with $-\infty \le a < b \le \infty$, and with a parametrization $\gamma(t) = (f(t), g(t))$ where $t \in [a,b]$. Thus, two possibilities arise: either $f'(t) \neq 0$ for any $t \in [a,b]$, or $f'(t) = 0$ and $g'(t) \neq 0$ for some $t \in [a,b]$. Why? Consider the case with $f'(t) \neq 0$ for any $t \in [a,b]$. Here, the curve C can be represented by $y = F(x)$. (See Exercise 5.) Then $g(t) = F(f(t))$. And applying the chain rule leads to

$$g'(t) = F'(f(t))f'(t) = F'(x)f'(t).$$

Thus, $F'(x) = \dfrac{g'(t)}{f'(t)}$. In the second case, if $f'(t_0) = 0$ for some $t_0 \in [a,b]$, then $g'(t_0) \neq 0$. Note that the derivative is sided if t_0 is at the endpoint. Hence, the curve C has a vertical tangent line at the point corresponding to t_0. Why? The above discussion is summarized next.

THEOREM 10.4.5. *If a smooth curve C is parametrized by $\gamma(t) = (f(t), g(t))$, then the slope $\dfrac{dy}{dx}$ of the tangent line to C at a point $P(x,y)$ is given by*

$$\frac{dy}{dx} = \frac{dy}{dt} \bigg/ \frac{dx}{dt},$$

provided $\dfrac{dx}{dt} \neq 0$.

Suppose the curve C is parametrized by $\gamma(t) = (f(t), g(t), h(t))$, where f, g, and h are differentiable functions. Then the vector $\dfrac{\vec{v}}{\|\vec{v}\|}$, with $\vec{v}(t) = \langle f'(t), g'(t), h'(t) \rangle$, is the *direction* vector pointing in the direction of the increasing values of t, thus determining the curve's *direction* (*orientation*). Again, if $h(t) \equiv 0$, then the curve C is in $\Re^2$. Section 10.6 discusses this idea in more depth. Theorem 10.4.5 is not that valuable if the parametrized curve C can be easily written as $y = F(x)$. However, if this is not the case, then Theorem 10.4.5 becomes valuable, as shall see later.

Example 10.4.6. Consider the curve C traced by a fixed point P on the circumference of a circle that rolls along a horizontal line without slipping. This curve

C is called a *cycloid*.[5] See Figure 10.4.1. Find a parametrization for this cycloid.

Answer. Let the circle of center Q and radius a roll along the x-axis beginning with P at the origin. A parameter t may be chosen to be the radian measure of the angle AQP, where A is where the circle touches the x-axis. See Figure 10.4.1. Since the circle rolls without slipping, the arc AP has the same length as the line segment $\overline{OA}$. Therefore, $at \equiv d(O,A)$. Also, the x-coordinate of the point P is given by

$$d(O,B) = d(O,A) - d(A,B) = at - a\sin t = a(t - \sin t).$$

Why? And the y-coordinate of the point P is given by

$$d(B,P) = d(A,Q) - d(Q,D) = a - a\cos t = a(1 - \cos t).$$

Thus, the parametric equations for the cycloid are

$$x(t) = a(t - \sin t), \text{ and } y(t) = a(1 - \cos t).$$

Note that the endpoint of each arch is called a *cusp*. ■

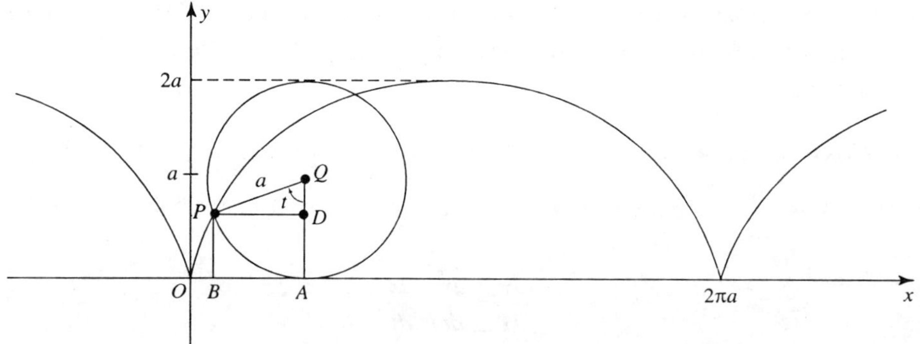

Figure 10.4.1

Consider the cycloid when $a < 0$. Then the graph parametrized by $x(t) = a(t - \sin t)$ and $y(t) = a(1 - \cos t)$ is called an *inverted cycloid* due to its similarity with the cycloid of Example 10.4.6 and with the exception that the circle of radius $|a|$ rolls below the x-axis. An inverted cycloid plays an important role in physics. In 1696, Johann Bernoulli posed the following problem. If a particle is to descend along a curve from the point A to the point B, with B not directly below A, and if the only force acting on

[5]The term cycloid was designated by Galileo in 1599. Galileo Galilei (1564–1642) was an Italian pioneer of modern mathematics, physics, and astronomy. Galileo is known for his study of free fall and for his use of the telescope. He was persecuted for supporting the Copernican theory of the solar system. Of worth noting is that Nicolaus Copernicus (1473–1543), whose theory was that the earth has a daily motion about its axis and a yearly motion around a stationary sun, was born in Poland under the name Mikołaj Kopernik.

the particle is gravity, what should be the shape of the curve so that the particle reaches B in the least possible amount of time? The problem was soon solved by Johann Bernoulli, his brother Jacob, Newton, Leibniz, and L'Hôpital. Their answer was an inverted cycloid, even though a straight line provides the shortest distance. Due to this fact, the curve of an inverted cycloid is known as the *brachistochrone*,[6] in Greek meaning the curve of "shortest time." In addition, the inverted cycloid is the only curve possessing the following property. If a particle slides down a curve towards the lowest point A with only gravity acting upon it, then the particle will reach A at the same time no matter where it starts along the curve. Due to this fact, the curve of an inverted cycloid is also known as the *tautochrone*,[7] in Greek meaning the curve of "equal time."

Definition 10.4.7. The curve C parametrized by $\gamma(t)$ is p*-periodic*, that is *periodic* with a period p, if and only if there exists a number $p>0$ such that $\gamma(t+p)=\gamma(t)$ for all $t\in\Re$. The smallest such number is called the *fundamental period* for γ. (Recall Definition 4.2.9.)

Definition 10.4.8. The curve C parametrized by $\gamma(t)$ with $t\in[a,b]$ is *closed* if and only if the initial point $\gamma(a)$ is the same as the terminal point $\gamma(b)$, that is $\gamma(a)=\gamma(b)$.

Exercises 10.4

1. Find an equation in rectangular coordinates that represents the curve C having the following indicated parametric equations. Sketch C and specify its direction.
   (a) $x(t)=t^3,\ y(t)=t^2$ (b) $x(t)=\sin^2 t,\ y(t)=\sin t$
   (c) $x(t)=-2\sin t,\ y(t)=3\cos t$, with $t\in[0,2\pi]$
   (d) $x(t)=e^t,\ y(t)=e^{-t}$
2. (a) Show that the functions $x(t)=x_0+t(x_1-x_0)$ and $y(t)=y_0+t(y_1-y_0)$ parametrize the line that passes through the points (x_0,y_0) and (x_1,y_1). See Example 10.4.2.
   (b) The equation $x(t)=x_0+t(x_1-x_0)$ is also commonly written as $x(t)=tx_1+(1-t)x_0$. What does this expression represent geometrically if t increases from 0 to 1? See Definition 5.6.22 and Definition 5.6.23.
3. Parametrize the line segment from the point $P(-1,3)$ to the point $Q(2,-5)$ for $t\in[0,4]$.
4. (a) Find a parametrization for some curve C which has a horizontal tangent line at some point in $\Re^2$.

---

[6]The brachistochrone problem was solved by Johann Bernoulli in 1697, and marked the beginning of the development of the calculus of variations.

[7]The tautochrone property of the inverted cycloid was known to the Dutch astronomer Christiaan Huygens, when in 1673, he published a description of an ideal pendulum clock in which the bob was to swing along a cycloid.

(b) Find a parametrization for some curve C which has a vertical tangent line at some point.
(c) Find a parametrization for some curve C which has a tangent line with a slope of one at some point.
(d) Find a parametrization for some curve C which has no tangent line at some point.
(e) Find a parametrization for some curve C which has two tangent lines at some point. See Exercise 8.

5. Suppose the curve C is parametrized by $\gamma(t)=(f(t),g(t))$ with $t\in[a,b]$, where $-\infty\le a<b\le\infty$. If f' is continuous on $[a,b]$, and $f'(t)\neq 0$ for any $t\in[a,b]$, show that C can be represented in rectangular coordinates by $y=F(x)$.

6. For $y^5=x^3$ give a parametrization that is
(a) smooth.
(b) not smooth.

7. Consider curves C_1 and C_2 given parametrically by $\gamma_1(t)=(3t+3,t+3,t)$ and $\gamma_2(t)=(t-2,t,2t-5)$, respectively.
(a) Do these curves intersect?
(b) If t represents time and particles travel on each curve with the positions determined by $\gamma_1(t)$ and $\gamma_2(t)$, will the particles ever collide?

8. Determine whether the curve $\gamma(t)=\left(t^2,t^3-t\right)$ intersects itself. If so, determine where, and find the equations for the tangent lines at that point(s). See Exercise 15 in Section 10.6.

9. Suppose the curve C is the graph of a nonnegative function F on the interval $[a,b]$. Let C be parametrized by $\gamma(t)=(f(t),g(t))$ with $t\in[c,d]$, where $f(c)=a$, $f(d)=b$, and f' and g are continuous on $[c,d]$. Then, show that the area between the curve C and the x-axis, and the lines $x=a$ and $x=b$, is given by $\int_c^d f'(t)g(t)\,dt$.

10. (a) Find the slope of the tangent line to the cycloid, given in Example 10.4.6, at $t=\frac{\pi}{4}$ for $a=1$ and $a=10$.
(b) Find an equation for all the tangent lines to the cycloid that are horizontal.
(c) Verify that $x'(t)$ and $y'(t)$ are both zero at the end of each arch of the cycloid.
(d) Verify that the cycloid is piecewise smooth.
(e) *(Torricelli's[8] Theorem)* Find the area under one arch of a cycloid. How does it compare to the area of the circle generating the cycloid?
(f) Find a rectangular equation of the cycloid in x and y.

---

[8]Evangelista Torricelli (1608–1647), born in Italy, discovered the principle of a barometer and was the first person to create a sustained vacuum.

11. Suppose C is the graph of the ellipse $\dfrac{x^2}{9}+\dfrac{y^2}{4}=1$.

(a) Verify that C can be parametrized by $x(t)=3\cos t$, and $y(t)=2\sin t$ with $t\in[0,2\pi]$.

(b) Use parametric equations to find $\dfrac{dy}{dx}$ at the point $\left(\dfrac{3\sqrt{3}}{2},1\right)$.

(c) Use implicit differentiation to find $\dfrac{dy}{dx}$ at the point $\left(\dfrac{3\sqrt{3}}{2},1\right)$.

(d) Use parametric equations to find $\dfrac{d^2y}{dx^2}$ at both $t=\dfrac{\pi}{6}$ and $t=\dfrac{\pi}{2}$. Why is one value larger than the other?

(e) Discuss the orientation of C for $t=0$, $\dfrac{\pi}{2}$, π, and $\dfrac{3\pi}{2}$.

12. Suppose C is an *astroid*[9] with the parametric equations $x(t)=a\cos^3 t$ and $y(t)=a\sin^3 t$, where a is a positive constant, and P is a point on C in the first quadrant. See Exercise 8 in Section 12.7.

(a) Find a rectangular equation in both x and y for C.

(b) Show that the length of the tangent line to the curve C at the point P, which is inside the first quadrant, is of constant length, independent of the position of P.

13. Suppose that the curve C_1 is parametrized by $\gamma_1(t)=(\cos t,\sin t,0)$, and that the curve C_2 is parametrized by $\gamma_2(t)=(0,\cos t,\sin t)$.

(a) Find points where C_1 intersects C_2.

(b) Show that C_1 and C_2 intersect each other at right angles.

14. Consider the *Witch of Agnesi*[10] curve C, constructed as follows. Draw a circle of radius a with center at the point $(0,a)$. Then draw the tangent line $y=2a$. Next, choose a point A on the line $y=2a$ and connect it with the origin O by the line segment AO. If the point B is where the line segment AO crosses the circle, then the point P is where the horizontal line through B and the vertical line through A intersect. The set of all points P makes up the curve C. See Figure 10.4.2.

---

[9] *Hypocycloid* curve is traced by a fixed point P on a circle of radius b as it rolls inside a fixed circle of radius a with $a>b$. If $a=4b$, then a hypocycloid has four cusps and is called an *astroid*. An astroid was first introduced by Johann Bernoulli in 1691.

[10] Maria Gaëtana Agnesi (1718–1799), an Italian linguist, was one of the first widely recognized woman mathematician. In 1748 Agnesi published the first mathematics text that included differential and integral calculus, infinite series, differential equations, and the bell-shaped curve, which at that time was called "versiera," meaning the "turning curve." When John Colson was translating the text into English in 1801, he must have confused "versiera" with "avversiera," meaning the "witch," and thus, the name "witch of Agnesi" stuck. The bell-shaped curve was discovered earlier by Fermat, and then studied by Grandi in 1703.

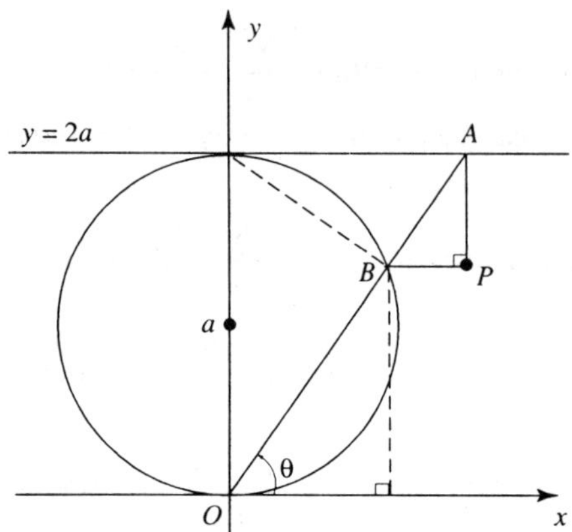

Figure 10.4.2

Find some parametric equations including an equation in Cartesian coordinates for the curve C. Also, graph C. See Exercises 1(m),1(n), and 1(o) in Section 8.1.

15. Are the curves given by $\gamma_1(t) = \left(\cot t, \sin^2 t\right)$ and $\gamma_2(t) = \left(\tan t, \cos^2 t\right)$ the same? Explain the difference between them, if any.

16. **(a)** The curve C parametrized by $\gamma(t) = (8 - 2\cos t, 4\cos t + 1)$ with $t \in [0, 2\pi]$ is closed. True or false?

(b) If the curve C given in part (a) is periodic, find a period p. Does C have a fundamental period? If so, what is it?

(c) Is there a difference in the shape of C when $t \in [0, \pi]$ versus $t \in [\pi, 2\pi]$?

17. What is the fundamental period for the curve C given by $\gamma(t) = (\sin t, \cos 2t, \sin 3t)$?

10.5* Lines and Planes in $\Re^3$

In Section 10.4 a special type of curve called a line was discussed. Suppose ℓ is a line which both passes through the point $P_0(x_0, y_0, z_0)$ and is parallel to the vector $\vec{u} = a\mathbf{i} + b\mathbf{j} + c\mathbf{k}$. To find an equation for the line ℓ, all points $P(x, y, z)$ for which the vector $\overrightarrow{P_0P}$ is parallel to the vector $\vec{u}$ need to be found. In terms of components, points that satisfy $\overrightarrow{P_0P} = t\vec{u}$ for some t real, must satisfy

$$(x - x_0)\mathbf{i} + (y - y_0)\mathbf{j} + (z - z_0)\mathbf{k} = t(a\mathbf{i} + b\mathbf{j} + c\mathbf{k}).$$

Now, equating the corresponding components on both sides of the equality and rearranging, we obtain that $x = x_0 + at$, $y = y_0 + bt$, and $z = z_0 + ct$. The following result has now been verified.

THEOREM 10.5.1. *If a line ℓ both passes through a point $P_0(x_0, y_0, z_0)$ and is parallel to a nonzero vector $\vec{u} = a\mathbf{i} + b\mathbf{j} + c\mathbf{k}$, then the parametric equations for ℓ are $x = x_0 + at$, $y = y_0 + bt$, and $z = z_0 + ct$, with $t \in \Re$.*

Naturally, t in Theorem 10.5.1 can be replaced by any multiple kt. Also, if a, b, and c are all different from 0, we can eliminate t and write an equation for the line ℓ in *symmetric form* as

$$\frac{x - x_0}{a} = \frac{y - y_0}{b} = \frac{z - z_0}{c}.$$

As we will see later, $\frac{x - x_0}{a} = \frac{y - y_0}{b}$ and $\frac{y - y_0}{b} = \frac{z - z_0}{c}$ are equations for two planes, and their intersection is in fact a line. Moreover, the line ℓ in Theorem 10.5.1 can be given by $\gamma(t) = (x_0 + at, y_0 + bt, z_0 + ct)$, which in turn can be represented by the vector-valued function

$$\begin{aligned} \vec{r}(t) &= (x_0 + at)\mathbf{i} + (y_0 + bt)\mathbf{j} + (z_0 + ct)\mathbf{k} \\ &= (x_0\mathbf{i} + y_0\mathbf{j} + z_0\mathbf{k}) + t(a\mathbf{i} + b\mathbf{j} + c\mathbf{k}) \equiv \vec{v} + t\vec{u}. \end{aligned}$$

Notice that $\vec{r}(t)$ resembles an equation for a line given in Section 1.2.

Example 10.5.2. Find parametric equations for the line ℓ passing through the points $P(1,-2,4)$ and $Q(-3,2,0)$.

Answer. The vector $\overrightarrow{PQ} = -4\mathbf{i} + 4\mathbf{j} - 4\mathbf{k}$ is parallel to the line ℓ, and thus, $\ell = (1 - 4t, -2 + 4t, 4 - 4t)$ for $t \in \Re$. ■

Note that if $t = 0$, then the particle traveling along ℓ is at the point P, and if $t = 1$, then the particle is at Q. Therefore, the directed line segment from P to Q is $(1 - 4t, -2 + 4t, 4 - 4t)$ for $t \in [0,1]$. Review Example 10.4.2. Also, observe that ℓ can be parametrized by $x = 1 - t$, $y = -2 + t$, and $z = 4 - t$ with $t \in \Re$. With this parametrization, ℓ is traveled by the particle four times slower than with the original parametrization given above. What if ℓ were parametrized by $\left(1 - \frac{s}{\sqrt{3}}, -2 + \frac{s}{\sqrt{3}}, 4 - \frac{s}{\sqrt{3}}\right)$? Then, if $s = 1$, the particle travels exactly 1 unit from P. Why? This special type of parametrization will be discussed in Section 10.7.

In the xy-plane, any relation of the form $ax + by = 0$, with $a, b \in \Re$, not both 0, represents a line through the origin. If $a \neq 0$, then the slope of a line perpendicular, that is normal, to $ax+by=0$ is $\frac{b}{a}$. Why? Therefore, the vector $a\mathbf{i}+b\mathbf{j}$ is perpendicular to the line $ax+by=0$. In fact, $a\mathbf{i}+b\mathbf{j}$ is perpendicular to any translation of this line. So, if the origin is translated to the point (x_0, y_0), then in the equation of the line x is replaced by $x-x_0$ and y by $y-y_0$ to obtain $a(x-x_0)+b(y-y_0)=0$. In $\Re^2$, this is an equation of a line passing through the point (x_0, y_0) with a normal vector $\vec{n}=a\mathbf{i}+b\mathbf{j}$. Naturally, an equation of the plane in $\Re^3$ would be similar, as the next result will show.

THEOREM 10.5.3. *An equation for the plane through the point* (x_0, y_0, z_0) *with a normal vector* $\vec{n}=\langle a,b,c\rangle$ *is given by*

$$a(x-x_0)+b(y-y_0)+c(z-z_0)=0.$$

Proof. Suppose that M is a plane and $P_0(x_0, y_0, z_0)$ is a point on M. M consists of all points $P(x,y,z)$ for which $\overrightarrow{P_0P}$ is perpendicular to the given vector $\vec{n}$. That is, P is on M if and only if $\vec{n}\cdot\overrightarrow{P_0P}=0$. In components, this is equivalent to

$$(a\mathbf{i}+b\mathbf{j}+c\mathbf{k})\cdot\left((x-x_0)\mathbf{i}+(y-y_0)\mathbf{j}+(z-z_0)\mathbf{k}\right)=0,$$

which simplifies to the desired expression. ■

Example 10.5.4. Find an equation for the plane passing through $P_0(-1,0,2)$, $P_1(1,1,3)$, and $P_2(0,2,-4)$.

Answer. Two vectors in the plane with a common point are $\overrightarrow{P_0P_1}=\langle 2,1,1\rangle$ and $\overrightarrow{P_0P_2}=\langle 1,2,-6\rangle$. To find a normal vector $\vec{n}$ to the plane, the cross product needs to be computed. Thus,

$$\vec{n}=\overrightarrow{P_0P_1}\times\overrightarrow{P_0P_2}=\begin{vmatrix}\mathbf{i} & \mathbf{j} & \mathbf{k}\\ 2 & 1 & 1\\ 1 & 2 & -6\end{vmatrix}=-8\mathbf{i}+13\mathbf{j}+3\mathbf{k}.$$

Using either P_0, P_1, or P_2, Theorem 10.5.3 gives an equation for the plane to be

$$-8x+13y+3k=-8(-1)+13(0)+3(2)=14.$$ ■

The graph of an equation involving three variables is called a *surface*. Planes are examples of surfaces. And future study will reveal other types of surfaces. *Cylinders* are formed by a curve in the plane M. This curve is moved through space in a direction perpendicular to M. Some other common surfaces are *quadric surfaces*. Examples of quadric surfaces centered at the origin are the following:

(a) the *sphere:* $x^2+y^2+z^2=r^2$,

(b) the *ellipsoid:* $\dfrac{x^2}{a^2}+\dfrac{y^2}{b^2}+\dfrac{z^2}{c^2}=1$,

(c) the *elliptic paraboloid:* $z=\dfrac{x^2}{a^2}+\dfrac{y^2}{b^2}$,

(d) the *elliptic cone:* $z^2=\dfrac{x^2}{a^2}+\dfrac{y^2}{b^2}$,

(e) the *hyperbolic paraboloid* (a *saddle*): $z=\dfrac{y^2}{b^2}-\dfrac{x^2}{a^2}$,

(f) the *hyperboloid of one sheet:* $\dfrac{x^2}{a^2}+\dfrac{y^2}{b^2}-\dfrac{z^2}{c^2}=1$, and

(g) the *hyperboloid of two sheets:* $-\dfrac{x^2}{a^2}-\dfrac{y^2}{b^2}+\dfrac{z^2}{c^2}=1$.

A comfortable awareness in graphing such surfaces using a calculator or computer is advisable.

Surfaces in comparison to curves are drastically different in nature. A projectile can move one of two ways on a curve, forward or backward. On a surface, however, a particle can move many different ways. For example, the surface of a right circular cylinder formed by moving the circle $x^2+y^2=4$ along the z-axis can be parametrized by $(\cos t,\sin t,z)$, where t and z are independent variables and vary through all real values. These types of functions will be studied in the next chapter.

Exercises 10.5

1. Find an equation for the line which both passes through $P_0(-1,2,3)$ and
   (a) is parallel to the line defined by $\ell(t)=(1,2t,-3+t)$.
   (b) is perpendicular to the line defined by $\ell(t)=(1,2t,-3+t)$.

2. Find the distance from the point $P(1,-1,2)$ to the line given by $\ell(t)=(1+t,2-2t,3+t)$.

3. Consider the lines $\ell_1(t)=(3t+3,t+3,t)$ and $\ell_2(t)=(t-2,t,2t-5)$. Review Exercise 7 in Section 10.4.
   (a) Do these lines intersect?
   (b) If t represents time and a particle travels on each line with its position determined by $\ell_1(t)$ and $\ell_2(t)$, will they ever collide?

4. Find a point where the line given by $\left(t+1,2t-1,\dfrac{1}{3}t\right)$ intersects the plane $2x-y+3z=6$.

5. Find the distance from the point $P(0,-1,2)$ to the plane $2x+3y-z=6$.

6. Find the (acute) angle between the planes $2x-y=7$ and $-x+y-3z=5$.

7. Find some parametric equations for the line in which the planes $-x+3y+z=3$ and $2x-y+2z=4$ intersect.

8. Two lines are *skew* if they are neither parallel nor intersect. Show that if ℓ_1 is the line through $P_1(1,2,7)$ and $Q_1(-2,3,-4)$, and ℓ_2 is the line through $P_2(2,-1,4)$ and $Q_2(5,7,-3)$, then ℓ_1 and ℓ_2 are skew.

9. Find an equation of the plane which contains the origin and the vectors $\vec{u}=2\mathbf{i}-\mathbf{j}+3\mathbf{k}$ and $\vec{v}=4\mathbf{j}-2\mathbf{k}$.

10. If the vectors $\vec{u}$, $\vec{v}$, and $\vec{w}$ have the same initial point, prove that $\vec{u}\cdot(\vec{v}\times\vec{w})=0$ if and only if these vectors lie in one plane.

10.6 Vector-Valued Functions

Definition 10.6.1. Suppose that f, g, and h are real-valued functions defined on $D\subseteq\Re$. Then the *vector-valued function*, or the *vector function* $\vec{r}$ on D is defined by

$$\vec{r}(t)=f(t)\mathbf{i}+g(t)\mathbf{j}+h(t)\mathbf{k}$$

with $t\in D$. The functions f, g, and h are the *component functions* of $\vec{r}$.

Preferably, $\vec{r}(t)$ is written as $f(t)\mathbf{i}+g(t)\mathbf{j}+h(t)\mathbf{k}$ rather than as $\langle f(t),g(t),h(t)\rangle$, since for example, in the first notation, if $h(t)\equiv 0$ then clearly $\vec{r}(t):D\subseteq\Re^1\to\Re^2$ instead of to $\Re^3$. In the vector notation $\langle f(t),g(t),h(t)\rangle$, however, $\vec{r}(t):D\subseteq\Re^1\to\Re^3$ no matter what h is, provided it is defined on D. Clearly, if h is not defined on D, neither is $\vec{r}$. For the case $g(t)\equiv 0$ and $h(t)\equiv 0$, $\vec{r}(t):D\subseteq\Re^1\to\Re^1$. Since the range of $\vec{r}$ are vectors $\langle f(t),g(t),h(t)\rangle$ determined by the points $(f(t),g(t),h(t))$, we may think of $\vec{r}$ as a function from $\Re^1$ to $\Re^3$. As values of t vary, the points $(f(t),g(t),h(t))$ *trace* the curve C in space. Hence, parametric equations for the curve C, that is, the image of γ given by $\gamma(t)=(f(t),g(t),h(t))$, and the range of the function $\vec{r}(t)=f(t)\mathbf{i}+g(t)\mathbf{j}+h(t)\mathbf{k}$, are one and the same. Again, if $h(t)\equiv 0$ and values of t vary, then $\vec{r}(t):D\subseteq\Re^1\to\Re^2$ and the image of γ given by $\gamma(t)=(f(t),g(t))$ represents the same parametrized curve in $\Re^2$. Next is a direct generalization of Definition 3.2.1 to vector-valued functions.

Definition 10.6.2. Suppose that $\vec{r}(t)=f(t)\mathbf{i}+g(t)\mathbf{j}+h(t)\mathbf{k}$ with $t\in D\subseteq\Re$, and t_0 is an accumulation point of D. Then, $\lim\limits_{t\to t_0}\vec{r}(t)=\vec{L}\equiv L_1\mathbf{i}+L_2\mathbf{j}+L_3\mathbf{k}$ if and only if for every $\varepsilon > 0$, there exists a real number $\delta > 0$, such that $\|\vec{r}(t) - \vec{L}\| < \varepsilon$, provided $0 < |t - t_0| < \delta$ and $t \in D$.

Note that $\|\vec{r}(t) - \vec{L}\| = \sqrt{[f(t) - L_1]^2 + [g(t) - L_2]^2 + [h(t) - L_3]^2}$ is the Euclidean distance in $\mathfrak{R}^3$, or in $\mathfrak{R}^2$ if $h(t) \equiv 0$. Thus, the limit in $\mathfrak{R}^3$corresponds to the limit in $\mathfrak{R}^2$. Definition 10.6.2 can be broken down into limits of individual components. Basically, if $\vec{r}(t) = f(t)\mathbf{i} + g(t)\mathbf{j} + h(t)\mathbf{k}$, then

$$\lim_{t \to t_0} \vec{r}(t) = \left[\lim_{t \to t_0} f(t) \right] \mathbf{i} + \left[\lim_{t \to t_0} g(t) \right] \mathbf{j} + \left[\lim_{t \to t_0} h(t) \right] \mathbf{k}.$$

This is formally recorded in the next theorem.

THEOREM 10.6.3. *Suppose that $\vec{r}(t) = f(t)\mathbf{i} + g(t)\mathbf{j} + h(t)\mathbf{k}$ with $t \in D \subseteq \mathfrak{R}$, and t_0 is an accumulation point of D. Then, $\lim_{t \to t_0} \vec{r}(t) = \vec{L} \equiv L_1\mathbf{i} + L_2\mathbf{j} + L_3\mathbf{k}$ if and only if $\lim_{t \to t_0} f(t) = L_1$, $\lim_{t \to t_0} g(t) = L_2$, and $\lim_{t \to t_0} h(t) = L_3$.*

Proof. ($\Rightarrow$) Suppose that $\lim_{t \to t_0} \vec{r}(t) = \vec{L}$. We will show only that $\lim_{t \to t_0} f(t) = L_1$. Other parts are left to the reader. Let $\varepsilon > 0$ be given. We need to find $\delta > 0$, such that whenever $0 < |t - t_0| < \delta$ and $t \in D$, then $|f(t) - L_1| < \varepsilon$. Since $\lim_{t \to t_0} \vec{r}(t) = \vec{L}$, there exists $\delta_1 > 0$ such that

$$\|\vec{r}(t) - \vec{L}\| = \sqrt{[f(t) - L_1]^2 + [g(t) - L_2]^2 + [h(t) - L_3]^2} < \varepsilon,$$

provided $0 < |t - t_0| < \delta_1$ and $t \in D$. But clearly, dropping two squared terms yields $\sqrt{[f(t) - L_1]^2} = |f(t) - L_1| < \varepsilon$. Hence, choosing $\delta = \delta_1$, the desired conclusion follows.

($\Leftarrow$) Let $\varepsilon > 0$ be given. Then, since $\lim_{t \to t_0} f(t) = L_1$, there exists $\delta_1 > 0$ such that $|f(t) - L_1| < \boxed{\frac{\varepsilon}{3}}$ for $0 < |t - t_0| < \delta_1$ and $t \in D$. Also, since $\lim_{t \to t_0} g(t) = L_2$, there exists $\delta_2 > 0$ such that $|g(t) - L_2| < \boxed{\frac{\varepsilon}{3}}$ for $0 < |t - t_0| < \delta_2$ and $t \in D$. And similarly, since $\lim_{t \to t_0} h(t) = L_3$, there exists $\delta_3 > 0$ such that $|h(t) - L_3| < \boxed{\frac{\varepsilon}{3}}$ for $0 < |t - t_0| < \delta_3$ and $t \in D$. Pick $\delta = \min\{\delta_1, \delta_2, \delta_3\}$. Then, if $t \in D$ and $0 < |t - t_0| < \delta$, by Exercise 11(c) of Section 10.3 we have that

$$\|\vec{r}(t)-\vec{L}\| \le |f(t)-L_1|+|g(t)-L_2|+|h(t)-L_3| < \frac{\varepsilon}{3}+\frac{\varepsilon}{3}+\frac{\varepsilon}{3}=\varepsilon .$$

Hence, $\lim_{t\to t_0} \vec{r}(t)=\vec{L}$. ■

Sequences are special types of functions, and using an extension of the notation from Chapter 2, with $\vec{a}_n = \left\langle \frac{\sin n}{n}, \frac{2n^2+1}{n^2}, \sqrt[n]{n} \right\rangle$, then $\lim_{n\to\infty} \vec{a}_n = \langle 0,2,1\rangle$. Why? Also, $\lim_{t\to 0}\left\langle \frac{\sin^2 t}{t}, \frac{\sin^2 t}{t^2}, \frac{t^2 \sin\frac{1}{t}}{\sin t}\right\rangle = \langle 0,1,0\rangle$, because

$$\lim_{t\to 0}\frac{\sin^2 t}{t} = \lim_{t\to 0} \sin t\cdot\frac{\sin t}{t} = 0\cdot 1 = 0,$$
$$\lim_{t\to 0}\frac{\sin^2 t}{t^2} = \left(\lim_{t\to 0}\frac{\sin t}{t}\right)^2 = 1^2 = 1, \text{ and}$$
$$\lim_{t\to 0}\frac{t^2\sin\frac{1}{t}}{\sin t} = \lim_{t\to 0}\left(\frac{t}{\sin t}\right)\left(t\sin\frac{1}{t}\right) = 1\cdot 0 = 0 .$$

L'Hôpital's rule could have been implemented.

Due to Theorem 10.6.3 a number of results can readily be presented with proofs left to Exercise 5.

THEOREM 10.6.4. *Suppose that $D \subseteq \Re$ is the domain for $\vec{u}(t)$ and $\vec{v}(t)$, $k: D \to \Re$ is a scalar function,*[11] *c is a scalar, and t_0 is an accumulation point of D. If $\lim_{t\to t_0}\vec{u}(t)$, $\lim_{t\to t_0}\vec{v}(t)$, and $\lim_{t\to t_0}k(t)$ are all finite, then*

(a) $\lim_{t\to t_0}\left[\vec{u}(t)\pm\vec{v}(t)\right] = \lim_{t\to t_0}\vec{u}(t) \pm \lim_{t\to t_0}\vec{v}(t)$.

(b) $\lim_{t\to t_0} c\vec{u}(t) = c\lim_{t\to t_0}\vec{u}(t)$.

(c) $\lim_{t\to t_0} k(t)\vec{u}(t) = \left[\lim_{t\to t_0} k(t)\right]\left[\lim_{t\to t_0}\vec{u}(t)\right]$.

(d) $\lim_{t\to t_0}\left[\vec{u}(t)\cdot\vec{v}(t)\right] = \left[\lim_{t\to t_0}\vec{u}(t)\right]\cdot\left[\lim_{t\to t_0}\vec{v}(t)\right]$.

(e) $\lim_{t\to t_0}\left[\vec{u}(t)\times\vec{v}(t)\right] = \left[\lim_{t\to t_0}\vec{u}(t)\right]\times\left[\lim_{t\to t_0}\vec{v}(t)\right]$, *provided $\vec{u},\vec{v}: D\to\Re^3$.*

---

[11] A function $k: D\subseteq\Re\to\Re$ in the context of vectors is called a *scalar function* since its image is a set of scalars.

Definition 10.4.4 expressed what is meant by a continuous curve. Employing Definition 10.6.2, the definition of continuity may be rewritten in terms of a limit.

Definition 10.6.5. A vector-valued function $\vec{r}(t)=f(t)\mathbf{i}+g(t)\mathbf{j}+h(t)\mathbf{k}$ with $t \in D \subseteq \Re$, is *continuous* at $t=t_0 \in D$ if and only if for every $\varepsilon>0$, there exists $\delta>0$ such that $\left\|\vec{r}(t)-\vec{r}(t_0)\right\|<\varepsilon$, for all $t \in D$ such that $|t-t_0|<\delta$.

As a consequence of Definition 10.6.5, the vector-valued function $\vec{r}$ is *continuous at* $t=t_0$, an accumulation point of its domain, if and only if $\lim_{t\to t_0} \vec{r}(t)=\vec{r}(t_0)$.

Definition 10.6.5 and Theorem 10.6.3 allow us to readily prove that sums and differences of continuous vector-valued functions are also continuous. See Exercise 7. Uniform continuity given in Definition 4.4.3 may also be extended to vector-valued functions. See Exercise 9. Now consider the concept of differentiability.

Definition 10.6.6. The vector-valued function $\vec{r}$ is *differentiable* if and only if $\vec{r}$ has a *derivative* at $t=t_0$ given by

$$\vec{r}'(t_0)=\lim_{t\to t_0}\frac{\vec{r}(t)-\vec{r}(t_0)}{t-t_0}$$

and this limit is finite.

As with single-valued functions, the above definition of $\vec{r}'(t_0)$ can be written as

$$\vec{r}'(t_0)=\lim_{h\to 0}\frac{\vec{r}(t_0+h)-\vec{r}(t_0)}{h}.$$

If $\vec{r}$ is differentiable at every value t in its domain D, that is if $\lim_{h\to 0}\frac{\vec{r}(t+h)-\vec{r}(t)}{h}$ is finite for all $t \in D$, then this limit represents the *derivative function* $\vec{r}'(t)$, and we say that $\vec{r}$ is *differentiable* (on D). Due to Theorem 10.6.3, $\vec{r}'(t)$ exists if and only if $f'(t)$, $g'(t)$, and $h'(t)$ exist, where $\vec{r}(t)=f(t)\mathbf{i}+g(t)\mathbf{j}+h(t)\mathbf{k}$, and $\vec{r}'(t)$ is given by $\vec{r}'(t)=f'(t)\mathbf{i}+g'(t)\mathbf{j}+h'(t)\mathbf{k}$. Why? As with limits and continuity of functions from $\Re^1$ to $\Re^1$, concepts of differentiability extend to vector-valued functions.

THEOREM 10.6.7. *Suppose that $\vec{u}$ and $\vec{v}$ are vector-valued functions with $t \in D \subseteq \Re$, and $k: D \to \Re$ is a scalar function, and all functions are differentiable. Then*

(a) $\frac{d}{dt}\left[\vec{u}(t)\pm\vec{v}(t)\right]=\vec{u}'(t)\pm\vec{v}'(t)$.

(b) $\frac{d}{dt}\left[\vec{u}(t)\cdot\vec{v}(t)\right]=\vec{u}(t)\cdot\vec{v}'(t)+\vec{u}'(t)\cdot\vec{v}(t)$.

(c) $\frac{d}{dt}\left[k(t)\vec{u}(t)\right]=k(t)\vec{u}'(t)+k'(t)\vec{u}(t)$.

(d) *(Chain Rule)* $\frac{d}{dt}\vec{u}(k(t)) = \vec{u}'(k(t))k'(t)$, *provided* $k(t) \in D$.

(e) $\frac{d}{dt}\|\vec{u}(t)\| = \frac{1}{\|\vec{u}(t)\|}[\vec{u}(t)\cdot\vec{u}'(t)]$, *if* $\vec{u}(t) \neq \vec{0}$.

(f) $\frac{d}{dt}[\vec{u}(t)\times\vec{v}(t)] = \vec{u}(t)\times\vec{v}'(t) + \vec{u}'(t)\times\vec{v}(t)$.

Proofs of Theorem 10.6.7 are left as Exercise 11. In view of Definition 10.4.4, the vector-valued function $\vec{r}(t)$, with $t \in [a,b]$, and $-\infty \le a < b \le \infty$, is *smooth* if it is *continuously differentiable* (that is, if all component functions of $\vec{r}(t)$ are continuously differentiable), and $\vec{r}'(t) \neq \vec{0}$ except perhaps when $t = a$ and/or $t = b$, in case they are finite values.

Suppose the curve C is determined by a smooth vector-valued function $\vec{r}$. And consider the functions $\vec{r}$ and $\vec{r}'$ graphically. From Figure 10.6.1, if $\vec{r}(t)$ represents the vector $\overrightarrow{OP}$, and $\vec{r}(t+h)$ represents the vector $\overrightarrow{OQ}$, then the vector $\overrightarrow{PQ} = \vec{r}(t+h) - \vec{r}(t)$.

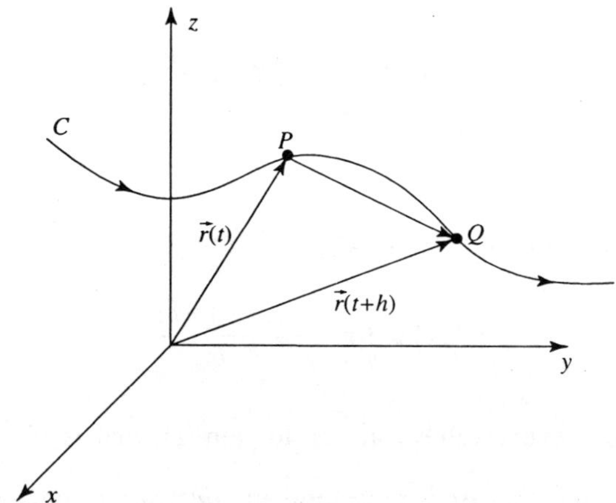

Figure 10.6.1

Why? As $h \to 0$, the vector $\overrightarrow{PQ}$ approaches the tangent to the curve C. Thus, a scalar multiple $\frac{1}{h}$ of $\overrightarrow{PQ}$ also approaches the tangent. Therefore, $\vec{r}'(t) = \lim_{h\to 0}\frac{1}{h}[\vec{r}(t+h) - \vec{r}(t)]$ is a *tangent vector* to the curve C. The *unit tangent vector* $\frac{\vec{r}'(t)}{\|\vec{r}'(t)\|}$, the normalized $\vec{r}'(t)$, will be denoted by $\vec{T}(t)$. Whether h is positive or negative, the vector $\frac{1}{h}[\vec{r}(t+h) - \vec{r}(t)]$ points in the direction of the increasing values of t. Why? Hence, taking limits as $h \to 0$, we find that $\vec{r}'(t)$ is both tangent to the curve C and points in the direction of motion. Since tangency represents a change of position, $\vec{r}'(t)$ is also called the *velocity vector* $\vec{V}(t)$. Recall that in Section 10.4 we called $f'(t)\mathbf{i}+g'(t)\mathbf{j}+h'(t)\mathbf{k}$ the direction vector for the curve C parametrized by $\gamma(t)=(f(t),g(t),h(t))$ or $\gamma(t)=(f(t),g(t))$ in the case when $h(t)\equiv 0$. The length of $\vec{r}'(t)$ is the *speed*, as Section 10.7 will verify. Similarly, if $\vec{r}$ is twice differentiable, then $\vec{r}''(t)=\dfrac{d}{dt}\vec{V}(t)\equiv\vec{A}(t)$ is the *acceleration vector.*

Example 10.6.8. Let $\vec{u}(t)=\langle\cos t,\sin t\rangle$ and $\vec{v}(t)=\langle\cos 2t,\sin 2t\rangle$.

(a) Find the image for each function.

(b) Find a value of t_0 for which $\vec{v}(t_0)$ gives the same point as $\vec{u}\left(\dfrac{\pi}{6}\right)$.

(c) Compare values of $\vec{u}'\left(\dfrac{\pi}{6}\right)$ with $\vec{v}'(t_0)$.

(d) Find an equation for the tangent line to the curve C determined by the function $\vec{u}$ at $t=\dfrac{\pi}{6}$.

(e) Find the acceleration vectors $\vec{u}''\left(\dfrac{\pi}{6}\right)$ and $\vec{v}''(t_0)$.

Answer to part (a). Since $\cos^2 z+\sin^2 z=1$, both functions have the unit circle $x^2+y^2=1$ for their image.

Answer to part (b). Since $\vec{u}\left(\dfrac{\pi}{6}\right)=\left\langle\dfrac{\sqrt{3}}{2},\dfrac{1}{2}\right\rangle$, we need to find a value t_0 so that $\langle\cos 2t_0,\sin 2t_0\rangle=\left\langle\dfrac{\sqrt{3}}{2},\dfrac{1}{2}\right\rangle$. Equating components, we have that $\cos 2t_0=\dfrac{\sqrt{3}}{2}$ and $\sin 2t_0=\dfrac{1}{2}$. Thus, one value of t_0 can be $\dfrac{\pi}{3}$.

Answer to part (c). Since both $\vec{u}$ and $\vec{v}$ are differentiable, we have $\vec{u}'(t)=\langle-\sin t,\cos t\rangle$ and $\vec{v}'(t)=\langle-2\sin 2t,2\cos 2t\rangle$. Thus, $\vec{u}'\left(\dfrac{\pi}{6}\right)=\left\langle-\dfrac{1}{2},\dfrac{\sqrt{3}}{2}\right\rangle$ and $\vec{v}'\left(\dfrac{\pi}{3}\right)=\langle-1,\sqrt{3}\rangle$, which are not the same values. However, $\vec{u}'$ at $\dfrac{\pi}{6}$ and $\vec{v}'$ at $\dfrac{\pi}{3}$ point in the same direction of travel, counterclockwise. It is obvious that the velocity of $\vec{v}$ is twice that of $\vec{u}$. $\vec{v}$ takes only π units of time to travel around the circle, whereas $\vec{u}$ takes 2π units of time.

Answer to part (d). The tangent line is the same for both curves at the point $\left(\dfrac{\sqrt{3}}{2},\dfrac{1}{2}\right)$. Since the tangent line is in $\Re^2$, its equation has a slope of $m=-\sqrt{3}$. Therefore, a desirable equation could be $y=-\sqrt{3}x+2$. Why?

Answer to part (e). Since $\vec{u}'(t) = \langle -\sin t, \cos t\rangle$ and $\vec{v}'(t) = \langle -2\sin 2t, 2\cos 2t\rangle$, and both vectors are differentiable, then $\vec{u}''(t) = \langle -\cos t, -\sin t\rangle$ and $\vec{v}''(t) = \langle -4\cos 2t, -4\sin 2t\rangle$. Therefore, $\vec{u}''\left(\frac{\pi}{6}\right) = \left\langle -\frac{\sqrt{3}}{2}, -\frac{1}{2}\right\rangle$ and $\vec{v}''\left(\frac{\pi}{3}\right) = \langle -2\sqrt{3}, -2\rangle$. Observe that both acceleration vectors point toward the center of the circle. Acceleration vectors are the forces that keep cars on the road while turning. Worth noting is that the acceleration of $\vec{v}$ is four times that of $\vec{u}$, even though the velocity of $\vec{v}$ is only twice that of $\vec{u}$, so wear your seat belt in either case. ■

As Example 10.6.8 points out, one cannot predict the behavior of a vector-valued function based on its looks. Next is another example why one "should not judge the book by its cover." Also see Exercise 17.

Example 10.6.9. Consider $\vec{u}, \vec{v}, \vec{w} : \Re^1 \to \Re^2$ defined by

$$\vec{u}(t) = \langle t, t^2\rangle,\ \vec{v}(t) = \langle t^3, t^6\rangle, \text{ and } \vec{w}(t) = \begin{cases} \langle t, t^2\rangle, & \text{if } t \le 0 \\ \langle t^3, t^6\rangle, & \text{if } t > 0. \end{cases}$$

All three functions are continuous, all have as curves the parabola $y = x^2$, are $(0,0)$ at $t = 0$, and trace the smooth-appearing curve in the same direction, that is from left to right. Why? Now, $\vec{u}$ is differentiable at the origin with the tangent vector $\vec{u}'(0) = \langle 1, 0\rangle$, and with the tangent line $y = 0$ in $\Re^2$. Why? The function $\vec{v}$ is also differentiable at the origin with the tangent vector $\vec{v}'(0) = \langle 0, 0\rangle$. Clearly, $\vec{v}$ has no tangent line. Why? The function $\vec{w}$ is not differentiable at the origin, and thus, $\vec{w}$ has no tangent line there. Why? ■

THEOREM 10.6.10. *Suppose $\vec{r}$ is a differentiable vector-valued function defined for $t \in D \subseteq \Re$. Then, $\|\vec{r}(t)\|$ is constant if and only if $\vec{r}$ and $\vec{r}'$ are orthogonal.*

What does Theorem 10.6.10 mean in $\Re^2$? And what does it mean in $\Re^3$? See Exercise 18. Theorem 10.6.10 has a nice application to the acceleration vector $\vec{A}$, where $\vec{A}(t) = \vec{r}''(t) = f''(t)\mathbf{i} + g''(t)\mathbf{j} + h''(t)\mathbf{k}$, provided $\vec{r}(t) = f(t)\mathbf{i} + g(t)\mathbf{j} + h(t)\mathbf{k}$ is twice differentiable. If $\vec{V}(t) \ne \vec{0}$, then we have the unit tangent vector $\vec{T}(t) = \frac{\vec{r}'(t)}{\|\vec{r}'(t)\|} = \frac{\vec{V}(t)}{\|\vec{V}(t)\|}$, where $v(t) = \|\vec{V}(t)\|$ is the *speed* of a particle traveling along the curve C determined by $\vec{r}(t)$. See Section 10.7. Thus, differentiating $\vec{V}(t) = v(t)\vec{T}(t)$, by Theorem 10.6.7, part (c), we obtain that

$$\vec{A}(t) = \vec{V}'(t) = v(t)\vec{T}'(t) + v'(t)\vec{T}(t).$$

Now, since $\left\|\vec{T}(t)\right\| = 1$, which is a constant, by Theorem 10.6.10 $\vec{T}'(t)$ is orthogonal to $\vec{T}(t)$. And since $\vec{T}(t)$ is tangent to the curve C at time t, we have that $\vec{T}'(t)$ is normal to C. The normalized vector $\dfrac{\vec{T}'}{\left\|\vec{T}'\right\|}$ is called the *unit normal vector* and is denoted by $\vec{N}$.

Thus, $\vec{A}_{\vec{T}}(t) \equiv v'(t)T(t)$ is tangent to the curve C, and $\vec{A}_{\vec{N}}(t) \equiv v(t)T'(t)$ is normal. Why? By writing $\vec{A}(t)$ as the sum of these vectors which are both tangent and normal to the curve C, the acceleration vector is being decomposed into the sum of the *tangential acceleration* $\vec{A}_{\vec{T}}(t)$, which is being directed along the path of motion, and the *normal acceleration* $\vec{A}_{\vec{N}}(t)$, which describes the rate at which the direction of motion is changing. See Figure 10.6.2.

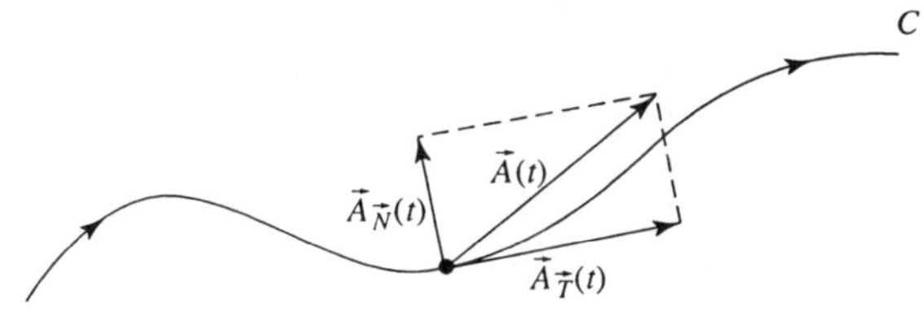

Figure 10.6.2

Example 10.6.11. If $\vec{r}(t) = t^2\mathbf{i} + t^3\mathbf{j}$, find the tangential and normal acceleration vectors $\vec{A}_{\vec{T}}(1)$ and $\vec{A}_{\vec{N}}(1)$.

Answer. As in the preceding discussion, the acceleration vector $\vec{A}$ can be written as

$$\vec{A}(t) = \vec{r}''(t) = v(t)\vec{T}'(t) + v'(t)\vec{T}(t),$$

where $\vec{T}(t)$ is the unit tangent vector and $v(t) \equiv \left\|\vec{V}(t)\right\| = \left\|\vec{r}'(t)\right\|$. To avoid lengthy calculations, see Exercise 20, $\vec{A}(t)$ is dotted with $\vec{T}(t)$ to obtain

$$\vec{A}(t)\cdot\vec{T}(t) = v(t)\vec{T}'(t)\cdot\vec{T}(t) + v'(t)\vec{T}(t)\cdot\vec{T}(t).$$

But, $\left\|\vec{T}(t)\right\| = 1$. Thus, $\vec{T}(t)\cdot\vec{T}(t) = 1$. Also, by Theorem 10.6.10, $\vec{T}'(t)\cdot\vec{T}(t) = 0$. Therefore, $\vec{A}(t)\cdot\vec{T}(t) = v'(t)$. Since $\vec{T}(t) = \dfrac{2}{\sqrt{4+9t^2}}\mathbf{i} + \dfrac{3t}{\sqrt{4+9t^2}}\mathbf{j}$ (why?), we have that $v'(t) = (2\mathbf{i} + 6t\mathbf{j})\cdot\vec{T}(t) = \dfrac{4+18t^2}{\sqrt{4+9t^2}}$. Thus, at $t = 1$, the tangential acceleration vector

$\vec{A}_{\vec{T}}(1) = v'(1)\vec{T}(1)$ becomes $\dfrac{44}{13}\mathbf{i} + \dfrac{66}{13}\mathbf{j}$. Why? Hence, the normal acceleration vector is

$$\vec{A}_{\vec{N}}(1) = \vec{A}_{\vec{T}}(1) - v'(1)\vec{T}(1) = (2\mathbf{i} + 6\mathbf{j}) - \left(\frac{44}{13}\mathbf{i} + \frac{66}{13}\mathbf{j}\right) = -\frac{18}{13}\mathbf{i} + \frac{12}{13}\mathbf{j}.$$

Figure 10.6.3 illustrates this situation. ■

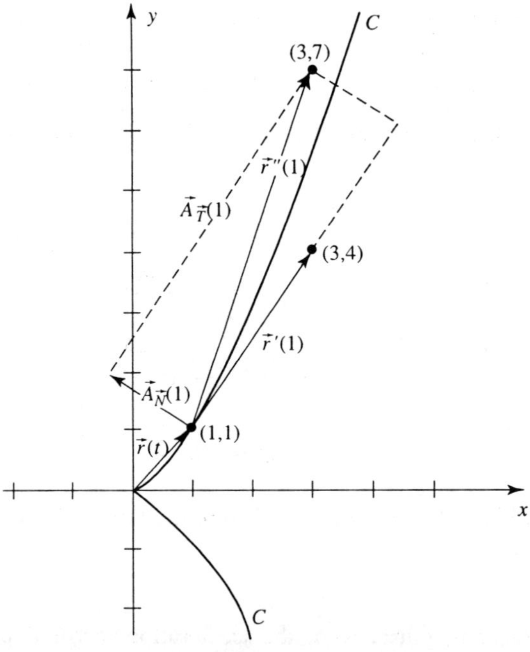

Figure 10.6.3

Before closing, an idea of the integration of vector-valued functions is necessary.

Definition 10.6.12. The vector-valued function $\vec{r}(t) = f(t)\mathbf{i} + g(t)\mathbf{j} + h(t)\mathbf{k}$ is *Riemann integrable* on the interval $[a,b]$ if and only if f, g, and h are all Riemann integrable on $[a,b]$, and

$$\int_a^b \vec{r}(t)\,dt = \left[\int_a^b f(t)\,dt\right]\mathbf{i} + \left[\int_a^b g(t)\,dt\right]\mathbf{j} + \left[\int_a^b h(t)\,dt\right]\mathbf{k}.$$

Also, if $\vec{R}'(t) = \vec{r}(t)$, then $\vec{R}(t)$ is an *antiderivative* of $\vec{r}(t)$, and

$$\int \vec{r}(t)\,dt = \left[\int f(t)\,dt\right]\mathbf{i} + \left[\int g(t)\,dt\right]\mathbf{j} + \left[\int h(t)\,dt\right]\mathbf{k} + \vec{C},$$

with $\vec{C}$ a constant vector.

THEOREM 10.6.13. *(Fundamental Theorem of Calculus) Suppose $\vec{r}(t)$ is a Riemann integrable vector-valued function on $[a,b]$. And suppose that $\vec{R}(t)$ is an antiderivative of $\vec{r}(t)$. Then*

$$\int_a^b \vec{r}(t)\,dt = \vec{R}(t)\Big|_a^b = \vec{R}(b) - \vec{R}(a).$$

See Exercise 27.

Exercises 10.6

1. Show that the following limits are not finite.

(a) $\lim_{x\to 0}\left\langle \frac{\sin^2 x}{x^4}, \frac{\sin x}{x^3}, \ln\frac{x+1}{x}\right\rangle$ **(b)** $\lim_{x\to 0}\left\langle \frac{x\sin\frac{1}{x}}{\sin x}, \frac{1-\cos x}{x^2}, \frac{\tan x}{x}\right\rangle$

2. **(a)** If D is the domain of $\vec{r}(t)$, and t_0 is an accumulation point of D, prove that $\lim_{t\to t_0}\vec{r}(t) = \vec{L}$ in $\Re^3$ if and only if $\lim_{t\to t_0}\left\|\vec{r}(t)-\vec{L}\right\| = 0$ in $\Re^1$. See Exercises 4 and 6 in Section 2.1.

(b) Demonstrate the idea of part (a) on the limit of the sequence $\{\vec{a}_n\}$, with $\vec{a}_n = \left\langle 1, \frac{n}{n+1}, \frac{1}{n}\right\rangle$ for $n \in N$.

3. If D is the domain of $\vec{r}(t)$, and t_0 is an accumulation point of D, prove that if $\lim_{t\to t_0}\vec{r}(t) = \vec{L}$, then $\lim_{t\to t_0}\|\vec{r}(t)\| = \|\vec{L}\|$. Is the converse true? Explain. See Exercise 5 in Section 2.1.

4. Rewrite Theorem 3.2.6 in a new setting for functions $\vec{r}:\Re^1 \to \Re^3$. Also, determine whether this new result is true or not.

5. Prove Theorem 10.6.4.

6. A sequence $\{\vec{a}_n\}$ is a *Cauchy sequence* if and only if for each $\varepsilon > 0$ there exists $n^* \in N$ such that $\|\vec{a}_m - \vec{a}_n\| < \varepsilon$, for all $m, n \geq n^*$. See Definition 2.5.6. Prove that Cauchy sequences are convergent. See Theorem 2.5.9.

7. Rewrite part (a) of Theorem 4.1.8 for vector-valued functions, and prove it.

8. Rewrite Exercise 4 in Section 4.1 for vector-valued functions, and prove it.

9. Suppose that $\vec{r}: D \subseteq \Re^1 \to \Re^3$ is continuous. Prove that the function $\|\vec{r}\|: D \subseteq \Re^1 \to \Re^1$ is continuous, where $\|\vec{r}\|(t) \equiv \|\vec{r}(t)\|$ for $t \in D$. Is the converse true? Explain.

10. Rewrite Definition 4.4.3 on *uniform continuity* of vector-valued functions. Also, prove that if a vector-valued function is continuous on a compact interval $[a,b]$, then it is uniformly continuous on $[a,b]$. See Theorem 4.4.6.

11. Prove Theorem 10.6.7.

12. If $\vec{u}$, $\vec{v}$, and $\vec{w}$ are differentiable vector-valued functions, prove that

$$\frac{d}{dt}[\vec{u}(t)\cdot\vec{v}(t)\times\vec{w}(t)] = \vec{u}(t)\cdot\vec{v}(t)\times\vec{w}'(t)+\vec{u}(t)\cdot\vec{v}'(t)\times\vec{w}(t)+\vec{u}'(t)\cdot\vec{v}(t)\times\vec{w}(t).$$

13. If $\vec{r}$ is *twice differentiable*, that is if both $\vec{r}'(t)$ and $\vec{r}''(t)$ exist, prove that

$$\frac{d}{dt}[\vec{r}(t)\times\vec{r}'(t)] = \vec{r}(t)\times\vec{r}''(t).$$

14. Suppose that $\vec{u}(t)=\langle\cos t,\sin t\rangle$ and $\vec{v}(t)=\langle\sin 2t,\cos 2t\rangle$. See Example 10.6.8.

(a) Show that the curve C for $\vec{u}$ is the same as that for $\vec{v}$.

(b) Find a value of t_0 for which $\vec{v}(t_0)$ gives the same point as $\vec{u}\left(\frac{\pi}{3}\right)$.

(c) Compare the values of $\vec{u}'\left(\frac{\pi}{3}\right)$ and $\vec{v}'(t_0)$.

(d) Find an equation for the tangent line to C at $\vec{u}\left(\frac{\pi}{3}\right)$.

(e) Find the tangential acceleration for both $\vec{u}$ at $\frac{\pi}{3}$ and $\vec{v}$ at t_0.

(f) Find the normal acceleration for both $\vec{u}$ at $\frac{\pi}{3}$ and $\vec{v}$ at t_0.

15. **(a)** Find equations of two tangent lines to the curve C traced by $\vec{u}(t)=\langle t^2-1,t^3-t\rangle$ at a point where the curve C intersects itself. See Exercise 8 in Section 10.4.

(b) Find $\vec{A}$, $A_{\vec{T}}$, and $A_{\vec{N}}$ at the points of intersection for each value of t.

(c) Find the vector $\vec{v}(t)$ which traces C in the opposite direction.

16. **(a)** Verify that for any curve C given by the smooth vector-valued function $\vec{r}$, $\vec{T}(t)$ points in the direction of the increasing values of t.

(b) Verify each statement in Example 10.6.9.

17. **(a)** Show that $\gamma(t)=\left(t|t|,t^2\right)$, with $t\in\Re$, is a parametrization for the curve given by $y=|x|$ in $\Re^2$.

(b) If $\vec{r}(t)=\left\langle t|t|,t^2\right\rangle$, show that $\vec{r}$ is differentiable everywhere, including the origin, even though $y=|x|$ is not differentiable at the origin.

(c) Find $\vec{r}'(0)$ and an equation for the tangent line at the origin, if possible.

18. Prove Theorem 10.6.10.

19. Suppose that a particle travels along the curve C given by $\vec{r}(t) = \langle \sin t, \sin 2t \rangle$ with time $t \in [0, 2\pi)$.[12]

(a) Sketch the curve C.

(b) Find the values of $t = t_0$ and $t = t_1$ at which the curve C intersects itself.

(c) Find $\vec{V}(t_0)$ and $\vec{V}(t_1)$.

(d) Find the equations of the two tangent lines to the curve C at t_0 and t_1.

(e) Find the unit tangent vectors at $t = t_0$ and $t = t_1$.

(f) Find the points on the curve at which the particle has locally the highest speed.

(g) Find $\vec{A}(t_0)$, $\vec{A}(t_1)$, $\vec{A}\left(\frac{\pi}{6}\right)$, and $\vec{A}\left(\frac{\pi}{2}\right)$.

(h) Find both the tangential and normal acceleration vectors at $t = t_0$ and $t = t_1$.

(i) Find a point in the first quadrant where the tangent to the curve C is horizontal.

(j) Compare the graph of C to the graphs of $\langle \sin 2t, \sin 3t \rangle$, $\langle \sin 3t, \sin 4t \rangle$, $\langle \sin t, \sin 4t \rangle$, $\langle \sin 3t, \sin 5t \rangle$, $\langle \cos 2t, \sin 5t \rangle$, and $\langle \cos 4t, \sin 5t \rangle$, each of which are Bowditch curves.

(k) Graph the curve $\left\langle \cos t, \frac{1}{2} \sin 2t \right\rangle$. How does it compare to $\langle \sin t, \sin 2t \rangle$? How does it compare to lemniscates $r^2 = \cos 2\theta$ given in polar coordinates? See Problem 10.9.15.

(l) The curve $\langle f(t), g(t) \rangle$ is p-*periodic* if and only if functions f and g are p-periodic. The value p is a period. Compare to Definition 4.2.9. Graph the curve C_1 given by $\gamma(t) = (\cos 2t, \sin 7t)$. Is C_1 periodic? Since the entire curve can be graphed with $t \in \left[\frac{\pi}{2}, \frac{3\pi}{2}\right]$, is π the *fundamental period* for C_1, that is, is π the smallest period? Is C_1 closed? Compare C_1 to the curve given by $\gamma_1(t) = (\cos(2t + 0.1), \sin(7t + 0.1))$.

20. If $\vec{r}(t) = t^2\mathbf{i} + t^3\mathbf{j}$, use direct calculations of $\vec{r}'(t)$, $\vec{T}(t)$, etc., to find the tangential and normal acceleration vectors at $t = 1$. See Example 10.6.11.

21. If $\vec{r}(t) = 3t\mathbf{i} + \left(9 - t^2\right)\mathbf{j}$, find the tangential and normal acceleration vectors at $t = 1$.

22. Write parametric equations for the tangent line to the curve C given by $\vec{r}(t) = 2\sqrt{t}\mathbf{i} - t^2\mathbf{j} + \frac{1}{t}\mathbf{k}$ at $t = 1$, $\left(1, -\frac{1}{16}, 4\right)$, and $\left(4, -8, \frac{1}{4}\right)$, if possible.

23. Suppose that a particle travels along the curve C parametrized by $\gamma(t) =$

---

[12]Nathaniel Bowditch (1773–1838) was a self-educated American astronomer, mathematician, and actuary. The curve $\langle \sin t, \sin 2t \rangle$ is an example of a *Bowditch curve.* The Bowditch curves are also referred to as *Lissajous figures,* named after Jules Antoine Lissajous (1822–1880), a French mathematician who contributed greatly to the study of optics.

$(8-2\cos t, 1+4\cos t)$, with $t\in[0,2\pi]$. Show that the particle starts at a standstill at the point $P(6,5)$, speeds up to reach its maximum speed at the midpoint, stops at the point $Q(10,-3)$, changes direction, and on the way to the terminal point P again reaches its maximum speed at $(8,1)$. Is γ smooth? See the beginning of Section 10.4.

24. Suppose that $\vec{r}$ is a differentiable vector-valued function and that $r=\|\vec{r}\|$. Show that if $r\neq 0$, then $\vec{r}\cdot\dfrac{d\vec{r}}{dt}=r\dfrac{dr}{dt}$.

25. Find all unit vectors that are tangent and all unit vectors that are normal to the curve $y=x^3-x$ at the point $(1,0)$.

26. The *scalar triple product,*[13] also known as the *box product,* of vector-valued functions $\vec{u}$, $\vec{v}$, and $\vec{w}$ is defined by $[\vec{u},\vec{v},\vec{w}]=\vec{u}\cdot(\vec{v}\times\vec{w})$. Show that

(a) $[\vec{u},\vec{v},\vec{w}]=[\vec{v},\vec{w},\vec{u}]=[\vec{w},\vec{u},\vec{v}]$.

(b) $[\vec{u},\vec{v},\vec{w}]=-[\vec{u},\vec{w},\vec{v}]$

(c) $[c\vec{u}+d\vec{v},\cdot,\cdot]=c[\vec{u},\cdot,\cdot]+d[\vec{v},\cdot,\cdot]$, where c and d are scalars and $\cdot$ represents any vector.

(d) $\dfrac{d}{dt}[\vec{u},\vec{v},\vec{w}]=[\vec{u}',\vec{v},\vec{w}]+[\vec{u},\vec{v}',\vec{w}]+[\vec{u},\vec{v},\vec{w}']$, provided $\vec{u}$, $\vec{v}$, and $\vec{w}$ are differentiable.

27. Prove Theorem 10.6.13.

28. If $\vec{u}$ and $\vec{v}$ are vector-valued functions Riemann integrable on $[a,b]$, and c is a scalar, prove that

(a) $\displaystyle\int_a^b[\vec{u}(t)+\vec{v}(t)]\,dt=\int_a^b\vec{u}(t)\,dt+\int_a^b\vec{v}(t)\,dt$. **(b)** $\displaystyle\int_a^b c\vec{u}(t)\,dt=c\int_a^b\vec{u}(t)\,dt$.

29. If $\vec{u}$ is a vector-valued function Riemann integrable on $[a,b]$ that has an antiderivative, and $\vec{c}$ is a fixed vector such that $\vec{c}\cdot\vec{u}$ is defined, prove that

$$\int_a^b\vec{c}\cdot\vec{u}(t)\,dt=\vec{c}\cdot\int_a^b\vec{u}(t)\,dt.$$

30. If $\vec{r}(t)=f(t)\mathbf{i}+g(t)\mathbf{j}+h(t)\mathbf{k}$ is a vector-valued function Riemann integrable on $[a,b]$, prove that $\|\vec{r}(t)\|$ is Riemann integrable on $[a,b]$, and that

$$\left\|\int_a^b\vec{r}(t)\,dt\right\|\le\int_a^b\|\vec{r}(t)\|\,dt.$$

See Exercise 5 in Section 6.3.

---

[13]For comparison, the *vector triple product* $\vec{u}\times\vec{v}\times\vec{w}$ has possibly different values depending on the grouping $(\vec{u}\times\vec{v})\times\vec{w}$ or $\vec{u}\times(\vec{v}\times\vec{w})$. Why? See Exercise 5 in Section 10.8.

10.7 Arc Length

Suppose that C is a smooth curve given by a smooth parametrization $\gamma(t)=\left(f(t),g(t)\right)$ for $t\in[a,b]$. Our goal is to define and compute the length of the curve C for $t\in[a,b]$. Since C is smooth, because f' and g' are continuous, the particle traveling on C does not suddenly change direction. And since $f'(t)$ and $g'(t)$ are not simultaneously 0 the particle does not stop or double back. Now, a partition $Q=\left\{t_0,t_1,\ldots,t_n\right\}$ of the interval $[a,b]$ is formed with $a=t_0<t_1<\ldots<t_n=b$, and with P_k corresponding to the point $\left(x(t_k),y(t_k)\right)$ on the curve for $k=0,1,2,\ldots,n$. The points $P_0,P_1,\ldots,P_n$ are connected by straight line segments whose finite length can be calculated. Let this length corresponding to the partition Q be denoted by L_Q. Clearly, the length of L_Q is shorter than the "length" of C. Thus, the *length* of C is defined by

$$L(\gamma)=\sup\left\{L_Q \mid Q \text{ is a partition of } [a,b]\right\},$$

if it is finite. $L(\gamma)$ is also referred to as the *arc length* of C. Though the next result should be almost obvious, a proof is required. See Exercise 14.

THEOREM 10.7.1. *Suppose that a smooth curve C is parametrized by $\gamma(t)=\left(f(t),g(t)\right)$ for $t\in[a,b]$. Then the length of C is $L(\gamma)=\lim\limits_{|Q|\to 0}L_Q$, where Q is a partition of $[a,b]$, and $|Q|$ is its norm.*

The formula for $L(\gamma)$, due to its difficulty, is not very useful in computations of arc length. The familiar integral formula for arc length presented in elementary calculus texts follows.

THEOREM 10.7.2. *If a curve C is given by a smooth parametrization $\gamma(t)=\left(f(t),g(t)\right)$ for $t\in[a,b]$, then the length of C is finite and equal to*

$$L(\gamma)=\int_a^b\sqrt{[f'(t)]^2+[g'(t)]^2}\,dt=\int_a^b\|\vec{r}\,'(t)\|\,dt.$$

Proof. Let $Q=\left\{t_0,t_1,\ldots,t_n\right\}$ be a partition of the interval $[a,b]$. And let C be represented by a vector-valued function $\vec{r}(t)=f(t)\mathbf{i}+g(t)\mathbf{j}$. Then for $a\le t_k<t_{k+1}\le b$, incorporating the distance formula, we have that

$$\sqrt{\left[f(t_{k+1})-f(t_k)\right]^2+\left[g(t_{k+1})-g(t_k)\right]^2}=\left\|\vec{r}(t_{k+1})-\vec{r}(t_k)\right\|=\left\|\int_{t_k}^{t_{k+1}}\vec{r}\,'(t)\,dt\right\|\le\int_{t_k}^{t_{k+1}}\|\vec{r}\,'(t)\|\,dt.$$

See Exercise 25 of Section 10.6 for a verification of this last inequality. Hence, on $[a,b]$ we have that $L_Q \le \int_a^b \|\vec{r}\,'(t)\|\,dt$.

Next we prove the reverse inequality. To this end, let $\varepsilon > 0$ be given. Since, by Exercise 10 of Section 10.6, $\vec{r}\,'$ is uniformly continuous on $[a,b]$, there exists $\delta > 0$ such that $\|\vec{r}\,'(t) - \vec{r}\,'(s)\| < \varepsilon$ for all $t, s \in [a,b]$ satisfying $|t-s| < \delta$. Therefore, if $Q = \{t_0, t_1, \dots, t_n\}$ is a partition of $[a,b]$ with $|Q| < \delta$, then $\|\vec{r}\,'(t)\| \le \|\vec{r}\,'(t_{k+1})\| + \varepsilon$ whenever $a \le t_k \le t \le t_{k+1} \le b$. Why? Hence, if $h \equiv \Delta t_{k+1} = t_{k+1} - t_k$, we have that

$$\begin{aligned}
\int_{t_k}^{t_{k+1}} \|\vec{r}\,'(t)\|\,dt &\le \int_{t_k}^{t_{k+1}} \left(\|\vec{r}\,'(t_{k+1})\| + \varepsilon\right) dt \\
&= \left\| \int_{t_k}^{t_{k+1}} \left[\vec{r}\,'(t) + \vec{r}\,'(t_{k+1}) - \vec{r}\,'(t)\right] dt \right\| + \varepsilon h \\
&\le \left\| \int_{t_k}^{t_{k+1}} \vec{r}\,'(t)\,dt \right\| + \left\| \int_{t_k}^{t_{k+1}} \left[\vec{r}\,'(t_{k+1}) - \vec{r}\,'(t)\right] dt \right\| + \varepsilon h \\
&\le \left\|\vec{r}(t_{k+1}) - \vec{r}(t_k)\right\| + \varepsilon h + \varepsilon h = \left\|\vec{r}(t_{k+1}) - \vec{r}(t_k)\right\| + 2\varepsilon h.
\end{aligned}$$

Therefore,

$$\sum_{k=0}^{n-1} \int_{t_k}^{t_{k+1}} \|\vec{r}\,'(t)\|\,dt \le \sum_{k=0}^{n-1} \left[\left\|\vec{r}(t_{k+1}) - \vec{r}(t_k)\right\| + 2\varepsilon h\right].$$

And thus, $\int_a^b \|\vec{r}\,'(t)\|\,dt \le L_Q + 2\varepsilon(b-a)$. But since ε was arbitrary, by Exercise 2 in Section 1.7, $\int_a^b \|\vec{r}\,'(t)\|\,dt \le L_Q$. Hence, the proof is complete. ∎

Often, authors of elementary calculus texts refer to *Duhamel's*[14] *Principle for Integrals* in completing a proof of Theorem 10.7.2. This method of proving, though somewhat incomplete, goes as follows. If $Q = \{t_0, t_1, \dots, t_n\}$ is a partition of the interval $[a,b]$, then the length L_Q is given by

---

[14]Jean Marie Constant Duhamel (1797–1872) was a French analyst and applied mathematician. He applied his results in partial differential equations to study heat flow, rational mechanics, and acoustics. Fresnel's work in optics and Duhamel's study of heat flow were mathematically similar.

$$L_Q = \sum_{k=0}^{n-1} \sqrt{\left[f(t_{k+1}) - f(t_k)\right]^2 + \left[g(t_{k+1}) - g(t_k)\right]^2}.$$

And, from the mean value theorem, Theorem 5.2.13, there exist points $\bar{t}_k$ and $\hat{t}_k$ in (t_k, t_{k+1}) for which $f(t_{k+1}) - f(t_k) = f'(\bar{t}_k)\Delta t_{k+1}$ and $g(t_{k+1}) - g(t_k) = g'(\hat{t}_k)\Delta t_{k+1}$, where $\Delta t_{k+1} = t_{k+1} - t_k$ with $k = 0, 1, \ldots, n-1$. Therefore, by Theorem 10.7.1 the length of C is given by

$$L(\gamma) = \lim_{|Q|\to 0} \sum_{k=0}^{n-1} \sqrt{\left[f'(\bar{t}_k)\right]^2 + \left[g'(\hat{t}_k)\right]^2}\,(\Delta t_{k+1}),$$

provided this limit is finite. The expression $L(\gamma)$ above resembles a Riemann sum with the problem being that $\bar{t}_k$ and $\hat{t}_k$ need not necessarily be the same point. Recall that in view of Exercise 6 in Section 6.2, if $c_k \in [t_k, t_{k+1}]$, we can write that

$$\lim_{|Q|\to 0} \sum_{k=0}^{n-1} \sqrt{\left[f'(c_k)\right]^2 + \left[g'(c_k)\right]^2}\,(\Delta t_{k+1}) = \int_a^b \sqrt{\left[f'(t)\right]^2 + \left[g'(t)\right]^2}\,dt \equiv L(\gamma).$$

Now, the proof is complete if we show that

$$\lim_{|Q|\to 0} \sum_{k=0}^{n-1} \sqrt{\left[f'(\bar{t}_k)\right]^2 + \left[g'(\hat{t}_k)\right]^2}\,(\Delta t_{k+1}) = \lim_{|Q|\to 0} \sum_{k=0}^{n-1} \sqrt{\left[f'(c_k)\right]^2 + \left[g'(c_k)\right]^2}\,(\Delta t_{k+1})$$

for some $c_k \in [t_k, t_{k+1}]$. Indeed, Duhamel's principle supports this result. Nonetheless, to complete the proof of Theorem 10.7.2 using this approach, Duhamel's principle must be proved. However, the proof is too complex to get involved in here.

Definition 10.7.3. If the curve C representing the function f has a finite length, then f is *rectifiable*.

Remark 10.7.4

(a) If a smooth parametrization for the curve C is $\gamma(t) = (f(t), g(t))$ with $t \in [a, b]$, then the formula for the arc length $L(\gamma)$ is

$$L(\gamma) = \int_a^b \sqrt{\left(\frac{dx}{dt}\right)^2 + \left(\frac{dy}{dt}\right)^2}\,dt.$$

Other familiar formulas from elementary calculus for $L(\gamma)$ are given in Exercise 2. Also, see Exercise 6 in Section 10.8.

(b) The formula for arc length already occurred in the discussion of functions of bounded variation in Part 3 of Section 5.6. Thus, connecting ideas and using Theorem 10.7.1, we have that a continuous function is of bounded variation if and only if it is rectifiable.

(c) In 3-space, if the curve C has a smooth parametrization $\gamma(t) = (x(t), y(t), z(t))$ for $t \in [a,b]$, then the length of C is given by

$$L(\gamma) = \int_a^b \sqrt{[x'(t)]^2 + [y'(t)]^2 + [z'(t)]^2}\, dt.$$

(d) Observe that from the proof of Theorem 10.7.2, if $\vec{r}$ is continuously differentiable on $[a,b]$, then $\vec{r}$ is rectifiable. The converse of this statement is not true. See Exercise 15. ■

Example 10.7.5. If

$$f(t) = t \text{ and } g(t) = \begin{cases} t\sin\dfrac{1}{t}, \text{ if } t \in (0,1] \\ 0, \text{ if } t = 0, \end{cases}$$

show that $\vec{r}(t) = f(t)\mathbf{i} + g(t)\mathbf{j}$ with $t \in [0,1]$ is not rectifiable. See Figure 3.3.4 and Problem 5.6.18.

Answer. Recall that the vector-valued function $\vec{r}(t)$ is continuous on $[0,1]$ and differentiable on $(0,1)$. Thus, $\vec{r}(t)$ is a "nice" function, though not rectifiable. Now, as pointed out in part (d) of Remark 10.7.4, though g' is unbounded on $(0,1)$, we may not conclude that $\vec{r}(t)$ is not rectifiable. We will need to show that L_Q tends to infinity. Thus, let $Q = \{0, t_1, \dots, t_n, 1\}$ be a partition of the interval $[0,1]$, where $t_k = \dfrac{2}{k\pi}$ for $k = 1, 2, \dots, n$ and $n \in N$ odd. If n is even, the discussion is similar. Observe that $\sin\dfrac{1}{t_k} = 0$ for k even, and that $\sin\dfrac{1}{t_k} = \pm 1$ for k odd. If $f(t) = t$, then

$$\begin{aligned} L_Q &= \sum_{k=0}^{n-1} \sqrt{[f(t_{k+1}) - f(t_k)]^2 + [g(t_{k+1}) - g(t_k)]^2} \\ &\geq \sum_{k=1}^{n-2} \sqrt{[f(t_{k+1}) - f(t_k)]^2 + [g(t_{k+1}) - g(t_k)]^2} \\ &\geq \sum_{k=1}^{n-2} |g(t_{k+1}) - g(t_k)| \\ &= \sum_{k=1}^{n-2} \left| t_{k+1} \sin\frac{1}{t_{k+1}} - t_k \sin\frac{1}{t_k} \right| \\ &= |t_1| + 2\sum_{\substack{k=3 \\ \text{odd}}}^{n-2} |t_k| = \frac{2}{\pi} + 2\sum_{\substack{k=3 \\ \text{odd}}}^{n-2} \left(\frac{2}{k\pi}\right). \end{aligned}$$

If $n \to \infty$, the series diverges. Why? Thus, $L_Q \to +\infty$ and the curve is not rectifiable. ■

A curve may be parametrized in many different ways as pointed out in Section 10.4. Though some parametrizations are very different, others are quite similar as we shall now see.

Definition 10.7.6. Suppose that a curve C has two parametrizations $\gamma_1(t)$ with $t \in [a,b]$ and $\gamma_2(t)$ with $t \in [c,d]$. These parametrizations for C are *equivalent* if and only if there exists a continuously differentiable function $u:[c,d] \to [a,b]$ which is both one-to-one and onto, with $u(c)=a$, $u(d)=b$, and $\gamma_2(t)=\gamma_1(u(t))$ for all $t \in [c,d]$.

Example 10.7.7. Consider the parametrizations $\gamma_1(t)=(t,-2t+17)$ with $t \in [6,10]$ and $\gamma_2(t)=(2t+6,-4t+5)$ with $t \in [0,2]$, both of which were presented in Section 10.4. These parametrizations are equivalent, since if we define $u:[0,2] \to [6,10]$ by $u(t)=2t+6$, then u satisfies all the conditions given in Definition 10.7.6. Why? ■

THEOREM 10.7.8. *Suppose that a curve C has two equivalent smooth parametrizations γ_1 and γ_2. If P is a point on the curve C, then γ_1 and γ_2 have the same orientation at P.*

Proof. Recall that γ_1 and γ_2 may be represented by vector-valued functions $\vec{r}_1$ and $\vec{r}_2$, respectively. And since γ_1 and γ_2 are equivalent, there exists a continuously differentiable function u such that both $u'(t)>0$ for all t and $\vec{r}_2(t)=\vec{r}_1(u(t))$. Why? If the point P is given by both $\vec{r}_1(t_1)$ and $\vec{r}_2(t_2)$, then $t_1=u(t_2)$. And the direction of $\vec{r}_1(t_1)$ is given by $\dfrac{\vec{r}_1'(t_1)}{\|\vec{r}_1'(t_1)\|}$. The direction of $\vec{r}_2(t_2)$ is of present interest. Since by Theorem 10.6.7, part (d), $\vec{r}_2'(t)=\vec{r}_1'(u(t))u'(t)$, we have that

$$\begin{aligned}\frac{\vec{r}_2'(t_2)}{\|\vec{r}_2'(t_2)\|} &= \frac{\vec{r}_1'(u(t_2))u'(t_2)}{\|\vec{r}_1'(u(t_2))u'(t_2)\|} = \frac{\vec{r}_1'(u(t_2))u'(t_2)}{|u'(t_2)|\|\vec{r}_1'(u(t_2))\|} \\ &= \frac{\vec{r}_1'(u(t_2))u'(t_2)}{u'(t_2)\|\vec{r}_1'(u(t_2))\|} \quad \text{(Why?)} \\ &= \frac{\vec{r}_1'(u(t_2))}{\|\vec{r}_1'(u(t_2))\|} = \frac{\vec{r}_1'(t_1)}{\|\vec{r}_1'(t_1)\|},\end{aligned}$$

which is the orientation of $\vec{r}_1$ at t_1. Hence, the proof is complete. ■

THEOREM 10.7.9. *If the curve C has two equivalent smooth parametrizations γ_1 and γ_2, then $L(\gamma_1)=L(\gamma_2)$.*

Proof. Let $\vec{r}_1$ and $\vec{r}_2$ be two vector-valued functions representing γ_1 and γ_2, respectively. Since γ_1 and γ_2 are equivalent, there exists a continuously differentiable function u such that $u'(t)>0$ for all $t\in[c,d]$, $\vec{r}_2(t)=\vec{r}_1(u(t))$, $u(c)=a$, and $u(d)=b$. Then, $\vec{r}_2'(t)=\vec{r}_1'(u(t))u'(t)$ for all $t\in[c,d]$. And since $u'(t)>0$, we have that

$$\begin{aligned} L(\gamma_2) &= \int_c^d \left\|\vec{r}_2'(t)\right\| dt = \int_c^d \left\|\vec{r}_1'(u(t))u'(t)\right\| dt \\ &= \int_c^d u'(t)\left\|\vec{r}_1'(u(t))\right\| dt = \int_a^b \left\|\vec{r}_1'(v)\right\| dv = L(\gamma_1), \end{aligned}$$

where $v=u(t)$. ■

Suppose that a smooth curve C is parametrized by $\gamma(t)=(f(t),g(t),h(t))$, where f, g, and h are continuously differentiable for $t\in[a,b]$. Then C is determined by a vector-valued function $\vec{r}(t)=f(t)\mathbf{i}+g(t)\mathbf{j}+h(t)\mathbf{k}$ with $t\in[a,b]$. Thus, $\vec{V}(t)=\vec{r}'(t)=f'(t)\mathbf{i}+g'(t)\mathbf{j}+h'(t)\mathbf{k}$, and so $\left\|\vec{V}(t)\right\|=\sqrt{[f'(t)]^2+[g'(t)]^2+[h'(t)]^2}$. Therefore, according to Theorem 10.7.2, the length of C on $[a,b]$ is given by $L(\gamma)=\int_a^b \left\|\vec{V}(t)\right\| dt$. Now, if $t\in[a,b]$ and a particle P travels along the curve C, then the position of P at time t, determined by the distance from the initial point $\vec{r}(a)=(f(a),g(a),h(a))$ to P, is given by

$$\int_a^t \left\|\vec{V}(\tau)\right\| d\tau = \int_a^t \left\|\vec{r}'(\tau)\right\| d\tau \equiv s(t).$$

Therefore, $s'(t)=\left\|\vec{V}(t)\right\|$, the speed of the particle at time t. And the distance traveled is obtained by integrating the speed. Also, the *direction* or *orientation* of C is determined by the increasing values of $s(t)$. Moreover, $s\equiv s(t)$ which is given by

$$s(t)=\int_a^t \sqrt{[f'(u)]^2+[g'(u)]^2+[h'(u)]^2}\, du$$

and is called the *arc length parameter* for the curve C.

Example 10.7.10. Suppose that a curve C is given parametrically by $\gamma(t)=(2t+1,3t-2)$ for $t\geq 0$. Give a parametrization $\gamma_1(s)$ for C where s is the arc length parameter of C.

Answer. Let $f(t)=2t+1$ and $g(t)=3t-2$. Thus, $f'(t)=2$ and $g'(t)=3$. And the arc length parameter for C is given by

$$s(t)=\int_0^t \sqrt{[f'(u)]^2+[g'(u)]^2}\,du=\int_0^t \sqrt{4+9}\,du=\sqrt{13}\,t,$$

for $t\geq 0$. Thus, $t=\frac{1}{\sqrt{13}}s$ with $s\geq 0$. Therefore,

$$\gamma(t)=(2t+1,3t-2)=\left(\frac{2}{\sqrt{13}}s+1,\frac{3}{\sqrt{13}}s-2\right)=\gamma_1(s). \qquad \blacksquare$$

Elaborating on this example, observe that the initial point P on the curve C occurs when $t=0$, that is P is $(1,-2)$. Also, the curve C is a line given in rectangular coordinates by $y=\frac{3}{2}x-\frac{7}{2}$, with $x\geq 1$. Why? In addition, note that $\gamma(1)$ gives the point $Q(3,1)$ which is $\sqrt{13}$ units away from P. But, $\gamma_1(1)$ gives the point $R\left(\frac{2}{\sqrt{13}}+1,\frac{3}{\sqrt{13}}-2\right)$ which is one unit away from P. Therefore, using the arc length parameter, for one unit of time a particle travels one unit of distance along the curve. That is, in general if $t=s$ is the arc length parameter for the curve C, then the curve C is traced by $\vec{r}(s)=f(s)\mathbf{i}+g(s)\mathbf{j}+h(s)\mathbf{k}$. Thus, $\|\vec{r}\,'(s)\|=\sqrt{[f'(s)]^2+[g'(s)]^2+[h'(s)]^2}$. But $s=\int_a^s \sqrt{[f'(u)]^2+[g'(u)]^2+[h'(u)]^2}\,du$ (why?), which upon differentiation gives $1=\sqrt{[f'(s)]^2+[g'(s)]^2+[h'(s)]^2}$. Hence, $\|\vec{r}\,'(s)\|=1$, and the curve is traced at a constant speed of 1.

Exercises 10.7

1. **(a)** Find the length of $(\ln\cos t, t)$ on the interval $\left[0,\frac{\pi}{3}\right]$.

(b) Find the length (*circumference*) of the circle given by $(\cos t,\sin t)$.

(c) Find the circumference of the circle given by $(\cos 2t,\sin 2t)$. See Example 10.6.8.

2. **(a)** Show that if $f:[a,b]\to\Re$ is continuously differentiable and $y=f(x)$, then the length L of f on $[a,b]$ is given by

$$L=\int_a^b \sqrt{1+\left(\frac{dy}{dx}\right)^2}\,dx.$$

(b) If the smooth curve C is given by $x = g(y)$ with $y \in [c,d]$, find the length of C.

3. Find the length of $f(x) = \frac{1}{6}x^3 + \frac{1}{2x}$ on the interval $[1,2)$.

4. *(Wren's[15] Theorem)* Find the length of one arch of a cycloid. See Example 10.4.6.

5. Find the length of the astroid given in Exercise 12 of Section 10.4.

6. Is the curve $\left(\sin 2\pi t, \cos 2\pi t, 2t - t^2\right)$ with $t \in [0,2]$ rectifiable? Explain.

7. Consider the curve C parametrized by $x = f(t)\cos t$ and $y = f(t)\sin t$, where f is continuously differentiable on $[a,b]$. Find the length of C.

8. Find the length of the curve C traced once by a parametrization $x = 2\cos^2 t$ and $y = \sin 2t$. See Problem 10.9.15, part (b), and Problem 10.9.20, part (a).

9. Determine if the parametrizations γ_1 and γ_2 are equivalent.

(a) $\gamma_1(t) = (2t + 6, -4t + 5)$ with $t \in [0,2]$, $\gamma_2(t) = \left(10 - \frac{1}{2}t, t - 3\right)$ with $t \in [0,8]$

(b) $\gamma_1(t) = \left(t, t^2\right)$ with $t \in [0,1]$, $\gamma_2(t) = \left(t^2, t^4\right)$ with $t \in [0,1]$

(c) $\gamma_1(t) = (\cos t, \sin t, t)$ with $t \in [0,\pi]$, $\gamma_2(t) = (\sin t, \cos t, t)$ with $t \in [0,\pi]$

(d) $\gamma_1(t) = (\cos t, \sin t)$ with $t \in [0,\pi]$, $\gamma_2(t) = \left(t, \sqrt{1 - t^2}\right)$ with $t \in [-1,1]$

10. **(a)** Find the length of a *circular helix*[16] defined by $\gamma(t) = (\cos t, \sin t, t)$ with $t \in [0, 4\pi]$.

(b) Find the parametrization using the arc length parameter for the helix in part (a).

(c) Find the speed of a particle traveling on the helix in parts (a) and (b) according to parametrizations from parts (a) and (b).

11. **(a)** Find the length of the curve C given by $\gamma(t) = \left(\cos\left(t^2\right), \sin\left(t^2\right)\right)$ on $[0,t]$.

(b) Find the speed of the particle traveling along the curve C given in part (a).

(c) Find the parametrization using the arc length parameter for the curve C given in part (a). Find the speed of the particle traveling according to this parametrization.

---

[15]Sir Christopher Wren (1632–1723), born in England, was an architect who both replanned the entire city of London following the fire of 1666 and supervised the rebuilding of 51 churches. As an architect, Wren's most famous design was that of Saint Paul's Cathedral. Wren was the first to find the length of a cycloid.

[16]*Helix* is a curve in 3-space which lies on a cylinder or cone, and whose tangent vector at every point has a constant angle. The thread of a bolt looks like a circular helix.

12. Suppose that a rectifiable curve C is given by $\gamma(t)$ with $t \in [a,b]$. And suppose that $c \in (a,b)$. If C_1 and C_2 are curves with C_1 given by $\gamma_1(t) = \gamma(t)$ for $t \in [a,c]$ and C_2 given by $\gamma_2(t) = \gamma(t)$ for $t \in [c,b]$, prove that both C_1 and C_2 are rectifiable, and that $L(\gamma) = L(\gamma_1) + L(\gamma_2)$.

13. Find the length of the curve C given by $\gamma(t) = (t, |t|)$ for $t \in [-1,1]$.

14. Prove Theorem 10.7.1.

15. Consider the sequence of squares $\{S_n\}$, with S_1, S_2, and S_3 is illustrated in Figure 10.7.1. In S_1, the sum of the lengths of

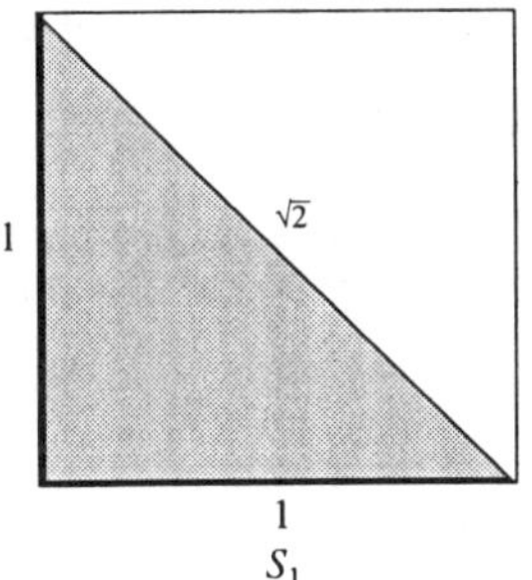

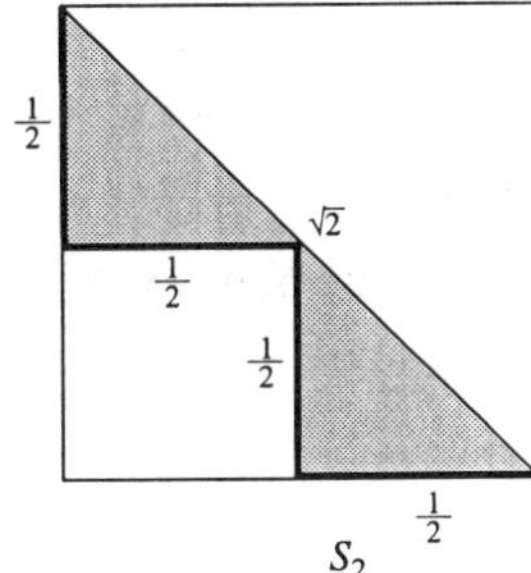

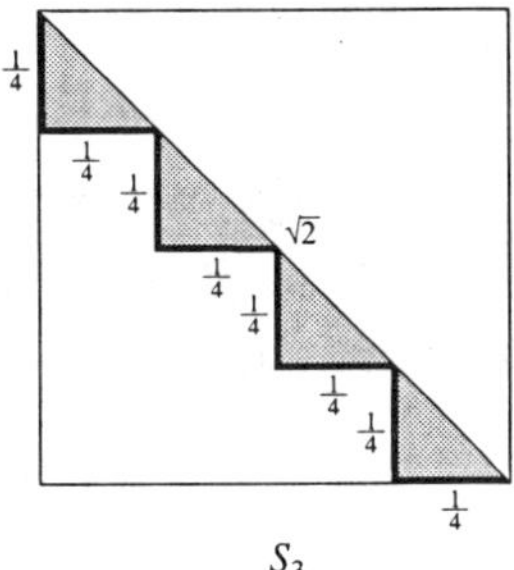

Figure 10.7.1

the thicker lines, that is, $1+1$, approximates the length of the diagonal, that is, $\sqrt{2}$. The diagonal is a curve with an arc length of $\sqrt{2}$. In S_2, the sum of the lengths of the thicker lines, that is, $\frac{1}{2}+\frac{1}{2}+\frac{1}{2}+\frac{1}{2}$ approximates $\sqrt{2}$. And in S_3, $\frac{1}{4}+\frac{1}{4}+\frac{1}{4}+\frac{1}{4}+\frac{1}{4}+\frac{1}{4}+\frac{1}{4}+\frac{1}{4}$ approximates $\sqrt{2}$. Observe that the shaded area between the diagonal and the stair-like curves tends to 0 as $k \to \infty$. The diagonal has an arc length of $\sqrt{2}$, whereas the length of the stair-like curve in each S_n that is to approximate the length of the diagonal is consistently 2. But the value of 2 is nowhere near that of $\sqrt{2}$. How can this be?

10.8* Review

Label each statement as true or false. If a statement is true, prove it. If not,

(i) give an example of why it is false, and
(ii) if possible, correct it to make it true, and then prove it.

1. If $\vec{u} \times \vec{v} = \vec{0}$, then $\vec{u} = \vec{0}$, or $\vec{v} = \vec{0}$, or both.

2. If $\vec{u}\cdot\vec{v}=0$, then $\vec{u}=\vec{0}$, or $\vec{v}=\vec{0}$, or both.

3. $(\vec{u}\times\vec{v})\cdot\vec{w}=\vec{u}\cdot(\vec{v}\times\vec{w})=[\vec{u},\vec{v},\vec{w}]$.

4. The direction of the vector $\vec{u}$ and the direction vector $\vec{u}$ have the same meaning.

5. $\vec{u}\times(\vec{v}\times\vec{w})=(\vec{u}\times\vec{v})\times\vec{w}$.

6. The length of the function f on the interval $[a,b]$ is given by $\int_a^b \sqrt{[f'(x)]^2+1}\,dx$, provided this integral is finite.

7. If $\vec{u}$ is a vector, then $\vec{u}=(\vec{u}\cdot\mathbf{i})\mathbf{i}+(\vec{u}\cdot\mathbf{j})\mathbf{j}+(\vec{u}\cdot\mathbf{k})\mathbf{k}$.

8. If $\vec{u}$ is a nonzero vector, with α being the angle between $\vec{u}$ and $\mathbf{i}$, β being the angle between $\vec{u}$ and $\mathbf{j}$, and γ being the angle between $\vec{u}$ and $\mathbf{k}$, then $\vec{u}=\|\vec{u}\|\left[(\cos\alpha)\mathbf{i}+(\cos\beta)\mathbf{j}+(\cos\gamma)\mathbf{k}\right]$.

9. If α, β, and γ are as in the previous exercise, then $\cos^2\alpha+\cos^2\beta+\cos^2\gamma=1$.

10. If $\gamma:[0,1]\to\Re^2$ is defined by

$$\gamma(t)=\begin{cases}\left(t,t^2\sin\dfrac{1}{t}\right), & \text{if } t\in(0,1]\\ (0,0), & \text{if } t=0,\end{cases}$$

then γ is rectifiable.

11. Suppose that $\gamma(t)=(f(t),g(t))$, with $t\in[a,b]$, is such that $f,g:[a,b]\to\Re$ are continuous and $f',g':(a,b)\to\Re$ are continuous, then γ parametrizes a rectifiable curve C.

12. $[\vec{u},\vec{u},\vec{v}]=0$ for all vectors $\vec{u}$ and $\vec{v}$.

13. $(\vec{u}+\vec{v})\times(\vec{u}-\vec{v})=2\vec{v}\times\vec{u}$.

14. $\vec{u}\times(\vec{v}\times\vec{w})+\vec{v}\times(\vec{w}\times\vec{u})+\vec{w}\times(\vec{u}\times\vec{v})=\vec{0}$.

15 If $\vec{u}$, $\vec{v}$, and $\vec{w}$ are mutually orthogonal, then $\vec{u}\times(\vec{v}\times\vec{w})=0$.

16. If $\vec{u}$, $\vec{v}$, and $\vec{w}$ are differentiable vector-valued functions, then

$$\frac{d}{dt}\{\vec{u}(t)\cdot[\vec{v}(t)\times\vec{w}(t)]\}=\frac{d}{dt}\{[\vec{u}(t)\cdot\vec{v}(t)]\times\vec{w}(t)\}.$$

17. $\dfrac{d}{dt}[\vec{u}\times(\vec{v}\times\vec{w})]=\vec{u}'\times(\vec{v}\times\vec{w})+\vec{u}\times(\vec{v}'\times\vec{w})+\vec{u}\times(\vec{v}\times\vec{w}')$.

18. If D is the domain of $\vec{r}(t)$ and t_0 is an accumulation point of D, then $\lim\limits_{t\to t_0}\vec{r}(t)=\vec{0}$ if and only if $\lim\limits_{t\to t_0}\|\vec{r}(t)\|=0$.

19. If x_i, y_i, z_i, with $i=1,2,3$, are differentiable functions of t, then

$$\frac{d}{dt}\begin{vmatrix} x_1 & y_1 & z_1 \\ x_2 & y_2 & z_2 \\ x_3 & y_3 & z_3 \end{vmatrix} = \begin{vmatrix} x_1' & y_1' & z_1' \\ x_2 & y_2 & z_2 \\ x_3 & y_3 & z_3 \end{vmatrix} + \begin{vmatrix} x_1 & y_1 & z_1 \\ x_2' & y_2' & z_2' \\ x_3 & y_3 & z_3 \end{vmatrix} + \begin{vmatrix} x_1 & y_1 & z_1 \\ x_2 & y_2 & z_2 \\ x_3' & y_3' & z_3' \end{vmatrix}.$$

20. If the curve C is given by a Lipschitz function $\vec{r}(t)$, then C is rectifiable.

21. The vectors $\vec{u}+\vec{v}$ and $\vec{u}-\vec{v}$ are orthogonal.

22. $(\vec{u}\times\vec{v})\cdot\vec{u}=\vec{0}$.

23. $\vec{u}\times\vec{v}=\vec{v}\times\vec{u}$.

24. $\vec{u}\times\vec{v}$ and $\vec{v}\times\vec{u}$ are never equal.

25. If $\vec{u}$ and $\vec{v}$ are differentiable, then $\frac{d}{dt}[\vec{u}(t)\times\vec{v}(t)]=\vec{u}(t)\times\vec{v}'(t)+\vec{v}(t)\times\vec{u}'(t)$.

26. If $\vec{r}$ is differentiable and $\vec{r}(t)\neq\vec{0}$ for any t, then $\frac{d}{dt}\|\vec{r}(t)\|=\frac{1}{\|\vec{r}(t)\|}[\vec{r}(t)\cdot\vec{r}'(t)]$.

27. If $\vec{r}$ is three times differentiable, then

$$\frac{d}{dt}\{\vec{r}(t)\cdot[\vec{r}'(t)\times\vec{r}''(t)]\}=\vec{r}(t)\cdot[\vec{r}'(t)\times\vec{r}'''(t)].$$

28. Two parametrizations for one curve can be equivalent even if one is smooth and the other is not.

29. $\vec{r}(t)=\left\langle t, t\cos\frac{1}{t}\right\rangle$ is rectifiable if $t\in(0,1]$.

30. Any vector $\vec{u}$ can be written as $\vec{u}=(\vec{u}\cdot\vec{v})\vec{v}+(\vec{v}\times\vec{u})\times\vec{v}$, where $\vec{v}$ is some unit vector. Moreover, $(\vec{u}\cdot\vec{v})\vec{v}$ is parallel to $\vec{v}$. And $(\vec{v}\times\vec{u})\times\vec{v}$ is orthogonal to $\vec{v}$.

31. The curve C given by $y=x^2-1$ can be parametrized by $\gamma_1(t)=(t,t^2-1)$ with $t\in\Re$, $\gamma_2(t)=(t^2,t^4-1)$ with $t\in\Re$, and $\gamma_3(t)=(t^3,t^6-1)$ with $t\in\Re$.

32. A circle with radius a may be parametrized by $\gamma(t)=(a\cos t, a\sin t)$ with $t\in[0,2\pi)$.

33. If $\vec{u}\cdot\vec{u}=1$, then $\vec{u}$ is a unit vector.

34. The lines $\ell_1=(2-3t,-1+t,1+2t)$ and $\ell_2=(-4+6t,1-2t,5-4t)$ are identical.

35. Suppose that $\vec{r}$ is a differentiable vector-valued function and $r=\|\vec{r}\|$. If $r\neq 0$, then

$$\frac{d}{dt}\frac{\vec{r}}{r}=\frac{1}{r^3}\left[\left(\vec{r}\times\frac{d\vec{r}}{dt}\right)\times\vec{r}\right].$$

36. If $\vec{u}$ and $\vec{v}$ are Riemann integrable vector-valued functions on $[a,b]$, then

$$\int_a^b[\vec{u}(t)\cdot\vec{v}(t)]\,dt=\left[\int_a^b\vec{u}(t)\,dt\right]\cdot\left[\int_a^b\vec{v}(t)\,dt\right].$$

37. If f is continuous on $[a,b]$ and $\vec{c}$ is a constant vector, then

$$\int_a^b \vec{c} f(t)\,dt = \vec{c}\int_a^b f(t)\,dt.$$

10.9* Projects

Part 1. Inner Product

Definition 10.9.1. Assume that $\vec{u}$, $\vec{v}$, and $\vec{w}$ are vectors and that c is a scalar. An *inner product* is a function that associates a real number $(\vec{u},\vec{v})$ with each pair of vectors $\vec{u}$ and $\vec{v}$ and satisfies the following properties:

- **(a)** $(\vec{u},\vec{v}) = (\vec{v},\vec{u})$
- **(b)** $(\vec{u},\vec{v}+\vec{w}) = (\vec{u},\vec{v}) + (\vec{u},\vec{w})$
- **(c)** $c(\vec{u},\vec{v}) = (c\vec{u},\vec{v})$
- **(d)** $(\vec{u},\vec{u}) \geq 0$, and $(\vec{u},\vec{u}) = 0$ if and only if $\vec{u} = \vec{0}$.

Problem 10.9.2. If $\vec{u}$ and $\vec{v}$ are vectors in $\Re^3$, show that $(\vec{u},\vec{v}) \equiv \vec{u}\cdot\vec{v}$ defines $(\vec{u},\vec{v})$ to be an inner product. Thus the dot product given in Definition 10.3.1 is a special case of an inner product.

Problem 10.9.3. Suppose that $\vec{p} \equiv p(x) = a_n x^n + a_{n-1}x^{n-1} + \cdots + a_1 x + a_0$ and $\vec{q} \equiv q(x) = b_n x^n + b_{n-1}x^{n-1} + \cdots + b_1 x + b_0$ are polynomials of degree n. Show that the function given by $(\vec{p},\vec{q}) = a_0 b_0 + a_1 b_1 + \cdots + a_n b_n$ is an inner product.

Problem 10.9.4. Suppose that f and g are continuous on an interval (a,b). Show that the function given by $(f,g) = \int_a^b f(x)g(x)\,dx$ is an inner product.

Problem 10.9.5. If f, g, and h are piecewise continuous on (a,b), show that

- **(a)** if the *norm* $\|f\|$ is defined by $\|f\| = \sqrt{(f,f)}$, then

$$\|f\| = \left\{ \int_a^b [f(x)]^2\,dx \right\}^{\frac{1}{2}}.$$

  See Definition 9.1.1 and Definition 10.2.1.
- **(b)** $(0,f) = (f,0) = 0$.
- **(c)** $(f+g,h) = (f,h) + (g,h)$.

(d) $(f, cg) = c(f, g)$, where $c \in \Re$.

(e) *(Cauchy–Schwarz inequality)* $|(f,g)| \le \|f\|\|g\|$. See Exercise 13 in Section 6.3 and Theorem 10.3.5.

(f) *(Triangle inequality* and *Minkowski inequality)* $\|f+g\| \le \|f\| + \|g\|$. See Exercise 15 in Section 6.3 and Theorem 10.3.6.

Definition 10.9.6. Two vectors $\vec{u}$ and $\vec{v}$ are *orthogonal* if and only if $(\vec{u}, \vec{v}) = 0$.

Orthogonality extends the concept of *perpendicular.* Recall that two vectors $\vec{u}$ and $\vec{v}$ in $\Re^3$ are perpendicular if $\vec{u} \cdot \vec{v} = 0$, that is, $(\vec{u}, \vec{v}) = 0$. See Section 10.3. If $\vec{u}$ and $\vec{v}$ represent piecewise continuous functions f and g on (a,b), then f and g are orthogonal on (a,b) if and only if $\int_a^b f(x)g(x)\,dx = 0$. See Definition 9.5.1.

Problem 10.9.7 *(Generalized Pythagorean Theorem)* If $\vec{u}$ and $\vec{v}$ are orthogonal and $\|f\| \equiv \sqrt{(f,f)}$, then show that $\|\vec{u}+\vec{v}\|^2 = \|\vec{u}\|^2 + \|\vec{v}\|^2$. Compare this result to Exercise 10(b) in Section 10.3.

Part 2. Polar Coordinates

Definition 10.9.8. Suppose that $P(x,y)$ is a point P in $\Re^2$ with rectangular coordinates, and 0 is the origin. Then $P[r,\theta]$ is P in *polar coordinates*[17] if $r = \left\|\overrightarrow{0P}\right\|$ and if θ is the angle between $\overrightarrow{0P}$ and the vector $\mathbf{i} = \langle 1, 0 \rangle$. Positive θ is measured counterclockwise from $\mathbf{i}$ and negative θ is measured clockwise from $\mathbf{i}$. If r is negative, then the point $P[r,\theta]$ may be represented by $P[-r, \theta+\pi]$, where $-r$ is clearly positive. The origin 0 is called the *pole.* And the positive x-axis is called the *polar axis.*

Remark 10.9.9

(a) The pole 0 may be represented by $[0, \theta]$ for any angle θ.

(b) Points $\left[1, \frac{2\pi}{3}\right]$, $\left[-1, \frac{5\pi}{3}\right]$, $\left[-1, -\frac{\pi}{3}\right]$, and $\left[1, -\frac{4\pi}{3}\right]$ represent the same point $P\left(-\frac{1}{2}, \frac{\sqrt{3}}{2}\right)$ in $\Re^2$.

(c) A point in polar coordinates is always labeled beginning with the value of r, even though when locating the point the angle θ is first to be sought out.

---

[17]Polar coordinates are a special case of a more general coordinate system called the *curvilinear coordinate system.*

(d) If $P(x,y)$ is a point P in Cartesian coordinates, and $P[r,\theta]$ is the point P in polar coordinates, then $x = r\cos\theta$, $y = r\sin\theta$, $x^2 + y^2 = r^2$, and $\tan\theta = \dfrac{y}{x}$ if $x \neq 0$. The first two expressions are used to change from polar coordinates to rectangular coordinates, while the last two expressions are used to change coordinates from rectangular to polar. ■

Problem 10.9.10

(a) Write $r = 2\sin\theta$ in rectangular coordinates. Observe that here $r = f(\theta)$, that is, r is a function of θ. Is the equivalent rectangular equation a function?

(b) Find rectangular coordinates for the points $\left[-\sqrt{2}, -\dfrac{\pi}{4}\right]$ and $\left[5, \arctan\dfrac{4}{3}\right]$ given in polar coordinates.

(c) Find polar coordinates for the point $\left(-1, \sqrt{3}\right)$.

(d) Verify that the equations for the x and y axes in polar coordinates are $\theta = 0$ and $\dfrac{\pi}{2}$, respectively.

(e) Verify that the equation for the line $y = mx$ in polar coordinates is $\theta = \arctan\dfrac{y}{x}$.

(f) Verify that the equation for the circle $x^2 + y^2 = a^2$ in polar coordinates is $r = a$.

Remark 10.9.11. Here are some tests for determining symmetry of a curve in polar coordinates. A curve in polar coordinates is called a *polar curve.*

(a) If r and θ are replaced by either r and $-\theta$, or $-r$ and $\pi - \theta$, and if the equation in polar coordinates remains unchanged, then this equation's curve is symmetric with respect to the x-axis, that is, $\theta = 0$.

(b) If r and θ are replaced by either $-r$ and $-\theta$, or r and $\pi - \theta$, and if the equation in polar coordinates remains unchanged, then this equation's curve is symmetric with respect to the y-axis, that is, $\theta = \dfrac{\pi}{2}$.

(c) If r and θ are replaced by either $-r$ and θ, or r and $\theta + \pi$, and if the equation in polar coordinates remains unchanged, then this equation's curve is symmetric with respect to the origin.

(d) If r and θ are replaced by r and $\dfrac{\pi}{2} - \theta$, and if the equation in polar coordinates remains unchanged, then this equation's curve is symmetric with respect to the line $y = x$, that is, $\theta = \dfrac{\pi}{4}$. ■

Problem 10.9.12

(a) Consider all symmetries with respect to $\theta = 0$, $\theta = \dfrac{\pi}{2}$, and the pole. If a polar curve has any two of these symmetries, prove that it must have the third.

(b) Test $r = \sin 2\theta$ for all symmetries with respect to $\theta = 0, \frac{\pi}{2}, \frac{\pi}{4}$ and the pole.

If $r = f(\theta)$ gives a smooth polar curve C, then we can find the slope $\frac{dy}{dx}$ of the tangent line to the curve at a point P without changing the equation into rectangular coordinates. For, if $\frac{dx}{d\theta} \neq 0$, then C is given parametrically as $x = r\cos\theta = f(\theta)\cos\theta$ and $y = r\sin\theta = f(\theta)\sin\theta$. Thus, by the chain rule and Theorem 10.4.5 we have that

$$\frac{dy}{dx} = \frac{dy}{d\theta}\Big/\frac{dx}{d\theta} = \frac{f'(\theta)\sin\theta + f(\theta)\cos\theta}{f'(\theta)\cos\theta - f(\theta)\sin\theta}.$$

At the origin, that is, at $[0, \theta^*]$ where $r = f(\theta^*) = 0$, the slope of the tangent line is $m = \left.\frac{dy}{dx}\right|_{\theta=\theta^*} = \tan(\theta^*)$, provided $f'(\theta^*) \neq 0$.

Remark 10.9.13

(a) Note that for several different values of θ^* a curve can pass through the origin. Thus, as in the case of other curves given parametrically, polar curves can have several tangent lines at one point.

(b) If $r = f(\theta)$, then the slope of the tangent line at $\theta = \theta^*$ is not necessarily given by $f'(\theta^*)$. Why?

(c) If $r = f(\theta)$ passes through the pole when $\theta = \theta^*$, and if $f'(\theta^*) \neq 0$, then the slope of a tangent line to the curve at the pole is $\tan(\theta^*)$. But, the slope of a line $\theta = \theta^*$ is also equal to $\tan(\theta^*)$. Why? Therefore, $\theta = \theta^*$ is an equation of a tangent line to $r = f(\theta)$ at the pole given by $[0, \theta^*]$. ■

Problem 10.9.14

(a) Find equations of the tangent lines to the curve $f(\theta) = 3\cos 2\theta$ at the pole.

(b) Find points on the curve $f(\theta) = 1 + \sin\theta$ at which a tangent line is horizontal or vertical.

(c) Find equation(s) of tangent line(s) to the curve $f(\theta) = 1 + \sin\theta$ at the pole, if possible. Explain.

(d) What do we find in solving $f'(\theta) = 0$ for θ? Explain.

Problem 10.9.15. Graph the following common polar curves for selected real values of a and b. Also, answer questions where asked.

(a) *Lines:* $\theta = a$, $r = a\sec\theta$, $r = a\csc\theta$

(b) *Circles:* $r = a$, $r = a\sin\theta$, $r = a\cos\theta$

(c) *Limacons:*[18] $r = a + b\sin\theta$, $r = a + b\cos\theta$ (Observe that if $|a| < |b|$, then the curve has an inner loop. If $|a| = |b|$, the curve is a *cardioid*.[19] If $a|b| > |a| > |b|$, the curve has a dimple. And if $|a| \geq 2|b|$, the curve is convex and tends to the shape of a circle as $|a|$ increases.)

(d) *Lemniscates:*[20] $r^2 = a\sin 2\theta$, $r^2 = a\cos 2\theta$ (How do these curves compare to the Bowditch curves discussed in Exercise 19 of Section 10.6?)

(e) *Petal curves:*[21] $r = a\sin(n\theta)$, $r = a\cos(n\theta)$, with n a nonzero integer (Note that if n is odd, the curve has n petals; if n is even, the curve has $2n$ petals, and if n is irrational, the curve has infinitely many petals. So how many petals does the curve $r = 2\sin 4\theta$ have? If θ starts at 0 and increases, in what order do the petals get traced?)

(f) *Spirals:* $r = f(\theta)$ where f is a monotonic function (Some examples of spirals are a *spiral of Archimedes* given by $r = a\theta$, a *logarithmic spiral*[22] given by $r = a\exp(\theta\cot a)$, and a *hyperbolic spiral*[23] given by $r = \dfrac{a}{\theta}$. Does a hyperbolic spiral have any asymptotes? Explain. Also, specify the direction of a hyperbolic spiral as θ increases.)

(g) $r = \cos\theta - \sin\theta$

---

[18]Étienne Pascal (1588–1651), a French lawyer born to wealth and father of Blaise Pascal, discovered the Limacon, which in Latin means "a snail." Gilles Personne de Roberval (1602–1675), a French mathematician and cartographer, named the limacon the *Limacon of Pascal.*

[19]The *cardioid*, a special case of the limacon, is a heart-shaped curve and an *epicycloid* of one loop. The cardioid is determined by a fixed point on a circle as the circle rolls on a stationary circle with the same radius. Observe that the cardioid $r = a(1 - \cos\theta)$ may also be written as $r = 2a\sin^2\left(\dfrac{1}{2}\theta\right)$.

[20]The lemniscate is a curve referred to as the *lemniscate of Bernoulli*, after Jacob Bernoulli. The term lemniscate comes from the Latin word *lemniscatus*, meaning "a pendant ribbon." The lemniscate is also a special case of a *Cassinian oval*, named after Italian mathematician and astronomer Giovanni Domenico Cassini (1625–1712).

[21]Petal curves were named between 1723 and 1728 by the Italian philosopher and mathematician Luigi Guido Grandi (1671–1742) who studied mechanics, astronomy, the "Witch of Agnesi" curve (1703), and experimented with a steam engine. Petal curves when centered at the origin are called *rhodonea curves*, the roses. See the footnote for Exercise 14 in Section 10.4.

[22]A logarithmic spiral, also called an *equiangular spiral* or a *logistic spiral*, was invented by Descartes in 1638, had its length computed by Torricelli, and was described as the coiling of a cone upon itself by D'Arcy Thompson. The spiral of Archimedes was described by Thompson as the coiling of a cylinder upon itself, such as a rope. If $\theta = \dfrac{\pi}{2}$, then the logarithmic spiral becomes a circle. A logarithmic spiral is carved on Jacob Bernoulli's tomb in Basel.

[23]A hyperbolic spiral, also called a *reciprocal spiral*, was introduced by Pierre Varignon in 1704, studied by Johann Bernoulli between 1710 and 1713, and later studied by Cotes in 1722. Pierre Varignon (1654–1722), a French mathematician and a priest, contributed mainly to differential equations and their applications in mechanics, fluid flow, and water and spring clocks.

THEOREM 10.9.16. *Suppose that a polar equation has the form* $r = \frac{de}{1 \pm e\cos\theta}$ *or* $r = \frac{de}{1 \pm e\sin\theta}$, *where* d *and* e *are positive real numbers. Then, this equation is an ellipse if* $0 < e < 1$, *a parabola if* $e = 1$, *or a hyperbola if* $e > 1$. *The number* e *is called the eccentricity.*

Problem 10.9.17. Prove Theorem 10.9.16 by finding the corresponding equations in rectangular coordinates.

Recall that intersection and collision are two different ideas, as was witnessed in Example 10.4.3. Curves may intersect, but particles traveling along those curves need not collide. For example, consider $r_1(\theta) = 1 - \cos\theta$ and $r_2(\theta) = 1 + \cos\theta$. Setting $r_1 = r_2$ we find $\left[1, \frac{\pi}{2}\right]$ and $\left[1, \frac{3\pi}{2}\right]$ are points of intersection for the two cardioids r_1 and r_2. But, in graphing the two cardioids, we see that both curves pass through the pole as well. This point was not obtained from solving $r_1 = r_2$ because curves pass through the pole for different values of θ.

Problem 10.9.18

(a) Show that $\left[2, \frac{\pi}{2}\right]$ is a point on the curve $r = 2\cos 2\theta$.

(b) Find all of the points of intersection for the curves $r^2 = \sin\theta$ and $r = 2 - \sin\theta$.

(c) Find all of the points of intersection for the curves $r = 6\sin\theta$ and $r = 2 + 2\sin\theta$.

(d) Find all of the points of intersection for the curves $r = 1 - \cos\theta$ and $r = 1 + \sin\theta$.

Recall from Exercise 7 of Section 10.7 that if $r = f(\theta)$ is continuously differentiable for $\theta \in [\alpha, \beta]$, then the length of the curve C in polar coordinates is

$$L = \int_\alpha^\beta \sqrt{[f(\theta)]^2 + [f'(\theta)]^2}\, d\theta.$$

Show that if the polar equation involves r and θ, where $r = r(t)$ and $\theta = \theta(t)$ are both continuously differentiable for $t \in [a, b]$, then $L = \int_a^b \sqrt{\left(\frac{dr}{dt}\right)^2 + r^2\left(\frac{d\theta}{dt}\right)^2}\, dt$.

Problem 10.9.19

(a) Find the length of the logarithmic spiral $r = e^\theta$ for $\theta \in [0, \pi]$.

(b) Use the formula for the length of a curve to verify that the circumference of a circle with radius a is given by $2\pi a$.

(c) Use the formula for the length of a curve to find the length of $r = -3\sin\theta$.

Concluding this project, we consider the area in polar coordinates. In Section 6.1 the definition of the Riemann integral in the plane for functions in rectangular coordinates was motivated by the search for the area under the curve. The area in polar coordinates is computed in a similar way. Consider the curve $r = f(\theta)$ which is continuous, nonnegative, and defined for $\theta \in [\alpha, \beta]$, where $\alpha < \beta \le \alpha + 2\pi$. The nonoverlapping region R between the graph of f and the lines $\theta = \alpha$ and $\theta = \beta$ will be found as follows. Partition the interval $[\alpha, \beta]$ into n subintervals to form a partition $P = \{\theta_0, \theta_1, \dots, \theta_n\}$. Next, choose $\phi_k \in [\theta_{k-1}, \theta_k]$ for $k = 1, 2, \dots, n$, and let $\Delta\theta_k = \theta_k - \theta_{k-1}$. The area of the subregion R_k bounded by f, $\theta = \theta_k$, and $\theta = \theta_{k-1}$, is to be approximated by the area of the sector of the circle having a radius $f(\phi_k)$ with $\phi_k \in [\theta_{k-1}, \theta_k]$, and having an angle $\Delta\theta_k$. The area of such a sector has the value of $\frac{1}{2}[f(\phi_k)]^2(\Delta\theta_k)$. Why? See Figure 10.9.1. The approximate area of the region R is the sum of the approximations of the subregions R_k, giving us $\sum_{k=1}^{n} \frac{1}{2}[f(\phi_k)]^2(\Delta\theta_k)$, which is a *Riemann sum* for the function $\frac{1}{2}[f(\theta)]^2$ on the interval $[\alpha, \beta]$. Now as n tends to ∞, the Riemann sum will tend to a definite integral. Hence, the area of the region R is given by $\frac{1}{2}\int_{\alpha}^{\beta} [f(\theta)]^2\, d\theta$.

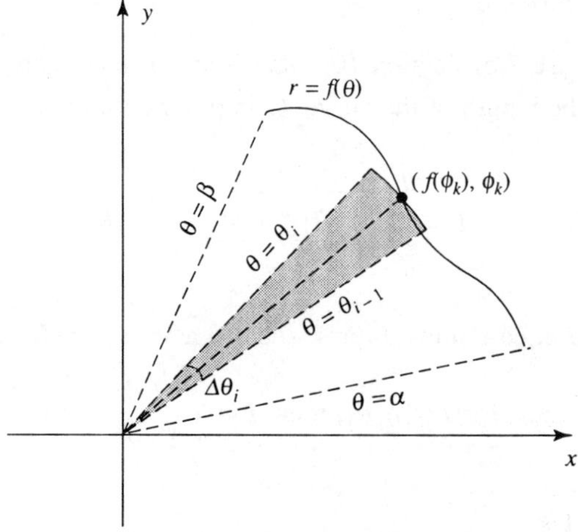

Figure 10.9.1

Problem 10.9.20. Find the area of the region R where R is

(a) inside the curve $r = 2\cos\theta$. See Exercise 8 in Section 10.7.
(b) inside the curve $r = \cos 2\theta$.
(c) the region inside both curves $r = 1$ and $r = 2\cos\theta$.
(d) inside $r = 2 + 2\sin\theta$ and outside $r = 6\sin\theta$.
(e) inside $r = 6\sin\theta$ and outside $r = 2 + 2\sin\theta$.

Problem 10.9.21. Suppose $r = f(\theta)$ is a cardioid C with its cusp at the pole.

(a) Find θ_0 for which the curve is at the pole.

(b) Compute $\lim_{\theta\to\theta_0^+} \frac{dy}{dx}$ and $\lim_{\theta\to\theta_0^-} \frac{dy}{dx}$.

(c) Find the length of C.[24]

(d) Find the area inside C.

(e) Find the parametric equations for C.

(f) Suppose a line ℓ cuts through C intersecting C at the cusp. Show that the length of the segment ℓ which is inside the cardioid is the same for all such lines ℓ.

(g) Suppose a line ℓ cuts through C intersecting C at the cusp. Show that the tangent lines to the cardioid at the points where ℓ intersects C are perpendicular.

Polar coordinates can be extended to *cylindrical coordinates* in $\Re^3$ where the point $P[r,\theta,z]$ is represented by a polar form in xy-plane and the usual z on z-axis. Can you guess why cylindrical coordinates are so called?

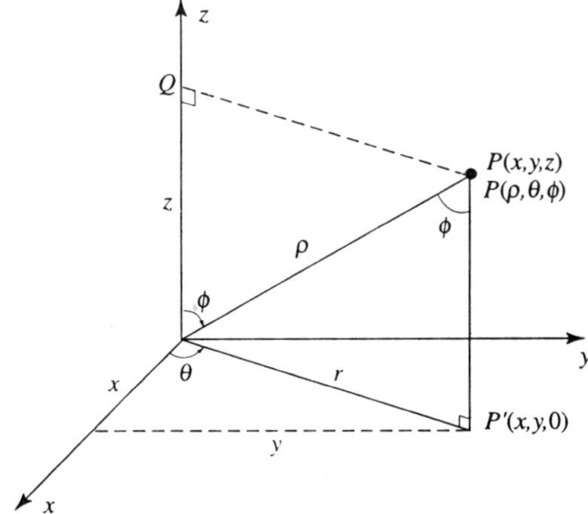

Figure 10.9.2

[24]Philippe de la Hire (1640–1718), a French artist, mathematician, and architect who contributed to astronomy, geodesy, and physics, was the first to calculate length of a cardioid in 1708.

Another popular coordinate system in $\Re^3$ is called the *spherical coordinate system,* where the point $P(\rho, \theta, \phi)$ in rectangular coordinates yields $x = \rho\cos\theta\sin\phi$, $y = \rho\sin\theta\sin\phi$, and $z = \rho\cos\phi$. Graphically, P is given as in Figure 10.9.2. From the distance formula we have that $\rho^2 = x^2 + y^2 + z^2$. Can you guess why the spherical coordinate system is so called?

Part 3. Cantor Function

Suppose the intervals $\left(a_k^{(n)}, b_k^{(n)}\right)$ are as defined in the discussion of the complement of the Cantor set in Part 4 of Section 7.7. The *Cantor function* $F(x)$ is defined to be

$$F(x) = \frac{2k-1}{2^n}, \text{ if } x \in \left(a_k^{(n)}, b_k^{(n)}\right).$$

Thus far, only a function in the complement of the Cantor set C has been defined. Next $F(x)$ for $x \in C$ is to be defined. If x_0 is any value in C, then there exists an increasing sequence $\{x_n\}$ of the endpoints of the intervals $\left(a_k^{(n)}, b_k^{(n)}\right)$ which converges to x_0, and a decreasing sequence $\{\bar{x}_n\}$ of the endpoints of the intervals $\left(a_k^{(n)}, b_k^{(n)}\right)$ which also converges to x_0. Define F at x_0 to be $F(x_0) = \lim_{n\to\infty} F(x_n) = \lim_{n\to\infty} F(\bar{x}_n)$. Hence, F is defined on the interval $[0,1]$.

Problem 10.9.22. Verify that

(a) F is defined on $[0,1]$.

(b) F is not constant.

(c) F is continuous on $[0,1]$.

(d) F is differentiable on $[0,1]$.

(e) $F'(x) = 0$ if $x \in \left(a_k^{(n)}, b_k^{(n)}\right)$.

(f) the length of the set $\{x \mid F'(x) = 0\}$ is 1, even though F is not constant on $[0,1]$.

(g) the area under the curve F on $[0,1]$ is $\frac{1}{2}$.

(h) the length of F on $[0,1]$ is 2.

(i) $\int_0^1 \sqrt{1 + [F'(x)]^2}\, dx = 1$. Does this contradict Theorem 10.7.2?

11

Functions of Two Variables

11.1 Basic Topology

Topology is an area of mathematics dealing with geometric properties that often do not depend on concepts such as length, size, or magnitude, and where the properties are preserved under continuous transformations. An introduction of this theory dates back to such names as Hausdorff,[1] Fréchet,[2] Riemann, and Cantor.

Definition 11.1.1. Suppose that $\vec{a} = \langle a_1, a_2, \ldots, a_n \rangle \in \Re^n$, $n \in N$, and r is a positive real number. A set $B(\vec{a};r)$, defined by $B(\vec{a};r) = \left\{ \vec{x} \,\middle|\, \vec{x} \in \Re^n \text{ and } \|\vec{x} - \vec{a}\| < r \right\}$, is called an *open ball* with *center* at $\vec{a}$ and *radius* r.

An open ball is also called a *basic open set* or a *neighborhood* of $\vec{a}$. In Definition

---

[1]Felix Hausdorff (1869–1942), born in Breslau, Germany (now Wrocław, Poland) created both the theory of topological spaces and the theory of metric spaces, introduced a partially ordered set, special types of ordinals, the notion of Hausdorff dimension, and the Hausdorff measure, and is given credit for the term *metric space*. In 1942, being a Jew, Hausdorff, his wife, and her sister, all committed suicide to avoid being taken to a Nazi camp.

[2]Maurice René Fréchet (1878–1973), a French topologist, founded the theory of abstract spaces, and made important contributions to calculus, probability, and statistics.

11.1.1, when $n=1$, $\vec{a} \equiv a$ and $\vec{x} \equiv x$. Also, the symbol $\|\vec{x}-\vec{a}\|$, called the *Euclidean distance* from $\vec{x}$ to $\vec{a}$, is sometimes denoted $d(\vec{x},\vec{a})$. In $\Re^1$, $\|\vec{x}-\vec{a}\|$ is an absolute value. Review Definition 2.5.1. In $\Re^1$, an open ball is an open interval, in $\Re^2$, it is an *open disk*, and in $\Re^3$, it is a solid *open sphere*.

Definition 11.1.2. A point $\vec{a} \in S \subseteq \Re^n$ with $n \in N$ is an *interior point* of S if and only if there exists an open ball centered at $\vec{a}$ that is entirely contained in S. The set of all interior points of S is called the *interior* of S, and is denoted by $\operatorname{int} S$.

Thus, $\vec{a} \in \operatorname{int} S$ if and only if there exists an $r>0$ such that $B(\vec{a};r) \subseteq S$.

Definition 11.1.3. A set $S \subseteq \Re^n$ with $n \in N$ is said to be *open* if and only if each point of S is an interior point of S.

Review Definition 4.1.16. $S \subseteq \Re^n$ is open, which means that S is open in $\Re^n$ for a particular n. For example, in Exercise 4 one must show that an interval $(0,1)$ is open in $\Re^1$, but not open in $\Re^2$. Also, in Exercise 8(a), one must show that "not open" does not mean closed. The formal definition of "closed" is given below.

THEOREM 11.1.4

(a) *An arbitrary union of open sets is open.*
(b) *A finite intersection of open sets is open.*

Proof of part (a). Let Δ be a given nonempty (*indexing*) set and for each element $\alpha \in \Delta$ there is a corresponding open set A_α. Assume that $S = \bigcup_{\alpha\in\Delta} A_\alpha \equiv \{x \mid x \in A_\alpha$ for at least one $\alpha \in \Delta\}$. We will prove that S is open. To this end, choose any $\vec{x} \in S$. Thus, there exists a particular α, call it α^*, for which $\vec{x} \in A_{\alpha^*}$. Since A_{α^*} is open, there exists $r>0$ such that $B(\vec{x};r) \subseteq A_{\alpha^*}$. But, $A_{\alpha^*} \subseteq S$. Therefore, $B(\vec{x};r) \subseteq S$. Hence, $\vec{x}$ is an interior point of S, so, S is open. ■

Proof of Theorem 11.1.4, part (b), is left as Exercise 6. In part (b), if the word "finite" is replaced by the word "arbitrary," then the statement is not true. This is Exercise 7.

Definition 11.1.5. A point $\vec{s}_0$ is an *accumulation point* of the set $S \subseteq \Re^n$ with $n \in N$ if and only if every open ball centered at $\vec{s}_0$ contains at least one point of S other than $\vec{s}_0$ itself.

Review Definition 2.5.2. As proven in Exercise 1 of Section 2.5, $\vec{s}_0$ is an accumulation point of S if and only if any open ball centered at $\vec{s}_0$ contains infinitely many points of S. Definition 4.1.15 and Theorem 4.1.17 may also be extended to $\Re^n$ for $n \in N$. That is, a set $S \subseteq \Re^n$ is *closed* if and only if the set S contains all of its accumulation points. And a set S is closed if and only if $\Re^n \setminus S$ for $n \in N$ is open. The *closure* $\overline{S}$ of S is defined by $\overline{S} = S \cup S'$, where S' is the set of all accumulation points of S, often called the *derived* set for S. $\overline{S}$ is closed. Why? Also, a set $S \subseteq \Re^n$ is *bounded* if and only if there exists $r > 0$ such that $S \subseteq B(\vec{0}; r)$.

THEOREM 11.1.6

(a) *A finite union of closed sets is closed.*
(b) *An arbitrary intersection of closed sets is closed.*

The proof of this theorem is left as an exercise.

THEOREM 11.1.7. *(Bolzano–Weierstrass Theorem for Sets) Every bounded infinite subset of $\Re^n$ with $n \in N$ has at least one accumulation point.*

For the case $n = 1$, Theorem 11.1.7 was proven in Section 2.5. The proof given for that theorem may not be directly extended to $\Re^2$ or $\Re^3$ since elements in $\Re^2$ and $\Re^3$ cannot be ordered, that is, arranged in an increasing fashion. A proof of Theorem 11.1.7 is left as the challenging Exercise 22. Instead, an application of Theorem 11.1.7 will be proven.

THEOREM 11.1.8. *(Cantor Intersection Theorem) Suppose that A_k with $k \in N$ are nonempty subsets of $\Re^n$ for $n \in N$ such that*

(a) $A_{k+1} \subseteq A_k$, *and*
(b) *each A_k is closed with A_1 bounded.*

Then, $\bigcap_{k=1}^{\infty} A_k$ is closed and nonempty.

Proof. If A_k is finite for some $k \in N$, then the proof is trivial. Why? Thus, assume that all of the A_ks are infinite. And define $S \equiv \bigcap_{k=1}^{\infty} A_k$. By Theorem 11.1.6, part (b), S is closed. Next, we need to show that $S \neq \phi$. Since the A_ks are infinite, define the infinite set $A = \{\vec{x}_1, \vec{x}_2, \ldots\}$, where $\vec{x}_k \in A_k$ and all of the points of A are distinct. Since $A \subset A_1$ and A_1 is bounded, so is A. By Bolzano–Weierstrass theorem for sets, A has an accumulation point, say $\vec{s}_0$. We will show that $\vec{s}_0 \in S$, and hence, that $S \neq \phi$. To show that $\vec{s}_0 \in S$, we will show that $\vec{s}_0 \in A_k'$ for all $k \in N$. Then, since the A_ks are closed, they must contain all of their accumulation points. And thus, $\vec{s}_0 \in A_k$ for all $k \in N$, which means that $\vec{s}_0 \in S$. Hence, consider the $B(\vec{s}_0; r)$ for any $r > 0$. Since $\vec{s}_0$ is an accumulation point of A, the $B(\vec{s}_0; r)$ contains infinitely many points of A. And since $A_{k+1} \subseteq A_k$, the $B(\vec{s}_0; r)$ will contain infinitely many points of A_k. Therefore, $\vec{s}_0 \in A_k'$. And thus, $\vec{s}_0 \in A_k$ for all $k \in N$. Hence, the intersection is nonempty. ■

Definition 11.1.9. The *boundary point* of a set S is a point $\vec{x}$ such that every open ball about $\vec{x}$ contains points of both S and the *complement* of S, that is, $\mathfrak{R}^n \setminus S$. The set of all boundary points of S, denoted by ∂S, is called the *boundary* of S.

Definition 11.1.10. Suppose that $n \in N$. Then,

(a) $A, B \subseteq \mathfrak{R}^n$ are *separated* if and only if $A \cap \overline{B} = \phi = \overline{A} \cap B$.

(b) a set $S \subseteq \mathfrak{R}^n$ is *disconnected* if and only if there exist open, disjoint, nonempty subsets A and B of $\mathfrak{R}^n$ such that $S = A \cup B$.

(c) a set $S \subseteq \mathfrak{R}^n$ is *connected* if and only if it is not disconnected.

(d) a set $S \subseteq \mathfrak{R}^n$ is a *region* if and only if S is a connected set that contains all of its interior points as well as any boundary points.

(e) a *closed region* is a region that contains all of its boundary points.

(f) an *open region* is a region that contains only interior points.

As we noted before, two sets A and B are *disjoint* if and only if $A \cap B = \phi$. Thus, two subsets A and B of $\mathfrak{R}^n$ are separated if they are disjoint, and neither contains an accumulation point of the other. That is, no point of A lies in $\overline{B}$ and vice versa. In mathematical literature, often, open regions are referred to as *domains*. These domains should not be confused with domains of functions. Without proof we will accept the fact that the only connected sets in $\mathfrak{R}^1$ are intervals and singletons.

Definition 11.1.11. A set $S \subseteq \mathfrak{R}^n$ with $n \in N$ is *arcwise connected* if and only if any two points in S can be connected by a path contained entirely in S.

In other words, the set S is arcwise or *pathwise connected* if and only if for any two points $\vec{a}$ and $\vec{b}$ in S, there exists a continuous function $f:[0,1] \to S$ such that $f(0) = \vec{a}$, $f(1) = \vec{b}$, and $f([0,1]) \subset S$. The image of f is a *path* from $\vec{a}$ to $\vec{b}$ that remains in S. In $\mathfrak{R}^1$, connected sets, that is, intervals, are arcwise connected and vice versa. However, unless a set S is open, it is not generally true that a nonempty connected set S must be arcwise connected. See Exercise 29 and Theorem 11.1.13. The converse is, however, true. The last two results are presented without proof.

THEOREM 11.1.12. *If S is arcwise connected, then S is connected.*

THEOREM 11.1.13. *If $S \subseteq \mathfrak{R}^2$ is a nonempty open connected set, then S is arcwise connected.*

Exercises 11.1

1. Prove that $\operatorname{int} S \subseteq S$.

2. Prove that a set S is open if and only if $S = \operatorname{int} S$.

3. Prove that an open ball, the empty set ϕ, and the *whole space* $\Re^n$ are open sets.

4. Prove that the interval $(0,1)$ is open in $\Re^1$ but not in $\Re^2$.

5. Prove that for real constants a, b, c, and d with $a < b$ and $c < d$, the Cartesian product $\{(x,y) | x \in (a,b), y \in (c,d)\}$ is an open set in $\Re^2$.

6. Prove part (b) of Theorem 11.1.4.

7. Find an infinite collection of open sets for which the intersection is
   (a) open. **(b)** not open.

8. **(a)** Give an example of a set that is neither open nor closed.
   (b) Give an example of a set that is both open and closed.

9. Prove that $\overline{S}$, S', the ϕ, and the whole space $\Re^n$ are closed.

10. Prove that S is closed if and only if $S = \overline{S}$.

11. Prove Theorem 11.1.6.

12. For each set S, find S'.
    (a) $S = \{(x,y) | y > 2x\}$ **(b)** Q, the rational numbers
    (c) $S = \left\{ \left(\frac{1}{n}, \frac{2}{n} \right) \middle| n \in N \right\}$

13. Prove that the $\operatorname{int} S$ is the largest open subset of S.

14. Prove that $\overline{S}$ is the smallest closed set containing S.

15. Assume that $A, B \subseteq \Re^n$.
    (a) Prove that $(\operatorname{int} A) \cap (\operatorname{int} B) = \operatorname{int}(A \cap B)$. Is this statement true for the union as well? Explain.
    (b) Prove that $A \subseteq B$ implies that $A' \subseteq B'$.
    (c) Prove that $A \subseteq B$ implies that $\overline{A} \subseteq \overline{B}$.
    (d) Prove that $(A \cup B)' = A' \cup B'$. Is this statement true for the intersection as well? Explain.
    (e) Prove that $A' = (\overline{A})'$.
    (f) Prove that $\overline{A \cup B} = \overline{A} \cup \overline{B}$. Is this statement true for the intersection as well? Explain.
    (g) Give an example where $\operatorname{int} A = \phi$, $\operatorname{int} B = \phi$, but $\operatorname{int}(A \cup B) = \Re$. Compare to part (a).

16. Show that the boundary of S is the same as the boundary of the complement of S.

17. Show that $\partial S = \overline{S} \cap \overline{\left(\Re^n \setminus S\right)}$ for $n \in N$ and thus, that ∂S is a closed set.

18. Show that the boundary of an open ball $B(\vec{a}; r)$ is the set of all points $\vec{x}$ such that $\|\vec{x} - \vec{a}\| = r$.

19. Show that the boundary of rational numbers Q in $\Re^1$ is $\Re^1$.

20. If $A \subseteq \Re^3$, prove that

(a) ∂A is closed.

(b) A is open if and only if $A \cap \partial A = \phi$.

(c) A is closed if and only if $\partial A \subseteq A$.

(d) $\partial A = \phi$ if and only if A is both open and closed.

21. Define distances between $\vec{x}$ and $\vec{y}$ in $\Re^n$ by

$$d_1(\vec{x},\vec{y}) = \max_{1 \le k \le n} |x_k - y_k| \text{ and } d_2(\vec{x},\vec{y}) = \sum_{k=1}^{n} |x_k - y_k|.$$

Suppose that $\vec{x} = (x_1, x_2)$ and $\vec{y} = (y_1, y_2)$ for $\vec{x}$ and $\vec{y}$ in $\Re^2$, and that $\vec{x} = (x_1, x_2, x_3)$ and $\vec{y} = (y_1, y_2, y_3)$ for $\vec{x}$ and $\vec{y}$ in $\Re^3$.

(a) What is the shape of $B_1(\vec{a};r) \equiv \left\{\vec{x} \,\middle|\, \vec{x} \in \Re^n \text{ and } d_1(\vec{x},\vec{a}) < r\right\}$ in $\Re^2$? How about in $\Re^3$?

(b) What is the shape of $B_2(\vec{a};r) \equiv \left\{\vec{x} \,\middle|\, \vec{x} \in \Re^n \text{ and } d_2(\vec{x},\vec{a}) < r\right\}$ in $\Re^2$? How about in $\Re^3$?

(c) Show that $d_1(\vec{x},\vec{y}) \le d(\vec{x},\vec{y}) \equiv \|\vec{x} - \vec{y}\| \le d_2(\vec{x},\vec{y})$ for all $\vec{x}, \vec{y} \in \Re^n$.

(d) Show that $d_2(\vec{x},\vec{y}) \le \sqrt{n}\, d(\vec{x},\vec{y}) \le n\, d_1(\vec{x},\vec{y})$ for all $\vec{x}, \vec{y} \in \Re^n$.

22. Prove Theorem 11.1.7.

23. Prove that $D \subseteq \Re^2$ is closed if and only if any sequence $\{(x_n, y_n)\}$ in D converges to a point in D. Note that a sequence $\{(x_n, y_n)\}$ *converges* to (a,b) if and only if $\langle x_n, y_n \rangle$ converges to $\langle a, b \rangle$.

24. *(Bolzano–Weierstrass Theorem for Sequences)* Prove that any bounded sequence in $\Re^2$ has at least one convergent subsequence. See Theorem 2.6.4. Also, note that a sequence $\{(x_n, y_n)\}$ is *bounded* if and only if there exists $M \in \Re$ such that $\|\langle x_n, y_n \rangle\| \le M$ for all $n \in N$, that is, the set $\left\{(x_n, y_n) \,\middle|\, n \in N\right\}$ is bounded.

25. Prove the following statements.

(a) In $\Re^1$, the set $S = \Re \setminus \{0\}$ is disconnected.

(b) In $\Re^1$, the set Q of rational numbers is disconnected.

(c) In $\Re^1$, a singleton is connected.

(d) In $\Re^2$, the point (a,b) is connected.

(e) ϕ is connected.

(f) Open balls are connected.

(g) Open regions are connected.

(h) $\Re^1$ is connected.

26. Prove that if A and B are nonempty separated sets, then $A \cup B$ is separated.

27. If sets A and B form a disconnection of the set S, and C is a connected subset of S, prove that $C \subseteq A$ or $C \subseteq B$.

28. If A is a connected set and B is a set such that $A \subseteq B \subseteq \overline{A}$, prove that B is connected, and hence, $\overline{A}$ is connected.

29. If $A \equiv \left\{(x,y) \middle| x \in [0,1] \text{ and } y = \frac{x}{n} \text{ with } n \in N\right\}$ and $B \equiv \left\{(x,0) \middle| x \in \left[\frac{1}{2},1\right]\right\}$, show that $S \equiv A \cup B$ is connected. Also note that S is not arcwise connected. Review Example 8.1.4 and Figure 8.1.2.

30. Prove that open ball in $\Re^2$ is arcwise connected.

31. A set $S \subseteq \Re^n$ with $n = 2$ or $n = 3$ is *convex* if and only if for any two points $\vec{x}, \vec{y} \in \Re^n$ and any $t \in [0,1]$, the line segment $(1-t)x + ty \in S$. Compare this statement to Definition 5.8.13. Show that open balls in $\Re^n$ for $n = 2$ and $n = 3$ are convex.

11.2 Limits and Continuity

The beta function in Section 6.6, and particular planes and surfaces in Section 10.5 touched on the idea of the domain of a function involving two independent variables. In this section and sections to follow, more will be learned about functions of two or more variables.

Consider the plane $z = 3$ in $\Re^3$. For any choice of x and y, the height is 3. Think of a calm water 3 feet deep, say. Since for any two *independent variables* x and y there exists one *dependent variable* z with a value of 3, we may call this surface a function and represent it by $f(x,y) = 3$. If the "water" was not calm, the function representing the wavy surface might be given by $f(x,y) = 3 + \sin x - \cos y$, equivalently, $f(x,y) = z$, where $z = 3 + \sin x - \cos y$.

In general, $f: D_f \subseteq \Re^2 \to \Re$ is a *real-valued function of two variables* with the *domain* $D_f \subseteq \Re^2$. And the set of all z values taken on by f is the function's *range*. Clearly, there is no more than one functional value z for each point (x,y) in D_f. If $f(x,y) = \sqrt{1 - x^2 - y^2}$, then the domain of f, D_f, consists of all points (x,y) satisfying $1 - x^2 - y^2 \geq 0$. Therefore, $D_f = \left\{(x,y) \middle| x^2 + y^2 \leq 1\right\}$, which is intuitively a closed unit disk. The range for this function is $R_f = \left\{z \middle| 0 \leq z \leq 1\right\}$, and the graph of f is a *surface* in $\Re^3$. Recall that the graphs of functions from $\Re^1$ to $\Re^1$ were called *curves* in $\Re^2$. Now, to graph $f(x,y) = \sqrt{1 - x^2 - y^2}$ by hand, one might first wish to know the *level curves*, that is, the curves in the xy-plane that represent the set of all points (x,y)

that are at a level c, for various real constants c. A level curve represents a solution to $f(x,y)=c$ in the xy-plane. Level curves with equally spaced levels are used to construct topographical maps. The closer the level curves are on such maps, the steeper the terrain. For contrast, *contour lines* are curves in space that represent the intersection of the surface $z=f(x,y)$ with horizontal planes $z=c$, where $c\in\Re$. Thus, level curves are projections of contour lines onto the xy-plane.

A sphere is an example of a surface that is not a function. Review Section 10.5 for other such surfaces. Consider a sphere with an equation involving three variables, say $x^2+y^2+z^2=9$, also written as $f(x,y,z)=9$ for $f(x,y,z)=x^2+y^2+z^2$. Along these lines, the previously given function $g(x,y)=\sqrt{1-x^2-y^2}\equiv z$ is a surface given by $f(x,y,z)=0$, where $f(x,y,z)=\sqrt{1-x^2-y^2}-z$. This notation will be seen again in Section 11.6.

Since there is a close relationship between a point (a,b) and the vector $\vec{v}=\langle a,b\rangle=a\mathbf{i}+b\mathbf{j}$, at times $f(a,b)$ will be represented by $f(\vec{v})$ in order to simplify the notation. If (a,b) represents a point P, then P and $\vec{P}$ will be used interchangeably. Also, at times sequences of points (a_n,b_n) will be represented by sequences of vectors $\vec{x}_n=\langle a_n,b_n\rangle$, previously discussed in Section 10.6. Moreover, to shorten the phrase if the point (a,b) is in the domain D, $\vec{v}=\langle a,b\rangle\in D$ will be written. Lastly, $\|(x,y)-(a,b)\|$ may be written for $\|\langle x,y\rangle-\langle a,b\rangle\|$.

Definition 11.2.1. Suppose that $f:D_f\subseteq\Re^2\to\Re$, and that (a,b) is an accumulation point of D. Then $\lim\limits_{(x,y)\to(a,b)} f(x,y)=L$, also written as $\lim\limits_{\substack{x\to a\\ y\to b}} f(x,y)=L$, if and only if for every $\varepsilon>0$ there exists $\delta>0$ such that $|f(x,y)-L|<\varepsilon$, provided that $0<\sqrt{(x-a)^2+(y-b)^2}<\delta$, and $(x,y)\in D$.

Recall that the set of points (x,y) in $\Re^2$ that satisfy $\sqrt{(x-a)^2+(y-b)^2}<\delta$ for some $\delta>0$ is called a *neighborhood* of (a,b), or an *open ball* about (a,b). Points (x,y) that satisfy $0<\sqrt{(x-a)^2+(y-b)^2}<\delta$ represent a *deleted (circular) neighborhood*, that is, an open ball with no center, also denoted by $B((a,b);\delta)\setminus\{(a,b)\}$, $0<\|(x,y)-(a,b)\|<\delta$, or $0<d((x,y),(a,b))<\delta$. Since

$$|x-a|=\sqrt{(x-a)^2}\le\sqrt{(x-a)^2+(y-b)^2}\text{ and}$$
$$|y-b|=\sqrt{(y-b)^2}\le\sqrt{(x-a)^2+(y-b)^2}$$

whenever $0<\sqrt{(x-a)^2+(y-b)^2}<\delta$, we also have that $0<|x-a|<\delta$ and

$0 < |y-b| < \delta$. The points (x,y) that satisfy $0 < |x-a| < \delta$ and $0 < |y-b| < \delta$ represent a *deleted rectangular neighborhood.* Compare this rectangular neighborhood to shapes presented in Exercise 21 of Section 11.1. Using vector notation, if $\vec{x} \equiv \langle x, y \rangle$ and $\vec{a} \equiv \langle a, b \rangle$, then $\lim_{\vec{x} \to \vec{a}} f(\vec{x}) = L$ if and only if for every $\varepsilon > 0$ there exists $\delta > 0$ such that $|f(\vec{x}) - L| < \varepsilon$ whenever $0 < d(\vec{x}, \vec{a}) < \delta$, and $\vec{x} \in D$. This is a generalization of Definition 3.2.1.

Intuitively speaking, $\lim_{(x,y) \to (a,b)} f(x,y) = L$ means that the value of $f(x,y)$ is within an arbitrary $\varepsilon > 0$ of L whenever points (x,y) from the domain of f, different from (a,b), are inside an open ball $B(\vec{a}; \delta)$. In some cases, the distance used to measure how far $\vec{x}$ is from $\vec{a}$ is not necessarily Euclidean. Review Exercise 21 in Section 11.1. As in Chapter 3, if $\lim_{(x,y) \to (a,b)} f(x,y) = L$, then the limit as (x,y) approaches (a,b) exists and is finite. However, (a,b) need not be in the domain of f and L need not be $f(a,b)$. For continuous functions L will be $f(a,b)$, as is discussed later.

Example 11.2.2. Show that $\lim_{(x,y) \to (a,b)} \left(x^2 + y^2\right)$ exists at any point $(a,b) \in \Re^2$. See Example 3.2.3.

Answer. In order to use the definition of a limit, we need to decide what the limiting value should be, and then prove that it was chosen correctly. Thus, we will prove that $\lim_{(x,y) \to (a,b)} \left(x^2 + y^2\right) = a^2 + b^2$. To this end, let $\varepsilon > 0$ be given. We need to find $\delta > 0$ such that if $0 < \|(x,y) - (a,b)\| < \delta$, then $\left|\left(x^2 + y^2\right) - \left(a^2 + b^2\right)\right| < \varepsilon$. Thus, we write

$$\begin{aligned} \left|\left(x^2 + y^2\right) - \left(a^2 + b^2\right)\right| &= |(x-a)(x+a) + (y-b)(y+b)| \\ &\le |x-a||x+a| + |y-b||y+b|. \end{aligned}$$

Now, the right-hand side above needs to be bounded by ε. Recall that $|x-a| \le \|(x,y) - (a,b)\|$, $|y-b| \le \|(x,y) - (a,b)\|$, and by Minkowski inequality, see Exercise 10(b) in Section 1.7, $\|(x,y) - (a,b)\| \le |x-a| + |y-b|$. Thus, the terms $|x+a|$ and $|y+b|$ need to be bounded above so that they can be eliminated. To do this we will restrict δ to be ≤ 1. Then

$$\begin{aligned} &|x+a| = |x-a+2a| \le |x-a| + |2a| \le 1 + |2a|, \text{ and} \\ &|y+b| = |y-b+2b| \le |y-b| + |2b| \le 1 + |2b|. \end{aligned}$$

Why? Since a and b are fixed, there exists $M \ge 1$ such that $|x+a| \le M$ and $|y+b| \le M$ for all (x,y) satisfying $0 < \|(x,y) - (a,b)\| < 1$. Thus,

$$\begin{aligned}\left|\left(x^2+y^2\right)-\left(a^2+b^2\right)\right| &\le |x-a||x+a|+|y-b||y+b| \\ &\le M\left(|x-a|+|y-b|\right), \text{ if } 0<\|(x,y)-(a,b)\|<1 \\ &\le M\left(\frac{\varepsilon}{2M}+\frac{\varepsilon}{2M}\right)=\varepsilon, \text{ if } 0<\|(x,y)-(a,b)\|<\delta\equiv\min\left\{1,\frac{\varepsilon}{2M}\right\}.\end{aligned}$$ ■

Remark 11.2.3. Suppose that $f:D\subseteq\Re^2\to\Re$. Then $f(x,y)$ has no limiting value at the point (a,b) if

(a) (a,b) is not an accumulation point of D.

(b) for every L there exists a particular $\varepsilon>0$ such that for every $\delta>0$ there exists some point $(x,y)\in D$ for which $0<d((x,y),(a,b))<\delta$ but $|f(x,y)-L|\ge\varepsilon$.

(c) there exist two sequences $\{\vec{x}_n\}$ and $\{\vec{y}_n\}$ in D converging to $\langle a,b\rangle$, but $\lim_{n\to\infty} f(\vec{x}_n)\ne\lim_{n\to\infty} f(\vec{y}_n)$. See Exercise 7. These sequences can be replaced by continuous curves (paths) that are defined on D and pass through the point (a,b).

(d) the *iterated limits* are not the same, that is, $\lim_{x\to a}\left[\lim_{y\to b} f(x,y)\right]\ne \lim_{y\to b}\left[\lim_{x\to a} f(x,y)\right]$. In other words, the limit is not independent of the method of approach to (a,b) used in part (c). See Exercise 1 in Section 11.7. Recall that Exercise 5(a) of Section 8.1 and part (c) of Remark 8.3.2 contained iterated limits. ■

Example 11.2.4. Show that $\lim_{(x,y)\to(0,0)}\dfrac{2x^2y}{x^4+y^2}$ does not exist.

Proof. In implementing part (c) of Remark 11.2.3, we will consider different continuous paths that pass through the origin and yet yield different values. Suppose that the points (x,y) approach $(0,0)$ via any straight line $y=mx$ for some $m\in\Re$. Then,

$$\lim_{\substack{(x,y)\to(0,0)\\ \text{with } y=mx}}\frac{2x^2y}{x^4+y^2}=\lim_{x\to0}\frac{2x^2(mx)}{x^4+(mx)^2}=\lim_{x\to0}\frac{2mx^3}{x^2\left(x^2+m^2\right)}=\lim_{x\to0}\frac{2mx}{x^2+m^2}=0.$$

If the line is vertical, then $x=0$ and $\lim_{\substack{(x,y)\to(0,0)\\ \text{with } x=0}}\dfrac{2x^2y}{x^4+y^2}=\lim_{y\to0}\dfrac{0}{y^2}=0$. Since we are to show that the limit does not exist, we need to find another path to $(0,0)$ that will give a different limiting value than 0 found from any of the lines above. Let us try a parabolic path, say, $y=x^2$. Then

$$\lim_{\substack{(x,y)\to(0,0)\\ \text{with } y=x^2}} \frac{2x^2y}{x^4+y^2} = \lim_{x\to 0} \frac{2x^2(x^2)}{x^4+(x^2)^2} = \lim_{x\to 0} \frac{2x^4}{2x^4} = 1 \neq 0.$$

Thus, the limit does not exist. ■

Observe that if we are to determine whether or not a given limit at (a,b) exists, then either we decide on the *limiting value* L and prove that it is indeed a correct limiting value, or we find two different paths that approach (a,b), but whose functional values do not approach the same value. Sometimes finding such paths may be difficult. Perhaps changing to a different coordinate system would produce quicker results. For example,

$$\lim_{(x,y)\to(0,0)} \frac{xy^2}{x^2+y^2} = \lim_{r\to 0} \frac{(r\cos\theta)(r\sin\theta)^2}{r^2} = \cos\theta\sin^2\theta \lim_{r\to 0} r = 0,$$

since $\cos\theta\sin^2\theta$ is bounded. See Exercise 2(a).

Many results obtained for functions of a single variable may be extended to functions of two or more variables. A few results are presented here, with proofs left to the reader. These results and others to follow will replace extended algebraic manipulations yielding quicker and more attractive conclusions.

THEOREM 11.2.5. *Suppose that* $f,g: D\subseteq \Re^2 \to \Re$, (a,b) *is an accumulation point of D,* $\lim_{(x,y)\to(a,b)} f(x,y) = A$, *and* $\lim_{(x,y)\to(a,b)} g(x,y) = B$. *Then*

(a) $\lim_{(x,y)\to(a,b)} (f\pm g)(x,y) = A\pm B.$

(b) $\lim_{(x,y)\to(a,b)} (fg)(x,y) = AB.$

(c) $\lim_{(x,y)\to(a,b)} \left(\frac{f}{g}\right)(x,y) = \frac{A}{B}$ *if* $B\neq 0$, *and if* $g(x,y)\neq 0$ *in a neighborhood of* (a,b).

(d) $\lim_{(x,y)\to(a,b)} |f(x,y)| = |A|.$

(e) $A \geq B$, *if* $f(x,y)\geq g(x,y)$ *on D.*

(f) $\lim_{(x,y)\to(a,b)} \sqrt{f(x,y)} = \sqrt{A}$ *if* $f(x,y)\geq 0$ *on D.*

Continuity Definitions 4.1.1 and 4.1.2 may also be extended to $\Re^2$.

Definition 11.2.6

(a) *(Local)* Suppose that $f: D\subseteq \Re^2 \to \Re$. Then f is *continuous at a point* $(a,b)\in D$ if and only if for any given $\varepsilon > 0$, there exists $\delta > 0$ such that $|f(x,y) - f(a,b)| < \varepsilon$ for all $(x,y)\in D$ with $\|(x,y)-(a,b)\| < \delta$.

(b) *(Global)* A function $f:D\subseteq\Re^2\to\Re$ is *continuous on a set* $E\subseteq D$ if and only if f is continuous at each point in E. If f is continuous at every point in its domain D, then f is *continuous*.

Definition 11.2.6 applies to the interior points as well as the boundary points of the domain D of a function f, provided that all points are in D. Also, observe that f is continuous at all *isolated points*, that is, *singletons* of the domain D. Isolated points are defined the same way in $\Re^2$ as in $\Re^1$. If (a,b) is an accumulation point of D, then f is continuous at (a,b) if and only if $\lim\limits_{(x,y)\to(a,b)} f(x,y)=f(a,b)$, equivalently, $\lim\limits_{(x,y)\to(a,b)}[f(x,y)-f(a,b)]=0$. In addition, note that if f is continuous at (a,b), then there exists an open ball $B((a,b);\delta)$, which is mapped by f into an open ball $B(f(a,b);\varepsilon)$. Thus, we can write that

$$f\big(B((a,b);\delta)\cap D\big)\subseteq B\big(f(a,b);\varepsilon\big).$$

Results for the single-variable functions in Chapter 4 may also be extended to $\Re^2$. The next theorem extends Exercises 5(k) and 5(l) from Section 4.1.

THEOREM 11.2.7. *A function $f:D\subseteq\Re^2\to\Re$ is continuous at $(a,b)\in D$ if and only if for any sequence $\{(x_n,y_n)\}$ in D converging to (a,b), the sequence $\{f(x_n,y_n)\}$ converges to $f(a,b)$.*

Proof. $(\Rightarrow)$ Suppose that f is continuous at $(a,b)\in D$. Let $\{(x_n,y_n)\}$ be a sequence in D which converges to (a,b). We will prove that $\{f(x_n,y_n)\}$ converges to $f(a,b)$. Thus, let $\varepsilon>0$ be given. Since f is continuous at (a,b), there exists $\delta>0$ such that $f\big(B((a,b);\delta)\cap D\big)\subseteq B\big(f(a,b);\varepsilon\big)$. Now, since $\{(x_n,y_n)\}$ converges to (a,b), there exists $n^*\in N$ such that $(x_n,y_n)\in B((a,b);\delta)$ for all $n\geq n^*$. Thus, $f(x_n,y_n)\in B\big(f(a,b);\varepsilon\big)$ for all $n\geq n^*$. Hence, since $\varepsilon>0$ is arbitrary, the sequence $\{f(x_n,y_n)\}$ converges to $f(a,b)$.

$(\Leftarrow)$ Suppose that for any sequence $\{(x_n,y_n)\}$ in D converging to (a,b), $\{f(x_n,y_n)\}$ converges to $f(a,b)$. We will prove that f is continuous at (a,b) by assuming to the contrary. Thus, assume that there exists $\varepsilon^*>0$ such that no $\delta>0$ gives $f\big(B((a,b);\delta)\cap D\big)\subseteq B\big(f(a,b);\varepsilon^*\big)$. In particular, if $\delta=\dfrac{1}{n}$ with $n\in N$, then

$$f\left(B\left((a,b);\frac{1}{n}\right)\cap D\right)\not\subseteq B\big(f(a,b);\varepsilon^*\big).$$

Thus, there exist points (x_n, y_n) in D such that $d((x_n, y_n),(a,b)) < \frac{1}{n}$. But $d(f(x_n, y_n), f(a,b)) \geq \varepsilon^*$. Hence, we have found a sequence $\{(x_n, y_n)\}$ converging to (a,b) for which $\{f(x_n, y_n)\}$ does not converge to $f(a,b)$. Hence, the proof is complete due to a contradiction to the assumption that f is not continuous. ■

THEOREM 11.2.8. *If functions $f, g : D \subseteq \Re^2 \to \Re$ are continuous at $(a,b) \in D$, then*

(a) *$f \pm g$ are continuous at (a,b).*

(b) *fg is continuous at (a,b).*

(c) *$\frac{f}{g}$ is continuous at (a,b), provided that $g(a,b) \neq 0$.*

(d) *$|f|$ is continuous at (a,b).*

(e) *$\sqrt{f}$ is continuous at (a,b), provided that $f \geq 0$.*

Proof of part (a). Incorporating Theorem 11.2.7, let $\{(x_n, y_n)\}$ be a sequence in D that converges to the point (a,b) in D. Since f and g are continuous at (a,b), $\{f(x_n, y_n)\}$ and $\{g(x_n, y_n)\}$ converge to $f(a,b)$ and $g(a,b)$, respectively. But the sequences $\{f(x_n, y_n)\}$ and $\{g(x_n, y_n)\}$ are in $\Re^1$. And by part (a) of Theorem 2.2.1,

$$\lim_{n\to\infty} \left[f(x_n, y_n) \pm g(x_n, y_n) \right] = \lim_{n\to\infty} f(x_n, y_n) \pm \lim_{n\to\infty} g(x_n, y_n) = f(a,b) \pm g(a,b).$$

Hence, by Theorem 11.2.7, $f \pm g$ are continuous at $(a,b) \in D$. ■

Proofs for the other parts of Theorem 11.2.8 are left as an exercise. Theorem 11.2.9, on continuity of the composition of functions, extends Theorem 4.1.9, proof of which is left to the reader.

THEOREM 11.2.9. *Consider functions $f : A \to \Re$ and $g : B \to \Re$ with $A, B \subseteq \Re^2$ such that $f(A) \subseteq B$. If f is continuous at $(a,b) \in A$ and g is continuous at $f(a,b) \in B$, then the function $g \circ f$ is continuous at (a,b).*

THEOREM 11.2.10. *The function $f : \Re^2 \to \Re$ is continuous if and only if $f^{-1}(S)$ is a nonempty open set in $\Re^2$, whenever S is a nonempty open set in $\Re$.*

Proof. $(\Rightarrow)$ Suppose that f is continuous and $S \subseteq \Re$ is a nonempty open set. Since we wish to prove that $f^{-1}(S)$ is open, we need to choose an arbitrary point $(a,b) \in f^{-1}(S)$, and show that (a,b) is an interior point of $f^{-1}(S)$. Now, $f(a,b) \in S$

and S is open. Thus, $B(f(a,b);r) \subseteq S$ for some $r \in \Re^+$. And since f is continuous at (a,b), there exists $\delta > 0$ such that $f(B((a,b);\delta)) \subseteq B(f(a,b);r) \subseteq S$. Why? Therefore, $B((a,b);\delta) \subseteq f^{-1}(S)$, and (a,b) is an interior point of $f^{-1}(S)$, meaning that $f^{-1}(S)$ is open.

($\Leftarrow$) Suppose $f^{-1}(S)$ is open in $\Re^2$ for an open set $S \subseteq \Re$. Let $(a,b) \in \Re^2$ and $\varepsilon > 0$ be given. Since $B(f(a,b);\varepsilon)$ is open, due to the given hypothesis, $f^{-1}(B(f(a,b);\varepsilon))$ is open in $\Re^2$. Therefore, $(a,b) \in f^{-1}(B(f(a,b);\varepsilon))$, making (a,b) an interior point of this set. Thus, there exists $\delta > 0$ such that $B((a,b);\delta) \subseteq f^{-1}(B(f(a,b);\varepsilon))$, and $f(B((a,b);\delta)) \subseteq B(f(a,b);\varepsilon)$. Hence, the continuity of f follows. Why? ■

Is Theorem 11.2.10 true if $f^{-1}(S)$ is empty? Why or why not?

COROLLARY 11.2.11. *The function $f:\Re^2 \to \Re$ is continuous if and only if $f^{-1}(E)$ is closed in $\Re^2$ for every E closed in $\Re^1$.*

To avoid extensive repetitions, we use definitions of the $\sup f$, the $\inf f$, the $\max f$, the $\min f$, and the function f being *bounded above*, *bounded below*, and *bounded*, extending from functions of a single variable to functions of two or more variables in a natural fashion. Definition for uniform continuity also extends to $\Re^n$. Compare the definition of following uniform continuity to Definition 4.4.3.

Definition 11.2.12. A function $f:D \subseteq \Re^2 \to \Re$ is *uniformly continuous on D* if and only if for any given $\varepsilon > 0$, there exists $\delta > 0$, such that $|f(\vec{x}) - f(\vec{t})| < \varepsilon$ for all $\vec{x}, \vec{t} \in D$ satisfying $\|\vec{x} - \vec{t}\| < \delta$.

The next theorem shows that Theorems 4.3.1, 4.3.2, and 4.4.6 are also true for functions of two variables.

THEOREM 11.2.13. *If $f:D \subseteq \Re^2 \to \Re$ is continuous, and if D is closed and bounded, then*

- ***(a)*** *f is bounded.*
- ***(b)*** *f attains both its maximum and minimum values.*
- ***(c)*** *f is uniformly continuous.*

All the three results of Theorem 11.2.13 are consequences of Bolzano–Weierstrass theorem for sequences presented in Theorem 2.6.4 and Exercise 24 in Section 11.1. Proofs of (a) and (c) can be obtained directly from proofs of Theorems 4.3.1 and 4.4.6, respectively. The proof of part (b), given next, indirectly proves that $f(D)$ is closed. See Exercise 2 in Section 4.3.

Proof of part (b). By part (a), f is bounded. Therefore, f has a finite supremum. Let $M = \sup\{f(x,y) \mid (x,y) \in D\}$, and let $\{z_n\}$ be a sequence in $f(D)$ that converges to M. We will prove that $M \in f(D)$. Let $\{\vec{x}_n\}$ be a sequence in D such that $z_n = f(\vec{x}_n)$. Now, since D is bounded, $\{\vec{x}_n\}$ has a subsequence $\{\vec{x}_{n_k}\}$ that converges to some point $\vec{x}_0 \in D$. Therefore, since f is continuous, $\{f(\vec{x}_{n_k})\}$ converges to $f(\vec{x}_0)$. But $\{f(\vec{x}_n)\}$ converges to M. Why? So $M = f(\vec{x}_0)$. Why? Therefore, $M \in f(D)$, and due to its nature, $M = \max f$. A proof for the minimum is similar. ■

THEOREM 11.2.14. *If $f: D \subseteq \Re^2 \to \Re$ is continuous, and if S is a connected subset of D, then $f(S)$ is connected; that is, $f(S)$ is an interval.*

Proof. We will prove the contrapositive of Theorem 11.2.14. That is, we will assume that f is continuous with $f(S)$ disconnected, and prove that S is not connected. To this end, let A and B be disjoint nonempty open sets such that $f(S) = A \cup B$. Since f is continuous, by Theorem 11.2.10, $f^{-1}(A)$ and $f^{-1}(B)$ are nonempty disjoint open subsets of $\Re^2$. Therefore, $f^{-1}(A) \cap f^{-1}(B) = \phi$. Also, $S = f^{-1}(A \cup B) = f^{-1}(A) \cup f^{-1}(B)$. Therefore, S is not connected, and the proof is complete. ■

Definition 11.2.15. The function $f: D \subseteq \Re^2 \to \Re$ is a *Lipschitz function* if and only if f satisfies the *Lipschitz condition*, namely, that there exists $L > 0$ (called a *Lipschitz constant*) such that $\left|f(\vec{x}) - f(\vec{t})\right| \le L\left\|\vec{x} - \vec{t}\right\|$, for all $\vec{x}, \vec{t} \in D$.

If in Definition 11.2.15, $L \in (0,1)$, then

$$d\left(f(\vec{x}), f(\vec{t})\right) = \left|f(\vec{x}) - f(\vec{t})\right| \le L\left\|\vec{x} - \vec{t}\right\| = Ld\left(\vec{x}, \vec{t}\right) < d\left(\vec{x}, \vec{t}\right).$$

Thus, the images of the two points are closer together than the two point themselves, resulting in a *contractive mapping*, that is, a *contractive function*. Is Theorem 4.3.7 true for the function $f: D \subseteq \Re^2 \to \Re$? Why or why not?

Exercises 11.2

1. For each function, specify the domain. Also, graph a few level curves.

(a) $f(x,y) = \sqrt{1 - x^2 - (y+1)^2}$ **(b)** $f(x,y) = x^2 + 4y^2$

(c) $f(x,y) = 2x - y$ **(d)** $f(x,y) = \sqrt{x - 2y}$

(e) $f(x,y) = \dfrac{y}{1 + x^2 + y^2}$ **(f)** $f(x,y) = \sqrt{36 - 4x^2 - 9y^2}$

2. Evaluate the given limits, and prove your conclusion using Definition 11.2.1.

(a) $\displaystyle\lim_{(x,y)\to(0,0)} \frac{xy^2}{x^2+y^2}$ **(b)** $\displaystyle\lim_{\substack{(x,y)\to(0,0)\\ x\neq y}} \frac{x^2-y^2}{x-y}$

(c) $\displaystyle\lim_{(x,y)\to(0,0)} \frac{xy\left(x^2-y^2\right)}{x^2+y^2}$ **(d)** $\displaystyle\lim_{(x,y)\to(0,0)} \frac{y}{1+x^2+y^2}$

(e) $\displaystyle\lim_{(x,y)\to(1,2)} \left(x^2+3y\right)$ **(f)** $\displaystyle\lim_{(x,y)\to(0,1)} \frac{x}{\sqrt{y}}$

3. Determine whether or not the given limits are finite. Explain yourself clearly.

(a) $\displaystyle\lim_{(x,y)\to(0,0)} \left(\frac{x^2-y^2}{x^2+y^2}\right)^2$ **(b)** $\displaystyle\lim_{(x,y)\to(0,0)} \frac{xy}{x^2+y^2}$

(c) $\displaystyle\lim_{(x,y)\to(0,0)} \frac{x^2-y^2}{x^2+y^2}$ **(d)** $\displaystyle\lim_{(x,y)\to(0,0)} \frac{1}{x^2+y^2}$

(e) $\displaystyle\lim_{(x,y)\to(0,0)} \frac{x^2}{x^2+y^2}$ **(f)** $\displaystyle\lim_{(x,y)\to(0,0)} \frac{xy(x-y)}{x^2+y^2}$

(g) $\displaystyle\lim_{(x,y)\to(0,0)} \frac{x^2-y^4}{x^2+y^4}$ **(h)** $\displaystyle\lim_{(x,y)\to(0,0)} \frac{x+y^2}{y^2}$

4. Consider the function

$$f(x,y)=\begin{cases}\dfrac{xy}{x^2+y^2}, \text{ if } (x,y)\neq(0,0)\\ 0, \text{ if } (x,y)=(0,0).\end{cases}$$

See Exercise 3(b).

(a) Use polar coordinates to show that $\displaystyle\lim_{(x,y)\to(0,0)} \frac{xy}{x^2+y^2}$ does not exist.

(b) Invoke part (c) of Remark 11.2.3 to show that the limit in part (a) does not exist.

(c) Show that f is not continuous at the origin.

(d) If $g(x)\equiv f(x,b)$, b any real constant, then show that g is continuous at $(0,0)$.

(e) If $h(y)\equiv f(a,y)$, a any real constant, then show that h is continuous at $(0,0)$.

(f) Show that $|f(x,y)|\leq\dfrac{1}{2}$.

5. Prove Theorem 11.2.5.

6. Find $f:\Re^2 \to \Re$ such that $\lim\limits_{\substack{(x,y)\to(0,0)\\ ax+by=0,\ a,b\in\Re}} f(x,y) = L$, but $\lim\limits_{(x,y)\to(0,0)} f(x,y)$ does not exist.

7. Prove that if $\lim\limits_{(x,y)\to(a,b)} f(x,y)$ is finite, then it is unique. See Theorem 2.1.9.

8. Determine whether or not the given functions are continuous. Prove your conclusions.

(a) $f(x,y) = x$ (This function is called a *projection function.*)

(b) $f(x,y) = \begin{cases} \dfrac{x-y}{x+y}, \text{ if } y \neq -x \\ 1, \text{ if } y = -x \end{cases}$

(c) $f(x,y) = \sqrt{x^2 + y^2}$ (This function is called a *norm function*. See Exercise 2(a) in Section 4.1.)

(d) $f(x,y) = \begin{cases} \dfrac{x^3 - xy^2}{x^2 + y^2}, \text{ if } (x,y) \neq (0,0) \\ 0, \text{ if } (x,y) = (0,0). \end{cases}$

9. Prove that $f(x,y) = \max\{x,y\}$ is continuous on $\Re^2$. This is another type of a *projection function*. See Exercise 8(a).

10. **(a)** Prove Theorem 11.2.8. **(b)** Prove Theorem 11.2.9.

(c) Prove that $\lim\limits_{(x,y)\to(a,b)} f(x,y) = 0$ if and only if $\lim\limits_{(x,y)\to(a,b)} |f(x,y)| = 0$. Compare to Exercise 4 in Section 2.1.

11. **(a)** Prove Corollary 11.2.11.

(b) Prove parts (a) and (b) of Theorem 11.2.13.

12. Write a negation of Definition 11.2.12, and use it to show that the functions $f:D \to \Re$ given below are not uniformly continuous on the open square $D \equiv (0,1)\times(0,1)$. See Exercise 1(a) of Section 4.4.

(a) $f(x,y) = \dfrac{1}{x+y}$ **(b)** $f(x,y) = \dfrac{1}{x} + y$

13. If $f:D \subseteq \Re^2 \to \Re$ is a Lipschitz function, prove that it is uniformly continuous on D.

14. Consider the function $f:\Re^2 \to \Re$ given in polar coordinates by

$$f(r,\theta) = \begin{cases} \dfrac{1}{2}\sin 2\theta, \text{ if } r \neq 0 \\ 0, \text{ if } r = 0. \end{cases}$$

(a) Is f a continuous function in polar coordinates?

(b) Write f in rectangular coordinates.

(c) Is f a continuous function in rectangular coordinates?

15. Suppose that D is an open region in $\Re^2$, $f: D \to \Re$ is continuous at $(a,b) \in D$, and $f(a,b) > 0$. Prove that there exists a neighborhood of (a,b) on which $f(x,y) > \frac{1}{2} f(a,b)$. See part (d) of Theorem 4.1.14.

16. **(a)** *(Intermediate Value Theorem)* Suppose that $f: D \subseteq \Re^2 \to \Re$ is continuous and S is a connected subset of D. If $\vec{a}, \vec{b} \in S$ such that $f(\vec{a}) < f(\vec{b})$, prove that for any k between $f(\vec{a})$ and $f(\vec{b})$, there exists a $\vec{c} \in S$ such that $f(\vec{c}) = k$. See Theorem 4.3.3.

(b) Suppose that $f: D \subseteq \Re^2 \to \Re$ is continuous and D is connected. If $f(x,y) > 0$ for at least one point in D, and $f(x,y) < 0$ for at least one point in D, prove that $f(x,y) = 0$ for at least one point in D.

17. *(Topologist's Sine Curve)* Let

$$A = \{(0,y) \mid y \in [-1,1]\} \text{ and}$$

$$B = \left\{(x,y) \,\middle|\, x \in (0,1] \text{ and } y = \sin\frac{1}{x}\right\}.$$

Show that $S \equiv A \cup B$ is connected. See Example 3.2.8 and Exercise 10 in Section 3.3. Note that S is not arcwise connected.

18. **(a)** Define what is meant by $\lim_{\|\vec{x}\| \to \infty} f(\vec{x}) = L$.

(b) Evaluate $\lim_{\|\vec{x}\| \to \infty} \frac{1}{x^2 + y^2}$.

(c) Evaluate $\lim_{\|\vec{x}\| \to \infty} \frac{x}{x^2 + y^2}$.

(d) Evaluate $\lim_{\|\vec{x}\| \to \infty} \frac{x^2}{x^2 + y^2}$. See Exercise 3(e).

11.3 Partial Derivatives

Definition 11.3.1. Suppose that $f: D \subseteq \Re^2 \to \Re$, where D is open and $(a,b) \in D$.

(a) *The partial derivative of f with respect to x at the point (a,b) is given by*

$$f_x(a,b) = \lim_{h \to 0} \frac{f(a+h,b) - f(a,b)}{h},$$

provided that this limit is finite.

(b) *The partial derivative of f with respect to y at the point (a,b)* is given by

$$f_y(a,b) = \lim_{k\to 0} \frac{f(a,b+k)-f(a,b)}{k},$$

provided that this limit is finite.

A few of the notations used for the partial derivative $f_x(a,b)$ are $\dfrac{\partial f}{\partial x}(a,b)$, $\dfrac{df}{dx}(x,b)\Big|_{x=a}$, $\dfrac{\partial f}{\partial x}\Big|_{(a,b)}$, $\dfrac{\partial}{\partial x}f(x,y)\Big|_{(a,b)}$, $D_x f\Big|_{(a,b)}$, and $z_x(a,b)$ if $z \equiv f(x,y)$. Thus, to evaluate $f_x(a,b)$, replace y by b in the definition of $f(x,y)$ to obtain a function of a single variable x, then differentiate f (with respect to x) as was done in Chapter 5, and then replace x by a. That is, define $g(x)=f(x,b)$ and compute $g'(a)$. Or, hold y constant, differentiate f with x as a single variable, and then replace x by a and y by b. This method, of course, does not always work and one must resort to Definition 11.3.1 or other methods. Note also that $f_x(a,b)$ is also given by

$$f_x(a,b) = \lim_{x\to a} \frac{f(x,b)-f(a,b)}{x-a},$$

provided that this limit is finite. A similar discussion holds for $f_y(a,b)$.

The partial derivative functions $f_x(x,y)$ and $f_y(x,y)$, also written as f_x and f_y, $\dfrac{\partial f}{\partial x}$ and $\dfrac{\partial f}{\partial y}$, are given by the limits

$$\frac{\partial f}{\partial x} = \lim_{h\to 0} \frac{f(x+h,y)-f(x,y)}{h} \text{ and } \frac{\partial f}{\partial y} = \lim_{k\to 0} \frac{f(x,y+k)-f(x,y)}{k},$$

provided that the limits are finite. The symbol $\dfrac{\partial}{\partial x}$ instructs us to find the partial derivative with respect to x of what follows the symbol. Thus, $\dfrac{\partial}{\partial x} f = \dfrac{\partial f}{\partial x}$.

Example 11.3.2. Compute $f_x(0,0)$ and $f_y(0,0)$ for each function, if possible.

(a) $f(x,y) = \sqrt{4-x^2-4y^2+8y}$ (the top half of an ellipsoid)

(b) $f(x,y) = \sqrt[3]{y}$

(c) $f(x,y) = \begin{cases} \dfrac{xy}{x^2+y^2}, & \text{if } (x,y) \neq (0,0) \\ 0, & \text{if } (x,y) = (0,0) \end{cases}$

(See Exercise 4 in Section 11.2 and Exercise 5 in Section 11.4.)

Answer to part (a). Since $f_x(x,0)=\frac{d}{dx}\sqrt{4-x^2}=-\frac{x}{\sqrt{4-x^2}}$, then $f_x(0,0)=0$. Similarly, $f(0,y)=\sqrt{4-4y^2+8y}$. And therefore, $f_y(0,y)=\frac{-4y+4}{\sqrt{4-4y^2+8y}}$, which gives $f_y(0,0)=2$. Observe that $f_x(0,0)\neq f_y(0,0)$.

Answer to part (b). Since $f(0,y)=\sqrt[3]{y}$, we have $f_y(0,y)=\frac{1}{3y^{\frac{2}{3}}}$. Therefore, $f_y(0,0)$ does not exist. Hence there is no partial derivative with respect to y at the origin. Since $f(x,0)=0$, then $f_x(0,0)=0$.

Answer to part (c). Here, Definition 11.3.1 is implemented. In an attempt to evaluate the desired limit at $(0,0)$, we write that

$$\lim_{h\to 0}\frac{f(0+h,0)-f(0,0)}{h}=\lim_{h\to 0}\frac{f(h,0)}{h}=\lim_{h\to 0}\frac{0}{h}=\lim_{h\to 0}0=0.$$

Thus, $f_x(0,0)=0$. Similarly, $f_y(0,0)=0$. Observe that as was seen in Exercise 4(a) of Section 11.2, f has no limit at the origin. And thus, as was seen in Exercise 4(c) of Section 11.2, f is not continuous at $(0,0)$. So, even though both partials exist and are equal at $(0,0)$, the function f is not continuous at $(0,0)$. Also, note that if $(x,y)\neq(0,0)$, then $f_x=\frac{y^3-x^2y}{\left(x^2+y^2\right)^2}$ and $f_y=\frac{x^3-xy^2}{\left(x^2+y^2\right)^2}$. But $f_x(0,0)$ and $f_y(0,0)$ cannot be computed from these formulas. Why not? ■

As mentioned earlier, often, in order to find $f_x(a,b)$, we define $g(x)\equiv f(x,b)$, compute $g'(x)$, and then evaluate $g'(a)$. This function g is a *cross-section* of f with $y=b$. In other words, $g(x)$ is a function representing the curve obtained by intersecting the surface f with the plane $y=b$. And $g'(a)$, also called $f_x(a,b)$, is the slope of the tangent line to the curve at the point $(a,b,f(a,b))\in\Re^3$. The partial $f_y(a,b)$ may be interpreted in a similar way.

Example 11.3.3. If g is a cross-section of the surface $z=\sqrt{36-4x^2-9y^2}$ with $x=1$, find parametric equations for the tangent line at

(a) $P\left(1,\sqrt{2},2\right)$. **(b)** $P\left(1,\sqrt{3},\sqrt{5}\right)$.

Answer to part (a). The point $P\left(1,\sqrt{2},2\right)$ does not lie on the surface. That is, if $f(x,y)\equiv\sqrt{36-4x^2-9y^2}$, then $f\left(1,\sqrt{2}\right)\neq 2$. Thus no tangent line exists at this point.

Answer to part (b). Since $f\left(1,\sqrt{3}\right)=\sqrt{5}$, the existence of an equation of the tangent line is possible. Indeed, $f_y(x,y)=-\dfrac{9y}{\sqrt{36-4x^2-9y^2}}$, which implies that the slope of the tangent line is given by $f_y\left(1,\sqrt{3}\right)=-\dfrac{9\sqrt{3}}{\sqrt{5}}$. Thus, if y changes by $\sqrt{5}$, then z drops by $9\sqrt{3}$. Therefore, the tangent line is parallel to the vector $\left\langle 0,\sqrt{5},-9\sqrt{3}\right\rangle$. Hence, the parametric equations for the tangent line at the point $P\left(1,\sqrt{3},\sqrt{5}\right)$ are $x=1$, $y=\sqrt{3}+\sqrt{5}\,t$, and $z=\ \sqrt{5}-9\sqrt{3}\,t$, with $t\in\Re$. ■

In general the functions f_x and f_y are also functions of x and y, so they must likely possess partial derivatives. If f_x has a partial derivative with respect to x, then a few representations of that *second partial derivative* of f with respect to x are denoted by f_{xx}, $\left(f_x\right)_x$, $\dfrac{\partial^2 f}{\partial x^2}$, and $\dfrac{\partial}{\partial x}\left(\dfrac{\partial f}{\partial x}\right)$. If f_x has a partial derivative with respect to y, called a *mixed partial derivative*, then a few representations of the resulting mixed partial may be denoted by f_{xy}, $\left(f_x\right)_y$, $\dfrac{\partial^2 f}{\partial y\,\partial x}$, and $\dfrac{\partial}{\partial y}\left(\dfrac{\partial f}{\partial x}\right)$. Partial derivatives with respect to x or y of f_y are given similarly. Partial derivatives of higher orders exist for many functions. If f_x can be differentiated with respect to y twice, then a few representations of the resulting mixed partial may be denoted by $((f_x)_y)_y=f_{xyy}=\dfrac{\partial}{\partial y}\left(\dfrac{\partial}{\partial y}\left(\dfrac{\partial f}{\partial x}\right)\right)=$ $\dfrac{\partial}{\partial y}\left(\dfrac{\partial^2 f}{\partial y\,\partial x}\right)=\dfrac{\partial^3 f}{\partial y^2\partial x}$. The partial derivatives, or simply the *partials*, denoted by $\dfrac{\partial^n f}{\partial x^n}$ and $\dfrac{\partial^n f}{\partial y^n}$ with $n\in N$, are called *pure partial derivatives of order n*. All other partial derivatives are called *mixed partials.* In particular, the partial derivatives f_{xx}, f_{xy}, f_{yx}, and f_{yy} are referred to as *partial derivatives of order two*, *partials of order two*, *second partial derivatives*, or *second partials*. Observe that

$$
\begin{aligned}
f_{xy} &= \frac{\partial}{\partial y}\left(\frac{\partial f}{\partial x}\right)=\frac{\partial}{\partial y}\left(\frac{\partial f}{\partial x}(x,y)\right)=\lim_{k\to 0}\frac{\dfrac{\partial f}{\partial x}(x,y+k)-\dfrac{\partial f}{\partial x}(x,y)}{k}\\
&= \lim_{k\to 0}\frac{\displaystyle\lim_{h\to 0}\frac{f(x+h,y+k)-f(x,y+k)}{h}-\lim_{h\to 0}\frac{f(x+h,y)-f(x,y)}{h}}{k}\\
&= \lim_{k\to 0}\lim_{h\to 0}\frac{f(x+h,y+k)-f(x,y+k)-f(x+h,y)+f(x,y)}{hk}.
\end{aligned}
$$

Similarly,

$$f_{yx} = \lim_{h\to 0}\lim_{k\to 0} \frac{f(x+h,y+k)-f(x+h,y)-f(x,y+k)+f(x,y)}{hk}.$$

Often, $f_{xy} = f_{yx}$. Sometimes, changing the order of these iterated limits produces different results, resulting in different mixed partials. Continuity guarantees that the iterated limits can be interchanged, which yields that f_{xy} and f_{yx} are equal. See part (d) of Remark 11.2.3 and Theorem 11.3.4, given next.

THEOREM 11.3.4. *(Clairaut's Theorem*[3] *) Suppose that D is an open region in $\Re^2$ and $f: D \to \Re$. If f_{xy} and f_{yx} are continuous on D, then $f_{xy} = f_{yx}$ on D.*

Proof. Let (a,b) be any point in D, and let

$$g(h,k) \equiv f(a+h,b+k) - f(a,b+k) - f(a+h,b) + f(a,b).$$

Also, define $p(x,y) = f(x+h,y) - f(x,y)$ and $q(x,y) = f(x,y+k) - f(x,y)$ for all $(x,y) \in D$. Then, $g(h,k) = p(a,b+k) - p(a,b)$ and $g(h,k) = q(a+h,b) - q(a,b)$. Now, in applying the mean value theorem, Theorem 5.2.14, to $g(h,k)$ above, we can write that

$$g(h,k) = kp_y(a,b+\theta_1 k) = k\left[f_y(a+h,b+\theta_1 k) - f_y(a,b+\theta_1 k)\right], \text{ for } \theta_1 \in (0,1),$$

and

$$g(h,k) = hq_x(a+\theta_2 h,b) = h\left[f_x(a+\theta_2 h,b+k) - f_x(a+\theta_2 h,b)\right], \text{ for } \theta_2 \in (0,1).$$

Apply the mean value theorem again to the last two expressions above to obtain that

$$g(h,k) = hkf_{yx}(a+\theta_3 h, b+\theta_1 k) \text{ and } g(h,k) = hkf_{xy}(a+\theta_2 h, b+\theta_4 k),$$

for $\theta_3, \theta_4 \in (0,1)$. Thus, $f_{yx}(a+\theta_3 h, b+\theta_1 k) = f_{xy}(a+\theta_2 h, b+\theta_4 k)$. Now let $h,k \to 0$. Since f_{xy} and f_{yx} are continuous at (a,b), $f_{xy}(a,b) = f_{yx}(a,b)$. Why? Does D need to be an open set? ■

For a shorter proof of Clairaut's theorem, see Section 12.2. Due to Theorem 11.3.4, $f_{xxyy} = f_{yyxx}$, provided that these partials are continuous. Why?

---

[3]Alexis Clairaut (1713–1765), a French mathematician, became at the age of eighteen the youngest person ever to be elected to the Paris Academy of Sciences. Clairaut is recognized for his study of differential equations, mechanics, and astronomy.

THEOREM 11.3.5. *Suppose that* D *is an open region in* $\Re^2$ *and* $f: D \to \Re$. *If* f, f_x, f_y, *and* f_{xy} *exist and are continuous on* D, *then* f_{yx} *also exists, and* $f_{xy} = f_{yx}$ *on* D.

Proof. Let (a,b) be any point in D, and let

$$g(x) \equiv f(x, b+k) - f(x,b),$$

where k is a constant. In applying the mean value theorem, Theorem 5.2.14, to the function g with x between a and $a+h$, with h a constant, we have that

$$g(a+h) - g(a) = hg'(a + \theta_1 h),$$

with $\theta_1 \in (0,1)$. Therefore, since $g'(x) = f_x(x, b+k) - f_x(x,b)$, we have that

$$g(a+h) - g(a) = h\left[f_x(a+\theta_1 h, b+k) - f_x(a+\theta_1 h, b)\right].$$

Now, in applying the mean value theorem to the function $p(y) \equiv f_x(a+\theta_1 h, y)$, with y between b and $b+k$, we have that $p(b+k) - p(b) = kp'(b+\theta_2 k)$, which is equivalent to

$$f_x(a+\theta_1 h, b+k) - f_x(a+\theta_1 h, b) = kf_{xy}(a+\theta_1 h, b+\theta_2 k),$$

with $\theta_2 \in (0,1)$. Therefore, from substitution, we obtain that

$$g(a+h) - g(a) = hkf_{xy}(a+\theta_1 h, b+\theta_2 k),$$

which in using the definition of g, is equivalent to

$$\left[f(a+h, b+k) - f(a+h, b)\right] - \left[f(a, b+k) - f(a,b)\right] = hkf_{xy}(a+\theta_1 h, b+\theta_2 k).$$

If we divide both sides by k and let $k \to 0$, from the definition of f_y on the left-hand side, and the continuity of f_{xy} on the right-hand side, we obtain

$$f_y(a+h, b) - f_y(a,b) = hf_{xy}(a+\theta_1 h, b).$$

Next, divide both sides by h and let $h \to 0$; then due to the definition of f_{yx} on the left-hand side and the continuity of f_{xy} on the right-hand side, we obtain $f_{yx}(a,b) = f_{xy}(a,b)$. Hence, the proof is complete. Where did we use the continuity of f, f_x, and f_y? Does D need to be open? ■

Before closing, we turn briefly to the concept of an operator. An *operator* is a mapping that transforms one function into another function. An operator defines an operation that is to be performed. The *differential operator* D, also denoted $D_x \equiv \frac{d}{dx}$ or $D_y \equiv \frac{d}{dy}$, is an example of an operator. If $f(x)$ is differentiable, then $Df = D_x f \equiv f'(x)$, a notation we have seen in Section 5.1. Thus, $D(x+3\sin x) = 1+3\cos x$. Other operators are the *Laplace operator* $\mathfrak{L}$, the *inverse Laplace operator* $\mathfrak{L}^{-1}$. The *forward* and *backward difference operators*, Δ and ∇, and the *central difference operator*, δ are among those briefly discussed next. L is a *linear operator* if $L(af+bg) = aLf + bLg \equiv aL(f) + bL(g)$ for any real constants a and b, and any functions f and g on which L can operate. For example, D is a linear operator. The Laplace operator $\mathfrak{L}$ and the inverse Laplace operator $\mathfrak{L}^{-1}$ are two linear operators from Sections 6.6 and 6.8, respectively. The forward and backward difference operators, Δ and ∇, also denoted by Δ_h and ∇_h, respectively, and the central difference operator, δ, are from Section 5.6. And, of course, $\frac{\partial}{\partial x}$ and $\frac{\partial}{\partial y}$ are operators introduced in this section. Another operator is the *shift operator*, denoted by E. Using the notation from Part 1 of Section 5.8, we define E_h by $E_h f(x) = f(x+h)$, with $h \in \Re$. The *unit operator*, 1, has the property that $1f = f$. And if two operators applied to the same function produce the same result, then they are *operationally equivalent*. For example, since

$$\Delta_h f(x) \equiv f(x+h) - f(x) = E_h f(x) - f(x) = (E_h - 1) f(x),$$

Δ_h and $E_h - 1$ are operationally equivalent. Thus, we may write that $\Delta_h = E_h - 1$. See Exercise 12 for more such operators.

Operators of higher orders can also be defined. For example,

$$\frac{d^2y}{dx^2} = \frac{d}{dx}\left(\frac{dy}{dx}\right) = \frac{d}{dx}\left(\frac{d}{dx}y\right) = D(Dy) \equiv D^2 y.$$

And in general, for any $n \in N$, $D^n y \equiv \frac{d^n y}{dx^n}$, provided that y is differentiable sufficiently many times. D^n is called the *differential operator of order* n. Also, $D^0 f \equiv f^{(0)} \equiv f$, meaning that $D^0 = 1$ is the unit operator. Observe that D^n can be written recursively as $D^n = D\left(D^{n-1}\right)$.

The operators $L_1 \equiv D^2 - 2$ and $L_2 \equiv 2D+1$ are linear. Why? Also, since $L_1 L_2 = L_2 L_1$, L_1 and L_2 are *commutative* . Computing $L_1 L_2$ yields

$$L_1 L_2 = \left(D^2 - 2\right)(2D+1) = 2D^3 + D^2 - 4D - 2.$$

$L_2 L_1$ is computed likewise. However, $L_1 L_2$ acts as a composition when operated on a function. For example, if $f(x) = x^3$, then

$$\begin{aligned} L_1L_2\left(x^3\right) &= L_1\left(L_2\left(x^3\right)\right) = \left(D^2-2\right)(2D+1)\left(x^3\right) = \left(D^2-2\right)\left(6x^2+x^3\right) \\ &= D^2\left(6x^2+x^3\right) - 2\left(6x^2+x^3\right) = 12+6x-12x^2-2x^3. \end{aligned}$$

Differential operators provide a nice method for solving certain types of differential equations. See Part 2 of Section 11.9. Further information on operators will be discussed later.

Now, if f is defined on $\Re^2$ and $f_x(x,y) = 2xy$, then by holding y constant, we can integrate with respect to x to obtain $f(x,y) = x^2y + C(y)$, where $C(y)$ is some function of y. This process is called a *partial antidifferentiation with respect to* x, and is written as

$$\int 2xy\,dx = x^2y + C(y), \text{ or as } \int 2xy\,\partial x = x^2y + C(y).$$

The function $x^2y + C(y)$ is called a *partial antiderivative with respect to* x. Similarly,

$$\int 2xy\,\partial y = xy^2 + c(x).$$

Partial integration is arrived at in a similar fashion as antidifferentiation. That is, if f is defined on $\Re^2$ with f_x and f_y Riemann integrable with respect to x and y, respectively, then

$$\int_{h_1(y)}^{h_2(y)} f_x(x,y)\,dx = f(x,y)\Big|_{x=h_1(y)}^{h_2(y)} = f\left(h_2(y),y\right) - f\left(h_1(y),y\right), \text{ and}$$

$$\int_{g_1(x)}^{g_2(x)} f_y(x,y)\,dy = f(x,y)\Big|_{y=g_1(x)}^{g_2(x)} = f\left(x,g_2(x)\right) - f\left(x,g_1(x)\right).$$

Exercises 11.3

1. Compute $f_x(0,0)$ and $f_y(0,0)$ for each function, if possible.

(a) $f(x,y) = \sqrt{1-x^2-(y+1)^2}$ (This is the top of a sphere. See Exercise 1(a) in Section 11.2.)

(b) $f(x,y) = x - y$ with $x, y \in [0,1]\times[0,1]$

(c) $f(x,y) = \begin{cases} \dfrac{xy}{x^2+y^2}, & \text{if } (x,y) \neq (0,0) \\ 1, & \text{if } (x,y) = (0,0) \end{cases}$

2. If $f(x,y) = 3x^2y^3 - \sin x$, find f_x, f_y, f_{xx}, f_{xy}, f_{yx}, f_{yy}, f_{xxy}, and f_{xyx}.

3. Suppose that

$$f(x,y)=\begin{cases}\dfrac{xy\left(x^2-y^2\right)}{x^2+y^2}, \text{ if } (x,y)\neq(0,0)\\ 0, \text{ if } (x,y)=(0,0).\end{cases}$$

(a) Use the definitions of partial derivatives to compute $f_x(0,y)$, $f_y(x,0)$, $f_{xy}(0,0)$, and $f_{yx}(0,0)$. Are the mixed partials at $(0,0)$ equal?

(b) Compute $f_x(x,y)$ and $f_y(x,y)$ for $(x,y)\neq(0,0)$. Are the values $f_x(0,y)$ and $f_y(x,0)$ the same as those found in part (a)?

4. List all of the third order partials of f.

5. Show that $f_{xx}+f_{yy}=0$ for each given function f.[4]

(a) $f(x,y)=x^2-y^2$ **(b)** $f(x,y)=\ln\dfrac{1}{\sqrt{x^2+y^2}}$

(c) $f(x,y)=e^x\cos y$ **(d)** $f(x,y)=x^3-3xy^2$

6. A function f, which has continuous partial derivatives of order four, such that $\nabla^2\left(\nabla^2 f\right)=0$, where ∇^2 is a Laplacian operator, is called a *biharmonic function*.[5] Often, we define $\nabla^4 f\equiv\nabla^2\left(\nabla^2 f\right)$. See the footnote to Exercise 5.

(a) Show that

$$\nabla^4 f=\frac{\partial^4 f}{\partial x^4}+2\frac{\partial^4 f}{\partial x^2\partial y^2}+\frac{\partial^4 f}{\partial y^4}.$$

See Exercise 12 in Section 11.8.

(b) Show that every harmonic function is biharmonic.

(c) Show that $f(x,y)=x^4-3x^2y^2$ is biharmonic but not harmonic.

7. Show that if $L_1\equiv D^2-2$, $L_2\equiv 2D+1$, $L\equiv L_1L_2$, and $f(x)=x^3$, then $L_1L_2f=L_2L_1f=Lf$.

---

[4]This exercise shows that the given functions satisfy *Laplace's equation* $u_{xx}+u_{yy}=0$. The term $u_{xx}+u_{yy}$ is often denoted by $\nabla^2 u$, pronounced "del squared u," or Δu, and is called the *Laplacian* of u. Also, $\nabla^2\equiv\dfrac{\partial^2}{\partial x^2}+\dfrac{\partial^2}{\partial y^2}$ is called a *Laplacian operator*. Laplace's equation is an important type of an *elliptic partial differential equation* with the theory of its solutions called the *potential theory*. The meaning of ∇^2 needs to be obtained from the context so that it will not be confused with difference operators. A function that both satisfies Laplace's equation and has continuous partial derivatives of order two, is called a *harmonic function*. See part (d) of Remark 11.4.4.

[5]Biharmonic functions are used to study elasticity.

8. If $f(x) = x^2 + \sin x$ and $Df(g(x)) \equiv f'(g(x))$, find $Df\left(\sin x^2\right)$.

9. **(a)** Show that $xD \neq Dx$.

(b) Give an example of two operators that are not commutative.

10. If $a, b \in \Re$ are constants, and $h = 1$, show that

(a) $f(x) = ae^{-3x} + bxe^{-3x}$ satisfies the equation $\left(D^2 + 6D + 9\right)y = 0$. That is, the operator $L \equiv D^2 + 6D + 9$ annihilates the function f. (An operator L *annihilates* a function f if $Lf(x) = 0$.) See part (b) of Problem 11.8.5.

(b) $f(x) = a\sin x + b\cos x - \dfrac{1}{4}x^2\cos x + \dfrac{1}{4}x\sin x$ satisfies the equation given by $\left(D^2 + 1\right)y = x\sin x$.

(c) $f(x) = a2^x + b3^x$ satisfies the equation $\left(E^2 - 5E + 6\right)y = 0$.

11. **(a)** Show that if $n \in N$, then the operator D^n annihilates any polynomial of degree $n - 1$.

(b) Show that if a is a real constant and $n \in N$, then the operator $(D - a)^n$ annihilates any function of the form $f(x) = p_{n-1}(x)e^{ax}$, where p_{n-1} is a polynomial of degree $n - 1$.

(c) Show that if $a, b \in \Re$ and $n \in N$, then the operator $\left[D^2 - 2aD + \left(a^2 + b^2\right)\right]^n$ annihilates the functions $y_1 = x^m e^{ax}\cos bx$ and $y_2 = x^m e^{ax}\sin bx$ for any $m = 0, 1, 2, \ldots, n$.

(d) Suppose that $L_1 = a_1D^2 + b_1D + c_1$ and $L_2 = a_2D^2 + b_2D + c_2$, where a_i, b_i, and c_i for $i = 1, 2$ are real constants. Also, suppose that L_1 annihilates $y_1(x)$ and L_2 annihilates $y_2(x)$, but that $L_1(y_2(x)) \neq 0$ and $L_2(y_1(x)) \neq 0$. Show that L_1 and L_2 are commutative, that L_1L_2 is linear, and that L_1L_2 annihilates $c_1y_1(x) + c_2y_2(x)$ for any constants c_1 and c_2.

(e) Show by example that a differential operator that annihilates a function need not be unique.

12. Show that

(a) $E_h = \Delta_h + 1$. **(b)** $1 = E_h - \Delta_h$.

(c) $\Delta_h = \delta E_{h+\frac{1}{2}}$. **(d)** $\delta = E_{h+\frac{1}{2}} - E_{h-\frac{1}{2}}$.

(e) $\Delta^2 = \left(E_h - 1\right)^2$, where Δ^2 is the second forward difference.

(f) $e^{ax}(D + a)y(x) = D\left(e^{ax}y(x)\right)$ for any differentiable function y and a real constant a.

13. Evaluate $\displaystyle\int_0^1\left[\int_1^{2x}\left(2xy^3 + 2xy^{-2}\right)dy\right]dx$.

11.4 Differentiation

In Section 5.1 a tangent line to a curve was discussed and given by a Taylor polynomial of degree 1. Also, the slope m of a nonvertical tangent line to a curve f at a point $P_0(x_0, f(x_0))$ was given by $f'(x_0)$. Now, in extending the idea of tangent lines to tangent planes and curves in $\Re^2$ to surfaces $z = f(x, y)$ in $\Re^3$, we wish to search for an equation of a nonvertical tangent plane to z at some point $P_0(x_0, y_0, f(x_0, y_0))$, provided that it exists. Hence, we will assume that the surface is sufficiently smooth near P_0, thus allowing a nonvertical tangent plane to exist at that point. To find such an equation for a nonvertical tangent plane, intersect a surface with a plane $y = y_0$ to produce a cross-section curve C_1 given by $f(x, y_0)$. Similarly, the intersection of a surface with a plane $x = x_0$ produces a cross-section curve C_2 given by $f(x_0, y)$. Section 11.3 held a similar discussion. The slopes of the tangent lines to C_1 and C_2 at P_0 are given by $f_x(x_0, y_0)$ and $f_y(x_0, y_0)$, respectively. These two tangent lines determine a tangent plane. And a normal vector to this plane is given by a cross product of any two vectors that are tangent to C_1 and C_2 at P_0. Since the slope of a tangent line to C_1 at P_0 is $f_x(x_0, y_0)$, also given by $\dfrac{f_x(x_0, y_0)}{1}$, an increase of 1 unit for x produces a change of $f_x(x_0, y_0)$ for z. Here, y does not change. Thus, a tangent vector to C_1 at P_0 can be given by

$$\vec{u} = 1 \cdot \mathbf{i} + 0 \cdot \mathbf{j} + f_x(x_0, y_0) \cdot \mathbf{k}.$$

Similarly, a tangent vector to C_2 at P_0 can be given by

$$\vec{v} = 0 \cdot \mathbf{i} + 1 \cdot \mathbf{j} + f_y(x_0, y_0) \cdot \mathbf{k}.$$

Also, a plane that contains both $\vec{u}$ and $\vec{v}$ is a tangent plane with a normal vector $\vec{n}$ given by

$$\vec{n} = \vec{v} \times \vec{u} = f_x(x_0, y_0) \cdot \mathbf{i} + f_y(x_0, y_0) \cdot \mathbf{j} - \mathbf{k}.$$

The above omitted calculations need to be verified. What would $\vec{u} \times \vec{v}$ produce? From Theorem 10.5.3, an equation of a nonvertical tangent plane is given by

$$f_x(x_0, y_0)(x - x_0) + f_y(x_0, y_0)(y, y_0) - 1(z - z_0) = 0.$$

And stated earlier, the above is an equation of a tangent plane to a curve $z = f(x, y)$ at a point $P_0(x_0, y_0, f(x_0, y_0))$ only if such a plane exists and is not vertical. For example, a circular cone given by $z = \sqrt{x^2 + y^2}$ has no tangent plane at the origin. Why? See Exercise 3. Furthermore, if a surface is given implicitly, then an equation for a tangent plane at P_0 may be rewritten as

$$\left.\frac{\partial z}{\partial x}\right|_{P_0}(x - x_0) + \left.\frac{\partial z}{\partial y}\right|_{P_0}(y - y_0) - (z - z_0) = 0,$$

where $\frac{\partial z}{\partial x}$ and $\frac{\partial z}{\partial y}$ may be found using implicit differentiation.

A generalization of the concept of differentiability from functions of a single variable to functions of two variables is now of interest. Recall that for $y = f(x)$ at the point $x = x_0$, we have that

$$f'(x_0) = \lim_{x \to x_0} \frac{f(x) - f(x_0)}{x - x_0} = \lim_{h \to 0} \frac{f(x_0 + h) - f(x_0)}{h},$$

provided that these limits exist. The temptation might arise to say that if $z = f(x, y)$ at the point $(x, y) = (x_0, y_0)$, then

$$f'(x_0, y_0) = \lim_{(x,y) \to (x_0, y_0)} \frac{f(x, y) - f(x_0, y_0)}{(x, y) - (x_0, y_0)}.$$

However, division by a vector is not defined. Thus, extension of the concept of differentiability requires a different point of view. This is done next.

Observe that $y = f(x)$ is differentiable at $x = x_0$ with $f'(x_0) = m$ if and only if

$$\left| \frac{f(x) - f(x_0)}{x - x_0} - m \right| = \varepsilon,$$

where $\varepsilon \to 0$ as $x \to x_0$. Equivalently, $|f(x) - f(x_0) - m(x - x_0)| = \varepsilon |x - x_0|$. Thus,

$$f(x) = f(x_0) + m(x - x_0) \pm \varepsilon |x - x_0|.$$

Therefore, f is differentiable at $x = x_0$ if there exists an m such that the above expression is satisfied with ε tending to 0 as x tends to x_0. Next, observe that if

$$T(x) \equiv f(x_0) + m(x - x_0),$$

which is a line through the point $(x_0, f(x_0))$, then

$$\frac{f(x)-T(x)}{|x-x_0|} \to 0 \text{ as } x \to x_0.$$

Thus, the line $y = T(x)$ is a tangent line to a curve f at $x = x_0$. An easy extension exists for functions of several variables. A surface $z = f(x,y)$ is differentiable at a point $(x,y) = (x_0, y_0)$ if and only if there exists a vector $\vec{m} = \langle m_1, m_2 \rangle$ such that

$$f(x,y) = f(x_0,y_0) + \langle m_1, m_2 \rangle \cdot \langle x - x_0, y - y_0 \rangle \pm \varepsilon \| x - x_0, y - y_0 \|,$$

where $\varepsilon \to 0$ as $(x,y) \to (x_0, y_0)$. Using a point and a vector notation interchangeably, if $\vec{P} = (x,y)$ and $\vec{P}_0 = (x_0, y_0)$, we have

$$f(\vec{P}) = f(\vec{P}_0) + \vec{m} \cdot \overrightarrow{PP_0} \pm \varepsilon \|\vec{P} - \vec{P}_0\|.$$

The tangent plane to $z = f(x,y)$ at the point (x_0, y_0) is given by

$$T(\vec{P}) \equiv f(\vec{P}_0) + \vec{m} \cdot \overrightarrow{PP_0}.$$

Thus,

$$z = f(x_0,y_0) + \langle m_1, m_2 \rangle \cdot \langle x - x_0, y - y_0 \rangle = z_0 + m_1(x - x_0) + m_2(y - y_0),$$

where $z_0 = f(x_0, y_0)$. Observe that $m_1 = f_x(x_0, y_0)$ and $m_2 = f_y(x_0, y_0)$. Why? Now, if $\vec{P} \equiv \vec{P}_0 + \vec{h}$, then $f(\vec{P}_0 + \vec{h}) = f(\vec{P}_0) + \vec{m} \cdot \vec{h} \pm \varepsilon \|\vec{h}\|$, where $\varepsilon \to 0$ as $\vec{h} \to (0,0)$. Hence, the following formal definition arises.

Definition 11.4.1. A function $f : D \subseteq \Re^2 \to \Re$, with D open, is *differentiable at* $\vec{P} \in D$ if and only if there exists a vector $\vec{m}$ such that

$$f(\vec{P} + \vec{h}) = f(\vec{P}) + \vec{m} \cdot \vec{h} \pm \varepsilon \|\vec{h}\|,$$

where $\varepsilon \to 0$ as $\vec{h} \to \vec{0}$.

Remark 11.4.2

(a) Definition 11.4.1 may be rephrased to say that f is differentiable at (x_0, y_0) if and only if there exists a vector $\vec{m} = \langle m_1, m_2 \rangle$ such that

$$\lim_{(x,y)\to(x_0,y_0)} \frac{f(x,y) - T(x,y)}{\|(x,y) - (x_0,y_0)\|} = \lim_{(x,y)\to(x_0,y_0)} \frac{f(x,y) - T(x,y)}{\sqrt{(x-x_0)^2 + (y-y_0)^2}} = 0,$$

where $T(x,y) \equiv f(x_0,y_0) + \langle m_1, m_2 \rangle \cdot \langle x - x_0, y - y_0 \rangle$ is the tangent plane. Here, the definition of the derivative of f at (x_0,y_0) may again be rephrased using the definition of a limit. Note that the condition $0 < \sqrt{(x-x_0)^2 + (y-y_0)^2} < \delta$ from Definition 11.2.1 will in the present case become $\sqrt{(x-x_0)^2 + (y-y_0)^2} < \delta$. Why?

(b) $\vec{m}$ is the *(total) derivative* of f at (x_0,y_0).

(c) From a previous discussion, if f is differentiable at (x_0,y_0), then the partial derivatives of f exist at (x_0,y_0). Also, $f_x(x_0,y_0) = m_1$ and $f_y(x_0,y_0) = m_2$. ■

The vector $\vec{m}$ is important, and thus, is given the special name gradient of f.

Definition 11.4.3. If $f: D \subseteq \Re^2 \to \Re$, with D open, then the *gradient* of f, denoted by grad f, ∇f, or $\nabla f(x,y)$, is the vector $\langle f_x, f_y \rangle$.

Remark 11.4.4

(a) A gradient ∇f is referred to as "del" f, and should not be confused with a backward difference operator.

(b) $\nabla f = \dfrac{\partial f}{\partial x}\mathbf{i} + \dfrac{\partial f}{\partial y}\mathbf{j}$, which is a vector even though it is commonly not written in bold or with an arrow above it.

(c) ∇ is a *vector differential operator* defined by $\nabla = \dfrac{\partial}{\partial x}\mathbf{i} + \dfrac{\partial}{\partial y}\mathbf{j}$. ∇ acquires meaning when operating on a function. Therefore,

$$\nabla f = \left(\frac{\partial}{\partial x}\mathbf{i} + \frac{\partial}{\partial y}\mathbf{j}\right) f = f_x\mathbf{i} + f_y\mathbf{j}.$$

(d) $\nabla^2 \equiv \nabla \cdot \nabla = \left(\dfrac{\partial}{\partial x}\mathbf{i} + \dfrac{\partial}{\partial y}\mathbf{j}\right) \cdot \left(\dfrac{\partial}{\partial x}\mathbf{i} + \dfrac{\partial}{\partial y}\mathbf{j}\right) = \dfrac{\partial^2}{\partial x^2} + \dfrac{\partial^2}{\partial y^2}$. So, $\nabla^2 f = f_{xx} + f_{yy}$ ∇^2 is called a *Laplacian operator.* See the footnote to Exercise 5 in Section 11.3. For contrast, see the footnote to Exercise 1 in Section 12.8.

(e) $\nabla f(x_0,y_0)$ may exist even if f is not differentiable at (x_0,y_0). That is, the existence of both partials at (x_0,y_0) is not sufficient to guarantee the differentiability at (x_0,y_0). See Exercise 4.

(f) If f is differentiable at (x_0,y_0), then the partials of f exist at (x_0,y_0), and the total derivative of f at (x_0,y_0) is given by $\nabla f(x_0,y_0)$. See part (c) of Remark 11.4.2. ■

Sufficient conditions for differentiability follow.

THEOREM 11.4.5. *If f_x and f_y exist in a neighborhood of the point (x_0, y_0), and f_x and f_y are continuous at (x_0, y_0), then f is differentiable at (x_0, y_0).*

Proof. We need to show that there exists a vector $\vec{m} = \langle m_1, m_2 \rangle$ such that

$$\lim_{(x,y)\to(x_0,y_0)} \frac{f(x,y) - f(x_0,y_0) - \langle m_1, m_2 \rangle \cdot \langle x - x_0, y - y_0 \rangle}{\|(x,y) - (x_0,y_0)\|} = 0.$$

That is, for any $\varepsilon > 0$ we need to find $\delta > 0$ so that

$$\frac{\left| f(x,y) - f(x_0,y_0) - \langle m_1, m_2 \rangle \cdot \langle x - x_0, y - y_0 \rangle \right|}{\sqrt{(x - x_0)^2 + (y - y_0)^2}} < \varepsilon,$$

or, equivalently, that

$$\left| f(x,y) - f(x_0,y_0) - \langle m_1, m_2 \rangle \cdot \langle x - x_0, y - y_0 \rangle \right| < \varepsilon \sqrt{(x - x_0)^2 + (y - y_0)^2},$$

whenever $0 < \sqrt{(x - x_0)^2 + (y - y_0)^2} < \delta$. See Definition 11.2.1. Now, by part (c) of Remark 11.4.2, the only candidates for m_1 and m_2 are $f_x(x_0, y_0)$ and $f_y(x_0, y_0)$, respectively. Therefore, we must show that there exists $\delta > 0$ so that

$$\left| f(x,y) - f(x_0,y_0) - f_x(x_0,y_0)(x - x_0) - f_y(x_0,y_0)(y - y_0) \right| < \varepsilon \sqrt{(x - x_0)^2 + (y - y_0)^2}$$

whenever $0 < \sqrt{(x - x_0)^2 + (y - y_0)^2} < \delta$. But since both partials are continuous at (x_0, y_0) and are defined in a neighborhood of (x_0, y_0), there exists $\delta_1 > 0$ such that

$$\left| f_x(x,y) - f_x(x_0,y_0) \right| < \boxed{\frac{\varepsilon}{2}} \text{ and } \left| f_y(x,y) - f_y(x_0,y_0) \right| < \boxed{\frac{\varepsilon}{2}},$$

whenever $\|(x,y) - (x_0,y_0)\| < \delta_1$. And, by the mean value theorem, Theorem 5.2.14, applied to the variable x, there exists a point (x_1, y), with x_1 between x_0 and x, such that

$$f(x,y) - f(x_0,y) = f_x(x_1,y)(x - x_0).$$

Similarly, there exists a point (x, y_1), with y_1 between y_0 and y, such that

$$f(x,y)-f(x,y_0)=f_y(x,y_1)(y-y_0).$$

Therefore, for $\delta=\delta_1$, and points (x,y) satisfying $0<\|(x,y)-(x_0,y_0)\|<\delta$, the previous expressions give

$$\begin{aligned} f(x,y)-f(x_0,y_0) &= [f(x,y)-f(x_0,y)]+[f(x_0,y)-f(x_0,y_0)] \\ &= f_x(x_1,y)(x-x_0)+f_y(x_0,y_1)(y-y_0). \end{aligned}$$

Hence, by combining the established expressions, when $0<\|(x,y)-(x_0,y_0)\|<\delta$, we have that

$$\begin{aligned} &|f(x,y)-f(x_0,y_0)-f_x(x_0,y_0)(x-x_0)-f_y(x_0,y_0)(y-y_0)| \\ &=|f_x(x_1,y)(x-x_0)+f_y(x_0,y_1)(y-y_0)-f_x(x_0,y_0)(x-x_0)-f_y(x_0,y_0)(y-y_0)| \\ &=|\langle f_x(x_1,y)-f_x(x_0,y_0), f_y(x_0,y_1)-f_y(x_0,y_0)\rangle\cdot\langle x-x_0, y-y_0\rangle| \\ &\le\|\langle f_x(x_1,y)-f_x(x_0,y_0), f_y(x_0,y_1)-f_y(x_0,y_0)\rangle\|\cdot\|\langle x-x_0, y-y_0\rangle\| \\ &\le\left[|f_x(x_1,y)-f_x(x_0,y_0)|+|f_y(x_0,y_1)-f_y(x_0,y_0)|\right]\cdot\|\langle x-x_0, y-y_0\rangle\| \\ &<\left(\frac{\varepsilon}{2}+\frac{\varepsilon}{2}\right)\cdot\|\langle x-x_0, y-y_0\rangle\|=\varepsilon\|(x,y)-(x_0,y_0)\|=\varepsilon\sqrt{(x-x_0)^2+(y-y_0)^2}. \end{aligned}$$

The first inequality above follows from the Cauchy–Schwarz inequality, Theorem 10.3.5, the second inequality, from Exercise 11(c) in Section 10.3. Thus, the proof is complete. ■

As discussed earlier, the conditions in Theorem 11.4.5 are only sufficient, but not necessary. We demonstrate this in exercise 6. Now, if f_x and f_y are defined in a neighborhood of (x_0,y_0) and are continuous at (x_0,y_0), which are the conditions of Theorem 11.4.5, then f is said to be *continuously differentiable* at (x_0,y_0). Thus, by Theorem 11.4.5, if f is continuously differentiable at (x_0,y_0), then f is differentiable at (x_0,y_0). Differentiability of f is not enough to yield continuity of f_x or f_y, but it is enough to yield the continuity of f.

THEOREM 11.4.6. *If f is differentiable at a point (x_0,y_0), then f is continuous at (x_0,y_0).*

Proof. Since f is differentiable at (x_0,y_0), by part (a) of Remark 11.4.2, there exists a vector $\vec{m}=\langle m_1,m_2\rangle$ such that

$$\lim_{(x,y)\to(x_0,y_0)} \frac{f(x,y)-T(x,y)}{\left\|(x,y)-(x_0,y_0)\right\|} = 0,$$

where $T(x,y) = f(x_0,y_0) + \langle m_1, m_2\rangle \cdot \langle x-x_0, y-y_0\rangle$, and since $(x,y)\to(x_0,y_0)$, we have that

$$\lim_{(x,y)\to(x_0,y_0)} \left\|(x,y)-(x_0,y_0)\right\| = 0.$$

The product of the above two limits yields that $\lim_{(x,y)\to(x_0,y_0)} [f(x,y)-T(x,y)] = 0$. Equivalently,

$$\lim_{(x,y)\to(x_0,y_0)} \left[f(x,y)-f(x_0,y_0)-m_1(x-x_0)-m_2(y-y_0)\right] = 0.$$

But $\lim_{(x,y)\to(x_0,y_0)} m_1(x-x_0) = 0 = \lim_{(x,y)\to(x_0,y_0)} m_2(y-y_0)$. Therefore,

$$\lim_{(x,y)\to(x_0,y_0)} \left[f(x,y)-f(x_0,y_0)\right] = 0,$$

which shows that f is continuous at (x_0,y_0). ■

In closing, Theorem 5.2.1 extends to functions of several variables. That is, the sum, difference, and product of differentiable functions are also differentiable, as well as the quotient, provided that the function in the denominator is not 0. See Exercise 14.

Exercises 11.4

1. Consider the point $P_0(1,-3,-2)$ on the sphere $x^2+y^2+z^2=14$.

(a) Find an equation of the tangent plane to the sphere at P_0 by solving the equation explicitly for z and then using the given formula.

(b) Find an equation of the tangent plane to the sphere at P_0 by computing $\frac{\partial z}{\partial x}$ and $\frac{\partial z}{\partial y}$ implicitly.

(c) At which points does this sphere have vertical tangent plane?

2. Find an equation of the tangent plane and a normal line to the given surface at the indicated point P_0.

(a) $x^2+y^2+z=9$, $P_0(1,-2,4)$ **(b)** $x^2+y^2-z^2=0$, $P_0(3,4,-5)$

3. Show that $f(x,y)=\sqrt{x^2+y^2}$ is not differentiable at the origin by showing that:

(a) there is no $\vec{m}$ as needed in Definition 11.4.1.
(b) $f_x(0,0)$ does not exist and using part (c) of Remark 11.4.2.

4. Consider the function $f(x,y)=\sqrt[3]{xy}$.

(a) Show that $f_x(0,0)=0=f_y(0,0)$. **(b)** Find $\nabla f\,(0,0)$.
(c) Show that f is not differentiable at $(0,0)$.
(d) Is f continuous at $(0,0)$? Explain.

5. Show that $f(x,y)=\begin{cases}\dfrac{xy}{x^2+y^2}, & \text{if } (x,y)\neq(0,0)\\ 0, & \text{if } (x,y)=(0,0)\end{cases}$ is not differentiable at $(0,0)$. See part (c) of Example 11.3.2.

6. Find a point (a,b) for which the function

$$f(x,y)=\begin{cases}(x-y)^2\sin\dfrac{1}{x-y}, & \text{if } x\neq y\\ 0, & \text{if } x=y\end{cases}$$

is differentiable at (a,b), but f_x and f_y are not continuous at (a,b).

7. If $f(x,y)=\begin{cases}\dfrac{xy}{\sqrt{x^2+y^2}}, & \text{if } (x,y)\neq(0,0)\\ 0, & \text{if } (x,y)=(0,0),\end{cases}$ show that

(a) f is continuous at $(0,0)$. **(b)** $f_x(0,0)$ and $f_y(0,0)$ both exist.
(c) f is not differentiable at $(0,0)$.

8. Consider

$$f(x,y)=\begin{cases}\left(x^2+y^2\right)\sin\dfrac{1}{x^2+y^2}, & \text{if } (x,y)\neq(0,0)\\ 0, & \text{if } (x,y)=(0,0).\end{cases}$$

(a) Show that $f_x(0,0)=0=f_y(0,0)$.
(b) Show that f_x and f_y are unbounded near $(0,0)$.
(c) Show that f_x and f_y are not continuous at $(0,0)$.

9. Show that the function f given by

(a) $f(x,y)=\dfrac{2x^2+y^2}{x-y}$ is continuously differentiable whenever $x\neq y$, and hence, f is differentiable.

(b) $f(x,y)=\begin{cases}\left(x^2+y^2\right)\sin\dfrac{1}{\sqrt{x^2+y^2}}, \text{ if } (x,y)\neq(0,0)\\ 0, \text{ if } (x,y)=(0,0)\end{cases}$

is differentiable at $(0,0)$, but not continuously differentiable at $(0,0)$.

(c) $f(x,y)=\begin{cases}\dfrac{xy\left(x^2-y^2\right)}{x^2+y^2}, \text{ if } (x,y)\neq(0,0)\\ 0, \text{ if } (x,y)=(0,0)\end{cases}$

is differentiable. Is f continuously differentiable? Is f_{xy} continuous in a neighborhood of $(0,0)$? See Exercise 3 in Section 11.3.

10. Show that ∇ is a linear operator.

11. If $f(x,y)=9-x^2-y^2$, find

(a) ∇f. (b) $\nabla^2 f$. (c) $\nabla^4 f$.

See Exercise 2(a).

12. If $u=u(x,y)$ and $v=v(x,y)$ have continuous second partial derivatives, prove that $\nabla^2(uv)=u\left(\nabla^2 v\right)+v\left(\nabla^2 u\right)+2\nabla u\cdot\nabla v$.

13. Prove that for two differentiable functions $f(x,y)$ and $g(x,y)$,

(a) $\nabla(fg)=f(\nabla g)+g(\nabla f)$.

(b) $\nabla\left(\dfrac{f}{g}\right)=\dfrac{g(\nabla f)-f(\nabla g)}{g^2}$, provided that $g\neq 0$ at that point.

14. Suppose that f and g are differentiable functions at (x_0,y_0). Prove that

(a) $f\pm g$ are differentiable at (x_0,y_0).

(b) fg is differentiable at (x_0,y_0).

(c) $\dfrac{f}{g}$ is differentiable at (x_0,y_0), provided $g(x_0,y_0)\neq 0$.

11.5 Directional Derivative

As commented earlier, the partial derivatives $f_x(a,b)$ and $f_y(a,b)$ measure the slope of a tangent line to a function $f(x,y)$ at a point $P(a,b)$ in the direction **i** and **j**, respectively. In this section, the slope of a tangent line to a function f at a point P in any direction is of interest. Before defining the extension of partial derivatives to directions other than **i** and **j**, observe that the limit expressions for $f_x(a,b)$ and $f_y(a,b)$ in part (a) of Definition 11.3.1 may be rewritten as

$$f_x(a,b) = f_x(\vec{p}) = \lim_{h\to 0} \frac{f(\vec{p}+h\mathbf{i}) - f(\vec{p})}{h}, \text{ and}$$
$$f_y(a,b) = f_y(\vec{p}) = \lim_{h\to 0} \frac{f(\vec{p}+h\mathbf{j}) - f(\vec{p})}{h},$$

where $\vec{p} = \langle a,b \rangle$. Why? Now, in order to define a directional derivative, we must replace $\mathbf{i}$ and $\mathbf{j}$ by another unit vector, say, $\vec{u}$.

Definition 11.5.1. Suppose that $f: D \subseteq \mathfrak{R}^2 \to \mathfrak{R}$, where D is open, and $(a,b) \in D$. If $\vec{u} = \langle u_1, u_2 \rangle$ is a unit vector and $\vec{p} = \langle a,b \rangle$, then the *directional derivative of f at (a,b) in the direction of $\vec{u}$* is given by

$$\frac{\partial f}{\partial \vec{u}}(a,b) \equiv \lim_{h\to 0} \frac{f(\vec{p}+\mathrm{h}\vec{u}) - f(\vec{p})}{h} = \lim_{h\to 0} \frac{f(a+hu_1, b+hu_2) - f(a,b)}{h},$$

provided this limit is finite.

Often, $\frac{\partial f}{\partial \vec{u}}$ is written as $D_{\vec{u}} f$. Also, $\frac{\partial f}{\partial \mathbf{i}} = f_x$ and $\frac{\partial f}{\partial \mathbf{j}} = f_y$. Thus, the partial derivatives f_x and f_y indeed measure the rate of change of f in the direction of $\mathbf{i}$ and $\mathbf{j}$, respectively. Certainly, in general, $f_x = D_{\mathbf{i}} f \neq D_{-\mathbf{i}} f$. See Exercise 2. Of importance is that $\vec{u}$ in Definition 11.5.1 be a unit vector. If $\vec{u}$ is not a unit vector, then before the directional derivative of f in the direction of $\vec{u}$ can be computed, $\vec{u}$ must be normalized. See Section 10.2.

THEOREM 11.5.2. *If f is differentiable at a point (a,b), then $D_{\vec{u}} f\ (a,b)$ exists in any direction $\vec{u}$, where $\vec{u}$ is a unit vector. Moreover,*

$$D_{\vec{u}} f\ (a,b) = \nabla f\ (a,b) \cdot \vec{u}.$$

Proof. Let $\varepsilon > 0$ be given. We will prove that if $\vec{u} = \langle u_1, u_2 \rangle$, with $\|\vec{u}\| = 1$, then

$$\lim_{h\to 0} \frac{f(a+hu_1, b+hu_2) - f(a,b)}{h} = \langle f_x(a,b), f_y(a,b) \rangle \cdot \langle u_1, u_2 \rangle.$$

Thus, we must show that there exists $\delta > 0$ such that

$$\left| \frac{f(a+hu_1, b+hu_2) - f(a,b)}{h} - \langle f_x(a,b), f_y(a,b) \rangle \cdot \langle u_1, u_2 \rangle \right| < \varepsilon$$

for all $0 < |h-0| = |h| < \delta$. See Definition 3.2.1. To this end, since f is differentiable at (a,b), there exists $\delta_1 > 0$, such that

$$\frac{\left|f(x,y)-f(a,b)-f_x(a,b)(x-a)-f_y(a,b)(y-b)\right|}{\|(x,y)-(a,b)\|}<\boxed{\varepsilon},$$

for all $(x,y)\in D$ with $\|(x,y)-(a,b)\|<\delta_1$. Why? In particular, we will choose only those points $(x,y)\in D$ for which $(x,y)=(a,b)+h(u_1,u_2)$ with $|h|<\delta_1$. Then, if $\delta=\delta_1$, we have that $\|(x,y)-(a,b)\|=|h|\|\vec{u}\|\le|h|\cdot 1<\delta_1=\delta$. And for these chosen points

$$\left|\frac{f(a+hu_1,b+hu_2)-f(a,b)-f_x(a,b)(hu_1)-f_y(a,b)(hu_2)}{h}\right|<\varepsilon.$$

Equivalently,

$$\left|\frac{f(a+hu_1,b+hu_2)-f(a,b)}{h}-f_x(a,b)u_1-f_y(a,b)u_2\right|<\varepsilon,$$

which is what we wanted to prove. ■

An important observation is that the differentiability of f at (a,b) is sufficient for the existence of a directional derivative in any direction, but not necessary. Even continuity is unnecessary for the existence of directional derivatives. See Exercise 3. A very nice consequence arises from Theorem 11.5.2.

COROLLARY 11.5.3. *If f is differentiable at (a,b), then*

(a) *f has a maximum rate of change at (a,b) in the direction of $\nabla f(a,b)$. Such maximum rate is $\|\nabla f(a,b)\|$.*

(b) *f has a minimum rate of change at (a,b) in the direction of $-\nabla f(a,b)$. Such minimum rate is $-\|\nabla f(a,b)\|$.*

Proof of part (a). If $\vec{u}$ is any unit vector, then, from Theorems 11.5.2 and 10.3.3,

$$\frac{\partial f}{\partial\vec{u}}(a,b)=\nabla f(a,b)\cdot\vec{u}=\|\nabla f(a,b)\|\ \|\vec{u}\|\cos\theta=\|\nabla f(a,b)\|\ \cos\theta,$$

where θ is the angle between $\nabla f(a,b)$ and $\vec{u}$. Therefore, this directional derivative is the largest when $\cos\theta=1$, that is, when $\theta=0$. But this is when $\vec{u}$ has the direction of $\nabla f(a,b)$. And in this case, the magnitude of the directional derivative is given by $\|\nabla f(a,b)\|$. ■

Part (a) of Corollary 11.5.3 says that the gradient vector $\nabla f(a,b)$ points in the direction in which f increases most rapidly. And the rate of increase in that direction is given by the magnitude of $\nabla f(a,b)$. As in Section 5.2, a differentiable function $f(x,y)$

attains its local maximum or its local minimum value at (a,b) if f has neither an increase nor a decrease at (a,b). Thus, both the largest and the smallest rate of change of a differentiable function f at (a,b) need to be 0. Hence, from Corollary 11.5.3, $\|\nabla f(a,b)\|$ must be zero. And hence, $\nabla f\,(a,b)=\vec{0}$. Why? The converse of this discussion is false. That is, there are differentiable functions $f(x,y)$ for which $\nabla f\,(a,b)=\vec{0}$, but which have no extremum at the point (a,b). See Exercise 8. The points that make $\nabla f=\vec{0}$ are called *critical points.*

Exercises 11.5

1. Find the directional derivative of each function at the specified point P in the direction of the vector $\vec{u}$.

(a) $f(x,y)=x+\sin y^2,\ P(1,0),\ \vec{u}=\mathbf{i}-2\mathbf{j}$

(b) $f(x,y)=\ln\left(x^2+y^2\right),\ P(-1,2),\ \vec{u}=\mathbf{j}$

2. **(a)** Use Definition 11.5.1 to show that $D_{\mathbf{i}}\,f=-D_{-\mathbf{i}}\,f$, provided that f_x exists.

(b) Use Theorem 11.5.2 to show that if f is differentiable at (a,b), then for any unit vector $\vec{u}$, $D_{-\vec{u}}\,f(a,b)=-D_{\vec{u}}\,f(a,b)$.

(c) If $D_{\vec{u}}f$ exists for a unit vector $\vec{u}=\langle u_1,u_2\rangle$, use Definition 11.5.1 to show that $D_{-\vec{u}}\,f=-D_{\vec{u}}\,f$.

3. If

$$f(x,y)=\begin{cases}\dfrac{x^2y}{x^6+2y^2}, & \text{if } (x,y)\neq(0,0)\\ 0, \text{ if } (x,y)=(0,0),\end{cases}$$

show that f is not continuous at $(0,0)$, but that it has a directional derivative in every direction at $(0,0)$.

4. Consider $f(x,y)=\begin{cases}0, \text{ if } xy\neq 0\\ 1, \text{ if } xy=0.\end{cases}$

(a) If g is defined by $g(x,y)=\begin{cases}0, \text{ if } (x,y)\neq(0,0)\\ 1, \text{ if } (x,y)=(0,0),\end{cases}$ is $g\equiv f$?

(b) Show that $f_x(0,0)=f_y(0,0)=0$, and that none of the other directional derivatives at $(0,0)$ exist.

(c) Show that f is not continuous at $(0,0)$.

(d) Using Theorem 11.4.6, show that f is not differentiable at $(0,0)$.

(e) Using Theorem 11.5.2, show that f is not differentiable at $(0,0)$.

5. If

$$f(x,y)=\begin{cases}\dfrac{xy}{x^2+y^2}, & \text{if } (x,y)\neq(0,0)\\ 0, \text{ if } (x,y)=(0,0),\end{cases}$$

show that $D_{\vec{u}}f\,(0,0)$ exists only if $\vec{u}=\langle 1,0\rangle$ or $\vec{u}=\langle 0,1\rangle$. See Exercise 5 in Section 11.4 and Exercise 13(d) in Section 11.6.

6. Use Theorem 11.5.2 to show that $f(x,y)=\sqrt{|xy|}$ is not differentiable at the origin.

7. Find the unit vector in the direction in which $f(x,y)=y^2\sin x$ increases most rapidly at the point $(0,-2)$. What is the maximum rate of change of f at $(0,-2)$?

8. Give an example of a differentiable function f on $\Re^2$ for which $\nabla f\,(a,b)=0$, but which has neither a local extremum nor an absolute extremum.

9. Find all of the critical points of $f(x,y)=x^3+3xy-y^3$.

10. Prove part (b) of Corollary 11.5.3.

11. Prove that, for two differentiable functions $f(x,y)$ and $g(x,y)$, with $a\in\Re$, and with a unit vector $\vec{u}$,

(a) $D_{\vec{u}}\,(af)=aD_{\vec{u}}\,f$. **(b)** $D_{\vec{u}}\,(f\pm g)=D_{\vec{u}}\,f\pm D_{\vec{u}}\,g$.

(c) $D_{\vec{u}}\,(fg)=f\,D_{\vec{u}}\,g+g\,D_{\vec{u}}\,f$.

(d) $D_{\vec{u}}\left(\dfrac{f}{g}\right)=\dfrac{g\,D_{\vec{u}}\,f+f\,D_{\vec{u}}\,g}{g^2}$, provided that $g\neq 0$ at that point.

11.6 Chain Rule

A similar analogy of the mean value theorem, Theorem 5.2.14, in which the derivative of f is replaced by ∇f, follows. Theorem 11.6.1 begins our assault on the chain rule.

THEOREM 11.6.1. *(Mean Value Theorem) Suppose that $f(x,y)$ is differentiable in an open ball B containing two points P and Q. Then there exists a point P_0 on the line segment connecting the points P and Q such that*

$$f(\vec{Q})-f(\vec{P})=\nabla f(\vec{P}_0)\cdot(\vec{Q}-\vec{P}).$$

The convexity of the ball B places the line segment connecting P and Q in B. The assumption to Theorem 11.6.1 could be rewritten to say that f is differentiable on an open set containing the points P, Q, and the line segment connecting these points. f

differentiable on an open and convex region R with the points P and Q in R could also have been a rewritten assumption. Rephrasing of the conclusion to Theorem 11.6.1, we could say that if $P(a,b)$ and $Q(a+h,b+k)$ are points in an open ball, then there exists $\theta \in (0,1)$ such that

$$f(a+h,b+k)-f(a,b)=hf_x(a+\theta h,b+\theta k)+kf_y(a+\theta h,b+\theta k).$$

Why?

In order to prove Theorem 11.6.1 we will define a function $g=g(t)$ whose graph connects the points $f(\vec{P})$ and $f(\vec{Q})$, and whose projection onto the xy-plane is a straight line segment. Then, since f is differentiable, so is g. Moreover, $g'(t)$ is precisely the directional derivative of f in the direction from P to Q. Then, since g is a function of a single variable, the mean value theorem, Theorem 5.2.14, can be applied. We implement this procedure in proving Theorem 11.6.1.

Proof of Theorem 11.6.1. Let P be a point (a,b), and let $\vec{u}$ be a unit vector given by $\vec{u}=\dfrac{\vec{Q}-\vec{P}}{\left\|\vec{Q}-\vec{P}\right\|}\equiv\langle u_1,u_2\rangle$. Next, we will define the curve $g:\Re\to\Re$ by

$$g(t)=f\left(\vec{P}+t\vec{u}\right)=f\left(a+tu_1,b+tu_2\right),$$

for $0\le t\le d\equiv\left\|\vec{Q}-\vec{P}\right\|$. We need to verify that g is differentiable. Thus, we write that

$$\begin{aligned}\lim_{h\to 0}\frac{g(t+h)-g(t)}{h} &= \lim_{h\to 0}\frac{f\left(\vec{P}+(t+h)\vec{u}\right)-f\left(\vec{P}+t\vec{u}\right)}{h}\\ &= \lim_{h\to 0}\frac{f\left(\left(\vec{P}+t\vec{u}\right)+h\vec{u}\right)-f\left(\vec{P}+t\vec{u}\right)}{h}\\ &= D_{\vec{u}}f\left(\vec{P}+t\vec{u}\right)=\nabla f\left(\vec{P}+t\vec{u}\right)\cdot\vec{u}.\end{aligned}$$

Why? Therefore, g is differentiable. And by Theorem 5.2.14, there exists $t_0\in(0,d)$ such that $g(d)-g(0)=g'(t_0)(d-0)$. This translates to

$$f\left(\vec{P}+\left\|\vec{Q}-\vec{P}\right\|\vec{u}\right)-f\left(\vec{P}\right)=\left[\nabla f\left(\vec{P}+t_0\vec{u}\right)\cdot\vec{u}\right]\left\|\vec{Q}-\vec{P}\right\|.$$

Equivalently,

$$f\left(\vec{Q}\right)-f\left(\vec{P}\right)=\nabla f\left(\vec{P}_0\right)\cdot\left(\vec{Q}-\vec{P}\right),$$

where $\vec{P}_0=\vec{P}+t_0\vec{u}=\vec{P}+\dfrac{t_0}{\left\|\vec{Q}-\vec{P}\right\|}\left(\vec{Q}-\vec{P}\right)$, a point on the line segment from P to Q. ■

Taylor's theorem, presented next, is a generalization of Theorems 11.6.1 and 5.3.8.

THEOREM 11.6.2. *(Taylor's Theorem) Suppose for any $n \in N$, a function $f(x,y)$ has continuous nth partial derivatives, and $(n+1)$st partial derivatives existing in an open ball $B(\vec{P};r)$. Then for $Q \in B(\vec{P};r)$, $f(\vec{Q}) = p_n(\vec{Q}) + R_n(\vec{Q})$, where*

$$p_n(\vec{Q}) = f(\vec{P}) + \left[(\vec{Q}-\vec{P})\cdot\nabla\right]f(\vec{P}) + \frac{1}{2!}\left[(\vec{Q}-\vec{P})\cdot\nabla\right]^2 f(\vec{P}) + \cdots + \frac{1}{n!}\left[(\vec{Q}-\vec{P})\cdot\nabla\right]^n f(\vec{P}),$$

and

$$R_n(\vec{Q}) = \frac{1}{(n+1)!}\left[(\vec{Q}-\vec{P})\cdot\nabla\right]^{n+1} f(\vec{P}_0),$$

with P_0 some point on the line segment connecting the points P and Q.

A proof of Theorem 11.6.2 would invoke Theorem 5.3.8 and be so similar to the proof of the mean value theorem, Theorem 11.6.1, that it is left as an exercise. Instead, an elaboration on the statement of Taylor's theorem is in order. Note that if Q is a point (x,y) and P is a point (a,b), then $(\vec{Q}-\vec{P})\cdot\nabla$ is an operator $(x-a)\frac{\partial}{\partial x} + (y-b)\frac{\partial}{\partial y}$.
Then, the formula in Theorem 11.6.2 becomes

$$\begin{aligned} f(x,y) &= f(a,b) + \left[(x-a)\frac{\partial}{\partial x} + (y-b)\frac{\partial}{\partial y}\right] f(a,b) \\ &+ \frac{1}{2!}\left[(x-a)\frac{\partial}{\partial x} + (y-b)\frac{\partial}{\partial y}\right]^2 f(a,b) + \cdots \\ &+ \frac{1}{n!}\left[(x-a)\frac{\partial}{\partial x} + (y-b)\frac{\partial}{\partial y}\right]^n f(a,b) + R_n(x,y), \end{aligned}$$

where $R_n(x,y)$ is the *remainder* given by

$$R_n(x,y) = \frac{1}{(n+1)!}\left[(x-a)\frac{\partial}{\partial x} + (y-b)\frac{\partial}{\partial y}\right]^{n+1} f(x_0,y_0),$$

with (x_0,y_0) some point on the line segment from P to Q. In the case when $\lim_{n\to\infty} R_n(x,y) = 0$, Taylor's theorem may be used to define an infinite series expansion for $f(x,y)$ in powers of $x-a$ and $y-b$.

A discussion of the differentiation of the function $z = f(x,y)$, where $x = x(t)$ and $y = y(t)$, is next.

THEOREM 11.6.3. *(Chain Rule) If $x = x(t)$ and $y = y(t)$ are differentiable at $t = t_0$, and if $z = f(x, y)$ is continuously differentiable on an open ball B containing the point $\left(x(t_0), y(t_0)\right)$, then*

$$\frac{dz}{dt} \equiv \frac{d}{dt} f\big(x(t), y(t)\big) = \frac{\partial z}{\partial x}\frac{dx}{dt} + \frac{\partial z}{\partial y}\frac{dy}{dt}.$$

Observe that a function f continuously differentiable means that both f_x and f_y are continuous, and so ∇f is continuous. Also, the functions $x(t)$ and $y(t)$ differentiable at $t = t_0$ means that $\vec{F}(t) \equiv \langle x(t), y(t)\rangle$ is differentiable at $t = t_0$. Therefore, by the chain rule

$$\begin{aligned}\frac{dz}{dt} &= \frac{d}{dt} f\big(x(t), y(t)\big) \equiv \frac{d}{dt}\left(f \circ \vec{F}\right)(t) = \left\langle \frac{\partial z}{\partial x}, \frac{\partial z}{\partial y}\right\rangle \cdot \left\langle \frac{dx}{dt}, \frac{dy}{dt}\right\rangle \\ &= \nabla f\big(x(t), y(t)\big) \cdot \left\langle \frac{dx}{dt}, \frac{dy}{dt}\right\rangle = \nabla f\left(\vec{F}(t)\right) \cdot \vec{F}'(t).\end{aligned}$$

Proof of Theorem 11.6.3. Note that $f\big(x(t), y(t)\big)$ is a function from $\Re$ to $\Re$. Thus,

$$\frac{d}{dt} f\big(x(t), y(t)\big) = \lim_{h \to 0} \frac{f\big(x(t+h), y(t+h)\big) - f\big(x(t), y(t)\big)}{h}.$$

Since $x(t)$ and $y(t)$ are continuous at $t = t_0$, there exists an h small enough so that $\big(x(t+h), y(t+h)\big) \in B$. Thus, by the mean value theorem, there exists a point P_0 on the line segment joining $\big(x(t), y(t)\big)$ and $\big(x(t+h), y(t+h)\big)$ so that

$$f\big(x(t+h), y(t+h)\big) - f\big(x(t), y(t)\big) = \nabla f\left(\vec{P}_0\right) \cdot \left[\langle x(t+h), y(t+h)\rangle - \langle x(t), y(t)\rangle\right].$$

Next, dividing both sides by h and taking the limit, we obtain

$$\begin{aligned}\frac{dz}{dt} &= \frac{d}{dt} f\big(x(t), y(t)\big) = \lim_{h \to 0}\left[\nabla f\left(\vec{P}_0\right) \cdot \frac{\langle x(t+h), y(t+h)\rangle - \langle x(t), y(t)\rangle}{h}\right] \\ &= \lim_{h \to 0} \nabla f\left(\vec{P}_0\right) \cdot \lim_{h \to 0} \frac{\langle x(t+h), y(t+h)\rangle - \langle x(t), y(t)\rangle}{h} \qquad \text{(Why?)} \\ &= \nabla f\big(x(t), y(t)\big) \cdot \vec{F}'(t). \quad \text{(Why?)}\end{aligned}$$

The proof is complete. ■

Example 11.6.4. If $z = e^{xy}$, where $x = t^2 - 1$ and $y = t^3 + t^2$, find $\dfrac{dz}{dt}$.

Answer. Let $f(x,y)=e^{xy}$. Then, $f_x = ye^{xy}$ and $f_y = xe^{xy}$. Thus, by the chain rule we have that

$$\begin{aligned}\frac{dz}{dt} &= \nabla f\left(t^2-1,t^3+t^2\right)\cdot\frac{d}{dt}\left\langle t^2-1,t^3+t^2\right\rangle\\ &= \left\langle\left(t^3+t^2\right)\exp\left[\left(t^2-1\right)\left(t^3+t^2\right)\right],\left(t^2-1\right)\exp\left[\left(t^2-1\right)\left(t^3+t^2\right)\right]\right\rangle\cdot\left\langle 2t,3t^2+2t\right\rangle\\ &= \left(5t^4+4t^3-3t^2-2t\right)\exp\left[\left(t^2-1\right)\left(t^3+t^2\right)\right].\ \text{Why?}\end{aligned}$$

Note that $\dfrac{dz}{dt}$ can also be computed by writing

$$\frac{dz}{dt}=\frac{\partial z}{\partial x}\frac{dx}{dt}+\frac{\partial z}{\partial y}\frac{dy}{dt}=\left(ye^{xy}\right)(2t)+\left(xe^{xy}\right)\left(3t^2+2t\right),$$

which yields the same result. ■

The chain rule, Theorem 11.6.3, may be extended to the assumptions $x = x(t,s)$ and $y = y(t,s)$, both in two variables. The partial differentiation symbol ∂ is used to indicate that s is considered to be fixed. Therefore, we may write

COROLLARY 11.6.5. *(Chain Rule) If $x = x(t,s)$ and $y = y(t,s)$ have continuous first partial derivatives at (t_0,s_0), and if $z = f(x,y)$ is continuously differentiable on an open ball B containing $\left(x(t_0,s_0),y(t_0,s_0)\right)$, then*

(a) $\dfrac{\partial z}{\partial t}=\dfrac{\partial z}{\partial x}\dfrac{\partial x}{\partial t}+\dfrac{\partial z}{\partial y}\dfrac{\partial y}{\partial t}$, *and*

(b) $\dfrac{\partial z}{\partial s}=\dfrac{\partial z}{\partial x}\dfrac{\partial x}{\partial s}+\dfrac{\partial z}{\partial y}\dfrac{\partial y}{\partial s}$.

Under suitable conditions of continuity and differentiability for functions $z = f(x,y)$ with $x = x(t)$ and $y = y(t)$, higher order derivatives can be computed. For example, using the product rule, we can write

$$\begin{aligned}\frac{d^2z}{dt^2} &= \frac{d}{dt}\left(\frac{dz}{dt}\right)=\frac{d}{dt}\left(\frac{\partial z}{\partial x}\frac{dx}{dt}+\frac{\partial z}{\partial y}\frac{dy}{dt}\right)\\ &= \left[\frac{\partial z}{\partial x}\frac{d}{dt}\left(\frac{dx}{dt}\right)+\frac{dx}{dt}\frac{d}{dt}\left(\frac{\partial z}{\partial x}\right)\right]+\left[\frac{\partial z}{\partial y}\frac{d}{dt}\left(\frac{dy}{dt}\right)+\frac{dy}{dt}\frac{d}{dt}\left(\frac{\partial z}{\partial y}\right)\right]\\ &= \frac{\partial z}{\partial x}\frac{d^2x}{dt^2}+\frac{dx}{dt}\left(\frac{\partial^2 z}{\partial x^2}\frac{dx}{dt}+\frac{\partial^2 z}{\partial y\,\partial x}\frac{dy}{dt}\right)+\frac{\partial z}{\partial y}\frac{d^2y}{dt^2}+\frac{dy}{dt}\left(\frac{\partial^2 z}{\partial x\,\partial y}\frac{dx}{dt}+\frac{\partial^2 z}{\partial y^2}\frac{dy}{dt}\right)\\ &= \frac{\partial z}{\partial x}\frac{d^2x}{dt^2}+\left(\frac{dx}{dt}\right)^2\frac{\partial^2 z}{\partial x^2}+2\frac{\partial^2 z}{\partial x\,\partial y}\frac{dx}{dt}\frac{dy}{dt}+\left(\frac{dy}{dt}\right)^2\frac{\partial^2 z}{\partial y^2}+\frac{\partial z}{\partial y}\frac{d^2y}{dt^2},\end{aligned}$$

where $\frac{d}{dt}\left(\frac{\partial z}{\partial x}\right)$ and $\frac{d}{dt}\left(\frac{\partial z}{\partial y}\right)$ were obtained from Theorem 11.6.3 with z replaced by $\frac{\partial z}{\partial x}$ and $\frac{\partial z}{\partial y}$, respectively.

Another application of the chain rule is in finding the derivative of a function defined implicitly. Implicitly defined functions were mentioned in Section 1.3, and implicit differentiation, in Section 5.2.

Definition 11.6.6. If $f: D \subseteq \Re \to \Re$, $F: \Re^2 \to \Re$, and $y \equiv f(x)$ on D, then the expression $F(x,y)=0$ defines f *implicitly*.

Observe that the expression $F(x,y)=0$ can be written as $G(x,y)=c$, where $F(x,y)=G(x,y)-c$ for any real constant c.

Remark 11.6.7. If a differentiable function $y=f(x)$ is defined implicitly by $F(x,y)=0$, where F is continuously differentiable, then Theorem 11.6.3 may be used in differentiating both sides of $F(x,y)=0$ to obtain

$$\frac{\partial F}{\partial x}\frac{dx}{dx}+\frac{\partial F}{\partial y}\frac{dy}{dx}=0,$$

which is equivalent to

$$\frac{\partial F}{\partial x}+\frac{\partial F}{\partial y}f'(x)=0.$$

And if $F_y \neq 0$, then $f'(x)=-\frac{F_x}{F_y}$. ■

In general, if $F(x,y)$ is continuously differentiable, it is not obvious that the expression $F(x,y)=0$ defines a differentiable function f, where $y=f(x)$. The existence of a differentiable function f may depend on the chosen point. For example, if $F(x,y)=x^2-y^2$ is defined on $\Re^2$, then the graph of $F(x,y)=0$ consists of two lines $y=x$ and $y=-x$. If we consider any point (x_0,y_0), different from the origin, which satisfies the equation $F(x,y)=0$, then (x_0,y_0) must be on one of the lines $y=x$ or $y=-x$. In either case, there exists an open ball about (x_0,y_0) for which y is a differentiable function of x. Note that this would not be the case if $(x_0,y_0)=(0,0)$. Why not? Conditions guaranteeing the existence of a differentiable function f, where $y=f(x)$ is implicitly defined by $F(x,y)=0$, are given in the following implicit function theorem whose challenging proof is left as an exercise.

THEOREM 11.6.8. *(Implicit Function Theorem also known as Dini's Theorem)* *Suppose that* $D \subseteq \Re^2$ *is an open set containing a point* (a,b) *and* $F: D \to \Re$. *Suppose further that* $F(a,b)=0$, $F_y(a,b) \neq 0$, *and that* F_x *and* F_y *are both continuous on* D, *making* F *continuously differentiable. Then there exists a neighborhood* N_r *of* $x=a$, *that is, an interval* $N_r=(a-r,a+r)$ *with* $r>0$, *and a function* $f: N_r \to \Re$ *for which*

(a) $(x, f(x)) \in D$ *whenever* $x \in N_r$,

(b) $F(x, f(x))=0$ *with* $x \in N_r$, *and*

(c) $f'(x)=\dfrac{dy}{dx}=-\dfrac{F_x}{F_y}$ *on* N_r.

Section 10.5 discusses an equation for the tangent plane to a surface. A search for such an equation may become easier through the use of the chain rule. First, finding an equation for a tangent line to a curve in the xy-plane using the chain rule will aid our search for an equation of the tangent plane to a surface. To this end, suppose that $f(x,y)$ has continuous partial derivatives in a neighborhood of a point $P_0(a,b)$. Consider its level curve $f(x,y)=c$, with c constant. See Section 11.2. This curve can be given parametrically by $x=x(t)$, $y=y(t)$, and the resulting vector-valued function $\vec{r}(t)=x(t)\mathbf{i}+y(t)\mathbf{j}$, where $a=x(t_0)$ and $b=y(t_0)$ for some $t=t_0$. Since $f(x(t),y(t))=c$, we have

$$\frac{d}{dt}f(x(t),y(t))=\frac{d}{dt}c=0.$$

And by the chain rule $\dfrac{\partial f}{\partial x}\dfrac{dx}{dt}+\dfrac{\partial f}{\partial y}\dfrac{dy}{dt}=0$, also written as $\nabla f \cdot \vec{r}'(t)=0$. In particular, at $t=t_0$, $\nabla f(P_0)\cdot \vec{r}'(t_0)=0$. But $\vec{r}'(t_0)$ is the tangent vector to this curve. See Section 10.6. Hence, the gradient is normal to the curve at P_0.

Extending this idea to a smooth level surface S given by $f(x,y,z)=c$, with c a constant, and ∇f, the *gradient* of f, defined to be $\langle f_x, f_y, f_z \rangle$, let C be any smooth curve on S given by

$$\vec{r}(t)=x(t)\mathbf{i}+y(t)\mathbf{j}+z(t)\mathbf{k},$$

with $P_0(x(t_0),y(t_0),z(t_0)) \equiv P_0(x_0,y_0,z_0)$, a point on C and S. Then, $\nabla f(P_0)\cdot \vec{r}'(t_0)=0$. But, $\vec{r}'(t_0)$ is the velocity (tangent) vector. Therefore, $\nabla f(P_0)$ is orthogonal to all velocity vectors at P_0. These velocity vectors at P_0 form the tangent plane. And $\nabla f(P_0)$ is normal to it. Now, since $\nabla f(P_0)=\langle f_x(P_0), f_y(P_0), f_z(P_0)\rangle$, from Theorem 10.5.3, an equation for the *tangent plane* to the level surface $f(x,y,z)=c$ at the point P_0 given by

$$f_x(P_0)(x-x_0)+f_y(P_0)(y-y_0)+f_z(P_0)(z-z_0)=0$$

exists. An equation of the *normal line* to this level surface is given parametrically by

$$x(t)=x_0+f_x(P_0)t,\ y(t)=y_0+f_y(P_0)t,\ \text{and } z(t)=z_0+f_z(P_0)t,$$

with $t\in\Re$.

Exercises 11.6

1. Prove Taylor's theorem, Theorem 11.6.2.

2. For $n=2$ and $n=3$, expand $\frac{1}{n!}\left[(x-a)\frac{\partial}{\partial x}+(y-b)\frac{\partial}{\partial y}\right]^n f(a,b)$, a term given in Theorem 11.6.2. Find the pattern.

3. Expand $y^2-x^2y-4x-6$ in powers of $x+1$ and $y-2$.

4. Write $p_3(x,y)$ in Taylor's theorem for the function $f(x,y)=x^2y^4-2x^3y^2+3xy^3$ about the point $P(2,-1)$.

5. Prove that if f is differentiable on an open ball B, and $\nabla f(x,y)=0$ for all $(x,y)\in B$, then $f(x,y)=c$, with c a constant. See Exercise 10 in Section 11.8.

6. Explain where continuous differentiability of the function f is used in the proof of Theorem 11.6.3.

7. If $z=f(x,y)$, $x=t\sin ts$, and $y=s\cos ts$, find both $\frac{\partial z}{\partial t}$ and $\frac{\partial z}{\partial s}$.

8. If $z=f(x,y)$, $x=x(t,s)$, $y=y(t,s)$, and $\frac{\partial^2 z}{\partial t^2}$ exists, find the formula for $\frac{\partial^2 z}{\partial t^2}$.

9. If $y=f(x)$ is a differentiable function, and $x=\sin t$, find $\frac{d^2y}{dt^2}$.

10. Consider the curve C given by the equation $4x^2-8x+y^2+4y=-4$.

(a) Show that the point $P_0\left(1+\frac{\sqrt{3}}{2},-3\right)$ is on C.

(b) Find the slope of a tangent line to C at P_0 by first solving the equation for y, and then computing the derivative.

(c) Find the slope of a tangent line to C at P_0 by implementing implicit differentiation without solving the equation for y.

(d) Find the slope of a tangent line to C at P_0 by implementing the method presented in Remark 11.6.7.

11. If $z = f(x,y)$ has continuous second partials and $x = r\cos\theta$ and $y = r\sin\theta$, show that

(a) $(z_x)^2 + (z_y)^2 = (z_r)^2 + r^{-2}(z_\theta)^2$.

(b) $z_{xx} + z_{yy} = z_{rr} + r^{-2}z_{\theta\theta} + r^{-1}z_r$. This expression gives the *Laplacian in polar coordinates*. See Exercise 5 in Section 11.3.

12. Suppose that $u = f(x,y)$ and $v = g(x,y)$ satisfy the *Cauchy–Riemann equations*

$$u_x = v_y \text{ and } u_y = -v_x.$$

Show that the *Cauchy–Riemann equations in polar coordinates* are

$$u_r = r^{-1}v_\theta \text{ and } v_r = -r^{-1}u_\theta.$$

13. Suppose that the function $f(x,y)$ is *homogeneous of degree* n, that is,

$$f(tx,ty) = t^n f(x,y),$$

for all x, y, and $t > 0$ such that $f(x,y)$ and $f(tx,ty)$ are defined, with n a real constant. See Exercise 34 in Section 5.2.

(a) Show that the functions $f_1(x,y) = \dfrac{xy}{x+y}$, $f_2(x,y) = \dfrac{2x}{x^2+y^2}$, $f_3(x,y) = 3x^{\frac{1}{3}} + x^{-\frac{2}{3}}y$, and $f_4(x,y) = \dfrac{x^2-y^2}{x^2+y^2}$ are all homogeneous. Find the degree of each.

(b) *(Euler's Theorem)* Prove that if $f(x,y)$ is differentiable and homogeneous of degree n at a point (x,y), then

$$xf_x(x,y) + yf_y(x,y) = nf(x,y),$$

which may be also be written as $\langle x,y\rangle \cdot \nabla f = nf$.

(c) Prove that if $f(x,y)$ has continuous second partial derivatives and is homogeneous of degree n, then

$$x^2 f_{xx} + 2xyf_{xy} + y^2 f_{yy} = n(n-1)f.$$

(d) Show that the formulas in parts (b) and (c) hold for the function

$$f(x,y) = \frac{xy}{x+y}.$$

See Exercise 5 in Section 11.5.

(e) Give a geometric interpretation of what is meant by a homogeneous function of degree n.

14. If $z = f\left(x^2 - y^2, y^2 - x^2\right)$, then show that $yz_x + xz_y = 0$.

15. If $u(x,t) = f(x - \alpha t) + g(x + \alpha t)$, and if the second partials of f and g exist, show that $u(x,t)$ satisfies the *one-dimensional wave equation* $u_{tt} = \alpha^2 u_{xx}$. The function $u(x,t)$ is called the *D'Alembert's complete solution,* obtained by D'Alembert in 1747. See Part 1 of Section 11.8.

16. Recall from the footnote to Exercise 5 in Section 11.3 that $f:\Re^2 \to \Re$ is *harmonic* if it has continuous second partial derivatives and satisfies

$$\nabla^2 f \equiv f_{xx} + f_{yy} = 0.$$

Prove that if f is harmonic and $g:\Re^2 \to \Re$ is defined by $g(x,y) = f\left(2xy, x^2 - y^2\right)$, then g is harmonic.

17. If $f(x,y) = \ln\left(x^2 + y^2\right)$, find ∇f at $(1,1)$. Also, graph the level curve that passes through $(1,1)$ together with $\nabla f\,(1,1)$. Is f a harmonic function? Explain.

18. Use a gradient to find an equation for a tangent line to the curve $4x^2 + y^2 = 8$ at the point $P(1,-2)$.

19. Find an equation for both the tangent plane and the normal line to the given surface at the given point P_0.

(a) $x^2 + y^2 + z^2 = 14$, $P_0(1,-3,-2)$ (See Exercise 1 in Section 11.4.)

(b) $z = 9 - x^2 - y^2$, $P_0(1,-2,4)$ (See Exercise 2 in Section 11.4.)

20. Show that the curve $\vec{r}(t) = t^2\mathbf{i} - \dfrac{2}{t}\mathbf{j} + 4t\mathbf{k}$ is tangent to the circular paraboloid $x^2 + y^2 + z = 9$ when $t = 1$.

21. Prove the implicit function theorem, Theorem 11.6.8.

11.7* Review

Label each statement as true or false. If a statement is true, prove it. If not,

(i) give an example of why it is false, and
(ii) if possible, correct it to make it true, and then prove it.

1. If $\lim_{x\to 0}\left[\lim_{y\to 0} f(x,y)\right] = L = \lim_{y\to 0}\left[\lim_{x\to 0} f(x,y)\right]$, then $\lim_{(x,y)\to(0,0)} f(x,y) = L$.

2. If $f:\Re^2 \to \Re$ is such that both $f(x,0)$ and $f(0,y)$ are continuous at $(0,0)$, then $f(x,y)$ is continuous at $(0,0)$.

3. An isolated point is a boundary point.
4. Any boundary point is also an accumulation point.
5. If $f: D \subseteq \Re^2 \to \Re$, D is open, and f is differentiable at a point $(a,b) \in D$, then f satisfies the Lipschitz condition.
6. If $z = f(x,y)$ has continuous second partials, $x = e^r \cos\theta$, and $y = e^r \sin\theta$, then $z_{xx} + z_{yy} = e^{-2r}(z_{rr} + z_{\theta\theta})$.
7. If f is both differentiable and homogeneous of degree n, then f_x and f_y are homogeneous of degree $n-1$.
8. If $\langle x,y\rangle \cdot \nabla f(x,y) = nf(x,y)$ for all (x,y) in the domain of f, then f is homogeneous of degree n.
9. If f has continuous first partials and $u = f(xy)$, then $xu_x + yu_y = 0$.
10. If f has continuous first partial derivatives on an open convex region, R, and if $f_x \equiv 0 \equiv f_y$ on R, then $f(x,y)$ is constant on R. See Section 11.6, Exercise 5.
11. If f is continuously differentiable and $z = f(x-y)$, then $z_x + z_y = 0$.
12. If f is a biharmonic function in rectangular coordinates x and y, then in the polar coordinates r and θ, we have that

$$\nabla^4 f = f_{rrrr} + \frac{2}{r^2} f_{rr\theta\theta} + \frac{1}{r^4} f_{\theta\theta\theta\theta} + \frac{2}{r} f_{rrr} - \frac{2}{r^3} f_{r\theta\theta} - \frac{1}{r^2} f_{rr} + \frac{4}{r^4} f_{\theta\theta} + \frac{1}{r^3} f_r .$$

See Exercise 6(a) in Section 11.3.
13. If $A, B \subseteq \Re^n$, then $\partial(A \cup B) = \partial A \cup \partial B$.
14. If $A, B \subseteq \Re^n$ such that A is closed and $\operatorname{int} A = \phi = \operatorname{int} B$, then $\operatorname{int}(A \cup B) = \phi$.
15. If $A \subseteq \Re^n$, then $\operatorname{int}(\Re^n \setminus A) = \Re^n \setminus \overline{A}$.
16. If f possesses all of its directional derivatives at (a,b), then f is differentiable at (a,b).
17. If f_x and f_y are continuous at (a,b), then f need not necessarily be continuous at (a,b).
18. If f_x and f_y both exist at (a,b), then all other directional derivatives exist at (a,b).
19. If f is not continuous at (a,b), then f cannot have all of its directional derivatives at (a,b).
20. If f has continuous first partial derivatives in a neighborhood of the point (a,b), and if $\vec{u}$ is any unit vector, then $\dfrac{\partial f}{\partial \vec{u}}(a,b) = \nabla f(a,b) \cdot \vec{u}$.
21. If m and n are positive integers, then $\displaystyle\lim_{(x,y)\to(0,0)} \frac{x^m y^n}{x^2 + y^2}$ exists if and only if $m + n > 2$.
22. If $f(x) = 2\sqrt[3]{x}$, then $Df(x^3) = 2$.

23. The function $f(x, y) = (\sin 2xy)\exp\left(x^2 - y^2\right)$ is harmonic.

24. There exists a closed subset of $\Re^2$ that has no boundary points.

25. A continuous image of an arcwise connected set is arcwise connected.

26. If A and B are connected sets that are not separated, then $A \cup B$ is a connected set.

27. If $f: D \subseteq \Re^2 \to \Re$ and $f_x(a,b) = f_y(a,b)$ for $(a,b) \in D$, then f is continuous at (a,b).

28. If $f: D \subseteq \Re^2 \to \Re$ is a contractive function defined on a closed and bounded set D, then f has a fixed point in D.

29. If $f(x,y) = \begin{cases} \dfrac{x^2 - xy}{x+y}, \text{ if } (x,y) \neq (0,0) \\ 0, \text{ if } (x,y) = (0,0), \end{cases}$ then $f_x(0,0) = 1$ and $f_y(0,0) = 0$.

30. If f is continuous on $\Re^2$ and $E \subset \Re^2$, then $f\left(\overline{E}\right) \subseteq \overline{f(E)}$.

31. If D is a bounded set in $\Re^2$ and $f: D \to \Re$ is uniformly continuous, then there exists a continuous function $F: \overline{D} \to \Re$ such that $F(x,y) = f(x,y)$ for all $(x,y) \in D$. See Theorem 4.4.7.

32. If $S \subset \Re^2$, then S is compact if and only if every sequence in S has a subsequence that converges to a point in S.

33. A function $f: \Re^2 \to \Re$ is continuous at a point (a,b) if and only if f is a continuous function in x at a, and f is a continuous function in y at b.

34. If $A \subseteq \Re^3$, then $A' = \partial A \cup (\text{int } A)$.

35. If f_x and f_y both exist and are equal at (a,b), then f is continuous at (a,b).

36. If f_x and f_y are both differentiable at (a,b), then $f_{xy}(a,b) = f_{yx}(a,b)$.

37. If f_x and f_y are both bounded in an open ball about (a,b), then f is continuous at (a,b).

38. The function $f(x,y) = \sqrt{|xy|}$ is continuous, but not differentiable at the origin.

39. If f and g are both differentiable functions in $\Re^1$, and if D is a differential operator, then $D(fg) = fDg + gDf$.

40. $\Delta^n = (E-1)^n$, $n \in N$.

41. If $A, B \subseteq \Re^3$, then $\overline{A \cap B} = \overline{A} \cap \overline{B}$. See Exercise 15(f) in Section 11.1.

42. If $A, B \subseteq \Re^3$, then $(\text{int } A) \cup (\text{int } B) = \text{int}(A \cup B)$. See Exercise 15(a) in Section 11.1.

43. If $A, B \subseteq \Re^3$, then $(A \cap B)' = A' \cap B'$. See Exercise 15(d) in Section 11.1.

44. If $A \subseteq \Re^3$, then $(A')' \subseteq A'$.

45. The differential operator transforms a differentiable function into a differentiable function.

46. There exists a function $f(x,y)$ such that f_{xy} exists for all (x,y), but f_y does not exist anywhere.

47. A convex subset of either $\Re^2$ or $\Re^3$ is connected.

48. If $u(r,\theta)$ satisfies $\nabla^2 u = 0$, then $v(r,\theta) \equiv u\left(\frac{1}{r},\theta\right)$ satisfies $\nabla^2 v = 0$.

49. If $A \subseteq \Re^n$, then $\overline{A} = (\operatorname{int} A) \cup (\partial A)$.

50. $\displaystyle\lim_{(x,y)\to(0,0)} \frac{\sin\left(x^2+y^2\right)}{x^2+y^2} = 1$.

51. If $\vec{N}$ is the outward unit normal to the curve C and $\dfrac{\partial f}{\partial \vec{N}} = 0$ on C, then f is constant on C.

11.8* Projects

Part 1. Application of Fourier Series to Partial Differential Equations

Vibrations of a violin string of length L are governed by the *one-dimensional wave equation*

$$\frac{\partial^2 u}{\partial t^2} = \alpha^2 \frac{\partial^2 u}{\partial x^2},$$

where $u(x,t)$ represents a displacement of the string in a plane, and the constant $\alpha^2 > 0$ depends on the linear density and tension of the string. Since the ends of the string are held fixed at $x = 0$ and $x = L$, u must satisfy the *boundary conditions*

$$u(0,t) = 0 \text{ and } u(L,t) = 0$$

for all t. Also, u must depend upon both the initial displacement and the initial velocity of each point on the string. Thus, for $t = 0$, the values of $u(x,0)$ and $\dfrac{\partial}{\partial t}u(x,0)$ must be specified. Therefore, we will assume that the *initial conditions* are

$$u(x,0) = f(x) \text{ and } u_t(x,0) = g(x)$$

for $x \in (0,L)$. For the initial and boundary conditions to be consistent,

$$f(0) = f(L) = 0 \text{ and } g(0) = g(L) = 0.$$

The following attempt to find the solution $u(x,t)$ to the wave equation, which must also satisfy both the boundary conditions and the initial conditions. An outline of this procedure, called *separation of variables,* is below with each of the steps requiring verification.

1. Assume that $u(x,t) = X(x)T(t)$, and show that the wave equation takes the form $\dfrac{T''(t)}{\alpha^2 T(t)} = \dfrac{X''(x)}{X(x)}$.

2. Now there exists a real constant k such that $\dfrac{X''(x)}{X(x)} = k$ and $\dfrac{T''(t)}{\alpha^2 T(t)} = k$.

3. The boundary conditions become $X(0)T(t) = 0$ and $X(L)T(t) = 0$.

4. Ignoring the trivial solutions, that is, $T(t) \equiv 0$, observe that $X(0) = X(L) = 0$.

5. Solve $X''(x) - kX(x) = 0$, with $X(0) = X(L) = 0$.

6. Show that $k = 0$ produces no *nontrivial,* that is, no identically zero, solution $u(x,t)$.

7. Show that $k > 0$ also does not produce a nontrivial solution.

8. If $k < 0$, then $X(x) = c_1 \cos\sqrt{-k}\, x + c_2 \sin\sqrt{-k}\, x$.

9. Show that $X(0) = 0$ implies that $X(x) = c_2 \sin\sqrt{-k}\, x$, and that $X(x) = c_2 \sin\sqrt{-k}\, x$, together with $X(L) = 0$, implies that $k = -\left(\dfrac{n\pi}{L}\right)^2$ for $n \in N$.

10. Show that $X_n(x) = d_n \sin\dfrac{n\pi}{L}x$ for $n \in N$ with d_n nonzero constants represent infinitely many solutions to the problem given in step 5.

11. Since $k = -\left(\dfrac{n\pi}{L}\right)^2$, solve $T''(t) + \dfrac{n^2\pi^2\alpha^2}{L^2}T(t) = 0$, for $n \in N$, to obtain $T_n(t) = k_n \cos\dfrac{n\pi\alpha}{L}t + \bar{k}_n \sin\dfrac{n\pi\alpha}{L}t$, with $n \in N$ and k_n and $\bar{k}_n$ constants.

12. Therefore, for all $n \in N$ show that

$$\begin{aligned} u_n(x,t) &= X_n(x)T_n(t) \\ &= \left(d_n \sin\frac{n\pi}{L}x\right)\left(k_n \cos\frac{n\pi\alpha}{L}t + \bar{k}_n \sin\frac{n\pi\alpha}{L}t\right) \\ &= \left(A_n \cos\frac{n\pi\alpha}{L}t + B_n \sin\frac{n\pi\alpha}{L}t\right)\sin\frac{n\pi}{L}x. \end{aligned}$$

13. Show that $u(x,t) = \displaystyle\sum_{k=1}^{\infty}\left(A_k \cos\frac{k\pi\alpha}{L}t + B_k \sin\frac{k\pi\alpha}{L}t\right)\sin\frac{k\pi x}{L}$ is also a solution for some constants A_k and B_k.

14. Since u needs to satisfy the initial conditions as well, show that for $x \in [0, L]$,

$$u(x,0) = \sum_{k=1}^{\infty} A_k \sin\frac{k\pi x}{L} = f(x), \text{ and } u_t(x,0) = \sum_{k=1}^{\infty} B_k \frac{k\pi\alpha}{L} \sin\frac{k\pi x}{L} = g(x).$$

15. Choose A_k and B_k so that $u(x,0)$ becomes a Fourier sine series for f and $u_t(x,0)$ becomes a Fourier sine series for g. Thus, let

$$A_n = \frac{2}{L}\int_0^L f(x)\sin\frac{n\pi x}{L}\,dx \text{ and } B_n\frac{n\pi\alpha}{L} = \frac{2}{L}\int_0^L g(x)\sin\frac{n\pi x}{L}\,dx$$

for $n \in N$.

16. Draw a final conclusion, and compare this problem to Problem 12.10.15.

Problem 11.8.1. Use separation of variables to solve the *one-dimensional heat equation*

$$u_t(x,t) = \beta u_{xx}(x,t),$$

which satisfies the boundary conditions $u(0,t) = 0$ and $u(L,t) = 0$ for $t > 0$, and the initial condition $u(x,0) = f(x)$ for $x \in [0, L]$.

Part 2. Operator Method for Solving Differential Equations

In Section 11.3, the differential operator D was introduced. Exploring differential operators further, consider the *second order linear differential equation with constant coefficients* of the form

$$(*) \qquad a_0\frac{d^2y}{dx^2} + a_1\frac{dy}{dx} + a_2 y = F(x),$$

where a_0, a_1, and a_2 are real constants, and the function F is continuous. If $F(x) \equiv 0$, then $(*)$ is said to be *homogeneous*. Also, $(*)$ may be rewritten using differential operators such as

$$\left(a_0 D^2 + a_1 D + a_2\right)y = F(x).$$

Note that the operator in parentheses is a linear operator. Why?

Problem 11.8.2. If y_1 and y_2 are *solutions* to the equation $(*)$ with $F(x) \equiv 0$, meaning that both $y = y_1$ and $y = y_2$ satisfy $(*)$, show that any *linear combination*, that is, $c_1 y_1 + c_2 y_2$, with c_1 and c_2 real constants, is also a solution to $(*)$. Demonstrate this idea on the differential equation $\dfrac{d^2 y}{dx^2} + y = 0$, assuming that $y_1 = \sin x$ and $y_2 = \cos x$ are two solutions.

Problem 11.8.3. *(Superposition Principle)* Suppose that $L \equiv a_0 D^2 + a_1 D + a_2$. If y_1 is a solution of $Ly = F_1(x)$ and y_2 is a solution of $Ly = F_2(x)$, show that for any real constants c_1 and c_2, the function $c_1 y_1(x) + c_2 y_2(x)$ is a solution of $Ly = c_1 F_1(x) + c_2 F_2(x)$.

Example 11.8.4. Use differential operators to find all functions y that satisfy

$$\frac{d^2 y}{dx^2} - \frac{dy}{dx} - 6y = 0,$$

that is, solve this differential equation.

Answer. Writing the given equation in an equivalent form, we obtain

$$\left(D^2 - D - 6\right)y = 0,$$

where the operator on the left side factors to give

$$(D-3)(D+2)y = 0.$$

Now, in setting $(D+2)y = u$, we obtain $(D-3)u = 0$. Then we multiply both sides of this equation by e^{-3x}, the so-called *integrating factor*. What is the relationship of e^{-3x} to the differential equation? And by Exercise 12(f) of Section 11.3, we obtain

$$e^{-3x}(D-3)u = D\left(e^{-3x}u\right) = 0,$$

which upon integration becomes $e^{-3x}u = c_1$, with c_1 some real constant of integration. Thus, $u = c_1 e^{3x}$. And since $(D+2)y = u = c_1 e^{3x}$, multiplication of both sides by the integrating factor e^{2x} gives $D\left(e^{2x}y\right) = c_1 e^{5x}$. Now, integration of both sides yields $e^{2x}y = \frac{1}{5}c_1 e^{5x} + c_2$, producing $y = \frac{1}{5}c_1 e^{3x} + c_2 e^{-2x} = C_1 e^{3x} + C_2 e^{-2x}$, with the constants $C_1 = \frac{1}{5}c_1$ and $C_2 = c_2$. ■

Problem 11.8.5. Use the operator method from Example 11.8.4 to solve

(a) $2\dfrac{d^2y}{dx^2}+\dfrac{dy}{dx}-y=0$.

(b) $\dfrac{d^2y}{dx^2}+6\dfrac{dy}{dx}+9y=0$. See Exercise 10(a) in Section 11.3.

(c) $\dfrac{d^2y}{dx^2}+3\dfrac{dy}{dx}+2y=4x$.

Problem 11.8.6. Consider the differential equation

$$a_0\frac{d^2y}{dx^2}+a_1\frac{dy}{dx}+a_2y=0,$$

where $a_0 \neq 0$, and suppose that the so-called *characteristic equation* $a_0m^2+a_1m+a_2=0$ for the given differential equation has two roots r_1 and r_2.

(a) Prove that if $r_1 \neq r_2$, then the most general solution for the given differential equation is of the form

$$y=c_1e^{r_1x}+c_2e^{r_2x},$$

where c_1 and c_2 are some constants. In the case where r_1 and r_2 are complex, show that they must be *complex conjugates* of each other, meaning that if $r_1=a+bi$, then $r_2=a-bi$. Moreover, using Euler's formula, show that in such a case a solution of the given differential equation for the given differential equation is of the form

$$y=e^{ax}\left(c_1\cos bx+c_2\sin bx\right).$$

(b) Prove that if $r_1=r_2$, meaning that if the characteristic equation has a double root r_1, then the most general solution is of the form

$$y=c_1e^{r_1x}+c_2xe^{r_1x}.$$

If $a\in\Re$ and $(D-a)y=f(x)$, then $\dfrac{dy}{dx}-ay=f(x)$. This special type of *first order linear differential equation* (See Problem 11.8.9.) may be solved as done previously by multiplying both sides of the differential equation by the integrating factor e^{-ax} to obtain $e^{-ax}(D-a)y=e^{-ax}f(x)$, which equivalently is $D\left(e^{-ax}y\right)=e^{-ax}f(x)$. Integration gives $y=e^{ax}\int e^{-ax}f(x)\,dx$, which allows us to define the operator $\dfrac{1}{D-a}$ by

$$\frac{1}{D-a}f(x) \equiv e^{ax}\int e^{-ax}f(x)\,dx,$$

where $\dfrac{1}{D-a}$ is the *inverse operator* for $D-a$, often denoted $(D-a)^{-1}$.

Problem 11.8.7. Solve the differential equations given in Problem 11.8.5 using inverse operators.

Operators may also be used to solve systems of differential equations. For example, let us solve the system

$$\begin{cases} \dfrac{dx}{dt} = 3x - 2y + e^t \\ \dfrac{dy}{dt} = 3y - 2x. \end{cases}$$

An equivalent representation for the given system is

$$\begin{cases} (D-3)x + 2y = e^t \\ 2x + (D-3)y = 0. \end{cases}$$

Eliminating y, we obtain $(D-5)(D-1)x = -2e^t$. Then, using some previously discussed operator methods in solving differential equation, obtain

$$x(t) = \frac{1}{2}te^t + \frac{1}{8}e^t + k_1e^t + k_2e^{5t}.$$

Upon substituting this result into the first equation in the system, we find that

$$y(t) = \frac{1}{2}te^t + \frac{3}{8}e^t + k_1e^t - k_2e^{5t}.$$

The solution to the system is given by both $x(t)$ and $y(t)$.

Problem 11.8.8. Use operators to solve the given systems.

(a) $$\begin{cases} \dfrac{dx}{dt} = x + y + 1 \\ \dfrac{dy}{dt} = x + y \end{cases}$$

(b) $$\begin{cases} \dfrac{dx}{dt} + \dfrac{dy}{dt} = x - 1 \\ \dfrac{dx}{dt} + \dfrac{dy}{dt} = 1 - y \end{cases}$$

Problem 11.8.9. If $p(x)$ and $q(x)$ are continuous functions and m is a real number, then the differential equation

$$\frac{dy}{dx}+p(x)y=q(x)y^m$$

is called a *Bernoulli equation*. Prove that in using the substitution $v=y^{1-m}$, where $m\neq 0$ and $m\neq 1$, the Bernoulli equation can be transformed into a *first order linear equation*

$$\frac{dy}{dx}+P(x)y=Q(x).$$ [6]

See Problem 12.10.25. If $m=0$ or $m=1$, then the Bernoulli equation is already a linear equation.

Problem 11.8.10. Solve $\dfrac{dy}{dx}+5y=-xy^2$.

Part 3. Separable and Homogeneous First Order Differential Equations

Consider a *separable differential equation* of the form

$$\frac{dy}{dx}=f(x)g(y).$$

See Part 3 of Section 12.10. Under suitable conditions, this equation can be solved as follows. First, divide both sides of the equation by $g(y)$ to obtain $\dfrac{1}{g(y)}\dfrac{dy}{dx}=f(x)$.

Then, if $G(y)$ is an antiderivative of $\dfrac{1}{g(y)}$ and $F(x)$ is an antiderivative of $f(x)$, the differential equation can be rewritten as

$$G'(y)\frac{dy}{dx}=F'(x),$$

or, equivalently, as

$$\frac{d}{dx}G(y(x))=\frac{d}{dx}F(x).$$

---

[6]The Bernoulli equation was presented by Jacob Bernoulli and then solved by his brother Johann in 1695. In 1696 Leibniz used the specified substitution to reduce the Bernoulli equation to a linear equation.

By Exercise 11 of Section 5.2, $G(y(x)) = F(x) + C$, where C is a real constant. Thus, if $y(x)$ satisfies this last expression, then y is an implicit solution for the given separable differential equation. See Part 3 of Section 12.10 about gaining or losing solutions.

Problem 11.8.11. Solve the given differential equations.

(a) $\dfrac{dy}{dx} = \dfrac{y+1}{2x-3}$

(b) $\dfrac{dy}{dx} = \dfrac{\tan y}{x^2+1}$

Definition 11.8.12. The differential equation $\dfrac{dy}{dx} = f(x,y)$ is said to be *homogeneous* if and only if there exists a function g such that $\dfrac{dy}{dx} = g\left(\dfrac{y}{x}\right)$.

Clearly, if $f(x,y) = -\dfrac{M(x,y)}{N(x,y)}$, and both functions M and N are homogeneous of the same degree m (see Exercise 13 in Section 11.6), then $\dfrac{dy}{dx} = f(x,y)$ is a homogeneous differential equation. An easy verification follows. Since $M(tx,ty)$ $t^m M(x,y)$, with $t = \dfrac{1}{x}$, we have $M\left(1,\dfrac{y}{x}\right) = \left(\dfrac{1}{x}\right)^m M(x,y)$. Similarly, $N\left(1,\dfrac{y}{x}\right) = \left(\dfrac{1}{x}\right)^m N(x,y)$. Thus,

$$\frac{dy}{dx} = -x^m M\left(1,\frac{y}{x}\right) \Big/ x^m N\left(1,\frac{y}{x}\right) \equiv g\left(\frac{y}{x}\right).$$

Problem 11.8.13. Prove that a homogeneous differential equation $\dfrac{dy}{dx} = f(x,y)$ can be transformed into a separable differential equation using the substitution $y = vx$.

Problem 11.8.14. Solve the given differential equations.

(a) $\dfrac{dy}{dx} = \dfrac{\sqrt{x+y}+\sqrt{x-y}}{\sqrt{x+y}-\sqrt{x-y}}$

(b) $\dfrac{dy}{dx} = \dfrac{x^2-xy}{y^2}$

12

Multiple Integration

12.1 Double Integral

The definition for double integral for functions of two variables is similar to that of integrals for functions of a single variable in Section 6.1. Suppose that a function $f: R \subset \Re^2 \to \Re$ is bounded on R, with R a rectangle given by $a \le x \le b$ and $c \le y \le d$. R may also be written as $[a,b] \times [c,d]$. Also, suppose that P_1 is a partition of the interval $[a,b]$ with n subintervals $[x_{i-1}, x_i]$ for $i = 0,1, \ldots, n$, and $a = x_0 < x_1 < \ldots < x_n = b$. Similarly, suppose that P_2 is a partition of the interval $[c,d]$ with m subintervals $[y_{j-1}, y_j]$ for $j = 0,1, \ldots, m$, and $c = y_0 < y_1 < \cdots < y_m = d$. Then, $P \equiv P_1 \times P_2$ is a *partition*, or a *grid* of the rectangle R, which subdivides R into *mn subrectangles* $R_{ij} = [x_{i-1}, x_i] \times [y_{j-1}, y_j]$ with the dimensions Δx_i by Δy_j, where $\Delta x_i \equiv x_i - x_{i-1}$ and $\Delta y_j \equiv y_j - y_{j-1}$. The maximum of the Δx_is and Δy_js is called the *norm* of a grid. And the area of a rectangle R_{ij} is given by $(x_i - x_{i-1})(y_j - y_{j-1}) = \Delta x_i \Delta y_j$, denoted by ΔA_{ij}. Also, $Q \equiv Q_1 \times Q_2$ is a *refinement* of $P_1 \times P_2$ if Q_1 is a refinement of P_1 and Q_2 is a refinement of P_2. And, if Q is a refinement of P, then $P \subseteq Q$.

Next, define

$$M_{ij}(f) = \sup\left\{ f(x,y) \,\middle|\, (x,y) \in R_{ij} \right\} \text{ and}$$
$$m_{ij}(f) = \inf\left\{ f(x,y) \,\middle|\, (x,y) \in R_{ij} \right\},$$

which give rise to *upper sums* $U(P,f)$, *lower sums* $L(P,f)$, and *Riemann sums* $S(P,f)$, defined by

$$U(P,f) = \sum_{i=1}^{n}\sum_{j=1}^{m} M_{ij}\Delta x_i \Delta y_j \equiv \sum_{i,j} M_{ij}\Delta A_{ij},$$
$$L(P,f) = \sum_{i=1}^{n}\sum_{j=1}^{m} m_{ij}\Delta x_i \Delta y_j \equiv \sum_{i,j} m_{ij}\Delta A_{ij}, \text{ and}$$
$$S(P,f) = \sum_{i=1}^{n}\sum_{j=1}^{m} f\left(c_i, d_j\right)\Delta x_i \Delta y_j \equiv \sum_{i,j} f\left(c_i, d_j\right)\Delta A_{ij},$$

with some arbitrary $c_i \in \left[x_{i-1}, x_i\right]$ and $d_j \in \left[y_{j-1}, y_j\right]$.

LEMMA 12.1.1. *Suppose that* $f : R \subset \Re^2 \to \Re$ *is bounded,* $R = [a,b] \times [c,d]$, *and* P *is a partition of* R.

(*a*) *If* $m = \min_R f(x,y)$ *and* $M = \max_R f(x,y)$, *then*

$$m(b-a)(d-c) \le L(P,f) \le S(P,f) \le U(P,f) \le M(b-a)(d-c).$$

(*b*) *If* Q *is a partition of* R *and* $P \subseteq Q$, *then* $L(P,f) \le L(Q,f)$ *and* $U(Q,f) \le U(P,f)$.

(*c*) *For any partition* Q *of* R, $L(P,f) \le U(Q,f)$.

(*d*) $\sup_P L(P,f) \le \inf_P U(P,f)$.

(*e*) $\sum_{i,j} A_{ij} = (b-a)(d-c)$, *the area of a rectangle* R.

Proofs for this lemma are omitted since proofs of Lemma 6.1.3 and Lemma 6.1.5 can easily be adapted to the present situation. Parts (d) and (e) are trivial.

Definition 12.1.2. A bounded function $f : R \subset \Re^2 \to \Re$, with R a rectangle given by $R = [a,b] \times [c,d]$, is *Riemann integrable* on R if and only if the *lower double integral*

$$\underline{\iint_R} f \equiv \sup\left\{ L(P,f) \,|\, P \text{ a partition of } R \right\} = \sup_P L(P,f) = I,$$

and the *upper double integral*

$$\overline{\iint_R} f \equiv \inf\{U(P,f) \mid P \text{ a partition of } R\} = \inf_P U(P,f) = I,$$

with I the value of the *Riemann integral* of f over R denoted by $\iint\limits_R f$, $\iint\limits_R f(x,y)\,dA$, or $\iint\limits_R f(x,y)\,dx\,dy$.

As we experienced in Chapter 6, it is not easy to compute the value of an integral using its definition. Other means of computation will be discussed later in this chapter. From Definition 12.1.2, if $f(x,y) \geq 0$ on a rectangle R, then $\iint\limits_R f$ represents the volume both under the surface and above the rectangle R in the xy-plane.

THEOREM 12.1.3. *A bounded function $f(x,y)$ on a rectangle $R=[a,b]\times[c,d]$ is Riemann integrable if and only if for any $\varepsilon > 0$, there exists a partition P of R such that $U(P,f) - L(P,f) < \varepsilon$.*

THEOREM 12.1.4. *Suppose that $f(x,y)$ is bounded on a rectangle $R=[a,b]\times[c,d]$. And suppose that there exists a sequence $\{P_n\}$ of partitions of R such that*

$$\lim_{n\to\infty} U(P_n, f) = I = \lim_{n\to\infty} L(P_n, f).$$

Then f is Riemann integrable and $\iint\limits_R f = I$.

Proofs of Theorem 12.1.3 and Theorem 12.1.4 are left as exercises. Many functions $f(x,y)$ are Riemann integrable, and certainly all continuous functions are Riemann integrable.

THEOREM 12.1.5. *Any continuous function $f(x,y)$ on a rectangle $R=[a,b]\times[c,d]$ is Riemann integrable.*

Proof. Let $\varepsilon > 0$ be given. Since f is continuous on a closed and bounded region R, by Theorem 11.2.13, f is bounded and uniformly continuous on R. And by Definition 11.2.12, if A is the area of a rectangle R, then there exists $\delta > 0$ such that

$$\left|f(\vec{x}) - f(\vec{t})\right| < \boxed{\frac{\varepsilon}{A}} \text{ for all } \vec{x},\vec{t} \in \Re \text{ satisfying } \left\|\vec{x} - \vec{t}\right\| < \delta.$$

Let P be any partition of R such that any two points in each subrectangle R_{ij} are within δ. Then, using previously established notation, we can write that

$$U(P,f)-L(P,f) = \sum_{i,j} M_{ij}\Delta A_{ij} - \sum_{i,j} m_{ij}\Delta A_{ij} = \sum_{i,j}\left(M_{ij}-m_{ij}\right)\Delta A_{ij}$$
$$< \sum_{i,j}\frac{\varepsilon}{A}\Delta A_{ij} = \frac{\varepsilon}{A}\sum_{i,j}\Delta A_{ij} = \frac{\varepsilon}{A}\cdot A = \varepsilon .$$

Therefore, f is Riemann integrable by Theorem 12.1.3. ■

The next theorem shows that a double integral satisfies the linear property.

THEOREM 12.1.6. *If functions f_1 and f_2 are Riemann integrable on a rectangle $R=[a,b]\times[c,d]$, then for c_1 and c_2 real constants, $c_1f_1+c_2f_2$ is Riemann integrable on R, and*

$$\iint_R c_1f_1+c_2f_2 = c_1\iint_R f_1 + c_2\iint_R f_2 .$$

The proof from Theorem 6.3.1 can easily be extended to a proof for Theorem 12.1.6. See Exercise 5. Our next result can also be proved by modifying the proof of Theorem 6.3.3.

THEOREM 12.1.7. *Suppose that $f(x,y)$ is bounded on a rectangle $R=[a,b]\times[c,d]$. Suppose further that $R=R_1\cup R_2$, where R_1 and R_2 are two rectangles. Then, f is Riemann integrable on R if and only if f is Riemann integrable on both R_1 and R_2. Moreover, if f is Riemann integrable on R, then*

$$\iint_R f = \iint_{R_1} f + \iint_{R_2} f .$$

Exercises 12.1

1. **(a)** Prove Lemma 12.1.1.
   (b) Prove Theorem 12.1.3.
   (c) Prove Theorem 12.1.4.
2. If $f(x,y)\equiv k$, with k a real constant, and if $R=[a,b]\times[c,d]$, show that $\iint_R f = k(b-a)(d-c)$. Note that $(b-a)(d-c)$ is the area of R, and $\iint_R f$ is the volume of a box with dimensions $b-a$ by $d-c$ by k.
3. Use Theorems 12.1.3 and 12.1.4 to compute the Riemann integral of $f(x,y)=x^2$ over a rectangle $R=[0,2]\times[0,1]$.

4. Suppose that $f, g: R \subset \Re^2 \to \Re$ are Riemann integrable functions on a rectangle $R = [a,b] \times [c,d]$.

(a) If $m \le f(x,y) \le M$ for all $(x,y) \in \Re$, show that

$$m(b-a)(d-c) \le \iint_R f \le M(b-a)(d-c).$$

(b) If $f(x,y) \ge 0$ for all $(x,y) \in \Re$, show that $\iint_R f \ge 0$.

(c) If $f(x,y) \le g(x,y)$ for all $(x,y) \in \Re$, show that $\iint_R f \le \iint_R g$.

5. Prove Theorem 12.1.6.

6. If $R = [0,1] \times [0,1]$ is a rectangle, and if f is defined on R by

$$f(x,y) = \begin{cases} 1, & \text{if } x \text{ is rational} \\ 0, & \text{otherwise,} \end{cases}$$

show that f is not Riemann integrable on R.

7. If $f, g: R \subset \Re^2 \to \Re$ are Riemann integrable on a rectangle $R = [a,b] \times [c,d]$, prove that

(a) $|f|$ is Riemann integrable and $\left| \iint_R f \right| \le \iint_R |f|$.

(b) fg is Riemann integrable.

(c) $\dfrac{f}{g}$ is Riemann integrable if there exists a constant $c > 0$ such that $|g(x,y)| \ge c$ for all $(x,y) \in \Re$.

8. Suppose that $f: D \subset \Re^2 \to \Re$ is continuous on an open region D. If $\iint_R f = 0$ for every rectangle $R \subset D$, prove that $f(x,y) \equiv 0$ on D.

9. Evaluate $\iint_R [x] + 2[y]$, where $R = [-1,3] \times [0,2]$.

10. *(Mean Value Theorem for Double Integrals)* If $f: R \subset \Re^2 \to \Re$ is continuous on a rectangle $R = [a,b] \times [c,d]$, prove that there exists $(x_0, y_0) \in R$ such that

$$\iint_R f = f(x_0, y_0)(b-a)(d-c).$$

See Corollary 6.3.9. Also, what are the values of $\iint_R 0$ and $\iint_R 1$? See Exercise 2.

12.2 Iterated Integrals

Suppose that $f: R \subset \mathfrak{R}^2 \to \mathfrak{R}$ is defined on a rectangle $R = [a,b] \times [c,d]$. If for each fixed x, f is an integrable function of y, then g can be defined by

$$g(x) = \int_c^d f(x,y)\,dy.$$

See the end of Section 11.3. If g is Riemann integrable on $[a,b]$, then

$$\int_a^b \left[\int_c^d f(x,y)\,dy \right] dx$$

called an *iterated integral*, whether written with or without brackets, is finite, and can be evaluated. Observe how closely an iterated integral resembles a double integral $\iint_R f$. The resemblance is more apparent in Riemann sums that, in using notation from Section 12.1, may be written as

$$S(P,f) = \sum_{i=1}^{n} \sum_{j=1}^{m} f\left(c_i, d_j\right) \Delta x_i \, \Delta y_j = \sum_{i=1}^{n} \left[\sum_{j=1}^{m} f\left(c_i, d_j\right) \Delta y_j \right] \Delta x_i .$$

Thus, under sufficient conditions the double integral and the iterated integral are one and the same. This is formally stated in Theorem 12.2.2 after the following preliminary lemma.

LEMMA 12.2.1. *Suppose that $f: R \subset \mathfrak{R}^2 \to \mathfrak{R}$ is bounded on a rectangle $R =$ $[a,b] \times [c,d]$. And suppose that for each x the function $h(y) \equiv f(x,y)$ is Riemann integrable (in y) on the interval $[c,d]$. Also, suppose that $g(x) \equiv \int_c^d f(x,y)\,dy$ on $[a,b]$. If P_1 is a partition of the interval $[a,b]$, and P_2 is a partition of the interval $[c,d]$, then, for a grid of the rectangle R given by $P \equiv P_1 \times P_2$, we must have*

$$L(P,f) \le L\left(P_1, g\right) \le U\left(P_1, g\right) \le U(P,f).$$

Proof. Let $P_1 = \left\{x_0, x_1, \ldots, x_n\right\}$ be a partition of $[a,b]$, and $P_2 = \left\{y_0, y_1, \ldots, y_m\right\}$ be a partition of $[c,d]$. In staying consistent with the notation given in Section 12.1, we have that

$$U(P,f)=\sum_{i=1}^{n}\sum_{j=1}^{m}M_{ij}\Delta x_i\Delta y_j,\quad L(P,f)=\sum_{i=1}^{n}\sum_{j=1}^{m}m_{ij}\Delta x_i\Delta y_j,$$

$$U(P_1,f)=\sum_{i=1}^{n}M_i\Delta x_i,\ \text{where } M_i=\sup\{g(x)\mid x\in[x_{i-1},x_i]\},\ \text{and}$$

$$L(P_1,f)=\sum_{i=1}^{n}m_i\Delta x_i,\ \text{where } m_i=\inf\{g(x)\mid x\in[x_{i-1},x_i]\}.$$

For any fixed i between 1 and n and any fixed $x\in[x_{i-1},x_i]$, $m_{ij}\le f(x,y)\le M_{ij}$ for all $y\in[y_{j-1},y_j]$ and any j between 1 and m. Then

$$m_{ij}[y_j-y_{j-1}]\le\int_{y_{j-1}}^{y_j}f(x,y)\,dy\le M_{ij}[y_j-y_{j-1}],$$

from which

$$\sum_{j=1}^{m}m_{ij}[y_j-y_{j-1}]\le\int_{c}^{d}f(x,y)\,dy\equiv g(x)\le\sum_{j=1}^{m}M_{ij}[y_j-y_{j-1}].$$

And since x is fixed and arbitrary, from the definition of m_i we have that

$$\sum_{j=1}^{m}m_{ij}[y_j-y_{j-1}]\le m_i.$$

Similarly, $\sum_{j=1}^{m}M_{ij}[y_j-y_{j-1}]\ge M_i$. Thus, since trivially $m_i\le M_i$, we have that

$$\sum_{j=1}^{m}m_{ij}[y_j-y_{j-1}]\le m_i\le M_i\le\sum_{j=1}^{m}M_{ij}[y_j-y_{j-1}].$$

Multiplying through by x_i-x_{i-1} and then taking the summation of the terms with i going from 1 to n completes the proof. ■

THEOREM 12.2.2. *(Fubini's[1] Theorem) Suppose that* $f:R\subset\Re^2\to\Re$ *is*

[1]Guido Fubini (1879–1943), born in Venice, Italy, is recognized for his work in differential geometry, differential equations, analytic functions, calculus of variations, nonlinear integral equations, and other areas. Fubini was a student of Dini and Bianchi. Luigi Bianchi (1856–1928) was an Italian mathematician whose work in non-Euclidean geometries was used by Einstein in his general theory of relativity. Fubini proved a more general version of Theorem 12.2.2 in 1907. Also, Cauchy knew of a less restrictive version for Theorem 12.2.2 almost a century earlier than Theorem 12.2.2 was presented. In 1938, escaping a Fascist government, Fubini emigrated to the United States.

Riemann integrable on a rectangle $R=[a,b]\times[c,d]$. *And suppose that for each* x *the function* $h(y)\equiv f(x,y)$ *is Riemann integrable (in* y*) on the interval* $[c,d]$. *Then the function* $g(x)\equiv\int_c^d f(x,y)dy$ *is Riemann integrable (in* x*) on the interval* $[a,b]$, *and*

$$\iint_R f=\int_a^b g(x)\,dx=\int_a^b\left[\int_c^d f(x,y)\,dy\right]dx.$$

Proof. Let $\varepsilon>0$ be given. Since f is Riemann integrable on R, by Theorem 12.1.3 there exists a partition $P\equiv P_1\times P_2$ of R such that $U(P,f)-L(P,f)<\varepsilon$. But by Lemma 12.2.1, $U(P_1,g)-L(P_1,g)<\varepsilon$. Thus, g is Riemann integrable on the interval $[a,b]$. And since $L(P,f)\le\int_a^b g(x)\,dx\le U(P,f)$ and ε is arbitrary, we must have that $\iint_R f=\int_a^b g(x)\,dx$, which proves the theorem. ■

COROLLARY 12.2.3. *Suppose that* $f:R\subset\Re^2\to\Re$ *is Riemann integrable on a rectangle* $R=[a,b]\times[c,d]$. *Suppose further that* $h(y)\equiv f(x,y)$ *is Riemann integrable (in* y*) on the interval* $[c,d]$, *and* $k(x)\equiv f(x,y)$ *is Riemann integrable (in* x*) on the interval* $[a,b]$. *Then*

$$\int_a^b\left[\int_c^d f(x,y)\,dy\right]dx=\int_c^d\left[\int_a^b f(x,y)\,dx\right]dy.$$

A proof of this result follows easily from Fubini's theorem, and is thus left as Exercise 1. Corollary 12.2.3 does not say that if both iterated integrals exist, then f is Riemann integrable on the rectangle R (see Exercise 4). Corollary 12.2.3 also does not say that if one iterated integral exists, then so must the other (see Exercise 5). For a discussion of the converse of Corollary 12.2.3, see Exercise 1 in Section 12.9. The contrapositive of Corollary 12.2.3 says that if two iterated integrals given in Corollary 12.2.3 are not equal, then the double integral does not exist, meaning that the function f is not Riemann integrable on a rectangle R. The moral to the story is that if we know that f is Riemann integrable on a rectangle R, then the value of the double integral can be computed using iterated integrals. And since both iterated integrals in Corollary 12.2.3 produce the same value, an evaluation of the easier of the two is desirable.

Example 12.2.4. If $R=[0,1]\times\left[0,\frac{\pi}{2}\right]$, evaluate $\iint_R x\sin xy$.

Answer. Since $f(x,y)\equiv x\sin xy$ is continuous on R, by Theorem 12.1.5, f is Riemann integrable on R. And by Fubini's theorem,

$$\iint_R x\sin xy = \int_0^1\left[\int_0^{\frac{\pi}{2}} x\sin xy\,dy\right]dx = \int_0^{\frac{\pi}{2}}\left[\int_0^1 x\sin xy\,dx\right]dy.$$

Now, since the middle integral is easier to compute than the last one, using the fundamental theorem of calculus, we can write that

$$\iint_R x\sin xy = \int_0^1\left[\int_0^{\frac{\pi}{2}} x\sin xy\,dy\right]dx = \int_0^1 (-\cos xy)\Big|_{y=0}^{\frac{\pi}{2}}\,dx = \int_0^1\left(1-\cos\frac{\pi}{2}x\right)dx = 1-\frac{2}{\pi}. \quad ■$$

Looking back at the proof for Clairaut's theorem, Theorem 11.3.4, applying Fubini's theorem would drastically shorten that proof giving the following.

Alternate proof of Clairaut's theorem. Since f_{xy} and f_{yx} are continuous on D, they are Riemann integrable on every rectangle $R \subset D$. We will prove that $\iint_R (f_{xy} - f_{yx}) = 0$ and then apply Exercise 8 from Section 12.1 to draw a final conclusion. Thus, to this end, if a rectangle $R = [a,b]\times[c,d]$, then using Fubini's theorem we can write that

$$\begin{aligned}
\iint_R (f_{xy} - f_{yx}) &= \iint_R f_{xy} - \iint_R f_{yx} = \int_a^b\int_c^d f_{xy}\,dy\,dx - \int_a^b\int_c^d f_{yx}\,dy\,dx \\
&= \int_a^b\int_c^d f_{xy}\,dy\,dx - \int_c^d\int_a^b f_{yx}\,dx\,dy = \int_a^b f_x(x,y)\Big|_{y=c}^{d}\,dx - \int_c^d f_y(x,y)\Big|_{x=a}^{b}\,dy \\
&= \int_a^b [f_x(x,d) - f_x(x,c)]\,dx - \int_c^d \left[f_y(b,y) - f_y(a,y)\right]dy \\
&= f(x,d)\Big|_a^b - f(x,c)\Big|_a^b - f(b,y)\Big|_c^d + f(a,y)\Big|_c^d = 0.
\end{aligned}$$

Since the rectangle R is arbitrary in D, the proof is complete. ■

Next, let us look at functions defined by a Riemann integral.

THEOREM 12.2.5. *If $f : R \subset \Re^2 \to \Re$ is continuous on a rectangle $R = [a,b]\times[c,d]$, and if $g(x) \equiv \int_c^d f(x,y)dy$, with x a parameter, then g is continuous on $[a,b]$.*

Proof. To show that g is continuous on $[a,b]$, we must show that for any $x_0 \in [a,b]$, we have $\lim_{x\to x_0} g(x) = g(x_0)$, that is,

$$\lim_{x\to x_0}\int_c^d f(x,y)\,dy = \int_c^d \lim_{x\to x_0} f(x,y)\,dy = \int_c^d f(x_0,y)\,dy.$$

Since f is continuous on a closed and bounded region R, f must be uniformly continuous on R. Hence, given any $\varepsilon > 0$, there exists $\delta > 0$, such that whenever (x,y) and (x_0,y) are in R with $\|(x,y)-(x_0,y)\| < \delta$, we must have

$$|f(x,y)-f(x_0,y)| < \boxed{\frac{\varepsilon}{d-c}}.$$

Therefore, since $|x-x_0| \le \|(x,y)-(x_0,y)\| < \delta$ and f is continuous on R, we can write

$$\begin{aligned}|g(x)-g(x_0)| &= \left|\int_c^d [f(x,y)-f(x_0,y)]\,dy\right| \le \int_c^d |f(x,y)-f(x_0,y)|\,dy \\ &< \int_c^d \frac{\varepsilon}{d-c}\,dy = \frac{\varepsilon}{d-c}(d-c) = \varepsilon.\end{aligned}$$

Why? Hence, g is continuous at x_0. ■

Theorem 12.2.5 can be used to prove Fubini's theorem. Theorem 12.2.5, as well as Theorem 12.2.6, which is to come, may be rewritten with y as the parameter. Theorem 12.2.6 permits interchanging of the integral with the derivative when differentiating with respect to the parameter.

THEOREM 12.2.6. *(Leibniz Rule) If $f(x,y)$ has continuous first partial derivatives on a rectangle $R=[a,b]\times[c,d]$, and if $g(x)\equiv\int_c^d f(x,y)\,dy$, then g is differentiable on (a,b), and $g'(x)\equiv\int_c^d f_x(x,y)\,dy$.*

Proof. Since f has continuous first partials, f is both differentiable on $(a,b)\times(c,d)$ and continuous on R. Thus, for each $x\in[a,b]$, $f(x,y)$ is Riemann integrable on $[c,d]$. Why? Choose $x_0\in(a,b)$. Then, if $x\ne x_0$ and k is between x and x_0, we can write that

$$\frac{g(x)-g(x_0)}{x-x_0} = \int_c^d \frac{f(x,y)-f(x_0,y)}{x-x_0}\,dy = \int_c^d f_x(k,y)\,dy,$$

Why? Since f_x is continuous, by Theorem 12.2.5, we can write

$$g'(x_0) = \lim_{x \to x_0} \frac{g(x) - g(x_0)}{x - x_0} = \lim_{x \to x_0} \int_c^d f_x(k, y)\, dy$$

$$= \int_c^d \lim_{x \to x_0} f_x(k, y)\, dy = \int_c^d f_x(x_0, y)\, dy,$$

which completes the proof. ■

Applications of Theorem 12.2.6 are left as Exercises 13 and 14, with generalizations of Theorems 12.2.5 and 12.2.6 left as Exercise 15.

In closing, consider the area between two curves. Suppose that g_1 and g_2 are continuous on an interval $[a, b]$ with $g_1(x) \le g_2(x)$ for all $x \in [a, b]$. Then the area of the region in the plane bounded by the curves $y = g_1(x)$, $y = g_2(x)$, $x = a$, and $x = b$, is given by the Riemann integral $A = \int_a^b [g_2(x) - g_1(x)]\, dx$. Written as an iterated integral, $A = \int_a^b \int_{g_1(x)}^{g_2(x)} 1\, dy\, dx$. For example, if a region is bounded by $y = x^2$, $y = x$, $x = 0$, and $x = 1$, then the area of this region is given by

$$A = \int_0^1 \int_{x^2}^{x} dy\, dx, \text{ or } A = \int_0^1 \int_{\sqrt{y}}^{y} dx\, dy.$$

Why? It seems reasonable to think of these two iterated integrals, which gave a value for the area of the region R, as double integrals yielding the value for the volume of a solid with the base R and the height to the surface being one, that is, $f(x, y) \equiv 1$. This idea is discussed in the next section.

Exercises 12.2

1. Prove Corollary 12.2.3.
2. Evaluate the given integrals on the specified rectangle R.
   (a) $\iint_R ye^{xy}$, where $R = [-1, 0] \times [-1, 1]$

   (b) $\iint_R x \sin xy$, where $R = [1, 2] \times [0, \pi]$

   (c) $\iint_R \sin x \cos y$, where $R = \left[0, \frac{\pi}{2}\right] \times \left[0, \frac{\pi}{2}\right]$

(d) $\iint_R \ln(\sin x)$, where $R = \left[0, \frac{\pi}{2}\right] \times [0,2]$ (See Exercise 9(h) in Section 6.5.)

3. Find the volume that is under the surface $f(x,y) = 2 - x^2 - y^2$ and above the rectangle $R = [0,1] \times [0,1]$.

4. If $f(x,y) = \dfrac{x-y}{(x+y)^3}$ is defined on the rectangle $R = [0,1] \times [0,1]$, show that

$$\int_0^1 \int_0^1 f(x,y)\,dy\,dx = \frac{1}{2} \text{ and } \int_0^1 \int_0^1 f(x,y)\,dx\,dy = -\frac{1}{2}.$$

Then conclude that f is not Riemann integrable on R.

5. Assume that the rectangle $R = [0,1] \times [0,2]$ and that f is defined on R by

$$f(x,y) = \begin{cases} 2, & \text{if } x \text{ is irrational} \\ 2y, & \text{if } x \text{ is rational.} \end{cases}$$

(a) Show that $\int_0^2 \int_0^1 f(x,y)\,dx\,dy$ does not exist.

(b) Show that $\int_0^1 \int_0^2 f(x,y)\,dy\,dx = 4$.

(c) Conclude that $\iint_R f$ does not exist.

6. (a) If f_{xy} is Riemann integrable on a rectangle $R = [a,b] \times [c,d]$, show that

$$\iint_R f_{xy} = f(a,c) - f(a,d) + f(b,d) - f(b,c).$$

(b) If $f_{xy} \equiv 0$ on $\Re^2$, use part (a) to show that $f \equiv 0$ on $\Re^2$.

7. (a) Suppose that both $f:[a,b] \to \Re$ and $g:[c,d] \to \Re$ are Riemann integrable, and that a rectangle $R = [a,b] \times [c,d]$. Prove that

$$\iint_R f(x)g(y) = \left[\int_a^b f\right]\left[\int_c^d g\right].$$

(b) Use part (a) to evaluate $\iint_R \left(x^2+1\right)\cos y$, where $R = [0,1] \times \left[0, \frac{\pi}{2}\right]$.

8. Suppose that $f,g:[a,b]\to\Re$ are Riemann integrable on $[a,b]$, and that a rectangle $R=[a,b]\times[a,b]$. Show that

(a) $\displaystyle \frac{1}{2}\int_a^b\int_a^b [f(x)g(y)-f(y)g(x)]^2\,dx\,dy=\left[\int_a^b f^2\right]\left[\int_a^b g^2\right]-\left[\int_a^b fg\right]^2.$

(b) $\displaystyle \frac{1}{2}\int_a^b\int_a^b [f(x)-f(y)][g(x)-g(y)]\,dx\,dy=(b-a)\int_a^b fg-\left[\int_a^b f\right]\left[\int_a^b g\right].$

9. Write two iterated integrals that represent the area of the region between the curves $y=x^2$ and $y=x^3$. Then evaluate the iterated integrals.

10. Write another iterated integral that is both equal to $\displaystyle\int_0^2\int_{y^2}^4 dx\,dy$ and represents the area of the same region. Then evaluate both iterated integrals.

11. Use iterated integrals to calculate the area of the region bounded by

(a) $y=x$ and $y=x^2-x$. **(b)** $y=x-1$ and $y^2=2x+6$.

12. Sketch the region whose area is given by

(a) $\displaystyle\int_{-2}^{1}\int_{\arctan x}^{2^x} dy\,dx.$ **(b)** $\displaystyle\int_{-1}^{1}\int_{\frac{1}{x^2+1}}^{\cosh x} dy\,dx.$

13. Prove that $\displaystyle\int_0^1 \frac{x-1}{\ln x}\,dx=\ln 2.$

14. Prove part (a) of Remark 6.6.7, that is, $\displaystyle\int_0^\infty \exp\left(-x^2\right)dx=\frac{\sqrt{\pi}}{2}$, using the following steps.

(a) If

$$f(x)=\left(\int_0^x e^{-t^2}\,dt\right)^2 \text{ and } g(x)=\int_0^1 \frac{e^{-x^2\left(t^2+1\right)}}{t^2+1}\,dt,$$

show that $f'(x)+g'(x)=0$.

(b) Show that $f(x)+g(x)=\dfrac{\pi}{4}$.

(c) Take limits and then draw the final conclusion.

See Problem 12.10.4 for an alternate method of proof.

15. Suppose that the functions $g(x)$ and $h(x)$ are continuously differentiable, that $g(x) \le h(x)$ on an interval $[a,b]$, and that $f(x,y)$ has continuous first partial derivatives in $\Re^2$. Then prove that $F(x) \equiv \int_{g(x)}^{h(x)} f(x,y)\,dy$ is continuous, and that

$$F'(x) \equiv \frac{d}{dx}\int_{g(x)}^{h(x)} f(x,y)\,dy = \int_{g(x)}^{h(x)} f_x(x,y)\,dy + f(x,h(x))h'(x) - f(x,g(x))g'(x).$$

This is a more general *Leibniz Rule* than the one given in Theorem 12.2.6. Also, see Exercise 6(b) in Section 6.4.

16. Suppose f is continuous on $[0,1]$ and

$$g(x,y) \equiv \begin{cases} x(y-1), & \text{if } x \le y \\ y(x-1), & \text{if } y < x \end{cases}$$

on $[0,1]\times[0,1]$. If $F(x) = \int_0^1 f(y)g(x,y)\,dy$, prove that $F''(x) = f(x)$ for $x \in (0,1)$.

17. From Leibniz rule we have that if f has continuous first partial derivatives on a rectangle $R = [a,b]\times[c,d]$, then

$$\frac{d}{dx}\int_c^d f(x,y)\,dy = \int_c^d f_x(x,y)\,dy \text{ and } \frac{d}{dy}\int_a^b f(x,y)\,dx = \int_a^b f_y(x,y)\,dx.$$

Are these formulas satisfied for the function $f(x,y) = \dfrac{-y}{x^2+y^2}$? How about for $f(x,y) = \dfrac{x}{x^2+y^2}$? See Exercise 6 in Section 12.6.

12.3 Integrals over General Regions

The Riemann integration of functions defined on a rectangle R has been discussed. An extension of integration to more general regions D, which might serve as domain for a function $f(x,y)$, is desirable. This extension of integration from a rectangle R to an arbitrary bounded region D is not trivial since there is no way to define a partition of D. Small rectangles may not cover D precisely, as viewed in Figure 12.3.1. Thus, we will enclose D in a rectangle R, as shown in Figure 12.3.1, then partition R, and extend the definition of the function $f(x,y)$ from D to R by

$$F(x,y) = \begin{cases} f(x,y), \text{ if } (x,y) \in D \\ 0, \text{ if } (x,y) \notin D. \end{cases}$$

Clearly, if F is Riemann integrable on R, then f should be Riemann integrable on D. And since there is no contribution to the upper and lower sums from the points outside of D the values of the integrals should be equal. For the sake of discussion, if f is continuous on D, then it should seem reasonable that f is Riemann integrable. Riemann integrability cannot however be concluded so readily since F need not necessarily be continuous on R. Thus, integrability does not follow from Theorem 12.1.5. Note that F is continuous at all points of R except perhaps along the boundary of D. From experience acquired in studying the Riemann integration of functions defined on subsets of $\Re^1$, recall that any finite number of discontinuities of a bounded function $f(x)$ do not effect the integrability of f. The reason is that these points could have been enclosed in intervals whose total length was so small that the upper and lower sums over these intervals differed only by a fraction of some arbitrary $\varepsilon > 0$. A generalization to $\Re^2$ is similar. That is, if the boundary of D can be covered by small enough rectangles so that the upper and lower sums over these rectangles differ only by a fraction of $\varepsilon > 0$, then integrability of f over D, or, equivalently, the integrability of F over R, will not be effected. These rectangles are shaded in Figure 12.3.1.

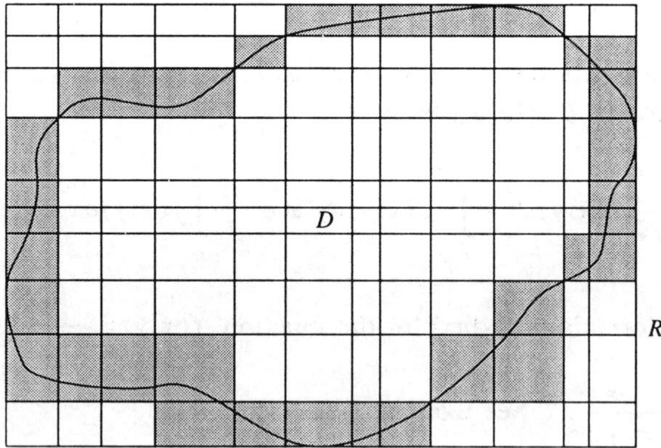

Figure 12.3.1

Definition 12.3.1. Suppose that for any $\varepsilon > 0$ there exists a finite number of rectangles R_{ij} whose union contains a set S. That is, if a finite number of rectangles *covers* a set S, such that the sum of the areas of these R_{ij}s is less than ε, then S is said to have a *Jordan*[2] *content zero*, or simply S, has *content zero*. Furthermore, if D is a bounded set, then D is *Jordan measurable* if and only if the boundary ∂D of D has content zero.

[2]Marie Ennemond Camille Jordan (1838–1922), a French mathematician, contributed greatly to algebra and group theory, introduced important concepts in topology, and was the first to introduce the concept of functions of bounded variation.

Intuitively, if a set S has content zero, then S has "zero area." Next, we will verify that many arbitrary bounded regions have a boundary of content zero.

THEOREM 12.3.2. *If C is a rectifiable curve, then C has content zero.*

Proof. Let s be the length of a curve C, and let n be some natural number. Suppose P_0 is an initial point of C, P_n is a terminal point of C, and $h = \frac{s}{n}$. Next, let P_1 be on C h units from P_0, let P_2 be on C $2h$ units from P_0, and so on. Construct a rectangle, which in this case is a square, with a center at each P_i and with sides parallel to axes each of length $3h$. If $P \equiv P(x, y)$ is any arbitrary point on C, then P is less than h units from one of the P_is. Hence, P is in at least one such rectangle. The area of each rectangle is $9h^2$, and there are n rectangles, so all of the points on the curve must be inside all n rectangles having a total area of $9h^2 n = \frac{9s^2}{n^2} \cdot n = \frac{9s^2}{n}$. Therefore, C can be covered by rectangles whose area is $\frac{9s^2}{n}$, which is less than ε if n is large enough. ■

Thus, in view of Theorem 10.7.1, if a curve C is smooth, it has content zero. Also, the union of a finite number of sets of content zero is again of content zero. Why? Therefore, if D is a bounded set whose boundary is piecewise continuously differentiable, then D is Jordan measurable. In fact, D is Jordan measurable if its boundary is made up of piecewise continuous curves (See Exercise 6). To rephrase, the set of discontinuities of a piecewise continuous function $f(x, y)$ has content zero.

THEOREM 12.3.3. *If f is continuous on a rectangle $R = [a, b] \times [c, d]$ except on a set of Jordan content zero, then f is Riemann integrable on R.*

Proof. Since f is bounded on R, there exists $M \in \Re$ such that $|f(x, y)| < M$ on R. Then for M_{ij} and m_{ij} as defined in Section 12.1, $M_{ij} - m_{ij} < 2M$. If S is a set of discontinuities of f, then S has content zero. Therefore, we may choose a grid P for a rectangle R so that the subrectangles R_{ij} that contain S have a total area less than $\boxed{\frac{\varepsilon}{4M}}$. Then, using the notation from Section 12.1, $U(P, f) - L(P, f)$ on the subrectangles that contain S is given by

$$\sum \left(M_{ij} - m_{ij} \right) \Delta A_{ij} < 2M \sum \Delta A_{ij} < 2M \cdot \frac{\varepsilon}{4M} = \frac{\varepsilon}{2}.$$

Let us consider the remaining rectangles in the grid P of R. On those rectangles, f is continuous. And since these rectangles are closed and bounded, f is uniformly continuous on them. By Theorem 12.1.5, f is Riemann integrable on each of these subintervals R_{ij}, and so by Theorem 12.1.3, there exists a refinement of the grid P so that $M_{ij} - m_{ij} < \dfrac{\varepsilon}{2A}$ on each of the refined subrectangles, with A the area of the rectangle R. Therefore, $U(P,f) - L(P,f)$ on the subintervals that do not contain S is given by

$$\sum \left(M_{ij} - m_{ij}\right)\Delta A_{ij} < \frac{\varepsilon}{2A}\sum \Delta A_{ij} < \frac{\varepsilon}{2A}\cdot A = \frac{\varepsilon}{2}.$$

Thus, on the entire rectangle R, $U(P,f) - L(P,f) < \varepsilon$. And by Theorem 12.1.5, f is Riemann integrable on R. ■

Compare Theorem 12.3.3 with Exercise 10 in Section 6.3. With all of these tools at hand we are ready to go back to the question of integrability of a bounded and continuous function f on a general bounded domain D. Assume that D has boundary of content zero, that is, D is Jordan measurable. If R is a rectangle containing D, and the extended function F of f is defined by

$$F(x,y) = \begin{cases} f(x,y), \text{ if } (x,y) \in D \\ 0, \text{ otherwise,} \end{cases}$$

then F is continuous except possibly on the ∂D that has content zero. Therefore, F is Riemann integrable on R. And thus, f is *Riemann integrable on* D. Moreover,

$$\iint_R F = \iint_D f.$$

Note that this definition is independent of the choice of a rectangle R that contains D. Also, note that all of the properties for Riemann integrable functions on a rectangle given thus far extend to the Jordan measurable sets.

So now, how are these integrals over more general regions computed? Since Fubini's theorem applies, computations using iterated integrals apply. Remember that an extended function to a rectangle R containing D brings no contribution to the value of the integral. Why? Therefore, the iterated integrals need only to cover the domain D. Thus, given a function f that is bounded and continuous on D, where $a \le x \le b$ and $g_1(x) \le y \le g_2(x)$, for g_1 and g_2 continuous on $[a,b]$,

$$\iint_D f = \int_a^b \int_{g_1(x)}^{g_2(x)} f(x,y)\,dy\,dx.$$

The domain D just described is often called a domain of the *first type*. On the other hand, if f is bounded and continuous on a domain of the *second type*, that is, on D, where $c \le y \le d$ and $h_1(y) \le x \le h_2(y)$, with h_1 and h_2 continuous on $[c,d]$, then

$$\iint_D f = \int_c^d \int_{h_1(y)}^{h_2(y)} f(x,y)\,dx\,dy.$$

Domains of either the first or second type are called *simple regions*. Simple regions are Jordan measurable regions (see Exercise 6), where no vertical or horizontal line intersects the boundary more than twice, unless the line itself is part of the boundary.

Example 12.3.4. Beware: Some iterated integrals do not correspond to a double integral. For instance, consider

$$\int_0^2 \int_1^{x^2} f(x,y)\,dy\,dx .$$

Since the curve $y=1$ is not below $y=x^2$ on an entire interval $[0,2]$, this iterated integral does not correspond to a double integral. However,

$$\int_1^{x^2} f(x,y)\,dy = \int_a^{x^2} f(x,y)\,dy - \int_a^{1} f(x,y)\,dy$$

holds for any value of a, so we can simply choose a so that $y=a$ is either below or on both $y=1$ and $y=x^2$ on the interval $[0,2]$, say, $a=0$. Then, the original iterated integral may be written as a combination of two double integrals as

$$\begin{aligned}\int_0^2 \int_1^{x^2} f(x,y)\,dy\,dx &= \int_0^2 \left[\int_0^{x^2} f(x,y)\,dy - \int_0^{1} f(x,y)\,dy \right] dx \\ &= \int_0^2 \int_0^{x^2} f(x,y)\,dy\,dx - \int_0^2 \int_0^{1} f(x,y)\,dy\,dx .\end{aligned}$$

These last two iterated integrals correspond to double integrals over domains called, say, D_1 and D_2. Can you describe D_1 and D_2? ■

As witnessed in Example 12.2.4, in deciding on the order of integration, it may be easier to evaluate one iterated integral than the other. In some cases, for functions over more general domains the order of integration may result in a fewer number of integrals need computation. Some of the exercises demonstrate this.

Theorem 12.3.3 demonstrated that if f is continuous on a rectangle $R=[a,b]\times[c,d]$, except perhaps on a set of Jordan content zero, then f is a Riemann integrable function on R. However, Riemann integrable functions on R do exist with discontinuities forming a set without Jordan content zero. Following is an example of such a function. Compare Example 12.3.5 to Dirichlet's function given in Example 3.2.10. Also, see exercise 8 in Section 6.2.

Example 12.3.5. Suppose that R is the rectangle $[0,1]\times[0,1]$, and suppose that S is a subset of R defined by

$$S = \{(x,y) \mid (x,y) \in R, \ x \text{ and } y \text{ are both rational}\}.$$

Now, define $f: R \to \Re$ by

$$f(x,y) = \begin{cases} \dfrac{1}{qs}, & \text{if } (x,y) = \left(\dfrac{p}{q}, \dfrac{r}{s}\right) \in S, \text{ in lowest terms} \\ 0, & \text{otherwise.} \end{cases}$$

Exercise 7 shows that S is not of content zero, and that f is Riemann integrable on R though not continuous on S. ■

Exercises 12.3

1. Compute $\iint_D f$ for each function f over the given domain D.

(a) $f(x,y) = 1$, where D is the region between the curves $y = x^2$ and $y = x^3$ (See Exercise 9 in Section 12.2.)

(b) $f(x,y) = 1$, where D is the region between the curves $y = \sqrt{x}$, $y = 0$, and $x = 4$ (See Exercise 10 in Section 12.2.)

(c) $f(x,y) = \dfrac{\sin x}{x}$, where D is the region between the curves $y = x$, $y = 0$, and $x = \dfrac{\pi}{2}$

(d) $f(x,y) = e^{y^2}$, where D is the region between the curves $y = 2x$, $x = 0$, and $y = 2$

2. Reverse the order of integration and evaluate the resulting iterated integral.

(a) $\displaystyle\int_{-2}^{4} \int_{\frac{1}{2}y^2 - 3}^{y+1} x \, dx \, dy$ (See Exercise 11(b) in Section 12.2.)

(b) $\displaystyle\int_{0}^{2} \int_{x^2}^{2x} (2x+3) \, dy \, dx$

(c) $\displaystyle\int_{0}^{1} \int_{-\sqrt{1-y^2}}^{\sqrt{1-y^2}} dx \, dy$

3. *(Dirichlet's Formula)* Verify that if $f(x,y)$ is continuous and $[a,b]$ is an interval, then

$$\int_a^b \int_a^x f(x,y)\,dy\,dx = \int_a^b \int_y^b f(x,y)\,dx\,dy.$$

What region do these iterated integrals represent?

4. Prove that if f is continuous on an interval $[a,b]$, then

$$2\int_a^b f(x)\,dx \cdot \int_x^b f(y)\,dy = \left[\int_a^b f(x)\,dx\right]^2.$$

5. Prove that if f is continuous on $\Re^1$, then

$$\int_0^x \int_0^y f(x)\,dx\,dy = \int_0^x (x-y)f(y)\,dy.$$

6. (a) Prove that simple regions are Jordan measurable.
   (b) Show that if

$$f(x) = \begin{cases} x\sin\dfrac{1}{x}, \text{ if } x \in (0,1] \\ 0, \text{ if } x = 0, \end{cases}$$

   then its image has a Jordan content zero. See Examples 3.3.8 and 10.7.5.

7. Consider the function f as defined in Example 12.3.5.
   (a) Show that S is not of Jordan content zero.
   (b) Show that f is not continuous at any point of S. And thus, that the set of discontinuities is not of content zero.
   (c) Prove that f is Riemann integrable on R.

12.4 Line Integrals

Suppose that $f(x,y)$ is continuous on a region D, containing a smooth curve C, in the xy-plane. C is smooth, so if A and B are points on C, then C must be rectifiable with a length of s between A and B. Suppose that $A = P_0$ is an initial point, $B = P_n$ is a terminal point, and $P_1, P_2, \ldots, P_{n-1}$ are points on C between A and B that divide C into n subarcs with lengths $\Delta s_1, \Delta s_2, \ldots, \Delta s_n$. On each subarc we can find the $\min f$, the $\max f$, or $f(c_i, d_i)$ for some arbitrary point (c_i, d_i) in the ith subarc. Multiplying these values by Δs_i and summing from the point A to the point B, we can define the lower, upper, and Riemann sums in a similar fashion to those given for single Riemann integrals. Omitting all of the repetitious details, we may immediately conclude that if

$$\sum_{i=1}^{n} f(c_i, d_i)\Delta s_i$$

as $n \to \infty$ has a finite limit L that is independent of the choice of points P_i on a curve C, then L is called the *line integral of f along C from A to B*, denoted $\int_C f\, ds$. The name "curve integral" seems more appropriate than the name line integral. At any rate,

$$\int_C f(x,y)\, ds = \lim_{n\to\infty} \sum_{i=1}^{n} f(c_i, d_i)\Delta s_i .$$

Recall that if C is smooth, then there exists a smooth parametrization $x = x(t)$, and $y = y(t)$ for C, with $t \in [a,b]$, where the initial point $A = (x(a), y(a))$, the terminal $B = (x(b), y(b))$, and points on C between A and B $P_i = (x(t_i), y(t_i))$. From Section 10.7, we have that the arc length parameter is

$$s(t) = \int_a^t \sqrt{[x'(u)]^2 + [y'(u)]^2}\, du .$$

Thus, $\dfrac{ds}{dt} = \sqrt{[x'(u)]^2 + [y'(u)]^2} = \left\|\vec{V}(t)\right\|$, where $\vec{V}(t) = \vec{r}\,'(t)$ with $\vec{r}(t) = x(t)\mathbf{i} + y(t)\mathbf{j}$. Therefore, using the mean value theorem, we may write that the *Riemann sum* is

$$\sum_{i=1}^{n} f(c_i, d_i)\Delta s_i = \sum_{i=1}^{n} f\left(x(\bar{t}_i), y(\bar{t}_i)\right)\frac{\Delta s_i}{\Delta t_i}\Delta t_i = \sum_{i=1}^{n} f\left(x(\bar{t}_i), y(\bar{t}_i)\right)s'(\hat{t}_i)\Delta t_i ,$$

where $\bar{t}_i$ and $\hat{t}_i$ are between t_{i-1} and t_i. Taking limits and reviewing a discussion involving Duhamel's principle presented in Section 10.7, we have that

$$\int_C f(x,y)\, ds = \int_a^b f(x,y)\frac{ds}{dt}\, dt = \int_a^b f\left(x(t), y(t)\right)\left\|\vec{V}(t)\right\| dt$$
$$= \int_a^b f\left(x(t), y(t)\right)\sqrt{[x'(t)]^2 + [y'(t)]^2}\, dt,$$

which is an ordinary Riemann integral in t.

Remark 12.4.1

(a) If $f(x,y) \ge 0$ for all (x,y) on a curve C, then the line integral for f may be interpreted as the area of a hanging curtain whose base is C and whose height at each point (x,y) is $f(x,y)$.

(b) Suppose that C is a line segment joining the point $(a,0)$ to the point $(b,0)$. If C is parametrized by $x = x$ and $y = 0$ with $x \in [a,b]$, then

$$\int_C f(x,y)\,ds = \int_a^b f(x,0)\,dx,$$

meaning that the line integral reduces to an ordinary Riemann integral.

(c) If $f(x,y) \equiv 1$ on C, then the line integral computes the length of C. See Theorem 10.7.2.

(d) Clearly, if two distinct points P and Q are connected by different curves, then each line integral may be different. See Exercise 5. However, two different smooth parametrizations of one curve C produce the same value as the corresponding line integral, which is certainly reasonable from the definition. See Exercise 7.

(e) If in the above development of the line integral, Δs_i is replaced by $\Delta x_i \equiv x_i - x_{i-1}$, then the *line integral of* f *along* C *with respect to* x would be obtained and given by

$$\int_C f(x,y)\,dx = \lim_{n\to\infty} \sum_{i=1}^{n} f(c_i, d_i)\Delta x_i.$$

Similarly, the *line integral of* f *along* C *with respect to* y would be given by

$$\int_C f(x,y)\,dy = \lim_{n\to\infty} \sum_{i=1}^{n} f(c_i, d_i)\Delta y_i.$$

Using this terminology, the line integral $\int_C f(x,y)\,ds$ is also called the *line integral with respect to arc length.*

(f) We can write that

$$\int_C f(x,y)\,dx = \int_a^b f(x(t),y(t))x'(t)\,dt, \text{ and that}$$

$$\int_C f(x,y)\,dy = \int_a^b f(x(t),y(t))y'(t)\,dt. \text{ Why?}$$

(g) $\int_C f(x,y)\,dx + g(x,y)\,dy$ is written to represent $\int_C f(x,y)\,dx + \int_C g(x,y)\,dy$.

(h) The linear property is satisfied by line integrals with respect to x, y, and the arc length parameter s. ■

Example 12.4.2

(a) Evaluate $\int_{C_1} xy\,ds$, where C_1 is given by $\gamma_1 = (-1+3t, 1+4t)$, for $t \in [0,1]$.

(b) Evaluate $\int_{C_2} xy\,ds$, where C_2 is given by $\gamma_2 = (2-3t, 5-4t)$, for $t \in [0,1]$.

Answer to part (a). If $x(t) = -1+3t$ and $y(t) = 1+4t$, then $x'(t) = 3$, $y'(t) = 4$, and $\sqrt{[x'(t)]^2 + [y'(t)]^2} = 5$. Thus,

$$\int_{C_1} xy\,ds = \int_0^1 (-1+3t)(1+4t)5\,dt = \frac{25}{2}.$$

Answer to part (b). If $x(t) = 2-3t$ and $y(t) = 5-4t$, then $x'(t) = -3$, $y'(t) = -4$, and $\sqrt{[x'(t)]^2 + [y'(t)]^2} = 5$. Thus, $\int_{C_2} xy\,ds = \frac{25}{2}$. ■

Note that in Example 12.4.2, C_1 and C_2 are identical except in their opposite orientations. In such a case, $C_1 = -C_2$. And since Δs_i is always positive, the line integrals with respect to arc length remain the same. Why? However, Δx_i and Δy_i change sign when orientation of the curve C is reversed. That is,

$$\int_{-C} f(x,y)\,dx = -\int_C f(x,y)\,dx,$$

and

$$\int_{-C} f(x,y)\,dy = -\int_C f(x,y)\,dy.$$

See Exercise 9. Also, note that the curve C_1 in Example 12.4.2 is a directed line segment from the point $(-1,1)$ to the point $(2,5)$ with a parametrization like that obtained from Example 10.4.2.

Suppose that the curve C is composed of a finite number of smooth curves $C_1, C_2, \ldots, C_n$, where the terminal point of C_i is the initial point of C_{i+1}. Thus C is *piecewise smooth.* See Definition 10.4.4. Then,

$$\int_C f\,ds = \sum_{i=1}^{n} \int_{C_i} f\,ds.$$

Why? The same holds for line integrals with respect to x and y.

Exercises 12.4

1. Evaluate the given line integrals with respect to arc length.

(a) $\int_C \left(x+\sqrt{y}\right)ds$, where C is parametrized by $\gamma(t)=\left(2t,t^2\right)$, for $t\in[0,1]$

(b) $\int_C \frac{3y+1}{(2x+y+4)\sqrt{x+2}}\,ds$, where C is parametrized by $\gamma(t)=\left(t^2-1,2t\right)$, for $t\in[0,1]$

(c) $\int_C \frac{10}{x^2+2y-26}\,ds$, where C is the line segment from the point $(0,1)$ to the point $(2,-1)$

(d) $\int_C xy\,ds$, where C is the parabola $y=x^2$ between the points $(0,0)$ and $(2,4)$.

2. Compute $\int_{C_1} xy\,dx$, $\int_{C_1} xy\,dy$, $\int_{C_2} xy\,dx$, and $\int_{C_2} xy\,dy$, where C_1 and C_2 are as given in Example 12.4.2.

3. If C is a smooth curve given by $y=g(x)$ with $x\in[a,b]$, show that

$$\int_C f(x,y)\,dx=\int_a^b f(x,g(x))\,dx, \text{ and that } \int_C f(x,y)\,dy=\int_a^b f(x,g(x))g'(x)\,dx.$$

Use the above expressions to compute $\int_C \frac{1}{x^2+1}\,dx$ and $\int_C \frac{1}{x^2+1}\,dy$, where C is the parabola $y=x^2$ between the points $(0,0)$ and $(2,4)$.

4. Evaluate $\int_{C_1} x^3y^2\,ds$ and $\int_{C_2} x^3y^2\,ds$, where C_1 is parametrized by $\gamma_1(t)=\left(t,\sqrt{1-t^2}\right)$ with $t\in[0,1]$, and C_2 is parametrized by $\gamma_2(t)=(\cos t,\sin t)$ with $t\in\left[0,\frac{\pi}{2}\right]$. Note that C_1 and C_2 are the same quarter of a unit circle. Are these line integrals equal? Should they be? See Exercise 7.

5. Evaluate $\int_{C_1} x\,ds$ and $\int_{C_2} x\,ds$, where C_1 and C_2 are curves connecting the points $(0,0)$ and $(1,1)$, with C_1 parametrized by $\gamma_1(t)=(2t,2t)$, with $t\in\left[0,\frac{1}{2}\right]$, and with C_2 parametrized by $\gamma_2(t)=\left(t,t^2\right)$ with $t\in[0,1]$. Are these line integrals equal? Should they be?

6. Evaluate $\int_C \left(x^3-y^3\right)dx$, where C is given parametrically by $\gamma(t)=\left(\sqrt{1-t^2},t\right)$ with $t\in[0,1]$.

7. Suppose that $f(x,y)$ is continuous on a region D in the xy-plane containing the two smooth curves C_1 and C_2 with equivalent parametrizations. Review Definition 10.7.6. Prove that $\int_{C_1} f\,ds=\int_{C_2} f\,ds$.

8. Suppose that $f(x,y)$ is continuous on a region D in the xy-plane which contains a smooth curve C. If $|f(x,y)|\le M$ for all (x,y) on C, and the length of C is bounded by L, prove that

$$\left|\int_C f(x,y)\,ds\right|\le ML.$$

9. Verify that $\int_{-C} f(x,y)\,dx=-\int_C f(x,y)\,dx$, and that $\int_{-C} f(x,y)\,dy=-\int_C f(x,y)\,dy$.

12.5 Vector Fields and Work Integrals

Consider a *vector field* $\vec{F}:D\subseteq\Re^2\to\Re^2$, where the domain is a subset of $\Re^2$ and the range consists of vectors. If at every point $P(x,y)$ in D we associate a unique vector having an initial point P, then a collection of such vectors represents a vector field. Examples of *velocity vector fields* are the flow of water in a river or air current. Sketch of the vector field defined by $\vec{F}(x,y)=\mathbf{i}+x\mathbf{j}$ is given in Figure 12.5.1. $\vec{F}(x,y)=\nabla f(x,y)$, a gradient of a differentiable function f, is another example of a vector field $\vec{F}$ that maps vectors to vectors (see Section 12.6). $\vec{F}(x,y)$ is a vector, so it can be written in terms of its *component functions* M and N as

$$\vec{F}(x,y)=M(x,y)\mathbf{i}+N(x,y)\mathbf{j}=\langle M(x,y),N(x,y)\rangle.$$

Note that M and N are scalar functions, also referred to as *scalar fields*.

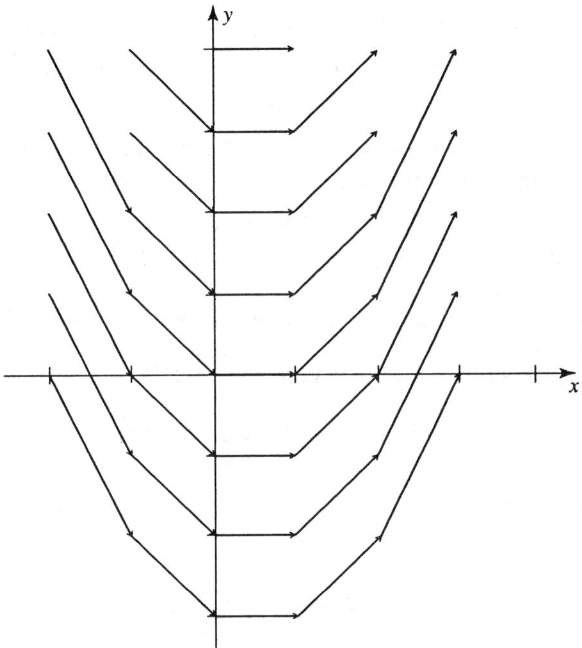

Figure 12.5.1

If a particle is in a velocity vector field, then the path it follows is called the *flow line*. Symbolically, if $\vec{V} = \vec{F}(\vec{r})$ is a velocity vector field, then the *flow line* is the path $\vec{r}(t)$ whose velocity vector $\vec{r}'(t) = \vec{V}(t)$. Thus, vectors in a vector field are tangent to flow lines. Flow lines are also called *streamlines*, and in differential equations, flow lines are known as *integral curves*. The *flow* of a vector field is the collection of its flow lines.

If $\vec{F}(x,y) = f_1(x,y)\mathbf{i} + f_2(x,y)\mathbf{j}$ forms a vector field, to find the flow lines we proceed as follows. If C is a flow line, and $x = x(t)$ and $y = y(t)$, with $t \in [a,b]$, a parametrization of C, then a position vector for C is given by $\vec{r}(t) = x(t)\mathbf{i} + y(t)\mathbf{j}$. Thus, we need to solve $\vec{r}'(\mathrm{t}) = \vec{F}(x(t), y(t))$, that is,

$$x'(t)\mathbf{i} + y'(t)\mathbf{j} = f_1(x(t), y(t))\mathbf{i} + f_2(x(t), y(t))\mathbf{j}.$$

Equating corresponding components,

$$x'(t) = f_1(x(t), y(t)) \text{ and } y'(t) = f_2(x(t), y(t)).$$

f_1 and f_2 are known functions, and thus, in order to find a flow line C parametrized by $\gamma(t) = (x(t), y(t))$, the above given system of differential equations must be solved.

Suppose that $\vec{F}(x,y) = M(x,y)\mathbf{i} + N(x,y)\mathbf{j}$ is some force field in an open region $D \subseteq \Re^2$, where the scalar functions M and N are continuous. Also, suppose that a smooth curve C in D is given by a smooth parametrization $\gamma(t) = (x(t), y(t))$ with $t \in [a,b]$. Then the curve C is determined by a vector-valued function $\vec{r}(t) = x(t)\mathbf{i} + y(t)\mathbf{j}$, and the motion along C is in a direction determined by the increasing values of t. The current objective is to determine the work done by $\vec{F}$ along C. For a constant force $\vec{F}$ moving along a straight line, the work W is given by

$$W = (\text{force in the direction of motion})(\text{distance moved}).$$

Thus, if θ is the angle between the force vector $\vec{F}$ and the direction of motion, then

$$W = \left\|\vec{F}\right\| \cos\theta \cdot (\text{distance moved}).$$

Therefore, the work done by a constant force $\vec{F}$ in moving an object from a point P to a point Q is given by

$$W = \vec{F} \cdot \overrightarrow{PQ} = \left\|\vec{F}\right\| \left\|\overrightarrow{PQ}\right\| \cos\theta,$$

where θ is the angle between $\vec{F}$ and $\overrightarrow{PQ}$. The value $\left\|\vec{F}\right\| \cos\theta$ is the force component in the direction of motion, and $\left\|\overrightarrow{PQ}\right\|$ is the distance moved. Next, we attempt to express W as an integral. A partition of the interval $[a,b]$ is made into n subintervals, which partitions C into n subarcs with $\vec{r}_0 = P$, $\vec{r}_n = Q$, and all other $\vec{r}_i$ being points P_i on C between P and Q, with the distance between the P_is given by $\Delta\vec{r}_i = \vec{r}_{i+1} - \vec{r}_i$. From the force field, there exists $\vec{F}(\vec{r}_i)$ acting on each point P_i and producing work $\vec{F}(\vec{r}_i) \cdot \Delta\vec{r}_i$. Now, summing over all of the subarcs, we obtain the Riemann sum

$$\sum_{i=1}^{n} \vec{F}(\vec{r}_i) \cdot \Delta\vec{r}_i .$$

If the limit is finite, then we have the *work integral*, also called the *line integral of a vector field* $\vec{F}$ *along the curve* C, with specified orientation given by

$$W = \int_C \vec{F} \cdot d\vec{r} = \lim_{n\to\infty} \sum_{i=1}^{n} \vec{F}(\vec{r}_i) \cdot \Delta\vec{r}_i .$$

For contrast, the line integral with respect to the arc length parameter s studied in Section 12.4 is referred to as the *line integral of a scalar field*. The above W yields the work done (as in elementary calculus) by a continuous force $F(x)$ directed along the x-axis from $x = a$ to $x = b$. That is,

$$W = \int_a^b F(x)\,dx\,.$$

If we express $\vec{F}$ and $\vec{r}$ in components, then $\vec{F} = M\mathbf{i} + N\mathbf{j}$ and $\Delta\vec{r} = \Delta x\mathbf{i} + \Delta y\mathbf{j}$, with $\vec{F}\cdot\Delta\vec{r} = M\Delta x + N\Delta y$. As a result, the work integral takes the form

$$W = \int_C M\,dx + N\,dy\,.$$

Review part (g) of Remark 12.4.1. Thus, if $\vec{F}$ is a continuous vector field, then there are a number of ways to express the work integral. For instance,

$$\begin{aligned}
\int_C \vec{F}\cdot d\vec{r} &= \int_C M(x,y)\,dx + N(x,y)\,dy = \int_C M(x,y)\,dx + \int_C N(x,y)\,dy \\
&= \int_a^b M\big(x(t),y(t)\big)x'(t)\,dt + \int_a^b N\big(x(t),y(t)\big)y'(t)\,dt \\
&= \int_a^b \big\langle M\big(x(t),y(t)\big), N\big(x(t),y(t)\big)\big\rangle\cdot\big\langle x'(t),y'(t)\big\rangle\,dt \\
&= \int_a^b \vec{F}\big(x(t),y(t)\big)\cdot\big\langle x'(t),y'(t)\big\rangle\,dt = \int_a^b \vec{F}\big(x(t),y(t)\big)\cdot r'(t)\,dt\,.
\end{aligned}$$

The work integral sums up the dot product of $\vec{F}$ and $\Delta\vec{r}$ along the curve C. Therefore, if $\vec{F}$ is facing in almost the same direction as $\Delta\vec{r}$, then the work integral is positive. And if $\vec{F}$ is facing in almost the complete opposite direction from $\Delta\vec{r}$, the work integral is negative. If $\vec{F}$ is perpendicular to C, then the integral is zero. The value of an integral may also be zero if positive and negative contributions along C cancel out. Summing up, we find that the work integral measures how much the curve C is going either with $\vec{F}$ or against it. The work integral is also independent of the parametrization of a curve C, provided that the orientation is not reversed. See Exercises 6 and 7. Other properties are presented in Exercises 8 - 10. The orientation of C given by $\vec{r}(t)$ is important in work integrals. A curve's orientation can easily be determined by the tangent vector obtained from $\vec{r}'(t)$.

Continuing with more observations, if L is the length of a curve, then we can rewrite the work integral containing an arc length of s as a parameter to obtain

$$\begin{aligned}
W &= \int_C M(x,y)\,dx + N(x,y)\,dy \\
&= \int_0^L \left[M\big(x(s),y(s)\big)\frac{dx}{ds} + N\big(x(s),y(s)\big)\frac{dy}{ds}\right] ds\,.
\end{aligned}$$

Why? $\dfrac{dx}{dt}\mathbf{i}+\dfrac{dy}{dt}\mathbf{j}=\vec{T}$, that is, the unit vector tangent to C in the direction of increasing s (see the end of Section 10.7), so we can write that

$$W=\int_{0}^{L}\left[M(x(s),y(s))i+N(x(s),y(s))j\right]\cdot\vec{T}\,ds=\int_{t=a}^{b}\vec{F}\cdot\vec{T}\,ds.$$

Conclusively, the work done by $\vec{F}$ involves only the tangential component of $\vec{F}$. The normal component of $\vec{F}$ does no work. Furthermore, if $\vec{F}$ represents the velocity field of fluid flow in a region in $\Re^2$ (such as a river), then $\int_{t=a}^{b}\vec{F}\cdot\vec{T}\,ds$ determines the flow and is referred to as a *flow integral*.

If a curve C is *closed*, that is, the initial point is the same as the terminal point, then the flow in the direction of a tangent to C is called the *circulation* of $\vec{F}$ around the curve C. And if the value of the integral is 0, then the fluid flow is said to be *irrotational*, meaning that the tendency of the fluid to rotate is 0. Also, suppose that s is an arc length parameter, C is a counterclockwise oriented closed smooth curve given by $\vec{r}(t)=x(t)\mathbf{i}+y(t)\mathbf{j}$, and $\vec{T}=\dfrac{dx}{ds}\mathbf{i}+\dfrac{dy}{ds}\mathbf{j}$. Then the outward unit normal vector to C is $\vec{N}=\dfrac{dy}{ds}\mathbf{i}-\dfrac{dx}{ds}\mathbf{j}$. Thus, using an integral with a circle on it to represent a closed curve oriented counterclockwise, we can write that

$$\oint_C\vec{F}\cdot\vec{N}\,ds=\oint_C\left[M(x(s),y(s))i+N(x(s),y(s))j\right]\cdot\vec{N}\,ds=\oint_C M(x,y)\,dy-N(x,y)\,dx.$$

These integrals represent an outward *flux*[3] across a closed curve C in the plane. Any parametrization of C in a counterclockwise direction may be used. Flow and flux will be further discussed in Section 12.8.

Exercises 12.5

1. Sketch the vector field given by

(a) $\vec{F}(x,y)=y\mathbf{i}+x\mathbf{j}$. **(b)** $\vec{F}(x,y)=-y\mathbf{i}+x\mathbf{j}$.

(c) $\vec{F}(x,y)=\mathbf{i}+\mathbf{j}$. **(d)** $\vec{F}(x,y)=\dfrac{x}{\sqrt{x^2+y^2}}\mathbf{i}+\dfrac{y}{\sqrt{x^2+y^2}}\mathbf{j}$.

---

[3]*Flux,* meaning *flow* in Latin, measures the rate at which a fluid either enters or leaves a region inside a closed curve in the xy-plane. Here $\vec{N}$ is an outward pointing normal vector. Curiously enough, the term flux is used in electric and magnetic fields even though there is no flow taking place there.

2. Consider the vector field given by $\vec{F}(x,y) = \mathbf{i} + x\mathbf{j}$, which was illustrated in Figure 12.5.1. Give an equation of the flow line that passes through the point $(1,1)$.

3. Consider the vector field given by $\vec{F}(x,y) = -x\mathbf{i} + y\mathbf{j}$. Sketch the flow of the vector field. Also, give an equation of the flow line that passes through the point $(1,2)$ when $t = 0$.

4. If Part 2 of Section 11.8 has been studied, find a flow line through the point $(-1,2)$ when $t = 0$ in the vector field given by $\vec{F}(x,y) = y\mathbf{i} + (6x + y)\mathbf{j}$.

5. Give an example of a vector field whose flow lines are circles with centers at the origin.

6. *(Independence of Parametrization)* If parametrizations for a smooth curve C are equivalent, then the values for $\int_C \vec{F} \cdot d\vec{r}$ are equal.

7. Suppose that $\vec{F}(x,y)$ is a continuous vector field containing a smooth curve C determined by $\vec{r}(t) = x(t)\mathbf{i} + y(t)\mathbf{j}$, with $t \in [a,b]$. Prove that

$$\int_{-C} \vec{F} \cdot d\vec{r} = -\int_C \vec{F} \cdot d\vec{r}.$$

8. Suppose that $\vec{F}(x,y)$ is a continuous vector field containing a smooth curve C determined by $\vec{r}(t) = x(t)\mathbf{i} + y(t)\mathbf{j}$, with $t \in [a,b]$. If $\vec{F}$ is *bounded* by M, meaning that $\left\|\vec{F}\right\| \le M$ for all points on a curve C, and the length of C is bounded by L, prove that

$$\left|\int_C \vec{F} \cdot d\vec{r}\right| \le ML.$$

See Exercise 8 in Section 12.4.

9. Suppose that both $\vec{F}$ and $\vec{G}$ are two continuous vector fields containing a smooth curve C determined by $\vec{r}(t) = x(t)\mathbf{i} + y(t)\mathbf{j}$, with $t \in [a,b]$. If α and β are two real scalars, prove the *linear property*

$$\int_C \left(\alpha\vec{F} + \beta\vec{G}\right) \cdot d\vec{r} = \alpha\int_C \vec{F} \cdot d\vec{r} + \beta\int_C \vec{G} \cdot d\vec{r}.$$

10. Suppose $\vec{F}$ is a continuous vector field containing a piecewise smooth curve C, where $\bigcup_{k=1}^{n} C_k = C$, with C_k parametrized by $\vec{r}(t) = x(t)\mathbf{i} + y(t)\mathbf{j}$ for $t \in \left[t_{k-1}, t_k\right]$. Prove that

$$\int_C \vec{F} \cdot d\vec{r} = \sum_{i=1}^{n} \int_{C_i} \vec{F} \cdot d\vec{r}.$$

11. Suppose that $\vec{F} = \frac{1}{x^2+1}\mathbf{i} + \frac{1}{x^2+1}\mathbf{j}$ and that a curve C is given by $y = x^2$ traced from the origin to the point $(2,4)$. Compute the value of $\int_C \vec{F} \cdot d\vec{r}$. See exercise 3 in Section 12.4. What would the value of this integral be if $\vec{F}$ was given instead by $\vec{F} = \frac{1}{x^2+1}\mathbf{i} + \frac{1}{y+1}\mathbf{j}$?

12. Suppose that $\vec{F}(x,y) = \sqrt{1+y^2}\,\mathbf{i} + x^2 y\,\mathbf{j}$ and that C is given by $y = 1 - x$ from $(0,1)$ to $(1,0)$, and by $y = \frac{x^2}{2} - \frac{1}{2x^2}$ from $(1,0)$ to $\left(2, \frac{15}{8}\right)$. Compute $\int_C \vec{F} \cdot d\vec{r}$.

13. Suppose that C is a vertical line segment in a continuous vector field $\vec{F}(x,y) = M(x,y)\mathbf{i} + N(x,y)\mathbf{j}$.

(a) Show that $\int_C \vec{F} \cdot d\vec{r}$ is independent of M.

(b) Show that if $N(x,y) \equiv 0$, then $\int_C \vec{F} \cdot d\vec{r} = 0$.

(c) Restate and then prove parts (a) and (b) for the case when C is a horizontal line segment.

14. **(a)** If a velocity field is $\vec{F} = x\mathbf{i} + y(1+y^2)\mathbf{j}$, find the circulation of a fluid around the ellipse $\vec{r}(t) = (3\cos t)\mathbf{i} - (2\sin t)\mathbf{j}$, with $t \in [0, 2\pi]$.

(b) If $\vec{F} = (x+y)\mathbf{i} - (x+y)\mathbf{j}$, find the circulation of a fluid around the ellipse $\vec{r}(t) = (2\sin t)\mathbf{i} - (\cos t)\mathbf{j}$, with $t \in [0, 2\pi]$.

15. Find the flux of a vector field $\vec{F} = x\mathbf{i} - y\mathbf{j}$ across the circle $x^2 + y^2 = 1$.

12.6 Gradient Vector Field

Definition 12.6.1. A vector field $\vec{F}$ defined on an open region $D \subseteq \Re^2$ is called a *gradient vector field*, or a *conservative*[4] *vector field*, if and only if $\vec{F}$ is the gradient for some scalar function (also called a scalar field) f defined on D, that is, $\vec{F} = \nabla f$ on D.

---

[4]In a conservative vector field, if a particle moves from one point to another, then the sum of the potential and kinetic energies remains constant, meaning that the total energy does not change. In physics, this is known as the *Law of Conservation of Energy*, from which the term "conservative" for the vector field comes.

A gradient vector field is a very special vector field with nice properties. We will explore some of them. Review Definition 11.4.3 and the fundamental theorem of calculus, Theorem 6.4.2, before going on.

THEOREM 12.6.2. *(Fundamental Theorem of Calculus for Line Integrals) Suppose that D is an open region in $\Re^2$, containing a smooth curve C determined by $\vec{r}(t) = x(t)\mathbf{i} + y(t)\mathbf{j}$ for $t \in [a,b]$. And suppose that $f: D \to \Re$ is continuously differentiable. If the starting point of C is $\vec{a} = \vec{r}(a)$ and the ending point of C is $\vec{b} = \vec{r}(b)$, then*

$$\int_C \nabla f \cdot d\vec{r} = f(\vec{b}) - f(\vec{a}).$$

Proof. $f(x,y)$, $x(t)$, and $y(t)$ are all differentiable and so is the composition $f(x(t), y(t))$. Why? By the chain rule, Theorem 11.6.3, since

$$\frac{d}{dt} f(x(t), y(t)) = \nabla f(x(t), y(t)) \cdot \langle x'(t), y'(t) \rangle,$$

we can write that

$$\begin{aligned}
\int_C \nabla f \cdot d\vec{r} &= \int_a^b \nabla f(x(t), y(t)) \cdot r'(t)\,dt = \int_a^b \nabla f(x(t), y(t)) \cdot \langle x'(t), y'(t) \rangle\,dt \\
&= \int_a^b \frac{d}{dt} f(x(t), y(t))\,dt = f(x(t), y(t))\Big|_a^b \\
&= f(x(b), y(b)) - f(x(a), y(a)) = f(\vec{b}) - f(\vec{a}).
\end{aligned}$$

■

$\nabla f = f_x \mathbf{i} + f_y \mathbf{j}$, if $M(x,y) \equiv f_x(x,y)$ and $N(x,y) \equiv f_y(x,y)$, so we can also write that

$$\int_C \nabla f \cdot d\vec{r} = \int_C M(x,y)\,dx + N(x,y)\,dy = \int_C f_x\,dx + f_y\,dy.$$

Observe that this integral is *path-independent*. That is, it depends only on the functional values at the endpoints of the curve, provided that we know which point is the starting point and which is the ending point, no matter how C is parametrized. If C is a closed smooth curve, then $\int_C \nabla f \cdot d\vec{r} = 0$, which may be written as $\oint_C \nabla f \cdot d\vec{r} = 0$.

In continuing, before computing $\int_C \vec{F} \cdot d\vec{r}$ for some continuous vector field $\vec{F}$ containing a smooth curve C, it is worthwhile to check whether or not $\vec{F}$ is a gradient of some function f. If such an f exists, then f is called a *potential function* for the vector field $\vec{F}$.

THEOREM 12.6.3. *If $\vec{F}(x,y) \equiv M(x,y)\mathbf{i} + N(x,y)\mathbf{j}$ is continuously differentiable on an open region $D \subseteq \Re^2$, and if $M(x,y) \equiv f_x(x,y)$ and $N(x,y) \equiv f_y(x,y)$ for some function f, then $M_y(x,y) = N_x(x,y)$.*

The value $N_x - M_y$ is called a *two-dimensional curl,* or *scalar curl,* of the vector field $\vec{F}$. In physics a curl is known as the *circulation density.* Theorem 12.6.3 says that if $\vec{F} = M\mathbf{i} + N\mathbf{j} = \nabla f$ for some f, then $M_y = N_x$. Thus, if $M_y \neq N_x$, that is, if the curl is not zero, then there is no function f such that $\vec{F} = \nabla f$. Proof of Theorem 12.6.3 is left as an exercise. Is the converse of Theorem 12.6.3 true? See Exercise 6. But, if $M_y = N_x$, then there is a chance that a potential function exists. Example 12.6.5 demonstrates how we can go about finding a potential function. First, however, in one-dimensional calculus, if f is continuous, then an antiderivative of f must exist, and in fact is given by $\int_a^x f(t)\,dt$. Why? In the present case, a vector field $\vec{F}$ plays the role of $f(t)$. However, $\vec{F}$ does not always have a potential function, the equivalent of an "antiderivative" for $\vec{F}$. Exercise 6 shows that even if $M_y = N_x$, it is not necessarily true that $\vec{F} \equiv M\mathbf{i} + N\mathbf{j}$ has a potential function. If, however, a region D is a *simply connected* open region, intuitively meaning that the open region D has no "holes," then the converse of Theorem 12.6.3 holds. Section 12.7 picks up this discussion. The following result was first proven by Euler in 1734.

THEOREM 12.6.4. *Suppose that $\vec{F}(x,y) \equiv M(x,y)\mathbf{i} + N(x,y)\mathbf{j}$ is continuously differentiable on a rectangle R. If $M_y = N_x$ on R, then there exists a potential function f for $\vec{F}$.*

Proof. We must find a function $f(x,y)$, with $(x,y) \in R$, such that $f_x(x,y) = M(x,y)$ and $f_y(x,y) = N(x,y)$. Now $f_x(x,y) = M(x,y)$ if

$$f(x,y) = \int M(x,y)\,\partial x + C(y),$$

as we have seen at the end of Section 11.3. Therefore,

$$f_y(x,y) = \frac{\partial}{\partial y}\int M(x,y)\,\partial x + C'(y).$$

If $f_y(x,y) = N(x,y)$, then we must have

$$N(x,y) = \frac{\partial}{\partial y}\int M(x,y)\,\partial x + C'(y),$$

which would give

$$C'(y) = N(x,y) - \frac{\partial}{\partial y}\int M(x,y)\,\partial x.$$

But this is true only if the right hand side is independent of x. A verification resides in solving Exercise 2. Then

$$C(y) = \int\left[N(x,y) - \frac{\partial}{\partial y}\int M(x,y)\,\partial x\right]dy.$$

Hence, $f(x,y) = \int M(x,y)\,\partial x + C(y)$, with $C(y)$ as given above, is a potential function for $\vec{F}$. ■

Theorem 12.6.4 is the converse of Theorem 12.6.3 but on a more restrictive region. The proof above for Theorem 12.6.4 outlines a procedure that can be followed to actually compute a potential function. See Exercise 12 for an alternate proof of Theorem 12.6.4. Also, where in the proof of Theorem 12.6.4 was it necessary that R be a rectangle? Would a proof still be valid if the rectangle R was replaced by an arbitrary convex region?

Example 12.6.5. Consider the vector field

$$\vec{F}(x,y) = M(x,y)\mathbf{i} + N(x,y)\mathbf{j} = \left(\sin x + 2xe^x y + x^2e^x y\right)\mathbf{i} + \left(x^2e^x + 1\right)\mathbf{j}.$$

Is $\vec{F}$ conservative on $\Re^2$? If so, find a potential function.

Answer. If $M(x,y) = \sin x + 2xe^x y + x^2e^x y$ and $N(x,y) = x^2e^x + 1$, then $M_y(x,y) = 2xe^x + x^2e^x = N_x(x,y)$. And therefore, since $\vec{F}$ is continuously differentiable on any rectangle R in $\Re^2$, according to Theorem 12.6.4, a potential function for $\vec{F}$ exists on $\Re^2$. We are looking for f such that $f_x = M$ and $f_y = N$ on $\Re^2$. Since integration of M with respect to x is lengthy, we will integrate N with respect to y and obtain

$$f(x,y) = \int N(x,y)\,\partial y + C(x) = x^2e^x y + y + C(x).$$

$f_x = M$, so we have

$$f_x(x,y) = 2xe^x y + x^2e^x y + C'(x),$$

which is equal to M if $C'(x) = \sin x$. Therefore, choose $C(x) = -\cos x$. Hence, a potential function for $\vec{F}$ is $f(x,y) = x^2e^x y + y - \cos x$, making $\vec{F}$ a conservative vector field on $\Re^2$. ■

By the fundamental theorem of calculus for line integrals, if $\vec{F}$ is a conservative vector field on $D \subseteq \Re^2$, then $\vec{F}$ can be integrated along most sophisticated curves simply by evaluating $\vec{F}$ at the curve's endpoints. However, the curve must not go outside the region D where $\vec{F}$ is not continuously differentiable. For example, a path in $\Re^2$ for the conservative vector field in Exercise 3(d) cannot cross the y-axis since $\vec{F}$ fails to be continuously differentiable at any of those points.

In correlation with the fundamental theorem for line integrals the concept of path independence was introduced.

Definition 12.6.6. Suppose that $\vec{F}(x,y)$ is a vector field defined on an open region $D \subseteq \Re^2$. Then $\vec{F}$ is *path independent* in D if and only if for any two smooth curves C_1 and C_2 in D connecting $\vec{a}$ to $\vec{b}$, we have

$$\int_{C_1} \vec{F}\cdot d\vec{r} = \int_{C_2} \vec{F}\cdot d\vec{r}.$$

Clearly gradient (conservative) vector fields are path independent, as the fundamental theorem for line integrals demonstrates. In fact, no other vector fields are path independent. The next theorem demonstrates how a potential function can be obtained given that the vector field $\vec{F}$ is path independent. In one-dimensional calculus, Theorem 12.6.7 is related to Theorem 6.4.4.

THEOREM 12.6.7. *Suppose that $\vec{F}(x,y) = M(x,y)\mathbf{i} + N(x,y)\mathbf{j}$ is a continuous vector field on an open region $D \subseteq \Re^2$. If $\vec{F}$ is path independent, then $\vec{F}$ has a potential function f. In fact,*

$$f(x,y) = \int_C \vec{F}\cdot d\vec{r} = \int_{(a,b)}^{(x,y)} \vec{F}\cdot d\vec{r},$$

where C is any curve connecting an arbitrary point (a,b) to (x,y) in D.

Proof. Let $P(a,b)$ be any point in D, and let

$$f(x,y) = \int_{(a,b)}^{(x,y)} \vec{F}\cdot d\vec{r} = \int_C \vec{F}\cdot d\vec{r}$$

for all $(x,y) \in D$. We will show that $\nabla f = \vec{F}$. $\vec{F}$ is path independent, so it does not matter which path from (a,b) to (x,y) is used to evaluate $\int_C \vec{F}\cdot d\vec{r}$, meaning $f(x,y)$. Thus, f is well-defined. Now, since D is open, there exists an open ball $B \subseteq D$ about (x,y). Let (x_1,y) be any point in B with $x_1 < x$. And define the curve C from (a,b) to (x,y) to be piecewise smooth and to be made up of paths C_1 and C_2, where C_1 is any path from (a,b) to (x_1,y), and C_2 is a horizontal line segment from (x_1,y) to (x,y). By Exercise 10 of Section 12.5, the value of the line integral along $C_1 + C_2$ is the same as along the curve C. Thus, we have that

$$f(x,y)=\int_{C_1}\vec{F}\cdot d\vec{r}+\int_{C_2}\vec{F}\cdot d\vec{r}=\int_{(a,b)}^{(x_1,y)}\vec{F}\cdot d\vec{r}+\int_{C_2}\vec{F}\cdot d\vec{r}.$$

The first of the integrals above is independent of x, so we have

$$f_x(x,y)=0+\frac{\partial}{\partial x}\int_{C_2}\vec{F}\cdot d\vec{r}.$$

If C_2 is parametrized by $\gamma_1(t)=(t,y)$ for $t\in[x_1,x]$, then

$$f_x(x,y)=\frac{\partial}{\partial x}\int_{C_2}\vec{F}\cdot d\vec{r}=\frac{\partial}{\partial x}\int_{x_1}^{x}\langle M(t,y),N(t,y)\rangle\cdot\langle 1,0\rangle\,dt=\frac{\partial}{\partial x}\int_{x_1}^{x}M(t,y)\,dt=M(x,y),$$

where the last equality follows from Theorem 6.4.7. Similarly, though using a vertical line segment instead of horizontal, we have $f_y(x,y)=N(x,y)$. For details see Exercise 9. Therefore, $\vec{F}=M\mathbf{i}+N\mathbf{j}=f_x\mathbf{i}+f_y\mathbf{j}=\nabla f$. And thus, f is a potential function for $\vec{F}$, meaning that $\vec{F}$ is conservative. ■

THEOREM 12.6.8. *If $\vec{F}$ is a continuous vector field on an open region $D\subseteq\Re^2$, then $\vec{F}$ is path independent if and only if $\int_C\vec{F}\cdot d\vec{r}=0$ for every closed curve C in D.*

Proof. ($\Rightarrow$) Suppose that $\vec{F}$ is path independent in D, and let C be an arbitrary curve in D. Choose any two points $\vec{a}$ and $\vec{b}$ on C. Then divide C into two parts, C_1 and C_2 which connect the points $\vec{a}$ and $\vec{b}$, where C_1 goes from $\vec{a}$ to $\vec{b}$ and C_2 goes from $\vec{b}$ to $\vec{a}$. Then, by Exercise 10 in Section 12.5,

$$\int_C\vec{F}\cdot d\vec{r}=\int_{C_1}\vec{F}\cdot d\vec{r}+\int_{C_2}\vec{F}\cdot d\vec{r}.$$

$\vec{F}$ is path independent, and both C_1 and $-C_2$ go from $\vec{a}$ to $\vec{b}$, so implementing Exercise 7 from Section 12.5, we have that

$$\int_{C_1}\vec{F}\cdot d\vec{r}=\int_{-C_2}\vec{F}\cdot d\vec{r}=-\int_{C_2}\vec{F}\cdot d\vec{r}.$$

Therefore, $\int_C\vec{F}\cdot d\vec{r}=0$.

Exercise 11 requests a proof of the converse to Theorem 12.6.8. ■

Remark 12.6.9. Combining Theorems 12.6.2, 12.6.7, and 12.6.8 for a continuous vector field $\vec{F}$ on an open region $D \subseteq \Re^2$, the following are equivalent.

(a) $\vec{F}$ is a conservative vector field.

(b) $\vec{F}$ is a gradient field.

(c) There exists a potential function f for $\vec{F}$.

(d) $\vec{F}$ is path independent.

(e) $\int_C \vec{F} \cdot d\vec{r} = 0$ for every closed curve C in D. C need not be smooth. ■

Exercises 12.6

1. Prove Theorem 12.6.3.

2. In the proof of Theorem 12.6.4, verify that $N(x,y) - \frac{\partial}{\partial y}\int M(x,y)\,\partial x$ is independent of x.

3. Determine whether or not the given vector field is conservative in the given region D. If it is, find a potential function.

(a) $\vec{F}(x,y) = y^2\mathbf{i} + 2xy\mathbf{j}$, with $D = \Re^2$

(b) $\vec{F}(x,y) = \left(3y + 4xy^2\right)\mathbf{i} + \left(3x + 3x^2y\right)\mathbf{j}$, with $D = \Re^2$

(c) $\vec{F}(x,y) = \left(3x^2y - \cos x + 4\right)\mathbf{i} + \left(x^3 - e^y\right)\mathbf{j}$, with $D = \Re^2$

(d) $\vec{F}(x,y) = \left(x^{-2} + y^3\right)\mathbf{i} + \left(3xy^2 - 1\right)\mathbf{j}$, with $D = \Re^2 \setminus (y\text{-axis})$

4. Compute $\int_C \vec{F} \cdot d\vec{r}$ for the given vector field $\vec{F}$ along the specified curve C.

(a) $\vec{F}(x,y) = y^2\mathbf{i} + 2xy\mathbf{j}$, with C a parabolic curve from the point $(0,0)$ to the point $(3,1)$. See Exercise 3(a).

(b) $\vec{F}(x,y) = x\mathbf{i} + (x+y)\mathbf{j}$, with C being the unit circle $x^2 + y^2 = 1$ traced once in a counterclockwise direction.

(c) $\vec{F}(x,y) = \left(e^x \sin x + 2xe^x y + x^2 e^x y\right)\mathbf{i} + \left(x^2 e^x + \sqrt{y^3 + 1}\right)\mathbf{j}$, with C being the ellipse $x^2 + xy + y^2 = 1$ traced clockwise.

(d) $\vec{F}(x,y) = \left(y^2 + 3x^2y - \cos x + 4\right)\mathbf{i} + \left(x^3 - e^y\right)\mathbf{j}$, with C being parametrized by $\gamma(t) = (\sin t, -\cos t)$, for $t \in [0, 2\pi]$.

5. **(a)** Find the work done by the force field $\vec{F}(x,y) = x^2y\mathbf{i} + xy^2\mathbf{j}$ in moving an object along a line segment from the point $(0,0)$ to the point $(2,3)$.

(b) Find the work done by the force field $\vec{F}(x,y) = xy^2\mathbf{i} + x^2y\mathbf{j}$ in moving an object along a line segment from the point $(0,0)$ to the point $(2,3)$.

6. Consider the vector field

$$\vec{F}(x,y) = \frac{-y}{x^2+y^2}\mathbf{i} + \frac{x}{x^2+y^2}\mathbf{j}$$

on a *punctured plane* $\Re^2 \setminus \{(0,0)\}$. See exercise 17 in Section 12.2, Exercise 7 in Section 12.7, and Example 12.7.4.

(a) If $\vec{F} = M\mathbf{i} + N\mathbf{j}$, show that $M_y = N_x$, that is, show that the curl of $\vec{F}$ is 0.

(b) Show that $\vec{F}$ is not path independent, by showing that $\int_C \vec{F}\cdot d\vec{r} \neq 0$ for C being the unit circle centered at $(0,0)$.

(c) Is $\vec{F}$ conservative? Explain.

(d) Is Theorem 12.6.4 contradicted? Explain.

(e) As in part (b), let C be the unit circle $x^2 + y^2 = 1$ traced counterclockwise. Is $\int_C \vec{F}\cdot d\vec{r} = \|\vec{F}\| \cdot (\text{length of } C)$? If so, what is this value?

7. Consider the vector field

$$\vec{F}(x,y) = \frac{x}{x^2+y^2}\mathbf{i} + \frac{y}{x^2+y^2}\mathbf{j}$$

on a punctured plane $\Re^2 \setminus \{(0,0)\}$.

(a) If $\vec{F} = M\mathbf{i} + N\mathbf{j}$, show that $M_y = N_x$, that is, show that the curl of $\vec{F}$ is 0.

(b) Show that $\int_C \vec{F}\cdot d\vec{r} = 0$, where C is the unit circle centered at $(0,0)$.

(c) Do parts (a) and/or (b) imply that $\vec{F}$ is conservative? Explain.

(d) Determine whether or not $f(x,y) = \frac{1}{2}\ln\left(x^2+y^2\right)$ is a potential function for $\vec{F}$.

(e) Is $\vec{F}$ conservative? Explain.

8. If $\vec{r} = x\mathbf{i} + y\mathbf{j} \neq \vec{0}$ and c is a scalar, a vector field $\vec{F}(x,y)$ defined by

$$\vec{F}(x,y) = \frac{c}{\|\vec{r}\|^3}\vec{r}$$

is called an *inverse square field.* Show that $f(x,y) = -\frac{c}{\|\vec{r}\|}$ is a potential function for $\vec{F}$.

9. Complete the proof of Theorem 12.6.7.

10. Consider $\vec{F}(x,y) = y^2\mathbf{i} + 2xy\mathbf{j}$. Find a potential function for $\vec{F}$ using Theorem 12.6.7. Compare your answer here with that given to exercise 3(a).

11. Complete the proof of Theorem 12.6.8.

12. Reprove Theorem 12.6.4 by showing that if $M_y = N_x$, then there exists a potential function f, which is given by $f(x,y) = \int_a^x M(u,b)\,du + \int_b^y N(x,v)\,dv$ with (a,b) any point in the rectangle.

12.7 Green's Theorem

We now turn our attention to the relationship of the line integral $\oint_C \vec{F}\cdot d\vec{r}$ of a vector field $\vec{F} = M\mathbf{i} + N\mathbf{j}$, around the boundary C of a closed region D in $\Re^2$, to the double integral $\iint_D N_x - M_y$, where $N_x - M_y$ is called the curl of $\vec{F}$, denoted $\operatorname{curl}\vec{F}$, as defined in Section 12.6. We will first consider the simple regions D described at the end of Section 12.3. More general regions will be discussed later in this section, where they will be subdivided into simple regions.

For line integrals $\oint_C \vec{F}\cdot d\vec{r}$, the orientation of a curve C is important. If C, being the boundary of a closed region D in $\Re^2$, is traced in such a way that the region is kept on the left, then C is *positively oriented.* Does this mean that C is always traced counterclockwise? What if D has a hole in it?

THEOREM 12.7.1. *(Green's[5] Theorem–circulation/curl form) Suppose that D is a simple region in $\Re^2$ with its boundary a closed positively oriented curve C. If $\vec{F}(x,y) = M(x,y)\mathbf{i} + N(x,y)\mathbf{j}$ is a continuously differentiable vector field on an open region containing D, then*

---

[5]George Green (1793–1841), a mathematically self-taught English windmill keeper made significant contributions to mathematics in the areas of analysis and applied mathematics. Green's first and most interesting paper, which went unnoticed for almost 25 years, was published at his own expense in 1828. Green's other papers included contributions to hydrodynamics, electricity, magnetism, and the reflection and refraction of light. Green's functions, Neumann's functions, and Robin's functions are often referred to as Green's functions of the first, second, and third kinds, respectively. Theorem 12.7.1, known as Green's theorem, was proved simultaneously by the Russian mathematician Mikhail Ostrogradsky in 1828.

$$\oint_C M\,dx + N\,dy = \iint_D \left(N_x - M_y\right) dx\,dy.$$

Recall that $\vec{F}$ continuously differentiable means that the first partials of M and N are continuous. Also, the line integral with a circle on the integral sign represents a closed positively oriented curve C.

Proof of Green's theorem. We need to consider two cases, one for each type of a simple region.

Case 1. Suppose that D is a Jordan measurable set of the form

$$D = \{(x,y) \mid a \le x \le b \text{ and } g_1(x) \le y \le g_2(x)\},$$

where g_1 and g_2 are continuous on $[a,b]$. D was called the region of the *first type* in Section 12.3. Due to the linearity of both line integrals and double integrals, Green's theorem can be proven by showing that

$$\oint_C M\,dx = -\iint_D M_y \text{ and } \oint_C N\,dy = \iint_D N_x.$$

Since each double integral can be written as an iterated integral, using the fundamental theorem of calculus, Theorem 6.4.2, we begin by writing

$$\iint_D M_y = \int_a^b \int_{g_1(x)}^{g_2(x)} M_y(x,y)\,dy\,dx = \int_a^b \left[M(x,g_2(x)) - M(x,g_1(x))\right] dx.$$

Now, consider $\oint_C M\,dx$. Since C may be piecewise smooth and composed of as many as four curves, we may need to write that

$$\oint_C M\,dx = \int_{C_1} M\,dx + \int_{C_2} M\,dx + \int_{C_3} M\,dx + \int_{C_4} M\,dx.$$

If this is the case, suppose that C_1 is traced by $y = g_1(x)$. Then C_2 and C_4, say, may be vertical line segments. And from Exercise 13 of Section 12.5,

$$\int_{C_2} M\,dx = 0 = \int_{C_4} M\,dx.$$

Now, parametrize the curves C_1 and $-C_3$ by $\gamma_1(t) = (t, g_1(t))$ with $t \in [a,b]$, and $\gamma_3(t) = (t, g_2(t))$ with $t \in [a,b]$, respectively. Keep in mind that C_3 must be positively oriented, and that γ_3 traces C_3 backwards. Why? Thus, we write

$$\oint_C M\,dx = \int_{C_1} M\,dx + \int_{C_3} M\,dx = \int_a^b M(t,g_1(t))\,dt - \int_a^b M(t,g_2(t))\,dt$$
$$= \int_a^b \left[M(t,g_1(t)) - M(t,g_2(t))\right]dt = -\iint_D M_y .$$

Exercise 1 requests a proof for $\oint_C N\,dy = \iint_D N_x$.

Case 2. Suppose that D is a Jordan measurable set of the form

$$D = \{(x,y) \mid c \le y \le d \text{ and } h_1(y) \le x \le h_2(y)\},$$

where h_1 and h_2 are continuous on $[c,d]$. Exercise 2 requests the details.

Thus, the proof is complete. ■

Remark 12.7.2

- **(a)** Green's theorem can at times be used to evaluate complicated line integrals by calculating easier iterated integrals. See Exercise 3.
- **(b)** A single iterated integral can, perhaps, be used to calculate a line integral along a piecewise smooth curve. See Exercise 4.
- **(c)** Evaluation of a double integral may be simplified using Green's theorem. See Exercises 5 and 6.
- **(d)** Green's theorem can be used to calculate the area of a region just by considering its boundary. See Remark 12.7.3. ■

Remark 12.7.3. Suppose that all of the conditions for Green's theorem are satisfied. $\iint_D 1\,dA$ is the area of D, so Green's theorem may be used to compute it. Observe that the curl $\vec{F} = 1$, that is, $N_x - M_y = 1$. There are a number of choices for such an M and N. For instance, we can choose $M(x,y) = 0$ and $N(x,y) = x$, or $M(x,y) = -y$ and $N(x,y) = 0$, or $M(x,y) = -\frac{1}{2}y$ and $N(x,y) = \frac{1}{2}x$. In view of Green's theorem, these three possibilities yield that the area A is given by

$$A = \iint_D 1\,dA = \oint_C x\,dy = -\oint_C y\,dx = \frac{1}{2}\oint_C x\,dy - y\,dx .$$ ■

Suppose that $\vec{G}$ is a vector field defined by $\vec{G}(x,y) = N(x,y)\mathbf{i} - M(x,y)\mathbf{j}$. Then, by Green's theorem we

$$\oint_C N\,dx - M\,dy = \iint_D \left(-M_x - N_y\right) dx\,dy.$$

Therefore, Green's theorem may be rewritten to have the conclusion

$$\oint_C M\,dy - N\,dx = \iint_D \left(M_x + N_y\right) dx\,dy.$$

If this is the case, Green's theorem is in the *flux/divergence form.* Forms for Green's theorem are called in this manner because of the following:

$\oint_C M\,dx + N\,dy = \oint_C \vec{F} \cdot \vec{T}\,ds$ gives the *counterclockwise circulation,*

$\oint_C M\,dy - N\,dx = \oint_C \vec{F} \cdot \vec{N}\,ds$ gives the *outward flux,*

$\iint_D \left(N_x - M_y\right) dA = \iint_D \operatorname{curl} \vec{F}\,dA$ is the *curl integral,* and

$\iint_D \left(M_x + N_y\right) dA = \iint_D \operatorname{div} \vec{F}\,dA$ is the *divergence integral.*

The concept of divergence is further discussed in Section 12.8. Also, integrals in vector form can easily be generalized to higher dimensions. Does Green's theorem hold for more general regions in $\Re^2$ rather than just for simple regions? It turns out that if a more general region can be decomposed into a finite number of simple regions, then Green's theorem will still hold.[6] As an example, suppose for now that a nonsimple region D with a smooth boundary $\partial D = C$ is as illustrated in Figure 12.7.1. D is subdivided into smaller regions whose union is D. In Figure 12.7.1, $D = D_1 \cup D_2$. Also, note that each D_1 and D_2 are simple, with top and bottom curves piecewise smooth, $C_1 \cup C_3$ the boundary of D_1, $C_2 \cup (-C_3)$ the boundary of D_2, and both positively oriented. Then, Green's theorem applies to both D_1 and to D_2, and therefore,

$$\oint_{C_1 \cup C_3} M\,dx + N\,dy = \iint_{D_1} \left(N_x - M_y\right) dA, \text{ and } \oint_{C_2 \cup (-C_3)} M\,dx + N\,dy = \iint_{D_2} \left(N_x - M_y\right) dA.$$

Now, since the line integrals along C_3 and $-C_3$ are opposite in sign, under addition they cancel to yield

[6]Green's theorem holds even for more general regions than given. All that is required is that the boundary of a bounded closed region consists of a finite number of rectifiable curves so that a line integral makes sense.

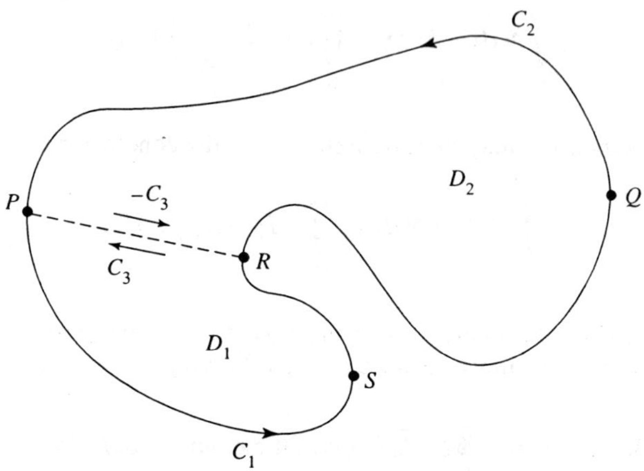

Figure 12.7.1

$$\oint_{C_1 \dot{\cup} C_2} M\,dx + N\,dy = \iint_{D_1 \cup D_2} \left(N_x - M_y\right) dA,$$

which is equivalent to

$$\oint_C M\,dx + N\,dy = \iint_D \left(N_x - M_y\right) dA,$$

the conclusion of Green's theorem. A similar argument may be applied to any region that can be subdivided into a finite number of simple regions, each with piecewise smooth boundaries. Note that not every simply connected bounded region can be subdivided into a finite number of simple regions. For example, the set S defined by

$$S = \left\{(x, y) \mid x \in [0,1], y \in \left[f(x), 2\right]\right\},$$

where

$$f(x) = \begin{cases} x \sin\dfrac{1}{x}, & \text{if } x \in (0,1] \\ 0, & \text{if } x = 0, \end{cases}$$

cannot be subdivided into a finite number of simple regions. Why not?

Example 12.7.4. Consider the vector field

$$\vec{F}(x, y) = \frac{-y}{x^2 + y^2}\mathbf{i} + \frac{x}{x^2 + y^2}\mathbf{j},$$

as given in Exercise 6 of Section 12.6. Show that $\oint_C \vec{F} \cdot d\vec{r} = 2\pi$ for every closed smooth curve C around the origin.

Answer. Let C be any closed smooth curve around the origin. Observe that $\vec{F}$ is not conservative and that Green's theorem does not apply to any region containing the origin. Why not? In addition, what is a parametrization for C? To get around all of these obstacles, consider the region D as in Figure 12.7.2, where C_1 is a circle with center $(0,0)$ and with a radius a small enough that it lies "inside" C.

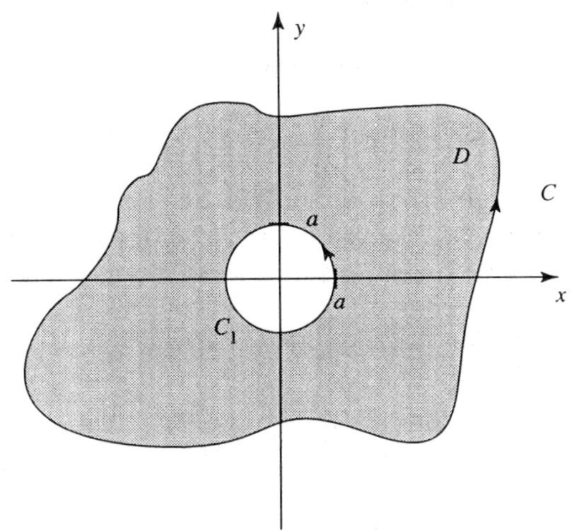

Figure 12.7.2

Then Green's theorem applies to D, which has as its boundary $C \cup (-C_1)$. And by Exercise 10 in Section 12.5 we have that

$$\oint_{C\cup(-C_1)} \vec{F}\cdot d\vec{r} = \oint_C M\,dx + N\,dy + \oint_{-C_1} M\,dx + N\,dy = \iint_D \left(N_x - M_y\right) dA$$

$$= \iint_D \left[\frac{y^2 - x^2}{\left(x^2+y^2\right)^2} - \frac{y^2 - x^2}{\left(x^2+y^2\right)^2}\right] dA = 0.$$

But, from Exercise 7 in Section 12.5,

$$\oint_C M\,dx + N\,dy = \oint_{C_1} M\,dx + N\,dy.$$

And in using Exercise 6 of Section 12.5, the second integral is easily computed using the parametrization $\gamma(t) = (a\cos t, a\sin t)$, for $t \in [0, 2\pi]$. Now the value is obtained by writing

$$\oint_C \vec{F}\cdot d\vec{r} = \oint_{C_1} \vec{F}\cdot d\vec{r} = \int_0^{2\pi} \left[M(x(t),y(t))x'(t) + N(x(t),y(t))y'(t)\right] dt = 2\pi. \qquad \blacksquare$$

Remark 12.7.5. Assume that $u(x, y)$ has continuous first and second partial derivatives in an open region R a subset of the xy-plane, containing a region D, which can be subdivided into simple regions, each with piecewise smooth boundaries. Define $M(x, y) = -u_y$ and $N(x, y) = u_x$. Then

$$\vec{F}(x, y) = M(x, y)\mathbf{i} + N(x, y)\mathbf{j} = -u_y\mathbf{i} + u_x\mathbf{j}$$

is a continuously differentiable vector field with a

$$\text{curl } \vec{F} = N_x - M_y = u_{xx} + u_{yy} \equiv \nabla^2 u,$$

where $\nabla^2 u$ is the *Laplacian* of u, defined in Section 11.3. Suppose that C is a positively oriented boundary of D given by $\vec{r}(t) = x(t)\mathbf{i} + y(t)\mathbf{j}$. And suppose that s is the arc length of C. Then, by Green's theorem we have that

$$\begin{aligned}\iint_D \nabla^2 u \, dx\, dy &= \oint_C M\, dx + N\, dy = \oint_C \left[Mx'(s) + Ny'(s)\right] ds \\ &= \oint_C \left(-\frac{\partial u}{\partial y}\frac{dx}{ds} + \frac{\partial u}{\partial x}\frac{dy}{ds}\right) ds = \oint_C \left(\frac{\partial u}{\partial x}\mathbf{i} + \frac{\partial u}{\partial y}\mathbf{j}\right)\cdot\left(\frac{dy}{ds}\mathbf{i} - \frac{dx}{ds}\mathbf{j}\right) ds \\ &= \oint_C \nabla u \cdot \vec{N}\, ds = \oint_C \frac{\partial u}{\partial \vec{N}}\, ds,\end{aligned}$$

where $\vec{N} = \dfrac{dy}{ds}\mathbf{i} - \dfrac{dx}{ds}\mathbf{j}$. $\vec{N}$ is clearly a unit vector since $\dfrac{d\vec{r}}{ds} = \dfrac{dx}{ds}\mathbf{i} + \dfrac{dy}{ds}\mathbf{j}$ is a unit tangent vector to C. Why? And $\vec{N}$ is normal since $\dfrac{d\vec{r}}{ds} \cdot \vec{N} = 0$. In fact, $\vec{N}$ is an outward unit normal vector to C, that is, $\vec{N}$ is directed to the exterior of C. Also, $\dfrac{\partial u}{\partial \vec{N}}$ represents a directional derivative. ■

Section 12.6 commented that the converse of Theorem 12.6.3 holds, but only if D is an open region with no holes. More formally, D must be an open simply connected region. A *simply connected* region D means that every simple closed curve in D also has its *interior* in D. A *simple curve* is a curve which does not intersect itself between its endpoints. This in turn means that D encloses a closed region that lies in D. In order to apply Green's theorem to smooth closed curves C, their interior must be in the domain of a vector field $\vec{F}$.

THEOREM 12.7.6. *(Curl Test) Suppose $\vec{F} = M\mathbf{i} + N\mathbf{j}$ is a continuously differentiable vector field on an open simply connected region $D \subseteq \Re^2$. If $M_y = N_x$, then $\vec{F}$ is conservative.*

Sketch of a proof. In view of Remark 12.6.9, we will show that $\int_C \vec{F}\cdot d\vec{r} = 0$ for every closed curve C in D. First, consider any simple closed curve C in D. Let R be the region enclosed by C. Since D is simply connected, $R \subseteq D$. Whether C is positively oriented or not, by Green's theorem

$$\oint_C \vec{F} \cdot d\vec{r} = \oint_C M\,dx + N\,dy = \iint_R \left(N_x - M_y\right) dA = \iint_R 0\,dA = 0.$$

Now, suppose that C is not simple and that it crosses itself a finite number of times at points $P_1, P_2, \dots, P_n$, with $n \in N$. Then C can be broken into n simple closed curves C_n. And in applying Green's theorem to each C_n, we can conclude that

$$\oint_C \vec{F} \cdot d\vec{r} = \oint_{C_1} \vec{F} \cdot d\vec{r} + \cdots + \oint_{C_2} \vec{F} \cdot d\vec{r} = 0.$$

The limiting process covers the case for a closed curve that intersects itself infinitely many times. ■

Exercises 12.7

1. Complete the proof of Case 1 in the proof of Theorem 12.7.1.
2. Prove Case 2 in the proof of Theorem 12.7.1.
3. Evaluate $\oint_C \vec{F} \cdot d\vec{r}$, where $\vec{F}(x, y) = \left(e^x \sin x + y\right)\mathbf{i} + \left(2x - \sqrt{y^3 + 1}\right)\mathbf{j}$, and C is the circle $x^2 + y^2 = 4$.
4. Evaluate $\oint_C 3x^2 y\,dx - 2xy^2 dy$, where C is a trapezoid with corners $(-2,2)$, $(-1,1)$, $(1,1)$, and $(2,2)$.
5. Suppose that $\vec{F}(x, y) = M(x, y)\mathbf{i} + N(x, y)\mathbf{j}$ is a continuous vector field. Suppose also that C is a closed smooth curve in $\Re^2$ that encloses a region D. Show that if $\vec{F}(x, y) = 0$ for all (x, y) on C, then $\iint_D \left(N_x - M_y\right) dA = 0$.
6. Suppose that D is the region bounded by the curves $x = y^2 - 1$ and $x = 3$. Also suppose that C is the boundary of D traversed clockwise. If $\vec{F}(x, y) = \left(\sin e^x\right)\mathbf{i} + \left(x - \arctan y^2\right)\mathbf{j}$, find $\int_C \vec{F} \cdot d\vec{r}$.
7. Consider the vector field

$$\vec{F}(x, y) = \frac{-y}{x^2 + y^2}\mathbf{i} + \frac{x}{x^2 + y^2}\mathbf{j},$$

as given in both Exercise 6 in Section 12.6 and Example 12.7.4.

(a) Show that $\iint\limits_D \operatorname{curl} \vec{F}\, dA = 0$, where D is a disk with boundary $x^2 + y^2 = 1$.

(b) Show that $\oint\limits_C \vec{F} \cdot d\vec{r} = 2\pi$, where C is the circle $x^2 + y^2 = 1$. Does this contradict Green's theorem? Explain.

(c) Show that $\oint\limits_C \vec{F} \cdot d\vec{r} = 0$ for every closed smooth curve C that neither passes through nor encloses the origin.

(d) Suppose that D is the *annular region* between the circles $x^2 + y^2 = 1$ and $x^2 + y^2 = 5$ with boundary C. Can Green's theorem be used to calculate $\oint\limits_C \vec{F} \cdot d\vec{r}$? If so, calculate this value.

8. Use Green's theorem to find the area of an astroid given by $\vec{r}(t) = \left(\cos^3 t\right)\mathbf{i} + \left(\sin^3 t\right)\mathbf{j}$, with $t \in [0, 2\pi]$. See Exercise 12 in Section 10.4.

9. Find the area inside the ellipse $\dfrac{x^2}{a^2} + \dfrac{y^2}{b^2} = 1$.

10. If the optional Section 10.9 on polar coordinates was covered, find the area of the cardioid $r = a(1 - \cos\theta)$, with $\theta \in [0, 2\pi]$, using a line integral formula given in Remark 12.7.3.

11. If $u(x, y) = x^2 + 2y^2$ and C is the boundary of a triangle whose vertices are $(1, 0)$, $(0, 2)$, and $(-1, 0)$, evaluate $\oint\limits_C \dfrac{\partial u}{\partial \vec{N}}\, ds$ as described in Remark 12.7.5.

12. Suppose that $u(x, y)$ is a harmonic function. See the footnote to Exercise 5 in Section 11.3. Let C be the boundary of a simple region $D \subseteq \Re^2$. If $\vec{N}$ is as given in Remark 12.7.5, prove that

(a) $\oint\limits_C \dfrac{\partial u}{\partial \vec{N}}\, ds = 0$, and that

(b) $\oint\limits_C u \dfrac{\partial u}{\partial \vec{N}}\, ds = \iint\limits_D (u_x)^2 + \left(u_y\right)^2 dA$.

13. Give an example of a curve that both is closed and intersects itself infinitely many times.

14. Evaluate $\int\limits_C \sin y\, dx + x \cos y\, dy$, where C is the boundary of the square $|x| + |y| = 1$.

15. Suppose that $D \subseteq \Re^2$ is a simple region, $\vec{F}$ is continuously differentiable on $D \cup \partial D$, and $\vec{F}$ is constant on ∂D. Prove that $\iint\limits_D \text{curl}\,\vec{F}\,dA = 0$.

12.8* Stokes' and Gauss' Theorems

In $\Re^3$, a vector field $\vec{F}$ is defined by

$$\vec{F}(x,y,z) = M(x,y,z)\mathbf{i} + N(x,y,z)\mathbf{j} + P(x,y,z)\mathbf{k},$$

where M, N, and P have partial derivatives. Also, a *three-dimensional curl* of $\vec{F}$ is a vector defined by

$$\text{curl}\,\vec{F} = \left(P_y - N_z\right)\mathbf{i} + \left(M_z - P_x\right)\mathbf{j} + \left(N_x - M_y\right)\mathbf{k}.$$

Thus, the $\text{curl}\,\vec{F} = \nabla \times \vec{F}$, where the *vector differential operator* ∇ is defined by

$$\nabla = \frac{\partial}{\partial x}\mathbf{i} + \frac{\partial}{\partial y}\mathbf{j} + \frac{\partial}{\partial z}\mathbf{k}.$$

See part (c) of Remark 11.4.4. Therefore, the $\text{curl}\,\vec{F}$ is a vector obtained by a formal expansion of the "determinant"

$$\begin{vmatrix} \mathbf{i} & \mathbf{j} & \mathbf{k} \\ \dfrac{\partial}{\partial x} & \dfrac{\partial}{\partial y} & \dfrac{\partial}{\partial z} \\ M & N & P \end{vmatrix}$$

about the first row. Also, $\left(\text{curl}\,\vec{F}\right)\cdot\vec{k} = \left(\nabla\times\vec{F}\right)\cdot\vec{k} = N_x - M_y$, which earlier was referred to as a scalar curl. Thus, if $\vec{F}(x,y) = M\mathbf{i} + N\mathbf{j} + 0\mathbf{k}$ and $\vec{T} = \frac{dx}{ds}\mathbf{i} + \frac{dy}{ds}\mathbf{j}$, where s is the arc length parameter, then Green's theorem in a circulation/curl form can be rewritten as

$$\oint\limits_C \vec{F}\cdot\vec{T}\,ds = \iint\limits_D \left(\text{curl}\,\vec{F}\right)\cdot\vec{k}\,dA = \iint\limits_D \left(\nabla\times\vec{F}\right)\cdot\vec{k}\,dA,$$

which is called *Stokes'*[7] *Theorem* in the plane. Now, if C is a circle C_r with a small radius r centered at a point (a,b), then

$$\oint_{C_r} \vec{F}\cdot\vec{T}\,ds = \iint_{\text{int}(C_r)} \left(\text{curl }\vec{F}\right)\cdot\vec{k}\,dA \approx \left[\text{curl }\vec{F}(a,b)\right]\cdot\vec{k}\left(\pi r^2\right),$$

meaning that the flow of, say, fluid in the direction of a tangent to C_r is determined by a curl $\vec{F}$. At the end of Section 12.5, the term used for this tendency of fluid to rotate about a point (a,b) was fluid's *circulation.*

Earlier, the vector $\nabla\times\vec{F}$ was defined. Now consider the

$$\nabla\cdot\vec{F} \equiv M_x + N_y + P_z,$$

which defines the *divergence* of $\vec{F}$, denoted div $\vec{F}$. If $\vec{T}$ and $\vec{F}$ are defined as earlier, and $\vec{N} = \frac{dy}{ds}\mathbf{i} - \frac{dx}{ds}\mathbf{j}$ is an outward normal vector, then Green's theorem, in flux/divergence form can be rewritten as

$$\oint_C \vec{F}\cdot\vec{N}\,ds = \iint_D \left(\text{div }\vec{F}\right)dA = \iint_D \nabla\cdot\vec{F}\,dA,$$

which is called *Gauss' Divergence Theorem* in the plane. The value for the above integrals is called the *flux* of $\vec{F}$ across C. Here also, if C_r is a small circle centered at (a,b) with a radius r, then

$$\oint_{C_r} \vec{F}\cdot\vec{N}\,ds = \iint_{\text{int}(C_r)} \left(\text{div }\vec{F}\right)dA \approx \left[\text{div }\vec{F}(a,b)\right]\left(\pi r^2\right),$$

which means that the flow of, say, a fluid through C_r is measured by div $\vec{F}$. Now, if div $\vec{F}(a,b)>0$, then fluid is diverging away from (a,b) to the exterior of C_r. And if div $\vec{F}=0$ inside C_r, then in view of Green's theorem, the net flow across the boundary is 0.

Parallel to Definition 12.6.1, a vector field $\vec{F}=\vec{F}(x,y,z)$ is said to be *conservative* if

---

[7]George Gabriel Stokes (1819–1903), an Irish mathematician, established the science of hydrodynamics, the science of geodesy, and made great advances in mathematical physics, elasticity, and in the wave theory of light. Stokes also discovered the principles of spectrum analysis, although credit is given to Bunsen and Kirchhoff. Gustav Robert Kirchhoff (1824–1887), born in Königsberg, Prussia (now Russia), was a German physicist who studied under Gauss. Kirchhoff made important contributions to the theory of circuits and elasticity. In 1852 Stokes explained the phenomenon of fluorescence. Stoke's theorem was first presented by William Thomson (1824–1907), an Irish physicist, who proposed the Kelvin absolute temperature scale, named after Thomson received the title of Baron Kelvin of Largs in 1892.

and only if $\vec{F} = \nabla f$ for some scalar field $f = f(x,y,z)$. If such an f exists, then f is called a *potential function* for $\vec{F}$. The equivalence statements in Remark 12.6.9 are true for $\vec{F} = \vec{F}(x,y,z)$.

Exercises 12.8

1. If $f = f(x,y,z)$ is a scalar field with continuous second partial derivatives, show that $\nabla \times \nabla f = \vec{0}$,[8] which is equivalent to the $\text{curl}(\text{grad}\, f) = \vec{0}$.

2. If $\vec{F} = \vec{F}(x,y,z)$ is a vector field with continuous second partial derivatives, show that $\nabla \cdot \left(\nabla \times \vec{F}\right) = 0$, which is equivalent to $\text{div}\left(\text{curl}\, \vec{F}\right) = 0$.

3. Suppose that a scalar field f and a vector fields $\vec{F}$ both possess partial derivatives.

(a) Show that $\nabla \cdot \left(f\vec{F}\right) = (f)\left(\nabla \cdot \vec{F}\right) + (\nabla f) \cdot \vec{F}$, which can be rewritten as

$$\text{div}\left(f\vec{F}\right) = (f)\left(\text{div}\, \vec{F}\right) + (\text{grad}\, f) \cdot \vec{F}.$$

(b) Show that $\nabla \times \left(f\vec{F}\right) = (f)\left(\nabla \times \vec{F}\right) + (\nabla f) \times \vec{F}$, which can be rewritten as

$$\text{curl}\left(f\vec{F}\right) = (f)\left(\text{curl}\, \vec{F}\right) + (\text{grad}\, f) \times \vec{F}.$$

(c) If $\vec{F} = \nabla g$ for some scalar function g possessing partial derivatives of order two, show that

$$\text{div}(f\nabla g) = (f)\left(\nabla^2 g\right) + \nabla f \cdot \nabla g.$$

4. Show that if $\vec{a}$ is a constant vector, then $\text{curl}\, \vec{a} = \vec{0}$.

5. Suppose that $\vec{F}$ and $\vec{G}$ are two vector fields with c a scalar.

(a) Show that both $\nabla \times \left(\vec{F} + \vec{G}\right) = \nabla \times \vec{F} + \nabla \times \vec{G}$ and $\nabla \times \left(c\vec{F}\right) = c\left(\nabla \times \vec{F}\right)$. That is, show that both

$$\text{div}\left(\vec{F} + \vec{G}\right) = \text{div}\, \vec{F} + \text{div}\, \vec{G} \text{ and } \text{div}\left(c\vec{F}\right) = c\, \text{div}\, \vec{F}.$$

Thus, a curl satisfies the linear property.

(b) Show that both $\nabla \cdot \left(\vec{F} + \vec{G}\right) = \nabla \cdot \vec{F} + \nabla \cdot \vec{G}$ and $\nabla \cdot \left(c\vec{F}\right) = c\left(\nabla \cdot \vec{F}\right)$. That is, show that divergence is linear.

---

[8]In view of this exercise, we define $(\nabla \times \nabla) f = \nabla \times (\nabla f)$. From this $\nabla \times \nabla = \vec{0}$, which is consistent with a vector cross product. For contrast, see part (d) of Remark 11.4.4.

(c) Show that $\nabla \cdot \left(\vec{F} \times \vec{G}\right) = \left(\nabla \times \vec{F}\right) \cdot \vec{G} - \left(\nabla \times \vec{G}\right) \cdot \vec{F}$.

6. If $\vec{F}(x,y,z) = (2x+z)\mathbf{i} - xy^2\mathbf{j} + xyz^2\mathbf{k}$, find both $\nabla \times \vec{F}$ and $\nabla \cdot \vec{F}$.

7. If $\vec{F}(x,y,z) = \frac{-y}{x^2+y^2}\mathbf{i} + \frac{x}{x^2+y^2}\mathbf{j} + z\mathbf{k}$, show that the curl $\vec{F}$ is 0. Also, if C is the circle $x^2 + y^2 = 1$ in the xy-plane, show that $\int_C \vec{F} \cdot d\vec{r}$ is not 0. Compare this problem to Exercise 6 in Section 12.6.

8. Prove that if $\vec{F}(x,y,z)$ is a continuously differentiable conservative vector field, then the curl $\vec{F} = \vec{0}$.

9. (a) If f has partial derivatives of order two, show that

$$\operatorname{div}(\operatorname{grad} f) = f_{xx} + f_{yy} + f_{zz}.$$

Note that $\operatorname{div}(\operatorname{grad} f) = \nabla \cdot (\nabla f) \equiv \nabla^2 f$, called the *Laplacian* of $f(x,y,z)$. See part (d) of Remark 11.4.4.

(b) If $f(x,y,z) = x^2yz^3$, find $\nabla^2 f$.

10. A function $f(x,y,z)$ is *harmonic* if it has continuous partial derivatives of order two and satisfies *Laplace's equation*

$$u_{xx} + u_{yy} + u_{zz} = 0.$$

Show that $f(x,y,z) = \left(x^2 + y^2 + z^2\right)^{-\frac{1}{2}}$ is harmonic. See Exercise 5 in Section 11.3.

11. If $\vec{r} = x\mathbf{i} + y\mathbf{j} + z\mathbf{k} \neq \vec{0}$ and c is a scalar, a vector field $\vec{F}$ defined by

$$\vec{F}(x,y,z) = \frac{c}{\|\vec{r}\|^3}\vec{r}$$

is called an *inverse square field*. See Exercise 8 in Section 12.6.

(a) Show that $f(x,y,z) = \frac{-c}{\|\vec{r}\|}$ is a potential function for $\vec{F}$, meaning that

$$\vec{F}(x,y,z) = \nabla \frac{-c}{\|\vec{r}\|}.$$

(b) Prove that both the curl $\vec{F}$ and the div $\vec{F}$ are 0.

(c) Show that f is harmonic.

12. If $\vec{F}(x,y) = 2y\mathbf{i} - 2x\mathbf{j}$ and if C is any closed curve in the xy-plane, find $\oint_C \vec{F} \cdot \vec{T}\, ds$

and $\oint\limits_C \vec{F}\cdot\vec{N}\,ds$.

13. Suppose that $\vec{r} = x\mathbf{i} + y\mathbf{j} + z\mathbf{k} \neq \vec{0}$.

(a) Show that $\nabla(\|\vec{r}\|) = \dfrac{\vec{r}}{\|\vec{r}\|}$. **(b)** Find the curl $\dfrac{\vec{r}}{\|\vec{r}\|}$.

14. Suppose that $D \subseteq \Re^2$ is a simple region and that $\vec{F} = \nabla f$ for some function $f(x, y)$ with continuous second partial derivatives on $D \cup \partial D$. Show that

$$\oint\limits_{\partial D} \vec{F}\cdot\vec{N}\,ds = \oint\limits_{\partial D} \nabla f\cdot\vec{N}\,ds = \iint\limits_D \nabla^2 f\,dA.$$

15. *(Green's First Identity)* Suppose $D \subseteq \Re^2$ is a simple region and $f, g: \Re^2 \to \Re$, where f is differentiable and g is twice continuously differentiable on $D \cup \partial D$. Prove that

$$\oint\limits_{\partial D} f\frac{\partial g}{\partial \vec{N}}\,ds = \iint\limits_D (f)\left(\nabla^2 g\right)dA + \iint\limits_D (\nabla f\cdot\nabla g)\,dA.$$

Conclude that

$$\oint\limits_{\partial D} f\frac{\partial f}{\partial \vec{N}}\,ds = \iint\limits_D \left[f\nabla^2 f + (f_x)^2 + \left(f_y\right)^2\right]dA.$$

16. *(Green's Second Identity)* Suppose $D \subseteq \Re^2$ is a simple region and $f, g: \Re^2 \to \Re$ where f and g have continuous second partial derivatives on $D \cup \partial D$. Prove that

$$\oint\limits_{\partial D} \left(f\frac{\partial g}{\partial \vec{N}} - g\frac{\partial f}{\partial \vec{N}}\right)ds = \iint\limits_D \left(f\nabla^2 g - g\nabla^2 f\right)dA.$$

17. *(Laplacian in Cylindrical Coordinates)* If $w = f(x, y, z)$ has continuous second partial derivatives, $x = r\cos\theta$ and $y = r\sin\theta$, show that

$$\nabla^2 w = w_{xx} + w_{yy} + w_{zz} = w_{rr} + r^{-2}w_{\theta\theta} + r^{-1}w_r + w_{zz}.$$

Compare results to Exercise 11(b) in Section 11.6.

18. *(Laplacian in Spherical Coordinates)* If $w = f(x, y, z)$ has continuous second partial derivatives, $x = \rho\cos\theta\sin\phi$, $y = \rho\sin\theta\sin\phi$, and $z = \rho\cos\phi$, show that

$$\nabla^2 w = w_{xx} + w_{yy} + w_{zz} = w_{\rho\rho} + \frac{1}{\rho^2}w_{\phi\phi} + \frac{1}{\rho^2\sin^2\phi}w_{\theta\theta} + \frac{2}{\rho}w_\rho + \frac{\cot\phi}{\rho^2}w_\phi.$$

12.9* Review

Label each statement as true or false. If a statement is true, prove it. If not,

(i) give an example of why it is false, and
(ii) if possible, correct it to make it true, and then prove it.

1. The converse of Fubini's theorem is false. That is, if

$$\int_a^b \int_c^d f(x,y)\,dy\,dx = \int_c^d \int_a^b f(x,y)\,dx\,dy,$$

then f need not be Riemann integrable on a rectangle $R=[a,b]\times[c,d]$.

2. If $f(x,y)=\begin{cases} 1, & \text{if } x \text{ is irrational} \\ 3y^2, & \text{if } x \text{ is rational} \end{cases}$ is defined on the rectangle $R=[0,1]\times[0,1]$, then f is Riemann integrable on R.

3. If $\vec{F}(x,y,z)=x\mathbf{i}+y\mathbf{j}+z\mathbf{k}$, then $\nabla\cdot\vec{F}=3$ and $\nabla\times\vec{F}=\vec{0}$ for all $(x,y,z)\in\Re^3$.

4. If $\vec{r}=x\mathbf{i}+y\mathbf{j}+z\mathbf{k}\neq\vec{0}$, then both $\nabla\dfrac{1}{\|\vec{r}\|}=\dfrac{-\vec{r}}{\|\vec{r}\|^3}$ and $\nabla^2\dfrac{1}{\|\vec{r}\|}=\vec{0}$.

5. The converse of Theorem 12.6.3 is true.

6. Any continuous curve is rectifiable.

7. If $\int_C \vec{F}\cdot d\vec{r}=0$, then C is a closed smooth curve.

8. If $f(x,y)$ is bounded on a rectangle $R=[a,b]\times[c,d]$, then

$$\underline{\iint_R} f \le \underline{\int_a^b}\left[\underline{\int_c^d} f\,dy\right]dx \le \underline{\int_a^b}\left[\overline{\int_c^d} f\,dy\right]dx \le \overline{\int_a^b}\left[\overline{\int_c^d} f\,dy\right]dx \le \overline{\iint_R} f.$$

9. If $f(x,y)$ is bounded on a rectangle $R=[a,b]\times[c,d]$, then

$$\underline{\iint_R} f \le \underline{\int_a^b}\left[\underline{\int_c^d} f\,dy\right]dx \le \overline{\int_a^b}\left[\underline{\int_c^d} f\,dy\right]dx \le \overline{\int_a^b}\left[\overline{\int_c^d} f\,dy\right]dx \le \overline{\iint_R} f.$$

10. If C is a unit circle centered at the origin and $\int_C \vec{F}\cdot d\vec{r}=0$, then $\vec{F}$ is conservative.

11. Suppose that $\vec{F}=M\mathbf{i}+N\mathbf{j}$ and C is a smooth closed curve enclosing a region D. If $\iint_D \left(N_x-M_y\right)dA=0$, then $\int_C \vec{F}\cdot d\vec{r}=0$.

12. If f and g are differentiable functions, then $\oint_C f(x)\,dx + g(y)\,dy = 0$ for every piecewise smooth simple closed curve C.

13. If both $f = f(x,y,z)$ and $\vec{F} = \vec{F}(x,y,z)$ have continuous second partial derivatives, then $\operatorname{curl}\left(\operatorname{curl}\vec{F}\right) = \operatorname{curl}\left(\operatorname{grad} f - \operatorname{curl}\vec{F}\right)$.

14. The $\operatorname{div}\vec{F}$ is a vector.

15. $\displaystyle\int_0^1 \frac{\ln(x+1)}{x^2+1}\,dx = \frac{\pi}{8}\ln 2$.

16. If $f(x) \le g(x)$ on $R = [a,b]\times[c,d]$, and if $\iint_R f(x)\,dx\,dy = 8$, then $\iint_R g(x)\,dx\,dy \ge 8$.

17. Suppose that f is an odd function in x, that is, $f(-x,y) = -f(x,y)$. And suppose that f is continuous on a rectangle $R = [a,b]\times[c,d]$. Then, $\iint_R f = 0$.

18. If f and g are continuous on a closed and bounded region D, and $g(x,y) \ge 0$ on D, then there exists a point $(x_0, y_0) \in D$ such that $\iint_D fg = f(x_0, y_0)\iint_D g$. See Theorem 6.3.8 and Exercise 10 in Section 12.1.

19. Suppose that $\vec{r}(t) = x(t)\mathbf{i} + y(t)\mathbf{j}$ determines a positively oriented boundary C on a simple region $D \subseteq \Re^2$. And suppose that $\vec{N} = \frac{dy}{ds}\mathbf{i} - \frac{dx}{ds}\mathbf{j}$ is an outward unit normal vector to C. If $u(x,y)$ has continuous second partial derivatives on $\Re^2$, then

$$\oint_C u\nabla u \cdot \vec{N}\,ds = \iint_D \left(u\nabla^2 u + \nabla u \cdot \nabla u\right) dA.$$

20. If $\int_a^b f = A$ and $\int_c^d g = B$, then $\iint_R f(x) + g(y) = A + B$, where $R = [a,b]\times[c,d]$.

21. The possibility of having two grids with neither a refinement of the other exists.

22. Every bounded function f on a rectangle $R = [a,b]\times[c,d]$ is Riemann integrable on R.

23. If $f(x,y)$ is continuous on an open rectangle, $R = (a,b)\times(c,d)$. And if $\int_a^b\int_c^d |f(x,y)|\,dy\,dx$ exists, then $\int_c^d\int_a^b |f(x,y)|\,dx\,dy$, $\int_a^b\int_c^d f(x,y)\,dy\,dx$, and $\int_c^d\int_a^b f(x,y)\,dx\,dy$ exist with $\int_a^b\int_c^d |f(x,y)|\,dy\,dx = \int_c^d\int_a^b |f(x,y)|\,dx\,dy$ and

$$\int_a^b \int_c^d f(x,y)\,dy\,dx = \int_c^d \int_a^b f(x,y)\,dx\,dy.$$

24. If $\vec{F}$ is a continuous vector field containing a smooth curve C, which is orthogonal to $\vec{F}$ at all points of C, then $\int_C \vec{F}\cdot d\vec{r} = 0$.

25. If $\int_C \vec{F}\cdot d\vec{r} = 0$ for some curve C in a vector space $\vec{F}$, then $\vec{F}$ is path independent.

26. If $\vec{F}$ is conservative and a closed smooth curve C is parametrized counterclockwise, then the flux of $\vec{F}$ is 0.

27. If $P(x,y)$ and $Q(x,y)$ are continuously differentiable on an open region containing a simple region D, then

$$\oint_{\partial D} PQ\,dx + PQ\,dy = \iint_D \left[Q\left(P_x - P_y\right) + P\left(Q_x - Q_y\right)\right] dA.$$

28. If a curve is positively oriented, then it is traced counterclockwise.

29. If C is a vertical line segment and $\vec{F} = M\mathbf{i} + N\mathbf{j}$ is a continuous vector field, then the value of $\int_C \vec{F}\cdot d\vec{r}$ is determined only by a scalar field M.

30. $\int_C \vec{F}\cdot d\vec{r}$ is a vector.

31. $\int_C \vec{F}\cdot d\vec{r} = \vec{F}\left(\vec{b}\right) - \vec{F}\left(\vec{a}\right)$, where $\vec{a}$ is the initial point on C and $\vec{b}$ is the ending point on C.

32. Suppose that C is a smooth closed curve and $\vec{F}$ is a continuous vector field. If $\hat{C}$ is the curve C traversed twice, then

$$\int_{\hat{C}} \vec{F}\cdot d\vec{r} = 2\int_C \vec{F}\cdot d\vec{r}.$$

33. Since $\vec{F}(x,y) = (2y - \ln x)\mathbf{i} + \left(2x + \sqrt[3]{y^4 + 7}\right)\mathbf{j}$ is conservative, $\int_C \vec{F}\cdot d\vec{r} = 0$ with C being any circle $x^2 + y^2 = r^2$, for $r \in \Re$.

34. Since $f(x,y) = 6\left[(xy)^{\frac{3}{2}} + (xy)^{\frac{4}{3}}\right]\ln xy$ is a potential function for $\vec{F}(x,y) = \dfrac{\sqrt{xy} + \sqrt[3]{xy}}{2x}\mathbf{i} + \dfrac{\sqrt{xy} + \sqrt[3]{xy}}{2y}\mathbf{j}$, then $\int_C \vec{F}\cdot d\vec{r} = 0$, with C being a quadrilateral with vertices $(-1,2)$, $(0,0)$, $(3,1)$, and $(3,5)$.

35. If C is a smooth curve given by the rectangular equation $y = g(x)$ with $x \in [a,b]$, then

$$\int_C f(x,y)\,ds = \int_a^b f(x, g(x))\sqrt{1+[y'(x)]^2}\,dx.$$

36. Every vector field has a potential function.

37. If a vector field $\vec{F}$ has a potential function f, then f is unique.

38. $\int_0^1\int_0^1 \frac{1}{x^2+y^2+4}\,dx\,dy < \frac{\pi}{4}$.

39. $\int_{-1}^1\int_{-1}^1 x\,dx\,dy = 4\int_0^1\int_0^1 x\,dx\,dy$.

12.10* Projects

Part 1. Change of Variables for Double Integrals

A change of variables is often helpful in simplifying the evaluation of a double integral. For the Riemann integral over $\Re^1$, the change of variable theorem appeared in Theorem 6.4.6. Looking at a change of variables in $\Re^2$ geometrically, suppose that $\int_D f(x,y)\,dA$ is under consideration, where f is continuous on a domain D. Let $x = g(u,v)$ and $y = h(u,v)$, where g and h have continuous first partial derivatives. This defines a *transformation* (a *mapping*) $T : S \subseteq \Re^2 \to \Re^2$, where $T(u,v) = (x,y)$ maps a region S from the uv-plane to a region D in the xy-plane. If $T(u_0, v_0) = (x_0, y_0)$, then (x_0, y_0) is the image of (u_0, v_0). And, if no two points have the same image, T is said to be *one-to-one*. Thus, T^{-1} exists. Now, suppose that P is a grid for the region S, which produces subrectangles R_{ij} in the uv-plane with dimensions Δu by Δv. Under the transformation T, these subrectangles R_{ij} are mapped to some subregions $T(R_{ij})$ in a domain D of the xy-plane. These subregions $T(R_{ij})$ are approximated by parallelograms. See Figure 12.10.1, where $T_u = \frac{\partial x}{\partial u}\mathbf{i} + \frac{\partial y}{\partial u}\mathbf{j}$ and $T_v = \frac{\partial x}{\partial v}\mathbf{i} + \frac{\partial y}{\partial v}\mathbf{j}$. Why? The area of the parallelogram is given by the absolute value of the cross product $|(\Delta u T_u) \times (\Delta v T_v)| = |T_u \times T_v|\Delta u\,\Delta v$, where

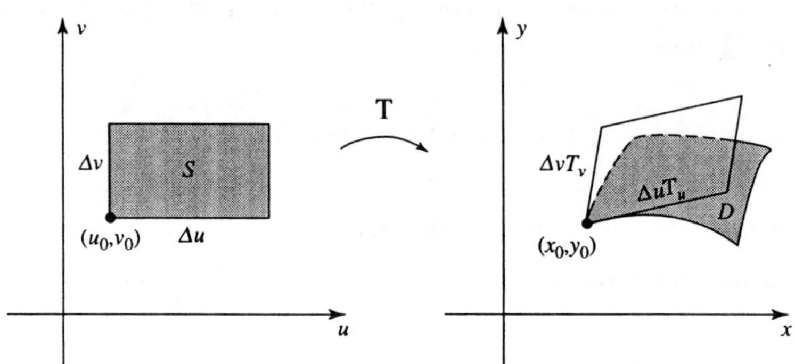

Figure 12.10.1

$$T_u \times T_v = \begin{vmatrix} \mathbf{i} & \mathbf{j} & \mathbf{k} \\ x_u & y_u & 0 \\ x_v & y_v & 0 \end{vmatrix} = \begin{vmatrix} x_u & y_u \\ x_v & y_v \end{vmatrix} \mathbf{k} \equiv \frac{\partial(x,y)}{\partial(u,v)} \mathbf{k},$$

where $\dfrac{\partial(x,y)}{\partial(u,v)} \equiv J(u,v)$, called the *Jacobian*[9] of the transformation T. Then, the area of the parallelogram is given by $|J(u,v)|\Delta u \Delta v$, which is approximately the area of the subregion $T\left(R_{ij}\right)$. $|J(u,v)|$ is called a *magnification factor*. To form a Riemann sum, multiply the area of each parallelogram by a functional value at any point in the subregion R_{ij}. Then let the norm of P go to 0. The result is the double integral

$$\iint_S f(g(u,v),h(u,v))|J(u,v)|\,du\,dv,$$

which equals $\iint_D f(x,y)\,dx\,dy$. Formally, we have

THEOREM 12.10.1. *(Change of Variables for Double Integrals) Suppose f is continuous and bounded on a Jordan measurable set D. Suppose further that T is a one to one transformation from S in the uv-plane to D in the xy-plane defined by the continuously differentiable functions $x = g(u,v)$ and $y = h(u,v)$. Moreover, assume that*

[9]Karl Gustav Jacob Jacobi (1804–1851), a German mathematician, who received his Ph.D. in 1825 from the University of Berlin, founded the theory of elliptic functions is known for his work in partial differential equations and elliptic functions. Jacobi polynomials are related to hypergeometric functions, Legendre polynomials, and Chebyshev polynomials. Jacobi was not the first to study the functional determinant, known as the Jacobian, which first appeared in 1815 in a paper written by Cauchy. Jacobi taught at the University of Königsberg until his death.

$\dfrac{\partial(x,y)}{\partial(u,v)} \neq 0$ *for any point in S. Then*

$$\iint_D f(x,y)\,dx\,dy = \iint_S f(g(u,v),h(u,v))|J(u,v)|\,du\,dv.$$

Problem 12.10.2. In a less friendly advanced calculus text, find a formal proof for the change of variables theorem which makes use of both line integrals and Green's theorem.

Problem 12.10.3.

(a) Set $x = r\cos\theta$ and $y = r\sin\theta$. Then, rewrite $\iint_R f(x,y)\,dx\,dy$ in polar coordinates as

$$\iint_S f(r\cos\theta, r\sin\theta)\,r\,dr\,d\theta.$$

(b) Observe that in part (a), $f(r\cos\theta, r\sin\theta)\cdot r$ is integrated over a region S in a rectangular coordinate plane $r\theta$.

(c) In part (a) it was shown that $J(r,\theta) = r$. What if $(0,0)$ is a point in the region R? Is the substitution used in part (a) still valid even though the Jacobian vanishes at $(0,0)$? Explain.

(d) Evaluate $\displaystyle\int_0^2 \int_0^{\sqrt{4-x^2}} \frac{xy}{\sqrt{x^2+y^2}}\,dy\,dx.$

(e) Evaluate $\displaystyle\int_{-1}^1 \int_0^{\sqrt{1-x^2}} \left(x^2+y^2\right)^{\frac{3}{2}}\,dy\,dx.$

(f) Evaluate the improper double integral $\displaystyle\int_0^\infty \int_0^\infty e^{-\left(x^2+y^2\right)}\,dx\,dy.$

(g) Find the area of a region inside the circle $r = 6\sin\theta$ but outside the cardioid $r = 2+2\sin\theta$. See part (e) of Problem 10.9.21.

(h) What does $\displaystyle\int_0^{\frac{\pi}{2}} \int_0^1 r\,dr\,d\theta$ represent geometrically?

Problem 12.10.4. Prove part (a) of Remark 6.6.7. That is, prove that $\Gamma\left(\dfrac{1}{2}\right) = \dfrac{\sqrt{\pi}}{2}$ by following the given steps.

(a) Since in using Definition 6.7.1, $\Gamma\left(\dfrac{1}{2}\right) = \displaystyle\int_0^\infty t^{-\frac{1}{2}} e^{-t}\,dt$, set $t = x^2$ and show that $\Gamma\left(\frac{1}{2}\right) = 2\int_0^\infty e^{-x^2}\,dx$.

(b) If $I = \Gamma\left(\frac{1}{2}\right)$, write $I^2 = \left[2\int_0^\infty e^{-x^2}\,dx\right]\cdot\left[2\int_0^\infty e^{-y^2}\,dy\right]$, and then apply Exercise 7(a) in Section 12.2 which needs to be extended to improper integrals. See Problem 12.10.3, part (f).

(c) Change to polar coordinates, compute, and then draw the final conclusion.

(d) Compare the above steps to those in Exercise 14 of Section 12.2.

Problem 12.10.5. Use Problem 12.10.4 to show that

(a) $\int_0^\infty e^{-2x^2}\,dx = \frac{\sqrt{2\pi}}{4}$.

(b) $\int_0^\infty e^{-3x^2}\,dx = \frac{\sqrt{3\pi}}{6}$.

(c) $\int_0^\infty e^{-4x^2}\,dx = \frac{\sqrt{\pi}}{4}$.

(d) $\int_0^\infty x^2 e^{-x^2}\,dx = \frac{\sqrt{\pi}}{4}$. See Exercise 12 in Section 6.6.

(e) $\int_0^\infty \sqrt{x}e^{-x}\,dx = \frac{\sqrt{\pi}}{2}$.

(f) $\int_0^1 \sqrt{-\ln x}\,dx = \frac{\sqrt{\pi}}{2}$.

Problem 12.10.6. If D is a trapezoid with vertices $(0,1)$, $(0,2)$, $(2,0)$, and $(1,0)$, find $\iint_D \sin\frac{y-x}{y+x}$.

Problem 12.10.7. Suppose that C is a positively oriented boundary of an annular region between the circles $x^2+y^2=4$ and $x^2+y^2=16$. If $\vec{F}(x,y) = -xy\mathbf{i} + x\mathbf{j}$ is a vector field, calculate both integrals in Green's theorem. Then show that their values are the same.

Problem 12.10.8. Prove that $\sum_{k=1}^{\infty}\frac{1}{k^2} = \frac{\pi^2}{6}$ by following the given steps.

(a) Use a geometric series to verify that

$$\int_0^1\int_0^1 \frac{1}{1-xy}\,dx\,dy = \int_0^1\int_0^1 \left(1 + xy + x^2y^2 + \cdots\right)dx\,dy.$$

(b) Pretend all is well and show that

$$\int_0^1\int_0^1 \left(1+xy+x^2y^2+\cdots\right)dx\,dy = \sum_{k=1}^{\infty}\frac{1}{k^2}.$$

(c) In order to show that $I \equiv \int_0^1\int_0^1 \frac{1}{1-xy}\,dx\,dy = \frac{\pi^2}{6}$, change the variables using the substitution $x = u\cos\theta - v\sin\theta$ and $y = u\sin\theta + v\cos\theta$ with a particular θ. Then show that $I = I_1 + I_2$, where

$$I_1 = 4\int_0^{\frac{\sqrt{2}}{2}}\int_0^{u} \frac{1}{2-u^2+v^2}\,dv\,du, \text{ and } I_2 = 4\int_{\frac{\sqrt{2}}{2}}^{\sqrt{2}}\int_0^{\sqrt{2}-u} \frac{1}{2-u^2+v^2}\,dv\,du.$$

(d) Show that $I_1 = \frac{\pi^2}{18}$ and $I_2 = \frac{\pi^2}{9}$. Then draw the final conclusion.

Part 2. Elliptic Equation

One of the simplest and most important partial differential equations in mathematical physics is an *elliptic equation,*

$$\nabla^2 u \equiv u_{xx} + u_{yy} = 0,$$

also called the *potential equation* or *Laplace's equation* in two dimensions. Solutions to $\nabla^2 u = 0$ that have continuous second partial derivatives are called *harmonic functions*. As witnessed in Exercise 5 of Section 11.3, many types of functions are solutions to Laplace's equation. We will search for functions that both satisfy Laplace's equation in a given domain D and satisfy certain conditions on the boundary C of D. In mathematical terms, a *boundary value problem of the first kind* will be studied, namely, the *Dirichlet's problem*,

$$\nabla^2 u = 0 \text{ in } D, \text{ with}$$
$$u = f \text{ on } C.$$

Here, assume that D can be decomposed into a finite number of simple regions. We also consider the *Neumann*[10] *problem*, or *boundary value problem of the second kind*, given by

$$\nabla^2 u = 0 \text{ in } D, \text{ with } \frac{\partial u}{\partial \vec{N}} = f \text{ on } C,$$

where $\frac{\partial u}{\partial \vec{N}} = \nabla u \cdot \vec{N}$, with $\vec{N}$ being an outward unit normal vector to C, as well as

[10]The Neumann problem is named after Karl Neumann. See the footnote in Section 6.6.

Robin's[11] *problem*, or a *boundary value problem of the third kind* involving the boundary condition with a linear combination of u and $\vec{N}$.

THEOREM 12.10.9. *If u_1 and u_2 are harmonic functions in $D \subseteq \Re^2$, and $u_1 = u_2$ on $C = \partial D$, then $u_1 = u_2$ in D.*

Theorem 12.10.9 states that a harmonic function is uniquely determined in D by its values on the ∂D. Also, Dirichlet's problem has a unique solution, provided it exists.

Proof of Theorem 12.10.9. If $w = u_1 - u_2$, then $\nabla^2 w = \nabla^2 u_1 - \nabla^2 u_2 = 0$ in D, and $w = 0$ on C. And from Exercise 15 in Section 12.8, Green's first identity yields that

$$\iint_D (w_x)^2 + (w_y)^2 \, dA = 0.$$

Now, since $(w_x)^2 + (w_y)^2 = \|\nabla w\|^2$ is continuous and nonnegative, $(w_x)^2 + (w_y)^2 = 0$ in D. Therefore, w must be constant on $D \cup \partial D$. Why? But, $w = 0$ on ∂D. Thus, $w \equiv 0$ on $D \cup \partial D$, and hence, $u_1 \equiv u_2$. ■

Problem 12.10.10. If u_1 and u_2 are harmonic solutions to the Neumann problem, show that u_1 and u_2 differ by a real constant.

Problem 12.10.11. Consider Robin's problem

$$\nabla^2 u = 0 \text{ in } D, \text{ with } \frac{\partial u}{\partial \vec{N}} + hu = g \text{ on } C,$$

where h and g are known functions. If $h \geq 0$, $h \not\equiv 0$, and u_1 and u_2 are harmonic functions that solve Robin's problem, then show that $u_1 \equiv u_2$.

If u is harmonic in D, then Green's first identity can be rewritten as

$$\oint_{\partial D} \frac{\partial u}{\partial \vec{N}} ds = \int_D \nabla^2 u \, dA.$$

Why? Therefore, if the Neumann problem has a solution u, $\nabla^2 u = 0$ in D, and $\dfrac{\partial u}{\partial \vec{N}} = f$ on ∂D, then the above yields

$$\oint_{\partial D} f \, ds = 0.$$

Without this last integral condition, called the *consistency condition*, the Neumann problem would have no solution. In the case where the partial differential equation is of

[11]Victor Gustave Robin (1855–1897), a French analyst and applied mathematician.

the form $\nabla^2 u = q$ in D, called *Poisson's*[12] *equation*, with the boundary condition $\dfrac{\partial u}{\partial \vec{N}} = f$ on ∂D, the consistency condition for the existence of solutions is

$$\oint_{\partial D} f\,ds = \iint_D q(x,y)\,dA.$$

Problem 12.10.12. Show that if u is a solution to the nonhomogeneous Robin's problem

$$\nabla^2 u = q(x,y) \text{ in } D, \text{ with } \frac{\partial u}{\partial \vec{N}} + hu = f(x,y) \text{ on } C,$$

then

$$\oint_C hu\,ds = \oint_C f\,ds - \iint_D q\,dA.$$

Problem 12.10.13. Prove that if u_1 and u_2 are harmonic solutions to the Robin's problem

$$\nabla^2 u + ku = q(x,y) \text{ in } D, \text{ with } \frac{\partial u}{\partial \vec{N}} + hu = f(x,y) \text{ on } C,$$

for $k < 0$ and $h > 0$, then $u_1 \equiv u_2$.

Problem 12.10.14

(a) *(Maximum Principle)* If u is harmonic on D and continuous on $D \cup \partial D$, then prove that the maximum value of u over $D \cup \partial D$ is attained on ∂D. That is, prove that $\max_{\partial D} u = M$ implies that $u \le M$ in D.

(b) Prove that a constant function u is the only harmonic function that can attain a maximum in the interior of D.

(c) Suppose u, v, and w are both harmonic on D and continuous on $D \cup \partial D$. If $u \le v \le w$ on ∂D, prove that $u \le v \le w$ in D.

Problem 12.10.15. Consider the Dirichlet's problem on a rectangle given by

$$\nabla^2 u = 0, \text{ if } 0 < x < a \text{ and } 0 < y < b,$$
$$u(x,0) = 0, \text{ and } u(x,b) = f(x), \text{ if } 0 \le x \le a, \text{ and}$$
$$u(0,y) = 0, \text{ and } u(a,y) = 0, \text{ if } 0 \le y \le b.$$

A solution to this boundary value problem may be obtained by implementing the

---

[12] Siméon Denis Poisson (1781–1840), a French mathematician whose most important works were in integration and Fourier series, published papers on electricity, magnetism, astronomy, and mechanics.

separation of variables method given below. Review Part 1 of Section 11.8. Verify each step in the following outline.

(a) Assume that $u(x,y) = X(x)Y(y)$ in order to obtain the two equations

$$(*) \qquad X'' - kX = 0 \text{ with } X(0) = 0 = X(a), \text{ and } k \text{ a constant, and}$$

$$(**) \qquad Y'' + kY = 0 \text{ with } Y(0) = 0, \text{ and } k \text{ the same constant.}$$

(b) For equation $(*)$ consider the case when $k = 0$, to obtain no nontrivial solutions to the equation.

(c) For equation $(*)$ consider the case when $k > 0$ to obtain no nontrivial solutions to the equation.

(d) Assume that $k < 0$, and solve equation $(*)$ to obtain $X_n(x) = \sin\dfrac{n\pi x}{a}$ for $k = -\left(\dfrac{n\pi}{a}\right)^2$, with $n \in N$.

(e) Show that if $k = -\left(\dfrac{n\pi}{a}\right)^2$ for $n \in N$, then equation $(**)$ has solutions of the form $Y_n(x) = \sinh\dfrac{n\pi y}{a}$.

(f) Conclude that $u_n(x,y) = c_n \sin\dfrac{n\pi x}{a}\sinh\dfrac{n\pi y}{a}$, for $n \in N$, satisfies Laplace's equation, and that u_n vanishes at $x = 0$, $x = a$, and $y = 0$.

(g) To obtain the solution to the given problem, let

$$u(x,y) = \sum_{k=1}^{\infty} c_k \sin\frac{k\pi x}{a}\sinh\frac{k\pi y}{a},$$

where the c_ks need to be determined. Then find the c_ks by requiring that $u(x,b) = f(x)$ and implementing the Fourier sine series.

(h) Conclude that the solution to the given problem is

$$u(x,y) = \sum_{k=1}^{\infty} b_k \sin\frac{k\pi x}{a}\left(\sinh\frac{k\pi y}{a}\Big/\sinh\frac{k\pi b}{a}\right),$$

where

$$b_k = \frac{2}{a}\int_0^a f(x)\sin\frac{k\pi x}{a}\,dx.$$

Problem 12.10.16. Consider the Dirichlet's problem on a disk, given by

$$\nabla^2 u = 0 \text{ if } x^2 + y^2 < a^2, \text{ with}$$
$$u = f \text{ on } x^2 + y^2 = a^2.$$

Follow the given outline to solve this problem:

(a) Let $x = r\cos\theta$ and $y = r\sin\theta$ for $0 \le \theta \le 2\pi$, and $0 < r \le a$. Then write $\nabla^2 u$ in polar coordinates. Review Exercise 11(b) in Section 11.6. Also, write the boundary condition as $u(a,\theta) = f(\theta)$ for $\theta \in [0, 2\pi]$, requiring that u be 2π-periodic.

(b) Apply separation of variables in assuming that $u(r,\theta) = R(r)T(\theta)$. Then obtain the two equations

$$(*) \qquad r^2 R'' + rR' - kR = 0, \text{ for } k \text{ constant, and}$$
$$(**) \qquad T'' + kT = 0.$$

(c) Since T is 2π-periodic, assume that both $T(-\pi) = T(\pi)$ and $T'(-\pi) = T'(\pi)$. Solve equation $(**)$ together with these conditions and conclude that $k = n^2$, for $n \in N$.

(d) Solve equation $(*)$ with $k = n^2$, $n \in N$.

(e) Implement a Fourier series to obtain the final solution to the given problem.

Problem 12.10.17

(a) Write the solution obtained in part (e) of Problem 12.10.16 as

$$u(r,\theta) = \frac{1}{2\pi}\int_{-\pi}^{\pi} \frac{a^2 - r^2}{a^2 - 2ar\cos(\theta - \phi) + r^2} f(\phi)\,d\phi,$$

known as *Poisson's integral formula* for the solution to the Dirichlet's problem on a disk.

(b) $K(a,r,\theta-\phi) \equiv \dfrac{1}{2\pi}\dfrac{a^2 - r^2}{a^2 - 2ar\cos(\theta-\phi) + r^2}$ with $0 \le r < a$ and $-\pi \le \theta \le \pi$ is called a *Poisson's kernel.* Show that $K(a,r,\theta-\phi) > 0$, and that

$$\int_{-\pi}^{\pi} K(a,r,\theta-\phi)\,d\phi = 1.$$

Problem 12.10.18. Find a solution to the Robin's problem on a rectangle given by

$$\nabla^2 u = 0, \text{ if } 0 < x < a \text{ and } 0 < y < b,$$
$$u(x,0) = f(x) \text{ and } u(x,b) = 0, \text{ if } 0 \le x \le a, \text{ and}$$
$$u_x(0,y) = 0 \text{ and } u_x(a,y) = 0, \text{ if } 0 \le y \le b.$$

Part 3. Exact Equations

Any first order differential equation $\frac{dy}{dx} = f(x,y)$ may be written as

$$(*)\qquad M(x,y) + N(x,y)\frac{dy}{dx} = 0.$$

If there exists $F = F(x,y)$ such that $F_x = M$ and $F_y = N$, then the equation becomes

$$\frac{\partial F}{\partial x} + \frac{\partial F}{\partial y}\frac{dy}{dx} = 0,$$

or, equivalently, $\frac{d}{dx}F(x, y(x)) = 0$. Thus, $F(x,y) = c$, for c a constant. Therefore, the function $y = y(x)$ is a solution to the given differential equation and is implicitly given by $F(x,y) = c$.

Definition 12.10.19. Consider the given differential equation $(*)$ on an open simply connected region R. If there exists $F = F(x,y)$ such that $F_x = M$ and $F_y = N$ for all $(x,y) \in R$, then equation $(*)$ is said to be *exact* on R, and F is a *potential function* for equation $(*)$.

THEOREM 12.10.20. *(Test for Exactness) Suppose that $M(x,y)$ and $N(x,y)$ are continuously differentiable on an open simply connected region R. Then equation $(*)$ is exact on R if and only if $M_y = N_x$ for all $(x,y) \in R$.*

Compare Theorem 12.10.20 to Theorems 12.6.3, 12.6.4, and 12.7.5. In fact, the proof of Theorem 12.6.4 outlines a procedure that can be followed in order to find a potential function, which leads to the solution of the differential equation. A potential function may also be obtained directly from the formula given in Exercise 12 of Section 12.6.

Problem 12.10.21. Determine whether or not the given equations are exact. Then find a solution to those which are exact.

(a) $y + 2x\frac{dy}{dx} = 0$

(b) $y^2 + 2xy\frac{dy}{dx} = 0$ (See Exercise 3(a) in Section 12.6.)

(c) $y\sin 2x + y^2 \sin x + \left(\sin^2 x - 2y\cos x\right)\frac{dy}{dx} = 0$

The differential equation in part (b) of Problem 12.10.21 is exact, but the simplified equation in part (a) is not. If the differential equation in part (a) is multiplied by $\mu(x,y) = y$, the resulting differential equation is exact. The nonzero function μ is called an *integrating factor* for the differential equation in part (a). The question as to how $\mu(x,y)$ is found in general to differential equations that are not exact is addressed next.

Suppose the differential equation $(*)$ is not exact on R. And suppose that $\mu(x,y)$ is its integrating factor. Then, the *essentially equivalent* equation $\mu M + \mu N \frac{dy}{dx} = 0$ is exact on R. And by Theorem 12.10.20, $\frac{\partial}{\partial y}(\mu M) = \frac{\partial}{\partial x}(\mu N)$ for all $(x,y) \in R$. Equivalently,

$$\mu M_y + M\mu_y = \mu N_x + N\mu_x.$$

This partial differential equation may be difficult to solve. If we knew that μ was a function only of x and not of y, then the above partial differential equation would become $\mu(M_y - N_x) = N\frac{d\mu}{dx}$. Why? And if $N \neq 0$, then this equation may be written as

$$\frac{1}{\mu}\frac{d\mu}{dx} = \frac{1}{N}\left(\frac{\partial M}{\partial y} - \frac{\partial N}{\partial x}\right),$$

which is separable if $\frac{1}{N}\left(\frac{\partial M}{\partial y} - \frac{\partial N}{\partial x}\right)$ is independent of y. If separable, then we have an easily obtainable solution for μ. Similarly, for an integrating factor $\mu = \mu(y)$, M and N must be so that $\frac{1}{M}\left(\frac{\partial N}{\partial x} - \frac{\partial M}{\partial y}\right)$ is independent of x. Can you verify this?

Problem 12.10.22. Solve the differential equation given by

$$xy^2 - y + \left(x + 2y^2\right)\frac{dy}{dx} = 0.$$

Problem 12.10.23. Consider the differential equation

$$(*) \qquad y(y+2x) - x^2\frac{dy}{dx} = 0.$$

(a) Show that equation $(*)$ is not exact.

(b) Show that $\mu(y) = y^n$ is an integrating factor for some integer n.

(c) Solve an essentially equivalent exact equation.

(d) Show that $y = 0$ is a solution of equation $(*)$, but not a solution of the essentially equivalent exact equation. Thus, a solution to equation $(*)$ was lost.

Problem 12.10.23. Consider the differential equation

$$(*) \qquad 2x + y^2 + \left(\frac{x^2}{y} + 3xy\right)\frac{dy}{dx} = 0.$$

(a) Show that the equation $(*)$ is not exact.

(b) Show that $\mu(y) = y^n$ is an integrating factor for some integer n.

(c) Solve the essentially equivalent exact equation.

(d) Show that $y = 0$ is a solution of the essentially equivalent exact equation, but not a solution of equation $(*)$. Thus, a solution to equation $(*)$ was gained.

Problem 12.10.25. Consider the equation $y - x\dfrac{dy}{dx} = 0$. Show that this equation

(a) is not exact.

(b) has an integrating factor $\mu = \mu(x)$.

(c) has an integrating factor $\nu = \nu(y)$.

(d) has an integrating factor $\eta(x, y) = x^r y^s$, where r and s are nonzero real constants.

Problem 12.10.26. Consider the differential equation

$$(*) \qquad M + N\frac{dy}{dx} = 0.$$

(a) If μ is an integrating factor for equation $(*)$, show that $c\mu$ is an integrating factor for $(*)$, with any nonzero constant c.

(b) If equation $(*)$ has a solution, show that $(*)$ has an integrating factor.

(c) If μ and ν are integrating factors for equation $(*)$ and $\mu \neq c\nu$ for any $c \in \Re$, then show that $\dfrac{\mu}{\nu} = C$, with C an arbitrary constant, defines an implicit solution to equation $(*)$.

(d) Use part (c) to solve the differential equation given in Problem 12.10.25.

Consider the *separable differential equation* $\dfrac{dy}{dx} = f(x)g(y)$, as given in Part 3 of Section 11.9. If we multiply it by $\dfrac{1}{g(y)}$ and rewrite, then we obtain

$$f(x) - \frac{1}{g(y)}\frac{dy}{dx} = 0,$$

which is an essentially equivalent exact equation. Therefore, separable differential equations have an integrating factor $\mu(y) = \dfrac{1}{g(y)}$. Watch for gaining or losing solutions.

Problem 12.10.27. Consider again the *linear differential equation*

$$(*) \qquad \frac{dy}{dx} + P(x)y = Q(x),$$

with P and Q continuous introduced in Part 2 of Section 11.8.

(a) Prove that equation $(*)$ has an integrating factor given by

$$\mu(x) = \exp\left(\int P(x)\,dx\right).$$

(b) Show that $(*)$ has a solution given by

$$y = \frac{1}{\mu(x)}\left(\int \mu(x)Q(x)\,dx + C\right),$$

with C an arbitrary constant.

(c) Solve $\dfrac{dy}{dx} - 5y = x$.

Hints and Solutions to Selected Exercises

As mentioned in the Preface, exercises often may be completed in a variety of ways. Thus, it is recommended that this part of the text be viewed after exercises have been attempted. Seeing a hint or the answer can discourage a correct and perhaps creative approach. Of course, if great difficulty arises in beginning a problem, then acknowledging this part of the text is encouraged.

Section 1.1

1. **(a)** $\{1\}$; **(b)** $\{1,3,5\}$; **(c)** $\{1,2,3,4,5\}$; **(d)** $\{(1,1),(1,8)\}$; **(e)** $\{(1,1),(1,8),(8,1),(8,8)\}$; **(f)** ϕ
2. $x=2, y=3$
6. **(a)** 6; **(b)** 23

Section 1.2

1. $\{(0,0),(0,1)\}$, $x=y^2+y$
2. **(a)** $\{x \mid x>-1\}$; **(b)** $\{x \mid x\geq 1 \text{ and } x\neq 2\}$; **(c)** $\{x \mid x\neq 0\}$; **(d)** $\{x \mid x\neq -2,1\}$
3. **(a)** bijection; **(b)** injection; **(c)** bijection; **(d)** none of these
4. **(a)** strictly increasing and bounded with no maximum, $\sup f=1, \inf f=\min f=0$;
   (b) strictly decreasing and bounded with no minimum, $\sup f=\max f=1, \inf f=0$;
   (c) $\min f=\inf f=-2, \sup f=\max f=\frac{3}{2}$; **(d)** bounded below with $\inf f=0$
5. Argue using definitions

6. (**a**) even; (**b**) neither; (**c**) even; (**d**) odd
7. (**a**) Assume that $f(-x)=f(x)$ and $g(-x)=g(x)$, and prove that $(f+g)(-x)=(f+g)(x)$ for all $x \in A$; (**c**) yes
9. Determine which function is higher and which is lower.
10. 6
11. Pick, for example, $f(x)=x^2$, $g(x)=x+1$.

Section 1.3

2. (**e**) Suppose $P(n)$ is the statement "$a-b$ is a factor of a^n-b^n." Then $P(1)$ is trivially true. Next, suppose $P(k)$ is true for some $k \in N$, i.e. $a-b$ is a factor of a^k-b^k. We will show that $P(k+1)$ is true, i.e. $a-b$ is a factor of $a^{k+1}-b^{k+1}$. To this end, we write $a^{k+1}-b^{k+1}=a^{k+1}-ab^k+ab^k-b^{k+1}=a\left(a^k-b^k\right)+b^k(a-b)$. Now, $a-b$ is a factor of a^k-b^k and $a-b$, so it is also a factor of the expression $a\left(a^k-b^k\right)+b^k(a-b)$. Hence $P(n)$ is true for all $n \in N$. (**h**) Suppose $P(n)$ is the statement "$2^n<n!$." Then, $P(4)$ means that $2^4<4!$, which is true. Next, suppose $P(k)$ is true for some integer $k \geq 4$, i.e. $2^k<k!$. We will show that $P(k+1)$ is true, i.e. $2^{k+1}<(k+1)!$. To this end, we write $2^{k+1}=2^k \cdot 2<(k!)\cdot 2<(k!)(k+1)=(k+1)!$. Hence, $P(n)$ is true for all $n \geq 4$.
   (**n**) Use the fact that $a \leq x_{k+1} \leq b$. (**s**) Use the fact that $\sqrt{k+1}>\sqrt{k}$ for $k \geq 1$.
3. (**a**) $\frac{x^{n+1}-2x+1}{1-x}$ when $x \neq 1$, and $1-n$ when $x=1$; (**b**) 6985
4. (**f**) -560; (**g**) Use the binomial theorem on $(1+1)^n$. (**i**) $2\left(2^{50}-1\right)$
5. (**a**) Suppose $P(n)$ is the statement "$(1+x)^n \geq 1+nx$." Then $P(1)$ is true. Next, suppose $P(k)$ is true for some $k \in N$, i.e. $(1+x)^k \geq 1+kx$. We will show that $P(k+1)$ is true, i.e. $(1+x)^{k+1} \geq 1+(k+1)x$. To this end, since $1+x>0$ and $kx^2 \geq 0$, we can write $(1+x)^{k+1}=(1+x)^k(1+x) \geq (1+kx)(1+x) \geq 1+(k+1)x$. Hence, $P(n)$ is true for all $n \in N$. $P(n)$ is also true if $x=-1$.
7. Write out a few terms and find the pattern, or use the method from Example 1.3.12. Then use induction. (**a**) $a_n=2n-1$; (**b**) $a_n=n$; (**c**) $a_n=2^n$; (**d**) $a_n=\frac{n(n+1)}{2}$; (**e**) $a_n=1$ if $n=1$ and $a_n=2^{n-2}$ if $n \geq 2$; (**f**) $a_n=\frac{n+1}{2n}$; (**g**) $a_n=2^n-1$; (**h**) $a_n=3^{1-n}$
8. Visit a library or see the *Instructor's Solutions Manual.*
9. Either use steps similar to those from Exercise 8, or use Exercise 8 with $f(x)=-\ln x$ and employ part (c) of Theorem 1.7.1.

Section 1.4

1. Consider $f(n)=2n$ for all $n \in N$.
2. To prove part (a) of Theorem 1.4.4, consider an identity function. To prove part (c), consider a composition function.
3. (**a**) $f(n)=n$ if n is even, and $f(n)=1-n$ if n is odd; (**b**) $f(n)=2^{n+2}$
4. Visit a library.

Section 1.5

1. (a) For all $p>0$ there exists a particular real number x such that $f(x+p)\neq f(x)$.
   (b) There exists $\varepsilon>0$ such that for all $\delta>0$, there are x and t in D such that $|x-t|<\delta$ and $|f(x)-f(t)|\geq\varepsilon$.
2. Prove the contrapositive.
3. Try a proof by contradiction.
5. (a) Let for example $r=\sqrt{2}, s=\sqrt{2}$. (b) P is false; take $r=\sqrt{2}, s=-\sqrt{2}$.
6. Suppose $S=\{n\in N\,|\,P(n)$ is false$\}$. Assume that $S\neq\phi$ to obtain a contradiction. Thus, if $S\neq\phi$, by the well-ordering principle, Axiom 1.3.1, S has a smallest member, say t. That is, $P(t)$ is false and $P(t-1)$ is true. But, from the hypothesis (b), if $P(i)$ is true for each $i=1,2,\dots,t-1$, then $P(t)$ is also true. We reached a contradiction. Therefore, $S=\phi$ and so $P(n)$ is true for all $n\in N$.
8. Use steps similar to those in the proof of Theorem 1.3.9 given in Exercise 6.

Section 1.6

1. (a) Assume that there are two and prove they must be equal. (d) Begin with $1+(-1)=0$. (f) Note that $-(-a)$ is the additive inverse of $-a$. (g) Suppose that $ab=0$ and $a\neq 0$. Show that $b=0$.
2. (a) Recall that $0=(-a)\cdot 0$. (b) Add something to both sides of the equality.
3. (b) Use Axiom (A9). (c) Use Axiom (A9).
4. (c) Consider the case in which $a<0$ or $a>0$. (d) Use a proof by contradiction.
5. Note that $r>0$ and $r-1\geq 0$ and use exercise 3(e).
6. Show that Q satisfies Axioms (A1) - (A12).
7. Use the set $T=\{-s\,|\,s\in S\}$ and the completeness axiom.
8. Suppose that there are two and show they must be equal.
9. Argue, using definitions.
10. Prove that (a) $\Rightarrow$ (b) $\Rightarrow$ (c) $\Rightarrow$ (d) $\Rightarrow$ (e) $\Rightarrow$ (a). To prove that (a) $\Rightarrow$ (b), let $z=\frac{y}{x}$. To prove that (b) $\Rightarrow$ (c), let $y=1$. To prove that (c) $\Rightarrow$ (d), use a contradiction. To prove that (d) $\Rightarrow$ (e), consider the set $S=\{m\in N\,|\,x<m\}$.
12. First, prove that (a,b) contains a rational number by considering the possibilities $0<a<b, a\leq 0<b$, or $a<b\leq 0$. Then employ Exercise 11.
13. (a) $\{\frac{1}{n}\,|\,n\in N\}$; (b) $\{1-\frac{1}{n}\,|\,n\in N\}$; (c) $\{x\in Q^+\,|\,x^2\leq 2\}$
14. See Exercise 13(c).
16. Let $x=\frac{p}{q}$ in the equation $f(x)=0$. Then multiply the result by q^n and solve for $a_n p^n$.
18. (a) $x=2,2,\sqrt{2},-\sqrt{2}$; (b) $x=2,\frac{2}{3},\frac{2}{3}$
19. (c) $p(x)=(x-1)^2(x+2)(2x-3)$; (d) $p(x)=(x-1)(x^2+4)$
20. (b) Factor and consider four cases. (c) Break it up into two inequalities.
21. Employ Theorem 1.6.11.
22. The converse is false.
23. Employ Theorem 1.6.11.
24. (a) $a=2, b=\frac{1}{2}$

Section 1.7

1. To prove part (b) of Theorem 1.7.1, first, prove that $a<b \Leftrightarrow a^2<b^2$. Then use this to prove that $a<b \Leftrightarrow \sqrt{a}<\sqrt{b}$. Use $\left(\sqrt{a}-\sqrt{b}\right)^2 \geq 0$ to prove part (c), and use $a^2+b^2 \leq a^2+2ab+b^2$ to prove part (d).
2. Try a proof by contradiction.
3. **(d)** Use Exercise 3(c). **(e)** First show that $\left|\frac{1}{b}\right|=\frac{1}{|b|}$ for $b \neq 0$.
4. **(a)** $x>\frac{1}{2}$; **(b)** $x \in\left[\frac{1}{3},9\right]$; **(c)** $x \in\left(-\infty,-\frac{1}{2}\right)\cup\left(\frac{7}{6},\infty\right)$; **(d)** $x \in[3,5)\cup(5,7]$; **(e)** no solution; **(f)** $x \in(-\infty,3)\cup(3,\infty)$; **(g)** $x=-\frac{7}{3},5$; **(h)** $x=5$
7. **(a)** Use the triangle inequality.
9. Try induction.
10. **(a)** Since $\sum_{k=1}^{n}\left(\alpha a_k+\beta b_k\right)^2 \geq 0$ for any real values α and β, pick $\alpha=\sum_{k=1}^{n}\left(b_k\right)^2$ and $\beta=-\sum_{k=1}^{n} a_k b_k$. **(b)** Use part (a) on $\sum_{k=1}^{n}\left(a_k+b_k\right)^2$.
15. center is $\left(-1,\frac{3}{4}\right)$, radius is $\frac{\sqrt{33}}{4}$
16. false
17. false

Section 1.8

2. **(a)** $\frac{x-b}{m}$; **(b)** $\frac{2-x}{1+x}$; **(c)** $\frac{x}{x-1}$; **(d)** $\frac{1}{3}\left(-1+\tan\frac{x}{2}\right)$, $x \in(-\pi,\pi)$;
3. $f(x)=3, g(x)=3$
4. $x \geq 0$ (or $x \leq 0$.) $f^{-1}(x)=\sqrt{\sqrt{x}-1}$, $x \geq 1$.
7. yes
9. **(a)** only if domain of f is $x \in\left[-\frac{\pi}{2},\frac{\pi}{2}\right]$; **(b)** $x \in[-1,1]$; **(c)** $x \in\left[-\frac{\pi}{2},\frac{\pi}{2}\right]$
10. **(a)** $\sqrt{1-x^2}$; **(c)** $\frac{x}{\sqrt{1+x^2}}$; **(d)** $2x\sqrt{1-x^2}$

Section 1.9

1. T	11. T	21. T	31. F	41. F	51. T
2. T	12. F	22. T	32. T	42. T	52. T
3. T	13. T	23. T	33. T	43. T	53. T
4. F	14. T	24. T	34. T	44. T	54. T
5. T	15. T	25. F	35. T	45. T	55. T
6. F	16. F	26. T	36. T	46. T	56. T
7. F	17. T	27. T	37. T	47. T	57. T
8. F	18. T	28. T	38. T	48. F	58. T
9. T	19. T	29. T	39. T	49. T	59. F
10. F	20. T	30. F	40. F	50. T	60. F

Section 2.1

1. $n^* = 2500$
2. (a) converges to 0; (b) Use Definition 2.1.2 and the fact that $n^2 - 2 \geq \frac{1}{2}n^2$, for all $n \geq 2$, to prove that the sequence converges to 0. (c) converges to 0; (d) converges to $\frac{1}{2}$; (e) converges to 0; (f) diverges; (g) converges to 0; (h) diverges; (i) converges to 0; (j) converges to 0; (k) diverges
3. (a) Employ Example 1.3.3. (b) Use Exercise 2(b) of Section 1.3 to prove that this sequence converges to $\frac{1}{3}$.
5. Use Exercise 6(b) of Section 1.7. The converse is false.
8. Use Definition 2.1.2 and Remark 2.1.8, part (f).
10. Use Definition 2.1.2 and Remark 2.1.8, part (f), to prove that $\{b_n\}$ converges to A.
11. $\{a_n\}$, where $a_n = (-1)^n$
12. Use Theorem 2.1.12 and the idea used in the proof of Theorem 2.1.11.
14. If $c > 1$, then $\sqrt[n]{c} = 1 + b_n$ with $b_n > 0$. If $0 < c < 1$, then $\sqrt[n]{c} = \frac{1}{1+d_n}$ with $d_n > 0$.
15. $\sqrt[n]{n} = 1 + b_n$ with $b_n > 0$, and from the binomial theorem, $n > 1 + \frac{n(n-1)}{2}(b_n)^2$, which can be used to prove this result.
18. Use Example 1.3.4.

Section 2.2

3. Factor $(a_n)^p - A^p$ and use Theorem 2.1.11.
4. Use the fact that $ab = \frac{1}{4}\left[(a+b)^2 - (a-b)^2\right]$.
6. Use the fact that $a_n b_n - AB = (a_n - A)(b_n - B) + A(b_n - B) + B(a_n - A)$.
7. Consider the cases $A = 0$ and $A > 0$ separately. In the second case, rationalize the numerators.
8. (a) Write $A - B = (A - a_n) + (b_n - B) + (a_n - b_n)$ and use Exercise 2 of Section 1.7.
9. Consider a proof by contradiction.
11. (a) 1; (b) 0; (c) 0; (d) Observe that $\frac{|r|}{n} < \frac{1}{2}$ if n is large to show that the limit is 0. (e) 0; (f) Rationalize the numerator to show that the limit is 0. (g) $\frac{1}{2}$; (h) 0; (i) Observe that $\sqrt[n]{n} < \sqrt[n]{n + \sqrt{n}} < \sqrt[n]{2n}$ to show that the limit is 1. (j) 2; (k) 0
12. $\{a_n b_n\}$ need not converge. If $\{b_n\}$ is bounded, $\{a_n b_n\}$ converges.
15. Use partial fraction decomposition on a_n.
18. (a) yes; (b) Try a proof by contradiction.
19. Use algebra to show that $a_n = \frac{b_n+1}{b_n-1}$, and thus $a_n + 1 = \frac{2b_n}{b_n-1}$.
20. Multiply a_n by $\frac{1}{n^q} \Big/ \frac{1}{n^q}$.
21. (a) 0; (b) $\frac{1}{2}$

Section 2.3

3. No. The sequence $\{a_n b_n\}$ may converge, diverge to $-\infty$, or oscillate.

4. **(a)** $b_n = 1$; **(b)** $b_n = \frac{a_n}{c}, c \neq 0$; **(c)** $b_n = (a_n)^2$; **(d)** $b_n = -1$; **(e)** $b_n = (-1)^n a_n$

5. **(b)** Observe that $n^3 - n + 1 > \frac{1}{2}n^3$ and $2n + 4 \leq 4n$, if $n \geq 2$.

10. $a_n = \frac{1}{n}$; $a_n = n$; $a_n = (-1)^n n$

11. **(a)** converges to 0; **(b)** diverges to $+\infty$; **(c)** converges to 0

14. Write $b_n = \frac{1}{a_n}(a_n b_n)$. Then use Theorem 2.3.6 and Exercise 5 of Section 2.2.

19. The rates are roughly the same.

Section 2.4

1. $a_n = n$ if n is odd, and $a_n = n + 2$ if n is even

2. $a_n = 1 - \frac{1}{n}$

4. **(a)** show $a_n \geq a_{n+1}$; **(b)** show $a_{n+1} < a_n$; **(c)** show $\frac{a_{n+1}}{a_n} < 1$; **(d)** show $a_n > a_{n+1}$

6. **(d)** Note that $\{a_n\}$ is increasing. Then use Example 1.3.4 to show that $\{a_n\}$ is bounded above. **(g)** $\{a_n\}$ is not monotone, but does converge to 0.

9. **(a)** Show that $0 < a_n < \frac{1}{2}$ and that $\{a_n\}$ is strictly decreasing. **(d)** Observe that since $a_n < b_n$ for all n, $A \leq B$.

10. **(a)** $a_1 = 1$ and $a_{n+1} = (n+1)a_n$ for all $n \in N$.

11. **(a)** Prove by induction that $\{a_n\}$ is strictly increasing and bounded above. Thus, it converges to, say, A. Then from the recursion formula, $A = \sqrt{6 + A}$, from which we can conclude that $A = 3$. **(e)** The sequence diverges. Thus, taking limits of the recursion formula yields a meaningless result. **(h)** Prove by induction that $\{a_n\}$ is decreasing and bounded below by 0. Thus, it converges to, say, A. Then from the recursion formula, we have no information as to what the limiting value is. See Exercise 7(f) of Section 1.3 to obtain an explicit formula for a_n. Then take limits. **(j)** Use the explicit formula for a_n from Exercise 7(h) of Section 1.3, and then take limits.

13. Write $B - (b_{n+1})^2 = \left[B - (b_n)^2\right]\left[\frac{B - (b_n)^2}{3(b_n)^2 + B}\right]^2$, and consider cases $B > 1$, $B = 1$, and $B < 1$. In each case, verify that $\{b_n\}$ converges so that limits of the recursion formula can be taken.

14. **(a)** converges to 1; **(b)** does not converge; **(c)** converges to 1; **(d)** converges to -1; **(e)** converges to 0; **(f)** oscillates

16. **(a)** Use part (c) of Theorem 1.7.1. **(b)** Use induction.

18. To prove that (b) implies (a), let S be an arbitrary bounded subset of $\Re$. To show that $\sup S$ exists and is finite, from an increasing sequence $\{a_n\}$ of points in S, and a decreasing sequence $\{b_n\}$ of upper bounds for S, that both converge to one value, namely, $\sup S$.

Section 2.5

2. (a) $\{x \mid x \in [-2,4]\}$; (b) $\{x \mid x \in (-\infty,2]\}$; (c) $\{1\}$; (d) ϕ; (e) $\{x \mid x \in [0,1]\}$
3. (a) $a_n = (-1)^n \frac{n}{n+1}$; (b) $S = \{x \mid x \in (0,1) \cup \{2\}\}$; (c)
6. (a) Suppose $\sup S \neq \max S$ and show that $s_0 \equiv \sup S$ is an accumulation point of S.
7. (b) Use partial fraction decomposition of $\frac{1}{n(n+1)}$ to show that $0 < a_m - a_n < \frac{1}{n+1}$ for $m > n$. (c) If $n > 1$, then $\frac{1}{n^2} < \frac{1}{n-1} - \frac{1}{n}$ and hence, $0 < a_m - a_n < \frac{1}{n} - \frac{1}{m} < \frac{1}{n}$ for $m > n$.
   (d) Consider $a_{2n} - a_n$. (f) Use Example 1.3.4.
11. Use an approach similar to the one used in proving Theorem 2.2.1, parts (a) and (b).
12. (a) To prove part (b) of Corollary 2.5.12, first, prove by induction that
    $\left|a_{n+p} - a_{n+p-1}\right| \leq k^p \left|a_n - a_{n-1}\right|$.
13. Consider subtracting $a_{n+1} = (a_n)^2$ from $a_{n+2} = (a_{n+1})^2$.
16. (a) Use the contraction principle to show that the sequence $\{a_n\}$ converges to, say, A. To find A, either write a_n explicitly using the method from Example 1.3.12, or use the fact that $a_{n+2} - a_{n+1} = \frac{1}{2}(a_n - a_{n+1})$.
19. $\{a_n\}$ converges to $\frac{\pi}{2}$.
20. The limiting value is approximately 0.73908133.

Section 2.6

1. (a) yes; (b) no; (c) no
2. (a) The subsequential limit points α are 0 and 1. (b) $\alpha = 0,1,-1$; (d) $\alpha = 1,-1$
3. Consider the sequence $\{a_{2^n}\}$.
8. Use Exercise 7 and Theorem 2.6.5.

Section 2.7

1. T	10. T	19. T	28. T	36. F	44. F
2. T	11. F	20. F	29. F	37. F	45. F
3. F	12. F	21. F	30. F	38. F	46. F
4. T	13. F	22. F	31. F	39. F	47. F
5. F	14. T	23. F	32. T	40. T	48. F
6. T	15. T	24. F	33. F	41. F	49. F
7. F	16. F	25. F	34. F	42. F	50. T
8. T	17. T	26. T	35. T	43. T	51. T
9. T	18. T	27. F			

Section 3.1

1. $\lim_{x \to -\infty} f(x) = -\infty$ if and only if for any real $K > 0$, there exists $M > 0$ such that $f(x) < -K$ provided $x \leq -M$ and x is in the domain of f.
4. Factor out x^n from the left-hand side.

5. (a) $-\frac{1}{2}$; (b) 0; (c) $-\frac{2}{3}$; (e) 0; (f) does not exist; (g) $+\infty$; (i) -1; (k) 0
6. (c) $+\infty$
8. (b) oblique asymptote $y=2x-2$; (c) no asymptotes; (e) horizontal asymptote $y=0$ as x tends to $-\infty$, and oblique asymptote $y=x^3$ as x tends to $+\infty$
10. (a) Since for any $x\in[1,\infty)$ there exists $n\in N$ such that $n\leq x\leq n+1$, deduce that $\left(1+\frac{1}{n+1}\right)^n\leq\left(1+\frac{1}{x}\right)^x\leq\left(1+\frac{1}{n}\right)^{n+1}$, and use the sandwich theorem. (b) $\frac{1}{e}$; (c) $+\infty$
13. Try a proof by contradiction.
15. $\sup S=1$, $\inf S=\min S=0$, $\max S$ does not exist.

Section 3.2

1. (a) 1; (b) 2; (c) 2; (d) 0; (e) $\frac{1}{\sqrt{2}}$; (f) 2
5. (b) Verify that $0\leq\left|x\sin\frac{1}{x}\right|\leq|x|$, and then use the sandwich theorem.
7. (a) 5; (b) $\frac{3}{2}$; (c) 1; (d) 0; (e) 1
8. (a) 1; (b) 1; (c) Limit does not exist.
14. (a) false; (b) false; (c) true
15. $\lim_{x\to 0}f(x)=1$, $\lim_{x\to 1}f(x)$ does not exist.

Section 3.3

2. Consider three possibilities: $f(a^+)=L$, $f(a^+)=+\infty$, or $f(a^+)=-\infty$.
4. (a) 1; (b) -2; (c) 0; (d) $+\infty$; (e) 0; (f) 0; (g) $+\infty$; (h) does not exist
5. (a) $x=0$ is an odd vertical asymptote. (b) $x=0$ is an odd vertical asymptote. (d) $x=-1$ is an even vertical asymptote.
9. (b) $f(0^+)=+\infty$, $f(0^-)=0$, $\lim_{x\to\pm\infty}f(x)=1$; (c) $f(0^+)=1$, $f(0^-)=0$; (d) $f(0^+)=0$, $f(0^-)=1$; (e) $f(0^+)=-\infty$; (g) $f(0^+)=+\infty$, $f(0^-)=-\infty$, $f(-1)=0$
10. (a) 0
11. (a) true; (b) false
13. (b) false
14. (a) 0; (b) 0; (c) 0; (d) 0
15. (a) 1; (b) 0; (c) 0; (d) 1
16. (a) correct; (b) incorrect
19. converges to e^{-1}
21. false

Section 3.4

1. T	9. F	17. F	25. T	33. F	41. F
2. T	10. T	18. T	26. T	34. F	42. T
3. F	11. T	19. T	27. F	35. F	43. T
4. F	12. T	20. T	28. F	36. F	44. T
5. F	13. F	21. T	29. F	37. T	45. T
6. T	14. F	22. F	30. F	38. F	46. T
7. T	15. F	23. F	31. F	39. F	47. F
8. T	16. F	24. F	32. F	40. F	

Section 4.1

1. $(-1,1)$
2. **(a)** on $\Re$; **(b)** when $x=\pm\frac{1}{n}$; **(c)** on $\Re$; **(d)** on $\Re$; **(f)** on $\Re\setminus\{1\}$; **(i)** on $\Re^+$; **(j)** on ϕ; **(k)** on $\Re$
3. **(a)** f and g are continuous at $x=0$. **(b)** f and g are continuous at $x=\frac{1}{2}$.
5. **(a)–(f)**, **(h)–(j)**, **(m)**, and **(n)** are false; others are true.
7. **(a)** Use Exercise 5(k). **(b)** no
9. Use Exercise 8.
10. **(b)** $f(x)=\begin{cases} x, & \text{if } x \text{ is rational} \\ 0, & \text{if } x \text{ is irrational;} \end{cases}$ **(d)** $f(x)=\begin{cases} 0, & \text{if } x \text{ is rational and } -2\le x\le 2 \\ \sin\frac{1}{x}, & \text{if } x \text{ is irrational and } -2\le x\le 2 \end{cases}$
11. **(a)** Consider cases: $b=1$, $b>1$ and $a=0$, $b>1$ and $a\neq 0$, and $0<b<1$.
12. **(a)** true; **(b)** false
15. converges to 0

Section 4.2

1. **(a)** not removable; **(b)** set $f(0)=1$; **(c)** set $f(0)=1$; **(d)** set $f(0)=0$; **(e)** set $f(0)=-1$; **(f)** no discontinuity at $x=0$; **(g)** set $f(0)=1$; **(h)** not removable; **(i)** not removable; **(j)** set $f(0)=0$; **(k)** set $f(0)=0$
2. **(b)** jump discontinuity at $x=0$ with jump of 2; **(c)** removable discontinuities at $x=-\frac{3}{2},1$ and jump discontinuities at $x=-1,0,1$; **(f)** removable discontinuities at $x=\pm\frac{1}{n}$, an oscillating discontinuity at $x=0$, and jump discontinuities at $x=1,-1$; **(h)** oscillating discontinuities at any $x\in(-\infty,0]$; **(i)** no discontinuities; **(j)** jump discontinuity at $x=0$ with the jump of 1; **(k)** infinite discontinuity at $x=0$; **(l)** oscillating discontinuity at $x=0$; **(p)** jump discontinuities at $x=1,-1$
3. all are false
4. **(d)** 4π; **(e)** $f(x)=\tan x$

Section 4.3

1. The three conditions are: f continuous, interval closed, interval bounded.
2. **(a)** If $\{y_n\}$ is a sequence in $f([a,b])$ converging to z_0, prove that $z_0\in f([a,b])$. Apply Theorem 2.6.4 to a sequence $\{x_n\}$ in $[a,b]$ for which $f(x_n)=y_n$.
3. **(b)** $f:[0,2]\to\Re$ defined by $f(x)=\begin{cases} \frac{x}{2}, & \text{if } 0\le x<1 \\ 3-x, & \text{if } 1\le x\le 2 \end{cases}$
5. Use Theorem 4.3.3.
6. Consider $f(x)=x^n-a$ for $x\ge 0$, and apply Theorem 4.3.3.
7. Argue by contradiction.
8. Try a proof by contradiction.
10. continuity
13. Verify that $\min\{f(x_1),f(x_2)\}\le\dfrac{k_1f(x_1)+k_2f(x_2)}{k_1+k_2}\le\max\{f(x_1),f(x_2)\}$, and apply

Theorem 4.3.3.

14. **(a)** $-2 \le x \le 1$; **(b)** $x < \sqrt{3}$ or $x > \sqrt{3.1}$; **(c)** $x \le 1$; **(d)** $x \in (-\infty, 0) \cup (0,1)$; **(e)** $x \in (-\infty, 0) \cup (0,1]$; **(g)** $x \in (-\infty, -3) \cup (-1,2)$; **(i)** $x \in \left(-\infty, -\frac{1}{2}\right] \cup (1, \infty)$

15. **(a)** Use $f(x) = 2^x - 3x$. **(b)** Use $g(x) = \frac{1}{3}\left(2^x\right)$.

Section 4.4

1. **(a)** no; **(b)** yes; **(c)** yes; **(d)** yes; **(e)** no; **(f)** yes

2. **(a)** false; **(b)** true; **(c)** true; **(d)** true; **(e)** false; **(f)** true; **(g)** false; **(h)** false; **(i)** false; **(j)** true

5. **(c)** Suppose f is unbounded and consider a sequence $\{x_n\}$ in D such that $|f(x_n)| \ge n$ for each $n \in N$. **(d)** Use $|(fg)(x) - (fg)(t)| \le |f(x)||g(x) - g(t)| + |g(t)||f(x) - f(t)|$.

6. **(a)** $f_1(x) = -x^2$; **(b)** $f_2(x) = \sin\left(x^2\right)$; **(d)** $f_4(x) = \sin x$

7. **(a)** yes; **(b)** no; **(c)** no; **(d)** no; **(e)** no; **(f)** no

8. $f(x) = \sqrt{x}$ with $x \in [0,1]$

9. Use Theorem 4.3.7 to obtain existence. To prove uniqueness, assume there are two.

11. **(a)** Define $g(x) = |x - f(x)|$ and show that g attains its maximum value, which is 0.

13. Use definition.

Section 4.5

1. T	10. F	19. T	27. T	35. F	43. T
2. T	11. F	20. T	28. F	36. T	44. F
3. F	12. F	21. F	29. F	37. F	45. T
4. F	13. F	22. F	30. F	38. F	46. T
5. T	14. T	23. T	31. T	39. F	47. T
6. T	15. F	24. F	32. T	40. T	48. F
7. T	16. F	25. T	33. F	41. T	49. F
8. F	17. T	26. F	34. F	42. T	50. T
9. T	18. T				

Section 5.1

2. A variety of formulas can be used.

3. **(a)** no; **(b)** yes; **(c)** no; **(d)** yes; **(e)** no; **(f)** no; **(g)** yes; **(h)** yes; **(i)** yes at $x = 0$ and no at $x = 1$; **(j)** no

4. **(a)** If $m = 0$, f is not continuous. If $m = 1$, f is continuous but not differentiable. If $m = 2$ or 3, f is differentiable. **(b)** If $x \ne 0$, f is not continuous, but $f'(0) = 1$.

5. **(a)** Determine when $f'(x) = 6$. **(b)** false

6. Use Theorem 5.1.8 to prove that f is not differentiable at any irrational value.

7. Split the fraction into two convenient parts.

8. **(a)** See Exercise 3(d); **(b)** $f(x) = \begin{cases} (x-1)^2(x+1)^2, \text{ if } x \text{ is rational} \\ 0, \text{ if } x \text{ is irrational;} \end{cases}$

(c) $f(x)=\begin{cases}(x+1)x^2, & \text{if } x \text{ is rational}\\ 0, & \text{if } x \text{ is irrational};\end{cases}$ (d) not possible;

(e) $f(x)=\begin{cases}x^2\sin\frac{1}{x}, & \text{if } x \text{ is irrational}\\ 0, & \text{if } x \text{ is rational};\end{cases}$ (f) $f(x)=\begin{cases}0, & \text{if } x \text{ is rational}\\ 1, & \text{if } x \text{ is irrational};\end{cases}$

(g) $f(x)=\sum_{k=0}^{\infty}\frac{1}{2^k}\cos\left(3^k x\right)$. Also, see Part 3 of Section 8.8.

9. (a) Use substitution in evaluating the limit. **(b)** $c=9$

10. (a) Write $2f(x+h)-f(x)-f(x-h)=[f(x)-f(x-h)]+2[f(x+h)-f(x)]$.

11. Use substitution in evaluating the limit.

12. First show that $f(0)=1$. Then, compute the limit that represents $f'(0)$ and use this limit when evaluating $f'(x)$.

13. Use a definition of the derivative and Exercise 10(a) of Section 3.1.

Section 5.2

1. **(b)** Write $|x|=\sqrt{x^2}$. Use it to get $\frac{d}{dx}\ln|x|=\frac{1}{x}$. **(d)** $\frac{-2xy^2-3y}{3x+2x^2y}$; **(e)** $y=-\frac{1}{4}x+\frac{3}{2}$; **(f)** $y=-\frac{4}{7}x+\frac{29}{7}$; **(g)** First show that $g(x)=x\sqrt[x]{x}$ is eventually increasing. **(h)** e^x; **(i)** $-\cos 1$; **(j)** $\frac{1}{\sqrt{1-x^2}}, \frac{1}{1+x^2}$; **(k)** $\frac{\cos x}{|\cos x|}$

2. To prove the second part, use the definition and the "trick" of adding a 0.

4. Use mathematical induction for both parts.

5. Change appropriate inequalities in the given proof of Theorem 5.2.8.

6. $f(x)=\frac{|x|}{x}$ if $x\in(-1,0)\cup(0,1)$, and $f(0)=0$.

7. $f(x)=x$ with $0\le x\le 1$ satisfies conditions (a) and (b) of Theorem 5.2.11 but not the conclusion. $f(x)=|x|$ with $-1\le x\le 1$ satisfies conditions (a) and (c) of Theorem 5.2.11 but not the conclusion. $f(x)=x$ with $0\le x<1$ and $f(1)=0$ satisfies conditions (b) and (c) of Theorem 5.2.11 but not the conclusion.

9. Use the mean value theorem with arbitrary x_1 and x_2 in the interval $[a,b]$.

10. **(a)** ($\Rightarrow$) Suppose that $a<x_1<x_2<b$ and use the mean value theorem. ($\Leftarrow$) Compute $\lim_{x\to a}\frac{f(x)-f(a)}{x-a}$ by computing the two-sided limits. **(b)** no; **(c)** yes

11. Consider the difference and use Exercise 9.

13. **(a)** Use the definition of uniform continuity and the mean value theorem. **(b)** Consider a function with a vertical tangent at an endpoint.

14. Use the mean value theorem on $g-f$. The converse is false.

15. **(a)** Use Exercise 10. Converse is false. **(b)** Use Exercise 10. Converse is false. **(d)** The relative and absolute minimum is $f\left(\frac{1}{e}\right)=\frac{1}{\sqrt[e]{e}}$. **(e)** Local minimum at $x=0$ and local maximum at $x=\frac{2}{3}$.

16. **(a)** Use Bolzano's intermediate value theorem to prove existence, and use Exercise 12 to prove uniqueness. **(b)** Try a proof by contradiction.

18. Use Rolle's theorem. It is possible for f' to have more roots than f.

19. Consider $g(x) = \exp(-\alpha x)f(x)$ and Rolle's theorem.

20. **(a)** Consider $f(t) = (1+t)^r$ and the mean value theorem. **(b)** Consider $f(x) = e^x$ and the mean value theorem. **(d)** Consider $f(x) = \arctan x$ and the mean value theorem. **(e)** Consider $f(x) = \sin x$ and the mean value theorem. **(f)** Consider $f(x) = \ln(x+1)$ and the mean value theorem. **(h)** Consider $f(x) = tx - x^t$ with $x \geq 0$, and use Exercise 15(a). Then, put $x = \frac{a}{b}$ with $a \geq 0$ and $b > 0$. The case when $b = 0$ is trivial. **(i)** Use $f(x) = \cos x$, $g(x) = 1 - \frac{1}{2}x^2$, and Exercise 14. **(j)** Determine where $f(x) = \frac{\ln x}{x}$ attains its maximum value.

21. **(a)** Consider $f(x) = \dfrac{a_n x^{n+1}}{n+1} + \dfrac{a_{n-1}x^n}{n} + \ldots + \dfrac{a_1 x^2}{2} + a_0 x$ and Rolle's theorem.

25. The mean value theorem does not guarantee the same c for both functions. Also, $g(a)$ could possibly equal $g(b)$, giving zero in the denominator.

26. To prove the first part, observe that $g(a) \neq g(b)$ and use the Cauchy's mean value theorem. For the second part, consider $h(x) = f(x)g(x) - f(a)g(x) - g(b)f(x)$ and Rolle's theorem.

28. Pick x, t to be any two distinct real numbers. Use the given condition to write a difference quotient. The sandwich theorem and Exercise 9 complete the proof.

29. Use the mean value theorem.

30. Use the mean value theorem on f in order to prove that $f'(x) \geq \frac{f(x)}{x}$, which in turn guarantees that $g'(x) \geq 0$.

31. Use Remark 5.2.15 to prove that $\frac{f(c+h)-f(c)}{h}$ approaches A when h approaches 0 from the right or from the left.

32. Use Exercise 1(b) of Section 5.1 with $\varepsilon = \frac{1}{2}f'(a)$.

33. **(a)** Consider $f(x) = 2\sqrt[11]{x^7}$. **(b)** $\frac{3}{2}$; **(c)** 0

35. Apply Theorem 5.2.17 to the functions $e^x f(x)$ and e^x on the interval (a,x).

36. Use $h(x) = f(x)\exp[g(x)]$ and Rolle's theorem.

37. Use $h(x) = \exp(-x) - \sin x$ and Rolle's theorem.

43. **(b)** $f(x) = \sqrt{2x}$

Section 5.3

1. Consider Exercise 10 of Section 5.2.

2. **(a)** f is concave up on $\left(-\sqrt{6},0\right) \cup \left(\sqrt{6},\infty\right)$. **(b)** Choose, say $f(x) = x^3$, $g(x) = \sin x$, $h(x) = \sqrt[3]{x}$. **(c)** $f'''(0)$ does not exist. **(d)** Try implicit differentiation.

4. Choose an appropriate value of n in the function $f(x) = \begin{cases} x^n \sin\frac{1}{x}, & \text{if } x > 0 \\ 0, & \text{if } x \leq 0. \end{cases}$

7. $f(x) = \frac{1}{x}\sin x^2$

8. Use mathematical induction.

10. Use the chain rule and the product rule.

11. **(a)** Use definitions. **(b)** no; **(c)** yes

14. Use Definition 5.3.6.

15. no

18. Define $F(t)=f(t)+\sum_{k=1}^{n}\frac{f^{(k)}(t)}{k!}(x-t)^k$ and $G(t)=\frac{(x-t)^{n+1}}{(n+1)(x-a)^n}$ on $[a,x]$, and use Cauchy's mean value theorem.

20. **(a)** $p_n(x)=1+x+x^2+...+x^n$; **(b)** $p_n(x)=1-\frac{x^2}{2!}+\frac{x^4}{4!}-\frac{x^6}{6!}+...+(-1)^{\frac{n+1}{2}}\frac{x^n}{n!}$, $n=1,3,5,...$; **(d)** $p_n(x)=(x-1)-\frac{1}{2}(x-1)^2+\frac{1}{3}(x-1)^3-...+\frac{(-1)^{n+1}}{n}(x-1)^n$

22. **(b)** Observe that the line segment joining the endpoints $(0,0)$ and $\left(\frac{\pi}{2},-1\right)$ is above the function $f(x)=-\sin x$. **(c)** The line segment joining the endpoints $(a,f(a))$ and $(b,f(b))$ is above the function $f(x)=\ln\frac{1}{x}$.

25. If $0<a\le b$, apply the mean value theorem to f over the intervals $[0,a]$ and $[b,a+b]$.

27. **(a)** Let $x_i\in(a,b)$ for $i=1,2,...,n$, let $A=\frac{x_1+x_2+...+x_n}{n}$, and define g on (a,b) by $g(x)=f(x)-xf'(A)$. Then, $g''(x)<0$ and g attains its maximum at $x=A$. This gives $g(x)\le g(A)$. Use this inequality to obtain the desired result. **(c)** Consider $f(x)=\ln x$.

28. Try a proof by contradiction.

29. Apply Rolle's theorem successively to $f,f',...,f^{(n-1)}$.

31. **(a)** Local minimum at $x=-\frac{3}{2},1$ and local maximum at $x=\frac{1}{2}$. **(c)** Local minimum at $x=0,1$ and local maximum at $x=\frac{1}{3}$.

Section 5.4

1. **(a)** $-\frac{1}{6}$; **(b)** 0; **(c)** $-\infty$; **(d)** 0; **(e)** k; **(f)** $\frac{5}{2}$; **(g)** ∞; **(h)** 0; **(i)** 0; **(j)** 0; **(k)** $\frac{1}{e}$; **(l)** $\frac{1}{e}$; **(m)** 0; **(n)** limit does not exist; **(o)** $\ln 2$; **(p)** 1; **(q)** 0; **(r)** 1; **(s)** 0; **(t)** $-\infty$

3. Set $t=\frac{1}{x}$, write $\lim_{x\to\infty}\frac{f(x)}{g(x)}=\lim_{t\to 0^+}\left[f\left(\frac{1}{t}\right)\div g\left(\frac{1}{t}\right)\right]$, and apply L'Hôpital's rule.

4. Differentiate with respect to h. The converse is false.

5. **(a)** ($\Rightarrow$) Apply the mean value theorem to f on the intervals $\left[a,\frac{a+b}{2}\right]$ and $\left[\frac{a+b}{2},b\right]$. ($\Leftarrow$) Let $a=x-h$, $\frac{a+b}{2}=x$, and $b=x+h$, and apply Theorem 5.4.4. **(b)** Use $f(t)\equiv\ln\frac{\sin t}{t}$ to prove that $2\ln\frac{\sin x}{x}\ge\ln\frac{\sin a}{a}+\ln\frac{\sin b}{b}$.

6. Apply L'Hôpital's rule and Exercise 10(g) of Section 5.1.

7. **(a)** 0; **(b)** 1; **(c)** f is increasing on $(0,e]$ and decreasing on $[e,\infty)$.

11. **(b)** 1; **(c)** -1

14. $\lim_{x\to 0}\frac{f'(x)}{g'(x)}=\frac{1}{2}$ and $\lim_{x\to 0}\frac{f(x)}{g(x)}=2$

15. **(a)** Use L'Hôpital's rule or Remark 3.2.11. **(b)** Use definition of the derivative and L'Hôpital's rule. **(c)** Simple roots are at $x=k\pi$, k a nonzero integer. **(d)** After differentiating $xf(x)=\sin x$ and adding the expressions, argue the fact that $f'(c)=0$ if the relative extremum occurs at $x=c$. **(e)** Use identities $\sin 2u=2\sin u\cos u$, $\cos^2\frac{x}{2}=\frac{1}{2}(1+\cos x)$, and the fact that $\lim_{t\to 0}\frac{\sin t}{t}=1$. **(f)** Use part (e) to show that the desired expression has the value of $\frac{3\sqrt{3}}{4\pi}$. **(g)** $\frac{\pi}{180}, \frac{\pi}{180}\cos x$

16. The result is not true if the condition that $\lim_{x\to c} f'(x)$ is finite is removed.

17. **(a)** yes; **(b)** $\{a_n\}$ converges to 1.

18. 3 times

20. For the case when $x=0$ use L'Hôpital's rule to show that $\lim_{x\to 0} g'(x)=g'(0)$.

Section 5.5

1. F	**10.** T	**19.** F	**28.** T	**37.** F	**45.** F
2. T	**11.** T	**20.** F	**29.** F	**38.** F	**46.** T
3. T	**12.** F	**21.** T	**30.** F	**39.** T	**47.** T
4. T	**13.** T	**22.** T	**31.** F	**40.** F	**48.** T
5. F	**14.** F	**23.** all T	**32.** T	**41.** F	**49.** T
6. F	**15.** T	**24.** all F	**33.** T	**42.** F	**50.** T
7. T	**16.** T	**25.** T	**34.** F	**43.** T	**51.** F
8. T	**17.** F	**26.** T	**35.** F	**44.** T	**52.** T
9. F	**18.** F	**27.** T	**36.** F		

Section 6.1

1. Use the fact that $\inf_{a\le x\le b} f(x)\le m_k \le f(c_k)\le M_k \le \sup_{a\le x\le b} f(x)$.

4. false

6. For any partition P, $m_k \ge 0$ for every k. Thus, $L(P,f)=\sum_{k=1}^{n} m_k \Delta x_k \ge 0$.

7. Pick $f,g:[-2,3]\to\Re$ given by $f(x)=\begin{cases}0, & \text{if } x \text{ is rational}\\ 4, & \text{if } x \text{ is irrational}\end{cases}$ and $g(x)=\begin{cases}4, & \text{if } x \text{ is rational}\\ 0, & \text{if } x \text{ is irrational.}\end{cases}$

8. Use Exercise 55 of Section 1.9

9. Consider the partition P_n with $|P_n|=\frac{2}{n}$ and formulas from Example 1.3.3 and Exercise 2 of Section 1.3.

10. For both parts (a) and (b) try $f:[0,2]\to\Re$ defined by $f(x)=\begin{cases}1, & \text{if } x \text{ is rational}\\ -1, & \text{if } x \text{ is irrational.}\end{cases}$

11. They are equal if $f(x)\ge 0$ and negatives of each other if $f(x)\le 0$.

12. $3, \frac{\pi}{4}$

13. Relate integrals to areas.

Section 6.2

1. Try $P=\{x_0=0, x_1=1-\frac{\varepsilon}{8}, x_2=1+\frac{\varepsilon}{8}, x_3=2-\frac{\varepsilon}{8}, x_4=2+\frac{\varepsilon}{8}, \ldots, x_8=4\}$.

3. Try a proof by contradiction.

4. Start with the definition of the limit and Theorem 6.2.1.

7. Use Exercise 8 of Section 4.1 when proving by contradiction.

8. Prove only that $f\in R[0,1]$.

10. **(b)** Use a function similar to the one given in Exercise 3 of Section 6.1. **(c)** true

11. **(a)** See Exercise 3 in Section 6.1. **(b)** Use $f(x)=x-[x]$, the Dirichlet's function, etc.

Section 6.3

4. (a) Write $fg = \frac{1}{4}(f+g)^2 - \frac{1}{4}(f-g)^2$. (b) Try the functions given as answers to Exercise 7 of Section 6.1. (c) $f, g:[0,1] \to \mathfrak{R}$ defined by $f(x)=0$ and $g(x) = \begin{cases} 0, & \text{if } x \text{ is rational} \\ 1, & \text{if } x \text{ is irrational.} \end{cases}$ (d) Each statement is false.
5. (a) Use Theorems 6.3.2 and 6.3.4, and Exercise 3(a) of Section 1.7.
6. Consider the Dirichlet's function f and $g(x) = \begin{cases} 1, & \text{if } 0 < x \le 1 \\ 0, & \text{if } x = 0. \end{cases}$
7. (a) Note that f is bounded away from 0. Now use Theorems 4.3.2 and 6.3.4.
8. Write $f \vee g = \frac{1}{2}\left[(f+g) + |f-g|\right]$ or use upper and lower sums.
9. Define $h = g - f$. Now use Exercise 13 of Section 6.2 and Theorem 6.3.1.
11. They all are Riemann integrable on $[0,1]$.
12. First show that $0 \le \int_{t-1}^{t} \frac{1}{x}dx - \frac{1}{t} \le \frac{1}{t-1} - \frac{1}{t}$, for $t > 1$. Use this to show that $\{a_n\}$ is a Cauchy sequence.
13. Consider $\int_a^b (\alpha f + \beta g)^2$ with $\alpha = \int_a^b g^2$ and $\beta = -\int_a^b fg$.
14. If $A = \left(\int_a^b |f|^p\right)^{\frac{1}{p}}$ and $B = \left(\int_a^b |g|^q\right)^{\frac{1}{q}}$, then in Exercise 20(h) of Section 5.2, let $a = \frac{|f|^p}{A^p}$, $b = \frac{|g|^q}{B^q}$, $\alpha = \frac{1}{p}$, and $1 - \alpha = \frac{1}{q}$.
15. For the case when $p = 2$, start with $\int_a^b (f+g)^2$ and use Exercise 13.
16. (b) Use Theorems 4.3.2 and 4.3.3.
17. Define $f:[0,\pi] \to \mathfrak{R}$ by $f(x) = \sin x$. **18.** $\frac{5}{3}$

Section 6.4

4. Choose, say, $f:[0,1] \to \mathfrak{R}$ defined by $f(x) = \begin{cases} x^2 \sin\frac{1}{x}, & \text{if } x \ne 0 \\ 0, & \text{if } x = 0. \end{cases}$
5. (a) $\frac{2}{3} - \frac{3\sqrt{3}}{8}$; (c) $\frac{x}{2}[\sin(\ln x) - \cos(\ln x)] + C$; (d) $\arctan 1 - \frac{1}{2}\ln 2$; (e) $1 - \cos(\ln 2)$; (f) 2; (g) $2e - 2$; (h) $\frac{1}{2}\arctan(2x+1) + C$; (i) $\frac{1}{x}e^x + C$; (j) Try $u = -\frac{1}{x}$; (k) $2\ln 2 - 1$; (l) not possible; (m) $-2\cos\sqrt{x} + C$; (n) $\frac{\pi}{4} - \frac{1}{2}$; (o) $\frac{1}{8}\frac{x}{x^2+4} + \frac{1}{16}\arctan\frac{x}{2} + C$; (p) $\ln\left[|x|\left(x^2+2x+2\right)\right] + \arctan(x+1) + C$
7. (a) $\sqrt{2+x^3}$; (b) $\sqrt{x^2+1}$; (c) $2x\arctan\left(x^4\right) - 2\arctan\left(4x^2\right)$
9. What is $f'(x)$ equal to?

10. (a) $\frac{1}{16}$; (b) $+\infty$; (c) 0; (d) $\frac{2}{3}$

12. Both statements are false.

14. Use induction to prove that $f(x)=f(a)+f'(a)(x-a)+\ldots+\frac{1}{n!}f^{(n)}(a)(x-a)^n+R_n(x)$, where $R_n(x)$ is as given.

17. $f(x)=\exp(x^2)$

18. Compare $\int_1^x \frac{1}{\sqrt{t}}dt$ with $\int_1^x \frac{1}{t}dt$.

21. (d) $\frac{1}{4}$; (e) π

22. Let $F(x)=\left[\int_0^x f(t)\,dt\right]^2-\int_0^x [f(t)]^3\,dt$ and show $F(x)\geq 0$ for all $x\in[0,1]$.

Section 6.5

2. Choose $f_1(x)=x=-f_2(x)$.

3. (a) f is improper Riemann integrable on $[1,\infty)$ if $p>1$. (c) infinite; (d) π

4. ($\Rightarrow$) Try a proof by contradiction.

7. (b) does not exist; (c) $-4\ln 2$

9. (a) -1; (b) π; (d) diverges; (e) π; (f) 1; (g) $n!$; (h) $-\frac{\pi}{2}\ln 2$; (i) 0

10. Integrals in parts (b), (c), (d) diverge, others converge.

11. (a) If $F(x)=\int_a^x f(t)dt,\ x\geq a$, then F is increasing and bounded above. Thus, its limit exists. (b) Use $h=g-f$.

12. Use integration by parts.

14. Use L'Hôpital's rule to evaluate the limit. The value is 2.

15. (a) Define $f:[0,\infty)\to\Re$ by $f(x)=\begin{cases}1, & \text{if } n-\frac{1}{n^2}\leq x\leq n \text{ with } n\in N\\ 0, & \text{otherwise.}\end{cases}$

16. 1

18. $f(x)=\frac{1}{x}$ and $g(x)=\frac{1}{x^2}$ on $[1,\infty)$.

20. (a) not improper Riemann integrable; (b) absolutely; (c) conditionally; (d) absolutely

22. Use integration by parts.

24. (a) $\int_0^{\frac{1}{2}} \frac{u^2}{u^4+1}du$; (b) $\int_0^{\frac{\pi}{2}} \cos^2\theta\, d\theta=\frac{\pi}{4}$; (c) $\frac{1}{3}\int_0^{\frac{\pi}{2}} du=\frac{\pi}{6}$

Section 6.6

2. Use induction.

5. (c) $\frac{4}{3}\sqrt{\pi}$

6. (a) In $\Gamma\left(\frac{5}{3}\right)$ use substitution $t=u^3$. (b) Let $x=\sqrt[3]{t}$. (c) $\frac{\pi}{\sqrt{2}}$; (e) $-\Gamma\left(\frac{4}{3}\right)$

7. Use Rolle's theorem.
8. **(a)** Take logarithms of both sides of the given expression and differentiate.
10. **(b)** $\frac{\sqrt{\pi}}{4}\operatorname{erf}^2 x + C$

11. **(a)** Show that $\frac{1}{\sqrt{2\pi}}\int_{-\infty}^{0}\exp\left(-\frac{t^2}{2}\right)dt = \frac{1}{2}$. **(b)** In part (a) let $u^2 = \frac{1}{2}t^2$.

12. **(a)** Use the definition of J_p. **(b)** Use the definition of J_p. **(c)** Use the product rule on part (a). **(d)** Use the product rule on part (b). **(e)** Add parts (c) and (d). **(f)** Subtract parts (c) and (d). **(g)** Set $k = 1$ in part (d).

13. **(a)** Use the definition of J_n. **(b)** Use the definition of J_{-m}. **(c)** Use Exercise 12(b) and Rolle's theorem. **(d)** Use Exercise 12(a) and Rolle's theorem. **(e)** $x^3 J_1 - 2x^2 J_2 + C$

14. **(a)** Replace z, z', and z'' into the given equation and draw the appropriate conclusion.

15. **(c)** $L_0(x) = 1, L_1(x) = 1 - x, L_2(x) = \frac{1}{2}\left(2 - 4x + x^2\right), L_3(x) = \frac{1}{6}\left(6 - 18x + 9x^2 - x^3\right)$

16. **(a)** $T_0(x) = 1, T_1(x) = x, T_2(x) = 2x^2 - 1, T_3(x) = 4x^3 - 3x, T_4(x) = 8x^4 - 8x^2 + 1$; **(b)** Let $\theta = \arccos x$ and show that the Chebyshev equation becomes $\frac{d^2y}{d\theta^2} + n^2 y = 0$. Then show that $T_n(\cos\theta)$ satisfies this equation. **(c)** Use the identity $(\cos k\theta)(\cos m\theta) = \frac{1}{2}\cos(k+m)\theta + \frac{1}{2}\cos(k-m)\theta$. **(d)** Let $u = \arccos x$ and employ the same identity as in part (c).

17. **(a)** $\frac{1}{x}, x > 0$; **(b)** $\frac{1}{x^2}, x > 0$; **(c)** $\frac{1}{x-a}, x > a$; **(d)** $\frac{k}{x^2 + k^2}, x > 0$; **(e)** $\frac{x}{x^2 + k^2}$, $x > 0$; **(f)** $\sqrt{\frac{\pi}{x}}, x > 0$; **(g)** does not exist; **(h)** $\frac{1}{x^2 - 1}, x > -1$

19. **(a)** yes; **(b)** no; **(c)** yes

20. **(a)** Use the inequality $0 \le e^{-xt}|f(t)| < Me^{-(x-a)t}$ and the comparison test.

21. **(d)** $\frac{6kx - 2k^3}{\left(x^2 + k^2\right)^3}$; **(e)** $\frac{1}{2}\ln\frac{x+1}{x-1}$, $\arctan\frac{1}{x}$

22. Break the integral at the multiples of p and, after suitable substitution, use the sum for the converging geometric series.

23. **(a)** Use Definition 6.6.13. **(b)** Use Definition 6.6.13.

24. **(a)** Prove that if $g'(t) = f(t)$, then $\mathcal{L}\{g(t)\}$ exists. Then, use Exercise 21(a).

28. Integrate by parts each integral in exercise 27.

Section 6.7

1. T	10. T	18. T	26. F	34. F	42. T
2. F	11. F	19. F	27. F	35. T	43. T
3. F	12. T	20. F	28. F	36. T	44. T
4. F	13. F	21. T	29. T	37. F	45. T
5. T	14. F	22. T	30. F	38. T	46. F
6. T	15. T	23. T	31. T	39. T	47. T
7. F	16. T	24. T	32. T	40. T	48. T
8. T	17. F	25. F	33. T	41. T	49. T
9. F					

Section 7.1

1. **(a)** converges to $\frac{1}{3}$; **(b)** converges to $-\frac{1}{6}$; **(c)** converges to $\frac{1}{2}$; **(d)** converges to $\frac{1}{2}$; **(e)** diverges; **(f)** diverges; **(g)** diverges; **(h)** converges to $\ln\frac{1}{2}$; **(i)** diverges; **(j)** diverges; **(k)** diverges; **(l)** diverges
3. **(a)** $\frac{2}{1}$; **(b)** $\frac{5}{1}$; **(c)** $\frac{1061}{330}$
4. Only (c) is true.
5. **(a)** $1 < x < 3$; **(b)** $|x| < 5$; **(c)** $|x| < \frac{\pi}{4}$; **(d)** $\frac{1}{e} < x < e$
8. Show that $\frac{1}{a_k}$ does not tend to 0.
10. **(a)** $\sum b_k$ converges $\Leftrightarrow \sum(a_k - a_{k+1})$ converges $\Leftrightarrow \{S_n\}$ converges where $S_n = \sum_{k=1}^{n}(a_k - a_{k+1})$. But, $S_n = a_1 - a_{n+1}$. So, $\{S_n\}$ converges if and only if $\{a_{n+1}\}$ converges.
11. $a_k = \dfrac{1}{2^k}$ and $b_k = \dfrac{1}{3^k}$
12. $a_k = \dfrac{1}{3^k}$ and $b_k = \dfrac{1}{2^k}$
13. Choose a_k and b_k to be the same as in Exercise 11.
14. Show that $S_n = b_1 + b_2 + \cdots + b_n \geq a_n\left(1 + \frac{1}{2} + \cdots + \frac{1}{n}\right)$, which is unbounded.
15. **(a)** $|x| < 1$; **(b)** $|x| < 1$ for both
17. Observe that $S_n = \frac{2}{n+1}$ if n is odd, and $S_n = 0$ if n is even.
18. **(a)** Bound the given series by a converging geometric series. **(b)** Use the fact that $n^2 \leq n!$ for $n \geq 4$.
20. true
21. **(a)** Use a triangle inequality and the Cauchy criterion.
23. Consider areas between the curves.

Section 7.2

1. Series given in parts (c), (e), (g), (h), (j), (m), (t), (v), (w), (x), and (z) all converge. Others diverge.
2. Series converges if and only if $p > 1$.
5. Use the fact that eventually $0 \leq (a_k)^2 \leq a_k < 1$.
6. Use the fact that $a_k + a_{k+1} \geq \sqrt{a_k a_{k+1}}$.
7. **(a)** Try a proof by contradiction. **(b)** Choose $a_n = \frac{1}{n}$ if $n = 1, 4, 9, \ldots$, and $a_n = \frac{1}{n^2}$ otherwise. **(c)** Choose $a_k = \frac{1}{k \ln k}$.
9. Each given series converges if and only if $p > 1$.
10. Try the limit comparison test.
11. **(a)** false; **(b)** true
12. $x > 1$
14. Use the Cauchy–Schwarz inequality to write $\left[\sum_{k=1}^{n}\sqrt{a_k}\left(\frac{1}{k}\right)\right]^2 \leq \left(\sum_{k=1}^{n} a_k\right)\left(\sum_{k=1}^{n}\frac{1}{k^2}\right)$, and then apply the comparison test.
15. **(a)** $f(x) = \sin \pi x$; **(b)** See Exercise 15(a) of Section 6.5.
16. Consider $\int_{\frac{1}{n+1}}^{\frac{1}{n}} f$.

Section 7.3

4. (a) diverges; (b) converges
5. (a) Use a_n as given in Example 7.3.7. (b) $b_n = n^n$
7. Series given in parts (a), (b), (d), (e), (f), (g), and (h) all converge. Others diverge.
8. Show that $\sum a_k$ converges.
9. $|x| < 1$
10. (a) diverges; (b) converges; (c) diverges
13. Note that we can write a_n as $\dfrac{1}{3^n}$ if n is odd, and $\dfrac{1}{2^n}$ if n is even.
18. (a) Use Exercise 9. (b) If $x \in (0,1)$ use Cauchy's root test.
20. Use Cauchy's ratio test.

Section 7.4

2. Series given in parts (a), (b), (d), and (e) all converge. Series in part (c) diverges.
3. The series converges for all p.
4. Choose a sequence similar to that in Example 7.3.10.
5. (a) $\sum \frac{(-1)^k}{k^2}$; (b) $\sum \frac{(-1)^k}{k}$; (c) $\sum \frac{1}{k}$
6. (a) converges absolutely; (b) diverges; (c) - (e) converge conditionally
7. $a_n = b_n = \frac{(-1)^n}{\sqrt{n}}$
8. $2 < x \le 4$
9. Use Leibniz's alternating series test.
11. Show that $\sum a_k{}^+$ diverges and $\sum a_k{}^-$ converges.
13. $\sum (-1)^{k+1}$
15. If $\{S_n\}$ is the sequence of partial sums for $\sum a_k$, show that $S_{3n} = \left(1 + \frac{1}{2} + \ldots + \frac{1}{3n}\right) >$ $\ln(3n+1) - \frac{2}{3}(1 + \ln n)$, which tends to $+\infty$ as $n \to \infty$.
16. (a), (c), (e), (f), (g) are true, and (b), (d) are false.
17. (a) Use the fact that $\frac{1}{k^2} = \frac{1}{(2k)^2} + \frac{1}{(2k-1)^2}$. (c) Use parts (a) and (b).
20. (e) no

Section 7.5

1. F	9. T	16. T	23. T	30. F	37. T
2. F	10. F	17. T	24. T	31. F	38. F
3. F	11. T	18. T	25. T	32. F	39. T
4. F	12. T	19. T	26. T	33. F	40. F
5. F	13. T	20. F	27. T	34. F	41. F
6. F	14. T	21. T	28. T	35. T	42. F
7. T	15. T	22. T	29. T	36. T	43. T
8. T					

Section 8.1

1. In parts (a), (b), (d), (e), (f), (h), (j), (k), (n), and (o), $f(x) = 0$. (c) $f(x) = 0$ if $0 \le x < 1$ or $x > 1$, and $f(1) = \frac{1}{2}$; (g) $f(x) = 0$ if $x \ne \frac{\pi}{2}$, and $f\left(\frac{\pi}{2}\right) = 1$; (i) $f(x) = 0$ if

$0 \le x < 1, f(1) = \frac{1}{2}$, and $f(x) = 1$ if $x > 1$; **(l)** $f(x) = 1$; **(m)** $f(x) = 0$ if $x \ne 0$ and $f(0) = 1$

2. **(a)** $f_n(x) = n$; **(c)** $f_n : (0, \infty) \to \mathfrak{R}$, where $f_n(x) = 0$ if $0 < x < \frac{1}{n}$, and $f_n(x) = \frac{1}{x}$ if $x \ge \frac{1}{n}$; **(e)** $f_n(x) = \frac{1}{nx}$ on $(0,1]$; **(g)** characteristic function of $[-n, n]$; **(i)** $f_n(x) = (-1)^n \left(\frac{n-1}{n}\right)$; **(j)** $f_n(x) = \frac{\sin nx}{\sqrt{n}}$ on $[0,1]$; **(l)** $f_n(x) = x^n$ on $[0,1]$
5. **(a)** $f_n(x) = x^n$ on $[0,1]$; **(b)** $f_n(x) = \frac{\sin nx}{\sqrt{n}}$ on $[0,1]$
7. Use the first derivative test to show that f_n attain their relative maximum values at $\frac{1}{n}$.

Section 8.2

1. **(a)** Functions in Exercise 1, parts (a), (d), and (f), of Section 8.1 all converge uniformly. Others do not. **(b)** not uniformly; **(c)** Convergence is uniform on (a, ∞) with $a > 0$. **(d)** If $D = [0, \infty)$ or $D = (0, \infty)$, then convergence is not uniform. Uniform if $D = [1, \infty)$.
4. $f_n(x) = g_n(x) = x + \frac{1}{n}$
5. Use the inequality $|f_n g_n - fg| \le |f_n||g_n - g| + |g||f_n - f|$.
7. $f_n(x) = \frac{\sin nx}{\sqrt{n}}$ on $[0,1]$
8. **(a)** $f_n(x) = \frac{1}{n}$ if $x \ge 0$, and $f_n(x) = -\frac{1}{n}$ if $x < 0$

Section 8.3

1. 0
4. Try a proof by contradiction.
5. If condition (a) does not hold, choose $f_n : [0,1] \to \mathfrak{R}$ defined by, $f_n(x) = 0$ if $x = 0$ or if $\frac{1}{n} \le x \le 1$, and $f_n(x) = 1$ if $0 < x < \frac{1}{n}$. If condition (b) does not hold, choose $f_n(x) = 1 - x^n$ on $[0,1]$. If condition (c) does not hold, choose $f_n : [0,2] \to \mathfrak{R}$ defined by, $f_n(x) = n^2x^2 - 2nx$ if $0 \le x \le \frac{2}{n}$, and $f_n(x) = 0$ if $\frac{2}{n} < x \le 2$ If condition (d) does not hold, choose $f_n(x) = 1 - x^n$ on $[0,1)$.
6. The converse is false. Consider $f_n(x) = \frac{x}{n}$ on $\mathfrak{R}$.
7. $f_n(x) = 1 + \frac{1}{n}$ if $x \ge 0$, and $f_n(x) = -\frac{1}{n}$ if $x < 0$
8. false

Section 8.4

1. Series given in parts **(a)**, **(b)**, **(c)**, **(g)**, **(i)**, and **(j)**, all converge uniformly. Series in parts **(d)**, **(e)**, **(f)**, and **(h)**, do not.
2. **(a)** The pointwise limit is $F(x) = 1$ if $x \ne 0$ with $F(0) = 0$. **(b)** The pointwise limit is $F(x) = 1$ if $x \in (0,1)$ with $F(0) = 0$ and $F(1) = 0$.
3. Use partial sums and refer to Exercise 2 of Section 8.2.
8. Note that $|a_n \sin nx| \le |a_n|$ for all $x \in \mathfrak{R}$.
9. Absolutely on $\left(\frac{1-\sqrt{5}}{2}, \frac{1+\sqrt{5}}{2}\right)$ and uniformly on any compact subset of $\left(\frac{1-\sqrt{5}}{2}, \frac{1+\sqrt{5}}{2}\right)$.

11. **(a)** $\sum_{k=0}^{\infty}(2x)^k$, if $|x|<\frac{1}{2}$; **(b)** $\sum_{k=0}^{\infty}(-1)^k(x-1)^k$, if $0<x<2$; **(c)** $-\frac{a}{b}\sum_{k=0}^{\infty}\left(\frac{x}{b}\right)^k$, if $|x|<b$;

(e) $\sum_{k=1}^{\infty}k2^k x^{k-1}$, if $|x|<\frac{1}{2}$

15. **(a)** Use Weierstrass M-test. **(b)** Use Cauchy's ratio test. **(c)** Choose $x_n=n$, $\varepsilon=1$.

16. If $x\in[0,1]$, then $|a_n x^n|\le|a_n|$ for all $n\in N$. Now use the Weierstrass M-test and Theorem 8.4.15.

17. If $a>1$, then for all $x\in[a,\infty)$, we have $n^{-x}\le n^{-a}$. Now use the Weierstrass M-test and Theorem 8.4.17.

18. In Example 8.4.16 use $x=\frac{1}{3}$ giving $\ln\frac{3}{2}$ as the desired result.

24. **(a)** Multiply F_n by $2\sin\frac{x}{2}$ and use identity $2\sin A\sin B=\cos(A-B)-\cos(A+B)$.

(b) Multiply G_n by $2\sin\frac{x}{2}$ and use identity for $2\cos A\sin B$.

25. Use Theorems 8.4.14 and 8.4.17.

26. Use Exercise 24(b) and Theorem 8.4.14.

Section 8.5

1. **(a)** yes; **(b)** no, unless $R=+\infty$

2. **(a)** $\sum\frac{x^k}{k^2}$; **(b)** $\sum_{k=0}^{\infty}\left(-\frac{1}{2}\right)^k\frac{(x-1)^k}{k+1}$; **(d)** $p(x)=3-(x-1)+(x-1)^2+0(x-1)^3+\cdots$

3. **(a)** $R=1, 1<x<3$; **(b)** $R=2, -2\le x<2$; **(c)** $R=3, -1<x<5$; **(d)** $R=\frac{e}{2}$, $-\frac{e}{2}<x<\frac{e}{2}$; **(e)** $R=3, -2<x<4$; **(f)** $R=4, -3\le x<5$; **(g)** $R=\infty$, $-\infty<x<\infty$

4. Try a proof by contradiction.

5. If $R\in(0,\infty)$, choose $x\in[a-r,a+r]$. Then, $|a_n(x-a)^n|\le|a_n|r^n$. Now apply Weierstrass M-test.

6. Use Cauchy's ratio test.

7. To show $R=1$, show that the series converges if $|x|<1$ and diverges if $|x|>1$.

10. Suppose that the series converges for some $x_0>1$ and reach a contradiction.

13. **(a)** $\frac{x}{(1-x)^2}$; **(b)** 2; **(c)** $\frac{2x^2}{(1-x)^3}$; **(d)** Let $x=\frac{1}{2}$ in part (c). The result is 4. **(e)** Use parts (b) and (d). The result is 6.

14. **(a)** $\ln x=\ln a+\sum_{k=1}^{\infty}\frac{(-1)^{k+1}}{k}\frac{(x-a)^k}{a^k}$, $0<x\le 2a$; **(b)** $\ln(x+1)=\sum_{k=0}^{\infty}\frac{(-1)^k}{k+1}x^{k+1}$. Interval of convergence includes $x=1$. **(c)** Interval of convergence does not include $x=1$.

(d) $2\sum\frac{x^{2k-1}}{2k-1}$

15. **(a)** $(-\infty,\infty)$; **(c)** Differentiate the power series term by term.

16. Differentiate the power series for J_n term by term.

Section 8.6

1. **(a)** $e^a\sum_{k=0}^{\infty}\frac{1}{k!}(x-a)^k$; **(b)** $\frac{\sqrt{2}}{2}-\frac{\sqrt{2}}{2}\left(x-\frac{\pi}{4}\right)-\frac{\sqrt{2}}{2}\frac{1}{2}\left(x-\frac{\pi}{4}\right)^2+\frac{\sqrt{2}}{2}\frac{1}{3!}\left(x-\frac{\pi}{4}\right)^3+\cdots$;

(c) $\sum_{k=0}^{\infty} x^{2k+1}$; (d) $1+\frac{1}{2}\sum_{k=1}^{\infty}(-1)^k\frac{(2x)^{2k}}{(2k)!}$; (e) $\sum_{k=0}^{\infty}(-1)^k\frac{x^{2k}}{(2k+1)!}$; (f) $\sum_{k=0}^{\infty}\frac{1}{(k+1)!}x^k$;

(g) $\sum_{k=0}^{\infty}\frac{(-1)^k}{k!}x^{2k}$; (i) not analytic; (j) $\sum_{k=0}^{\infty}\frac{1}{(2k)!}x^{2k}$

3. (a) Let $\left(e^x\right)^{-1}=a_0+a_1x+a_2x^2 \ldots$. Find a_i's by equating coefficients in $\left(e^x\right)\left(e^x\right)^{-1}=1$.

4. (a) 1; (b) $2\ln 2$; (c) $\frac{2}{5}$; (d) $\frac{1}{2}\ln 2$

5. (a) $f^{(40)}(0)=0$ and $f^{(5)}(0)=24$

6. The limits in (a) - (d) have the value 1. The limit in (e) is 0.

7. (a) and (b) Use power series representation for e^x. (d) To prove the second inequality show the remainder $R_2(x)\le 0$. (e) Show the remainder $R_1(x)\le 0$.

8. Find n for which $|R_n|<0.00005$ and then compute n terms in the series for e.

13. (a) $\tan x = x+\frac{1}{3}x^3+\frac{2}{15}x^5+\frac{17}{315}x^7+\cdots$

Section 8.7

1. T	9. T	17. T	24. F	31. F	38. F
2. T	10. F	18. T	25. F	32. T	39. F
3. F	11. F	19. T	26. T	33. T	40. T
4. F	12. F	20. T	27. F	34. T	41. T
5. T	13. T	21. T	28. F	35. F	42. T
6. T	14. F	22. T	29. T	36. F	43. F
7. T	15. T	23. T	30. T	37. F	44. T
8. T	16. T				

Section 9.1

1. $\sqrt{2L}$

2. For proof of part (c) use the identity $\sin A\cos B=\frac{1}{2}\sin(A-B)+\frac{1}{2}\sin(A+B)$.

6. The Fourier series for f_1, f_2, and f_3 is $\sum(-1)^{k+1}\frac{2}{k}\sin kx$.

7. (a) $\frac{4}{\pi}\sum_{k=1}^{\infty}\frac{1}{2k-1}\sin[(2k-1)x]$; (b) $\frac{\pi}{4}+\sum_{k=1}^{\infty}\left\{\frac{-2}{(2k-1)^2\pi}\cos[(2k-1)x]+\frac{(-1)^{k+1}}{k}\sin kx\right\}$; (c) 5;

(d) $\frac{\pi^2}{3}+4\sum_{k=1}^{\infty}\frac{(-1)^k}{k^2}\cos kx$; (e) $2\sum_{k=1}^{\infty}\frac{(-1)^{k+1}}{k}\sin kx$; (f) $\sin\frac{x}{2}$; (g) $\frac{8}{\pi}\sum_{k=1}^{\infty}\frac{(-1)^{k+1}k}{k^2-1}\sin kx$;

(h) $\frac{2}{\pi}-\frac{4}{\pi}\sum_{k=1}^{\infty}\frac{\cos 2kx}{4k^2-1}$; (i) $\frac{1}{2}-\frac{4}{\pi^2}\sum_{k=1}^{\infty}\frac{1}{(2k-1)^2}\cos[(2k-1)\pi x]$;

(j) $-\frac{1}{2}-\frac{10}{\pi}\sum_{k=1}^{\infty}\frac{1}{2k-1}\sin[(2k-1)x]$; (k) $\frac{1}{2}+\frac{1}{2}\cos 2x$

Section 9.2

2. Show that $f(x)=\sum_{k=1}^{\infty}\frac{(-1)^{k+1}}{k^2}\cos kx$. Then use $x=\pi$.

3. (a) Show that $f \sim 2\sum_{k=1}^{\infty}\frac{(-1)^{k+1}}{k}\sin kx$, and use $x=\frac{\pi}{2}$. (b) Show $f \sim \frac{4}{\pi}\sum_{k=1}^{\infty}\frac{\sin[(2k-1)x]}{2k-1}$, and use $x=\frac{\pi}{2}$.

4. Show that $f(x) \sim \frac{8}{3}+\frac{16}{\pi^2}\sum_{k=1}^{\infty}\frac{(-1)^{k+1}}{k^2}\cos\frac{k\pi x}{2}$. Then use $x=0$.

5. (a) pointwise but not uniform

6. (a) Find the Fourier series for $f(x)=\frac{|x|}{x}$ on $[-\pi,\pi]$. (b) no; (c) yes

9. (b) Use part (a).

10. (a) Integrate expression in Exercise 8.

11. (a) Use Exercises 9(b) and 10(a).

12. Multiply the Fourier series given in Definition 9.1.4 by $f(x)$ and integrate term by term.

Section 9.3

1. (a) $\frac{4}{\pi}\sum_{k=1}^{\infty}\frac{1}{2k-1}\sin[(2k-1)x]$, 1; (b) $\frac{4}{\pi}\sum_{k=1}^{\infty}\frac{(-1)^{k+1}}{k}\sin\frac{k\pi x}{2}$,

$1-\frac{8}{\pi^2}\sum_{k=1}^{\infty}\frac{1}{(2k-1)^2}\cos\frac{(2k-1)\pi x}{2}$; (c) $\sin x$, $\frac{2}{\pi}-\frac{4}{\pi}\sum_{k=1}^{\infty}\frac{\cos 2kx}{4k^2-1}$; (d) $\sum_{k=1}^{\infty}\frac{2}{k}\sin\frac{kx}{2}$,

$\frac{2}{\pi}\sum_{k=1}^{\infty}\frac{1}{k^2}\left[1-(-1)^k\right]\cos\frac{kx}{2}$; (e) $\frac{2}{\pi^2}\sum_{k=1}^{\infty}\frac{1}{k^2}\left[1-(-1)^k\right]\sin k\pi x$,

$\frac{1}{6}+\sum_{k=1}^{\infty}\frac{2}{k^2\pi^2}\left[(-1)^{k+1}-1\right]\cos k\pi x$, (f) $\sin x$,

$\frac{8}{3\pi}\cos\left(\frac{1}{2}x\right)-\frac{8}{\pi}\sum_{k=1}^{\infty}\frac{1}{(2k+1)^2-4}\cos\frac{[(2k+1)x]}{2}$

3. $2+\frac{4}{\pi}\sum_{k=1}^{\infty}\frac{(-1)^{k+1}}{k}\sin k\pi x$

4. No difference

6. x if $x\in(-1,1)$, and 0 if $x=-1,1$

7. (a) $\frac{4}{\pi}\sum_{k=1}^{\infty}\frac{\sin(2k-1)x}{2k-1}$; (b) Set $x=\frac{\pi}{2}$

12. $3+\sum_{k=1}^{\infty}\frac{9}{k^2\pi^2}\sqrt{1+k^2\pi^2}\cos\left(\frac{2k\pi x}{3}+\arctan k\pi\right)$

13. (d) $2+\sin 2x+\frac{16}{\pi}\sum_{k=1}^{\infty}(-1)^{k+1}\frac{k}{4k^2-1}\sin kx$

15. pointwise and uniformly

Section 9.4

1. (a) Let $x=0$ in the Fourier series for f. (b) Let $x=\pi$ in the Fourier series for f. (c) Use Theorem 9.4.3 and then let $x=0$.

2. $\pi^2-2+12\sum_{k=1}^{\infty}\frac{(-1)^k}{k^2}\cos kx$

4. (a) Let $x=\frac{1}{2}$ in the Fourier sine series for f.

5. Try a verification by contradiction.

Section 10.1

2. right triangle

4. $2x-2y-6z=23$

5. (a) $(x-1)^2+(y+2)^2+(z-3)^2=6$; (b) $(x-1)^2+(y+2)^2+(z-3)^2=4$

6. $\left(x-\frac{1}{2}\right)^2+(y-2)^2+(z+1)^2=\frac{57}{4}$

7. Center is $\left(\frac{1}{2},-1,-\frac{3}{2}\right)$, radius is $\frac{3}{\sqrt{2}}$

8. interior of the sphere given in Exercise 7.

Section 10.2

8. (a) $\left(\sqrt{29}\right)\left(\frac{2}{\sqrt{29}}\mathbf{i}+\frac{3}{\sqrt{29}}\mathbf{j}-\frac{4}{\sqrt{29}}\mathbf{k}\right)$; (c) $(1)\langle 1,1,0\rangle+(-1)\langle -1,1,-2\rangle+(3)\langle 0,1,-2\rangle$

Section 10.3

2. (a) $\|\vec{u}\|\|\vec{v}\|$; (b) $-\|\vec{u}\|\|\vec{v}\|$

5. false

6. $\vec{u}=\vec{0}$

7. (a) If $\vec{u}$ and $\vec{v}$ are adjacent sides, show that $\|\vec{u}+\vec{v}\|^2+\|\vec{u}-\vec{v}\|^2=2\|\vec{u}\|^2+2\|\vec{v}\|^2$.
   (b) If $\vec{u}$ and $\vec{v}$ are adjacent sides, show that $(\vec{u}+\vec{v})\cdot(\vec{u}-\vec{v})=0$.

9. (a) In a triangle inequality $\|\vec{u}+\vec{v}\|\le\|\vec{u}\|+\|\vec{v}\|$, replace $\vec{u}$ by $\vec{u}-\vec{v}$. (c) In part (a) replace $\vec{v}$ by $-\vec{v}$. (d) In Theorem 10.3.6 replace $\vec{v}$ by $-\vec{v}$. (e) Use Theorem 10.3.5.

10. (a) When $\theta=0, \theta=\pi, \vec{u}=\vec{0}, \vec{v}=\vec{0}$. (b) When $\theta=0, \theta=\pi, \vec{u}=\vec{0}, \vec{v}=\vec{0}$.

14. $\vec{u}=\left(\frac{4}{5}\mathbf{i}+\frac{8}{5}\mathbf{k}\right)+\left(\frac{6}{5}\mathbf{i}-3\mathbf{j}-\frac{3}{5}\mathbf{k}\right)$

15. $\vec{u}\times\vec{v}=5\mathbf{i}+8\mathbf{j}-6\mathbf{k}$, $\vec{u}\times\vec{u}=\vec{0}$

16. To show that $\vec{u}\times\vec{v}$ is orthogonal to $\vec{u}$, show $(\vec{u}\times\vec{v})\cdot\vec{u}=0$.

20. (a) $5\mathbf{i}+8\mathbf{j}-6\mathbf{k}$; (b) 0; (c) -11

21. (c) no

22. $\vec{u}=\vec{0}$

24. $\frac{21}{2}$

25. (a) Use Exercise 23(a); (b) Use Exercise 23(b)

27. (a) no; (b) yes

Section 10.4

1. (a) $x^2=y^3, x\in\Re$; (b) $x=y^2$ between the points $(1,1)$ and $(1,-1)$;
   (c) $9x^2+4y^2=36$; (d) $xy=1, x>0$

2. (b) directed line segment on the x-axis from x_0 to x_1.

3. $x(t)=\frac{3}{4}t-1,\ y(t)=-2t+3,\ t\in[0,4]$

4. (a) $x(t)=t^3, y(t)=t^5$; (b) $x(t)=t^5, y(t)=t^3$; (c) $x(t)=t^3-1, y(t)=t^3$;
   (d) $x(t)=t,\ y(t)=\sin\frac{1}{t}$ if $t\neq 0, y(0)=0$; (e) $x(t)=t^2, y(t)=t^3-t$

6. (a) $x(t)=t^{\frac{5}{3}}, y(t)=t$; (b) $x(t)=t^5, y(t)=t^3$

7. (a) $(0,2,-1)$ is the point of intersection. (b) no

8. The curve intersects itself at the point $(1,0)$ for the values $t=-1,1$. Tangent lines are $y=x-1$ and $y=-x+1$.

10. (a) For $a=1$ or $a=10$, at $t=\dfrac{\pi}{4}$, the slope of the tangent line is $m=\frac{\sqrt{2}}{2-\sqrt{2}}$. (b) $y=2a$

(e) $3\pi a^2$; (f) $a\arcsin\frac{\sqrt{2ay-y^2}}{a} = \sqrt{2ay-y^2}+x$

11. (b) $\frac{dy}{dx} = -\frac{2\sqrt{3}}{3}$; (d) $\frac{d^2y}{dx^2} = -\frac{16}{9}$ at $t=\frac{\pi}{6}$, and $\frac{d^2y}{dx^2} = -\frac{2}{9}$ at $t=\frac{\pi}{2}$

12. (a) $x^{\frac{2}{3}}+y^{\frac{2}{3}}=a^{\frac{2}{3}}$

13. (a) $(0,-1,0), (0,1,0)$

14. $x = 2a\cot\theta$ and $2a\sin^2\theta$ for $\theta\in(0,\pi)$, or $x=2a\tan\theta$ and $2a\cos^2\theta$ for $\theta\in\left(-\frac{\pi}{2},\frac{\pi}{2}\right)$, $y=\frac{8a^3}{x^2+4a^2}$

16. (a) true; (b) Fundamental period is 2π; (c) none, but the orientation is opposite

17. 2π

Section 10.5

1. (a) $\ell(t)=(-1,2+2t,3+t), t\in\Re$; (b) $\ell(t)=(-1+3t,2+t,3-2t), t\in\Re$

2. $\sqrt{\frac{35}{6}}$

4. $(4,5,1)$

5. $\frac{11}{\sqrt{14}}$

6. $\approx 66°$

7. $x=3+7t, y=2+4t, z=-5t, t\in\Re$

9. $5x-2y-4z=0$

Section 10.6

1. (a) $\lim_{x\to 0}\frac{\sin^2 x}{x^4}=+\infty$ and $\lim_{x\to 0}\frac{\sin x}{x^3}=+\infty$; (b) $\lim_{x\to 0}\frac{x\sin\frac{1}{x}}{\sin x}$ does not exist

2. (b) $\lim_{n\to\infty}\vec{a}_n=\langle 1,1,0\rangle$

3. Converse is false.

9. Converse is false.

14. (b) $t_0=\frac{\pi}{12}$; (c) $\vec{u}'\left(\frac{\pi}{3}\right)=-\frac{\sqrt{3}}{2}\mathbf{i}+\frac{1}{2}\mathbf{j}, \vec{v}'\left(\frac{\pi}{12}\right)=\sqrt{3}\mathbf{i}-\mathbf{j}$, (d) $\sqrt{3}x+3y=2\sqrt{3}$; (e) $\vec{0}$, $\vec{0}$; (f) $\left\langle-\frac{1}{2},-\frac{\sqrt{3}}{2}\right\rangle, \left\langle -2,-2\sqrt{3}\right\rangle$

15. (a) $y=x$ and $y=-x$; (b) $\vec{A}(1)=2\mathbf{i}+6\mathbf{j}, \vec{A}_{\vec{T}}(1)=4\mathbf{i}+4\mathbf{j}, \vec{A}_{\vec{N}}(1)=-2\mathbf{i}+2\mathbf{j}$, and $\vec{A}(-1)=2\mathbf{i}-6\mathbf{j}, \vec{A}_{\vec{T}}(-1)=4\mathbf{i}-4\mathbf{j}, \vec{A}_{\vec{N}}(-1)=-2\mathbf{i}-2\mathbf{j}$; (c) $\left\langle t^2-1,t-t^3\right\rangle$

17. (c) $\vec{r}'(0)=\langle 0,0\rangle$. No tangent line.

19. (b) $t_0=0, t_1=\pi$; (c) $\vec{V}(0)=\langle 1,2\rangle, \vec{V}(\pi)=\langle -1,2\rangle$; (d) $y=2x, y=-2x$; (e) $\vec{T}(0)=\left\langle\frac{1}{\sqrt{5}},\frac{2}{\sqrt{5}}\right\rangle, \vec{T}(\pi)=\left\langle-\frac{1}{\sqrt{5}},\frac{2}{\sqrt{5}}\right\rangle$; (f) Absolute maximum speed is when $t=0$, $\pi, 2\pi$, i.e. at the origin. The other local highest speed is when $t=\frac{\pi}{2},\frac{3\pi}{2}$, i.e. at the points $(1,0)$ and $(-1,0)$; (g) $\vec{A}(0)=\langle 0,0\rangle=\vec{A}(\pi), \vec{A}\left(\frac{\pi}{6}\right)=\left\langle-\frac{1}{2},-2\sqrt{3}\right\rangle, \vec{A}\left(\frac{\pi}{2}\right)=\langle -1,0\rangle$

20. $\vec{A}_{\vec{T}}(1)=\frac{44}{13}\mathbf{i}+\frac{66}{13}\mathbf{j}, \vec{A}_{\vec{N}}(1)=-\frac{18}{13}\mathbf{i}+\frac{12}{13}\mathbf{j}$,

21. $\vec{A}_{\vec{T}}(1)=-\frac{12}{5}\mathbf{i}+\frac{8}{5}\mathbf{j}, \vec{A}_{\vec{N}}(1)=\frac{12}{5}\mathbf{i}-\frac{18}{5}\mathbf{j}$,

22. At $t=1, x=2+\tau, y=-1-2\tau, z=1-\tau$, with $\tau\in\Re$. No tangent at the point $\left(4,-8,\frac{1}{4}\right)$.

25. Unit tangent vectors are $\frac{1}{\sqrt{5}}\mathbf{i}+\frac{2}{\sqrt{5}}\mathbf{j}$ and $-\frac{1}{\sqrt{5}}\mathbf{i}-\frac{2}{\sqrt{5}}\mathbf{j}$. Unit normal vectors are $-\frac{2}{\sqrt{5}}\mathbf{i}+\frac{1}{\sqrt{5}}\mathbf{j}$ and $\frac{2}{\sqrt{5}}\mathbf{i}-\frac{1}{\sqrt{5}}\mathbf{j}$.

Section 10.7

1. **(a)** $\ln(2+\sqrt{3})$; **(b)** 2π; **(c)** 2π
2. **(b)** $\int_c^d \sqrt{1+\left(\frac{dx}{dy}\right)^2}\,dy$
3. $\frac{17}{12}$
4. $8a$
5. $6a$
6. yes
7. $L=\int_a^b \sqrt{[f(t)]^2+[f'(t)]^2}\,dt$
8. 2π
9. **(a)** no; **(b)** yes; **(c)** no; **(d)** no
10. **(a)** $4\sqrt{2}\pi$; **(b)** $\gamma_1(s)=\left(\cos\frac{s}{\sqrt{2}},\sin\frac{s}{\sqrt{2}},\frac{s}{\sqrt{2}}\right)$, $s\in[0,4\sqrt{2}\pi]$; **(c)** $\sqrt{2},1$
11. **(a)** t^2; **(b)** $2t$; **(c)** $\gamma_1(s)=(\cos s,\sin s)$, $s\geq 0$

Section 10.8

1. F	8. T	14. T	20. T	26. T	32. T
2. F	9. T	15. F	21. F	27. T	33. T
3. T	10. T	16. T	22. F	28. T	34. T
4. F	11. F	17. T	23. F	29. F	35. T
5. F	12. T	18. T	24. F	30. T	36. F
6. F	13. T	19. T	25. F	31. F	37. T
7. T					

Section 11.1

7. **(a)** $A_n=(-n,n)$; **(b)** $A_n=\left(-\frac{1}{n},\frac{1}{n}\right)$
8. **(a)** interval $[1,2)$; **(b)** ϕ
12. **(a)** $\{(x,y)\,|\,y\geq 2x\}$; **(b)** $\Re$; **(c)** interval $[0,2]$
15. **(g)** $A=$all rationals and $B=$all irrationals

Section 11.2

2. **(a)** 0; **(b)** 0; **(c)** 0; **(d)** 0; **(e)** 7; **(f)** 0
3. In **(d)** the limit is infinite. In **(f)** the value is 0. In other parts limits do not exist.
6. $\frac{x^2y}{x^4+y^2}$
8. **(a)** yes; **(b)** no; **(c)** yes; **(d)** yes
14. **(a)** no; **(b)** $f(x,y)=xy$; **(c)** yes
18. **(b)** 0; **(c)** 0; **(d)** does not exist

Section 11.3

1. **(a)** $f_x(0,0)=0$, $f_y(0,0)$ does not exist; **(b)** $f_x(0,0)=1$, $f_y(0,0)=-1$; **(c)** $f_x(0,0)$ and $f_y(0,0)$ do not exist

3. **(a)** $f_x(0,y)=-y$, $f_y(x,0)=x$, $f_{xx}(0,0)=-1$, $f_{yx}(0,0)=1$; **(b)** $f_x(x,y)=xy\frac{4xy^2}{(x^2+y^2)^2}+y\frac{x^2-y^2}{x^2+y^2}$, $f_y(x,y)=xy\frac{-4x^2y}{(x^2+y^2)^2}+x\frac{x^2-y^2}{x^2+y^2}$

4. $f_{xxx}, f_{xxy}, f_{xyx}, f_{yxx}, f_{yyx}, f_{yxy}, f_{xyy}, f_{yyy}$

8. $2\sin x^2+\cos\left(\sin x^2\right)$

9. **(b)** $L_1=xD+1$, $L_2=D-1$

13. $\frac{11}{20}$

Section 11.4

1. **(a)** $x-3y-2z=14$; **(c)** when $z=0$

2. **(a)** $2x-4y+z=14$, and $x=1-2t$, $y=-2+4t$, $z=4-t$, with $t\in\Re$;
(b) $3x+4y+5z=0$, and $x=3+3t$, $y=4+4t$, $z=-5+5t$, with $t\in\Re$

4. **(b)** $\vec{0}$; **(d)** yes

9. **(c)** yes, no

11. **(a)** $\langle -2x,-2y\rangle$; **(b)** -4; **(c)** 0

Section 11.5

1. **(a)** $\frac{1}{\sqrt{5}}$; **(b)** $\frac{4}{5}$

4. **(a)** no

5. Start with any unit vector $\vec{u}=\langle u_1,u_2\rangle$ and show $D_{\vec{u}}f(0,0)$ exists only if $u_1=0$ or $u_2=0$ but not both.

6. Find $\vec{u}$ so that $\|\vec{u}\|=1$ but $\frac{\partial f}{\partial \vec{u}}(0,0)$ does not exist.

7. **i, 4**

8. $f(x,y)=xy$ at $(0,0)$

11. Use Exercise 14 of Section 11.4.

Section 11.6

1. Use Theorem 5.3.8 on the function $g(t)\equiv f\left(\vec{P}+t\vec{u}\right)$, where $\vec{u}=\frac{\vec{Q}-\vec{P}}{\left\|\vec{Q}-\vec{P}\right\|}$. Follow similar steps to those given in the proof of Theorem 11.6.1.

2. $=\frac{1}{2!}\left[h^2 f_{xx}(a,b)+2hkf_{xy}(a,b)+k^2 f_{yy}(a,b)\right]$, if $n=2$

3. $3(y-2)-2(x+1)^2+2(x+1)(y-2)+(y-2)^2-(x+1)^2(y-2)$

7. Use $\frac{\partial z}{\partial t}=\frac{\partial z}{\partial x}\frac{\partial x}{\partial t}+\frac{\partial z}{\partial y}\frac{\partial y}{\partial t}$.

9. $-(\sin t)\frac{dy}{dx}+\left(\cos^2 t\right)\frac{d^2y}{dx^2}$

10. **(b)** $2\sqrt{3}$

12. Set $x=r\cos\theta$ and $y=r\sin\theta$. Then use $\frac{\partial v}{\partial \theta}=\frac{\partial v}{\partial x}\frac{\partial x}{\partial \theta}+\frac{\partial v}{\partial y}\frac{\partial y}{\partial \theta}$. Similarly, compute $\frac{\partial u}{\partial r}$.

13. **(a)** f_1 is of degree 1, f_2 is of degree -1, f_3 is of degree $\frac{1}{3}$, f_4 is of degree 0;
(b) Differentiate both sides of the identity with respect to t and then set $t=1$;

(c) Differentiate both sides of the identity with respect to t twice and then set $t=1$;

14. Let $u=x^2-y^2$, $v=y^2-x^2$ and use $z_x=f_u u_x+f_v v_x$

16. Let $u=2xy$, $v=x^2-y^2$.

17. $\nabla f(1,1)=\mathbf{i}+\mathbf{j}$

18. $y=2x-4$

Section 11.7

1. F	10. T	19. F	28. F	36. T	44. T
2. F	11. T	20. T	29. T	37. T	45. F
3. T	12. T	21. T	30. T	38. T	46. T
4. F	13. F	22. F	31. T	39. T	47. T
5. T	14. T	23. T	32. T	40. T	48. T
6. T	15. T	24. T	33. F	41. F	49. T
7. T	16. F	25. T	34. F	42. F	50. T
8. T	17. F	26. T	35. F	43. F	51. F
9. F	18. F	27. F			

Section 12.1

1. Extend proofs from Chapter 6 to the current context.
4. (a) Use Lemma 12.1.1. (b) Use part (a) with $m=0$. (c) Consider $h\equiv g-f\geq 0$.
6. Show $L(P,f)=0$ and $U(P,f)=1$.
7. See proofs of similar results in Chapter 6.
8. Consider a proof by contradiction.
9. 12

Section 12.2

1. Argue that each iterated integral is equal to the double integral.
2. (a) $2+e^{-1}-e$; (b) 1; (c) 1; (d) $-\pi\ln 2$
3. $\frac{4}{3}$
4. Consider writing $\frac{x-y}{(x+y)^3}=\frac{2x}{(x+y)^3}-\frac{1}{(x+y)^2}$.
5. (a) Show that $\int_0^1 f(x,y)\,dx$ does not exist; (b) $\int_0^2 f(x,y)\,dy=4$
7. (a) Use Fubini's theorem. (b) $\frac{4}{3}$
8. In both parts, use Exercise 7(a).
9. $\int_0^1\int_{x^3}^{x^2} dy\,dx=\int_0^1\int_{\sqrt{y}}^{\sqrt[3]{y}} dx\,dy=\frac{1}{12}$

10. $\int_0^4\int_0^{\sqrt{x}} dy\,dx=\frac{16}{3}$

11. (a) $\int_0^2\int_{x^2-x}^{x} dy\,dx=\frac{4}{3}$; (b) Evaluate $\int_{-2}^{4}\int_{\frac{1}{2}y^2-3}^{y+1} dx\,dy$.

13. Use Leibniz rule.
16. Use Exercise 15.

Section 12.3

1. (a) $\frac{1}{12}$; (c) 1; (d) $\frac{1}{4}\left(e^4-1\right)$

2. (b) $\int_0^4 \int_{\frac{y}{2}}^{\sqrt{y}} (2x+3)\,dx\,dy$; (c) $\int_{-1}^{1} \int_0^{\sqrt{1-x^2}} dy\,dx$

3. Graph the region.

4. Write the left hand side as two iterated integrals, apply Dirichlet's formula and combine.

5. Let $g(x)$ equal the left hand side and $h(x)$ equal the right hand side. Show $g'=h'$ and $g(0)=h(0)=0$ to obtain the desired conclusion.

7. (a) Note that every subrectangle of R contains a point of S. (b) $f=0$ as close to $(a,b)\in S$ as we wish. (c) Use the fact that lower sums are zero to show that if the integral exists, its value must be 0. Thus, show $U(P,f)<\varepsilon$.

Section 12.4

1. (a) $2\left(2\sqrt{2}-1\right)$; (c) $2\sqrt{2}\ln\frac{2}{3}$; (d) Use the parametrization $x=t,\ y=t^2$, with $0\le t\le 2$.

2. $\int_{C_1} xy\,dx=\frac{15}{2}$, $\int_{C_1} xy\,dy=10$

3. Use part (f) of Remark 12.4.1 with parametrization $x=t,\ y=g(t)$, for $t\in[a,b]$. Also, $\int_C \frac{1}{x^2+1}dx=\arctan 2$ and $\int_C \frac{1}{x^2+1}dy=\ln 5$.

4. $\frac{2}{15}$

5. $\frac{\sqrt{2}}{2}$, $\frac{1}{12}\left(5\sqrt{5}-1\right)$

6. Use the substitution $t=\sin u$.

9. Argue that in Riemann sums Δx_i and Δy_i change sign when the orientation of the curve is reversed.

Section 12.5

2. $y=\frac{1}{2}x^2+\frac{1}{2}$

3. $y=\frac{2}{x}$

4. $y=-2x$

5. $\vec{F}(x,y)=-y\mathbf{i}+x\mathbf{j}$

8. Use Cauchy–Schwarz inequality, Theorem 10.3.5.

9. Use the definition of the work integral and linearity properties of Riemann integrals.

11. $\arctan 2+\ln 5$

12. $\frac{\sqrt{2}}{2}+\frac{1}{2}\ln\left(1+\sqrt{2}\right)+\frac{4}{3}+\frac{81}{16}$

13. (a) If C connects points (a,b) and (a,c), show that $\int_C \vec{F}\cdot d\vec{r}=\int_0^1 N(a,y(t))y'(t)\,dt$.

    (b) Use part (a).

14. (a) Show that $\int_C \vec{F}\cdot d\vec{r}=0$; (b) -4π

15. 2π

Section 12.6

1. Since $M = f_x$ and $N = f_y$, then $M_y = f_{xy}$ and $N_x = f_{yx}$. By Clairaut's theorem, $M_y = N_x$.
2. Show that partial of this expression with respect to x is zero.
3. **(a)** $f(x,y) = xy^2 + C$; **(b)** does not exist; **(c)** $f(x,y) = x^3y - e^y + 4x - \sin x$; **(d)** $f(x,y) = xy^3 - y - \frac{1}{x}$
4. **(a)** 3; **(b)** π; **(c)** 0; **(d)** 0
5. **(a)** 19.5; **(b)** 18
6. **(b)** $\int_C \vec{F} \cdot d\vec{r} = 2\pi$; **(c)** no; **(e)** 2π
7. **(c)** no; **(d)** Show that $f_x\mathbf{i} + f_y\mathbf{j} = \vec{F}$. **(e)** yes
8. Show that $\nabla f = \vec{F}$.

Section 12.7

3. 4π
4. $-\frac{45}{2}$
5. Since $M = 0$ and $N = 0$ for all (x,y) on C, by Green's theorem we have that $\iint_D (N_x - M_y)\, dA = \oint_C M dx + N dy = 0$.
6. $-\frac{32}{3}$
7. **(a)** Show $\operatorname{curl} \vec{F} = 0$; **(d)** 0
8. $\frac{3}{8}\pi$
9. $ab\pi$
10. $\frac{3\pi}{2}a^2$
11. 12
12. **(a)** Follows from Remark 12.7.5, since u is harmonic. **(b)** Let $M = -uu_y$, $N = uu_x$. Then $\vec{F} = M\mathbf{i} + N\mathbf{j}$ with $\operatorname{curl} \vec{F} = (u_x)^2 + (u_y)^2$. Now apply Green's theorem.
14. 0
15. We cannot sat that $\operatorname{curl} \vec{F} = 0$ since we only know $\vec{F}$ on ∂D.

Section 12.8

7. $\int_C \vec{F} \cdot d\vec{r} = 2\pi$
9. **(b)** $2yz^3 + 6x^2yz$
12. $\oint_C \vec{F} \cdot \vec{T}\, ds = (-4)(\text{area inside } C)$, $\oint_C \vec{F} \cdot \vec{N}\, ds = 0$
13. **(b)** $\vec{0}$

Section 12.9

1. T	8. T	15. T	22. F	28. F	34. F
2. F	9. T	16. F	23. T	29. F	35. T
3. T	10. F	17. F	24. T	30. F	36. F
4. F	11. F	18. T	25. F	31. F	37. F
5. F	12. T	19. T	26. F	32. T	38. T
6. F	13. F	20. F	27. T	33. F	39. F
7. F	14. F	21. T			

Index of Symbols

Greek Alphabet

A	α	alpha	N	ν	nu
B	β	beta	Ξ	ξ	xi
Γ	γ	gamma	O	o	omicron
Δ	δ	delta	Π	π	pi
E	ε	epsilon	P	ρ	rho
Z	ζ	zeta	Σ	σ	sigma
H	η	eta	T	τ	tau
Θ	θ, ϑ	theta	Y	υ	upsilon
I	ι	iota	Φ	ϕ, φ	phi
K	κ	kappa	X	χ	chi
Λ	λ	lambda	Ψ	ψ	psi
M	μ	mu	Ω	ω	omega

Index

A

B

C

D

E

F

G

H

I

J

K

L

M

N

O

P

Q

R

S

T